# COMBINATORICS OF SET PARTITIONS

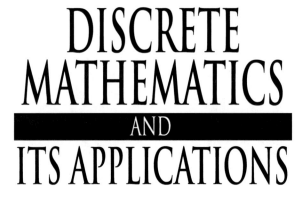

Series Editor
**Kenneth H. Rosen, Ph.D.**

R. B. J. T. Allenby and Alan Slomson, How to Count: An Introduction to Combinatorics, Third Edition

Juergen Bierbrauer, Introduction to Coding Theory

Katalin Bimbó, Combinatory Logic: Pure, Applied and Typed

Donald Bindner and Martin Erickson, A Student's Guide to the Study, Practice, and Tools of Modern Mathematics

Francine Blanchet-Sadri, Algorithmic Combinatorics on Partial Words

Miklós Bóna, Combinatorics of Permutations, Second Edition

Richard A. Brualdi and Dragoš Cvetković, A Combinatorial Approach to Matrix Theory and Its Applications

Kun-Mao Chao and Bang Ye Wu, Spanning Trees and Optimization Problems

Charalambos A. Charalambides, Enumerative Combinatorics

Gary Chartrand and Ping Zhang, Chromatic Graph Theory

Henri Cohen, Gerhard Frey, et al., Handbook of Elliptic and Hyperelliptic Curve Cryptography

Charles J. Colbourn and Jeffrey H. Dinitz, Handbook of Combinatorial Designs, Second Edition

Martin Erickson, Pearls of Discrete Mathematics

Martin Erickson and Anthony Vazzana, Introduction to Number Theory

Steven Furino, Ying Miao, and Jianxing Yin, Frames and Resolvable Designs: Uses, Constructions, and Existence

Mark S. Gockenbach, Finite-Dimensional Linear Algebra

Randy Goldberg and Lance Riek, A Practical Handbook of Speech Coders

Jacob E. Goodman and Joseph O'Rourke, Handbook of Discrete and Computational Geometry, Second Edition

Jonathan L. Gross, Combinatorial Methods with Computer Applications

**Titles (continued)**

Jonathan L. Gross and Jay Yellen, Graph Theory and Its Applications, Second Edition

Jonathan L. Gross and Jay Yellen, Handbook of Graph Theory

David S. Gunderson, Handbook of Mathematical Induction: Theory and Applications

Richard Hammack, Wilfried Imrich, and Sandi Klavžar, Handbook of Product Graphs, Second Edition

Darrel R. Hankerson, Greg A. Harris, and Peter D. Johnson, Introduction to Information Theory and Data Compression, Second Edition

Darel W. Hardy, Fred Richman, and Carol L. Walker, Applied Algebra: Codes, Ciphers, and Discrete Algorithms, Second Edition

Daryl D. Harms, Miroslav Kraetzl, Charles J. Colbourn, and John S. Devitt, Network Reliability: Experiments with a Symbolic Algebra Environment

Silvia Heubach and Toufik Mansour, Combinatorics of Compositions and Words

Leslie Hogben, Handbook of Linear Algebra

Derek F. Holt with Bettina Eick and Eamonn A. O'Brien, Handbook of Computational Group Theory

David M. Jackson and Terry I. Visentin, An Atlas of Smaller Maps in Orientable and Nonorientable Surfaces

Richard E. Klima, Neil P. Sigmon, and Ernest L. Stitzinger, Applications of Abstract Algebra with Maple™ and MATLAB®, Second Edition

Richard E. Klima and Neil P. Sigmon, Cryptology: Classical and Modern with Maplets

Patrick Knupp and Kambiz Salari, Verification of Computer Codes in Computational Science and Engineering

William Kocay and Donald L. Kreher, Graphs, Algorithms, and Optimization

Donald L. Kreher and Douglas R. Stinson, Combinatorial Algorithms: Generation Enumeration and Search

Hang T. Lau, A Java Library of Graph Algorithms and Optimization

C. C. Lindner and C. A. Rodger, Design Theory, Second Edition

Nicholas A. Loehr, Bijective Combinatorics

Toufik Mansour, Combinatorics of Set Partitions

Alasdair McAndrew, Introduction to Cryptography with Open-Source Software

Elliott Mendelson, Introduction to Mathematical Logic, Fifth Edition

Alfred J. Menezes, Paul C. van Oorschot, and Scott A. Vanstone, Handbook of Applied Cryptography

Stig F. Mjølsnes, A Multidisciplinary Introduction to Information Security

Jason J. Molitierno, Applications of Combinatorial Matrix Theory to Laplacian Matrices of Graphs

Richard A. Mollin, Advanced Number Theory with Applications

Richard A. Mollin, Algebraic Number Theory, Second Edition

**Titles (continued)**

Richard A. Mollin, Codes: The Guide to Secrecy from Ancient to Modern Times

Richard A. Mollin, Fundamental Number Theory with Applications, Second Edition

Richard A. Mollin, An Introduction to Cryptography, Second Edition

Richard A. Mollin, Quadratics

Richard A. Mollin, RSA and Public-Key Cryptography

Carlos J. Moreno and Samuel S. Wagstaff, Jr., Sums of Squares of Integers

Goutam Paul and Subhamoy Maitra, RC4 Stream Cipher and Its Variants

Dingyi Pei, Authentication Codes and Combinatorial Designs

Kenneth H. Rosen, Handbook of Discrete and Combinatorial Mathematics

Douglas R. Shier and K.T. Wallenius, Applied Mathematical Modeling: A Multidisciplinary Approach

Alexander Stanoyevitch, Introduction to Cryptography with Mathematical Foundations and Computer Implementations

Jörn Steuding, Diophantine Analysis

Douglas R. Stinson, Cryptography: Theory and Practice, Third Edition

Roberto Togneri and Christopher J. deSilva, Fundamentals of Information Theory and Coding Design

W. D. Wallis, Introduction to Combinatorial Designs, Second Edition

W. D. Wallis and J. C. George, Introduction to Combinatorics

Lawrence C. Washington, Elliptic Curves: Number Theory and Cryptography, Second Edition

DISCRETE MATHEMATICS AND ITS APPLICATIONS

Series Editor KENNETH H. ROSEN

# COMBINATORICS OF SET PARTITIONS

Toufik Mansour

University of Haifa
Israel

BOWLING GREEN STATE
UNIVERSITY LIBRARIES

CRC Press
Taylor & Francis Group
Boca Raton   London   New York

CRC Press is an imprint of the
Taylor & Francis Group, an **informa** business

A CHAPMAN & HALL BOOK

CRC Press
Taylor & Francis Group
6000 Broken Sound Parkway NW, Suite 300
Boca Raton, FL 33487-2742

© 2013 by Taylor & Francis Group, LLC
CRC Press is an imprint of Taylor & Francis Group, an Informa business

No claim to original U.S. Government works

Printed in the United States of America on acid-free paper
Version Date: 20120622

International Standard Book Number: 978-1-4398-6333-6 (Hardback)

This book contains information obtained from authentic and highly regarded sources. Reasonable efforts have been made to publish reliable data and information, but the author and publisher cannot assume responsibility for the validity of all materials or the consequences of their use. The authors and publishers have attempted to trace the copyright holders of all material reproduced in this publication and apologize to copyright holders if permission to publish in this form has not been obtained. If any copyright material has not been acknowledged please write and let us know so we may rectify in any future reprint.

Except as permitted under U.S. Copyright Law, no part of this book may be reprinted, reproduced, transmitted, or utilized in any form by any electronic, mechanical, or other means, now known or hereafter invented, including photocopying, microfilming, and recording, or in any information storage or retrieval system, without written permission from the publishers.

For permission to photocopy or use material electronically from this work, please access www.copyright.com (http://www.copyright.com/) or contact the Copyright Clearance Center, Inc. (CCC), 222 Rosewood Drive, Danvers, MA 01923, 978-750-8400. CCC is a not-for-profit organization that provides licenses and registration for a variety of users. For organizations that have been granted a photocopy license by the CCC, a separate system of payment has been arranged.

**Trademark Notice:** Product or corporate names may be trademarks or registered trademarks, and are used only for identification and explanation without intent to infringe.

**Visit the Taylor & Francis Web site at**
http://www.taylorandfrancis.com

**and the CRC Press Web site at**
http://www.crcpress.com

إلى حِكْمَتي ... وَعِلْمي

إلى مَن عَلَّمَني عِلْمَ الْحَياةِ

إلى أدَبي ... وَحِلْمي

إلى رُوْحِ أبي أهْدي هَذا الْكِتابَ الْمُتَواضِعَ

# Contents

| | |
|---|---|
| Preface | xix |
| Acknowledgment | xxv |
| Author Biographies | xxvii |

**1 Introduction**   1
- 1.1 Historical Overview and Earliest Results . . . . . . . . . . . 1
- 1.2 Timeline of Research for Set Partitions . . . . . . . . . . . 10
- 1.3 A More Detailed Book . . . . . . . . . . . . . . . . . . . . 20
  - 1.3.1 Basic Results . . . . . . . . . . . . . . . . . . . . . 20
  - 1.3.2 Statistics on Set Partitions . . . . . . . . . . . . . . 21
  - 1.3.3 Pattern Avoidance in Set Partitions . . . . . . . . . . 23
  - 1.3.4 Restricted Set Partitions . . . . . . . . . . . . . . . . 25
  - 1.3.5 Asymptotics Results on Set Partitions . . . . . . . . . 25
  - 1.3.6 Generating Set Partitions . . . . . . . . . . . . . . . 26
  - 1.3.7 Normal Ordering and Set Partitions . . . . . . . . . . 27
- 1.4 Exercises . . . . . . . . . . . . . . . . . . . . . . . . . . . . 27

**2 Basic Tools of the Book**   31
- 2.1 Sequences . . . . . . . . . . . . . . . . . . . . . . . . . . . . 31
- 2.2 Solving Recurrence Relations . . . . . . . . . . . . . . . . . 39
  - 2.2.1 Guess and Check . . . . . . . . . . . . . . . . . . . . 39
  - 2.2.2 Iteration . . . . . . . . . . . . . . . . . . . . . . . . 40
  - 2.2.3 Characteristic Polynomial . . . . . . . . . . . . . . . 40
- 2.3 Generating Functions . . . . . . . . . . . . . . . . . . . . . 45
- 2.4 Lagrange Inversion Formula . . . . . . . . . . . . . . . . . 63
- 2.5 The Principle of Inclusion and Exclusion . . . . . . . . . . 64
- 2.6 Generating Trees . . . . . . . . . . . . . . . . . . . . . . . . 67
- 2.7 Exercises . . . . . . . . . . . . . . . . . . . . . . . . . . . . 71

**3 Preliminary Results on Set Partitions**   75
- 3.1 Dobiński's Formula . . . . . . . . . . . . . . . . . . . . . . 75
- 3.2 Different Representations . . . . . . . . . . . . . . . . . . . 84
  - 3.2.1 Block Representation . . . . . . . . . . . . . . . . . 85
  - 3.2.2 Circular Representation and Line Diagram . . . . . . 89

|       |       | 3.2.3 | Flattened Set Partitions | 90 |
|---|---|---|---|---|

|  |  | 3.2.3 | Flattened Set Partitions . . . . . . . . . . . . . . . . . | 90 |
|---|---|---|---|---|
|  |  | 3.2.4 | More Representations . . . . . . . . . . . . . . . . . . | 95 |
|  |  |  | 3.2.4.1 Canonical Representation . . . . . . . . . . | 95 |
|  |  |  | 3.2.4.2 Graphical Representation . . . . . . . . . . | 96 |
|  |  |  | 3.2.4.3 Standard Representation . . . . . . . . . . | 96 |
|  |  |  | 3.2.4.4 Rook Placement Representation . . . . . . . | 98 |
|  | 3.3 | Exercises . . . . . . . . . . . . . . . . . . . . . . . . . . . . . . | | 99 |
|  | 3.4 | Research Directions and Open Problems . . . . . . . . . . . | | 103 |
| **4** | **Subword Statistics on Set Partitions** | | | **107** |
|  | 4.1 | Subword Patterns of Size Two: Rises, Levels and Descents | . | 109 |
|  |  | 4.1.1 | Number Levels . . . . . . . . . . . . . . . . . . . . . . | 114 |
|  |  | 4.1.2 | Nontrivial Rises and Descents . . . . . . . . . . . . . | 117 |
|  | 4.2 | Peaks and Valleys . . . . . . . . . . . . . . . . . . . . . . . . | | 119 |
|  |  | 4.2.1 | Counting Peaks in Words . . . . . . . . . . . . . . . | 120 |
|  |  | 4.2.2 | Counting Peaks . . . . . . . . . . . . . . . . . . . . . | 123 |
|  |  | 4.2.3 | Counting Valleys . . . . . . . . . . . . . . . . . . . . | 126 |
|  | 4.3 | Subword Patterns: $\ell$-Rises, $\ell$-Levels, and $\ell$-Descents . . . . . | | 131 |
|  |  | 4.3.1 | Long-Rise Pattern . . . . . . . . . . . . . . . . . . . | 131 |
|  |  | 4.3.2 | Long-Level Pattern . . . . . . . . . . . . . . . . . . . | 136 |
|  |  | 4.3.3 | Long-Descent Pattern . . . . . . . . . . . . . . . . . | 139 |
|  | 4.4 | Families of Subword Patterns . . . . . . . . . . . . . . . . . . | | 141 |
|  |  | 4.4.1 | The Patterns $122\cdots 2, 11\cdots 12$ . . . . . . . . . . . . . | 141 |
|  |  | 4.4.2 | The Patterns $22\cdots 21, 211\cdots 1$ . . . . . . . . . . . . | 143 |
|  |  | 4.4.3 | The Pattern $m\rho m$ . . . . . . . . . . . . . . . . . . . | 146 |
|  |  | 4.4.4 | The Pattern $m\rho(m+1)$ . . . . . . . . . . . . . . . . | 148 |
|  |  | 4.4.5 | The Pattern $(m+1)\rho m$ . . . . . . . . . . . . . . . . | 151 |
|  | 4.5 | Patterns of Size Three . . . . . . . . . . . . . . . . . . . . . . | | 152 |
|  | 4.6 | Exercises . . . . . . . . . . . . . . . . . . . . . . . . . . . . . . | | 158 |
|  | 4.7 | Research Directions and Open Problems . . . . . . . . . . . | | 159 |
| **5** | **Nonsubword Statistics on Set Partitions** | | | **165** |
|  | 5.1 | Statistics and Block Representation . . . . . . . . . . . . . . | | 167 |
|  | 5.2 | Statistics and Canonical and Rook Representations . . . . . | | 169 |
|  | 5.3 | Records and Weak Records . . . . . . . . . . . . . . . . . . . | | 175 |
|  |  | 5.3.1 | Weak Records . . . . . . . . . . . . . . . . . . . . . . | 176 |
|  |  | 5.3.2 | Sum of Positions of Records . . . . . . . . . . . . . . | 179 |
|  |  | 5.3.3 | Sum of Positions of Additional Weak Records . . . . | 181 |
|  | 5.4 | Number of Positions Between Adjacent Occurrences of a Letter | | 182 |
|  |  | 5.4.1 | The Statistic dis . . . . . . . . . . . . . . . . . . . . | 185 |
|  |  | 5.4.2 | The Statistic $m$-Distance . . . . . . . . . . . . . . . | 190 |
|  |  | 5.4.3 | Combinatorial Proofs . . . . . . . . . . . . . . . . . . | 193 |
|  | 5.5 | The Internal Statistic . . . . . . . . . . . . . . . . . . . . . . | | 196 |
|  |  | 5.5.1 | The Statistic int . . . . . . . . . . . . . . . . . . . . | 198 |
|  |  | 5.5.2 | The Statistic $int_1$ . . . . . . . . . . . . . . . . . . . | 200 |

|     | 5.6  | Statistics and Generalized Patterns | 202 |
| --- | --- | --- | --- |
|     | 5.7  | Major Index | 209 |
|     | 5.8  | Number of Crossings, Nestings and Alignments | 215 |
|     | 5.9  | Exercises | 216 |
|     | 5.10 | Research Directions and Open Problems | 219 |
| **6** | **Avoidance of Patterns in Set Partitions** | | **223** |
|     | 6.1 | History and Connections | 223 |
|     | 6.2 | Avoidance of Subsequence Patterns | 225 |
|     |     | 6.2.1 Pattern-Avoiding Fillings of Diagrams | 226 |
|     |     | 6.2.2 Basic Facts and Patterns of Size Three | 236 |
|     |     | 6.2.3 Noncrossing and Nonnesting Set Partitions | 237 |
|     |     | 6.2.4 The Patterns $12\cdots(k+1)12\cdots k$, $12\cdots k12\cdots(k+1)$ | 239 |
|     |     |     6.2.4.1 The Pattern $12\cdots(k+1)12\cdots k$ | 240 |
|     |     |     6.2.4.2 The Pattern $12\cdots k12\cdots(k+1)$ | 241 |
|     |     |     6.2.4.3 The Equivalence | 242 |
|     |     | 6.2.5 Patterns of the Form $1(\tau+1)$ | 243 |
|     |     | 6.2.6 The Patterns $12\cdots k1$, $12\cdots k12$ | 244 |
|     |     | 6.2.7 Patterns Equivalent to $12\cdots m$ | 250 |
|     |     | 6.2.8 Binary Patterns | 250 |
|     |     | 6.2.9 Patterns Equivalent to $12^k13$ | 256 |
|     |     | 6.2.10 Landscape Patterns | 259 |
|     |     | 6.2.11 Patterns of Size Four | 264 |
|     |     |     6.2.11.1 The Pattern 1123 | 264 |
|     |     |     6.2.11.2 Classification of Patterns of Size Four | 267 |
|     |     | 6.2.12 Patterns of Size Five | 268 |
|     |     |     6.2.12.1 The Equivalence $12112 \sim 12212$ | 268 |
|     |     |     6.2.12.2 Classification of Patterns of Size Five | 270 |
|     |     | 6.2.13 Patterns of Size Six | 270 |
|     |     | 6.2.14 Patterns of Size Seven | 272 |
|     | 6.3 | Generalized Patterns | 275 |
|     |     | 6.3.1 Patterns of Type $(1,2)$ | 276 |
|     |     | 6.3.2 Patterns of Type $(2,1)$ | 280 |
|     | 6.4 | Partially Ordered Patterns | 289 |
|     |     | 6.4.1 Patterns of Size Three | 290 |
|     |     | 6.4.2 Shuffle Patterns | 294 |
|     | 6.5 | Exercises | 298 |
|     | 6.6 | Research Directions and Open Problems | 300 |
| **7** | **Multi Restrictions on Set Partitions** | | **307** |
|     | 7.1 | Avoiding a Pattern of Size Three and Another Pattern | 308 |
|     |     | 7.1.1 The Patterns 112, 121 | 309 |
|     |     | 7.1.2 The Pattern 123 | 312 |
|     |     | 7.1.3 The Pattern 122 | 314 |
|     | 7.2 | Pattern Avoidance in Noncrossing Set Partitions | 315 |

|  |  | 7.2.1 | CC-Equivalences | 320 |
|---|---|---|---|---|
|  |  | 7.2.2 | Generating Functions | 322 |
|  |  | 7.2.3 | CC-Equivalences of Patterns of Size Four | 326 |
|  |  | 7.2.4 | CC-Equivalences of Patterns of Size Five | 326 |
|  | 7.3 |  | General Equivalences | 327 |
|  | 7.4 |  | Two Patterns of Size Four | 330 |
|  | 7.5 |  | Left Motzkin Numbers | 334 |
|  |  | 7.5.1 | The Pairs $(1222, 1212)$ and $(1112, 1212)$ | 335 |
|  |  | 7.5.2 | The Pair $(1211, 1221)$ | 336 |
|  |  | 7.5.3 | The Pair $(1222, 1221)$ | 337 |
|  | 7.6 |  | Sequence A054391 | 339 |
|  |  | 7.6.1 | The Pairs $(1212, 12221), (1212, 11222), (1212, 11122)$ | 340 |
|  |  | 7.6.2 | The Pair $(1221, 12311)$ | 340 |
|  |  | 7.6.3 | The Pairs $(1221, 12112), (1221, 12122)$ | 342 |
|  | 7.7 |  | Catalan and Generalized Catalan Numbers | 346 |
|  | 7.8 |  | Pell Numbers | 347 |
|  |  | 7.8.1 | Counting $\mathcal{P}_n(1211, 1212, 1213)$ by inv | 348 |
|  |  | 7.8.2 | Counting $\mathcal{P}_n(1212, 1222, 1232)$ by Comaj | 351 |
|  | 7.9 |  | Regular Set Partitions | 354 |
|  |  | 7.9.1 | Noncrossing Regular Set Partitions | 357 |
|  |  | 7.9.2 | Nonnesting Regular Set Partitions | 358 |
|  | 7.10 |  | Distance Restrictions | 359 |
|  | 7.11 |  | Singletons | 362 |
|  | 7.12 |  | Block-Connected | 364 |
|  | 7.13 |  | Exercises | 367 |
|  | 7.14 |  | Research Directions and Open Problems | 372 |
| **8** | **Asymptotics and Random Set Partition** |  |  | **379** |
|  | 8.1 |  | Tools from Probability Theory | 380 |
|  | 8.2 |  | Tools from Complex Analysis | 387 |
|  | 8.3 |  | Z-statistics | 394 |
|  | 8.4 |  | Set Partitions as Geometric Words | 401 |
|  | 8.5 |  | Asymptotics for Set Partitions | 405 |
|  |  | 8.5.1 | Asymptotics for Bell and Stirling Numbers | 405 |
|  |  |  | 8.5.1.1 On the Number of Blocks | 409 |
|  |  |  | 8.5.1.2 On the Number of Distinct Block Sizes | 411 |
|  |  | 8.5.2 | On Number of Blocks in a Noncrossing Set Partition | 416 |
|  |  | 8.5.3 | Records | 417 |
|  | 8.6 |  | Exercises | 418 |
|  | 8.7 |  | Research Directions and Open Problems | 420 |

| 9 | Gray Codes, Loopless Algorithms and Set Partitions | 423 |
|---|---|---|
| | 9.1 Gray Code and Loopless Algorithms | 424 |
| | 9.2 Gray Codes for $\mathcal{P}_n$ | 428 |
| | 9.3 Loopless Algorithm for Generating $\mathcal{P}_n$ | 434 |
| | 9.4 Exercises | 436 |
| | 9.5 Research Directions and Open Problems | 437 |

| 10 | Set Partitions and Normal Ordering | 439 |
|---|---|---|
| | 10.1 Preliminaries | 440 |
| | 10.2 Linear Representation and $\mathcal{N}((a^\dagger a)^n)$ | 446 |
| | 10.3 Wick's Theorem and $q$-Normal Ordering | 449 |
| | 10.4 $p$-Normal Ordering | 452 |
| | 10.5 Noncrossing Normal Ordering | 457 |
| |     10.5.1 Some Preliminary Observations | 459 |
| |     10.5.2 Noncrossing Normal Ordering of $(a^r(a^\dagger)^s)^n$ | 460 |
| |     10.5.3 Noncrossing Normal Ordering of $(a^r+(a^\dagger)^s)^n$ | 462 |
| |     10.5.4 $k$-Ary Trees and Lattice Paths | 465 |
| | 10.6 Exercises | 469 |
| | 10.7 Research Directions and Open Problems | 471 |

| A | Solutions and Hints | 473 |
|---|---|---|
| B | Identities | 501 |
| C | Power Series and Binomial Theorem | 503 |
| D | Chebychev Polynomials of the Second Kind | 507 |
| E | Linear Algebra and Algebra Review | 511 |
| F | Complex Analysis Review | 513 |
| G | Coherent States | 517 |
| H | C++ Programming | 519 |
| I | Tables | 537 |
| J | Notation | 543 |

| Bibliography | 547 |
|---|---|
| Index | 573 |

# List of Tables

| | | |
|---|---|---|
| 1.1 | Timeline of first research steps on set partitions | 10 |
| 1.2 | Timeline research for statistics on set partitions | 12 |
| 1.3 | Timeline research for pattern avoidance on set partitions | 14 |
| 1.4 | Timeline research for restricted set partitions | 15 |
| 1.5 | Timeline research for asymptotics on set partitions | 17 |
| 1.6 | Timeline research for set partitions and algebra | 18 |
| 1.7 | Timeline research for generating set partitions | 19 |
| 1.8 | Timeline research for the normal ordering and set partitions | 20 |
| 2.1 | The set $A(n)$ | 35 |
| 2.2 | Set partitions in $A(n)$ with exactly $k$ blocks | 37 |
| 2.3 | Particular solutions for certain $q(n)$ | 43 |
| 3.1 | Set partitions avoiding a pattern of [3] | 87 |
| 3.2 | Set partitions avoiding two patterns of [3] | 88 |
| 3.3 | Even set partitions avoiding a pattern of [3] | 88 |
| 3.4 | The cardinality $|\text{Flatten}_n(\tau)|$, where $\tau \in S_3$ and $1 \leq n \leq 10$ | 91 |
| 3.5 | Three letter subsequence pattern | 92 |
| 3.6 | The cardinality $|\text{Flatten}_n(\tau)|$, where $\tau \in S_4$ and $1 \leq n \leq 10$ | 103 |
| 4.1 | Generating functions for 2-letter patterns | 119 |
| 4.2 | $\{\mathcal{P}_n(\tau)\}_{n=1}^{12}$ for four letter subword patterns $\tau$ | 161 |
| 4.3 | $\{\mathcal{P}_n(\tau)\}_{n=1}^{12}$ for five letter subword patterns $\tau$ | 161 |
| 6.1 | Number of partitions in $\mathcal{P}_n(\tau)$, where $\tau \in \mathcal{P}_3$ | 237 |
| 6.2 | The cardinality $|\mathcal{P}_n(12 \cdots k12)|$ for $2 \leq k \leq 5$ and $1 \leq n \leq 10$ | 244 |
| 6.3 | Number of set partitions of $\mathcal{P}_n(\tau)$, where $\tau \in \mathcal{P}_4$ | 267 |
| 6.4 | Nonsingleton equivalence classes of subsequence patterns of $\mathcal{P}_5$ | 270 |
| 6.5 | Nonsingleton equivalence classes of subsequence patterns of $\mathcal{P}_6$ | 271 |
| 6.6 | Nonsingleton equivalence classes of subsequence patterns of $\mathcal{P}_7$ | 272 |
| 6.7 | Three-letter generalized patterns of type $(1,2)$ | 276 |
| 6.8 | Three-letter generalized patterns of type $(2,1)$ | 285 |
| 6.9 | The number $a_\tau(n)$ of set partitions of $[n]$ that contain exactly once a subsequence pattern of size three, where $n = 3, 4, \ldots, 10$ | 302 |
| 6.10 | The number $a_\tau(n)$ of set partitions of $[n]$ that contain a subsequence pattern of size four, where $n = 4, 5, \ldots, 10$ | 302 |

| | | |
|---|---|---|
| 6.11 | The number $a_\tau(n)$ of set partitions of $[n]$ that contain a subsequence pattern of size five, where $n = 5, 6, \ldots, 11$ . . . . . . . | 303 |
| 7.1 | Generating functions $F_{\mathbf{a}}(x, y)$, where $\mathbf{a}$ is a nonempty composition of size at most three . . . . . . . . . . . . . . . . . . . . . | 312 |
| 7.2 | The equivalence classes of $(3, 4)$-pairs . . . . . . . . . . . . . . | 315 |
| 7.3 | Nonsingleton cc-equivalences of patterns of size four . . . . . | 326 |
| 7.4 | Nonsingleton cc-equivalences of patterns of size five . . . . . . | 327 |
| 7.5 | Nonsingleton equivalences of pair patterns of size four . . . . | 333 |
| 7.6 | Number of set partitions of $[n]$ with exactly $r$ circular connectors . . . . . . . . . . . . . . . . . . . . . . . . . . . . . . . . . . . | 365 |
| 7.7 | Number of set partitions of $[n]$ with exactly $r$ connectors. . . | 365 |
| 7.8 | Number of sparse set partitions of $\mathcal{P}_n(\tau)$ with $\tau \in \mathcal{P}_4$ . . . . . | 373 |
| 7.9 | Number of sparse set partitions of $\mathcal{P}_n(\tau, \tau')$ . . . . . . . . . . | 373 |
| 8.1 | Values for $P_{n,k}$ for all $1 \leq k \leq n \leq 3$ . . . . . . . . . . . . . . | 404 |
| 8.2 | Number of set partitions of $[n]$ and estimation . . . . . . . . . | 407 |
| 9.1 | Listing $\mathcal{P}_4$ . . . . . . . . . . . . . . . . . . . . . . . . . . . . . . . | 428 |
| I.1 | The numbers $P_n(\tau)$ where $\tau$ is a subsequence pattern of size six and $n = 6, 7, \ldots, 11$ . . . . . . . . . . . . . . . . . . . . . . . | 537 |
| I.2 | The numbers $P_n(1212, \tau)$ where $\tau$ is a subsequence pattern of size four and $n = 4, 5, \ldots, 12$ . . . . . . . . . . . . . . . . . . . . | 540 |
| I.3 | The numbers $P_n(1212, \tau)$ where $\tau$ is a subsequence pattern of size five and $n = 6, 7, \ldots, 12$ . . . . . . . . . . . . . . . . . . . | 541 |
| I.4 | The numbers $P_n(1212, \tau)$ where $\tau$ is a subsequence pattern of size six and $n = 7, 8, \ldots, 12$ . . . . . . . . . . . . . . . . . . . | 541 |

# List of Figures

| | | |
|---|---|---|
| 1.1 | Diagrams used to represent set partitions in 16th century Japan | 1 |
| 1.2 | Stirling numbers of the second kind . . . . . . . . . . . . . . | 6 |
| 1.3 | Rhyming patterns for 5-line verses. . . . . . . . . . . . . | 7 |
| 1.4 | Rhyming patterns for 6-line verses. . . . . . . . . . . . . | 28 |
| 2.1 | Different ways of drawing non-intersecting chords on a circle between 4 points . . . . . . . . . . . . . . . . . . . . . . . . | 62 |
| 2.2 | Generating tree for $[2]^n$. . . . . . . . . . . . . . . . . . . . | 69 |
| 2.3 | Generating tree for set partitions. . . . . . . . . . . . . . . | 70 |
| 3.1 | The circular representation and the line diagram of the set partition 18/235/47/6 . . . . . . . . . . . . . . . . . . . . . . | 90 |
| 3.2 | A graphical representation of the set partition $\pi = 1231242$ . | 96 |
| 3.3 | The standard representation of 1357/26/4/89 . . . . . . . . . | 96 |
| 3.4 | The standard representation of 4/235/167/89 and its corresponding Dyck path . . . . . . . . . . . . . . . . . . . . . . | 97 |
| 3.5 | A rook placement . . . . . . . . . . . . . . . . . . . . . . . | 99 |
| 3.6 | A rook placement representation of 1357/26/4/89 . . . . . . . | 100 |
| 5.1 | A rook placement . . . . . . . . . . . . . . . . . . . . . . . | 171 |
| 5.2 | The vectors $\mathrm{csw}(r), \mathrm{cnw}(r),$ and $\mathrm{rnw}(r)$ . . . . . . . . . . . . | 171 |
| 5.3 | The rook placement $f(112132242)$ . . . . . . . . . . . . . . | 172 |
| 5.4 | The rook placement $g(11213114233)$ . . . . . . . . . . . . . | 173 |
| 5.5 | The standard representation of the partition 15/27/34/68. . . | 211 |
| 5.6 | The $\mathrm{cr}_2(\pi)$ for $\pi \in \mathcal{P}_n(\{1,2\},\{4,6\})$ . . . . . . . . . . . . . | 212 |
| 5.7 | The $\mathrm{pmaj}(\pi)$ for $\pi \in \mathcal{P}_n(\{1,2\},\{4,6\})$ . . . . . . . . . . . | 214 |
| 6.1 | Three diagrams . . . . . . . . . . . . . . . . . . . . . . . . | 226 |
| 6.2 | A diagram and its filling . . . . . . . . . . . . . . . . . . . | 227 |
| 6.3 | A Ferrers diagram . . . . . . . . . . . . . . . . . . . . . . | 228 |
| 6.4 | A stack polyomino . . . . . . . . . . . . . . . . . . . . . . | 228 |
| 6.5 | A moon polyomino . . . . . . . . . . . . . . . . . . . . . . | 239 |
| 7.1 | Generating tree for set partitions . . . . . . . . . . . . . . . | 317 |
| 7.2 | Rook placement of the set partition 19/26/348/5/7 . . . . . . | 356 |
| 7.3 | The mapping $\psi_3$ . . . . . . . . . . . . . . . . . . . . . . | 357 |

| | | |
|---|---|---|
| 7.4 | The bijection $\Phi$ maps 19/25/36/48/7 to 18/247/35 . . . . . . | 357 |
| 7.5 | The linear representation of a linked set partition . . . . . . . | 368 |
| 10.1 | The linear representations of the contractions of the word $aaa^\dagger a^\dagger$ . . . . . . . . . . . . . . . . . . . . . . . . . . . . . . | 442 |
| 10.2 | The linear representation of the Feynman diagram $\gamma$. . . . . . | 443 |
| 10.3 | The linear representations of the contractions of the word $(a^\dagger a)^3$ . . . . . . . . . . . . . . . . . . . . . . . . . . . . . . . . | 448 |
| 10.4 | The set $\mathcal{L}_2$ of $L$-lattices paths . . . . . . . . . . . . . . . . . . | 468 |
| 10.5 | 2-Motzkin paths of length 2 . . . . . . . . . . . . . . . . . . . | 470 |

# *Preface*

This book gives an introduction to and an overview of the methods used in the combinatorics of pattern avoidance and pattern enumeration in set partitions, a very active area of research in the last decade. The first known application of set partitions arose in the context of tea ceremonies and incense games in Japanese upper class society around A.D. 1500. Guests at a Kado ceremony would be smelling cups with burned incense with the goal to either identity the incense or to identity which cups contained identical incense. There are many variations of the game, even today. One particular game is named genji-ko, and it is the one that originated the interest in set partitions. Five different incense were cut into five pieces, each piece put into a separate bag, and then five of these bags were chosen to be burned. Guests had to identify which of the five incense were the same. The Kado ceremony masters developed symbols for the different possibilities, so-called genji-mon. Each such symbol consists of five vertical bars, some of which are connected by horizontal bars. Fifty-two symbols were created, and for easier memorization, each symbol was identified with one of the fifty-four chapters of the famous *Tale of Genji* by Lady Murasaki. In time, these genji-mon and two additional symbols started to be displayed at the beginning of each chapter of the *Tale of Genji* and in turn became part of numerous Japanese paintings. They continued to be popular symbols for family crests and Japanese kimono patterns in the early 20th century and can be found on T-shirts sold today.

Until the late 1960s, individual research papers on various aspects of set partitions appeared, but there was no focused research interest. This changed after the 1970s, when several groups of authors developed new research directions. They studied set partitions under certain set of conditions, and not only enumerated the total number of these objects, but also certain of their characteristics. Other focuses were the relation between the algebra and set partitions such as the noncrossing set partitions, the appearance of set partitions in physics where the number of set partitions is a good language to find explicit formula for normal ordering form of an expression of boson operators, and the study of random set partition to obtain asymptotics results.

We wrote this book to provide a comprehensive resource for anybody interested in this new area of research. It will combine the following in one place:

- provide a self-contained, broadly accessible introduction to research in this area,

- present an overview of the history of research on enumeration of and pattern avoidance in set partitions,

- present several links between set partitions and other areas of mathematics, and in normal ordering form of expressions of boson operators in physics,

- describe a variety of tools and approaches that are also useful to other areas of enumerative combinatorics.

- suggest open questions for further research, and

- provide a comprehensive and extensive bibliography.

Our book is based on my own research and on that of my collaborators and other researchers in the field. We present these results with consistent notation and have modified some proofs to relate to other results in the book. As a general rule, theorems listed without specific references give results from articles by the author and their collaborators, while results from other authors are given with specific references. Many results of my own, with or without collaborators, articles are omitted, so we refer the reader to all these articles to find the full details in the subject.

## Audience

The book is intended primarily for advanced undergraduate and graduate students in discrete mathematics with a focus on set partitions. Additionally, the book serves as a one-stop reference for a bibliography of research activities on the subject, known results, and research directions for any researcher who is interested to study this topic. The main chapter of the book is based on the research papers of the author and his coauthors.

## Outline

In Chapter 1 we present a historical perspective of the research on set partitions and give very basic definitions and early results. This is followed by an overview of the major themes of the book: Dobiński's formula and representations of set partitions, subword statistics on set partitions, nonsubword statistics on set partitions, pattern avoidance in set partitions, multi-pattern avoidance in set partitions, asymptotics and random set partitions,

Gray codes, loopless algorithm and set partitions, and normal ordering and set partitions. We also provide a time line for the articles in these major areas of research.

In Chapter 2 we introduce techniques to solve recurrence relations, which arise naturally when counting set partitions. This chapter contains several basic examples to illustrate out techniques on solving recurrence relations. We also provide the basic definitions and basic combinatorial techniques that are used throughout the book such as Dyck paths, Motzkin paths, generating trees, Lagrange inversion formula, and the principle of inclusion and exclusion.

In Chapter 3 we focussed on basic steps for set partitions. We start by presenting Dobiński's formula and its extensions. Then we provide different representations of set partitions, where some of them are nonattractive representations and other are attractive representations in the set partitions research. For the nonattractive representations, we provide the full results on our area. But for the attractive representations we only give very basic definitions and results, leaving the full discussions to the rest of the book chapters.

Chapter 4 deals with statistics on set partitions, where we use the term *statistic* exclusively for subword statistics (counting occurrences of a subword pattern), for instance, the number times two adjacent terms in a set partition are the same. We connect early research on the rises, levels, and falls of statistics with the enumeration of subword patterns. Also we consider subword pattern of length three and general longer subword patterns.

Chapter 5 completes the study of the previous chapter by dealing with statistics on set partitions, where our statistics have not been introduced "naturally" by subword patterns. Here, we focus on different types of statistics, where the research on set partitions has taken interest in them. For instance, we deal with major index, inversions, records, weak records, sum of positions of records on set partitions, crossings, nestings, alignments, etc. We also consider the case of generalized patterns in set partitions.

In Chapter 6 we present results on pattern avoidance for subsequence, generalized patterns and partially ordered patterns. We provide a complete classification with respect to Wilf-equivalence for subsequence patterns with sizes three to seven. We also provide exact formulas for patterns of sizes three and four, and some formulas for other patterns. Our basic tools is based on pattern-avoidance fillings of diagrams, where it has received a lot of attention, for example, in the research articles of Backelin, Krattenthaler, de Mier, Rubey West, and Xin. We conclude the chapter with results on avoidance of generalized patterns and partially ordered patterns.

Chapter 7 deals with multi-restrictions on set partitions. Indeed, this chapter complements the previous chapter by providing a complete classification with respect to Wilf-equivalence for two (or more) subsequence patterns with sizes four, and for two subsequence patterns with one of size three and another for sizes up to 20. Noncrossing set partitions also have received a lot of attention in this chapter, where we provide a full classification with respect to Wilf-equivalence for two subsequence patterns $(1212, \tau)$ with patterns $\tau$ of sizes four and five. We conclude the chapter with links between set partitions and other combinatorial sequences such as left Moztkin numbers, sequence A054391, generalized Catalan numbers, Catalan numbers, and Pell numbers. At the end of this chapter we present other types of multi-restrictions that have not been created from subsequence/generalized patterns such as $d$-regular set partitions, distances and singletons and set partitions, and block-connected set partitions.

In Chapter 8 we focus on random set partitions, after first presenting tools from probability theory. Then, we describe asymptotic results for various statistics on random set partitions. Finally, we derive asymptotic results using tools from complex analysis.

In Chapter 9 we focus on generating set partitions to provide a Gray code algorithm and generate them looplesly. After presenting tools for generating algorithms, Gray codes, and loopless algorithms, we provide our generating Gray code and then a loopless algorithm for set partitions.

In Chapter 10 we provide several results that connect the set partitions and the problem of normal ordering of boson expressions. After we present a modern review of normal ordering, we focus on normal ordering of the operator $(aa^*)^n$. Here, we establish two bijections between the set of contractions of this operator and the set of set partitions of $[n]$. Also, we provide an extension of Wick's theorem to the case of $q$-normal ordering. Again, the noncrossing set partitions have take a part in our research, where the language of noncrossing set partitions have been used.

The chapters build upon each other, with exception of the last three chapters, which can be read (almost) independently of previous chapters. Most chapters start with a section describing the history of the particular topic and its relations to other pervious chapters. New methods and definitions are illustrated with worked examples, and (sometimes) we provide Maple code. At the end of each chapter we present a list of exercises that correspond to the topics in the chapter, where some of them are related to published research papers. Then we provide research directions that extend the results given in the chapter. Questions posed in the research directions can be used for projects, thesis topics, or research.

The appendix gives basic background from different area of mathematics. We also provide a number of C++ programs that are used in the book (classification of pattern avoidance according to Wilf-equivalence) and a number of tables that containing the output of such programs. In addition, we present a detailed description of the relevant Maple functions.

## Support Features

We have included hints and answers to exercises in Appendix A. In addition, the C++ programs and their outputs can be downloaded from the author's web page http://math.haifa.ac.il/manbook.html. We will regularly update the author's web page to post references to new papers in this research area as well as any corrections that might be necessary. (No matter how hard we try, we know that there will be typos.) We invite readers to give us feedback at *tmansour@univ.haifa.ac.il*.

# *Acknowledgment*

Every book has a story of how it came into being and the people that supported the author(s) along the way. Ours is no exception. The idea for this book came when Mark Shattuck emailed the author asking some enumeration problems on set partitions, after collaborating via the Internet on a few papers. Toufik had been working on several combinatorial problems on set partitions and had a background in pattern avoidance in different combinatorial families such as permutations, words, and compositions. Combining these lines of inquiry, he started to work on pattern avoidance in set partitions.

Several years later, after a mostly long-distance collaboration that has made full use of modern electronic technology (e-mail and Skype[1]), this book has taken its final shape. Along the way we have received encouragement for our endeavor from William Y.C. Chen, Mark Dukes, Vít Jelínek, Silvia Heubach, Arnold Knopfmacher, Augustine Munagi, Matthias Schork, Simone Severini, Armend Sh. Shabani, Mark Shattuck, and Chunwei Song. A special thanks to Armend Sh. Shabani for reading the previous version of this book.

And then there are the people in our lives who supported us on a daily basis by giving us the time and the space to write this book. Their moral support has been very important. Toufik thanks his wife Ronit for her support and understanding when the work on the book took him away from spending time with her and from playing with their children Itar, Atil, and Hadil. Also, Toufik thanks our larger families for cheering us on and supporting us even though they do not understand the mathematics that we have described.

---

[1] Skype$^{TM}$ is a registered trademark of Skype Limited.

# *Author Biographies*

**Toufik Mansour** obtained his Ph.D. degree in mathematics from the University of Haifa in 2001. He spent one year as a postdoctoral researcher at the University of Bordeaux (France) supported by a Bourse Chateaubriand scholarship, and a second year at the Chalmers Institute in Gothenburg (Sweden) supported by a European Research Training Network grant. He has also received a prestigious MAOF grant from the Israeli Council for Education. Toufik has been a permanent faculty member at the University of Haifa since 2003 and was promoted to associate professor in 2008. He spends his summers as a visitor at institutions around the globe, for example, at the Center for Combinatorics at Nankai University (China) where he was a faculty member from 2004 to 2007, and at The John Knopfmacher Center for Applicable Analysis and Number Theory, University of the Witwatersrand (South Africa).

Toufik's area of specialty is enumerative combinatorics, and more generally, discrete mathematics and its applications to physics, biology, and chemistry. Originally focusing on pattern avoidance in permutations, he has extended his interest to colored permutations, set partitions, words, and compositions. Toufik has authored or co-authored more than 200 papers in this area, many of them concerning the enumeration of set partitions. He has given talks at national and international conferences and is very active as a reviewer for several journals, including *Advances in Applied Mathematics*, *Ars Combinatorica*, *Discrete Mathematics*, *Discrete Applied Mathematics*, *European Journal of Combinatorics*, *Journal of Integer Sequences*, *Journal of Combinatorial Theory Series A*, *Annals of Combinatorics*, and the *Electronic Journal of Combinatorics*.

# Chapter 1

# Introduction

## 1.1 Historical Overview and Earliest Results

The first known application of set partitions arose in the context of tea ceremonies and incense games in Japanese upper-class society around A.D. 1500. Guests at a Kado ceremony would be smelling cups with burned incense with the goal to either identify the incense or to identify which cups contained identical incense. There are many variations of the game, even today. One particular game is named *genji-ko*, and it is the one that originated the interest in $n$-set partitions. Five different incense sticks were cut into five pieces, each piece put into a separate bag, and then five of these bags were chosen to be burned. Guests had to identify which of the five were the same. The Kado ceremony masters developed symbols for the different possibilities, so-called *genji-mon*. Each such symbol consists of vertical bars, some of which are connected by horizontal bars. For example, the symbol ||⊓ indicates that incense 1, 2, and 3 are the same, while incense 4 and 5 are different from the first three and also from each other (recall that the Japanese write from right to left). Fifty-two symbols were created, and for easier memorization, each symbol was identified with one of the chapters of the famous *Tale of Genji* by Lady

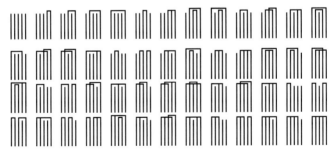

**FIGURE 1.1**: Diagrams used to represent set partitions in 16th century Japan

Murasaki. Figure 1.1 shows the diagrams[1] used in the tea ceremony game. In

---
[1] www.viewingjapaneseprints.net/texts/topictexts/artist_varia_topics/genjimon7.html

time, these genji-mon and two additional symbols started to be displayed at the beginning of each chapter of the *Tale of Genji* and in turn became part of numerous Japanese paintings. They continued to be popular symbols for family crests and Japanese kimono patterns in the early 20th century, and can be found on T-shirts sold today.

How does the tea ceremony game relate to set partitions? Before making the connection, let us define what we mean by a set partition in general, and by a set partition in particular.

**Definition 1.1** *A set partition $\pi$ of a set $S$ is a collection $B_1, B_2, \ldots, B_k$ of nonempty disjoint subsets of $S$ such that $\cup_{i=1}^k B_i = S$. The elements of a set partition are called* blocks, *and the size of a block $B$ is given by $|B|$ the number of elements in $B$. We assume that $B_1, B_2, \ldots, B_k$ are listed in increasing order of their minimal elements, that is, $\min B_1 < \min B_2 < \cdots < \min B_k$. The set of all set partitions of $S$ is denoted by $\mathcal{P}(S)$.*

Note that an equivalent way of representing a set partition is to order the blocks by their maximal element, that is, $\max B_1 < \max B_2 < \cdots < \max B_k$. Unless otherwise noted, we will use the ordering according to the minimal element of the blocks.

**Example 1.2** *The set partitions of the set $\{1, 3, 5\}$ are given by*

$$\{1,3,5\};\ \{1,3\},\{5\};\ \{1,5\},\{3\};\ \{1\},\{3,5\}\ and\ \{1\},\{3\},\{5\}.$$

There is another representation of a set partition, which arises from considering them as words that satisfy certain set of conditions.

**Definition 1.3** *Let $\pi$ be any set partition of the set $[n] = \{1, 2, \ldots, n\}$. We represent $\pi$ in either* sequential *or* canonical *form. In the sequential form, each block is represented as sequence of increasing numbers and different blocks are separated by the symbol $/$. In the canonical representation, we indicate for each integer the block in which it occurs, that is, $\pi = \pi_1 \pi_2 \cdots \pi_n$ such that $j \in B_{\pi_j}$, $1 \leq j \leq n$. We denote the set of all set partitions of $[n]$ by $\mathcal{P}_n = \mathcal{P}([n])$, and the number of all set partitions of $[n]$ by $p_n = |\mathcal{P}_n|$, with $p_0 = 1$ (as there is only one set partition of the empty set). Also, we denote the set of all set partitions of $[n]$ with exactly $k$ blocks by $\mathcal{P}_{n,k}$.*

**Example 1.4** *The set partitions of $[3]$ in sequential form are $1/2/3$, $1/23$, $12/3$, $13/2$, and $123$, while the set partitions of $[3]$ in canonical representation are $123$, $122$, $112$, $121$, and $111$, respectively. Thus, $p_3 = 5$.*

**Example 1.5** *The set partition $14/257/3/6$ has canonical form $1231242$. We have that $\pi_1 = \pi_4 = 1$, as both $1$ and $4$ are in the first block. Likewise, $\pi_2 = \pi_5 = \pi_7 = 2$, as $2$, $5$, and $7$ are in the second block.*

The two representations can easily be distinguished due to the vertical bars, except in the single case when all elements of the set $[n]$ are in a single block. In this case, $\pi = 12345\cdots n$, and its corresponding canonical form is $11\cdots 1$. On the other hand, the set partition $12345\cdots n$ in canonical form represents the partition $1/2/\cdots/n$ in sequential form. The canonical representations can be formulated in terms of words under certain conditions. At first, we explain what we mean by the concept of a word, and then we characterize what word presents a canonical representation of set partition.

**Definition 1.6** *Let $A$ be a (totally ordered) alphabet on $k$ letters. A word $w$ of size $n$ on the alphabet $A$ is an element of $A^n$ and is also called word of size $n$ on alphabet $A$. In the case $A = [k]$, an element of $A^n$ is called $k$-ary word of size $n$. Words with letters from the set $\{0,1\}$ are called binary words or binary strings, and words with letters from the set $\{0,1,2\}$ are called ternary words or ternary strings.*

**Example 1.7** *The 2-ary words of size three are 111, 112, 121, 122, 211, 212, 221, and 222, the binary strings of size two are given by 00, 01, 10, and 11, while the ternary strings of size two are given by 00, 01, 02, 10, 11, 12, 20, 21, and 22.*

As we have shown, any set partition of $[n]$ can be given by its canonical representation that, which is a word, under certain conditions can be formulated by the following fact.

**Fact 1.8** *A (canonical representation of a) set partition $\pi = \pi_1\pi_2\cdots\pi_n$ of $[n]$ is a word $\pi$ such that $\pi_1 = 1$, and the first occurrence of the letter $i \geq 1$ precedes that of $j$ if $i < j$.*

Now we can make the connection between genji-ko and set partitions: each of the possible incense selections corresponds to a set partition of $[5]$, where the partition is according to flavor of the incense. Thus, ||𝍷 could be written as the set partition $123/4/5$ of $[5]$. More details about the connections of genji-ko to the history of Japanese mathematics can be found in two article by Tamaki Yano [370, 371] (in Japanese).

Takakazu Seki

According to Knuth [196], a systematic investigation of the mathematical question, namely finding the number of set partitions of $[n]$ for any $n$, was first undertaken by Takakazu Seki and his students in the early 1700s. Takakazu Seki was born into a samurai warrior family, but was adopted at an early age by a noble family named Seki Gorozayemon whose name he carried.

Seki[2,3], who was an infant prodigy in mathematics and was self-educated, became known as "The Arithmetical Sage" (a term which is carved on his tombstone) and soon had many pupils. One of his pupils, Yoshisuke Matsunaga found a basic recurrence relation for the number of set partitions of $[n]$, as well as a formula for the number of set partitions of $[n]$ with exactly $k$ blocks of sizes $n_1, n_2, \ldots, n_k$ with $n_1 + \cdots + n_k = n$.

**Theorem 1.9** *Let $p_n$ be the number of set partitions of $[n]$. Then $p_n$ satisfies the recurrence relation*
$$p_n = \sum_{j=0}^{n-1} \binom{n-1}{j} p_j$$
*with initial condition $p_0 = 1$.*

**Proof** Assume that the first block contains $j+1$ elements from the set $[n]$, where $0 \leq j \leq n-1$. Since the first block contains the minimal element of the set, namely 1, we need to choose $j$ elements from the set $\{2, 3, \ldots, n\}$ to complete the first block. Thus, the number of set partitions of $[n]$ with exactly $j+1$ elements in the first block is given by $\binom{n-1}{j} p_{n-1-j}$. Summing over all possible values of $j$, we obtain
$$p_n = \sum_{j=0}^{n-1} \binom{n-1}{j} p_{n-1-j} = \sum_{j=0}^{n-1} \binom{n-1}{n-1-j} p_{n-1-j} = \sum_{j=0}^{n-1} \binom{n-1}{j} p_j,$$
with $p_0 = 1$. □

**Theorem 1.10** *The number of set partitions of $[n]$ with exactly $k$ blocks of sizes $n_1, n_2, \ldots, n_k$ with $n_1 + n_2 + \cdots + n_k = n$ is given by*
$$\prod_{j=1}^{k} \binom{n - 1 - n_1 - \cdots - n_{j-1}}{n_j - 1}.$$

**Proof** The proof is similar to the one for Theorem 1.9. For the first block, we choose $n_1 - 1$ elements from the set $\{2, 3, \ldots, n\}$. From the $n - n_1$ available elements, we place the minimal element into the second block and then choose $n_2 - 1$ elements from the $n - n_1 - 1$ remaining elements, and so on, until we have placed all elements. Thus, the number of set partitions of $[n]$ with exactly $k$ blocks of sizes $n_1, n_2, \ldots, n_k$ with $n_1 + n_2 + \cdots + n_k = n$ is given by
$$\binom{n-1}{n_1 - 1}\binom{n-1-n_1}{n_2-1}\cdots\binom{n-1-n_1-\cdots-n_{s-1}}{n_s-1},$$
which completes the proof. □

---

[2] http://www.gap-system.org/~history/Biographies/Seki.html
[3] http://apprendre-math.info/anglais/historyDetail.htm?id=Seki

A more generalized formula for the number of set partitions of $[n]$ into $k_j$ blocks of sizes $n_1, n_2, \ldots, n_m$ with $k_1 n_1 + \cdots + k_m n_m = n$ can be obtained directly from Theorem 1.10 (see Exercise 1.5). These results were not published by Matsunaga himself, but were mentioned (with proper credit given) in Yoriyuki Arima's book *Shūki Sanpō*,[4] which was published in 1769. One of the questions posed in this text was to find the value of $n$ for which the number of set partitions of $[n]$ is equal to $678,570$ (the answer is $n = 11$).

Additional results were derived by Masanobu Saka in 1782 in his work *Sanpō-Gakkai*. Saka established a recurrence for the number of set partitions of $[n]$ into $k$ subsets, and using this recurrence, he computed the values for $n \leq 11$.

**Definition 1.11** *The number of set partitions of $[n]$ into $k$ blocks is denoted by $\left\{ {n \atop k} \right\}$ or $\mathrm{Stir}(n,k)$. The values $\mathrm{Stir}(n,k)$ are called Stirling numbers of the second kind (Sequence A008277 in [327]).*

**Theorem 1.12** *The number of set partitions of $[n]$ into $k$ blocks satisfies the recurrence*
$$\mathrm{Stir}(n+1, k) = k\,\mathrm{Stir}(n, k) + \mathrm{Stir}(n, k-1),$$
*with $\mathrm{Stir}(1,1) = 1$, $\mathrm{Stir}(n,0) = 0$ for $n \geq 1$, and $\mathrm{Stir}(n,k) = 0$ for $n < k$.*

**Proof** For any set partition of $[n+1]$ into $k$ blocks, there are two possibilities: either $n+1$ forms a single block, or the block containing $n+1$ has more than one element. In the first case, there are $\mathrm{Stir}(n, k-1)$ such set partitions, while in the second case, the element $n+1$ can be placed into one of the $k$ blocks of a set partition of $[n]$ into $k$ blocks, that is, there are $k\,\mathrm{Stir}(n,k)$ such set partitions. □

**Remark 1.13** *Note that by definition,*
$$p_n = \sum_{k=1}^{n} \mathrm{Stir}(n, k).$$

*The sequence $\{p_n\}_{n \geq 0}$ is also known as the* Bell numbers[5] *and denoted by $\{\mathrm{Bell}_n\}_{n \geq 0}$ (see sequence A000110 in [327] and [123]).*

Saka was not the first one to discover these sequences. James Stirling,[6] on the other side of the globe in England, had found these sequences in a purely algebraic setting in his book *Methodus Differentialis* [345] in 1730. Stirling's interest was in speeding up convergence of series, and the values $\mathrm{Stir}(n,k)$ arise as coefficients when expressing monomials in terms of falling polynomials.

---

[4] http://en.wikipedia.org/wiki/Yoriyuki_Arima
[5] In honor of famous mathematician Eric Temple Bell (1883–1960), for example see http://en.wikipedia.org/wiki/Bell_number
[6] http://www-history.mcs.st-andrews.ac.uk/Biographies/Stirling.html

**Definition 1.14** *Polynomials of the form*

$$z(z-1)\cdots(z-n+1)$$

*are called* falling polynomials *and are denoted by* $(z)_n$.

**Example 1.15** *The first three monomials can be expressed as follows in terms of falling polynomials as*

$$z = z = (z)_1,$$
$$z^2 = z + z(z-1) = (z)_1 + (z)_2,$$
$$z^3 = z + 3z(z-1) + z(z-1)(z-2) = (z)_1 + 3(z)_2 + (z)_3.$$

The values for the coefficients in the falling polynomials were given as the first table in the introduction of *Methodus Differentialis*, reproduced here as Figure 1.2, where columns correspond to $n$, and rows correspond to $k$.

| 1 | 1 | 1 | 1 | 1 | 1 | 1 | 1 | 1 | 1 | &c. |
|---|---|---|---|---|---|---|---|---|---|---|
|   | 1 | 3 | 7 | 15 | 31 | 63 | 127 | 255 | &c. |   |
|   |   | 1 | 6 | 25 | 90 | 301 | 966 | 3025 | &c. |   |
|   |   |   | 1 | 10 | 65 | 350 | 1701 | 7770 | &c. |   |
|   |   |   |   | 1 | 15 | 140 | 1050 | 6951 | &c. |   |
|   |   |   |   |   | 1 | 21 | 266 | 2646 | &c. |   |
|   |   |   |   |   |   | 1 | 28 | 462 | &c. |   |
|   |   |   |   |   |   |   | 1 | 36 | &c. |   |
|   |   |   |   |   |   |   |   | 1 | &c. |   |
|   |   |   |   |   |   |   |   |   | &c. |   |

**FIGURE 1.2**: Stirling numbers of the second kind

The description given by Stirling on how to compute these values makes it clear that he did not use the recurrence given by Saka given in Theorem 1.12. To read more about how James Stirling used the falling polynomials for series convergence, see the English translation of *Methodus Differentialis* with annotations Tweedle [345]. Despite Stirling's earlier discovery of the values Stir(n, k), Saka receives credit for being the first one to associate a combinatorial meaning to these values, which are now named after James Stirling.

**Theorem 1.16** *For all $n \geq 1$,*

$$z^n = \sum_{k=1}^{n} \text{Stir}(n,k)(z)_k.$$

**Proof** We proceed the proof by induction on $n$. For $n = 1, 2$ we have $\text{Stir}(1,1)(z)_1 = z$ and $\text{Stir}(2,1)(z)_1 + \text{Stir}(2,2)(z)_2 = z + z(z-1) = z^2$. Now assume that the claim holds for $n$, and let us prove it for $n+1$. By Theorem 1.12 and induction hypothesis we have that

$$\begin{aligned}
z^{n+1} = z \sum_{k=1}^{n} \text{Stir}(n,k)(z)_k &= \sum_{k=1}^{n} \text{Stir}(n,k)(z)_k(z-k+k) \\
&= \sum_{k=1}^{n} \text{Stir}(n,k)(z)_{k+1} + \sum_{k=1}^{n} k\text{Stir}(n,k)(z)_k \\
&= \sum_{k=1}^{n+1} \text{Stir}(n,k-1)(z)_k + \sum_{k=1}^{n+1} k\text{Stir}(n,k)(z)_k \\
&= \sum_{k=1}^{n+1} (\text{Stir}(n,k-1) + k\text{Stir}(n,k))(z)_k = \sum_{k=0}^{n} \text{Stir}(n+1,k)(z)_k,
\end{aligned}$$

which completes the induction step. □

**FIGURE 1.3**: Rhyming patterns for 5-line verses.

While set partitions were studied by several Japanese authors and Toshiaki Honda devised algorithms to generate a list of all set partitions of $[n]$, the problem did not receive equal interest in Europe. There were isolated incidences of research, but no systematic study. The first known occurrence of set partitions in Europe also occurred outside of mathematics, in the context of the structure of poetry. In the second book of *The Arte of English Poesie* [282], George Puttenham[7] in 1589 compares the metrical form of verses to arithmetical, geometrical, and musical patterns. Several diagrams, which are in essence the same as the genji-mon, were given on Page 101 of [282]. Figure 1.3 is drawn after Puttenhams illustration for the 5-line verses.

In Puttenham's illustration rhyming lines are connected, just like bars representing incense with the same scent were connected in the genji-mon. For example, the leftmost pattern describes a 5-line verse in which the first, third and fifth line rhyme, and also the second and fourth. Puttenham placed additional conditions on the structure of the verses – each line had to rhyme with at least one other line and the rhyming scheme should not result in two separate smaller verses. For example, a rhyming scheme corresponding to 12/345 would not be allowed. It is not clear whether Puttenham only listed the rhyming schemes that were used in English poetry of the time, or whether he

---

[7]http://en.wikipedia.org/wiki/George_Puttenham

set out to find all possible patterns. He does certainly have preferences for some of the structures over others, as he references Figure 1.3 with the following remark: "The staffe (=stanza) of five hath seven proportions ... whereof some of them be harsher and unpleasaunter to the eare then some be."

The first mathematical investigation of set partitions was conducted by Gottfried Leibniz[8] in the late 1600s. An unpublished manuscript shows that he tried to enumerate the number of ways to write $a^n$ as a product of $k$ factors, which is equivalent to the question of partitioning a set of $n$ elements into $k$ blocks. He enumerated the cases for $n \leq 5$, and unfortunately double-counted the case for $n = 4$ into two blocks of size 2 and the case for $n = 5$ the case into three blocks of sizes one, two, and two. These two mistakes prevented him from seeing that $\text{Stir}(n, 2) = 2^{n-1} - 1$, and also the recursion given in Theorem 1.12. Further details can be found in the commentary by Knobloch [187, Pages 229–233] and the reprint of Leibnitz's original manuscripts [188, Pages 316–321].

The second investigation was made by John Wallis,[9] who asked a more general question in the third chapter of his *Discourse of Combinations, Alternations, and Aliquot Parts* in 1685 (For example, see Jordan [154], Riordan [292], Goldberg et al. [117], and Knuth [194]). He was interested in questions relating to proper divisors (=*aliquot parts*) of numbers in general and integers in particular. The question of finding all the ways to factor an integer is equivalent to finding all partitions of the multiset consisting of the prime factors of the integer (with multiplicities). He devised an algorithm to list all factorizations of a given integer, but did not investigate special cases. For example, factorization of a squarefree integer answers the question of set partitions.

Back to Japan, the modification of (1.9) was given by Masanobu Saka in 1782, where he showed that the number of set partitions of $[n]$ with exactly $k$ blocks is given by $\text{Stir}(n, k)$, the Stirling number of the second kind. After 1782, Bell numbers has received more attention. It seems that the first occurrence in print of Bell number has never been traced, but this number has been attributed to Euler (see Bell [23]), where there is no exact reference verification of this fact. Following Eric Temple Bell, we shall denote them by $\epsilon_n$ and call them exponential numbers [23, 24]. On the other hand, Touchard [342, 343] used the notation $a_n$ to celebrate the birth of his daughter Anne, and later Becker and Riordan [22] used the notation $B_n$ in honor of Bell. Throughout this book we will use the notation $\text{Bell}_n$. Since 1877, an explicit expression for the $n$-th Bell number has an attractive question, where Dobiński [92] gave an explicit formula for the $n$-th Bell number and d'Ocagne [93] studied the generating function for the sequence $\{\text{Bell}_n\}_{n \geq 0}$. In 1902, Anderegg [3] showed

$$2e = \sum_{j \geq 1} \frac{j^2}{j!}, \quad 5e = \sum_{j \geq 1} \frac{j^3}{j!}, \quad 15e = \sum_{j \geq 1} \frac{j^4}{j!},$$

---
[8] http://www-history.mcs.st-andrews.ac.uk/Biographies/Leibniz.html
[9] http://www-history.mcs.st-and.ac.uk/Biographies/Wallis.html

## Introduction

(see Exercise 3.3), and in particular he obtained Dobiński's formula. In the 1930s, Becker and Riordan [22] studied several arithmetic properties of Bell and Stirling numbers, and Bell [23, 24] recovered the Bell numbers. Later, Epstein [104] studied the exponential generating function for the Bell numbers (also, see Williams [359] and Touchard [342, 343]). In the 1960s, Cohn, Even, Menger, Karl, and Hooper [79] presented several basic properties for the Bell and Stirling numbers, and Rota [297] presented the first formal and modern attention to the set partitions. Rota [297] said

*"A great many problems of enumeration can be interpreted as counting the number of partitions of a finite set",*

for instance:

The number of rhyme schemes for $n$ verses. A *rhyme scheme* for $n$ verses is the pattern of rhyme on $n$ lines of a poem or song. It is usually referred to by using letters or numbers to indicate which lines rhyme. In other words, it is the pattern of end rhymes or lines. A rhyme scheme gives the scheme of the rhyme; a regular pattern of rhyming words in a poem (the end words). For instance, the rhyme scheme 1121 for 4 verses indicates four-line stanza in which the first, second and fourth lines rhyme. Here it is an example of this rhyme scheme:

| | |
|---|---|
| Combinatorics of set partitions | 1 |
| Combinatorics of compositions | 1 |
| Combinatorics of words | 2 |
| Combinatorics of permutations | 1 |

Actually, rhyme scheme $\pi = \pi_1\pi_2\cdots\pi_n$ for $n$ verses it also represents a set partition. Since $\pi$ satisfies that conditions $\pi_1 = 1$ and $\pi_i \leq 1 + \max_{j \in [i-1]} \pi_j$, we have that $\pi$ is the canonical representation (see Definition 1.3) of a set partition.

The number of ways of distributing $n$ distinct elements into $n$ boxes such that the minimal element of the $i$-th box is less than the minimal element of the $(i+1)$-st box (empty boxes permitted and all the empty boxes are on the left side of the nonempty boxes ). Actually, if we identify the boxes as blocks, then we have a trivial bijection between the such ways of distributing $n$ distinct elements and set partitions of $[n]$. For example, when $n = 3$ we have five such ways, namely $123/\emptyset/\emptyset$, $12/3/\emptyset$, $13/2/\emptyset$, $1/23/\emptyset$, and $1/2/3$.

The number of equivalence relations among $n$ elements (see [292]). Let $B_1/B_2/\cdots/B_k$ be any set partition of $[n]$, we says $a_i$ and $a_j$ in the same equivalence class if and only if $i$ and $j$ are in the same block. Thus, the number of equivalence relations among $n$ elements is given by the $n$-th Bell number. For example, if we have three elements, $a, b, c$, then we have the following five possibilities: $a, b, c$ are in the same class, or $a, b$ in one class and $c$ another class, or $a, c$ in one class and $b$ another class, or $a$ in one class and $b, c$ another class, or $a, b, c$ in different classes.

The number of decompositions of an integer into coprime ($a, b$ are said to be coprime if there is no $k > 1$ dividing both $a$ and $b$) factors when $n$ distinct primes are concerned (see Bell [23, 24]). To see this let $n = a_1 a_2 \cdots a_n$ be any integer such that $a_1, \ldots, a_n$ are $n$ distinct prime numbers. Then each decomposition of $n$ into coprime factors says $n = n_1 n_2 \cdots n_k$ can be coded as a set partitions by $B_1/\cdots/B_k$, where the element $j$ in the block $B_i$, if and only if $a_j$ divides $n_i$. Clearly, this defines bijection between the such decompositions and set partitions of $[n]$. For instance, if $n = 2 \cdot 3 \cdot 5$, then the decompositions of $n$ into coprime factors are $n = (2)(3)(5)$, $n = (2)(15)$, $n = (6)(5)$, $n = (10)(3)$, and $n = (30)$.

For other examples see Binet and Szekeres [340] and Riordan [292]. It seems to us that the first published explicit formula for the $n$-Bell number was returned to 1877, where Dobiński [92] showed Theorem 3.2. Later, in 1902, Anderegg [3] recovered once more the same formula (actually he generalized it; see Exercise 3.3).

## 1.2 Timeline of Research for Set Partitions

True, the study of set partitions was begun in 1782 by Saka, but it wasn't until almost 100 years later when people started to focus on this combinatorial object. From 1877 until 1974 the research was focused on finding explicit formulas for the Bell and Stirling numbers and some arithmetic properties of these numbers. We allocate this research in Table 1.1 which presents the years research was first research undertaken on set partitions: The first steps in finding recurrence relations, explicit formulas and generating functions were investigated in [3, 21, 22, 23, 24, 79, 80, 81, 82, 92, 93, 193, 288, 292, 297, 298, 345, 342, 343]. Then the researchers were interested on properties of the Bell number and Stirling numbers (see [62, 104, 359]) and on extensions, generalizations and behavior of these numbers (see [19, 26, 66, 67, 222, 249]).

Table 1.1: Timeline of first research steps on set partitions

| Year | References |
|------|------------|
| 1730 | Stirling [345] |
| 1782 | Saka |
| 1877 | Dobiński [92]; d'Ocagne [93] |
| 1902 | Anderegg [3] |

| 1933 | Touchard [342] |
|------|----------------|
| 1934 | Becker and Riordan [22]; Bell [23, 24] |
| 1939 | Epstein [104] |
| 1945 | Williams [359] |
| 1952 | Becker [21] |
| 1956 | Touchard [343] |
| 1958 | Riordan [292] |
| 1962 | Cohn, Even, Menger, Karl, and Hooper [79]; Rényi [288] |
| 1964 | Rota [297, 298] |
| 1968 | Comtet [80] |
| 1970 | Baróti [19] |
| 1972 | Comtet [81] |
| 1973 | Knuth [193]; Menon [249] |
| 1974 | Comtet [82]; Bender [26] |
| 1976 | Carlitz [66, 67] |
| 1995 | Canfield [62] |
| 2011 | Mansour, Munagi, and Shattuck [222]; Mansour and Shattuck [234] |

As in all branches of science, the first research steps were very critical to development of the field, leading to the huge number of researches and hundreds of research papers behind this book. After the research cited in Table 1.1, the interest on set partitions has been devoted to wide range of topics: set partitions satisfy certain set of conditions, counting set partitions according to a statistic, asymptotic enumerations, pattern avoidance, generating algorithms, and application in science in general and in physics in particular. Since it is hard to present one time line table that contains all our data research, we present several tables such that each of them focuses on particular points of the research on set partitions.

The first interest shown in set partitions was the counting of set partitions of $[n]$ according to one (or more) parameter/statistic (see definitions in Chapters 4 and 5) as summarized in Table 1.2. Actually, we can look at Saka's result on counting set partitions according to the number of blocks, whereas in 1782 he studied the number of set partitions with exactly $k$ blocks. More precisely, he showed that the number of set partitions of $\mathcal{P}_{n,k}$ is given by Stir(n, k). It took several decades until the research really became active and rich in statistics on set partitions, but we can see that the study of statistics on set partitions actually began in 1982 with Milne's paper[252], who seems to have pioneered the study of set partitions statistics with distribution given by the $q$-Stirling numbers. It took 200 years until Milne published his paper, and then followed several papers on set partition statistics. These statistics can be formulated either on set partitions that are given by the block representation (see Definition 1.1) or by the canonical representation (see Definition 1.3). Long after Saka's paper, Gould [118] gave combinatorial definitions of the $q$-

Stirling numbers of the first and second kind, where $n$ and $k$ are nonnegative integers. When $q = 1$, these numbers reduce to the ordinary Stirling numbers.

Table 1.2: Timeline research for statistics on set partitions

| Year | References |
|---|---|
| 1782 | Saka |
| 1890 | Cayley |
| 1961 | Gould [118] |
| 1968 | Comtet [80] |
| 1970 | Kreweras [203] |
| 1974 | Comtet [82] |
| 1982 | Milne [252]; Gessel [113]; Devitt and Jackson [91] |
| 1986 | Garsia and Remmel [111] |
| 1991 | Wachs and White [349]; Sagan [305] |
| 1994 | White [354]; Simion [322] |
| 1996 | Johnson [150, 151] |
| 1997 | Biane [35] |
| 1999 | Bóna [47]; Constantine [83] |
| 2000 | Klazar [178, 179] |
| 2003 | Deodhar and Srinivasan [89] |
| 2004 | Sapounakis and Tsikouras [311]; Wagner [350] |
| 2005 | Shattuck [319] |
| 2006 | Kasraoui and Zeng [159] |
| 2007 | Mansour and Munagi [218]; Yano and Yoshida [369] |
| 2008 | Goyt [120]; Chen, Gessel, Yan, and Yang [71]; Ishikawa, Kasraoui, and Zeng [142, 143]; Chen, Gessel, Yan, and Yang [71] |
| 2009 | Mansour and Munagi [220]; Goyt and Sagan [122]; Kasraoui and Zeng [160]; Poznanović and Yan [277]; Mansour and Shattuck [238] |
| 2010 | Mansour and Shattuck [230]; Mansour, Shattuck, and Yan [242]; Knopfmacher, Mansour, and Wagner [190]; Mansour, Shattuck, and Wagner [241]; Mansour and Munagi [221]; Shattuck [320]; Sun and Wu [338] |
| 2011 | Josuat-Vergés and Rubey [155]; Kasraoui, Stanton, and Zeng [158]; Mansour, Munagi, and Shattuck [222]; Sun and Wu [337]; Mansour and Shattuck [236] |
| 2012 | Mansour and Shattuck [233, 239, 235] |

After that several $q$-Stirling numbers and $q$-Bell numbers, that is, several statistics on set partitions have been defined and studied (see [35, 80, 91, 113, 118, 120, 150, 151, 203, 252, 305, 319, 349, 354, 350]). Let us give some examples we will present in detail in the next chapters: Milne [252] showed that the $q$-analogue of Stirling numbers of the second kind of Gould is the distribution generating function for a set partition analogue of the inv statistic (see also [89]). Then Sagan [305] (see also [71, 120, 122]) introduced a set partition analogue of the maj statistic, where he showed that this statistic has $q$-Stirling numbers as its distribution generating function and that the joint distribution of the maj and the inv statistics is exactly the $p, q$-stirling numbers of Wachs and White [349, 354] (see also [111]). In 1994, Simion [322] extended the statistics of Wachs and White [349, 354] to the set of noncrossing set partitions (see also [369, 277]) and ordered set partitions by Ishikawa, Kasraoui, and Zeng [142, 143] (see also [160]).

Other statistics that come from permutation patterns have been based to a large extent on the research of set partition statistics. The problem of counting occurrences of a pattern in a set partition has been introduced by Klazar [178, 179]. For instance, Bóna [47] studied the generating function for the number of set partitions of $[n]$ with exactly $k$ crossings (see the definitions in next chapters). Note that the case $k = 1$ has been considered by Cayley in 1890. Mansour, Shattuck, and Yan [242] studied the generating functions for the number of set partitions of $[n]$ according to a subword pattern (see Chapter 4). Also, Kasraoui and Zeng [159] studied the distribution of crossings, nestings, and alignments of two edges in set partitions. The rises, levels, and drops statistics were considered by Mansour and Munagi [218, 220] (see also Yang [368]). As the set of noncrossing set partitions has been considered according to several statistics, so also have other statistics on subsets of set partitions have studied and considered; see for example [190, 238, 230, 241, 222, 236, 233, 239, 235, 311, 320, 338, 337] We allocate these researches in Table 1.2, which presents the time research for statistics on set partitions.

In Chapter 6, we will focus on a very specific type of restrictions, namely, restrictions characterized by pattern avoidance. Permutation patterns or avoiding patterns research becomes an important interest in enumerative combinatorics, as evidenced in the book by Kitaev [172]. Pattern avoidance was first studied for $S_n$ that avoid a subsequence pattern in $S_3$. The first known explicit solution seems to reach back to Hammersley [132], who found the number of permutations in $S_n$ that avoid the subsequence pattern of 321. In [194, Chapter 2.2.1] and [193, Chapter 5.1.4] Knuth shows that for any $\tau \in S_3$, we have $|S_n(\tau)| = \frac{1}{n+1}\binom{2n}{n}$ (see [327, Sequence A000108]). Other researchers considered restricted permutations in the 1970s and early 1980s (see, for example, [295], [299], and [300]) but the first systematic study was not undertaken until 1985, when Simion and Schmidt [325] found the number of permutations in $S_n$ that avoid any subset of subsequence patterns in $S_3$. Burstein [55] extended the study of pattern avoidance in permutations to the study of pattern avoidance in words, where he determined the number of words over an alphabet $[k]$ of length $n$ that would avoid any subset of subsequence patterns in $S_3$. Later, Burstein and Manosur [58] considered subsequence patterns with repeated letters. Recently, subsequence pattern avoidance has been studied for compositions (see [138] and references therein). These results were followed by several works on set partitions that avoid a fixed (subsequence, generalized, ordered partially) pattern. In 1996, Klazar [176] defined new types of restrictions, which led to the studying of pattern avoidance on set partitions as described in Chapter 6. He said two blocks, $X, Y$, of a set partition are *crossing* (respectively, *nesting*) if there are four elements $x_1 < y_1 < x_2 < y_2$ (respectively, $x_1 < y_1 < y_2 < x_2$) such that $x_1, x_2 \in X$ and $y_1, y_2 \in Y$. Then he showed that the number of noncrossing (respectively, nonseting) set partitions in $\mathcal{P}_n$ is given by the $n$-th Catalan number. The definition of crossing and nesting on set partitions has been considered by different meanings in [68, 69] and [146] (see also [9, 57]). Later, Sagan [306] enumerated the number of set partitions of $[n]$ that avoid a pattern of size three. This result was extended by Goyt [120], where he found formulas for the number of set partitions of $[n]$ that avoid a subset of patterns in $\mathcal{P}_3$ (see also [121, 122]). The year 2008 was a very critical year in the research of pattern avoidance on set partitions, where Klazar [185] studied the growth rates of set partitions and Jelínek and Mansour [146] (see also [148, 149, 231]) have classified the patterns of sizes at most seven according to the Wilf-equivalent on set partition. Some specific enumeration on pattern-avoiding set partitions have been considered by several authors. For instance, Bousquet-Mélou and Xin [50] gave a recurrence relation for the number of 3-noncrossing (123123-avoiding) set partitions of $[n]$. For the case of $k$-noncrossing set partitions we refer the reader to [69]. For other examples, see [229, 363, 366, 251]. We allocate these researches in Table 1.3, which presents the time research for pattern avoidance on set partitions.

**Table 1.3**: Timeline research for pattern avoidance on set partitions

| Year | References |
|---|---|
| 1958 | Miksa, Moser, and Wyman [251] |
| 1996 | Klazar [176]; Yang [368] |
| 2005 | Bousquet-Mélou and Xin [50] |
| 2006 | Burstein, Elizalde, and Mansour [57] |
| 2007 | Chen, Deng, Du, Stanley, and Yan [69] |
| 2008 | Goyt [120]; Jelínek and Mansour [146]; Klazar [185] Mansour and Severini [229] |
| 2009 | Goyt and Sagan [122]; Xin and Zhang [363]; Yan [366] |
| 2010 | Goyt and Pudwell [121]; Jelínek and Mansour [148]; Sagan [306] |
| 2011 | Armstrong, Stump, and Thomas [9] Jelínek, Mansour, and Shattuck [149]; Mishna and Yen [254] |
| 2012 | Mansour and Shattuck [231] |

The research on set partitions does not stop only on studying the determination the number of set partitions in $\mathcal{P}_n$ or in $\mathcal{P}_{n,k}$ satisfying a certain set of conditions characterized by permutation patterns, but it extends to general conditions. Thus, in next timeline table we describe the research on set partitions that satisfy a certain set of conditions, where the conditions are not characterized in terms of patterns. For instance, Prodinger [278] asked and determined the number of *d-regular* set partitions in $\mathcal{P}_n$, that is, set partitions in $\mathcal{P}_n$ such that any two distinct elements, $a, b$, in the same block satisfy $|b - a| \geq d$ for given $d$. The study of $d$-regular set partitions followed by several research works via Chen, Deng, and Du [68], Kasraoui [156], Mansour and Mbarieky [217], and Deng, Mansour, and Mbarieky [88]. In another example, we refer the reader to [205], where Erdös-Ko-Rado theorems on set partitions are considered. More precisely, they studied set of set partitions where each two set partitions have $t$ blocks in common. Sun and Wu [337] studied the set partitions without singletons (blocks of size one) and set partitions with fixed largest singleton. For other examples, see [61, 120, 168, 258, 337].

Here, we present several examples of research papers where each links between set partitions that satisfy a certain set of conditions and other combinatorial families such as trees and lattice paths (for definitions, see next chapter). Poupard [276] linked the set of noncrossing set partitions to the set of noncrossing dissection convex polygons. Prodinger [279] found a correspondence between ordered trees and noncrossing partitions, and Dershowitz and Zaks [90] established another bijection between ordered trees with prescribed numbers of nodes and leaves and noncrossing partitions with prescribed numbers of elements and blocks. Němeček and Klazar [261] constructed a bijection between the set of noncrossing set partitions of $[n]$ and the set of words over $\{-1, 0, 1\}$ of size $m - 2$, which have every initial sum nonnegative. In another example, Mansour [216] studied the number of smooth set partitions (a set partition $B_1/B_2 \cdots /B_k$ is said to be *smooth* if $i \in B_s$ implies that $i + 1 \in B_{s-1} \cup B_s \cup B_{s+1}$) and constructed a bijection between the set of smooth set partitions of $\mathcal{P}_n$ and the set of symmetric Dyck paths of size $2n - 1$ with no *peaks at even level*. For other examples, see [74, 76, 77, 155, 166, 177, 208, 232, 260, 269, 367, 369, 372, 374].

The concept of set partitions has been extended to the more general combinatorial family. If we try to copy the problem to the new combinatorial family and study it, then we will find general results. For instance, Bender [26] studied set partitions of multisets. and Lindquist and Sierksma [210] studied partitioning labeled $p$ cards into $m$ line boxes.

With the refinement order, the set of noncrossing set partitions becomes a *lattice* (for definitions, see next chapters), denoted by $\mathcal{NC}(n)$. Via an order-reversing involution, it is proved that NC($n$) is *self-dual*. The study of the lattice $\mathcal{NC}(n)$ has received a lot of attention, for example, see [324, 326]. We review the restricted set partitions research in Table 1.4.

Table 1.4: Timeline research for restricted set partitions

| Year | References |
|---|---|
| 1972 | Poupard [276] |
| 1974 | Bender [26] |
| | Continued on next page |

| Year | References |
|---|---|
| 1981 | Prodinger [278]; Lindquist and Sierksma [210] |
| 1983 | Prodinger [279] |
| 1986 | Dershowitz and Zaks [90] |
| 1991 | Simion and Ullman [326] |
| 1998 | Klazar [177]; Liaw, Yeh, Hwang, and Chang [208] |
| 2000 | Simion [324] |
| 2003 | Klazar [181]; Němeček and Klazar [261]; Panayotopoulos and Sapounakis [269] |
| 2004 | Natarajan [260] |
| 2005 | Chen, Deng, and Du [68] |
| 2007 | Yano and Yoshida [369] |
| 2008 | Chu and Wei [77]; Chen, Wu, and Yan [76]; Goyt [120]; Ku and Renshaw [205]; Munagi [258]; Yeung, Ku, and Yeung [372] |
| 2009 | Kasraoui [156]; Callan [61]; Mansour [216]; Mansour and Mbarieky [217]; Mansour and Munagi [219] |
| 2010 | Chen and Wang [74] |
| 2011 | Deng, Mansour, and Mbarieky [88]; Jakimczuk [144]; Kim [168, 166]; Mansour and Shattuck [232]; Sun and Wu [337]; Yan and Xu [367]; Zhao and Zhong [374] |
| 2012 | Josuat-Vergés and Rubey [155], Mansour and Shattuck [237], Marberg [244] |

In 1928, appear the first results on asymptotics for the number of set partitions appeared, where Knopp obtained only the dominant term in the asymptotic expansion of $Bell_n$. Later, Moser and Wyman [256] used contour integration to obtain the complete asymptotic formula for $Bell_n$ then generalized by Hayman [134]. In 1957, Moser and Wyman [257] presented asymptotic formula for the Stirling number of the second kind (see also [46, 85, 86, 103, 154, 364]). Several research papers have fixed a statistic on set partitions and then studied the asymptotic behavior of the average of these statistics in a random set partition of $[n]$, or have fixed a subset of set partitions of $[n]$ and then studied asymptotically the number of elements of this subset when $n$ grows to infinity. For instance, Bender, Odlyzko, and Richmond [27] studied the number of irreducible set partitions of $[n]$, where a set partition of $[n]$ is said to be *irreducible* if no proper subinterval of $[n]$ is a union of blocks. In [191], the asymptotic behavior of the number of distinct block sizes in a set partition of $[n]$ (see also [356]) is considered. In [190], the mean and the variance of two parameters on set partitions of $[n]$ is asymptotically considered, which are related to the records of words. For other examples, see [84, 157, 272, 308].

Other results on set partitions obtained by means of probability theory in general and random variables in particular. For instance, Port [275] studied the connection between Bell and Stirling numbers and the moments of a Poisson random variable. For a good review paper we refer to [270], where it examines some properties of set partitions with emphasis on a statistical viewpoint. The paper cited 58 references focused on proofs of Dobiński's formula and some variants, equality of Bell numbers with moments of the Poisson distribution, derivation of the 2-variable generating function for Stirling numbers, and how to generate a set partition at random. For other examples, see [253, 271, 274].

We allocate these researches in Table 1.5, which presents the time research for asymptotics on set partitions.

**Table 1.5**: Timeline research for asymptotics on set partitions

| Year | References |
| --- | --- |
| 1928 | Knopp [192] |
| 1950 | Jordan [154] |
| 1955 | Moser and Wyman [256] |
| 1956 | Hayman [134] |
| 1957 | Moser and Wyman [257] |
| 1958 | de Bruijn [85] |
| 1974 | Bleick and Wang [46] |
| 1981 | de Bruijn [86] |
| 1983 | Wilf [356] |
| 1985 | Bender, Odlyzko, and Richmond [27] |
| 1995 | Yakubovich [364] |
| 1997 | Mingo and Nica [253]; Pitman [270]; Pittel [271] |
| 1998 | Port [275] |
| 1999 | Knopfmacher, Odlyzko, Pittel, Richmond, Stark, Szekeres, and Wormald [191] |
| 2000 | Pittel [272] |
| 2001 | Elbert [103] |
| 2006 | Salvy [308] |
| 2010 | Popa [274]; Knopfmacher, Mansour, and Wagner [190] |
| 2011 | Czabarka, Erdős, Johnson, Kupczok, and Székely [84], Kasraoui [157], Mansour and Shattuck [240], Oliver and Prodinger [266] |
| 2012 | Arizmendi [6] |

In 1972, Kreweras [204] showed that noncrossing set partitions form a lattice, ordered by refinement, and obtained several interesting properties of this lattice. After Kreweras's paper, published many research papers on the connection between noncrossing set partitions and algebra in general and *lattices* and *posets* in particular. Note that some of the authors gave algebraic properties for several results on noncrossing set partitions and others extended the set of noncrossing set partitions to noncrossing set partitions in other Weyl groups such us noncrossing set partitions of type $B$.

Here, in our book, the focus will be only on the enumerative combinatorics side of set partitions. Hence, we will not consider the algebraic properties or algebraic applications of the concept set partitions in general and noncrossing set partitions in particular. Indeed, the main reason for that is the huge number of research papers that focus on the connections between the algebra and set partitions.

In the next table, Table 1.6, we present some relevant references that describe these connections.

**Table 1.6**: Timeline research for set partitions and algebra

| Year | References |
|---|---|
| 1972 | Kreweras [204] |
| 1980 | Edelman [97] |
| 1982 | Edelman [98] |
| 1991 | Simion and Ullman [326] |
| 1993 | Montenegro [255] |
| 1994 | Canfield and Harper [64]; Edelman and Simion [99]; Speicher [328]; Bennett, Dempsey, and Sagan [30] |
| 1996 | Nica [262] |
| 1997 | Bożejko, Kümmerer, and Speicher [51]; Nica and Speicher [264]; Reiner [287]; Stanley [334]; Biane [35] |
| 1998 | Athanasiadis [10] |
| 2000 | Bóna and Simion [49]; Oravecz [267]; Simion [323] |
| 2001 | Canfield [63] |
| 2003 | Biane, Goodman, and Nica [36] |
| 2004 | Athanasiadis and Reiner [12]; Halverson and Lewandowski [131] |
| 2005 | Barcucci, Bernini, Ferrari, and Poneti [16]; Krattenthaler [200] |
| 2006 | Bessis and Corran [33]; Thomas [341]; Zoque [375] |
| 2007 | Athanasiadis, Brady, and Watt [11]; Bessis and Reiner [34]; Ehrenborg and Readdy [101]; Hankin and West [133]; McCammond [247]; Rey [289] |
| 2008 | Armstrong and Eu [8]; Reading [284] |
| 2009 | Armstrong [7]; Ingalls and Thomas [141]; Mamede [213]; Nica and Oancea [263]; Savitt [313] |
| 2010 | Fink and Giraldo [106]; Krattenthaler and Müller [202]; Reading [285]; Rubey and Stump [302] |
| 2011 | Armstrong, Stump, and Thomas [9]; Chen and Wang [75]; Kim [167, 169]; Reading [286]; Rhoades [290] |
| 2012 | Marberg [245] |

In 1973, the first papers on generating algorithms for set partitions appeared, where Ehrlich [102] presented a loopless (loop-free) algorithm for generating the set partitions of $[n]$. More precisely, Ehrlich found a loopless algorithm for generating $\mathcal{P}_n$ with the following property: successive partitions differ in one letter with the difference being 1, except if the change is in the largest letter of the set partition, which can change to 1. The algorithm of Ehrlich is similar to Knuth's algorithm. Both algorithms successively append the values $1, 2, \ldots, 1 + \max_j \pi_j$ to set partitions in $\mathcal{P}_{n-1}$, but in a different order.

**Table 1.7**: Timeline research for generating set partitions

| Year | References |
|------|------------|
| 1973 | Ehrlich [102] |
| 1976 | Kaye [165] |
| 1983 | Stam [330] |
| 1984 | Semba [317] |
| 1988 | ER [105] |
| 1990 | Ruskey [303] |
| 1994 | Ruskey and Savage [304] |
| 1997 | Grebinski and Kucherov [125]; Savage [312] |
| 2002 | Orlov [268] |
| 2005 | Knuth [195] |
| 2008 | Mansour and Nassar [223] |
| 2009 | Huemer, Hurtado, Noy, and Omaña-Pulido [140] |
| 2011 | Mansour, Nassar, and Vajnovszki [224] |

Kaye [165] used the block representation for set partitions to describe an algorithm for producing all the set partitions of $[n]$ such that each set partition is obtained from its predecessor by shifting only one element from one block to another (distance 1). Ruskey [303] presented a modification of Knuth's algorithm in which the distance is at most 2. Ruskey and Savage [304] generalized Ehrlich's results to the set of strings of nonnegative integers satisfying $1 \leq a_1 \leq k$ and $a_i \leq 1 + \max\{a_1, a_2, \ldots, a_{i-1}, k-1\}$. Note that the case $k = 1$ gives the set partitions $\mathcal{P}_n$. In particular, they showed two different Gray codes for $\mathcal{P}_n$ [304]. Later, Mansour and Nassar [223] constructed an algorithm that creates a Gray code with distance 1 and a loopless algorithm that generates the set partitions of $[n]$. For other generating algorithms for set partitions, see [105, 195, 268, 312, 317]. Later, generating algorithms for noncrossing set partitions were constructed, for example see [140]. We note these researches in Table 1.7, which presents the time research generating set partitions.

The normal ordering form of a "boson operator" (see Chapter 10) is a well known problem in physics, which has been investigated by many research papers. For the physics motivations and the physics results, we refer the reader to a thesis of Blasiak [38] (and references therein), which provides a comprehensive resource for anybody interested in this area. The connection between normal ordering and set partitions was discovered by Katriel and Kibler [164] from 1974–2002. Indeed, they showed a normal ordering of a boson operator is related to Stirling numbers of the second kind and Bell numbers. Until 2007 it was no combinatorial connection between set partitions and normal ordering problem, then several papers were published by Mansour, Schork and Severini [226, 227, 225, 228]. They established bijections between the set of the contractions of a boson operator and set partitions. They generalized this study to include several types of statistics on set partitions that have or do not have physical meaning on the normal ordering problem of a boson operator. These statistics are covering, the noncrossing normal ordering form, $q$-normal ordering form, $q$-Wick theorem, and $p$-normal ordering form. We allocate these researches in Table 1.8, which presents the timeline for research generating set partitions.

**Table 1.8**: Timeline research for the normal ordering and set partitions

| Year | References |
|---|---|
| 1974 | Katriel [161] |
| 1992 | Katriel and Kibler [164] |
| 2000 | Katriel [162] |
| 2002 | Katriel [163] |
| 2003 | Blasiak, Penson, and Solomon [42, 43]; Schork [315] |
| 2004 | Blasiak, Horzela, Penson, and Solomon [40]; Blasiak, Penson, and Solomon [44] |
| 2005 | Blasiak [38]; Blasiak, Horzela, Penson, Duchamp, and Solomon [39]; Blasiak, Penson, Solomon, Horzela, and Duchamp [45]; Méndez, Blasiak, and Penson [248]; Varvak [346] |
| 2006 | Schork [316] |
| 2007 | Mansour, Schork, and Severini [226, 227]; Blasiak, Horzela, Penson, Solomon, and Duchamp [41] |
| 2008 | Mansour and Schork [225]; Mansour, Schork, and Severini [228] |

## 1.3  A More Detailed Book

In this section we will have a closer look at the major research activities for set partitions area, by presenting relevant definitions and basic examples, as well as an overview of the structure of the book.

### 1.3.1  Basic Results

Our basic results chapter, Chapter 3, will be used as a first step to consider the research on set partitions. The first part of this chapter deals with the Dobiński's formula and its $q$-analog, where we discus the Dobiński's formula that discovered in 1887 and its elegant proof that was given by Rota [297] in 1964. Then it took only 18 years to discover a very beautiful contribution for set partitions, where Milne [252] succeeded in giving a $q$-analog of results of Rota. Indeed, we believe that the formal study of set partition statistics was initiated by Milne in 1982. The second part focuses on various representations of set partitions. As in almost all the combinatorial structures, the representation of an element plays a critical rule in introducing statistics and information on the structure, we define several equivalence representations of set partitions. Each of these representations has its value and its results, some of them being more attractive and receiving more attention than the others.

## 1.3.2 Statistics on Set Partitions

Our next step is to look at the research on set partitions following the contribution of Milne [252]. In Chapters 4 and 5, we consider the enumeration of set partitions according to a fixed statistic.

**Definition 1.17** *A statistic on set partitions $\mathcal{P}_n$ is a function $s : \mathcal{P}_n \to \mathbb{N} \cup \{0\} = \{0, 1, 2, \ldots\}$.*

**Example 1.18** *Up to now we met several statistics on set partitions. For instance, the number of blocks, the number elements in the first block, the number of singletons (blocks of size one), and number of distinct blocks (blocks unequal sizes). Other statistics have also introduced and studied, such as rises, levels, and descents (falls).*

Results on the statistics rise, level, and descent can be generalized by thinking of them as two-letter subword patterns. In order to define what we mean by a pattern we need to introduce the following terminology.

**Definition 1.19** *Let $S$ be any finite subset (or collection of subsets, or sequence) of real number with cardinality $|S| = n$. Then the* reduce form *of $S$ is the unique order-preserving bijection $S \to [n]$. We denote the reduce form of $S$ by* Reduce(S).

**Example 1.20** *The reduced form of the sequence $S = 43676$ is Reduce(S) = 21343, as the terms of the sequence are in order $3 < 4 < 6 < 7$, and therefore, 3 is the smallest element, 4 is the second smallest element, 6 is the third smallest element, and 7 is the fourth smallest element. The reduced form of the collection $\{1,3,8\}/\{2,7\}$ (set partition on $\{1,2,3,7,8\}$) is Reduce(S) = $\{1,3,5\}/\{2,4\}$ and the reduced form of the set partition $\pi = 235/4$ of $\{2,3,4,5\}$ is Reduce($\pi$) = 124/3.*

**Definition 1.21** *A* permutation *of size $n$ is a word over alphabet $[n]$ such that each letter occurs exactly once.*

*A* composition *of size $n$ is a word over alphabet $\mathbb{N} = \{1, 2, \ldots\}$ (positive* integer numbers) such that the sum of its letters is $n$, the number of the letters in the word is called the *length of the composition*.

A sequence (permutation, composition, word or canonical representation of a set partition) $\pi$ *contains a substring* $\sigma$ if $\sigma$ occurs in $\pi$, that is, there exist $\pi'$ and $\pi''$ such that $\pi$ has the form $\pi'\sigma\pi''$. Otherwise, we say that $\pi$ *avoids the substring* $\sigma$ or is *$\sigma$-avoiding*.

A sequence (permutation, composition, word or canonical representation of a set partition) $\pi$ *contains a subword pattern* $\sigma$ if $\pi$ can be written as $\pi = \pi'\pi''\pi'''$ such that $\operatorname{Reduce}(\pi'') = \sigma$. Otherwise, we say that $\pi$ *avoids the subword* $\sigma$ or *$\sigma$-avoiding*.

**Example 1.22** *The sequence* $31432213$ *avoids the substring* $12$, *but has two occurrences of the subword pattern* $12$ *(namely,* $14$ *and* $13$*).*

*The set partition* $\pi = 121322415$ *avoids the substring* $31$, *but has three occurrences of the subword pattern* $21$ *(namely,* $21$, $32$ *and* $41$*).*

Note that for general concept of pattern avoidance, we refer the reader Chapter 6.

For longer sequences, it is often useful to write the sequence in standard short-hand notation as follows.

**Definition 1.23** *The notation* $a^j$ *in a sequence stands for a subsequence of* $j$ *consecutive occurrences of the letter* $a$. *Also, the notation* $S^j$ *(respectively,* $S^*$*) denotes the set of all the words that can be obtained from* $j$ *letters (respectively, a finite number of letters) in the set* $S$.

**Example 1.24** *The sequence* $111222232222$ *can be written as* $1^3 2^4 3 2^4$ *in short-hand notation. The set* $\{0,1\}^n$ *means all the binary words of length* $n$, *and the set* $\{0,1\}^*$ *contains all the binary words of finite size.*

After we defined the notion of a subword pattern, it is clear that the subwords patterns $11$, $12$ and $21$ correspond to level, rise and descent. Mansour and Munagi [218] enumerated set partitions according to subword patterns of length two. This result was generalized by the same authors [220] (for some combinatorial proofs, see Shattuck [320]), where they studied the number of set partitions of $[n]$ according to statistic $m$-levels ($m-1$ consecutive levels), $m$-rises ($m-1$ consecutive rises) and $m$-descents ($m-1$ consecutive descents). Some of their results were further extended to 3-letter patterns by Mansour and Shattuck [230] and to families of $\ell$-letter patterns by Mansour, Shattuck, and Yan [242], as we will discuss in Chapter 4.

The statistic language of subword patterns is not always a good language to formulate our statistics on set partitions. Sometimes, our statistics have complicated characterization in the language of subword patterns, but they have simple characterization in another language. In Chapter 5, we will consider several such statistics, called *nonsubword statistics*.

# Introduction

**Example 1.25** *For instance, let $\pi$ be any canonical representation of a set partition, we define $f_i(\pi)$ to be the difference between the position of the rightmost occurrence of the letter $i$ and the leftmost occurrence of the letter $i$ in $\pi$. Note that if the letter $i$ does not occur in $\pi$ we define $f_i(\pi)$ to be 0. Then we define the total statistic on $\pi$ to be $f(\pi) = \sum_{i \geq 1} f_i(\pi)$. As an example, if $\pi = 121341233212 \in \mathcal{P}_{12,4}$ then $f_1(\pi) = 11 - 1 = 10$, $f_2(\pi) = 12 - 2 = 10$, $f_3(\pi) = 9 - 4 = 5$ and $f_4(\pi) = 5 - 5 = 0$, which gives*

$$f(\pi) = 10 + 10 + 5 + 0 = 25.$$

*This statistic can be defined in different way if we use the block representation, namely, $f(\pi) = \sum_{i \geq 1} \max B_i - \min B_i$ where $B_i$'s are the blocks of the set partition $\pi$. As an example, the block representation of the set partition $\pi = 121341233212 \in \mathcal{P}_{12,4}$ is $\{1,3,6,11\}/\{2,7,9,12\}/\{4,8,9\}/\{5\}$, which leads to $f(\pi) = 11 - 1 + 12 - 2 + 9 - 4 + 5 - 5 = 25.$*

More precisely, in Chapter 5 we will enumerate the number of set partitions of $[n]$ according to several nonsubword statistics. Followed Milne [252] several authors considered such enumerations. For instance, in 2004, Wagner [350] studied the following three nonsubword statistics on set partitions $\pi = B_1/B_2 \cdots /B_k$:

$$\sum_{i=1}^{k} i|B_i|, \quad \sum_{i=1}^{k}(i-1)|B_i| \text{ and } \sum_{i=1}^{k}(i-1)(|B_i|-1).$$

For other examples and deep details we refer the reader to Chapter 5.

## 1.3.3 Pattern Avoidance in Set Partitions

In Chapter 6, we will change our focus from enumeration of set partitions according to the number of times a certain pattern occurs to the avoidance of such patterns. Clearly, our results for pattern avoidance of subword patterns can be obtained from the previous chapters by looking at zero occurrences. In this chapter, we will consider more general patterns, namely subsequence, generalized, and partially ordered patterns. Subsequence patterns are those where the individual parts of the pattern do not have adjacency requirements, while generalized patterns have some adjacency requirements.

**Definition 1.26** *A sequence (permutation, word, composition or set partition) $\pi = \pi_1 \pi_2 \cdots \pi_n$ contains a subsequence pattern $\tau = \tau_1 \tau_2 \cdots \tau_m$ if there exist a m-term subsequence of $\pi$ such that its reduced form equals $\tau$. Otherwise, we say that $\pi$ avoids the subsequence pattern $\tau$ or is $\tau$-avoiding.*

**Example 1.27** *The set partition $1213211445621$ in $\mathcal{P}_{13}$ contains the subsequence pattern $2134$,*

**Definition 1.28** *Let $\tau$ be any pattern. We denote the set of set partitions of $[n]$ (respectively, set partitions of $[n]$ with exactly $k$ blocks) that avoid the pattern $\tau$ by $\mathcal{P}_n(\tau)$ (respectively, $\mathcal{P}_{n,k}(\tau)$).*

Note that the above definition presents only one meaning of avoiding and containing a pattern. Actually, in the literature several meanings are known for avoiding and containing a subsequence pattern in set partitions. For more details, we refer the reader to Chapter 6.

More general patterns are the partially ordered patterns, which are further generalizations in the sense that some of the letters of the pattern are not comparable. Here, we present only an example, leaving the full details to Section 6.4.

**Example 1.29** *An occurrence of the (subword) pattern $\tau = 1'1''2''$ in a set partition $\pi = \pi_1\pi_2\cdots\pi_n \in \mathcal{P}_n$ means that there is a subword $\pi_i\pi_{i+1}\pi_{i+2}$ with $\pi_{i+1} < \pi_{i+2}$ and no restrictions placed on $\pi_i$. For instance, the set partition $12132 \in \mathcal{P}_5$ has one occurrence of the subword pattern $1'1''2''$, namely $213$.*

*An occurrence of the subsequence pattern $\tau = 1'1''2''$ in a set partition $\pi = \pi_1\pi_2\cdots\pi_n \in \mathcal{P}_n$ means that there is a subsequence $\pi_i\pi_j\pi_k$ with $\pi_j < \pi_k$, and no restrictions placed on $\pi_i$. For instance, the set partition $12132 \in \mathcal{P}_5$ has two occurrences of the subsequence pattern $1'1''2''$, namely $113$ and $213$.*

Chapter 6 is divided into three main sections, which are corresponding to three type of patterns: subsequence patterns, generalized patterns, and partially ordered patterns. In Section 6.2, based on result of Jelínek and Mansour [146], we determine all the equivalence classes of patterns of size at most seven, where our classification is largely based on several new infinite families of pairs of equivalent patterns. For example, we prove that there is a bijection between $k$-noncrossing $(12\cdots k12\cdots k$-avoiding) and $k$-nonnesting $(12\cdots kk\cdots 21$-avoiding) set partitions. In Section 6.3 we present several known and new results on generalized 3-letter patterns (those that have the some adjacency requirements). We derive results for permutation and multi-permutation patterns of types $(1,2)$ and $(2,1)$, which are the only generalized patterns not investigated in pervious section. Finally, in Section 6.4, we discuss the results on partially ordered patterns.

The discussion of pattern avoidance on set partitions does not stop on avoiding one pattern, also, it is extended to cover avoiding of two or more patterns in the same time.

**Definition 1.30** *A sequence (permutation, word, composition or set partition) $\pi = \pi_1\pi_2\cdots\pi_n$ avoids a list $T$ of patterns or $T$-avoiding if it avoids each pattern in the list $T$. We denote the set of set partitions of $[n]$ (respectively, set partitions of $[n]$ with exactly $k$ blocks) that avoid $T$ by $\mathcal{P}_n(T)$ (respectively, $\mathcal{P}_{n,k}(T)$.*

In Chapter 7, we extend the equivalence classes of set partitions, avoiding a subsequence pattern of size three and four to determine the equivalence classes of set partitions, avoiding either two patterns of sizes three and $k$ or two patterns of size four. In the case of set partitions of $[n]$ that avoid two patterns such that one of the patterns has size three, we include enumeration results of the generating functions for such number of such set partitions. Moreover, in Chapter 7 we study other restrictions, as we will discuss in our next paragraph.

### 1.3.4 Restricted Set Partitions

In Chapter 7, we will discuss different types of restrictions on set partitions. A first type of restrictions is to make certain conditions on the size of the elements in the same block of a set partition. For example, a set partition $B_1/B_2/\cdots/B_k$ of $[n]$ is said to be *d-regular*, if for any two distinct elements, $a, b$, in the same block, we have $|b-a| \geq d$. The study of $d$-regular set partitions has been receiving a lot of attention, and it seems that the first consideration of $d$-regular set partitions goes back to Prodinger [278] who called them *d-Fibonacci partitions*. Another example is to introduce (see [77]) certain set of conditions on the distances among the elements in each block of set partition of $A$, where $A$ is given by a set of $n$ nonnegative integers. For instance, we consider set partitions of $A$ such that the distance between any two elements, $x_i, x_j$, in the same block satisfies $|i - j| \neq \ell$. A second type of restriction is to avoid a certain set of conditions, where these our conditions may or may not be characterized in terms of pattern avoidance language. In the case when our restrictions are characterized by pattern avoidance, avoiding more than one pattern, we will classify the equivalence classes among pair of patterns of several general types (see [149]). More precisely, we will classify pairs of patterns $(\sigma, \tau)$, where $\sigma \in \mathcal{P}_3$ is a pattern with at least two distinct letters and $\tau \in \mathcal{P}_k(\sigma)$. Then we will provide an upper bound for the number of equivalence classes, and provide an explicit formula for the generating function of all such avoidance classes, showing that in all cases this generating function is rational. Also, we will consider the case of two patterns of size four and we provide an explicit formula for the generating function of all such avoidance classes. We will focus on several subsets of set partitions, where they linked to different combinatorial structures such as Dyck paths and Motzkin paths (for definitions see the next Chapter). Another type of restriction is to introduce conditions on the size of the blocks, for instance to determine the number of set partitions of $[n]$, such that each block has at least two elements, or to determine the number of set partitions of $[n]$, such that the blocks have different sizes. We will study restricted set partitions, as well as other variations of set partitions in Chapter 7.

### 1.3.5 Asymptotics Results on Set Partitions

In the study of statistics on set partitions, sometimes one is not able to get explicit results or nice generating functions. Then the question becomes what can be said about the quantity of interest as $n$ tends to infinity. Tools to obtain such asymptotic results come from probability theory and complex analysis, or a combination of the two.

In order to utilize the probability approach, one considers the set partitions to be random, with each set partition of $[n]$ occurring equally likely. With this approach, several authors have derived interesting asymptotic results. For instance, [27] studied asymptotically the number of irreducible set partitions of $[n]$, and [191] considered the asymptotic behavior of the number of distinct block sizes in a set partition of $[n]$ (see also [356]). Chapter 8 presents an introduction to the tools from complex analysis and probability theory and then presents several asymptotics problems in set partitions.

### 1.3.6 Generating Set Partitions

*Gray codes* are named after Frank Gray who patented the binary reflected Gray code (BRGC) in 1953 for use in pulse code communications. Many authors were interested in combinatorial Gray codes of different *combinatorial classes*, such as a Gray code for set partitions.

**Definition 1.31** *A class of combinatorial objects is a pair $(U, |\cdot|)$ where $U$ is a finite or denumerable set and the mapping $|\cdot| : U \to \mathbb{N}$ is such that the inverse image of any integer is finite.*

For instance, the set of permutations, set of compositions, set of words over finite alphabet, and set of set partitions are combinatorial classes, our focus being combinatorial class set partitions. The area of combinatorial Gray codes was popularized by Wilf in his invited address at the SIAM Conference on Discrete Mathematics in 1988 and his subsequent SIAM monograph [357].

**Definition 1.32** *A Gray code for a combinatorial class of objects is a listing of the objects as words of letters so that the transition from an object to its successor takes only a "small change" or a small number of different letters. The definition of "small change" or small distance, depends on the particular family.*

**Example 1.33** *One can list the set partitions in $\mathcal{P}_3$ as a Gray code of distance 1 as follows:*
$$111,\ 112,\ 121,\ 122\ and\ 123.$$

*Here, we used the canonical representation to represent each set partition.*

Loopless generation has a history which dates back to 1973, when Ehrlich [102] formulated explicit criterions for loop-free concepts. This concept presents an interesting challenge in the field of combinatorial generation. The loopless algorithms must generate each combinatorial object from its predecessor in no more than a constant number of operations. Hence, each object is generated in constant time. This means that powerful programming structures such as recursion and looping cannot be used in the code for generating successive objects, although we always need one loop to generate all objects of some given class.

**Definition 1.34** *An algorithm which generates the objects of a combinatorial family is said to be* loopless *or* loop-free *if it takes no more than a constant amount of time between successive objects. A loopless algorithm which generates a given combinatorial class in a Gray code order is called a loopless Gray code or a loop-free Gray code algorithm.*

In Chapter 9, we give an introduction to Gray codes and loopless algorithm tools and then present several results on generating set partitions in Chapter 9. Generating set partitions has been initiated on 1973 by Ehrlich [102]. From that time, a huge number of research papers on generating combinatorial families has been published and some of them have been focused on our combinatorial class, namely set partitions. For a good survey of combinatorial generation we refer the reader to [312] and [357].

### 1.3.7 Normal Ordering and Set Partitions

Since Katriel's [161] seminal work, the combinatorial aspects of *boson normal ordering* have been receiving a lot of attention, considered intensively, have also been investigated for many meanings, for example, see [39, 42, 43, 45, 44, 110, 162, 163, 227, 228, 248, 250, 315, 346, 361] and the references given therein. We refer the reader to Wilcox [355] for the earlier literature on normal ordering of noncommuting operators. Since Katriel and Kibler's [164] results, the combinatorics of normal ordering have been linked to the set partitions. Mansour, Schork, and Severini [226] defined generalizations of boson normal ordering. These are based on the number of *contractions*, where they shed further light onto the combinatorics of set partitions arising from boson normal ordering problem. Also, they presented a $q$-version of Wick's theorem on normal ordering in [227]. The links between noncrossing set partitions and noncrossing normal ordering of an expression have been considered by Mansour, Schork and Severini in [228]. In Chapter 10 we present a modern review of normal ordering with focusing on the connections between boson normal ordering problem and set partitions.

---

## 1.4 Exercises

**Exercise 1.1** *List the block representations of all the set partitions of* [4]*.*

**Exercise 1.2** *List the canonical representations of members of $\mathcal{P}_4$.*

**Exercise 1.3** *List the block representation of members of $\mathcal{P}(\{1,3,4\})$.*

**Exercise 1.4** *List the block representations of the set partitions of $[5]$ for which each block contains an odd number of elements.*

**Exercise 1.5** *Use Theorem 1.10 to obtain an explicit formula for the number of set partitions of $[n]$ with $n_j$ subsets of size $k_j$, where $k_1 n_1 + \cdots + k_m n_m = n$.*

**Exercise 1.6** *Find an explicit formula for the number of set partitions of $[n]$ such that the first block has exactly either one or two elements.*

**Exercise 1.7** *Puttenham seemed to have been interested primarily in rhyme schemes that satisfied the following three properties:*
*(1) every line should rhyme with at least one other line (completeness);*
*(2) the scheme should not consist of two smaller stanzas (indecomposable); and*
*(3) the poem should not be trivial, that is, should not have all lines rhyme with each other.*
*Calling such a rhyme scheme* desirable, *answer the following questions:*
*(a) Did Puttenham listed all the desirable 5-line rhyme schemes in Figure 1.3? If not, draw the missing one(s).*

*Puttenham also listed rhyme schemes for 6-line verses. He claims: "The sixaine or staffe of sixe hath ten proportions, whereof some be usuall, some not usuall, and not so sweet as one another." Figure 1.4 below gives the schemes as drawn by Puttenham for the 6-line verses.*

**FIGURE 1.4**: Rhyming patterns for 6-line verses.

*(b) Are all the listed schemes desirable? If not, indicate which scheme(s) do not fit.*
*(c) Sixaines can occur in three types: three blocks of size two, two blocks of size three, or one block of size four together with a block of size two. Determine how many schemes there are of each type, and list any schemes missing from Puttenham's list either in sequence notation or draw them.*
*(d) Write a program that produces the desirable partitions of $[n]$, and compute the number of such partitions for $n = 1, 2, \ldots, 7$.*

**Exercise 1.8** *Given the set partition* $\pi = 12/378/49/56$ *of* $[9]$. *Find the canonical form of* $\pi$ *and then find number rises, descents and levels in* $\pi$.

**Exercise 1.9** *Prove that* $\text{Stir}(n,2) = 2^{n-1} - 1$, *for all* $n \geq 2$.

**Exercise 1.10** *Prove that the sequence* $\left\{ \frac{\text{Stir}(n,k+1)}{\text{Stir}(n,k)} \right\}_{k=1}^{n-1}$ *is strictly decreasing.*

**Exercise 1.11** *List canonical representations of all the set partitions of* $[5]$ *that avoid the subword pattern* 11.

**Exercise 1.12** *List canonical representations of all the set partitions of* $[5]$ *that avoid the subsequence pattern* 11.

**Exercise 1.13** *Write a Gray code for* $\mathcal{P}_4$ *with distance one.*

**Exercise 1.14** *Write a Gray code for* $\mathcal{AP}_4(111)$ *with distance one, where* 111 *is a subsequence pattern.*

**Exercise 1.15** *Write a Maple code to produce all the set partitions of* $[5]$ *with exactly three blocks.*

# Chapter 2

## Basic Tools of the Book

Counting elements of a set is the heart of mathematics and embodies the magic of the combinatorics. Usually, we are interested in counting objects that depend on a parameter (or two, or three ...), for example finding the number of set partitions of $[n]$ (here, the parameter is $n$) or finding the number of set partitions of $[n]$ with exactly $k$ blocks (here, the parameters are $n$ and $k$). By exhibiting all possibilities of set partitions for small values of $n$, we find the number of these objects, but as $n$ increases, we need to be smarter about counting. The main goal of this chapter is to provide and to overview of some basic facts and ideas of counting without proofs, where we will illustrate several techniques such as the use of recurrence relations, generating functions, and combinatorial bijections with very basic examples. The proofs of these basic facts and ideas can be founded in any basic book on combinatorics; for example, see the second chapter in [138] to find such proofs and other examples related to words and compositions.

## 2.1 Sequences

The number of combinatorial objects that depend on a parameter $n$ for each value $n$ can be expressed as a sequence.

**Definition 2.1** *A sequence with values in a set $A$ or a sequence in $A$, can be expressed formally as a function $a: I \to A$, where $I$ is a subset of the integers $\mathbb{Z}$, and, for the objects we are interested in, $A \subseteq \mathbb{N}_0$, the set of nonnegative integers. The set $I$ is called the* index set, *and the set $A$ consists of the values of the sequence. The full sequence is usually given as an ordered list. For example, when $I = \mathbb{N}$, then $a = a_1 a_2 \cdots$, or $a = \{a_n\}_{n=1}^{\infty} = \{a_n\}_{n \geq 1}$ for short. If $I = [m]$, then $a$ is called a* finite sequence of size $m$ *or $m$-term sequence. A sequence $\{b_n\}_{n \in J}$ is a* subsequence of the sequence $\{a_n\}_{n \in I}$ *if $J \subseteq I$.*

**Example 2.2** *The sequence $1, 2, 3, 1, 8, 8$ is a 6-term sequence where the first term is 1, the second term is 2, the third term is 3, and so on. The sequence*

$\{3n-2\}_{n\in\mathbb{N}}$ *is a subsequence of the sequence* $\{n\}_{n\in\mathbb{N}}$ *of positive integer numbers.*

When we have two sets with the same cardinality, that is, the same number of objects, then we may ask if there is a one-to-one correspondence between the objects of the first set and the objects of the second set. The one-to-one correspondence idea is important and necessary when we classify our sets according to the cardinalities of the sets (Wilf-equivalence) in the next chapters.

**Definition 2.3** *A function* $f: A \to B$ *is* surjective *or* onto *if for all* $y \in B$ *there is an element* $x \in A$ *such that* $f(x) = y$. *A function* $f: A \to B$ *is* injective *or* one-to-one *if for all* $x, y \in A$, $x \neq y$ *implies that* $f(x) \neq f(y)$. *A function is* bijective *or* one-to-one and onto *if and only if it is both injective and surjective. A* bijection *or a* one-to-one correspondence *is a bijective function. A function* $f: A \to A$ *is* involution *on* $A$ *if* $f(f(a)) = a$ *for all* $a \in A$.

**Example 2.4** *Let* $f: \mathbb{N} \to \mathbb{N}$ *be a function defined by* $f(n) = n^2$ *for all* $n$. *For any* $n, m$ *in* $\mathbb{N}$, $f(n) = f(m)$ *implies that* $n^2 = m^2$ *which is equivalent to* $n = m$. *Thus,* $f$ *is injective. But* $f$ *is not surjective, since there no* $n \in \mathbb{N}$ *such that* $f(n) = n^2 = 2$.

**Example 2.5** *We will use a bijection to establish that the number of set partitions of* $[n]$ *such that the first block contains exactly one element equals the number of set partitions of* $[n-1]$, $n \geq 1$. *Let* $A(n)$ *denote the set partitions of* $[n]$ *such that the first block contains exactly one element. We describe a bijection between* $A(n)$ *and* $\mathcal{P}_{n-1}$ *as follows: For any set partition* $\pi = \pi_1 \pi_2 \cdots \pi_n \in A(n)$ *we define* $f(\pi) = \text{Reduce}(\pi_2 \pi_3 \cdots \pi_n)$. *From the fact that* $\pi \in A(n)$ *we have that* $\pi_1 = 1$, $\pi_2 = 2$ *and* $\pi_j > 1$ *for all* $j > 1$. *Thus,* $f(\pi) \in \mathcal{P}_{n-1}$. *The function* $f$ *is injective: since if* $f(\pi) = f(\theta)$, *then* $\pi_j = \theta_j$ *for all* $j > 1$, *and* $\pi_1 = \theta_1 = 1$, *which obtains that* $\pi = \theta$. *The function* $f$ *is surjective: if* $f(\pi) = \text{Reduce}(\pi) = \pi'_1 \pi'_2 \cdots \pi'_{n-1}$ *then* $\pi_1 = 1$ *and* $\pi_j = \pi'_{j-1} + 1$ *for all* $j = 2, 3, \ldots, n-1$. *For example,* $f(122324) = 11213$. *Hence,* $f$ *is a bijection, that is,* $|A(n)| = |\mathcal{P}_{n-1}|$. *Moreover, since we know that* $|B(n)| = \text{Bell}_{n-1}$, *we obtain that the number of set partitions in* $A(n)$ *is given by* $\text{Bell}_{n-1}$.

**Example 2.6** *Let* $a_n$ *be the number of elements (other than 1) in the first block in all set partitions of* $[n]$, *and let* $b_n$ *be the number of singleton blocks (blocks with exactly one element) in all set partitions of* $[n]$, *excluding any singleton blocks equal to* $\{1\}$. *Since there is only one set partition of* $[1]$, *namely* $\{1\}$, *then* $a_1 = b_1 = 0$. *Also, there are two set partitions of* $[2]$, *namely,* $\{1,2\}$ *and* $\{1\}, \{2\}$, *so* $a_1 = b_1 = 1$.

*Now let us prove combinatorially that* $a_n = b_n$ *for all* $n$ *by presenting a bijection* $f$ *between the number of elements (other than 1) in the first block and the number of singleton blocks in all set partitions of* $[n]$, *excluding any singleton blocks equal to* $\{1\}$. *Let* $B_1, B_2, \ldots, B_k$ *be a set partition with* $k$ *blocks* such that $B_{i_j}$ contains only one element for $j = 1, 2, 3, \ldots s$. We construct a new set partition of $[n]$ by doing the following: All the elements in $B_{i_1} \cup \cdots \cup B_{i_s}$, together with 1, define a block (the first block), and each element of $B_1$ other than the element 1 defines a singleton block, where the rest of the blocks remain unchanged. This mapping is an involution (see Definition 2.3) on $\mathcal{P}_n$ and thus it is a bijection. For example, $1/2345 \to 1/2345$, $12/34/5 \to 15/2/34$ and $123/4/5 \to 145/2/3$.

**Definition 2.7** *A permutation of $[n]$ or of size $n$ is a one-to-one function from $[n]$ to itself; that is, is a bijection from $[n]$ to itself. Why? We denote permutations of $[n]$ by $\pi = \pi_1 \pi_2 \cdots \pi_n$, and the set of all permutations of size $n$ by $\mathcal{S}_n$ ($\mathcal{S}$ stands for symmetric group).*

*An involution $\pi$ of size $n$ is a one-to-one involution from $[n]$ to itself, that is, $\pi_{\pi_i} = i$ for all $i = 1, 2, \ldots, n$. We denote the set of all involutions of size $n$ by $\mathcal{I}_n$.*

The definition easily generalizes to any set of $n$ elements, as any such set can be associated with the set $[n]$.

**Example 2.8** *There are six permutations of size 3, namely 123, 132, 213, 231, 312, and 321. Among these permutations there are four involutions of size 3, namely, 123, 132, 213, and 321.*

Now, let us return to sequences. They can be defined either by an explicit formula or via a recurrence relation. An explicit formula for a sequence allows for direct computation of any term of the sequence by knowing just the value of $n$, without the need to compute any other term(s) of the sequence.

**Example 2.9** (Words) *Given a set $A$ of $k$ symbols, say $A = [k]$, we want to count the number of ways to write a word of size $n$ over alphabet $A$ (see Definition 1.6), which equals the number of functions from $[n]$ to $A$. There are $k$ choices for each symbol in the word, therefore,*

$$|[k]^n| = \underbrace{k \cdot k \cdots k}_{n \text{ times}} = k^n.$$

*With this explicit formula it is easy to compute the number of words of size $n = 10$ over alphabet $[2]$ as $2^{10} = 1024$.*

**Example 2.10** (Permutations) *Given a set of $n$ symbols, we want to count the number of ways these $n$ symbols can be arranged, which is equal to the number of permutations of size $n$ (see Definition 2.7). There are $n$ choices for the first element in the arrangement, $n-1$ choices for the next element, $n-2$ choices for the third element, $\ldots$, and one choice for the last element. Therefore,*

$$|\mathcal{S}_n| = n \cdot (n-1) \cdot (n-2) \cdots 2 \cdot 1 = n!.$$

Note that $n!$ reads "$n$ factorial" and $0! = 1$ by definition. With this explicit formula it is easy to compute the number of permutations of size 9 as $9! = 362880$.

On the other hand, a recurrence relation defines the value of the general term of the sequence in terms of the preceding value(s) of the sequence, together with an initial condition or a set of initial conditions. The initial conditions are necessary to ensure a uniquely defined sequence.

**Example 2.11** (Set Partitions) *By Theorem 1.9 we have that the sequence $\{\text{Bell}_n\}_{n \geq 0}$ satisfies the recurrence relation*

$$\text{Bell}_n = \sum_{j=0}^{n-1} \binom{n-1}{j} \text{Bell}_j = \sum_{j=0}^{n-1} \frac{(n-1)!}{j!(n-1-j)!} \text{Bell}_j$$

*with the initial condition $\text{Bell}_0 = 1$. From the initial conditions we can easily compute $\text{Bell}_1 = \text{Bell}_0 = 1$ and $\text{Bell}_2 = \text{Bell}_1 + \text{Bell}_0 = 2$. The first sixteen terms of the sequence are 1, 2, 5, 15, 52, 203, 877, 4140, 21147, 115975, 678570, 4213597, 27644437, 190899322, 1382958545, and 10480142147 (see Sequence A000110 in [327]). To obtain these (and, with appropriate modification, more) values we can use the following Maple codes:*

```
Bell:=proc(n)
  if n<0 then
    return "sequence not defined for negative indices";
  elif n=0 then return 1;
    else return(sum(binomial(n-1,j)*Bell(j),j=0..n-1));
  end if;
end proc:
seq(Bell(n),n=1..16);
```

*or*

```
with(combinat,bell);
seq(bell(n),n=1..16);
```

**Example 2.12** (Recursion for the Fibonacci Sequence) *The Fibonacci sequence is given by the following recurrence relation*

$$\text{Fib}_n = \text{Fib}_{n-1} + \text{Fib}_{n-2}$$

*with the initial conditions $\text{Fib}_0 = 0$ and $\text{Fib}_1 = 1$. From the initial conditions we can easily compute $\text{Fib}_2 = \text{Fib}_1 + \text{Fib}_0 = 1$, $\text{Fib}_3 = 2$, ... The first sixteen terms of the sequence are given by 0, 1, 1, 2, 3, 5, 8, 13, 21, 34, 55, 89, 144, 233, 377, and 610 (see Sequence A000045 in [327]). To obtain these (and, with appropriate modification, more) values we can use the following Maple codes:*

# Basic Tools of the Book 35

```
Fib:=proc(n)
  if n<0 then
    return "seq not defined for negative indices";
  elif n=0 then return 0;
    elif n=1 then return 1;
      else Fib(n-1)+Fib(n-2);
  end if;
end proc:
seq(Fib(n),n=0..15);
```

*or*

```
with(combinat,fibonacci):
seq(fibonacci(n),n=0..15);
```

*Note that if we change the initial conditions for the Fibonacci recurrence we obtain another famous sequence. The Lucas sequence is defined by*

$$\text{Luc}_n = \text{Luc}_{n-1} + \text{Luc}_{n-2}$$

*with the initial conditions* $\text{Luc}_0 = 2$ *and* $\text{Luc}_1 = 1$.

For a history of these two sequences, as well as applications where these sequences occur, see [198]. It is customary to use the subword Fib and Luc for these two sequences. Another sequence that is closely related to these two is the shifted Fibonacci sequence, defined as $\text{fib}_n = \text{Fib}_{n+1}$ for $n \geq 0$, which occurs naturally in many different contexts.

**Example 2.13** *For instance, let $A(n)$ be the set of all set partitions of $[n]$ such that each block contains either one element or two consecutive elements. For the first few values of $n$, we can easily make a list of such set partitions, as shown in Table 2.1, and count their number directly.*

**Table 2.1**: *The set $A(n)$*

| n | Elements of $A(n)$ |
|---|---|
| 1 | 1 |
| 2 | 11, 12 |
| 3 | 112, 122, 123 |
| 4 | 1122, 1123, 1223, 1233, 1234 |

The sequence $\{|A(n)|\}_{n \geq 0}$ looks very much like the shift Fibonacci sequence. We also have to check whether this Fibonacci pattern continues beyond the first few values. Let's think about a systematic way to create the set partitions in $A(n)$ recursively. Each such set partition its last block is either $\{n\}$ or $\{n-1, n\}$. We can create those with a last block $\{n\}$ by appending a set partition of $[n-1]$, and those with a last block $\{n-1,n\}$ by appending a set partitions of $[n-2]$. It is customary to define the number of set partitions of $[0]$ to be 1 (for the empty set partition). If we denote the number of set partitions in $A(n)$ by $a_n$, then we obtain the following recursion:

$$a_n = a_{n-1} + a_{n-2}$$

with the initial conditions $a_0 = a_1 = 1$, which, by induction, shows that $a_n = \text{fib}_n$ for all $n$.

**Example 2.14** (Restricted Ternary Words) *Let's determine $a_n$, the number of ternary words (words over alphabet $\{0,1,2\}$) of size $n$ which do not start with 0 and do not contain the substrings 22, 01, and 02. If the ternary word starts with a 1, we can append any of the $a_{n-1}$ such words of size $n-1$. If the ternary word starts with a 2, then the second letter has to be a 1 or a 0, that is, the word starts with 21 or with 20. In the first case, there are no restrictions on how the word can continue, so we can append any of the $a_{n-2}$ such words. If the word starts with 20, then the only possible words are of the form $200\cdots 0$, and there is exactly one word. The initial conditions are $a_0 = 1$ and $a_1 = 2$. Overall,*

$$a_n = a_{n-1} + a_{n-2} + 1, \qquad a_0 = 1, a_1 = 2.$$

In the above two examples, the recurrence relations are obtained easily, whereas explicit formulas are very difficult to derive directly (but not always!!). Very often, recursive relation can be derived naturally, which describes how a sequence evolves from one step to the next. Note how we can divide the objects to be counted into small classes, each of which is counted separately. In Example 2.13, we considered the last block of the set partition. In that case, we could have just as easily focused on the first block, as each first block is either $\{1\}$ or $\{1,2\}$. Focusing on the first or the last block are common methods for obtaining small classes, but we will see other possibilities in the examples throughout the book.

Recurrence relation can be used to find the value of a specific term of the sequence, say $\text{Bell}_{100}$, where all preceding values $\text{Bell}_j$ for $j = 0, 1, \ldots, 99$ have to be determined, unless an explicit formula can be derived. In Section 2.2 we will present several methods of obtaining an explicit formula from a recurrence relation.

Several times we are interested in more than one parameter, for instance, in addition to the number of set partitions of $[n]$ such that each block contains either one element or two elements we may want to keep track of the number of blocks. In this case, we obtain a sequence with several indices.

**Definition 2.15** *Let $k \geq 1$. A sequence with $k$ indices is a function $a: I^k \to A$, denoted by $\{a_{n_1,\ldots,n_k}\}_{n_1,\ldots,n_k \in I}$ or $\{a_{\vec{n}}\}_{\vec{n} \in I^k}$, where $I \subseteq \mathbb{Z}$. The element $a_{\vec{n}}$ of a sequence $\{a_{\vec{n}}\}_{\vec{n} \in I^k}$ is called the $\vec{n}$-th term, and the vector $\vec{n}$ of integers is the sequence vector of indices.*

If a sequence has only two indices, it is easy to list the sequence values in a two-dimensional table.

**Example 2.16** (Continuation of Example 2.13) *We denote the number of set partitions in $A(n)$ with exactly $k$ blocks by $a_{n,k}$. From Table 2.1, we can read off that $a_{1,1} = 1$, $a_{2,1} = 1$, and $a_{2,2} = 1$. In order to obtain the recurrence relation for this sequence, we now also keep track of what happens to the number of blocks when we create the set partitions recursively. Whenever our last block is $\{n\}$ or $\{n-1,n\}$, the number of blocks increases by one. Thus,*

$$a_{n,k} = a_{n-1,k-1} + a_{n-2,k-1}, \quad a_{0,0} = 1, a_{1,1} = 1, a_{n,0} = 0 \text{ if } n \neq 0.$$

*From the definition, $a_{n,k} = 0$ if $k > n$, and we therefore present the values as a triangular array in Table 2.2, where only the values corresponding to $0 \leq k \leq n$ are listed. Note that summing the elements across a row gives the number of set partitions in $A(n)$.*

**Table 2.2**: *Set partitions in $A(n)$ with exactly $k$ blocks*

| $n \backslash k$ | 0 | 1 | 2 | 3 | ... |
|---|---|---|---|---|---|
| 0 | $a_{0,0} = 1$ | | | | |
| 1 | $a_{1,0} = 0$ | $a_{1,1} = 1$ | | | |
| 2 | $a_{2,0} = 0$ | $a_{2,1} = 1$ | $a_{2,2} = 1$ | | |
| 3 | $a_{3,0} = 0$ | $a_{3,1} = 0$ | $a_{3,2} = 2$ | $a_{3,3} = 1$ | ... |
| 4 | $a_{4,0} = 0$ | $a_{4,1} = 0$ | $a_{4,2} = 1$ | $a_{4,3} = 3$ | ... |
| ⋮ | ⋮ | ⋮ | ⋮ | ⋮ | ⋱ |

In order to compute the values for $a_{i,j}$ given in Table 2.2, we can use the following Maple code:

```
a:=Matrix(5,5):
a[1,1]:=1:
for k from 2 to 5 do
  a[1,k]:=0:
od:
for n from 2 to 5 do
  a[n,1]:=0;
  for k from 2 to 5 do
    if n=k then a[n,k]:=1;
    elif n<k then a[n,k]:=0;
      else a[n,k]:=a[n-1,k-1]+a[n-2,k-1];
    end if;
  od:
od:
"print/rtable"(a);
```

**Example 2.17** *How many possible arrangements to set partition of $[n]$ into $k$ nonempty subsets? We already know the answer which is $\mathrm{Stir}(n,k)$ the Stirling number of the second kind, as described in Theorem 1.12. These numbers satisfy the recurrence relation*

$$\mathrm{Stir}(n+1,k) = k\mathrm{Stir}(n,k) + \mathrm{Stir}(n,k-1)$$

*with the initial conditions $\mathrm{Stir}(1,1) = \mathrm{Stir}(n,0) = 1$ for all $n \geq 1$, and $\mathrm{Stir}(n,k) = 0$ for all $n < k$.*

Other examples of sequences with two indices are *k-element permutations* and *k-element combinations* of $n$ objects.

**Example 2.18** (Permutations Revisited) *How many possible arrangements to order $k$ objects from a set of $n$ objects in one line? We already know the answer when $k = n$ which is $n!$ as it is shown in Example 2.10. Again, there are $n$ choices for the first element, $n-1$ choices for the second element, $\ldots$, and $n-k+1$ choices for the last element, in the arrangement. Thus, if $a_{n,k}$ denotes the number of $k$-element permutations of $n$ objects, then*

$$a_{n,k} = n \cdot (n-1) \cdots (n-k+1) = \frac{n!}{(n-k)!}.$$

**Example 2.19** (Combinations) *Instead of counting the $k$-combinations directly, we make a connection with $k$-element permutations of $n$ objects by creating the $k$-element permutations in the following two steps: First, let us select the $k$ elements, and second arrange them in all possible orders. If we denote the number of $k$-element combinations of $n$ objects by $b_{n,k}$, then the first step can be done in $b_{n,k}$ ways, and each such selection can be arranged in $k!$ ways (see Example 2.10). Therefore, $a_{n,k} = b_{n,k} \cdot k!$, which leads to $b_{n,k} = \frac{a_{n,k}}{k!} = \frac{n!}{(n-k)!k!} = \binom{n}{k}$.*

The values $b_{n,k}$ occur in the binomial theorem, and $\binom{n}{k}$ reads "$n$ choose $k$". These values also appear in the famous Pascal's Triangle (Pascal's Triangle is named after Blaise Pascal,[1] but was known about 500 years earlier both in the Middle East by Al-Karaji[2] and in China by Jia Xian.[3]) There exist lot of interesting properties for the binomial coefficients. For example, we can state the following properties.

**Fact 2.20** *For all $n, k \geq 1$,*
 (i) $\binom{n}{k} = \binom{n-1}{k-1} + \binom{n-1}{k}$.
 (ii) $\sum_{i=0}^{n} \binom{n}{i} = 2^n$.
 (iii) $\sum_{i=k}^{n} \binom{i}{k} = \binom{n+1}{k+1}$.

**Proof** See Exercise 2.2. □

---
[1] www-history.mcs.st-andrews.ac.uk/Mathematicians/Pascal.html
[2] turnbull.mcs.st-and.ac.uk/~history/Mathematicians/Al-Karaji.html
[3] www-history.mcs.st-andrews.ac.uk/Mathematicians/Jia_Xian.html

## 2.2 Solving Recurrence Relations

It is well known that there is no general method for solving recurrence relations, which is why it is an art. Here, we will present several methods for solving recurrence relations, some of which can be used only for special types of these recurrence relations. Therefore, we start by classifying recurrence relations.

**Definition 2.21** *A linear recurrence relation of order $r$ is of the form*

$$g_0(n)a_n + g_1(n)a_{n-1} + \cdots + g_r(n)a_{n-r} = h(n)$$

*for all $n \geq r$, where $g_r \neq 0$. The recurrence relation is called P-recursive if $g_0, g_1, \ldots, g_r$ and $h$ are polynomials in $n$. When all the polynomials $g_i$ are constant, the recurrence relation reduces to a* linear recurrence relation of order $r$ with constant coefficients. *If $h(n) = 0$, then the recurrence relation is called* homogeneous. *Otherwise, it is called* nonhomogeneous.

**Example 2.22** (Set partitions with each block has either one, two, or three consecutive elements) *Let $a_n$ be the number of set partitions of $[n]$ such that each block contains either one, two, or three consecutive elements. Then, $a_0 = 1$ (the empty set partition), $a_1 = 1$ (the set partition 1), and $a_2 = 2$ (the set partitions 11 and 12). In general, any such set partition of $[n]$ its last block is either $\{n\}$, $\{n-1, n\}$ or $\{n-2, n-1, n\}$. Thus (see Example 2.13),*

$$a_n = a_{n-1} + a_{n-2} + a_{n-3}, \quad a_0 = a_1 = 1, a_2 = 2.$$

*In this case, the order of the recurrence is 3.*

There are several methods for solving recurrence relations: (1) guess and check, (2) iteration (repeated substitution), (3) characteristic polynomial, and (4) generating functions.

### 2.2.1 Guess and Check

The first method is based on guessing a solution and then proving by induction that it is correct. Usually, finding the right guess is a problem of magic!

**Example 2.23** (Total Number of Arrangements of $[n]$) *Consider the recurrence relation*

$$a_n = n \cdot a_{n-1} + 1, \quad n \geq 1$$

*with the initial condition $a_0 = 1$. Computing the first few values gives $a_0 = 1$, $a_1 = 1 + 1 = 2$, $a_2 = 2 + 2 + 1 = 5$, $a_3 = 6 + 6 + 3 + 1 = 16$ and $a_4 =$*

$24 + 24 + 12 + 4 + 1 = 65$. A reasonable guess is that $a_n = n! \sum_{k=0}^{n} \frac{1}{k!}$ for all $n \geq 0$, which follows by induction:

$$a_{n+1} = (n+1)a_n + 1 = (n+1)! \sum_{k=0}^{n} \frac{1}{k!} + 1 = (n+1)! \sum_{k=0}^{n+1} \frac{1}{k!}.$$

Note that this recurrence relation is P-recursive. The explicit formula for $a_n$ can be obtained by Maple code:

```
rsolve({a(n)=n*a(n-1)+1,a(0)=1},a(n));
```

### 2.2.2 Iteration

The iteration method does not require guessing the solution. The idea is to iterate the recurrence, simplify the result and then to recognize a pattern for the general term of the sequence. Again, by induction we can show that the pattern is correct.

**Example 2.24** (Number Perfect Matchings on $[2n]$) A prefect matching is a set partition where each block contain exactly two elements. Let $a_n$ be the number of perfect matching on $[2n]$. Then, by considering the elements in the first block, we obtain the following recurrence relation

$$a_n = (2n-1) \cdot a_{n-1}, \quad n \geq 1$$

with the initial condition $a_0 = 1$. Repeated substitution gives

$$\begin{aligned}
a_n &= (2n-1) \cdot a_{n-1} \\
&= (2n-1)(2n-3) \cdot a_{n-2} \\
&= (2n-1)(2n-3)(2n-5) \cdot a_{n-3} \\
&= \cdots = (2n-1)(2n-3) \cdots 5 \cdot 3 \cdot a_0 \\
&= (2n-1)!!
\end{aligned}$$

The explicit formula for $a_n$ can be obtained by Maple:

```
rsolve({a(n)=(2*n-1)*a(n-1),a(0)=1},a(n));
```

Note that $n!!$ reads "$n$ double factorial", and $0!! = 1!! = 1$ by definition.

### 2.2.3 Characteristic Polynomial

Now we describe how to obtain an explicit solution for a linear recurrence relation with constant coefficients $c_1, c_2, \ldots, c_r$ of order $r$ of the form

$$a_n + c_1 a_{n-1} + c_2 a_{n-2} + \cdots + c_r a_{n-r} = h_n \quad \text{for all } n \geq r, \qquad (2.1)$$

where $h_n$ is a polynomial in $n$, not depending on $a_n$. When possible to solve a nonhomogeneous recurrence relation, we need to solve the corresponding homogeneous recurrence relation as part of the process, so at first we solve the linear homogeneous recurrence relations with constant coefficients of the form

$$a_n + c_1 a_{n-1} + c_2 a_{n-2} + \cdots + c_r a_{n-r} = 0 \quad \text{for all } n \geq r. \quad (2.2)$$

Following the guess and check method, one might guess that the solution has the form $\xi^n$, give or take a multiplicative constant. Checking the guess by substituting it into the recurrence relation 2.2, we obtain that $\xi$ satisfies $\Delta(\xi) = 0$, where $\Delta$ is the polynomial

$$\Delta(x) = x^r + c_1 x^{r-1} + c_2 x^{r-2} + \cdots + c_r.$$

**Definition 2.25** *The polynomial $\Delta(x)$ is called the* characteristic polynomial *of the recurrence relation 2.2.*

The fundamental theorem of algebra gives that the characteristic polynomial $\Delta(x)$ has $r$ (complex) roots, counting multiplicities. The next theorem describes how the solutions look in the case of roots with multiplicity bigger than one, and presents that there are $r$ independent solutions for a linear recurrence relation with constant coefficients of order $r$.

**Theorem 2.26** *Let $\xi \in \mathbb{C}$ be any root of the characteristic polynomial $\Delta(x)$ of the recurrence relation 2.2 with multiplicity $m$. Then, the basic solutions $n^i \xi^n$, $i = 0, 1, \ldots, m-1$, satisfy (2.2).*

**Proof** See Exercise 2.3. $\square$

In order to find the general solution of (2.3), we need only to take a linear combination of the $r$ different basic solutions given in Theorem 2.26.

**Theorem 2.27** *If $a_1(n), \ldots, a_r(n)$ are different sequences satisfying (2.2), then for any constants $k_1, \ldots, k_r \in \mathbb{C}$, the sequence $\sum_{j=1}^{n} k_j \cdot a_j(n)$, called the* general solution, *satisfies (2.2).*

The above theorems describe an algorithm for finding explicit formula for any linear homogeneous recurrence relation as presented in (2.2) with $r$ initial conditions:

- Find all the roots of the characteristic polynomial $\Delta(x)$ and their multiplicity.

- Determine the general solution as described in Theorem 2.27.

- By the initial conditions we obtain a linear system of $r$ equations in $r$ unknown constants, and then we solve it to derive a specific solution.

Note that in order to apply our algorithm for finding an explicit formula for a homogeneous linear recurrence relation with constant coefficients of order $r$, we need to deal with a polynomial equation of order $r$ for which we do not have formulas, when $r \geq 4$, to find the exact roots unless we can factor the characteristic polynomial.

**Example 2.28** *Consider the recurrence relation*

$$a_n = 8a_{n-1} - 21a_{n-2} + 18a_{n-3} \tag{2.3}$$

*with initial conditions* $a_0 = 1$, $a_1 = 1$ *and* $a_2 = 2$. *The characteristic polynomial of* (2.3) *is given by*

$$\Delta(x) = x^3 - 8x^2 + 21x - 18 = (x-2)(x^2 - 6x + 9) = (x-2)(x-3)^2.$$

*Thus, the characteristic polynomial has two roots:* $\xi = 2$ *with multiplicity one, and* $\xi = 3$ *with multiplicity two. From Theorem 2.27, we obtain that the general solution for* (2.3) *is given by*

$$a_n = k_1 \cdot 2^n + (k_2 + k_3 \cdot n) \cdot 3^n.$$

*Using the initial conditions results in the equations* $k_1 + k_2 = 1$, $2k_1 + 3(k_2 + k_3) = 1$, *and* $4k_1 + 9(k_2 + 2k_3) = 2$. *The values of* $k_1, k_2, k_3$ *can be computed by using the Maple code*

```
solve({k_1*2^0+(k_2+k_3*0)*3^0=1,k_1*2^1+(k_2+k_3*1)*3^1=1,
k_1*2^2+(k_2+k_3*2)*3^2=2},{k_1,k_2,k_3});
```

*We obtain that* $k_1 = 5$, $k_2 = -4$ *and* $k_3 = 1$, *which gives that the explicit formula is given by* $a_n = 5 \cdot 2^n - (4-n)3^n$.

**Example 2.29** (Fibonacci and Lucas sequences) *We can now find explicit solutions for the Fibonacci and the Lucas sequence. Recall that the two sequences have the same recurrence relation*

$$a_n = a_{n-1} + a_{n-2},$$

*with initial conditions* $a_0 = \mathrm{Fib}_0 = 0$, $a_1 = \mathrm{Fib}_1 = 1$ *for the Fibonacci sequence and* $a_0 = \mathrm{Luc}_0 = 2$, $a_1 = \mathrm{Luc}_1 = 1$ *for the Lucas sequence. Since the recurrence relation is the same, they both have the same characteristic polynomial* $\Delta(x) = x^2 - x - 1$, *and hence the same general solution:*

$$a_n = k_1 \cdot \left(\frac{1+\sqrt{5}}{2}\right)^n + k_2 \cdot \left(\frac{1-\sqrt{5}}{2}\right)^n.$$

*Note that it is customary in the context of the Fibonacci sequence to define* $\alpha = \frac{1+\sqrt{5}}{2}$, *the Golden Ratio*[4] $\Phi$, *and* $\beta = \frac{1-\sqrt{5}}{2}$. *The initial conditions for*

---
[4] http://mathworld.wolfram.com/GoldenRatio.html

the Fibonacci sequence give $0 = k_1 + k_2$ and $1 = k_1 \cdot \alpha + k_2 \cdot \beta$, which implies $k_1 = 1/\sqrt{5}$ and $k_2 = -1/\sqrt{5}$, and thus

$$\text{Fib}_n = \frac{1}{\sqrt{5}} \left[ \left( \frac{1+\sqrt{5}}{2} \right)^n - \left( \frac{1-\sqrt{5}}{2} \right)^n \right]. \qquad (2.4)$$

This formula was first derived by Abraham De Moivre,[5] and independently by Daniel Bernoulli.[6] For the derivation of the explicit formula for the Lucas sequence, see Exercise 2.4. We can find the explicit formula for $\text{Fib}_n$ by using the following Maple code:

```
rsolve({a(n)=a(n-1)+a(n-2),a(0)=0,a(1)=1},a(n));
```

Now we indicate how to solve a linear nonhomogeneous recurrence relation with constant coefficients.

**Theorem 2.30** *The general solution of the nonhomogeneous linear recurrence relation with constant coefficients of order $r$ as given in (2.1) is of the form $p(n) + q(n)$, where $q(n)$ is the general solution of the associated homogeneous recurrence relation (2.2), and $p(n)$ is any solution of (2.1).*

**Proof** See Exercise 2.5. $\square$

The solution $p(n)$ in Theorem 2.30 is called the *particular solution*. In order to find the general solution for a nonhomogeneous recurrence relation, we need to

- find the general solution for the associated homogeneous recurrence relation;
- guess a particular solution;
- use the initial conditions to determine the specific solution.

**Table 2.3**: Particular solutions for certain $q(n)$

| $q(n)$ | $p(n)$ |
|---|---|
| $cn^r d^n, r \in \mathbb{N}$ | $d^n(c_r n^r + \cdots + c_1 n + c_0)$ |
| $\sin(sn)$ or $\cos(sn)$ | $c_1 \sin(sn) + c_2 \cos(sn)$ |
| $d^n \sin(sn)$ or $d^n \cos(sn)$ | $c_1 d^n \sin(sn) + c_2 d^n \cos(sn)$ |

The difficult part is to guess the particular solution. There is no approach that works for all types of functions that may occur on the right-hand side

---
[5] http://www-groups.dcs.st-and.ac.uk/~history/Mathematician/De_Moivre.html
[6] http://www-groups.dcs.st-and.ac.uk/~history/Mathematicians/Bernoulli_Daniel.html

of (2.1). However, for certain common functions, the form of the particular function is known. If $p(n)$ is a constant multiple of a function listed on the left-hand side of Table 2.3, then the function given on the right-hand side is a particular solution. The constants $d$, $r$, and $s$ are given, while the constants $k_i$ have to be determined from the given recurrence relation. If $q(n)$ is a linear combination of functions given on the left-hand side of Table 2.3, and none of these functions is a solution of the homogeneous recurrence relation, then the particular solution is a linear combination of the respective functions given on the right-hand side of Table 2.3, see [127, 334, 348]. For example, if $q(n) = d^n + \sin(sn)$, then $p(n) = k_1 \cdot d^n + k_2 \sin(sn) + k_3 \cos(sn)$.

In order to illustrate how to solve a nonhomogeneous recurrence relation, we present the following example.

**Example 2.31** (Restricted Set Partitions) *A block (subset) is said to be poor if it contains either one, or two consecutive elements. A set partitions of $[n]$ is called almost-poor if each of its block is either a poor or right-most block. For example, the almost-poor set partitions of $[n]$ are given by* 1; 11, 12; 111, 112, 122, 123; 1111, 1122, 1123, 1222, 1223, 1233, 1234, *where* $n = 1, 2, 3, 4$. *The recurrence relation for the number of almost-poor set partitions of $[n]$ is given by*

$$a_n = a_{n-1} + a_{n-2} + 1, \qquad (2.5)$$

*with initial conditions $a_1 = 1$ and $a_2 = 2$. (Why? See Example 2.13.) The associated homogeneous recurrence relation is the same as the one for the Fibonacci sequence, and therefore, the general solution is given by $a_n = k_1 \alpha^n + k_2 \beta^n$ (see Example 2.29) with $\alpha = (1+\sqrt{5})/2$ and $\beta = (1-\sqrt{5})/2$. The next step is to find a particular solution. In this case, the polynomial $h(n)$ is a constant, so the particular solution is constant, too, say $p(n) = c$. Substituting this particular solution into (2.5) leads to $c = c + c + 1$, which is equivalent to $c = -1$. Thus, the general solution for the nonhomogeneous recurrence relation has the form*

$$a_n = k_1 \alpha^n + k_2 \beta^n - 1. \qquad (2.6)$$

*By substituting the initial conditions results in (2.6) we obtain*

$$1 = k_1 \alpha + k_2 \beta - 1 \quad \text{and} \quad 2 = k_1 \alpha^2 + k_2 \beta^2 - 1,$$

*which gives $k_1 = \frac{3+\sqrt{5}}{2\sqrt{5}}$ and $k_2 = \frac{-3+\sqrt{5}}{2\sqrt{5}}$. Hence, (2.6) can be written as*

$$a_n = \frac{3+\sqrt{5}}{2\sqrt{5}} \alpha^n - \frac{3-\sqrt{5}}{2\sqrt{5}} \beta^n - 1 = \text{Fib}_{n+1} - 1.$$

*Once more we can find the explicit formula for $a(n)$ by using the Maple code*

```
rsolve({a(n)=a(n-1)+a(n-2)+1,a(1)=1,a(2)=2},a(n));
```

## 2.3 Generating Functions

In this section, we discus another tool to derive explicit formulas from recurrence relation, namely generating functions. This tool not only helps us to derive formulas from recurrence relation, but can be used directly to count number of objects in a sequence of sets. Once you get the hang of this tool, I guarantee you will see how fun and addicting these problems can be. Generating functions are formal power series, and we will discuss the theory of formal power series after stating the definitions and some basic examples. Throughout the book we will use two types of generating functions, ordinary and exponential. Also, we will present several examples that describe how to convert an ordinary generating function to exponential generating function.

**Definition 2.32** *The* ordinary *(respectively,* exponential*) generating function for the sequence* $\{a_n\}_{n \geq 0}$ *is given by* $A(x) = \sum_{n \geq 0} a_n x^n$ *(respectively, $E(x) = \sum_{n \geq 0} a_n \frac{x^n}{n!}$).*

Since we will primarily deal with ordinary generating functions, we will omit "ordinary" in the remainder of the book. We start by computing the generating functions for some basic examples.

**Example 2.33** *Let $a_n = c^n$ for all $n \geq 0$, then the generating function (the ordinary) for the sequence $\{a_n\}_{n \geq 0}$ is given by*

$$A(x) = \sum_{n \geq 0} a_n x^n = \sum_{n \geq 0} c^n x^n = \frac{1}{1 - cx},$$

*and the exponential generating function for the same sequence is given by*

$$E(x) = \sum_{n \geq 0} a_n \frac{x^n}{n!} = \sum_{n \geq 0} \frac{c^n x^n}{n!} = e^{cx}.$$

**Example 2.34** (Fibonacci sequence) *Now we will derive the generating function for the Fibonacci sequence and from it the explicit formula given in Example 2.29. We start with the recurrence relation for the Fibonacci numbers, $\text{Fib}_n = \text{Fib}_{n-1} + \text{Fib}_{n-2}$ for $n \geq 2$, with initial conditions $\text{Fib}_0 = 0$ and $\text{Fib}_1 = 1$. Since the generating function for the sequence $\{\text{Fib}_n\}_{n \geq 0}$ is defined as $\text{Fib}(x) = \sum_{n \geq 0} \text{Fib}_n x^n$, we multiply the recurrence relation by $x^n$, and then sum over the values of $n$ for which the recurrence is valid:*

$$\sum_{n \geq 2} \text{Fib}_n x^n = \sum_{n \geq 2} \text{Fib}_{n-1} x^n + \sum_{n \geq 2} \text{Fib}_{n-2} x^n,$$

which is equivalent to

$$\text{Fib}(x) - \text{Fib}_1 x - \text{Fib}_0 = x \sum_{n \geq 2} \text{Fib}_{n-1} x^{n-1} + x^2 \sum_{n \geq 2} \text{Fib}_{n-2} x^{n-2}$$

$$= x \sum_{n \geq 1} \text{Fib}_n x^n + x^2 \sum_{n \geq 0} \text{Fib}_n x^n$$

$$= x(\text{Fib}(x) - \text{Fib}_0) + x^2 \text{Fib}(x).$$

Using the above initial conditions and solving for $\text{Fib}(x)$ obtains

$$\text{Fib}(x) = \frac{x}{1 - x - x^2} = \frac{-x}{(x + \alpha)(x + \beta)},$$

where $\alpha = \frac{1+\sqrt{5}}{2}$ and $\beta = \frac{1-\sqrt{5}}{2}$. In order to derive the coefficients of $x^n$, we need the Taylor series expansion of the generating function $\text{Fib}(x)$. Instead of using the definition of the Taylor series expansion which involves derivatives of the function, we make use of functions whose Taylor series we know (see Example 2.33). By partial fraction decomposition, we can rewriting the denominator as product of factors in the correct form for the geometric series:

$$\frac{1}{-(x+\alpha)(x+\beta)} = \frac{1}{\frac{1}{\beta}(-\beta x - \beta \alpha)\frac{-1}{\alpha}(-\alpha x - \alpha \beta)}$$

$$= \frac{1}{(1 - \beta x)(1 - \alpha x)}$$

$$= \frac{1}{\sqrt{5}(1 - \alpha x)} - \frac{1}{\sqrt{5}(1 - \beta x)}.$$

Thus,

$$\text{Fib}(x) = \frac{x}{\sqrt{5}} \left( \frac{1}{1 - \alpha x} - \frac{1}{1 - \beta x} \right).$$

Using Example 2.33 gives

$$\text{Fib}(x) = \frac{x}{\sqrt{5}} \left( \sum_{n \geq 0} (\alpha x)^n - \sum_{n \geq 0} (\beta x)^n \right) = \sum_{n \geq 0} \frac{1}{\sqrt{5}} (\alpha^n - \beta^n) x^n.$$

Now we can read off $\text{Fib}_n$ as the coefficient of $x^n$ to obtain

$$\text{Fib}_n = \frac{1}{\sqrt{5}} (\alpha^n - \beta^n)$$

as in Example 2.29. In Maple, the generating function can be obtained from the recurrence relation by giving an optional argument to rsolve:

```
rsolve({Fib(n)=Fib(n-1)+Fib(n-2),
Fib(0)=0,Fib(1)=1},Fib,'genfunc'(x));
```

Basic Tools of the Book    47

**Example 2.35** (Fibonacci sequence) *Now we will derive the exponential generating function for the Fibonacci sequence and from it the explicit formula given in Example 2.29. Again, we start with the recurrence relation for the Fibonacci numbers, $\text{Fib}_n = \text{Fib}_{n-1} + \text{Fib}_{n-2}$ for $n \geq 2$, with initial conditions $\text{Fib}_0 = 0$ and $\text{Fib}_1 = 1$. Since the exponential generating function for the sequence $\{\text{Fib}_n\}_{n \geq 0}$ is defined as $E(x) = \sum_{n \geq 0} \text{Fib}_n \frac{x^n}{n!}$, we multiply the recurrence relation by $\frac{x^n}{n!}$, and then sum over the values of $n$ for which the recurrence is valid:*

$$\sum_{n \geq 2} \text{Fib}_n \frac{x^n}{n!} = \sum_{n \geq 2} \text{Fib}_{n-1} \frac{x^n}{n!} + \sum_{n \geq 2} \text{Fib}_{n-2} \frac{x^n}{n!},$$

*which is equivalent to*

$$E(x) - \text{Fib}_1 x - \text{Fib}_0$$
$$= \int_0^x \sum_{n \geq 2} \text{Fib}_{n-1} \frac{t^{n-1}}{(n-1)!} dt + \int_0^x \int_0^x \sum_{n \geq 2} \text{Fib}_{n-2} \frac{x^{n-2}}{(n-2)!} dt dt$$
$$= \int_0^t (E(t) - \text{Fib}_0) dt + \int_0^x \int_0^x E(t) dt dt.$$

*Thus,*

$$\frac{d^2}{dx^2} E(x) = \frac{d}{dx} E(x) + E(x).$$

*Solving this differential equation for $E(x)$ gives*

$$E(x) = k_1 e^{\alpha x} + k_2 e^{\beta x},$$

*where $\alpha = \frac{1+\sqrt{5}}{2}$ and $\beta = \frac{1-\sqrt{5}}{2}$. Using the initial conditions $E(0) = 0$ (that is, $\text{Fib}_0 = 0$) and $\frac{d}{dx} E(x)|_{x=0} = 1$ (that is, $\text{Fib}_1 = 1$) give $k_1 + k_2 = 0$ and $k_1 \alpha + k_2 \beta = 1$. Solving for the constants $k_1$ and $k_2$, we obtain that $k_1 = \frac{1}{\sqrt{5}}$ and $k_2 = \frac{-1}{\sqrt{5}}$. Hence,*

$$E(x) = \frac{1}{\sqrt{5}} \left( e^{\alpha x} - e^{\beta x} \right) = \sum_{n \geq 0} \frac{1}{\sqrt{5}} (\alpha^n - \beta^n) \frac{x^n}{n!}.$$

*Now we can read off $\text{Fib}_n$ as the coefficient of $x^n$ to obtain*

$$\text{Fib}_n = \frac{1}{\sqrt{5}} (\alpha^n - \beta^n)$$

*as in Example 2.29.*

**Example 2.36** *The exponential generating function for $|\mathcal{S}_n|$, the number of permutations of size $n$ (of which there are $n!$) is given by*

$$E(x) = \sum_{n \geq 0} |\mathcal{S}_n| \frac{x^n}{n!} = \sum_{n \geq 0} x^n = \frac{1}{1-x}.$$

**Example 2.37** *Modifying Example 2.36 slightly, we derive the exponential generating function for the sequence $a_n = 2^n n!$. Then*

$$E(x) = \sum_{n \geq 0} a_n \frac{x^n}{n!} = \sum_{n \geq 0} \frac{2^n n!}{n!} x^n = \sum_{n \geq 0} 2^n x^n = \frac{1}{1-2x},$$

*We can find the generating function $E(x)$ by using the Maple function* sum:

```
sum(2^n*n!/n!*x^n,n=1..infinity);
```

The above examples show that it is convenient to use the exponential generating function rather than the ordinary generating function when the sequence terms involve factorial terms. Note that we have considered the generating function in two different ways: If we already have an explicit formula, then we just use the definition to obtain the generating function. In Example 2.34 presents how to obtain the generating function from a recurrence relation and then an explicit formula. Do we always have these two choices? The answer is yes; any linear recurrence relation with constant coefficients can also be solved using the generating function approach, as we will demonstrate now. But we can say that the generating function approach has the advantage that we can solve the nonhomogeneous recurrence relation directly, without first solving the associated homogeneous relation, see Example 2.40.

**Lemma 2.38** *Given a linear recurrence relation with constant coefficients of the form*

$$a_n + c_1 a_{n-1} + c_2 a_{n-2} + \cdots + c_r a_{n-r} = h_n \quad \text{for all } n \geq d,$$

*where $d \geq r$, $c_1, \ldots, c_r$ are constants, and $h_n$ is any function. The associated generating function is given by*

$$A(x) = \frac{\sum_{j=0}^{r} c_j x^j \sum_{i=0}^{d-j-1} a_i x^i + \sum_{n \geq d} h_n x^n}{\sum_{j=0}^{r} c_j x^j}. \qquad (2.7)$$

**Proof** We proceed the proof as in Example 2.29. By multiplying the recurrence relation by $x^n$ and summing over $n \geq d$, we have

$$\sum_{n \geq d} \left( \sum_{j=0}^{r} c_j a_{n-j} \right) x^n = \sum_{n \geq d} h_n x^n.$$

Separating terms that are independent of $n$, and splitting off $x^j$ obtains

$$\sum_{j=0}^{r} c_j x^j \sum_{n \geq d} a_{n-j} x^{n-j} = \sum_{n \geq d} h_n x^n,$$

or equivalently,

$$\sum_{j=0}^{r} c_j x^j \left( A(x) - \sum_{i=0}^{d-j-1} a_i x^i \right) = \sum_{n \geq d} h_n x^n.$$

Solving for $A(x)$ completes the proof. □

The version of the above lemma in terms of exponential generating functions is given by the following lemma.

**Lemma 2.39** *Given a linear recurrence relation with constant coefficients of the form*

$$a_n + c_1 a_{n-1} + c_2 a_{n-2} + \cdots + c_r a_{n-r} = h_n \quad \text{for all } n \geq d,$$

*where $d \geq r$, $\alpha_1, \ldots, \alpha_r$ are constants, and $h_n$ is any function. The associated exponential generating function $E(x) = \sum_{n \geq 0} a_n \frac{x^n}{n!}$ satisfies*

$$\sum_{j=0}^{r} c_j \frac{d^{d-j}}{dx^{d-j}} A(x) = \frac{d^d}{dx^d} H(x), \tag{2.8}$$

*where $H(x)$ is the exponential generating function for the sequence $\{h_n\}_{n=d}^{\infty}$, that is, $H(x) = \sum_{n \geq d} h_n \frac{x^n}{n!}$.*

**Proof** We proceed the proof as in Example 2.35. By multiplying the recurrence relation by $\frac{x^n}{n!}$ and summing over $n \geq d$, we have

$$\sum_{n \geq d} \left( \sum_{j=0}^{r} c_j a_{n-j} \right) \frac{x^n}{n!} = \sum_{n \geq d} h_n \frac{x^n}{n!}.$$

Exchanging the sums and separating terms that are independent of $n$ gives

$$\sum_{j=0}^{r} c_j \underbrace{\int_0^x \cdots \int_0^x}_{j \text{ times}} \sum_{n \geq d} a_{n-j} \frac{x^{n-j}}{(n-j)!} \underbrace{dt \cdots dt}_{j \text{ times}} = \sum_{n \geq d} h_n \frac{x^n}{n!},$$

or, equivalently,

$$\sum_{j=0}^{r} c_j \underbrace{\int_0^x \cdots \int_0^x}_{j \text{ times}} \left( A(x) - \sum_{i=0}^{d-j-1} a_i \frac{x^i}{i!} \right) \underbrace{dt \cdots dt}_{j \text{ times}} = \sum_{n \geq d} h_n \frac{x^n}{n!},$$

which gives (2.8) as required. □

We can apply this formula to the nonhomogeneous recurrence relation of Example 2.31.

**Example 2.40** Let $a_n = a_{n-1} + a_{n-2} + 2$ with initial conditions $a_0 = a_1 = 1$ and $a_2 = 2$. According to Lemma 2.38, the relevant parameters are $r = 2$, $d = 3$, $c_0 = 1$, $c_1 = c_2 = -1$, $h_n = 2$ for all $n \geq 3$, and therefore, $\sum_{n \geq 3} h_n x^n = 2x^3 \sum_{n \geq 0} x^n = \frac{2x^3}{1-x}$. By substituting this expression and the parameter values into (2.7), we obtain that

$$A(x) = \frac{1 + \frac{2x^3}{1-x}}{1 - x - x^2} = \frac{1 - x + 2x^3}{(1-x)(1-x-x^2)}.$$

Maple ables to find the generating function also in this case:

```
rsolve({a(n)=a(n-1)+a(n-2)+1,a(0)=1,a(1)=1,a(2)=2},
  a, 'genfunc'(x));
```

Now let us use the exponential generating function techniques. The exponential generating function for the sequence $\{2\}_{n \geq 3}$ is given by $\sum_{n \geq 3} h_n \frac{x^n}{n!} = 2e^x - 1 - 2x - x^2$. Substituting this expression and the parameter values into (2.8), we obtain

$$\frac{d^2}{dx^2} E(x) - \frac{d}{dx} E(x) - AEx) = 2e^x.$$

Maple is able to solve this differential equation:

```
dsolve(diff(diff(E(x),x),x)-diff(E(x),x)-E(x)=exp(x),E(x));
```

This gives

$$E(x) = k_1 e^{\alpha x} + k_2 e^{\beta x} - 2e^x,$$

where $\alpha = \frac{1+\sqrt{5}}{2}$ and $\beta = \frac{1-\sqrt{5}}{2}$. In order to find the constants $k_1$ and $k_2$, we use the initial conditions $E(0) = 1$ and $\frac{d}{dx} E(x) \mid_{x=0} = 1$. This leads to $k_1 + k_2 - 2 = 1$ and $\alpha k_1 + \beta k_2 - 2 = 1$. Solving for $k_1$ and $k_2$ we obtain that

$$E(x) = \frac{3(1-\beta)}{\sqrt{5}} e^{\alpha x} - \frac{3(1-\alpha)}{\sqrt{5}} e^{\beta x} - 2e^x.$$

In order for generating functions to be useful, we need to be able to add and multiply them. However, we will not be concerned with questions of convergence at this point, but rather work with *formal power series*. (We will change our point of view when we derive results on the asymptotic behavior of the coefficients.) Let $\{a_{\vec{n}}\}_{\vec{n} \in \mathbb{N}^k}$ be any sequence with $k$ indices, $\mathbb{L}$ be a ring (see Definition E.4), and $\mathbb{L}[x_1, \ldots, x_k] = \mathbb{L}[\vec{x}]$ be the set of all polynomials in $k$ indeterminates $\vec{x} = (x_1, \ldots, x_k)$ with coefficients in $\mathbb{L}$.

**Definition 2.41** *The* ordinary *and* exponential generating function *for the sequence $\{a_{\vec{n}}\}_{\vec{n} \in \mathbb{N}^k}$ are given by*

$$A(\vec{x}) = \sum_{\vec{n} \in \mathbb{N}^k} a_{\vec{n}} \vec{x}^{\vec{n}} \quad \text{and} \quad E(\vec{x}) = \sum_{\vec{n} \in \mathbb{N}^k} a_{\vec{n}} \frac{\vec{x}^{\vec{n}}}{\vec{n}!},$$

respectively, where $\vec{x}^{\vec{n}} = \prod_{j=1}^{k} x_j^{n_j}$ and $\vec{n}! = \prod_{j=1}^{k} n_j!$. In this case, we say that $A$ (respectively, $E$) is the ordinary (respectively, exponential) generating function such that $x_i$ marks $n_i$ for all $i = 1, 2, \ldots, k$.

Two formal power series $A(\vec{x})$ and $B(\vec{x})$ are equal if $a_{\vec{n}} = b_{\vec{n}}$ for all $\vec{n} \in \mathbb{N}^k$, and we write $A(\vec{x}) = B(\vec{x})$. The set of formal power series or generating functions in $\vec{x} = (x_1, \ldots, x_k)$ is denoted by $\mathbb{L}[[\vec{x}]]$. The addition and subtraction operations of $A$ and $B$ are given by

$$A(\vec{x}) \pm B(\vec{x}) = \sum_{\vec{n} \in \mathbb{N}^k} (a_{\vec{n}} \pm b_{\vec{n}}) \vec{x}^{\vec{n}},$$

and the multiplication of $A$ and $B$ or the convolution of $A$ and $B$ is defined by the Cauchy product rule:

$$C(\vec{x}) = A(\vec{x}) \cdot B(\vec{x}) = \sum_{\vec{n} \in \mathbb{N}^k} \sum_{\vec{m}' + \vec{m}'' = \vec{n}} a_{\vec{m}'} b_{\vec{m}''} \vec{x}^{\vec{n}},$$

where $\vec{m}', \vec{m}'' \in \mathbb{N}^k$ and $\vec{m}' + \vec{m}'' = (m_1' + m_1'', \ldots, m_k' + m_k'')$. Note that it can be shown that the set $\mathbb{L}[[\vec{x}]]$ of all formal power series in $\vec{x}$ over $\mathbb{L}$ is a ring with the addition and multiplication as defined above.

**Example 2.42** Let $A(x; q)$ be the generating function for the number of words of size $n$ over alphabet $[k]$, where $q$ marks the number of $k$'s. Since each word can be decomposed as $\pi = \pi^{(1)} k \pi^{(2)} k \cdots \pi^{(m)} k \pi^{(m+1)}$ such that $\pi^{(j)}$ is a word over alphabet $[k-1]$, we have that

$$A(x; q) = \sum_{m \geq 0} x^m q^m \left( \frac{1}{1 - (k-1)x} \right)^{m+1}$$

$$= \frac{\frac{1}{1-(k-1)x}}{1 - \frac{xq}{1-(k-1)x}} = \frac{1}{1 - (k-1)x - xq}.$$

**Example 2.43** Let $a_n$ be the number words of size $n$ over alphabet $[k]$ with no two $k$'s adjacent. Let us find the generating function $A(x) = \sum_{n \geq 0} a_n x^n$. The letter $k$ does not occur in the word. In this case the generating function for such words is given by $\frac{1}{1-(k-1)x}$. The letter $k$ occurs at least once, say, $m$ times. In this case, such a word as $\pi$ can be decomposed as $\pi^{(1)} k \pi^{(2)} \cdots k \pi^{(m+1)}$, where $\pi^{(j)}$ is a nonempty word over alphabet $[k-1]$ for all $j = 2, 3, \ldots, m$, and $\pi^{(1)}$ and $\pi^{(m+1)}$ are words over alphabet $[k-1]$. Thus, the generating function for such words is given by

$$x^m \frac{1}{1-(k-1)x} \left( \frac{1}{1-(k-1)x} - 1 \right)^{m-1} \frac{1}{1-(k-1)x}$$

$$= \frac{x^{2m-1}(k-1)^{m-1}}{(1-(k-1)x)^{m+1}}.$$

Hence, the generating function $A(x)$ is given by

$$A(x) = \frac{1}{1-(k-1)x} + \sum_{m\geq 1} \frac{x^{2m-1}(k-1)^{m-1}}{(1-(k-1)x)^{m+1}}$$

$$= \frac{1}{1-(k-1)x} + \frac{x/(1-(k-1)x)^2}{1-(k-1)x^2/(1-(k-1)x)}$$

$$= \frac{1}{1-(k-1)x} + \frac{x}{(1-(k-1)x)(1-(k-1)x-(k-1)x^2)}.$$

Throughout the book we will be interested in the following families of generating functions.

**Definition 2.44** *A generating function $A(\vec{x}) \in \mathbb{L}[[x]]$ is called* rational *if there exist two polynomials $p(\vec{x}), q(\vec{x}) \in \mathbb{L}[\vec{x}]$ such that $A(\vec{x}) = \frac{p(\vec{x})}{q(\vec{x})}$. It is* algebraic *over a field $\mathbb{L}[\vec{x}]$ if there exists a nontrivial polynomial $R \in \mathbb{L}[\vec{x}]$ such that $A(\vec{x})$ satisfies a polynomial equation $R(A; x_1, \ldots, x_k) = 0$.*

Note that every rational generating function is algebraic.

**Example 2.45** *The generating function* $\mathrm{Fib}(x) = \frac{x}{1-x-x^2}$ *for the Fibonacci sequence is rational, whereas the generating function $A(x)$ satisfying the polynomial equation $A(x) = 1 + xA(x) + x^2 A(x)^2$ (see Example 2.62) is algebraic, but not rational.*

Klazar [180] showed that the ordinary generating function for the sequence $\{\mathrm{Bell}_n\}_{n\geq 0}$, that is, $\mathrm{Bell}(x) = \sum_{n\geq 0} \mathrm{Bell}_n x^n$ satisfies no algebraic equation over the field of rational functions $\bar{\mathbb{C}}(x)$ (see Exercise 2.21).

Now, we define two other operations on the ring $\mathbb{L}[[\vec{x}]]$ that will assist us in manipulating generating functions.

**Definition 2.46** *The* derivative *and the* integral *of $A(\vec{x})$ with respect to $x_i$ are defined by*

$$\frac{\partial}{\partial x_i} A(\vec{x}) = \sum_{\vec{n} \in \mathbb{N}^k} a_{\vec{n}} n_i x_i^{n_i - 1} \prod_{j \neq i} x_j^{n_j}$$

*and*

$$\int A(\vec{x}) dx_i = \sum_{\vec{n} \in \mathbb{N}^k} a_{\vec{n}} \frac{1}{n_i + 1} x_i^{n_i + 1} \prod_{j \neq i} x_j^{n_j},$$

*respectively.*

**Example 2.47** *The generating function for the sequence $\{nm + 2\}_{n,m\geq 0}$ is* *given by*

$$\sum_{n,m\geq 0}(nm+2)x^n y^m = 2\sum_{n,m\geq 0}x^n y^m + \sum_{n,m\geq 0}mnx^n y^m$$
$$= \frac{2}{(1-x)(1-y)} + xy\frac{\partial^2}{\partial x \partial y}\sum_{n,m\geq 0}x^n y^m$$
$$= \frac{2}{(1-x)(1-y)} + xy\frac{\partial^2}{\partial x \partial y}\left(\frac{1}{(1-x)(1-y)}\right)$$
$$= \frac{2}{(1-x)(1-y)} + \frac{xy}{(1-x)^2(1-y)^2}$$
$$= \frac{2-2x-2y+3xy}{(1-x)^2(1-y)^2}.$$

Now, let us see how we can obtain the ordinary, exponential generating function of a new sequence from the ordinar, exponential generating function of a related series. For example, if $A(x)$ is the ordinary, exponential generating function for the sequence $\{a_n\}_{n\geq 0}$, what is the ordinary, exponential generating function of the sequences $\{a_{n-1}\}_{n\geq 1}$ and $\{a_{n+1}\}_{n\geq 0}$? We use the definition of the ordinary, exponential generating function to obtain that

$$\sum_{n\geq 1}a_{n-1}x^n = x\sum_{n\geq 1}a_{n-1}x^{n-1} = xA(x),$$

$$\sum_{n\geq 1}a_{n-1}\frac{x^n}{n!} = \int_0^x \sum_{n\geq 1}a_{n-1}\frac{t^{n-1}}{(n-1)!}dt = \int_0^x A(t)dt$$

and

$$\sum_{n\geq 0}a_{n+1}x^n = \frac{1}{x}\sum_{n\geq 1}a_n x^n = \frac{1}{x}(A(x)-a_0),$$

$$\sum_{n\geq 0}a_{n+1}\frac{x^n}{n!} = \frac{d}{dx}\sum_{n\geq 0}a_{n+1}\frac{x^{n+1}}{(n+1)!} = \frac{d}{dx}A(x).$$

These prove that the ordinary generating functions for the sequences

$$\{a_{n-1}\}_{n\geq 1} \text{ and } \{a_{n+1}\}_{n\geq 0}$$

are given by $xA(x)$ and $\frac{1}{x}(A(x)-a_0)$, respectively, and the exponential generating functions for the sequences $\{a_{n-1}\}_{n\geq 1}$ and $\{a_{n+1}\}_{n\geq 0}$ are given by $\int_0^x A(t)dt$ and $\frac{d}{dx}A(x)$, respectively. By applying this rule repeatedly we obtain a general rule for computing the generating function of a shifted sequence from the ordinay/exponential generating function of the original sequence.

**Rule 2.48** (Wilf [358, Rule 1, Chapter 2]) *If $A(x)$ is the generating function for the sequence $\{a_n\}_{n\geq 0}$, then the generating functions for the sequences*

$\{a_{n+k}\}_{n\geq 0}$ and $\{a_{n-k}\}_{n\geq k}$, for any integer $k \geq 0$, are given by

$$\frac{A(x) - a_0 - a_1 x - \ldots - a_{k-1} x^{k-1}}{x^k} \text{ and } x^k A(x),$$

respectively.

**Proof** See Exercise 2.6. □

The above rule can be written in terms of exponential generating function as follows.

**Rule 2.49** (Wilf [358, Rule 1', Chapter 2]) *If $A(x)$ is the exponential generating function for the sequence $\{a_n\}_{n\geq 0}$, then the exponential generating functions for the sequences $\{a_{n+k}\}_{n\geq 0}$ and $\{a_{n-k}\}_{n\geq k}$, for any integer $k \geq 0$, are given by*

$$\underbrace{\int_0^x \int_0^{t_1} \cdots \int_0^{t_{k-1}}}_{k \text{ times}} A(t_{k-1}) dt_{k-1} \cdots dt_1 dx \text{ and } \frac{d^k}{dx^k} A(x),$$

respectively.

**Proof** See Exercise 2.6. □

Another rule is the relation between a sequence and its partial sum sequence. The *partial sum sequence* of sequence $\{a_n\}_{n\geq 0}$ is the sequence $\{a_0 + a_1 + \cdots + a_n\}_{n\geq 0}$.

**Rule 2.50** *If $A(x)$ is the generating function for the sequence $\{a_n\}_{n\geq 0}$, then the generating function for the sequence $\{a_0 + a_1 + \cdots + a_n\}_{n\geq 0}$ is given by $\frac{A(x)}{1-x}$.*

**Proof** See Exercise 2.7. □

Another very common modification of a sequence is multiplication by $n$, or in general, by a polynomial in $n$. Let us start again with the simplest case and determine the generating function of $\{n\, a_n\}_{n\geq 0}$ from the generating function $A(x)$ of $\{a_n\}_{n\geq 0}$. Again, we use the definition of the generating function and use algebraic transformations to express it in terms of $A(x)$:

$$\sum_{n\geq 0} n\, a_n x^n = x \sum_{n\geq 0} n\, a_n x^{n-1} = x A'(x).$$

Thus, multiplying a sequence by $n$ results in differentiating the generating function and then multiplying by $x$.

**Definition 2.51** *We denote the differential operator by $D$, and write $Df$ to indicate $f'$ for any function $f$. The second derivative of $f$ will be denoted by $D^2 f$, and in general, $D^k f$ denotes the k-th derivative of $f$.*

Therefore, the generating function for the sequence $\{na_n\}_{n\geq 0}$ is given by $xDA(x)$. Repeatedly applying this fact shows that the generating function for the sequence $\{n^k a_n\}_{n\geq 0}$ is given by $(xD)^k A(x)$. Together with the linearity of the generating function, we obtain the following general rule.

**Rule 2.52** (Wilf [358, Rule 2, Chapter 2]) *Let $P(n)$ be a polynomial in $n$. Then the generating function for the sequence $\{P(n)a_n\}_{n\geq 0}$ is given by $P(xD)A(x)$.*

We illustrate this rule with an example.

**Example 2.53** *What is the generating function of $\{\binom{n+1}{3}\}_{n\geq 0}$?* Looking at this sequence through the generating functions, we identify this as the sequence $\{1\}_{n\geq 0}$, multiplied by the polynomial $P(n) = \binom{n+1}{3} = \frac{1}{6}n^3 - \frac{1}{6}n$. Thus, the generating function for the sequence $\{\binom{n+1}{3}\}_{n\geq 0}$ is given by

$$\begin{aligned} A(x) &= \frac{1}{6}(xD)^3 \frac{1}{1-x} - \frac{1}{6}(xD)\frac{1}{1-x} \\ &= \frac{1}{6}(xD)^2 \frac{x}{(1-x)^2} - \frac{x}{6(1-x)^2} \\ &= \frac{1}{6}(xD)\frac{x(1+x)}{(1-x)^3} - \frac{x}{6(1-x)^2} \\ &= \frac{x(1+4x+x^2)}{6(1-x)^4} - \frac{x}{6(1-x)^2} \\ &= \frac{x^2}{(1-x)^4}. \end{aligned}$$

We can find the generating function $A(x)$ by using the following Maple code

```
simplify(sum(binomial(n+1,3)*x^n,n=2..infinity));
```

and we can simplify the generating function $A(x)$ by using the following Maple code

```
simplify(1/6*x*diff(x*diff(x*diff(1/(1-x),x),x),x)
-1/6*x*diff(1/(1-x),x));
```

So in this instance, the rule for the exponential generating function of a shifted sequence is simpler than the corresponding one for the ordinary generating function. If we look at the impact of multiplying a sequence with a polynomial, we obtain exactly the same result as in the case of ordinary generating functions.

**Rule 2.54** (Wilf [358, Rule 2', Chapter 2]) *Let $P(n)$ be a polynomial in $n$ and let $E(x)$ be the exponential generating function for the sequence $\{a_n\}_{n\geq 0}$. Then the exponential generating function for the sequence $\{P(n)a_n\}_{n\geq 0}$ is given by $P(xD)E(x)$.*

Now, let us see how we can derive the ordinary/exponential generating function equation of a new sequence which is defined by another sequence.

**Theorem 2.55** *Let $\{a_n\}_{n\geq 0}$ and $\{b_n\}_{n\geq 0}$ be any two sequences such that $b_n = \sum_{j=0}^n \binom{n}{j} a_j$ for all $n \geq 0$. Then the generating functions $B(x) = \sum_{n\geq 0} b_n x^n$ and $A(x) = \sum_{n\geq 0} a_n x^n$ satisfy*

$$B(x) = \frac{1}{1-x} A(x/(1-x)).$$

**Proof** Multiplying $b_n = \sum_{j=0}^n \binom{n}{j} a_j$ by $x^n$ and summing over all $n \geq 0$, we obtain

$$B(x) = \sum_{n\geq 0} \left( \sum_{j=0}^n \binom{n}{j} a_j \right) x^n = \sum_{j\geq 0} a_j \sum_{n\geq j} \binom{n}{j} x^n.$$

Thus, by Exercise 2.10, $B(x) = \sum_{j\geq 0} a_j \frac{x^j}{(1-x)^{j+1}} = \frac{1}{1-x} A(x/(1-x))$, as claimed. □

**Example 2.56** *In order to obtain the generating function for the number of set partitions of $[n]$, we state that the number $\text{Bell}_n$ of set partitions of $[n]$ satisfies the recurrence relation (see Example 2.11)*

$$\text{Bell}_n = \sum_{j=0}^{n-1} \binom{n-1}{j} \text{Bell}_j$$

*with the initial condition $\text{Bell}_0 = 1$. If we transform this recurrence in terms of generating functions by Theorem 2.55, we have*

$$B(x) - 1 = \frac{x}{1-x} B(x/(1-x)).$$

*Iterating the equation infinity times we obtain*

$$B(x) = \sum_{n\geq 0} \frac{x^n}{(1-x)(1-2x)\cdots(1-nx)}.$$

*If we would like to count those set partitions of $[n]$ with a fixed number of blocks, then we use the recurrence relation*

$$\text{Stir}(n, k) = k\text{Stir}(n-1, k) + \text{Stir}(n-1, k-1).$$

Multiplying the above recurrence by $x^n$ and summing over $n \geq k$, we obtain that the generating function $S_k(x) = \sum_{n \geq k} \text{Stir}(n,k) x^n$ satisfies

$$S_k(x) = kx S_k(x) + x S_{k-1}(x)$$

with $S_0(x) = 1$. Iterating the equation $k$ times we derive that

$$\sum_{n \geq k} \text{Stir}(n,k) x^n = \frac{x^k}{\prod_{i=1}^{k}(1-ix)}, \tag{2.9}$$

for all $k \geq 0$.

The above rule can be stated in terms of exponential generating functions as follows.

**Theorem 2.57** *Let $\{a_n\}_{n \geq 0}$ and $\{b_n\}_{n \geq 0}$ be any two sequences such that $b_n = \sum_{j=0}^{n} \binom{n}{j} a_j$ for all $n \geq 0$. Then the exponential generating functions $B(x) = \sum_{n \geq 0} b_n \frac{x^n}{n!}$ and $A(x) = \sum_{n \geq 0} a_n \frac{x^n}{n!}$ satisfy*

$$B(x) = e^x A(x).$$

**Proof** Multiplying $b_n = \sum_{j=0}^{n} \binom{n}{j} a_j$ by $\frac{x^n}{n!}$ and summing over all $n \geq 0$ we obtain

$$B(x) = \sum_{n \geq 0} \left( \sum_{j=0}^{n} \binom{n}{j} a_j \right) \frac{x^n}{n!} = \sum_{n \geq 0} \sum_{j=0}^{n} \frac{x^{n-j}}{(n-j)!} \frac{a_j x^j}{j!} = e^x A(x),$$

as claimed. □

**Example 2.58** *In order to obtain the exponential generating function for the number of set partitions of $[n]$, we state that the number $\text{Bell}_n$ of set partitions of $[n]$ satisfies the recurrence relation (see Example 2.11)*

$$\text{Bell}_{n+1} = \sum_{j=0}^{n} \binom{n}{j} \text{Bell}_j$$

with the initial condition $\text{Bell}_0 = 1$. If we transform this recurrence in terms of exponential generating functions by Theorem 2.57, then we have

$$\sum_{n \geq 0} \text{Bell}_{n+1} \frac{x^n}{n!} = e^x B(x),$$

which is equivalent to $\frac{d}{dx} B(x) = e^x B(x)$, and hence $B(x) = e^{e^x + c}$. Using the initial condition $B(0) = 1$ we obtain $B(x) = e^{e^x - 1}$.

Looking at Examples 2.56 and 2.58 ones can ask if there is a direct method to move from ordinary generating function $A(x)$ to exponential generating function $E(x)$ without doing direct enumeration from the beginning. The answer is yes; the technique is to find the coefficient of $x^n$ in $A(x)$, say $a_n$, and then write explicit formula for the exponential generating function for the sequence $\{a_n\}_{n\geq 0}$. Note that it seems at first glance this that procedure is very simple, but actually this is not correct as we will show in Chapter 5. To illustrate our technique, we consider the following example.

**Example 2.59** *Example 2.56 gives that the ordinary generating function for the number of set partitions of $[n]$ is given by $B(x) = \sum_{j\geq 0} \frac{x^j}{(1-x)(1-2x)\cdots(1-jx)}$. We expand it into partial fractions:*

$$B(x) = 1 + \sum_{m\geq 1} \frac{b_m}{1/x - m}.$$

The coefficient $b_m$ can be found by multiplying by $\frac{1}{x} - m$ and setting $x = \frac{1}{m}$:

$$b_m = \sum_{j\geq 0} \frac{1/x - m}{\prod_{i=1}^{j}(1/x - i)}\Big|_{x=1/m} = \sum_{j\geq m} \frac{1}{\prod_{i=1, i\neq m}^{j}(1/x - i)}\Big|_{x=1/m}$$

$$= \sum_{j\geq m} \frac{(-1)^{j-m}}{(m-1)!(j-m)!} = \frac{1}{(m-1)!} \sum_{j\geq m} \frac{(-1)^{j-m}}{(j-m)!}$$

$$= \frac{e^{-1}}{(m-1)!}.$$

Therefore,

$$\sum_{j\geq 1} \frac{x^j}{\prod_{i=1}^{j}(1-ix)} = \sum_{m\geq 1} \frac{e^{-1}x}{(m-1)!(1-mx)}.$$

Now we expand $\frac{x}{1-mx}$ into a geometric series:

$$\sum_{j\geq 1} \frac{x^j}{\prod_{i=1}^{j}(1-ix)} = \sum_{m\geq 1} \frac{e^{-1}}{m!} \sum_{n\geq 1} (mx)^n.$$

Since we would like to work with the exponential generating function rather than the ordinary generating function, we introduce a factor $\frac{1}{n!}$:

$$\sum_{m\geq 1} \frac{e^{-1}}{m!} \sum_{n\geq 1} \frac{(mx)^n}{n!} = \sum_{m\geq 1} \frac{e^{-1}}{m!} (e^{mx} - 1)$$

$$= e^{-1} \sum_{m\geq 1} \frac{e^{mx}}{m!} - e^{-1} \sum_{m\geq 1} \frac{1}{m!} = e^{-1}(e^{e^x} - 1) - e^{-1}(e^1 - 1)$$

$$= e^{e^x - 1} - 1.$$

## Basic Tools of the Book

Counting the empty set partition we obtain that the exponential generating function for the number of set partitions of $[n]$ is given by $e^{e^x-1}$, as discussed in Example 2.58.

Now we give two enumeration examples in lattice paths.

**Definition 2.60** *A lattice path of size $n$ is a sequence of points $z_1 z_2 \cdots z_n$ with $n \geq 1$ such that each point $z_i$ belongs to the plane integer lattice, and consecutive points $z_i$ and $z_{i+1}$ are connected by a line segment. We will consider lattice paths in $\mathbb{Z}^2$ whose permitted step types are up-steps $U = (1,1)$, down-steps $D = (1,-1)$, and horizontal (level) steps $H = (1,0)$.*

**Example 2.61** (Dyck paths and Catalan numbers) *A Dyck path of size $2n$ is a lattice path consisting of $n$ $U$s and $n$ $D$s such that no initial segment of the path has more $D$s than $U$s. For example, the Dyck paths of size six are*

$$UDUDUD,\ UDUUDD,\ UUDDUD,\ UUDUDD,\ \text{and}\ UUUDDD.$$

*Dyck paths of size $2n$ can be used to encode valid arrangements of parentheses in a mathematical expression, with a $U$ representing a left parenthesis and a $D$ representing a right parenthesis. The above Dyck paths correspond to the following parentheses arrangements:*

$$()()(),\quad ()(()),\quad (())(),\quad (()()),\quad \text{and}\quad ((())).$$

*We now like to count the number of Dyck paths using generating functions. In order to do that, we need to set up a recurrence relation of some form. Similar to how we proceeded in Example 2.56, we decompose the Dyck path into smaller Dyck paths by using the first return decomposition: For each position $j$, compute the difference between the number of $U$s and the number of $D$s in the first $j$ positions. Clearly, $j \geq 0$. We consider the first position $j$ where there is an equal number of $U$s and $D$s. Then we can write the Dyck path $P$ of size $2n$ as $P = U P' D P''$, where the $D$ is at position $j = 2k + 2$. (Note that $P'$ and $P''$ may be empty paths.) By construction, $P'$ is a Dyck path of size $2k$, and since the number of $U$s and $D$s balance at position $2k + 2$, the word $P''$ is also a Dyck path, and it has size $2(n - k - 1)$. This representation is unique, and we can therefore count the Dyck paths according to this decomposition to get the following recurrence relation, where $d_n$ counts the Dyck paths of size $2n$:*

$$d_n = \sum_{k=0}^{n-1} d_k\, d_{n-k-1} \quad \text{for } n \geq 1.$$

*Let $D(x)$ be the generating function for the sequence $\{d_n\}_{n \geq 0}$. Adjusting for the fact that the convolution sum has upper index $n - 1$ instead of $n$ we get*

$$D(x) - 1 = x D(x)^2$$

which implies
$$D(x) = \frac{1 - \sqrt{1-4x}}{2x}.$$

We can solve the above equation by using the following Maple code

```
solve(dd-1=x*dd^2,x);
```

In order to extract the coefficients of $x^n$ in $D(x)$, we make use of the following well-known identity (see (C.1))

$$\sqrt{1+t} = \sum_{m \geq 0} \binom{1/2}{m} t^m = 1 + \sum_{m \geq 1} (-1)^{m-1} \frac{(2m-2)!}{2^{2m-1} m!(m-1)!} t^m.$$

Thus,

$$\begin{aligned} D(x) &= -\frac{1}{2x} \sum_{n \geq 1} (-1)^{n-1} \frac{(2n-2)!}{2^{2n-1} n!(n-1)!} (-4x)^n \\ &= \sum_{n \geq 1} \frac{1}{n} \binom{2n-2}{n-1} x^{n-1} = \sum_{n \geq 0} \frac{1}{n+1} \binom{2n}{n} x^n. \end{aligned} \qquad (2.10)$$

This implies that $d_n$ is given by the n-th Catalan number $\text{Cat}_n = \frac{1}{n+1}\binom{2n}{n}$. The Catalan sequence[7] $1, 1, 2, 5, 14, 132, 429, 1430, 4862, 16796, \ldots$, which occurs as Sequence A000108 in [327] counts an enormous number of different combinatorial structures (see Stanley [334, Page 219 and Exercise 6.19]). It was first described in the 18th century by Leonard Euler,[8] who was interested in the number of different ways of dividing a polygon into triangles. The sequence is named after Eugène Charles, Catalan[9] who also worked on the problem and discovered the connection to parenthesized expressions.

**Example 2.62** (Motzkin paths and Motzkin numbers) *A Motzkin path of size n is a lattice path consisting of n letters $U, D, H$, such that no initial segment of the path has more $D$s than $U$s. For example, the Motzkin paths of size four are $HHHH, HHUD, HUDU, HUHD, UDHH, UDUD, UHDU, UHHD$, and $UUDD$. We now like to count the number of Motzkin paths using generating functions. To do so, we need to set up a recurrence relation of some form. Similar to how we proceeded in Example 2.61, we decompose the Motzkin word into smaller Motzkin paths by using the first return decomposition: For each position $j$, compute the difference between the number of $U$s and the number of $D$s in the first $j$ positions. Clearly, $j \geq 0$. We consider the first position $j$ where there is an equal number of $U$s and $D$s. Then we can write the Motzkin path $P$ of size $n$ as either $P = HP'$ or $P = UP''DP'''$, where the $D$ is at position $j \geq 2$. (Note that $P', P'',$ and $P'''$ may be empty paths.) By*

---

[7] http://mathworld.wolfram.com/CatalanNumber.html
[8] http://turnbull.mcs.st-and.ac.uk/ history/Mathematicians/Euler.html
[9] http://turnbull.mcs.st-and.ac.uk/ history/Mathematicians/Catalan.html

construction, $P'$ is a Motzkin path of size $n-1$, $P''$ is a Motzkin path of size $j-2$, and since the number of $U$s and $D$s balance at position $j$, the lattice path $P'''$ is also a Motzkin path, and it has size $n-j$. This representation is unique, and we can therefore count the Motzkin paths according to this decomposition to get the following recurrence relation, where $m_n$ counts the Motzkin paths of size $n$:

$$m_n = m_{n-1} + \sum_{k=2}^{n} m_{k-2}\, m_{n-k} \quad n \geq 1.$$

Let $M(x)$ be the generating function for the sequence $\{m_n\}_{n\geq 0}$. Adjusting for the fact that the convolution sum has upper index $n-1$ and instead of $n$ we get

$$M(x) - 1 = xM(x) + x^2(M(x))^2,$$

which implies

$$M(x) = \frac{1 - x - \sqrt{1 - 2x - 3x^2}}{2x^2}.$$

We can solve the above equation by using the following Maple code

```
solve(mm-1=x*mm+x*mm^2,x);
```

In order to extract the coefficient $x^n$ in $M(x)$, we make use of the following well-known identity (see (C.1)):

$$\sqrt{1+t} = \sum_{k\geq 0} \binom{1/2}{k} t^k = 1 + \sum_{k\geq 1}(-1)^{k-1}\frac{(2k-2)!}{2^{2k-1}k!(k-1)!} t^k.$$

Thus,

$$M(x) = \frac{-1}{2x} + \frac{1}{2x^2}\sum_{k\geq 1}\frac{(2k-2)!}{2^{2k-1}k!(k-1)!} x^k(2+3x)^k$$

$$= \frac{-1}{2x} + \sum_{k\geq 1}\sum_{j=0}^{k}\binom{k}{j}\frac{2^{k-j}3^j(2k-2)!}{2^{2k}k!(k-1)!} x^{k+j-2},$$

which implies that the coefficient of $x^n$ in $M(x)$ is given by

$$\sum_{k\geq 1}\binom{k}{n+2-k}\frac{3^{n+2-k}(2k-2)!}{2^{n+2}k!(k-1)!}$$

$$= \frac{3^{n+2}}{2^{n+2}}\sum_{k\geq 1}\frac{3^{-k}}{k}\binom{2k-2}{k-1}\binom{k}{n+2-k},$$

for all $n \geq 1$. This implies that $m_n$ is given by the m-th Motzkin number $\mathrm{Mot}_n = \frac{3^{n+2}}{2^{n+2}}\sum_{k\geq 1}\frac{3^{-k}}{k}\binom{2k-2}{k-1}\binom{k}{n+2-k}$. The Motzkin sequence[10] 1, 1, 2, 4, 9, 21, 51, 127, 323, 835, ..., which occurs as Sequence A001006 in [327] counts

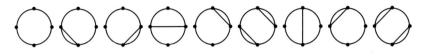

**FIGURE 2.1**: Different ways of drawing non-intersecting chords on a circle between 4 points

an enormous number of different combinatorial structures (see Stanley [334, Page 238 and Exercise 6.37]).

It was first described by Theodore Samuel Motzkin,[11] who was interested in the number of different ways of drawing nonintersecting chords on a circle between $n$ points (see Figure 2.1 when $n = 4$).

**Definition 2.63** *We define $[\sigma]$ to be a subsequence of a longer sequence $\pi$, where $[\sigma]$ refers to the sequence $\sigma + k$ with $k$ is the number of distinct symbols of $\pi$ that appear before the first symbol of $\sigma$ in $\pi$.*

**Example 2.64** *The sequence $1[1122]1$ refers to the sequence $122331$, and $11[1221][1122]$ should be understood as $1123324455$.*

We can ask the following question: Does there exist a bijection between the set of different ways of drawing nonintersecting chords on a circle between $n$ points and the set of Motzkin paths of size $n$? (See Exercise 2.11)

A connection between set partitions and Motzkin paths can be given as follows.

**Example 2.65** *Let $A_n$ be the set partitions $\pi = \pi_1 \pi_2 \cdots \pi_n$ of $[n]$ such that there are no $i < j < k$ with $\pi_i = \pi_j = \pi_k$ and there is are no $i < j < k < \ell$ such that $\pi_i = \pi_k < \pi_j = \pi_\ell$ (later we call this properties by avoiding 1212 and 111). Each set partition $\pi \in A_n$ can be decomposed as either $\pi = 1[\pi']$ or $\pi = 1[\pi']1[\pi'']$.*

*Now we define $\rho$ a map between $A_n$ and the set of Motzkin paths of size $n$ as follows. We read the letters of a set partition $\pi$ in $A_n$ from left to right and in each step we add steps to our path $P$. When we see a letter occurs exactly once in $\pi$ we add to the path a step $H$. When a letter occurs twice (clearly, any letter can occur at most twice, otherwise there exists $i < j < k$ such that $\pi_i = \pi_j = \pi_k$), we add $U$ if it is the leftmost letter and $D$ if it is the rightmost letter in $\pi$. Clearly, at the end, $P$ is a Motzkin path of size $n$, and $\rho$ is a bijection.*

---

[10] http://en.wikipedia.org/wiki/Motzkin_number
[11] http://en.wikipedia.org/wiki/Theodore_Motzkin

## 2.4 Lagrange Inversion Formula

Here, we present a strong tool for solving certain kinds of functional equations: the Lagrange inversion formula. It is useful for deriving explicit formulas and studying asymptotic analysis. The Lagrange inversion formula applies to functional equations of the form $z = x\,\phi(z)$, where $\phi(z)$ is a given function of $z$, when our goal is to determine $z$ as a function of $x$. More precisely, assume we are given the power series expansion of the function $\phi = \phi(z)$, convergent in some neighborhood of the origin. Thus, the question is to find the power series expansion of the solution of $z = x\phi(z)$, say, $z = z(x)$, in some neighborhood of the origin. The answer is given by the following theorem.

**Theorem 2.66** (Wilf [358, Theorem 5.1.1], Lagrange inversion formula) *Let $f(z)$ and $\phi(z)$ be formal power series in $z$, with $\phi(0) = 1$. Then there is a unique formal power series $z = z(x)$ that satisfies $z = x\phi(z)$. Furthermore, the value $f(z(x))$ of $f$ at that root $z = z(x)$, when expanded in a power series in $x$ about $x = 0$, satisfies*

$$[x^n]f(z(x)) = \frac{1}{n}[z^{n-1}](f'(z)\phi^n(z)). \tag{2.11}$$

Note that the function $f$ in (2.11) allows us to compute the coefficients of the unknown function $z(x)$ (in which case $f(z) = z$), or any function of $z(x)$, for example, the $k$-th power (in which case $f(z) = z^k$). For instance, let's give an example of the Lagrange inversion formula in action: we will derive once more that Dyck paths of size $2n$ are counted by the Catalan numbers, this time with the help of the Lagrange inversion formula.

**Example 2.67** *In Example 2.61 we have derived that the generating function $C(x)$ for the number of Dyck paths of size $2n$ satisfies the functional equation $C(x) = 1 + xC^2(x)$. Now, we define $z(x) = C(x) - 1 = xC^2(x)$. This implies that $z(x) = C(x) - 1$, $\phi(z) = (1 + z)^2$ and $\phi(0) = 1$. If we choose $f(z) = z$ and apply Theorem 2.66, then we obtain*

$$[x^n]C(x) = \frac{1}{n}[z^{n-1}](1+z)^{2n} = \frac{1}{n}\binom{2n}{n-1} = \frac{1}{n+1}\binom{2n}{n},$$

*for all $n \geq 1$. Thus, we have once more proved that the number of Dyck paths of size $2n$ for $n \geq 1$ is given by the $n$-th Catalan number $\mathrm{Cat}_n = \frac{1}{n+1}\binom{2n}{n}$.*

This example shows that the Lagrange inversion formula can give results very easily; earlier, we had to work a little harder and use a special identity to obtain the coefficients in Example 2.61.

## 2.5 The Principle of Inclusion and Exclusion

In this section we deal with another useful and powerful tool in enumeration, which is the principle of inclusion and exclusion, see [82]. To do so, we need the following definition.

**Definition 2.68** *For two sets $A$ and $B$ we define*

1. *Union of the sets $A$ and $A'$, denoted $A \cup A'$, is the set of all elements that are members of $A$, $A'$, or both.*

2. *Intersection of the sets $A$ and $A'$, denoted $A \cap A'$, is the set of all elements that are members of both $A$ and $A'$.*

3. *Cartesian product of the sets $A$ and $A'$, denoted $A \times A'$, is the sets whose members are all possible ordered pairs $(a, a')$, where $a \in A$ and $a' \in A'$.*

Also, let $B \subset A$, we define the complement of the set $B$, denoted $\overline{B}$, to be the set of all elements that are members of $A$ and not members of $B$. This operation is also called the set difference of $A$ and $B$, denoted by $A \backslash B$. Let $A$ and $B$ be any two sets. We say that $A$ and $B$ are disjoint sets if $A \cap B = \emptyset$, or in words $A$ and $B$ have no element in common, where we denote the empty set by $\emptyset$.

**Example 2.69** *Let $A = [2] = \{1,2\}$, $B = \{-1,3,7\}$ and $C = \{-1,1\}$ be three subsets of the set of $\{-1,0,1,2,3,4,5,6,7\}$. Then $2 \in A$ but $3$ is not element of $A$, that is, $3 \notin A$. Also, the union of $A$ and $B$ is the set $A \cup B = \{-1,1,2,3,7\}$ and union of $A$ and $C$ is the set $A \cup C = \{-1,1,2\}$. The intersection of $B$ and $C$ is the set $B \cap C = \{-1\}$. The complement of $A \cup B$ is the set $\{0,4,5,6\}$. The set difference of $A$ and $C$ is the set $\{2\}$.*

There several properties of the operations union, intersection, complement and set difference. Here we summarize some of these properties, where we omit the proof.

**Fact 2.70** *For any set $A$,*

$$A \cap \emptyset = \emptyset, \quad A \cup \emptyset = A, \quad A \backslash \emptyset = A,$$

*and for any subset $B$ of $A$*

$$\overline{B} \cap B = \emptyset, \quad \overline{B} \cup B = A, \quad \overline{\overline{B}} = B.$$

*For any two sets $A$ and $B$,*

$$A \cap B = B \cap A, \quad A \cup B = B \cup A$$

and for any two subsets $B$ and $C$ of $A$,

$$\overline{B \cap C} = \overline{B} \cup \overline{C} \text{ and } \overline{B \cup C} = \overline{B} \cap \overline{C} \text{ (de Morgan laws)}.$$

For any three sets $A$, $B$ and $C$,

$$A \cap (B \cap C) = (A \cap B) \cap C, \qquad A \cup (B \cup C) = (A \cup B) \cup C,$$
$$A \cap (B \cup C) = (A \cap B) \cup (A \cap C).$$

Our first step for stating the principle of inclusion and exclusion is to introduce the addition principle which can be expressed in terms of sets as follows: For any two disjoint sets $A$ and $B$,

$$|A \cup B| = |A| + |B|,$$

where $|A|$ denotes the number of elements of the set $A$. In this statement, we require that the sets $A$ and $B$ are disjoint. Can you express $|A \cup B|$ when $A$ and $B$ are not necessary disjoint sets?. To do so, any element in $A \cap B$ (if there is) is counted exactly twice when we count elements of $A$ and elements of $B$. Thus, to obtain the exact count of $|A \cup B|$, the number $|A| + |B|$ should be corrected. It follows that

$$|A \cup B| = |A| + |B| - |A \cap B|. \tag{2.12}$$

**Example 2.71** *Find the number of positive integers from the set* $[100]$ *which divisible by 4 or 5? The positive integers which we are looking for are 4, 5, 8, 10, 12, 15, 16, 20, ..., 100, how many integers here?. In order to present the solution more formally, let $A$ be the set of all integers in $[100]$ which divisible by 4 and $B$ be the set of integers in $[100]$ which are divisible by 5. Clearly, our task is to find $|A \cup B|$. By (2.12) we have that*

$$|A \cup B| = 25 + 20 - |A \cap B| = 25 + 20 - 5 = 40,$$

*where $A \cap B$ is the set of all integers in $[100]$ which are divisible by 4 and 5, that is divisible by 20.*

Identity (2.12) is the simplest form of a principle called the *principle of inclusion and exclusion*, which is a very useful and powerful tool in enumeration.

**Theorem 2.72** (The Principle of Inclusion and Exclusion) *Let $A_1, \ldots, A_k$ be any subsets in a set $A$. Then*

$$|A_1 \cap \cdots \cap A_k| = \sum_{j=1}^{k} (-1)^{j-1} \left( \sum_{1 \leq i_1 < \cdots < i_j \leq k} |A_{i_1} \cap \cdots \cap A_{i_j}| \right).$$

**Proof** We define the characteristic function respect to $A$, denoted by $I_A$, to be the function $I_A : A \to \{0, 1\}$ with $I_A(a) = 1$ if $a \in A$, and $I_A(a) = 0$ otherwise. Note that the function $I_A$ satisfies the following properties

1. $I_{\bar{B}} = I_A - I_B$, for any subset $B$ of a set $A$.

2. $I_{A \cap B} = I_A \cdot I_B$, for any two sets $A$ and $B$.

Thus, by Fact 2.70, 1 and 2 we have

$$\begin{aligned} I_{\overline{A_1 \cup \cdots \cup A_k}} &= I_{\overline{A_1} \cap \cdots \cap \overline{A_k}} \\ &= I_{\overline{A_1}} \cdots \cdots I_{\overline{A_k}} \\ &= (I_A - I_{A_1}) \cdots \cdots (I_A - I_{A_k}) \\ &= I_A + \sum_{j=1}^{k}(-1)^j \left( \sum_{1 \leq i_1 < \cdots < i_j \leq k} I_{A_{i_1} \cap \cdots \cap A_{i_j}} \right). \end{aligned}$$

Therefore, by summing the values $I_{\overline{A_1 \cup \cdots \cup A_k}}(s)$ over all $s \in A$ we obtain

$$|\overline{A_1 \cup \cdots \cup A_k}| = |A| + \sum_{j=1}^{k}(-1)^j \left( \sum_{1 \leq i_1 < \cdots < i_j \leq k} |A_{i_1} \cap \cdots \cap A_{i_j}| \right),$$

which implies that

$$\begin{aligned} |A_1 \cup \cdots \cup A_k| &= |A| - |\overline{A_1 \cup \cdots \cup A_k}| \\ &= \sum_{j=1}^{k}(-1)^{j-1} \left( \sum_{1 \leq i_1 < \cdots < i_j \leq k} |A_{i_1} \cap \cdots \cap A_{i_j}| \right), \end{aligned}$$

as required. □

**Example 2.73** (Theorem 1 in [77]) *We say that a set partition of* $A = \{x_1, x_2, \ldots, x_n\}$ *has distance greater than $\ell$ if the distance between any two elements in the same block is at least $\ell$, that is, $|i-j| \geq \ell$ for any two different elements $x_i, x_j$ in the same block. Let $n \geq \max\{1, \ell\}$, then, the number of set partitions of $A$ have distance greater than $\ell$ is given by*

$$S_\ell(n,k) = \sum_{j=0}^{k-\ell} \frac{(-1)^{k-\ell-j}}{(k-\ell)!} \binom{k-\ell}{j} j^{n-\ell}.$$

*We proceed the proof by the principle of inclusion and exclusion. Fix $n > \ell$. Let $M$ be the set of all distributions of $A$ into $k$ distinguishable blocks $M_1, \ldots, M_k$ with $|i - j| > \ell$ for any two elements $x_i, x_j$ in the same block. We denote the subset $M$ that contains of those distributions for which the block $M_k$ is empty by $B_k$. Let $S_\ell^*(n, k)$ be the number of the distributions in $M$ such that there no empty block. Then we have $S_\ell^*(n, k) = k! S_\ell(n, k)$ and $S_\ell^*(n,k) = |\cap_{j=1}^{k} \overline{B_j}|$. Let $\theta = \theta_1 \cdots \theta_j$ be any increasing sequence with terms in $[n]$. The cardinality $|\cap_{j=1}^{k} \overline{B_j}|$ can be found as follows: We distribute the first $\ell$ elements of $A$ into $\ell$*

blocks from $k-j$ ones (excluding $M_i$ with $i$ as term of $\theta$), which equals $\binom{k-j}{\ell}\ell!$. On the other hand, we have $(k-\ell-j)^{n-\ell}$ ways to distribute the remaining $n-\ell$ elements $x_{\ell+1},\ldots,x_n$ into the $k-j$ blocks such that $x_i$ with $\ell < i \le n$ would not meet of the precedent $\ell$ elements $x_{i-1},\ldots,x_{i-\ell}$. Thus,

$$\left|\cap_{i=1}^{j} B_{\theta_i}\right| = \binom{k-j}{\ell}\ell!(k-\ell-j)^{n-\ell}.$$

According to Theorem 2.72 we obtain that

$$\begin{aligned}
S_\ell^*(n,k) &= \sum_{j=0}^{k}\sum_{\theta}(-1)^j \left|\cap_{i=1}^{j} B_{\theta_i}\right| \\
&= \sum_{j=0}^{k}(-1)^j \binom{k}{j}\binom{k-j}{\ell}\ell!(k-\ell-j)^{n-\ell} \\
&= \frac{k!}{(k-\ell)!}\sum_{j=0}^{k-\ell}(-1)^j\binom{k-\ell}{j}(k-\ell-j)^{n-\ell} \\
&= \frac{k!}{(k-\ell)!}\sum_{j=0}^{k-\ell}(-1)^{k-\ell-j}\binom{k-\ell}{j}j^{n-\ell},
\end{aligned}$$

which implies

$$S_\ell(n,k) = \sum_{j=0}^{k-\ell}\frac{(-1)^{k-\ell-j}}{(k-\ell)!}\binom{k-\ell}{j}j^{n-\ell},$$

as required.

---

## 2.6 Generating Trees

In this section we deal with another useful and powerful tool in enumeration, which is the method of generating trees, see [353]. To do so, we start with the following definition.

**Definition 2.74** *A plane tree $T$ can be defined recursively as a finite set of nodes, such that one distinguished node $r$ is called the* root *of $T$, and the remaining nodes form an ordered partition $(T_1,T_2,\ldots,T_m)$ of $m$ disjoint nonempty sets, each of which is a plane tree. We will draw plane trees with the root on the top level. The edges connecting the root of the tree to the roots of $T_1,T_2,\ldots,T_m$, which will be drawn from left to right on second level. For each node $v$, the nodes in the next lower level adjacent to $v$ are called the* children *of $v$, and $v$ is said to be their* parent. *Clearly, each node other than*

$r$ has exactly one parent. A node of $T$ is called a leaf if it has no children (by convention, we assume that the empty tree, formed by a single node, has no leaves), otherwise it is said to be an internal node.

It is well known that the number of plane trees with $n$ edges equals $\operatorname{Cat}_n = \frac{1}{n+1}\binom{2n}{n}$; see [335, Exercise 6.19e].

Generating trees were introduced by West in [353] as a method to capture the structure of a recurrence. He described the $n$-th term of the recurrence as number of nodes in the $n$-th level of the associated generating tree. Since this has become standard terminology in any paper on combinatorics, we will introduce this technique here.

**Definition 2.75** *A generating tree is a labeled plane tree (the vertices have labels) such that if $v_1$ and $v_2$ are two nodes with the same label, and $\ell$ is any label, then $v_1$ and $v_2$ have exactly the same number of children with label $\ell$. To specify a generating tree it therefore suffices to specify:*

1. *the label of the root, and*

2. *a set of succession rules explaining how to derive from the label of a parent the labels of all of its children.*

**Example 2.76** (The Complete Binary Tree) *The complete binary tree is a plane tree in which each vertex has exactly two children. Since all the nodes in the complete binary tree are identical, it is enough to use only one label, which we choose to be (2). So we get the following description:*

**Root**: (2)

**Rule**: (2) $\rightsquigarrow$ (2)(2).

**Example 2.77** (The Fibonacci Tree) *The Fibonacci tree has many uses in computer science applications. It is a variant of the binary tree and can be visualized as follows. The root is a red node. A red node has a blue offspring, and a blue node has a blue and a red offspring. The two different types of nodes are thus distinguished by the number of offspring, so we label the nodes according to their number of children and obtain these rules:*

**Root**: (1)

**Rules**: (1) $\rightsquigarrow$ (2), (2) $\rightsquigarrow$ (1)(2).

West [353] introduced this object to study the number of permutations of $[n]$ satisfying certain set of conditions. Here we extend this to words and set partitions as follows.

**Example 2.78** We can count the words in $[k]^n$ by inserting a letter from $[k]$ to the right of the rightmost letter of a word. For the purpose of enumeration, we only keep track of labels on words in the case when members of $[k]^n$ are formed from the empty word. For example, the generating tree for the words in $[k]^n$ is given by

**Root:** $(k)$

**Rules:** $(k) \rightsquigarrow (k)^k$,

where $(i)^k$ is shorthand for $k$ labels $(i)$. The generating tree for the words is given in Figure 2.2, showing both the words and their respective labels.

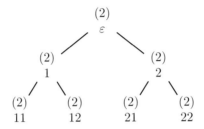

**FIGURE 2.2**: Generating tree for $[2]^n$.

**Example 2.79** In setting of the pervious example, set partitions of $[n]$ will be regarded as words of the form $\pi_1 \pi_2 \cdots \pi_{n-1} j$ over the alphabet $\{1, 2, \ldots\}$, where the letter $j$ belongs to the set $\{1, 2, \ldots, 1 + \max_j \pi_j\}$. Thus, if we label each set partition by a label $(a)$, where $a$ is the maximum element of the partition, we see that the generating tree for the set partitions of $[n]$ is given by

**Root:** $(1)$

**Rules:** $(a) \rightsquigarrow (a)^a (a+1)$,

where $(i)^k$ is shorthand for $k$ labels $(i)$. The generating tree for the words is given in Figure 2.3, showing both the set partitions and their respective labels.

The next step is to derive generating trees for set partitions that satisfy certain set of conditions. Given a set $S$ of conditions, we define $\mathcal{P}_n(S)$ to be set of all set partitions of $[n]$ that do not satisfy any condition in the set $S$. One can define the rooted tree for set partitions $\mathcal{P}_n(S)$ as follows. The nodes at level $n$ are precisely the elements of $\mathcal{P}_n(S)$. The parent of a set partition $\pi = \pi_1 \pi_2 \cdots \pi_{n+1} \in \mathcal{P}_{n+1}(S)$ is the unique set partition $\pi' = \pi_1 \pi_2 \cdots \pi_n$. Note that here we define the parent via the child, which ensures that the

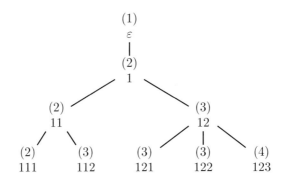

**FIGURE 2.3**: Generating tree for set partitions.

conditions in $S$ are preserved. If we were to define the child via the parent, then we would have to delete those children for which the insertion of the value $\pi_{n+1}$ leads to that $\pi$ satisfies one of the conditions of $S$ or $\pi$ is not a set partition.

**Definition 2.80** *The generating tree whose vertices at level $n$ are the set partitions of $\mathcal{P}_n(S)$ is denoted by $\mathcal{T}(S)$.*

The difficult work, which is truly an art, consists of translating this backward structural description of the generating tree $\mathcal{T}(S)$ into succession rules. Examples of pattern avoidance in permutations, words and compositions can be found in [48, 138] and the references cited therein. Here we give an example of pattern avoidance in set partitions.

**Example 2.81** *Let $\mathcal{P}_n(121)$ be the set of partitions $\pi = \pi_1\pi_2\cdots\pi_n \in \mathcal{P}_n$ such that there no $i,j,k$ with $1 \leq i < j < k \leq n$ and $\pi_i = \pi_k < \pi_j$. If $\pi = \pi_1\pi_2\cdots\pi_n \in \mathcal{P}_n(121)$ then its children are exactly $\pi j \in \mathcal{P}_{n+1}(121)$ such that either $j = \max_i \pi_i$ or $j = 1 + \max_i \pi_i$. Thus, the tree $\mathcal{T}(121)$ can be expressed as*

    **Root**: $(1)$

    **Rules**: $(2) \rightsquigarrow (2)$.

*Since each node in the tree has two children but the root has only one child, we get that the number of nodes at level $n$ is given by $2^{n-1}$. Hence, $|\mathcal{P}_n(121)| = 2^{n-1}$ for all $n \geq 1$.*

It is not always easy to move from the generating tree $\mathcal{T}(S)$ to explicit formula for $\mathcal{P}_n(S)$ as shown in the above example. In this case, maybe it is helpful to translate our generating tree to a functional equation.

Basic Tools of the Book                                              71

**Example 2.82** *The generating tree $\mathcal{T}(23\text{-}1)$ for the set partitions $\pi = \pi_1\pi_2\cdots\pi_n$ of $[n]$ that there is no $i,j$ such that $i+1 < j$ and $\pi_j < \pi_i < \pi_{i+1}$ is given by*

**Root:** $(1,1,1)$

**Rules:** $(a,b,c) \rightsquigarrow (a,b,a)\cdots(a,b,c)(c,b,c+1)\cdots(c,b,b)(c,b+1,b+1).$

*To see this, we label each set partition $\pi = \pi_1\cdots\pi_n \in \mathcal{P}_n$ by $(a,b,c)$, where*

$$c = \pi_n,\ b = \max_{1 \le i \le n} \pi_i,\ \text{and}\ a = \begin{cases} 1, & \text{if } \pi = 11\cdots 1; \\ \max\{\pi_i : \pi_i < \pi_{i+1}\}, & \text{otherwise}. \end{cases}$$

*Clearly, the set partition $\{1\}$ of $[1]$ is labelled by $(1,1,1)$ and $c \ge a$ (for otherwise, $\pi$ would contain the pattern 23-1). If we have a set partition $\pi$ associated with a label $(a,b,c)$, then each child of $\pi$ is a set partition of the form $\pi' = \pi c'$, where $c' = a, a+1, \ldots, b+1$, for if $c' < a$, then $\pi'$ would contain 23-1, which is not allowed. Thus, we have three cases:*
*(1) if $b \ge c' > c$, then $\pi'$ is labelled by $(c,b,c')$,*
*(2) if $c' = b+1$, then $\pi'$ is labelled by $(c,b+1,b+1)$,*
*(3) if $c \ge c' \ge a$, then $\pi'$ is labelled by $(a,b,c')$.*
*Combining the above cases yields our generating tree.*
*Looking at our generating tree $\mathcal{T}(23\text{-}1)$, we see it is not easy to find explicit formula for the number of nodes at the n-th level.*

To end this, we note that most of the cases, in general, we do not know how to deal with finding an explicit formula for the number of nodes at level $n$ in the generating tree $\mathcal{T}(S)$. But some cases can be done as we will show in the next chapters. Also, note that if the two trees $\mathcal{T}(S)$ and $\mathcal{T}(S')$ are isomorphic, that means there a bijection between the generating tree $\mathcal{T}(S)$ and $\mathcal{T}(S')$, then the number of nodes at the $n$-th level in $\mathcal{T}(S)$ is the same number of nodes at the $n$-th level in $\mathcal{T}(S')$, namely $|\mathcal{P}_n(S)| = |\mathcal{P}_n(S')|$.

---

## 2.7 Exercises

**Exercise 2.1** *Write an explicit formula for the n-th term of the sequence $a_n$, where*

  i. $a_n$ *is the square of $n$.*

 ii. $a_n$ *is the sum of the first $n$ positive integer numbers.*

iii. $a_n$ *is the number of subsets of a set of $n$ elements.*

iv. $a_n$ *is the number of functions from the set $[n]$ to the set $[3]$.*

**Exercise 2.2** For all $n, k \geq 1$,

i. $\binom{n}{k} = \binom{n-1}{k-1} + \binom{n-1}{k}$.

ii. $\sum_{i=0}^{n} \binom{n}{i} = 2^n$.

iii. $\sum_{i=k}^{n} \binom{i}{k} = \binom{n+1}{k+1}$.

**Exercise 2.3** Prove Theorem 2.26.

**Exercise 2.4** Find explicit formula for $L_n$, the n-th Lucas number.

**Exercise 2.5** Prove Theorem 2.30.

**Exercise 2.6** Prove Rule 2.48 and Rule 2.49.

**Exercise 2.7** Prove Rule 2.50.

**Exercise 2.8** Prove the identity $\sum_{i=0}^{n} \binom{n}{i} \text{Fib}_i = \text{Fib}_{2n}$ for $n \geq 0$, (where $F_n$ is the n-th Fibonacci number) by using rules for generating functions.

**Exercise 2.9** The n-th Pell number, denoted by $\text{Pell}_n$, is defined via the following recurrence relation (see [327, Sequence A000129])

$$\text{Pell}_n = 2\text{Pell}_{n-1} + \text{Pell}_{n-2}, \quad \text{Pell}_0 = 0 \text{ and } \text{Pell}_1 = 1.$$

i. Find the generating function $F(x) = \sum_{n \geq 0} \text{Pell}_n x^n$.

ii. Show that $\text{Pell}_n = \sum_{j=0}^{\lfloor \frac{n-1}{2} \rfloor} \binom{n-1-j}{j} 2^{n-1-2j}$.

iii. Show that $\text{Pell}_n = \frac{\sqrt{2}}{4} \left( (1+\sqrt{2})^n - (1-\sqrt{2})^n \right)$.

**Exercise 2.10** Prove that for all $k \geq 1$,

$$\frac{1}{(1-x)^k} = \sum_{j \geq 0} \binom{j+k-1}{k-1} x^j.$$

**Exercise 2.11** Find a bijection between the set of different ways of drawing nonintersecting chords on a circle between $n$ points and the set of Motzkin paths of size $n$.

**Exercise 2.12** Let $A_n$ be the set of all Dyck paths of size $2n$ with no maximal sequence of contiguous down-steps of size one. Find a formula for the cardinality of the set $A_n$.

**Exercise 2.13** Let $\mathcal{P}_n^\ell$ be the set of all set partitions $\pi = \pi_1\pi_2\cdots\pi_n$ of $[n]$ that satisfies
$$\pi_{i+\ell} - \pi_i \leq 1, \text{ for all } i = 1, 2, \ldots, n - \ell.$$
Establish a bijection between the set partitions in $\mathcal{P}_n^1$ and the set of Dyck paths of size $2n$.

**Exercise 2.14** Following Exercise 2.13, show that the generating function for the number of set partitions in $\mathcal{P}_n^1$ with at most $m$ blocks is given by
$$\frac{U_m\left(\frac{1}{2\sqrt{x}}\right)}{\sqrt{x}U_{m+1}\left(\frac{1}{2\sqrt{x}}\right)},$$
where $U_n(t)$ is the $n$-th Chebyshev polynomial of the second kind (see Appendix D).

**Exercise 2.15** Following Exercise 2.13, show that the number of set partitions in $\mathcal{P}_n^\ell$, $n \geq \ell$, is given by
$$\sum_{\sigma=\sigma_1\sigma_2\cdots\sigma_\ell} \prod_{j=1}^\ell \frac{\sigma_j + 1}{\sigma_j + 1 + \lfloor\frac{n-j}{\ell}\rfloor}\binom{2\lfloor\frac{n-j}{\ell}\rfloor + \sigma_j}{\lfloor\frac{n-j}{\ell}\rfloor},$$
where $\sigma = \sigma_1\cdots\sigma_\ell \in \mathcal{P}_\ell$.

**Exercise 2.16** Prove that the generating function of the number of all set partitions $\pi = \pi_1\pi_2\cdots\pi_n \in \mathcal{P}_{n,k}$ such that $\pi_{i+1} - \pi_i \leq 1$ is given by
$$\frac{1}{\sqrt{x}U_{k+1}\left(\frac{1}{2\sqrt{x}}\right)U_k\left(\frac{1}{2\sqrt{x}}\right)},$$
where $U_n(t)$ is the $n$-th Chebyshev polynomial of the second kind.

**Exercise 2.17** Fix $m, k \geq 1$. Then the generating function for the number of all set partitions $\pi = \pi_1\pi_2\cdots\pi_n \in \mathcal{P}_{n,k}$ such that $\pi_{i+1} - \pi_i \geq -m$, $i = 1, 2, \ldots, n-1$, is given by
$$P_{k,m}(x) = x^k \prod_{s=1}^k \left(1 + \sum_{p=\max\{s+1-m,1\}}^s \frac{x\sum_{j\geq 0}(-1)^j\binom{p-1-mj}{j}x^j}{\sum_{j\geq 0}(-1)^j\binom{s+m-mj}{j}x^j}\right).$$

**Exercise 2.18** Establish bijections between
(1) set of all set partitions $\pi = \pi_1\pi_2\cdots\pi_n \in \mathcal{P}_{n,k}$ such that $\pi_{i+1} - \pi_i \geq -1$, for all $i = 1, 2, \ldots, n-1$.
(2) set of all set partitions $\pi = \pi_1\pi_2\cdots\pi_n \in \mathcal{P}_{n,k}$ such that $\pi_{i+1} - \pi_i \leq 1$, for all $i = 1, 2, \ldots, n-1$.
(3) set of all Dyck paths of size $2n$ with no steps above the line $y = k$.

**Exercise 2.19** A symmetric Dyck path of size $n$ is a Dyck path which is symmetric to the vertical line $x = n$. Find the generating function for the number of symmetric Dyck paths of size $n$.

**Exercise 2.20** Solve the recurrence relation $a_n = 2^{n-1} + \sum_{k=0}^{n-1} \binom{n-1}{k} a_k$ with the initial condition $a_0 = 2$.

**Exercise 2.21** Show that the ordinary generating function
$$\text{Bell}(x) = \sum_{n \geq 0} \text{Bell}_n x^n$$
satisfies no algebraic equation over the field of rational functions $\mathbb{C}(x)$.

**Exercise 2.22** Show the following sequences are strictly decreasing for any $n \geq 3$,
$$\left\{ \frac{\frac{k-1}{n-k+1} \text{Stir}(n,k)}{\text{Stir}(n, k-1)} \right\}, \quad k = 2, 3, \ldots, n.$$

**Exercise 2.23** A sequence $\{a_n\}_{n \geq 0}$ os nonegative numbers is log-concave if $a_n^2 \geq a_{n+1} a_{n-1}$, for $n \geq 1$. Show that the Stirling numbers of the second kind $\text{Stir}(n, k)$ with $n$ fixed are log-concave sequences in $m$.

**Exercise 2.24** Let $\{c_n(a, b, c, s)\}_{n \geq 0}$ be a sequence defined by
$$c_n(a, b, c, d) = abc_{n-1}(a, b, c, d) + cc_{n-1}(a+d, b, c, d)$$
with $c_0(a, b, c, d) = 1$. Prove that the generating function
$$\sum_{n \geq 0} c_n(a, b, c, d) \frac{x^n}{n!} = e^{abx + \frac{c}{bd}(e^{bdx} - 1)},$$
and as consequence of this show
$$c_n(a, b, c, d) = b^n \sum_{j=0}^{n} a^{n-j} d^j \binom{n}{j} \text{Bell}_j(c/(bd)).$$

See [234] to find extra properties for this sequence.

**Exercise 2.25** A 2D shape is a two-dimensional grid of unit squares whose vertices are the lattice points and any finite set of it having left-justified with no restrictions on the row sizes. A two-dimensional set partition of $[n]$ is a distribution of the elements of $[n]$ among the cells of 2D shape such that (1) if a row has empty cell, then all the cells on its right are empty cells, (2) if the first column has empty cell, then the cells that below it are empty, and (3) in each row, and in the first column, the cells appear in order of their increasing smallest elements. Show that the exponential generating function for the number of two-dimensional set partitions of $[n]$ is given by $e^{e^{e^x-1}-1}$.

# Chapter 3

# Preliminary Results on Set Partitions

As we discussed in Chapter 1, the Bell numbers were discovered by Yoshisute Matsunaga before 1769, James Stirling in 1730, and Masanobu Saka in 1782. It took more than 100 years after Stirling and Saka to derive an explicit formula for the $n$-th Bell number. It seems to us that the first formula for the $n$-th Bell number was derived by Dobiński in 1887. Later, several published results studied several properties of Bell numbers. In the first part of this chapter we will present the Dobiński's formula for the $n$-th Bell number as discovered in 1887. Later, Rota [297] in 1964 presented a very elegant proof for Dobiński's formula. After, it took only 18 years for another significant contribution to set partition when Milne [252] presented the $q$-analog of the results of Rota. Actually, we think that the formal study of statistics on set partitions was initiated by Milne in 1982.

To provide result on set partitions, as on all combinatorial classes, we need to represent our set partitions in different representations. For the set partitions there are several well known representations, and each of them has its aim value and its result, but some of them are more attractive than the others. From a historical point of view, the set partition is defined by block representation (see Definition 1.1). Also, as discussed in the introduction each set partition can be represented as a word, the canonical representation (see Definition 1.3). In this chapter we focus not only on blocks and canonical representations of set partitions, but will consider the other four representations as well, namely, flattened, graphical, standard, and rook placement representations. We will discuss some basic results, each of these to smooth way for a discussion of the results of our book.

## 3.1 Dobiński's Formula

As discussed in Chapter 1, Matsunaga found the first recurrence relation for the sequence $\{\text{Bell}_n\}_{n\geq 0}$, where $\text{Bell}_n$ is the number of set partitions of $[n]$ (see Examples 2.56 and 2.58, and see [327, Sequence A000110]). More precisely, he showed that

$$\text{Bell}_n = \sum_{j=0}^{n-1} \binom{n-1}{j} \text{Bell}_j \quad \text{with} \quad \text{Bell}_0 = 1 \qquad (3.1)$$

(see Theorem 1.9). The modification of this recurrence relation was given by Saka in 1782, where he showed that the number of set partitions of $[n]$ with exactly $k$ blocks is given by $\text{Stir}(n,k)$, the Stirling number of the second kind.

**Definition 3.1** *Let $\pi$ be any set partition; we denote the number of blocks in $\pi$ by $\text{blo}(\pi)$.*

It seems to us that the first published explicit formula for the $n$-th Bell number dates back to 1877, where Dobiński [92] showed it in Theorem 3.2. Later, in 1902, Anderegg [3] recovered once more the same formula (actually he generalized it, see Exercise 3.3). More recently, a variation Dobiński's formula is given by Wilf [358], and a simple proof it is given by Pitman [270, 271] (see Exercise 3.2).

**Theorem 3.2** (Dobiński [92]) *Let $n \geq 1$. Then the number of set partitions of $[n]$ is given by*

$$\text{Bell}_n = \frac{1}{e} \sum_{j \geq 0} \frac{j^n}{j!}.$$

**Proof** In order to show this, we use the exponential generating function for the sequence $\{\text{Bell}_n\}_{n \geq 0}$, which is given by $B(x) = e^{e^x - 1}$ as shown in Example 2.58. Let us find the coefficient of $x^n$ in the generating function $B(x)$ as follows:

$$\sum_{n \geq 0} \text{Bell}_n \frac{x^n}{n!} = B(x) = \frac{1}{e} e^{e^x} = \frac{1}{e} \sum_{j \geq 0} \frac{(e^x)^j}{j!} = \frac{1}{e} \sum_{j \geq 0} \frac{e^{jx}}{j!}$$

$$= \frac{1}{e} \sum_{j \geq 0} \sum_{i \geq 0} \frac{j^i x^i}{j! i!}.$$

Thus, by comparing the coefficient of $x^n$ on both sides, we obtain $\text{Bell}_n = \frac{1}{e} \sum_{j \geq 0} \frac{j^n}{j!}$, as claimed. $\square$

The above proof for Dobiński's formula can be formulated by using Theorem 1.16 (see Exercise 3.2). In 1964, Rota [297] found a very elegant proof for Theorem 3.2. He gave a new formula for the $n$-th Bell number, which differs from Theorem 3.2. In order to present the details of Rota's proof, we need the following definitions.

**Definition 3.3** *An* operator *(not to be confused with operation) is a mapping from one vector space to another. A* linear operator *is operator that preserves the operations of vector space: addition ($A(x+x') = A(x) + A(x')$) and scalar* multiplication $(A(\alpha x) = \alpha x)$. Let $L$ be an operator on the vector space $\mathbb{R}[x]$ of all polynomials in single variable $x$ with the basis $\{(x)_0 = 1, (x)_1, (x)_2, \ldots\}$ such that $L$ is defined by $L((x)_j) = 1$ for all $j \geq 0$, where $(x)_j$ are falling polynomials.

**Theorem 3.4** (Rota [297]) *For all $n \geq 0$,*
$$\mathrm{Bell}_n = L(x^n).$$

**Proof** Note that the cardinality of the set of all functions $f : [n] \to [m]$ is given by $m^n$ (Explain!). Let us count such functions in a different way. Let $f : [n] \to [m]$ be any function and let $k = |\{f(j)|j \in [n]\}|$; we define $\pi_f$ to be a set partition in $\mathcal{P}_{n,k}$ such that $a, b \in [n]$ are in the same block if and only if $f(a) = f(b)$. Each function $f$ with set partition $\pi_f$ takes distinct values on distinct blocks, so the number of the functions $f : [n] \to [m]$ with a set partition $\pi_f$ is the same as the number of one-to-one functions from the set $[\mathrm{blo}(\pi_f)]$ to the set $[m]$, which it is given by $m(m-1)\cdots(m-\mathrm{blo}(\pi)+1) = (m)_{\mathrm{blo}(\pi)}$ (Explain!). Thus, we obtain the following identity

$$\sum_{\pi \in \mathcal{P}_n} (m)_{\mathrm{blo}(\pi)} = m^n. \tag{3.2}$$

Now we are ready to prove the theorem. Let $V = \mathbb{R}[x]$ be the vector space of all polynomials in single variable $x$ with the basis $\{(x)_j \mid j \geq 0\}$. Applying the operator $L$ (see Definition 3.3) on both sides of (3.2), we obtain $\sum_{\pi \in \mathcal{P}_n} L((x)_{\mathrm{blo}(\pi)}) = L(x^n)$, which, by $L((x)_j) = 1$, implies that $\mathrm{Bell}_n = L(x^n)$ as claimed. $\square$

Rota [297] showed that the above theorem can be used to rederive the recurrence relation for the Bell numbers (see (3.1) and Theorem 1.9).

**Theorem 3.5** (Rota [297]) *For all $n \geq 0$,*
$$\mathrm{Bell}_{n+1} = \sum_{j=0}^{n} \binom{n}{j} \mathrm{Bell}_j$$

*with $\mathrm{Bell}_0 = 1$.*

**Proof** By the proof of Theorem 3.4 and the fact that $x(x-1)_n = (x)_{n+1}$, we have
$$L(x(x-1)_n) = L((x)_{n+1}) = 1 = L((x)_n).$$
Since $\{(x)_j \mid j \geq 0\}$ is a basis for the vector space $V = \mathbb{R}[x]$ and the linearity of the operator $L$, we obtain
$$L(xp(x-1)) = L(p(x))$$
for any polynomial $p(x) \in V$. In particular, if $p(x) = (1+x)^n$, then $L(x^{n+1}) =$

$L((1+x)^n)$, which, by the linearity of $L$, gives $L(x^{n+1}) = \sum_{j=0}^{n} \binom{n}{j} L(x^j)$. Hence, by Theorem 3.4, we have

$$\text{Bell}_{n+1} = \sum_{j=0}^{n} \binom{n}{j} L(x^j) = \sum_{j=0}^{n} \binom{n}{j} \text{Bell}_j$$

as required. □

Rota [297] continued his work to recover Dobiński's formula.

**Theorem 3.6** (Recovering Dobiński's formula by Rota [297]) *The number of set partitions of $[n]$ is given by*

$$\text{Bell}_n = \frac{1}{e} \sum_{j \geq 0} \frac{j^n}{j!}.$$

**Proof** By the Taylor expansion of $e^x$, we have $e^x = \sum_{j \geq 0} \frac{x^j}{j!}$. Then, taking the $n$-th derivative, we obtain $e = \sum_{j \geq 0} \frac{(j)_n}{j!}$ for all $n \geq 0$. Thus, Theorem 3.4 gives

$$L((x)_n) = 1 = \frac{1}{e} \sum_{j \geq 0} \frac{L((j)_n)}{j!}.$$

Since $\{(x)_j \mid j \geq 0\}$ is a basis of the vector space $V = \mathbb{R}[x]$, we have

$$L(p(x)) = \frac{1}{e} \sum_{j \geq 0} \frac{L(p(j))}{j!}$$

for any polynomial $p(x) \in V$. Substituting $p(x) = x^n$ and using Theorem 3.4, we complete the proof. □

As we said at the beginning of this chapter, the first step on the study of the $q$-analog of Bell and Stirling numbers was initiated by Milne [252] in 1982. Indeed, Milne extended the idea of Rota (Theorems 3.4–3.6) to derive the $q$-analog. More precisely, he generalized the vector space $V$ and the operator $L$ in Definition 3.3 to derive the $q$-Bell and $q$-Stirling numbers. We start with the following definitions.

**Definition 3.7** Let $q > 1$ and $V = \mathbb{R}[x]$ be the vector space of all polynomials in single variable $x$. For $p(x) \in V$, we define the $q$-analog $n$-th difference operator $\Delta_q^n$ as follows:

$$\Delta_q^0(p(x)) = I(p(x)) = p(x),$$
$$\Delta_q^1(p(x)) = (E - I)(p(x)) = p(x+1) - p(x),$$
$$\Delta_q^{n+1}(p(x)) = \Delta_q^n((E - q^n I)(p(x))) = \Delta_q^n(p(x+1)) - q^n \Delta_q^n(p(x)),$$

where $E : p(x) \to p(x+1)$ is the shift operator.

Clearly, $\Delta_q^n(p(x)) = (E - q^{n-1}I)(E - q^{n-2}I)\cdots(E - I)(p(x))$, and $\Delta_1^n = \Delta^n$.

**Definition 3.8** *The q-analog (or q-number, q-bracket) of n is defined by*

$$[n]_q = \frac{1-q^n}{1-q} = 1 + q + \cdots + q^{n-1}.$$

*We define the q-analog of the factorial or q-factorial to be*

$$[n]_q! = [n]_q[n-1]_q \cdots [1]_q \text{ with } [0]_q! = 1.$$

*More generally, we define* $[n]_{j;q} = [n]_q[n-1]_q \cdots [n-j+1]_q$, *with* $[n]_{j;q} = 0$ *for* $j > n$. *Thus, we can define the q-binomial coefficients, also known as Gaussian coefficients, Gaussian polynomials, or Gaussian binomial coefficients, as*

$$\begin{bmatrix} n \\ j \end{bmatrix} = \frac{[n]_{j;q}}{[j]_q!}.$$

**Example 3.9** *We have* $[3]_q = 1 + q + q^2$, $[2]_q = 1 + q$, *and* $[1]_q = 1$. *Thus* $[3]_q! = (1+q)(1+q+q^2)$, $[2]_q! = 1 + q$, *and* $[1]_q! = 1$. *This gives* $\begin{bmatrix} 3 \\ 2 \end{bmatrix} = \frac{[3]_q!}{[2]_q![1]_q!} = 1 + q + q^2$.

By using the q-binomial theorem (see Appendix C), we have

$$\Delta_q^n(p(x)) = \sum_{j=0}^{n}(-1)^j \begin{bmatrix} n \\ j \end{bmatrix} q^{\binom{j}{2}} E^{n-i}(p(x)).$$

**Definition 3.10** *Let $V_q$ be the vector space consisting all polynomials in one variable $q^x$ with coefficients of rational functions in $q$.*

Clearly, the set $\{[x]_{j;q} \mid j \geq 0\}$ is a basis for $V_q$. Let $p(x) \in V_q$ be a polynomial in $q^x$ of degree at most $n$; then we can write

$$p(x) = \sum_{j=0}^{n} \alpha_j [x]_{j;q}, \qquad (3.3)$$

where $\alpha_j$ is a polynomial in $q$. Now let us find an explicit formula for the coefficients $\alpha_j$. By induction on $n$ (do this!) we have

$$\Delta_q^n[x]_{j;q} = [j]_{n;q}[x]_{j-n;q}q^{n(x-j+n)}.$$

It follows $\alpha_j = \frac{\Delta_q^j p(0)}{[j]_q!}$. Note that for $p(x) = x^n$ and $q = 1$, we have $\alpha_j = \frac{\Delta^j p(0)}{j!} = \text{Stir}(n,j)$ and $[x]_{j;q} = (x)_j$. That is, when we substitute $q = 1$ in

(3.3), we obtain (3.2), namely, $x^n = \sum_{j=0}^{n} \text{Stir}(n,j)(x)_j$. In order to generalize this identity, we set $p(x) = ([x]_q)^n \in V_q$ in (3.3). This gives

$$([x]_q)^n = \sum_{j=0}^{n} \frac{\Delta_q^j ([x]_q)^n \mid_{x=0}}{[j]_q!} [x]_{j,q},$$

which invites the following definition of $q$-analog of Stirling numbers of the second kind.

**Definition 3.11** *The $q$-Stirling number of the second kind is defined by*

$$\text{Stir}_q(n,k) = \frac{\Delta_q^k ([x]_q)^n \mid_{x=0}}{[k]_q!}.$$

Therefore, the $q$-analog of (3.2) is given by

$$([x]_q)^n = \sum_{j=0}^{n} \text{Stir}_q(n,j)[x]_{j,q}, \qquad (3.4)$$

which implies

$$([x]_q)^{n+1} = \sum_{j=0}^{n+1} \text{Stir}_q(n+1,j)[x]_{j,q} = \sum_{j=0}^{n+1} \text{Stir}_q(n,j)([x-j]_q q^j + [j]_q)[x]_{j,q})$$

$$= \sum_{j=0}^{n+1} (q^{j-1} \text{Stir}_q(n,j-1) + [j]_q \text{Stir}(n,j))[x]_{j,q}.$$

Hence, $q$-Stirling numbers of the second kind can be defined by the recurrence relation

$$\text{Stir}_q(n+1,j) = q^{j-1} \text{Stir}_q(n,j-1) + [j]_q \text{Stir}(n,j).$$

Thus we can define the $q$-analog of the Bell numbers as follows.

**Definition 3.12** *For all $n \geq 0$,*

$$\text{Bell}_{n;q} = \sum_{k=0}^{n} \text{Stir}_q(n,k).$$

Now we indicate how we use the finite operator methods of [297] to derive the $q$-analog the Dobiński's formula (Theorem 3.2). In order to do that, we extend the operator $L$ that is defined by Rota in Definition 3.3 to a linear operator on the vector space $V_q$.

**Definition 3.13** *Let $L_q$ be a linear operator on the vector space $V_q$ defined by $L_q(1) = 1$ and $L_q([x]_{j;q}) = 1$, for all $j \geq 1$.*

The next theorem needs to use the $q$-exponential function.

**Definition 3.14** *The $q$-exponential function is defined by $e_q(x) = \sum_{j \geq 0} \frac{x^j}{[j]_q!}$ (see Appendix C).*

**Theorem 3.15** (Milne [252, Corollary 1.1]) *For all $n \geq 0$,*

$$\text{Bell}_{n+1;q} = \frac{1}{e_q(1)} \sum_{j \geq 0} \frac{[j]_q^n}{[j]_q!}.$$

**Proof** By applying the operator $L_q$ to both sides of (3.4), we obtain

$$L_q([x]_q)^n = \sum_{j=0}^{n} \text{Stir}_q(n,j) = \text{Bell}_{n;q},$$

which implies

$$1 = \frac{1}{e_q(1)} \sum_{j=0}^{n} \frac{1}{[j]_q!} = \frac{1}{e_q(1)} \sum_{j \geq n} \frac{1}{[j-n]_q!} = \frac{1}{e_q(1)} \sum_{j \geq 0} \frac{[j]_{n;q}}{[j]_q!}.$$

Since $L_q([x]_{n;q}) = 1$, we have

$$L_q([x]_{n;q}) = \frac{1}{e_q(1)} \sum_{j \geq 0} \frac{[j]_{n;q}}{[j]_q!}.$$

By the fact that $L_q$ is linear operator on $V_q$ and the set $\{[x]_{j;q} \mid j \geq 0\}$ is a basis for the vector space $V_q$, we obtain

$$L_q(p(x)) = \frac{1}{e_q(1)} \sum_{j \geq 0} \frac{p(j)}{[j]_q!}.$$

By setting $p(x) = [x]_q^{n+1}$, we complete the proof. □

Now we shift our focus to the generating function for the Bell and Stirling numbers. As shown in Example 2.58, the exponential generating function for the number of set partitions of $[n]$ is given by $e^{e^x - 1}$, that is,

$$\sum_{n \geq 0} \text{Bell}_n \frac{x^n}{n!} = e^{e^x - 1}.$$

This can be extended as follows.

**Theorem 3.16** (see Comtet [82]) *The exponential generating function for the number of set partitions of $[n]$ according to the number of blocks is given by $e^{q(e^x - 1)}$. Moreover, for any given $k$, the exponential generating function for the number of set partitions of $[n]$ with exactly $k$ blocks is given by*

$$\sum_{n \geq k} \text{Stir}(n,k) \frac{x^n}{n!} = \frac{1}{k!}(e^x - 1)^k.$$

**Proof** Let $E_n(q)$ be the exponential generating function for the number of set partitions of $[n]$ according to the number of blocks. Then the proof of Theorem 1.9 shows

$$E_n(q) = q \sum_{j=0}^{n-1} \binom{n-1}{j} E_j(q)$$

with $E_0(q) = 1$. Theorem 2.57 and Rule 2.49 give $E(x;q) = \sum_{j \geq 0} E_j(q) \frac{x^j}{j!} = e^{q(e^x - 1)}$. Therefore, the coefficient of $q^k$ in $E(x;q)$ is given by $\frac{1}{k!}(e^x - 1)^k$, as required. $\square$

**Corollary 3.17** *The mean of the number blocks in all set partitions of $[n]$ is given by $\frac{\text{Bell}_{n+1}}{\text{Bell}_n} - 1$.*

**Proof** By Theorem 3.16 we have

$$\frac{d}{dq} e^{q(e^x - 1)} \big|_{q=1} = (e^x - 1)e^{e^x - 1} = \frac{d}{dx}(e^{e^x - 1}) - e^{e^x - 1},$$

which, by Rule 2.49, implies that the number of blocks in all set partitions of $[n]$ is given by $\text{Bell}_{n+1} - \text{Bell}_n$. Thus, the mean of the number blocks in all set partitions of $[n]$ is given by $\frac{\text{Bell}_{n+1} - \text{Bell}_n}{\text{Bell}_n} = \frac{\text{Bell}_{n+1}}{\text{Bell}_n} - 1$, as claimed. $\square$

In 1960, Klein-Barmen [186] found an explicit formula for the Stirling number of the second kind.

**Theorem 3.18** (Klein-Barmen [186]) *For all $1 \leq k \leq n$,*

$$\text{Stir}(n,k) = \frac{1}{(k-1)!} \sum_{j=0}^{k-1} (-1)^j \binom{k-1}{j} (k-j)^{n-1}.$$

**Proof** By Theorem 3.16 we have

$$\left[\frac{x^n}{n!}\right] \frac{(e^x - 1)^k}{k!} = \left[\frac{x^n}{n!}\right] \sum_{j=0}^{k} \frac{(-1)^j}{k!} \binom{k}{j} e^{(k-j)x}$$

$$= \frac{1}{k!} \sum_{j=0}^{k} (-1)^j \binom{k}{j} (k-j)^n$$

$$= \frac{1}{(k-1)!} \sum_{j=0}^{k-1} (-1)^j \binom{k-1}{j} (k-j)^{n-1},$$

as required. $\square$

The above theorem together with the fact that $\text{Bell}_n = \sum_{j=0}^{n} \text{Stir}(n,j)$ gives the following result.

**Corollary 3.19** (Cohn, Even, Menger, Karl, and Hooper [79]) *For all $n \geq 1$,*

$$\text{Bell}_n = \sum_{k=1}^{n} \left( \frac{1}{(k-1)!} \sum_{j=0}^{k-1} (-1)^j \binom{k-1}{j} (k-j)^{n-1} \right).$$

In Chapter 2 we showed that $\sum_{n \geq k} \text{Stir}(n,k) x^n = \frac{x^k}{\prod_{i=1}^{k}(1-ix)}$ (see (2.9)). Comtet [81] extended this results as follows.

**Theorem 3.20** (Comtet [81]) *Let $\{a_{n,k}\}_{n,k \geq 0}$ be a sequence. Then the following are equivalent characterizations of the sequence $\{a_{n,k}\}_{n,k}$:*

$$a_{n,k} = a_{n-1,k-1} + u_k a_{n-1,k}, \quad a_{n,0} = u_0^n, \quad a_{0,k} = \delta_{0,k}, \tag{3.5}$$

$$a_{n,k} = \sum_{d_1+d_2+\cdots+d_k=n-k,\ d_i \geq 0} \prod_{j=0}^{k} u_j^{d_j}, \tag{3.6}$$

$$\sum_{n \geq 0} a_{n,k} x^n = \frac{x^k}{(1-u_0 x)(1-u_1 x)\cdots(1-u_k x)}. \tag{3.7}$$

**Proof** Let us assume that (3.5) holds, and prove (3.6) by induction on $n$:

$a_{n,k}$
$= a_{n-1,k-1} + u_k a_{n-1,k}$

$$= \sum_{\substack{\sum_{i=1}^{k-1} d_i = n-k,\\ d_i \geq 0}} \prod_{j=0}^{k-1} u_j^{d_j} + u_k \sum_{\substack{\sum_{i=1}^{k} d_i = n-1-k,\\ d_i \geq 0}} \prod_{j=0}^{k} u_j^{d_j}$$

$$= \sum_{\substack{\sum_{i=1}^{k} d_i = n-k,\\ d_i \geq 0,\ d_k = 0}} \prod_{j=0}^{k} u_j^{d_j} + \sum_{\substack{\sum_{i=1}^{k-1} d_i + (d_k+1) = n-k,\\ d_i \geq 0}} u_k^{d_k+1} \prod_{j=0}^{k-1} u_j^{d_j}$$

$$= \sum_{\substack{\sum_{i=1}^{k} d_i = n-k,\\ d_i \geq 0}} \prod_{j=0}^{k} u_j^{d_j},$$

which completes the proof. From the step of the induction, we obtain that (3.6) implies (3.5).

Let $A_k(x) = \sum_{n \geq 0} a_{n,k} x^n$. By multiplying (3.5) by $x^n$ and summing over all $n$, we obtain $A_k(x) = x A_{k-1}(x) + u_k x A_k(x)$, which gives

$$A_k(x) = \frac{x}{1-u_k x} A_{k-1}(x) = \frac{x^2}{(1-u_k x)(1-u_{k-1} x)} A_{k-2}(x)$$

$$= \cdots = \frac{x^k}{(1-u_k x)\cdots(1-u_0 x)} A_0(x),$$

where $A_0(x) = 1$, which holds immediately from the initial conditions. Thus, (3.5) implies (3.7). On the other hand, if (3.7) holds, then

$$A_k(x) = xA_{k-1}(x) + u_k x A_k(x),$$

which gives (3.5). □

For instance, the above theorem for $u_k = k$ gives the ordinary generating function for the Stirling numbers of the second kind (see Theorem 1.12 and Example 2.56).

## 3.2 Different Representations

To provide results on set partitions, as on all combinatorial structures, we need different representations for our set partitions. For the set partitions, are well known several representation, and each of them has its aim value and its results, but some of them more attractive than the others. From the historical point of view, the set partitions is defined by block representation (see Definition 1.1). Also, as discussed in the introduction, each set partition can be represented as a word, the canonical representation (see Definition 1.3). In this chapter we focus not only on block and canonical representations os set partitions, but will consider as well other four representations, namely, flattened, graphical, standard, and rook placement representations. The rook placement and flattened representations are combinatorial objects studied in their own right, see, for example, [333] for rook placement representation and [61] for flattened representation. In [241], Mansour, Shattuck, and Wagner provided the graphical representation to study the distance between elements of the set partitions, and in [176, 178, 179], Klazar defined the standard representation to restrict the set partitions to a small class that has combinatorial interpretation with Dyck paths. In [306], Sagan investigated another type of restriction on block representation of set partitions, where the restrictions are defined directly on the blocks. Throughout this book we will deal with the following representations:

- Block representation
- Circular representation
- Line diagram
- Flattened set partitions
- Canonical representation
- Graphical representation

- Standard representation

- Rook placement representation

Each of these representations has its own magic on set partitions. Also, framing questions about set partitions in terms of any of these representation have the potential to lead to new combinatorial questions. At the end of this chapter we conclude with dimensional set partitions, especially the two dimensional set partitions introduced by Mansour, Munagi, and Shattuck [222]. Rewriting the research on set partitions according to these representations, we discover that block and flattened representations are less attractive compared to other representation on the set partitions and pattern avoidance research. Thus, in this chapter, we will consider detail the block and flattened representations. But the other representation will be considered shortly, namely, basic definitions and examples. We separate the results in chapters and then into small sections since it will be tough to write all the results in the book (see Chapters 4–7) in one chapter.

As we said, in the literature there exist several equivalent ways of representing a set partition, and we already discussed two of theses representations in Chapter 1 (see Definitions 1.1 and 1.3). For each of these representations we will present the necessary definitions and examples to illustrate them. More precisely, on each representation we will define the representation itself and the concept of "pattern" or "statistic". This leads to several variations of the definition of the pattern avoidance problem or the counting pattern problem on set partitions. Here, we present only the very basic results on the attractive representations, while the other representations will be investigated in full detail in the next few chapters. Actually, the book is devoted to those attractive representations.

### 3.2.1 Block Representation

From a historical point of view (see Definition 1.1), a set partition $\pi$ of $S$ is defined as a collection $B_1, \ldots, B_k$ of nonempty disjoint subsets, which are called blocks, of $S$ such that $\cup_{i=1}^{k} B_i = S$, where the blocks satisfy the following order property $\min B_1 < \min B_2 < \cdots < \min B_k$. Throughout the book we will call this representation the *block representation* of $\pi$ and we will write $\pi = B_1/B_2/\cdots/B_k$. In order to define the pattern in set partitions via this representation, we will need the notion of a subpartition.

**Definition 3.21** *A set subpartition of a set partition* $\pi = B_1/B_2/\cdots/B_k$ *of $S$ is a set partition $\pi'$ of $S' \subseteq S$ such that each block of $\pi'$ is contained in a different block of $\pi$.*

**Example 3.22** *Let $\pi$ be the set partition $135/26/4$. Then $\pi$ has a set subpartition $35/4$, but not $13/5$, since $1, 3,$ and $5$ are in the same block of $\pi$.*

**Definition 3.23** Let $\pi$ and $\sigma$ be any two set partitions; we say that $\pi$ contains $\sigma$ as a pattern if there exists a set subpartition $\pi'$ of $\pi$ such that $\mathrm{Reduce}(\pi') = \sigma$. In this case, $\pi'$ is called an occurrence of $\sigma$ in $\pi$. If $\pi$ has no occurrence of $\sigma$, then we say that it avoids $\sigma$. In this context, we denote the set partitions of $[n]$ that avoid the pattern $\sigma$ by $\mathcal{P}_n(\pi)$.

**Example 3.24** The set partition $\pi = 135/26/4$ contains four occurrences of the pattern $12/3$, namely, $13/6$, $13/4$, $15/6$, and $35/6$. On the other hand, the set partition $\pi$ avoids the pattern $12/34$.

From the above definition we can obtain the following fact, where we leave the proof to the reader.

**Fact 3.25** (Sagan [306, Theorem 2.2]) The set partitions in $\mathcal{P}_n(1/2/\cdots/k)$ ($\mathcal{P}_n(12\cdots k)$) equals the set partitions of $[n]$ with at most $k-1$ blocks (at most $k-1$ elements in each block). Moreover, the exponential generating function for the number of set partitions in $\mathcal{P}_n(1/2/\cdots/k)$ ($\mathcal{P}_n(12\cdots k)$) is given by

$$\sum_{j=0}^{k-1} \frac{(e^x - 1)^j}{j!} \quad \left( e^{\sum_{j=1}^{k-1} \frac{x^j}{j!}} \right).$$

**Definition 3.26** Given a set partition $\pi = B_1/B_2/\cdots/B_k$ of $[n]$, we define the complement of $\pi$ to be the set partition $\overline{\pi}$ where its blocks are given by $\overline{B_i} = \{n + 1 - b \mid b \in B_i\}$ for all $i = 1, 2, \ldots, k$.

**Example 3.27** The complement of the set partition $13/26/457$ is $134/26/47$.

**Fact 3.28** (Sagan [306, Lemma 2.4]) For any set partition $\tau$,

$$\mathcal{P}_n(\overline{\tau}) = \{\overline{\pi} \mid \pi \in \mathcal{P}_n(\tau)\}.$$

Moreover, for all $n \geq 0$,

$$|\mathcal{P}_n(\overline{\tau})| = |\mathcal{P}_n(\tau)|.$$

In 2006, Sagan [306] found exact formulas for the number of set partitions, which avoid certain specific patterns. In particular, he enumerated and characterized those set partitions avoiding any set partitions of [3] (see Table 3.1).

## Preliminary Results on Set Partitions

**Table 3.1**: Set partitions avoiding a pattern of [3]

| $\tau$ | $\{|\mathcal{P}_n(\tau)|\}_{n\geq 1}$ | Reference |
|---|---|---|
| 1/2/3, 13/2 | $2^{n-1}$ | Fact 3.25 and Proposition 3.31 |
| 12/3, 1/23 | $\binom{n}{2}+1$ | Fact 3.25 and Exercise 3.8 |
| 123 | $\sum_{i=0}^{n/2}\binom{n}{2i}(2i)!!$ | Fact 3.25 |

All the results of Table 3.1 except the case 13/2 are easy consequences of Facts 3.25 and 3.28. To count the number of set partitions in $\mathcal{P}_n(13/2)$, we need the following definition.

**Definition 3.29** *We say that set partition $\pi$ of $[n]$ it layered if it has the form*

$$\pi = 12\cdots j_1/(j_1+1)(j_1+2)\cdots j_2/\cdots/(j_m+1)(j_m+2)\cdots j_{m+1},$$

*where $j_{m+1}=n$.*

**Example 3.30** *Among the five set partitions of [3] there are four layered set partitions, namely, 1/2/3, 1/23, 12/3, and 123.*

**Proposition 3.31** (Sagan [306, Theorem 2.5]) *For all $n \geq 1$,*

$$|\mathcal{P}_n(13/2)| = 2^{n-1}.$$

**Proof** By Definition 3.29 we have that each layered set partition does not contain an occurrence of the pattern 13/2. Now, assume $\pi$ avoids 13/2 and let $B$ be the nonlast block of $\pi$. If $\max B > \min B$, then from the fact that $\pi$ avoids 13/2, we have that all the numbers $\min B+1, \min B+2, \ldots, \max B-1$ are in $B$. Hence $B = \{\min B, \min B+1, \ldots, \max B\}$. Thus $\pi$ is a layered set partition. If we denote the number of layered set partitions of $[n]$ by $a_n$, then $a_n$ satisfies the recurrence relation $a_n = a_{n-1}+a_{n-2}+\cdots+a_0$ with the initial condition $a_0 = 1$. Hence, $a_n = 2^{n-1}$ (prove it!), which completes the proof. $\square$

**Definition 3.32** *Let $T$ be any set subpartition patterns. We denote the set of set partitions in $\mathcal{P}_n$ that avoids every pattern in $T$ by $\mathcal{P}_n(T)$.*

In 2008, Goyt [120] extended the results in Table 3.1 to cover the cardinalities of the set partitions in $\mathcal{P}_n$ that avoid a subset of $\mathcal{P}_3$. Table 3.2 (see Section 2 in [120]) presents the cardinalities of the set $\mathcal{P}_n(T)$, where $T$ contains two set partition patterns and the set $\mathcal{P}_n(T)$ is not empty for larger $n$.

**Table 3.2**: Set partitions avoiding two patterns of [3]

| $T$ | $\{|\mathcal{P}_n(T)|\}_{n\geq 1}$ |
|---|---|
| $\{12/3, 123\}, \{1/23, 123\}$ <br> $\{12/3, 1/2/3\}, \{1/23, 1/2/3\}$ <br> $\{1/23, 13/2\}, \{12/3, 13/2\}$ <br> $\{1/2/3, 13/2\}$ | $n$ |
| $\{1/2/3, 12/3\}, \{1/23, 12/3\}$ | 3 |
| $\{13/2, 123\}$ | Fib$_{n+1}$ |

Actually, all the results in Table 3.1 can be proved easily. For instance, applying the proof of Proposition 3.31 to the case of avoiding both the patterns $13/2$ and $1/2/3$ we obtain that each set partition in $\mathcal{P}_n(13/2, 1/2/3)$ has the form $12\cdots i/(i+1)(i+2)\cdots n$ for some $1 \leq i \leq n$, which implies $|\mathcal{P}_n(13/2, 1/2/3)| = n$ for all $n \geq 1$. Another example is

$$|\mathcal{P}_n(13/2, 123)| = \text{Fib}_{n+1}.$$

Again, by Proposition 3.31, each set partition $\pi$ in $\mathcal{P}_n(13/2, 123)$ can be written as either $\pi = \pi'/n$ or $\pi = \pi''/(n-1)n$, where $\pi' \in \mathcal{P}_{n-1}(13/2, 123)$ and $\pi'' \in \mathcal{P}_{n-2}(13/2, 123)$. Hence, if $a_n = |\mathcal{P}_n(13/2, 123)|$, then $a_n$ satisfies the recurrence relation $a_n = a_{n-1} + a_{n-2}$ with $a_0 = a_1 = 1$. Hence, by Example 2.12, we have $a_n = \text{Fib}_{n+1}$.

Goyt [120] considered also sets $T$ of three patterns, and as excepted from Table 3.1, the results are simple and the set $\mathcal{P}_n(T)$ contains only few elements. Moreover, he introduced odd and even set partitions and considered enumeration of odd (even) set partitions that avoid a pattern (a set of patterns).

**Definition 3.33** *The sign of a set partition $\pi$ of $n$ with exactly $k$ blocks is defined by* $\text{Sign}(\pi) = (-1)^{n-k}$. *Even (Odd) set partitions $\pi$ satisfy* $\text{Sign}(\pi) = 1$ *(*$\text{Sign}(\pi) = -1$*). We denote the set of even (odd) set partitions of $[n]$ that avoid a pattern $\tau$, a set of pattern $T$ by $E\mathcal{P}_n(\tau)$, $E\mathcal{P}_n(T)$ ($O\mathcal{P}_n(\tau)$, $O\mathcal{P}_n(T)$).*

**Table 3.3**: Even set partitions avoiding a pattern of [3]

| $\tau$ | $|E\mathcal{P}_{2n}(\tau)|$ | $|E\mathcal{P}_{2n+1}(\tau)|$ | Reference |
|---|---|---|---|
| $1/2/3$ | $2^{2n-1} - 1$ | $1$ | Proposition 3.34 |
| $1/23$ | $n^2 - n + 1$ | $n^2 + 1$ | Exercise 3.10 |
| $13/2$ | $2^{2n-2}$ | $2^{2n-1}$ | Proposition 3.34 |
| $123$ | $\sum_{i=0}^{\frac{n-1}{2}} \binom{2n}{4i+2}(4i+2)!!$ | $\sum_{i=0}^{\frac{2n-1}{4}} \binom{2n+1}{4i+2}(4i+2)!!$ | Exercise 3.10 |

Goyt [120] extended the results of Table 3.1 to the case of even/odd set partitions as describe in Table 3.3. Note that the case of odd set partitions can be achieved by using the following simple relation $E\mathcal{P}_n(T) + O\mathcal{P}_n(T) = \mathcal{P}_n(T)$ for all set $T$ of patterns and for all $n \geq 0$.

**Proposition 3.34** (Goyt [120]) *Let $n \geq 1$. Then*

(i) *for $n$ odd, $|E\mathcal{P}_n(1/2/3)| = 1$; and for $n$ even, $|E\mathcal{P}_n(1/2/3)| = 2^{n-1} - 1$.*

(ii) *for $n \geq 1$, $|E\mathcal{P}_n(1/23)| = \lfloor (n-1)^2/4 \rfloor + 1$.*

(iii) *for $n \geq 2$, $|E\mathcal{P}_n(13/2)| = 2^{n-2}$.*

(iv) *for $n \geq 1$, $|E\mathcal{P}_n(123)| = \sum_{i=0}^{(n-2)/4} \binom{n}{4i+2}(4i+2)!!$.*

**Proof** (i) Any set partition $\pi$ of $[n]$ that avoids $1/2/3$ has at most two blocks. If $n$ is odd, then there is only one even (odd) set partition with exactly one (two) blocks. Since $|\mathcal{P}_n(1/2/3)| = 2^{n-1}$, we obtain the result. The proof for even $n$ is very similar.

(ii) See Exercise 3.10.

(iii) By Proposition 3.31 it suffices to give a sign reversing involution $f : \mathcal{P}_n(13/2) \to \mathcal{P}_n(13/2)$. By the proof of Proposition 3.31, each set partition $\pi = B_1/B_2/\cdots/B_k$ of $[n]$ with exactly $k$ blocks that avoids $13/2$ is a layered set partition, and thus, $n \in B_k$. We define $f(\pi) = B_1/B_2/\cdots/B_{k-1} \cup n$ if $B_k = n$, and $f(\pi) = B_1/B_2/\cdots/B_k\setminus\{n\}/n$ otherwise. Clearly, $f(\pi)$ is a layered set partition of $[n]$ and $\text{Sign}(f(\pi)) = -\text{Sign}(\pi)$, which completes the proof.

(ii) See Exercise 3.10. □

Goyt, also, studied even (odd) set partitions that avoid a set of either two or three patterns. The most interesting case, as we can guess from Table 3.3, is the case of avoiding $T = \{13/2, 123\}$. In this case, Goyt [120, Proposition 4.7] showed that the number of even set partitions of $[n]$ that avoid $T$ is given by $\text{Fib}_{n+1}/2$, for all $n \geq 1$ (Prove it!).

### 3.2.2 Circular Representation and Line Diagram

Here we will suggest other two representation for a set partition, namely, the circular representation and the line diagram.

**Definition 3.35** *Let $\pi$ be any set partition. We present the elements of the set $[n]$ as $n$ points around a circle, and two (cyclically) successive elements of the same block of $\pi$ are joined by a segment. We call this representation by the name* circular representation.

**Definition 3.36** Let $\pi = B_1/B_2 \cdots /B_k$ be any set partition of $[n]$ with exactly $k$ blocks. We present the elements of the $i$-th block $B_i$ as $|B_i|$ points $(j, i)$, $j \in B_i$, on the segment between the the points $(\min B_i, i)$ and $(\max B_i, i)$. We call this representation by the name line diagram.

**Example 3.37** Figure 3.1 represents the circular representation and line diagram of the set partition $18/235/47/6$ of $[8]$.

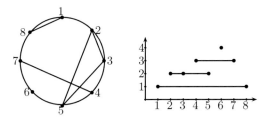

**FIGURE 3.1**: The circular representation and the line diagram of the set partition $18/235/47/6$

### 3.2.3 Flattened Set Partitions

In 2009, Callan [61] represented set partitions as permutations. He enumerated the number of set partitions of $[n]$ that their corresponding permutations satisfy under certain set of conditions. Up to now, there exist only a few results in this direction. Actually, there is only one reference due to Callan.

**Definition 3.38** Let $B_1/B_2/\cdots/B_k$ be any set partition of $[n]$. We define Flatten($\pi$) to be the permutation (see Definition 2.7) obtained by concatenating the blocks of $\pi$ (each block is written as an increasing word). In this context Flatten($\pi$) is called the flattened set partition of $\pi$.

**Example 3.39** If $\pi$ is the set partition $145/268/37$, then the flattened set partition of $\pi$ is the permutation $14526837$.

Now we will present several results and techniques of [61] to study the number of set partitions $\pi$ such that Flatten($\pi$) avoids a pattern (see next definition).

**Definition 3.40** Let $\pi$ be any set partition of $[n]$, and let $\tau$ be any subsequence pattern. We say that $\pi$ avoids $\tau$ or $\pi$ is $\tau$-avoiding if the permutation Flatten($\pi$) avoids $\tau$. Otherwise, we say that $\pi$ contains $\tau$. We denote the set of all set partitions in $\mathcal{P}_n$ ($\mathcal{P}_{n,k}$) that avoid the pattern $\tau$ by Flatten$_n(\tau)$ (Flatten$_{n,k}(\tau)$).

Coding this definition as C++ program (see Appendix H), we obtain the cardinalities of the set $\text{Flatten}_n(\tau)$ for small $n$ and patterns $\tau$ of size three and four (see Tables 3.4 and 3.6).

**Example 3.41** (Callan [61, Section 2.3]) Let $\pi = B_1/B_2/\cdots/B_k$ and $\pi' = \text{Reduce}(B_2/\cdots/B_k)$. It is not hard to check that $\pi \in \text{Flatten}_n(213)$ if and only if (Expalin?)

(1) $B_1 = \{1, 2, \ldots, i\} \cup \{j, j+1, \ldots, n\}$ with $1 \leq i < j$, and

(2) $\text{Flatten}(\pi')$ avoids the subsequence pattern 213.

Thus, if we denote the number of set partitions in $\text{Flatten}_n(213)$ such that the first block has size $k$ by $f_{n,k}$ and $f_n = |\text{Flatten}_n(213)|$, then $f_{n,k} = k f_{n-k}$, $f_{n,n} = 1$, and $f_n = \sum_{k=1}^n f_{n,k}$. Therefore, the sequence $\{f_n\}_{n \geq 0}$ satisfies the recurrence relation

$$f_n = 1 + \sum_{k=1}^{n-1} k f_{n-k} = 1 + \sum_{k=1}^{n-1} (n-k) f_k = 1 + n \sum_{k=1}^{n-1} f_k - \sum_{k=1}^{n-1} k f_k,$$

which implies

$$f_n - 2f_{n-1} + f_{n-2} = f_{n-1}.$$

By the definitions we can state the following initial conditions $f_0 = f_1 = 1$ and $f_2 = 2$. Using the tools of Chapter 2 for solving recurrence relations, we obtain $f_n = \text{Fib}_{2n-1}$ for all $n \geq 1$.

**Table 3.4**: The cardinality $|\text{Flatten}_n(\tau)|$, where $\tau \in S_3$ and $1 \leq n \leq 10$

| $\tau \backslash n$ | 1 | 2 | 3 | 4 | 5 | 6 | 7 | 8 | 9 | 10 |
|---|---|---|---|---|---|---|---|---|---|---|
| 123 | 1 | 2 | 1 | 0 | 0 | 0 | 0 | 0 | 0 | 0 |
| 132 | 1 | 2 | 4 | 8 | 16 | 32 | 64 | 128 | 256 | 512 |
| 213, 312 | 1 | 2 | 5 | 13 | 34 | 89 | 233 | 610 | 1597 | 4181 |
| 231 | 1 | 2 | 5 | 14 | 42 | 132 | 429 | 1430 | 4862 | 16796 |
| 321 | 1 | 2 | 5 | 15 | 51 | 188 | 731 | 2950 | 12235 | 51822 |

**Example 3.42** (Callan [61, Section 2.5]) Let $\pi = B_1/B_2/\cdots/B_k$ and $\pi' = \text{Reduce}(B_2/\cdots/B_k)$. It is easy to verify that $\pi \in \text{Flatten}_n(312)$ if and only if (Why?)

(1) $B_1 = \{1, 2, \ldots, i\} \backslash \{j\}$ with $i \geq j \geq 2$, and

(2) $\text{Flatten}(\pi')$ avoids the subsequence pattern 312.

Thus, if the first block of $\pi$ has size $k \leq n-1$, then there are exactly $k$ choices for $j$, namely, $2, 3, \ldots, k+1$. Let $f_n = |\text{Flatten}_n(312)|$, so

$$f_n = f_{n-1} + 2f_{n-2} + \cdots + (n-1)f_1 + 1,$$

where 1 counts the case that the first block of $\pi$ is $[n]$. This leads to $f_n - 2f_{n-1} - f_{n-2} = f_{n-1}$ with $f_0 = f_1 = 1$ and $f_2 = 2$. Solving this recurrence relation as we discussed in Chapter 2, we obtain $f_n = \text{Fib}_{2n-1}$ for all $n \geq 1$.

The flatten of a set partition was introduced by Callan [61]. Actually, Callan introduced flatten as a set partition to study a different meaning of avoiding (in this book we will introduce several meaning of avoiding and restriction on set partitions. Indeed, we already saw one meaning given in the previous subsection) on set partitions as given in Definitions 3.38 and 3.40. He found explicit formulas for the cardinalities $|\text{Flatten}_n(\tau)|$, where $\tau$ is a subsequence pattern of size three (see Table 3.5). In order to show the results of this table, we need to consider the cases 231 and 321 (see Exercise 3.6 and Examples 3.41 and 3.42). To do so, we need the following definition.

Table 3.5: Three letter subsequence pattern

| $\tau$ | $\{|\text{Flatten}_n(\tau)|\}_{n \geq 1}$ | Reference |
|---|---|---|
| 123 | $1, 2, 1, 0, 0, 0, \ldots$ | Exercise 3.6 |
| 132 | $2^{n-1}$ | Exercise 3.6 |
| 213, 312 | $\text{Fib}_{2n-1}$ | Example 3.41, Example 3.42 |
| 231 | $\text{Cat}_n = \frac{1}{n+1}\binom{2n}{n}$ | Theorem 3.45 |
| 321 | $\sum_{k=0}^{n-1} \binom{n-1}{k} \text{Cat}_k$ | Theorem 3.46 |

**Definition 3.43** Let $\pi = \pi_1 \pi_2 \cdots \pi_n$ be any permutation of size $n$. A descent terminator is either $\pi_1$ or $\pi_i$ such that $\pi_i < \pi_{i-1}$. A right-to-left minima or RL-minima of $\pi$ is an element $\pi_i$ such that $\pi_i < \pi_j$ for all $j = i+1, i+2, \ldots, n$. A left-to-right maxima or LR-maxima of $\pi$ is an element $\pi_i$ such that $\pi_i > \pi_j$ for all $j = 1, 2, \ldots, i-1$.

**Example 3.44** For example, if $\pi = 1213214132$, then the descent terminators of $\pi$ are given by $1, 1, 2, 1, 1, 2$, the right-to-left minima of $\pi$ are given by $2, 1$, and the left-to-right-maxima of $\pi$ are given by $1, 2, 3, 4$.

**Theorem 3.45** (Callan [61, Section 2.4]) *The number of set partitions in* $\text{Flatten}_n(231)$ *is given by* $\text{Cat}_n$ *the $n$-th Catalan number.*

**Proof** Let $M_\pi$ be the set of all the RL-minima of $\pi$ that are not descent terminators. Define $f_{n,k}$ to be the number of set partitions $\pi \in \text{Flatten}_n(231)$ such that $|M_\pi| = k$. We proceed to obtain the proof in three steps, where in the first step we investigate the case $k = 0$, in the second step we reduce the general case to the case $k = 0$, and in the last step we combine the first two steps together to complete the proof.

**Step 1.** Let $\pi$ be any set partition in $Flatten_n(231)$ such that $|M_\pi| = 0$. Clearly, the $RL$-minima and descent terminators of $\pi$ coincide. Since the last entry of $\pi$ is an $RL$-minima, it is also a descent terminator, which implies that the last block of $\pi$ contains exactly one element. Now, let $B = \{b_1, b_2, b_3, \ldots\}$ be the nonlast block of $\pi$, and then its size is at most two. Otherwise, $b < b_2 < b_3$ or $b_2 < b < b_3$, where $b$ is the first entry of the next block, that is, $b_2$ is an $RL$-minima and it is not a descent terminator – a contradiction. Additionally, if $B$ is a nonlast singleton block, then the first element of the next block of $B$ is $RL$-minima, which is not a descent terminator. Therefore, each nonlast block has size two and the last block of $\pi$ has size one, which implies that $n$ is an odd number and $\pi$ can be expressed as

$$\{b_1 = 1, c_1\}/\ldots/\{b_m, c_m\}/\{c_{m+1}\}.$$

If $b_2 > 2$, then $c_1 = 2$ and $b_2$ is $RL$-minima that is not descent terminator, which leads to a contradiction, and thus $b_2 = 2$. Next, we show that $b_i \leq 2i-2$ for all $i = 2, 3, \ldots, m+1$. If $b_i > 2i-2$ for some $i \geq 2$, then all the elements of the set $[2i-2]$ cannot occur after $b_i$ because $b_i$ is $RL$-minima. This makes the first $i-1$ blocks to consist all the elements of the set $[2i-2]$ leaving $c_{i-1}$ an $RL$-minima, which gives a contradiction. Hence the elements $b_1, b_2, \ldots, b_{m+1}$ satisfy

$$3 \leq b_3 < b_4 < \cdots < b_{m+1} \text{ and } b_i \leq 2i-2 \text{ for all } i = 3, 4, \ldots, m+1. \quad (3.8)$$

Hence, there is a bijection between the set partitions $\pi \in Flatten_n(231)$ with $|M_\pi| = 0$ and the set of sequences $\{b_i\}_{i=3}^{m+1}$ satisfying (3.8). To see its inverse, let $A = [n]\setminus\{b_1, b_2, \ldots, b_{m+1}\}$ and let us define $c_i$ as follows. Let $c_m$ be the smallest element in $A$ such that $c_m > b_{m+1}$ and define $c_i$ to be the smallest element in $A\setminus\{c_{i+1}, \ldots, c_m\}$ such that $c_i > b_{i+1}$. For instance, if $n = 9$ and $b_1 = 1$, $b_2 = 2$, $b_3 = 3$, $b_4 = 6$, and $b_5 = 8$, then $c_4 = 9$, $c_3 = 7$, $c_2 = 4$, and $c_1 = 5$. The number of the sequences $\{b_i\}_{i=1}^{m+1}$ satisfying (3.8) is given by $Cat_m$ (see Exercise 3.11).

**Step 2.** Let $\pi \in Flatten_n(231)$ with $|M_\pi| = k$. Clearly, $M_\pi \subseteq \{2, 3, \ldots, n\}$. Let $L_\pi$ be the set of all elements of $M_\pi$ that is a minimal element of a block in $\pi$. Clearly, $L_\pi \subseteq M_\pi$. Define $\pi'$ to be the set partition obtained from $\pi$ by removing each element $i \in M_\pi$ from its block and, if $i \in L_\pi$, concatenating this block with the currently preceding block. Thus $\pi'$ is a set partition in $Flatten_{n-k}(231)$. For instance, if $\pi = \{1\}/\{2, 3, 8\}/\{4, 7\}/\{5, 6\}$, then $M_\pi = \{2, 3, 6\}$, $L_\pi = \{2\}$, and $\pi' = \{1, 2, 6\}/\{3, 5\}/\{4\}$. We claim that there is a bijection from the set partitions in $Flatten_n(231)$ with $|M_\pi| = k$ to the triplets $(M, L, \pi')$ such that $|M| = k$, $L \subseteq M \subseteq \{2, 3, \ldots, n\}$ and $\pi' \in Flatten_{n-k}(231)$ with $|M_{\pi'}| = 0$. In order to establish this bijection, let $(M, L, \pi')$ be such triple and let us define $\pi$ as follows. For each $i \in M$ (from smallest to largest), locate the first block $B$ of $\pi'$ such that its first element less than $i$, then insert $i$ in

94     Combinatorics of Set Partitions

$B$ such keeping an increasing order in $B$, where we increase each element $j$ by 1 if $j \geq i$. After inserting all elements of $M$, we will obtain a set partition $\pi \mathcal{P}_n$ in which the descent terminators are the block initiators and there is no initiator element in $M$. Hence, the map $\pi' \mapsto \pi$ is a bijection. Therefore, $f_{n,k} = \binom{n-1}{k} 2^k f_{n-k;0}$, where $\binom{n-1}{k}$ counts the number of choices of $k$ elements of $M$ from the set $\{2, 3, \ldots, n\}$, and $2^k$ counts the number of choices to divide the $k$ elements to two disjointed sets $M'$ and $L$, where $M = M' \cup L$.

**Step 3.** Combining Steps 1 and 2, we obtain

$$f_{n,k} = \binom{n-1}{k} 2^k \mathrm{Cat}_{(n-k-1)/2},$$

where $\mathrm{Cat}_m$ is defined to be 0 when $m$ is not an integer. Summing over all possible values of $k$, we derive that the number of set partitions in $\mathrm{Flatten}_n(231)$ is given by

$$|\mathrm{Flatten}_n(231)| = \sum_{k=0}^{n-1} \binom{n-1}{k} 2^k \mathrm{Cat}_{(n-k-1)/2}$$

$$= \sum_{k=0}^{(n-1)/2} \binom{n-1}{2k} 2^{n-1-2k} \mathrm{Cat}_k.$$

By Touchard's identity (see B.1), we have $|\mathrm{Flatten}_n(231)| = \mathrm{Cat}_n$, which completes the proof. □

The techniques that were used in the proof of the previous theorem can be extended to obtain one more result.

**Theorem 3.46** (Callan [61, Section 2.6]) *The number of set partitions in $\mathrm{Flatten}_n(321)$ is given by $\sum_{k=0}^{n} \binom{n}{k} \mathrm{Cat}_k$.*

**Proof** Again, let $M_\pi$ be the set of all the $RL$-minima of $\pi$ that are not descent terminators. Let $f_{n,k}$ be the number of set partitions $\pi \in \mathrm{Flatten}_n(321)$ with $|M_\pi| = k$. We proceed with the proof by three steps. In the first step we investigate the case $k = 0$, in the second step we reduce the general case to the case $k = 0$, and in the last step we combine the first two steps together to complete the proof.

**Step 1.** We show that $f_{n,0} = \mathrm{Riord}_{n-1}$ by establishing a bijection between the set partitions $\pi$ in $\mathrm{Flatten}_n(321)$ with $|M_\pi| = 0$ and the set of Dyck paths of size $2n - 2$ with no maximal sequence of contiguous down steps of size one. This bijection can be described as follows (actually, it is another formalization of Krattenthaler's [199] bijection). Let $\pi = \pi_1 \cdots \pi_n \in \mathrm{Flatten}_n(321)$ with $|M_\pi| = 0$; we define

$$\pi' = (n+1-\pi_n)(n+1-\pi_{n-1})\cdots(n+1-\pi_2).$$

Clearly, $\pi'$ is a permutation in $\mathcal{S}_n(321)$. Let $M = \{m_1, \ldots, m_k\}$ be the list of $LR$-maxima of $\pi'$ and $L = \{\ell_1, \ldots, \ell_k\}$ be the list of the positions of $LR$-maxima of $\pi'$. Our Dyck path can be defined as

$$U^{m_1-m_0} D^{\ell_2-\ell_1} \cdots U^{m_k-m_{k-1}} D^{\ell_{k+1}-\ell_k},$$

where $m_0 = 0$ and $\ell_{k+1} = n$. Obviously, our Dyck path has no maximal sequence of contiguous down steps of size one, and thus, the map is a bijection.

**Step 2.** By using the same bijection $\pi \mapsto (M, L, \pi')$ as given in the proof of Theorem 3.42, we obtain that the number of set partitions $\pi \in \text{Flatten}_n(321)$ with $|M_\pi| = k$ is given by $f_{n,k} = \binom{n-1}{k} 2^k f_{n-k,0}$.

**Step 3.** Steps 1 and 2 obtain

$$|\text{Flatten}_n(321)| = \sum_{k=0}^{n-1} \binom{n-1}{k} 2^k \text{Riord}_{n-k-1},$$

which, by Appendix B.2, gives $|\text{Flatten}_n(321)| = \sum_{k=0}^{n} \binom{n}{k} \text{Cat}_k$, as required. □

### 3.2.4 More Representations

In last two subsection we investigated two representations of set partitions, namely, block and flattened representations. As we discussed in the introduction to the current chapter, there are other representations. Actually, the other representations are very attractive in set partitions research. This invites us to separate the results regarding the other representations to several chapters and then to small sections since it will be tough to write all the results of the book (see Chapters 4, 5, 6, and 7) in one or two chapters. In any case, here we present the elementary definitions of each of four popular representations: canonical representation, graph representation, standard representation, and Rook placement representation.

#### 3.2.4.1 Canonical Representation

Canonical representation, as discussed in Definition 1.3, plays the most popular representation in this book. Note that the definitions of the concept of pattern-avoidance and pattern-counting in set partitions are obtained directly from definitions on set of words and set of permutations (see [48, 138]).

**Definition 3.47** *Let $\pi = \pi_1 \pi_2 \cdots \pi_n$ be a canonical representation of a set partition of $[n]$. Let $\tau = \tau_1 \cdots \tau_k$ be any word of size $k$. We say that $\pi$ contains $\tau$ as a subsequence pattern if there exist a subsequence $\pi' = \pi_{i_1} \pi_{i_2} \cdots \pi_{i_k}$ in $\pi$ such that $\pi'$ is order isomorphic to $\tau$, that is, $\pi_{i_a} < \pi_{i_b}$ (respectively, $\pi_{i_a} = \pi_{i_b}$, $\pi_{i_a} > \pi_{i_b}$) if and only if $\tau_a < \tau_b$ (respectively, $\tau_a = \tau_b$, $\tau_a > \tau_b$). Otherwise, we say that $\pi$ avoids the subsequence $\tau$ or is $\tau$-avoiding.*

**Example 3.48** *The canonical representation of the set partition $146/235/7$*

is $\pi = 1221213$. Here, $\pi$ contains the patterns 1212 and 1213, but it avoids the pattern 1231.

#### 3.2.4.2 Graphical Representation

Another representation that is used in our book is the graphical representation.

**Definition 3.49** *A graphical representation of a set partition $\pi = \pi_1\pi_2\cdots\pi_n$ of $[n]$ is given by the set of points $(\pi_i, i)$, $i = 1, 2, \ldots, n$, in the lattice $\mathbb{Z}^2$.*

Actually, a set $A$ of points in the first quarter of the lattice $\mathbb{Z}^2$ is a graphical representation for a member of $\mathcal{P}_{n,k}$ if $A$ contains only points of the form $(j, i)$ such that $j \leq i$, $j = 1, 2, \ldots, k$ and $i = 1, 2, \ldots, n$, with at least one point on each vertical line and no two points on the same horizontal line.

For example, the graphical representation of the set partition $\pi = 1231242 \in \mathcal{P}_{7,4}$ is given below in Figure 3.2.

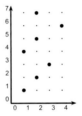

**FIGURE 3.2**: A graphical representation of the set partition $\pi = 1231242$

#### 3.2.4.3 Standard Representation

Now we are interested to represent a set partition as a linear graph.

**Definition 3.50** *Let $\pi = B_1/B_2\cdots/B_k$ be any set partition of $[n]$. The graph on the vertex set $[n]$ whose edge set consists of arcs connecting the elements of each block in numerical order is called the* standard representation *of $\pi$. We always write an arc $e$ as a pair $(i, j)$ with $i < j$, and say that $i$ is the* left-hand endpoint *of $e$ and $j$ is the* right-hand endpoint *of $e$.*

For example, the standard representation of the set partition $1357/26/4/89$ has the arc set $\{(1, 3), (3, 5), (5, 7), (2, 6), (8, 9))\}$; see Figure 3.3.

**FIGURE 3.3**: The standard representation of $1357/26/4/89$

The very basic result that can be mentioned here is the enumeration of noncrossing set partitions.

**Definition 3.51** *A noncrossing set partition $\pi$ is a set partition such that any two edges in its standard representation does not cross. In other words, a noncrossing set partition $\pi$ is a set partition such that its canonical form avoids the subsequence pattern 1212.*

The study of noncrossing set partitions goes back at least to Becker [21], where they are called "planar rhyme schemes". The first systematic study of noncrossing set partitions began by Kreweras [203], Poupard [276], Prodinger [279], and Simion [324].

**Theorem 3.52** *There is a bijection between the set of noncrossing set partitions of $[n]$ and the set of Dyck paths of size $2n$. Moreover, the number of noncrossing set partitions of $[n]$ is given by $\mathrm{Cat}_n$, the n-th Catalan number.*

**Proof** Let $\pi$ be a standard representation of a noncrossing set partition of $[n]$. We label the vertices of $\pi$ from 1 to $n$ from left to right. Let $v_1 = 1, v_2, \ldots, v_s$ be a maximal connected path in $\pi$, that is, there is an arc between the vertices $v_i$ and $v_{i+1}$ for all $1 \leq i \leq s-1$, where $s$ maximal. Define the induced graph on the vertices $v_i + 1, v_i + 2, \ldots, v_{i+1} - 1$ by $\pi^{(i)}$. Now we define a map $f$ recursively:
$$f(\pi) = U^s Df(\pi^{(1)}) Df(\pi^{(2)}) \cdots Df(\pi^{(s)}),$$
where $f(\pi)$ is the empty path when $\pi$ is an empty set partition. Clearly, $f(\pi) = UD$ if $\pi$ has exactly one vertex. By induction on the number vertices in $\pi$ we have that $f$ is a bijection. For instance, see Figure 3.4. Hence, Example 2.61 completes the proof. □

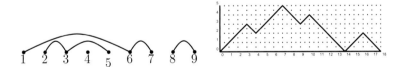

**FIGURE 3.4**: The standard representation of $4/235/167/89$ and its corresponding Dyck path

**Corollary 3.53** *The number of noncorssing set partitions of $\mathcal{P}_{n,k}$ is given by*
$$\frac{1}{n}\binom{n}{k}\binom{n}{k-1}.$$

**Proof** Theorem 3.52 shows that the number of noncorssing set partitions of $\mathcal{P}_{n,k}$ equals the number of blocks in a Dyck paths of length $2n$, which is well known by the Narayana number[1] $\frac{1}{n}\binom{n}{k}\binom{n}{k-1}$ (for example, see [207]). □

A simple and direct derivation for the number of noncrossing set partitions of $[n]$ has been considered by Liaw, Yeh, Hwang, and Chang [208], where they proved the following result (see [90, 203]).

**Theorem 3.54** (Liaw, Yeh, Hwang, and Chang [208]) *The number of noncrossing set partitions of $[n]$ with exactly $k$ blocks such that the $i$-th block has size $a_i$ is given by*
$$\frac{n(n-1)(n-2)\cdots(n-k+2)}{a_1!a_2!\cdots a_k!}.$$

**Proof** The proof is based on the following claim:

- There is a bijection between the set $\mathcal{NC}$ of noncrossing set partitions of $[n]$ with exactly $k$ blocks of sizes $a_1, a_2, \ldots, a_k$ and the set $\mathcal{V}$ of vectors $(i_1, \ldots, i_{k-1})$ where $1 \leq i_j \leq n$ and the $i_j$'s are distinct for $1 \leq j \leq k-1$.

At first let us see how this claim yields the theorem. Assume that the blocks are distinguishable, so by our claim we have $|\mathcal{NC}| = |\mathcal{V}| = n(n-1)\cdots(n-k+2)$. However, when $a_i = a_j$, then interchanging the elements does not lead to a different set partition, since blocks can be identified only through their sizes. Thus we must divide the number of such set partitions by $a_1!a_2!\cdots a_k!$.

Now we are ready to prove the claim. Let $\pi$ be a set partition given by its standard representation. Label each component of $\pi$ by the label of the leftmost of its vertex. Define the vector $I_\pi = (i_1, i_2, \ldots, i_{k-1})$ to be the order of the components labels of $\pi$. For instance, if $\pi$ is given in Figure 3.4, then the vector $I_\pi = (1, 2, 4)$. The map $\pi \to I_\pi$ is well defined. On the other side, for a given vector $J = (j_1, \ldots, j_{k-1})$, for each $s = k-1, k-2, \ldots, 1$, we connect the vertex $j_s$ with the next $a_s$ nonmarked vertices and each vertex we marked. At the end we connect all the other nonmarked vertices. Hence, the map $\pi \to I_\pi$ is a bijection from the set $\mathcal{NC}$ to the set $\mathcal{V}$. □

#### 3.2.4.4 Rook Placement Representation

Another interest representation of set partitions can be formulated as follows.

**Definition 3.55** *The $n$-th triangular board is the board consisting of $n-1$ columns with $n-i$ cells in the $i$-th column and $i-1$ cells in the $i$-th row, (first column is the leftmost column and first row is the top row). For convenience, we also join pending edges at the right of the first row and at the top of the first column. A rook placement is a way of placing nonattacking rooks on such*

---
[1] see http://en.wikipedia.org/wiki/Narayana_number.

a board, that is, putting no two rooks in the same row or column. Let $R_{n,k}$ be the set of all rook placements of $n - k$ rooks on the n-triangular shape, where a rook is indicated by a black disk.

For instance, Figure 3.5 illustrates a 9-th triangular board and an element of $R_{9,5}$.

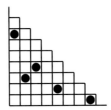

**FIGURE 3.5**: A rook placement

Our basic result to present here is the correspondence between the rook placements and set partitions.

**Theorem 3.56** (Stanley [333, Page 75]) *There exists a bijection between the set of set partitions $\mathcal{P}_{n,k}$ and the set of rook placements $R_{n,k}$. Moreover, $|R_{n,k}| = \text{Stir}(n, k)$.*

**Proof** Define $\rho : \mathcal{P}_{n,k} \to R_{n,k}$ as follows. First, label the rows (including the pending edge) of the $n$-th triangular board from top to bottom in increasing order by $1, 2, \ldots, n$, and the columns (including the pending edge) from left to right in increasing order by $1, 2, \ldots, n$. Then, if $\pi \in \mathcal{P}_{n,k}$ given by its linear graph representation (respectively, by its canonical representation), we construct $\rho(\pi)$ by placing a rook in the cell on the column labeled by $i$ and the row labeled by $j$ if and only if $(i, j)$ is an arc of $\pi$ (respectively, $i$ and $j$ belong to the same block $B$ of $\pi$, where $i < j$ and there are no elements in $B$ between $i$ and $j$). It is not hard to show that the map $\rho$ is well defined and bijective. The rest follows from Theorem 1.12. For example, Figure 3.6 presents the rook placement representation of the set partition $1357/26/4/89$. □

## 3.3 Exercises

**Exercise 3.1** *Solve the recurrence relation $a_n = 1 + \sum_{j=1}^{n-1} \sum_{i=0}^{j-1} \binom{n-1}{j}\binom{j-1}{i} a_i$ with $a_0 = 1$.*

**Exercise 3.2** *Use Theorem 1.16 to obtain a proof for Theorem 3.2.*

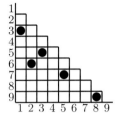

**FIGURE 3.6**: A rook placement representation of 1357/26/4/89

**Exercise 3.3** *Show*

$$\begin{vmatrix} 1 & -\binom{m-1}{0} & -\binom{m-1}{1} & \cdots & -\binom{m-1}{m-2} \\ 1 & 1 & -\binom{m-2}{0} & \cdots & -\binom{m-2}{m-3} \\ 1 & 0 & 1 & \cdots & -\binom{m-3}{m-4} \\ 1 & 0 & 0 & 1 & \cdots & -\binom{m-4}{m-5} \\ \vdots & & & & \vdots \\ 1 & 0 & 0 & 0 & \cdots & 1 & -\binom{1}{0} \\ 1 & 0 & 0 & 0 & \cdots & 0 & 1 \end{vmatrix} e = \sum_{j\geq 1} \frac{j^m}{j!}.$$

**Exercise 3.4** *Let $n \geq 0$. Show*

$$\mathrm{Bell}_{n+1} = \frac{1}{e} \sum_{j+k \geq n} \frac{(j+k-n)^k}{(j+k-n)!(n-k)!k!}.$$

**Exercise 3.5** *Prove $k!\mathrm{Stir}(n,k) = \sum \frac{n!}{i_1! i_2! \cdots i_k!}$, where the sum is over all possible $(i_1, i_2, \ldots, i_k)$ of nonnegative integers with $i_1 + i_2 + \cdots + i_k = n$.*

**Exercise 3.6** *Characterize the set*

i. $\mathrm{Flatten}_{n,k}(123)$.

ii. $\mathrm{Flatten}_n(132)$.

**Exercise 3.7** *A singleton block in a set partition is a block with exactly one element. Let $F(x;p,q)$ be the exponential generating function for the number of set partitions of $[n]$ according to the number of singleton blocks and number of blocks, that is,*

$$F(x;p,q) = \sum_{n\geq 0} \frac{x^n}{n!} \left( \sum_{\pi \in \mathcal{P}_n} p^{\mathrm{blo}(\pi)} q^{\text{number singleton blocks in } \pi} \right).$$

*Show that*

i. $F(x;p,q) = e^{p(e^x - 1 - x + qx)}$.

ii. The exponential generating function for the number of singleton blocks in all the set partitions of $[n]$ according to the number of blocks is given by $pxe^{p(e^x-1)}$.

iii. The exponential generating function for the number of singleton blocks, excluding any singleton block that equals $\{1\}$, in all the set partitions of $[n]$ according to the number of blocks, is given by $p^2xe^xe^{p(e^x-1)}$.

**Exercise 3.8** Show that the number of set partitions in $\mathcal{P}_n(1/23)$ is given by $\binom{n}{2}+1$ for all $n \geq 1$.

**Exercise 3.9** Find a bijection between the set $\mathcal{P}_n(13/2, 123)$ and the set of all permutations in $\mathcal{S}_n$ that avoid all patterns 123, 132, and 213.

**Exercise 3.10** Show that $|E\mathcal{P}_n(1/23)| = \lfloor(n-1)^2/4\rfloor + 1$ and $|E\mathcal{P}_n(123)| = \sum_{i=0}^{(n-2)/4} \binom{n}{4i+2}(4i+2)!!$, for $n \geq 1$.

**Exercise 3.11** Let $A_m$ be the set of all sequences $b_1b_2\cdots b_{m+1}$ such that $b_1 = 1 < b_2 = 2 < b_3 < b_4 < \cdots < b_{m+1}$ and $b_i \leq 2i-2$ for all $i = 3, 4, \ldots, m+1$. Prove that $|A_m| = \text{Cat}_m$ for all $m \geq 0$.

**Exercise 3.12** A noncrossing cycle set partition of $[n]$ whose elements arranged into a circle in its natural order, is a set partition of $[n]$ such that there are no four elements $a < b < c < d$, where $a$ and $c$ are in one block and $b, d$ are in another. Show that the number of noncrossing cycle set partitions of $[n]$ with exactly $k \geq 2$ blocks of sizes $n_1, n_2, \ldots, n_k$ is given by

$$\frac{(k-2)!\binom{n}{k-2}}{\prod_{j=1}^k n_j!}.$$

**Exercise 3.13** Following [121], we define a colored set partition of $[n]$ as a set partition of $[n]$ where each element in the set partition is given one of $c$ colors. We will denote by $\mathcal{P}_n \wr C_c$ the set of all colored set partitions, borrowing notation from the notion of the wreath product of groups. Find a formula for the number of colored set partitions of $[n]$ that avoid the pattern $1^a1^b$, that is, there is no subsequence 11 in the set partition such that the first element in the subsequence has color $a$, and the second element in the subsequence has color $b$ (for other cases of patterns, see [121]).

**Exercise 3.14** An involution set partition of $[n]$ is a set partition of size $[n]$ such that the size of each block is at most two. We denote the set of all involution set partitions of $[n]$ by $\mathcal{IP}_n$. Find an explicit formula for the cardinality of the set $\mathcal{IP}_n$.

**Exercise 3.15** *Following Exercise 3.14, show that the number of noncrossing involution set partitions of $[n]$ with exactly $k \geq n/2$ blocks is given by*

$$\frac{1}{k}\binom{k}{n-k}\binom{n}{k-1}.$$

**Exercise 3.16** *Let $t(n,k)$ be the number of set partitions of $[n]$ ($n$ even) with exactly $k$ blocks such that $2i$ and $2i-1$ are not both minimum elements of blocks. Show*

i. *the numbers $t(n,k)$ satisfy the recurrence relation $t(n,k) = k^2 t(n-2,k) + (2k-1)t(n-2,k-1)$*

ii. *the exponential generating function $\sum_{n,k\geq 0} t(n,k)\frac{x^n y^k}{n!}$ is given by $e^{y(\cosh(x)-1)}$.*

**Exercise 3.17** *Show that the number of set partitions of $[n]$ with exactly one crossing is given by $\binom{2n-5}{n-4}$. (Note that, in [47] it has been proved that the generating function for the number of set partitions of $[n]$ with exactly $k$ crossings is a rational function of $x$ and $\sqrt{1-4x}$).*

**Exercise 3.18** *A Dumont permutation of the second kind of size $2n$ is a permutation $\pi = \pi_1 \pi_2 \cdots \pi_{2n} \in S_{2n}$ such that, for every $i = 1, 2, \ldots, n$, $\pi_{2i} < 2i$ and $\pi_{2i-1} \geq 2i-1$ (see [95]). Find a bijection between the set 3142-avoiding Dumont permutation of the second kind of size $2n$ and the set of noncrossing set partitions of $[n]$.*

**Exercise 3.19** *Let $\rho : \mathcal{P}_{n,k} \to R_{n,k}$ be the bijection that was described in Theorem 3.56. Show that $\pi \in \mathcal{P}_{n,k}$ does not contain $i, i+1, i+2$ in the same block for some $i$ if and only if $\rho(\pi)$ has no adjacent two nonempty cells in any diagonal.*

**Exercise 3.20** *Use the exponential generating function for the sequence $\{\text{Bell}_n\}_{n\geq 0}$, namely, $\text{Bell}(x) = e^{e^x - 1}$ (see Example 2.58), to show that $\text{Bell}_p \equiv 2 \pmod p$, for any prime $p$.*

**Exercise 3.21** *Use the previous exercise to show that*

$$\sum_{j=1}^{p-1}(-1)^k \text{Bell}_j \equiv 1 \pmod{p}$$

*for any prime $p$.*

**Exercise 3.22** *Let $M_{n,r}$ denote, for $0 \leq r \leq n$, the $n$-multiset*

$$\{1, 1, \ldots, 1, 2, 3, \ldots, n-r+1\},$$

where the element 1 appears with multiplicity $r$, and the remaining $n - r$ elements each appear with multiplicity 1. We define $\text{Bell}_{n,r} = |M_{n,r}|$. Show that

$$\text{Bell}_{n,r} = \sum_{k=0}^{n-1-r} \binom{n-1-r}{k} \text{Bell}_{n-1-k,r} + \text{Bell}_{n-1,r-1},$$

for all $n \geq r + 1$.

## 3.4  Research Directions and Open Problems

We now suggest several research directions that are motivated both by the results and exercises of this and earlier chapters.

**Research Direction 3.1** *Section 3.2.1 presents explicit formulas for the number of set partitions of $[n]$ that avoid a subpartition of size three. Fix $\tau$ to be a subpartition of $[k]$. For instance, $\tau = 1/23 \cdots k$ or $\tau = 1/2/34 \cdots k$. Find the number of set partitions of $[n]$ that avoid the subpartition $\tau$ according to Definition 3.23, where $\tau = 1/2/\cdots \ell/(\ell+1)(\ell+2)\cdots k$. In the case $k = 3$, see Table 3.1.*

**Research Direction 3.2** *Section 3.2.3 gives the full classification for the cardinalities of the sets $\text{Flatten}_n(\tau)$ for subsequence patterns $\tau$ of size three. This leads to the larger question of classifying the cardinalities of the sets $\text{Flatten}_n(\tau)$ for subsequence patterns $\tau$ of size four. Using a computer program (see Appendix H), we have obtained the sequences of the number of set partitions in $\text{Flatten}_n(\tau)$ for $n = 1, 2, \ldots, 10$ and subsequence pattern $\tau$ of size four (see Table 3.6).*

**Table 3.6**: The cardinality $|\text{Flatten}_n(\tau)|$, where $\tau \in S_4$ and $1 \leq n \leq 10$.

| $\tau \backslash n$ | 1 | 2 | 3 | 4 | 5 | 6 | 7 | 8 | 9 | 10 |
|---|---|---|---|---|---|---|---|---|---|---|
| 1234 | 1 | 2 | 5 | 7 | 7 | 3 | 0 | 0 | 0 | 0 |
| 1243 | 1 | 2 | 5 | 13 | 33 | 81 | 193 | 449 | 1025 | 2305 |
| 1324, 1423 | 1 | 2 | 5 | 13 | 34 | 89 | 233 | 610 | 1597 | 4181 |
| 1342 | 1 | 2 | 5 | 14 | 42 | 132 | 429 | 1430 | 4862 | 16796 |
| 1432, 3142, 4132 | 1 | 2 | 5 | 15 | 51 | 188 | 731 | 2950 | 12235 | 51822 |
| 2134 | 1 | 2 | 5 | 15 | 48 | 157 | 521 | 1751 | 5951 | 20424 |
| 2143 | 1 | 2 | 5 | 15 | 51 | 187 | 716 | 2812 | 11222 | 45280 |
| 2314, 2413, 3412 | 1 | 2 | 5 | 15 | 50 | 178 | 663 | 2552 | 10071 | 40528 |
| 2341 | 1 | 2 | 5 | 15 | 51 | 190 | 759 | 3203 | 14129 | 64645 |

*Continued on next page*

| $\tau\backslash n$ | 1 | 2 | 3 | 4 | 5 | 6 | 7 | 8 | 9 | 10 |
|---|---|---|---|---|---|---|---|---|---|---|
| 2431, 3241 | 1 | 2 | 5 | 15 | 52 | 202 | 858 | 3909 | 18822 | 94712 |
| 3124 | 1 | 2 | 5 | 15 | 48 | 156 | 507 | 1643 | 5313 | 17163 |
| 3214, 4213, 4312 | 1 | 2 | 5 | 15 | 52 | 201 | 840 | 3709 | 17035 | 80551 |
| 3421 | 1 | 2 | 5 | 15 | 52 | 202 | 858 | 3910 | 18846 | 95058 |
| 4123 | 1 | 2 | 5 | 15 | 48 | 156 | 509 | 1663 | 5436 | 17772 |
| 4231 | 1 | 2 | 5 | 15 | 52 | 202 | 858 | 3910 | 18846 | 95059 |
| 4321 | 1 | 2 | 5 | 15 | 52 | 203 | 876 | 4114 | 20737 | 110836 |

To extend the results of Section 3.2.3, find an explicit formula for the cardinalities $|\text{Flatten}_n(\tau)|$ for any subsequence pattern $\tau \in \mathcal{S}_4$.

**Research Direction 3.3** *The above research direction can also be extended as follows. We say a sequence $\pi = \pi_1\pi_2\cdots\pi_n$ avoids xy-z (x-yz) if there is no subsequence $\pi_a\pi_{a+1}\pi_b$ ($\pi_a\pi_b\pi_{b+1}$) in $\pi$ such that its reduced form equals with the reduced form of xyz. For instance, the set partition 1213114112 avoids 1-23 but does not avoid 12-3.*

*Our suggestions to study the cardinalities of the sets* $\text{Flatten}_n(xy\text{-}z)$ *and* $\text{Flatten}_n(x\text{-}zy)$, *where xyz is any permutation of size at least three. Note that our calculations suggest the following:*

$$|\text{Flatten}_n(1\text{-}32)| = |\text{Flatten}_n(13\text{-}2)| = 2^{n-1},$$

$$|\text{Flatten}_n(23\text{-}1)| = \sum_{k=0}^{[n/2]} \frac{2^{n-3k}n!}{(n-2k)!k!} = \text{Sequence } A005425 \text{ in } [327],$$

$$|\text{Flatten}_n(31\text{-}2)| = |\text{Flatten}_n(2\text{-}13)| = \text{Fib}_{2n-1},$$

$$|\text{Flatten}_n(2\text{-}31)| = \text{Cat}_n,$$

$$|\text{Flatten}_n(31\text{-}2)| = \text{Sequence } A005773 \text{ in } [327].$$

*Moreover, if we define* $\text{Flatten}_n(\tau^{(1)}, \tau^{(2)})$ *to be the of all set partitions $\pi$ of $[n]$ such that $\text{Flatten}(\pi)$ avoids both $\tau^{(1)}, \tau^{(2)}$, then our calculations suggest the following:*

$$|\text{Flatten}_n(2\text{-}31, 2\text{-}13)| = |\text{Flatten}_n(2\text{-}31, 3\text{-}12)| = \text{Pell}_n,$$

*where* $\text{Pell}_n$ *is the n-th Pell number (defined by* $\text{Pell}_0 = 0$, $\text{Pell}_1 = 1$ *and* $\text{Pell}_{n+2} = 2\text{Pell}_{n+1} + \text{Pell}_n$; *see [327, Sequence A000129]). We can give more and more examples that connect our flattened set partitions to other combinatorial structures, but we consider the above starting point of research.*

**Research Direction 3.4** *Following [275] (see Exercise 3.16), it will be interesting to study the number of set partitions of $[n]$ with exactly $k$ blocks such that the numbers of $a_{i,1}, \ldots, a_{i,\ell}$ are not all minimum elements of blocks. For* instance, when $a_{i,1} = 2i$, $a_{i,2} = 2i - 1$, and $\ell = 2$, the problem is solved by Port [275] in 1998, where Port showed that the number of set partitions of $[n]$ (n even) with exactly $k$ blocks such that $2i$ and $2i - 1$ are both not minimum elements of blocks relate to Touchard numbers (see [343]). Thus the question is to find (if there is) any relation between our general problem and combinatorial structures.

**Research Direction 3.5** *Following Exercise 3.14 and results of Section 3.2.1, find the number of involution set partitions of $[n]$ that avoid a subpartition of $\tau$ (see Definition 3.23).*

**Research Direction 3.6** *One question that can be asked is to study the number of rook placements $R_{n,k}$ (the proof of Theorem 3.56 gives a bijection between the set of set partitions $\mathcal{P}_{n,k}$ and the set of rook placements $R_{n,k}$) according to ceratin set of statistics. Here we suggest the following statistic. In [20], the X-ray of a permutation is studied, which is defined as the sequence of antidiagonal sums in the associated permutation matrix. Inspired by this, we define the X-ray of a set partition $\pi$, denoted by $X_\pi$, to be the sequence of the number of nonempty cells in the antidiagonals in the associated rook placement of $\pi$. For instance, the X-ray of the set partition $\pi = 1357/26/4/89$ of [9] is given by $X_\pi = 010002000100001$. Clearly, the X-ray of a set partition in $\mathcal{P}_n$ is a sequence of nonnegative integer numbers of size $2n-3$. Now, one can ask, for example, the following questions: For a fixed sequence $a = a_1 a_2 \cdots a_{2n-3}$ characterize the set of all $\pi \in \mathcal{P}_n$ with $X_\pi = a$, and count such set partitions. Find the set of $\pi \in \mathcal{P}_n$ such that the X-ray of $\pi$ is a binary sequence (ternary sequence)?*

# Chapter 4

# Subword Statistics on Set Partitions

A statistic (see Definition 1.17) on a set partition is a characteristic such as the number of blocks, rises, levels, descents, odd letters, etc. In this chapter we will focus on "word-statistics", a special type of statistics. A word-statistic is a statistic that can be expressed in terms of occurrences of subword patterns. The current chapter will present several results and techniques to obtain generating functions for word-statistics on set partitions, where (mostly) we identify a set partition with the corresponding canonical sequence (see Definition 1.3) and employ this representation to define patterns among set partitions.

The earliest word-statistics were rises, levels, and drops, considered by Mansour and Munagi [218] as we will see in Section 4.1 (also, see Section 4.2). Rises, levels, and drops (descents) can be regarded as the simplest of patterns, namely, two letter subword patterns (see Definition 4.3). A rise corresponds to the subword pattern 12, a descent corresponds to the subword pattern 21, and a level corresponding to the subword pattern 11. In this chapter we use the term *pattern* exclusively for subword patterns, enumerating set partitions that contain or avoid them. More general patterns will be investigated in Section 4.2.

In next section we will present results on the enumeration of set partitions according to the statistics blocks, rises, levels, and descents. We obtain our goal by deriving a generating function for several statistics of interest simultaneously, and then obtain results for individual statistics as special cases.

Following the research on words [59] and compositions [137, 138, 243], Mansour and Munagi [220] (for some combinatorial proofs, see Shattuck [320]) studied the number of set partitions of $[n]$ according to word-statistics $\ell$-levels ($\ell-1$ consecutive levels), $\ell$-rises ($\ell-1$ consecutive rises), and $\ell$-descents ($\ell-1$ consecutive descents). These results will be presented in Section 4.3. Some of their results were further extended to three letter subword patterns by Mansour and Shattuck [230] and to families of $\ell$-letter subword patterns by Mansour, Shattuck, and Yan [242]. These results will be presented in Section 4.2, starting with three letter subword patterns and follows by the results for more general $\ell$-letter subword patterns. Moreover, in [220], [242], and [320] results were presented on the total number of occurrences of a fixed subword pattern. This can be done by using generating function techniques or by combinatorial proofs. The generating function technique can be used as follows.

**Definition 4.1** Let $\alpha_1, \ldots, \alpha_s$ be any set of statistics on the set partitions.

We denote the generating function for the number of set partitions of $\mathcal{P}_n$ according to the statistics $\alpha_1, \ldots, \alpha_s$ by

$$P_{\alpha_1,\ldots,\alpha_s}(x; q_1, \ldots, q_s) = \sum_{n \geq 0} x^n \sum_{\pi \in \mathcal{P}_n} \prod_{j=1}^{s} q_j^{\alpha_j(\pi)}.$$

More generally, We denote the generating function for the number of set partitions of $\mathcal{P}_{n,k}$ according to the statistics $\alpha_1, \ldots, \alpha_s$ by

$$P_{\alpha_1,\ldots,\alpha_s}(x, y; q_1, \ldots, q_s) = \sum_{n,k \geq 0} x^n y^k \sum_{\pi \in \mathcal{P}_{n,k}} \prod_{j=1}^{s} q_j^{\alpha_j(\pi)}.$$

Clearly, $P_{\alpha_1,\ldots,\alpha_s}(x, y; q_1, \ldots, q_s) = P_{\text{blo},\alpha_1,\ldots,\alpha_s}(x; y, q_1, \ldots, q_s)$, where $\text{blo}(\pi)$ is the number of the blocks in a set partition $\pi$.

In the proofs, we also need to consider generating functions for set partitions that start with specified letters (as in the proof of Lemma 4.6).

**Definition 4.2** Let $\alpha_1, \ldots, \alpha_s$ be any set of statistics on the set partitions. We denote the generating function for the number of set partitions $\pi = \pi_1 \pi_2 \cdots \pi_n \in \mathcal{P}_n$ according to the statistics $\alpha_1, \ldots, \alpha_s$ such that $\pi_1 \cdots \pi_m = \theta_1 \cdots \theta_m$ by

$$P(\theta_1 \cdots \theta_m) = P(\theta_1 \cdots \theta_m | x, y)$$
$$= P_{\alpha_1,\ldots,\alpha_s}(\theta_1 \cdots \theta_m | x, y; q_1, \ldots, q_s)$$
$$= \sum_{n \geq 0} \sum_{\pi = \theta_1 \cdots \theta_m \pi' \in \mathcal{P}_n} x^n y^{\text{blo}(\pi)} \prod_{j=1}^{s} q_j^{\alpha_j(\pi)}.$$

Let $P_\tau(x; q)$ be any generating function that counts number of objects parameterized with parameter $n$ according a fixed statistic, say $\tau$. In order to find the total number of occurrence of the statistic $\tau$ in all of the set objects parameterized $n$, we use the fact

$$\sum_{n \geq 0} \left( \sum_\pi \tau(\pi) \right) x^n = \frac{d}{dq} P_\tau(x; q) \big|_{q=1}.$$

This trick will be used several times in this chapter. But on the combinatorial proofs, there is no specific trick that works in all our proofs. Thus, we give several combinatorial proofs to illustrate our method, while we leave a few of them as exercises.

## 4.1 Subword Patterns of Size Two: Rises, Levels and Descents

In this section we derive general results involving number of statistics (blocks, rises, levels, and descents) on set partitions, which then can be applied to special cases. To do so, we start by fixing our statistics.

**Definition 4.3** *Let $\pi = \pi_1\pi_2\cdots\pi_n$ be any set partition of $[n]$. We say that $\pi$ has a rise, level and descent at $i$ if $\pi_i < \pi_{i+1}$, $\pi_i = \pi_{i+1}$ and $\pi_i > \pi_{i+1}$, respectively. We denote the number of blocks, rises, levels, and descents in $\pi$ by $\mathrm{blo}(\pi)$, $\mathrm{ris}(\pi)$, $\mathrm{lev}(\pi)$, and $\mathrm{des}(\pi)$, respectively. The generating function for the number of set partitions of $[n]$ according to the number of blocks, rises, levels, and descents is given by*

$$F(x,y) = F(x,y;r,\ell,d) = P_{\mathrm{ris,lev,des}}(x,y;r,\ell,d).$$

**Example 4.4** *The set partition $\pi = 12311242 \in \mathcal{P}_8$ has four rises (at $i = 1$, $i = 2$, $i = 5$ and $i = 6$), one level (at $i = 4$), and two descents (at $i = 3$ and $i = 7$). Thus, $\mathrm{blo}(\pi) = 4$, $\mathrm{ris}(\pi) = 4$, $\mathrm{lev}(\pi) = 1$, and $\mathrm{des}(\pi) = 2$.*

In our proofs, we also need to consider generating functions for the number of set partitions that start with fixed letters.

**Definition 4.5** *We denote the generating function for the number of set partitions $\pi = \pi_1\pi_2\cdots\pi_n \in \mathcal{P}_n$ according to the number of blocks, rises, levels and descents such that $\pi_1\cdots\pi_m = \theta_1\cdots\theta_m$ by*

$$F(\theta_1\cdots\theta_m|x,y) = P(\theta_1\cdots\theta_m|x,y;r,\ell,d)$$
$$= \sum_{n\geq 0}\sum_{\pi=\theta_1\cdots\theta_m\pi'\in\mathcal{P}_n} x^n y^{\mathrm{blo}(\pi)} r^{\mathrm{ris}(\pi)} \ell^{\mathrm{lev}(\pi)} d^{\mathrm{des}(\pi)}.$$

*Also, we denote the generating function*

$$F(12\cdots m|x,y) - F(12\cdots m+1|x,y)$$

*by $F_m(x,y) = F_m(x,y|r,\ell,d)$. Moreover, we define*

$$F_m^*(x,y) = \sum_{j\geq m} F_j(x,y).$$

Clearly, the generating function $F_m(x,y)$ enumerates the number of set partitions of $[n]$ according to number of blocks, rises, levels, and descents that start by $12\cdots mj$ with $1 \leq j \leq m$. From Definitions 4.3 and 4.5, we immediately have

$$F(x,y) = 1 + \sum_{k\geq 1} F_k(x,y), \qquad (4.1)$$

where 1 counts the empty set partition, and

$$F_m(x,y) = x^m y^m r^{m-1} + \sum_{j=1}^{m} F(12\cdots mj|x,y), \qquad (4.2)$$

where $x^m y^m r^{m-1}$ counts the set partition $12\cdots m$. Now, our plan is to find an explicit formula for the generating function $F_m(x,y)$, which will allow us to write an explicit formula for the generating function $F(x,y)$. To do so, we need the following two lemmas.

**Lemma 4.6** *For all $1 \leq j \leq m-2$,*

$$F(12\cdots m(j+1)|x,y) = \frac{1+xd-x\ell}{1+xr-x\ell} F(12\cdots mj|x,y) \qquad (4.3)$$

*with $F(12\cdots mm|x,y) = x\ell F_m^*(x,y)$ and*

$$F(12\cdots m1|x,y) = \frac{x^{m+1} y^m r^{m-1}(d-r)}{1+x(r-\ell)}$$
$$+ \frac{x(xdr - xr\ell + d)F_m^*(x,y) + x(r-d)F_m(x,y)}{1+x(r-\ell)}. \qquad (4.4)$$

**Proof** Let $1 \leq j \leq m-1$. From Definitions 4.3 and 4.5 we have

$$F(12\cdots mj|x,y)$$
$$= x^{m+1} y^m r^{m-1} d + \sum_{k=1}^{j-1} F(12\cdots mjk|x,y) + F(12\cdots mjj|x,y)$$
$$+ \sum_{k=j+1}^{m} F(12\cdots mjk|x,y) + F(12\cdots mj(i+1)|x,y) \qquad (4.5)$$
$$= x^{m+1} y^m r^{m-1} d + xd \sum_{k=1}^{j-1} F(12\cdots mk|x,y) + x\ell F(12\cdots mj|x,y)$$
$$+ xr \sum_{k=j+1}^{m-1} F(12\cdots mk|x,y) + x^2 dr F_m^*(x,y) + xd F_{m+1}^*(x,y).$$

Therefore,

$$F(12\cdots m(j+1)|x,y) - F(12\cdots mj|x,y)$$
$$= (xd - x\ell)F(12\cdots mj|x,y) + (x\ell - xr)F(12\cdots m(j+1)|x,y),$$

which implies (4.3). Also, (4.5) gives

$$(1 - x\ell)F(12\cdots m1|x,y) = x^{m+1} y^m r^{m-1} d$$
$$+ xr \sum_{k=2}^{m-1} F(12\cdots mk|x,y) + x^2 dr F_m^*(x,y) + xd F_{m+1}^*(x,y).$$

By using (4.2), we obtain

$$(1 + xr - x\ell)F(12\cdots m1|x,y)$$
$$= x^{m+1}y^m r^{m-1}d + xr\left(F_m(x,y) - x^m y^m r^{m-1} - F(12\cdots mm|x,y)\right)$$
$$+ x^2 dr F_m^*(x,y) + xd F_{m+1}^*(x,y),$$

which is equivalent to

$$(1 + xr - x\ell)F(12\cdots m1|x,y)$$
$$= x^{m+1}y^m r^{m-1}d + xr\left(F_m(x,y) - x^m y^m r^{m-1} - x\ell F_m^*(x,y)\right)$$
$$+ x^2 dr F_m^*(x,y) + xd F_{m+1}^*(x,y),$$

and by the fact that $F_{m+1}^*(x,y) = F_m^*(x,y) - F_m(x,y)$ we obtain (4.4). It is obvious that $F(12\cdots mm|x,y) = x\ell F_m^*(x,y)$, as required. (Why?) □

**Lemma 4.7** *For all $m \geq 1$, the generating function $F_m(x,y)$ satisfies*

$$F_m(x,y) = x^m y^m r^{m-1}$$
$$+ \left(1 - \frac{r(1+x(d-\ell))}{r-d} + \frac{d(1+x(d-\ell))}{r-d}\frac{(1+x(r-\ell))^m}{(1+x(d-\ell))^m}\right) F_m^*(x,y) \quad (4.6)$$

*with $F_0(x,y) = 1$.*

**Proof** Lemma 4.6 together with (4.2) gives

$$F_m(x,y) = x^m y^m r^{m-1} + x\ell F_m^*(x,y)$$
$$+ \frac{x^{m+1}y^m r^{m-1}(d-r) + x(xdr - xr\ell + d)F_m^*(x,y) + x(r-d)F_m(x,y)}{1+x(r-\ell)}$$
$$\cdot \sum_{j=1}^{m-1}\left(\frac{1+x(d-\ell)}{1+x(r-\ell)}\right)^{j-1}.$$

By using the fact that $\sum_{j=0}^{m-2} a^j = \frac{1-a^{m-1}}{1-a}$, we obtain the result after several simple algebraic operations. □

In order to solve the recurrence in the above lemma, we consider the following general type of recurrence relation:

$$f_\ell = a_\ell + b_\ell \sum_{j \geq \ell} f_j, \quad \ell = 0, 1, \ldots. \quad (4.7)$$

By this definition we have

$$b_{\ell+1} f_\ell - b_\ell f_{\ell+1} = b_{\ell+1} a_\ell - b_\ell a_{\ell+1} + b_\ell b_{\ell+1} f_\ell,$$

which implies that (4.7) is equivalent to

$$f_\ell = \frac{a_\ell b_{\ell+1} - a_{\ell+1} b_\ell}{b_{\ell+1}(1-b_\ell)} + \frac{b_\ell}{b_{\ell+1}(1-b_\ell)} f_{\ell+1}, \quad \ell = 0, 1, \ldots. \quad (4.8)$$

The following lemma solves this recurrence relation, where the proof is left as Exercise 4.1.

**Lemma 4.8** *Let $\{g_\ell\}_{\ell \geq 0}$ be any sequence satisfying the recurrence relation*

$$g_\ell = \alpha_\ell + \beta_\ell g_{\ell+1},$$

*for $\ell \geq 0$. Assume that $g_\ell \to 0$ as $\ell \to \infty$. Then $g_\ell = \sum_{i \geq \ell} \left( \alpha_i \prod_{j=\ell}^{i-1} \beta_j \right)$.*

Define $f = \sum_{\ell \geq 0} f_\ell$. By applying Lemma 4.8 on (4.8), we obtain an explicit formula for $f$, where $f_\ell$ satisfies (4.7).

**Proposition 4.9** *Let $f_\ell$ be any sequence satisfying (4.7) and $f_\ell \to 0$ as $\ell \to \infty$. Then $f = \sum_{i \geq 0} \frac{a_i}{\prod_{j=0}^i (1-b_j)}$.*

**Proof** By Lemma 4.6 and (4.8) we obtain that for all $\ell \geq 0$,

$$f_\ell = \sum_{i \geq \ell} \left( \frac{a_i b_{i+1} - a_{i+1} b_i}{b_{i+1}(1-b_i)} \prod_{j=\ell}^{i-1} \frac{b_j}{b_{j+1}(1-b_j)} \right).$$

By summing over all possible values of $\ell$, we have

$$f = \sum_{i \geq 0} \frac{a_i b_{i+1} - a_{i+1} b_i}{b_{i+1}(1-b_i)} \left[ 1 + \sum_{j=0}^{i-1} \left( \frac{b_j}{b_i \prod_{k=j}^{i-1}(1-b_k)} \right) \right]$$

$$= \frac{a_0 b_1 - a_1 b_0}{b_1 (1-b_0)} + \sum_{i \geq 1} \frac{a_i b_{i+1} - a_{i+1} b_i}{b_{i+1}} \left[ \frac{1}{1-b_i} + \sum_{j=0}^{i-1} \left( \frac{b_j}{b_i \prod_{k=j}^{i}(1-b_k)} \right) \right]$$

$$= \frac{a_0 b_1 - a_1 b_0}{b_1(1-b_0)} + \sum_{i \geq 1} \left( \frac{a_i b_{i+1} - a_{i+1} b_i}{b_{i+1}} \sum_{j=0}^{i} \frac{b_j}{b_i \prod_{k=j}^{i}(1-b_k)} \right)$$

$$= \frac{a_0 b_1 - a_1 b_0}{b_1(1-b_0)} + \sum_{i \geq 1} \left( \frac{a_i b_{i+1} - a_{i+1} b_i}{b_{i+1} b_i \prod_{j=0}^{i}(1-b_j)} \sum_{j=0}^{i} b_j \prod_{k=0}^{j-1}(1-b_k) \right).$$

So, by using the fact that $1 - \sum_{j=0}^i b_j \prod_{k=0}^{j-1}(1-b_k) = \prod_{j=0}^i (1-b_j)$ (prove it!!), we obtain

$$f = \frac{a_0 b_1 - a_1 b_0}{b_1(1-b_0)} + \sum_{i \geq 1} \left( \frac{a_i b_{i+1} - a_{i+1} b_i}{b_{i+1} b_i \prod_{j=0}^{i}(1-b_j)} \left( 1 - \prod_{k=0}^{i}(1-b_k) \right) \right).$$

Therefore,

$$f = \frac{a_0 b_1 - a_1 b_0}{b_1(1-b_0)} + \sum_{i\geq 1} \frac{a_i b_{i+1} - a_{i+1} b_i}{b_{i+1} b_i \prod_{j=0}^{i}(1-b_j)} - \sum_{i\geq 1}\left(\frac{a_i}{b_i} - \frac{a_{i+1}}{b_{i+1}}\right)$$

$$= \frac{a_0 b_1 - a_1 b_0}{b_1(1-b_0)} - \frac{a_1}{b_1} + \sum_{i\geq 1} \frac{a_i b_{i+1} - a_{i+1} b_i}{b_{i+1} b_i \prod_{j=0}^{i}(1-b_j)}$$

$$= \frac{a_0 - \frac{a_1}{b_1}}{1-b_0} + \sum_{i\geq 1} \frac{\frac{a_i}{b_i} - \frac{a_{i+1}}{b_{i+1}}}{\prod_{j=0}^{i}(1-b_j)}$$

$$= \frac{a_0 - \frac{a_1}{b_1}}{1-b_0} + \sum_{i\geq 1} \frac{a_i}{b_i \prod_{j=0}^{i}(1-b_j)} - \sum_{i\geq 2} \frac{a_i}{b_i \prod_{j=0}^{i-1}(1-b_j)}$$

$$= \frac{a_0}{1-b_0} + \sum_{i\geq 1} \frac{a_i}{b_i \prod_{j=0}^{i}(1-b_j)} - \sum_{i\geq 1} \frac{a_i}{b_i \prod_{j=0}^{i-1}(1-b_j)},$$

which implies

$$f = \frac{a_0}{1-b_0} + \sum_{i\geq 1} \frac{a_i}{\prod_{j=0}^{i}(1-b_j)} = \sum_{i\geq 0} \frac{a_i}{\prod_{j=0}^{i}(1-b_j)},$$

as claimed. □

Now, we state an explicit formula for the generating function $F(x, y; r, \ell, d)$ for the number set partitions of $[n]$ according to the statistics number blocks, number rises, number levels, and number descents.

**Theorem 4.10** *We have*

$$F(x, y; r, \ell, d) = 1 + \sum_{i\geq 1} \frac{x^i y^i r^{i-1}(r-d)^i}{(1+x(d-\ell))^i \prod_{j=1}^{i}\left(r - d\left(\frac{1+x(r-\ell)}{1+x(d-\ell)}\right)^j\right)}.$$

*Moreover, the generating function for the number of set partitions of $[n]$ according to number of rises, levels and descents with exactly $k \geq 1$ blocks is given by*

$$\frac{x^k r^{k-1}(r-d)^k}{(1+x(d-\ell))^k \prod_{j=1}^{k}\left(r - d\left(\frac{1+x(r-\ell)}{1+x(d-\ell)}\right)^j\right)}.$$

**Proof** Lemma 4.7 gives that $F(x, y; r, \ell, d) = 1 + \sum_{i\geq 1} f_i$, where $f_i$ satisfies the recurrence relation $f_i = a_i + b_i \sum_{j\geq i} f_j$ with

$$a_i = x^i y^i r^{i-1},$$

$$b_i = 1 - \frac{r(1+x(d-\ell))}{r-d} + \frac{d(1+x(d-\ell))}{r-d}\left(\frac{1+x(r-\ell)}{1+x(d-\ell)}\right)^i,$$

for all $i \geq 1$, $a_0 = 1$ and $b_0 = 0$. Therefore, by Proposition 4.9 we have

$$F(x,y,r,\ell,d) = 1 + \sum_{i\geq 1} \frac{a_i}{\prod_{j=1}^{i}(1-b_j)}$$

$$= 1 + \sum_{i\geq 1} \frac{x^i y^i r^{i-1}}{\prod_{j=1}^{i}\left[\frac{r(1+x(d-\ell))}{r-d} - \frac{d(1+x(d-\ell))}{r-d}\left(\frac{1+x(r-\ell)}{1+x(d-\ell)}\right)^j\right]}$$

$$= 1 + \sum_{i\geq 1} \frac{x^i y^i r^{i-1}(r-d)^i}{(1+x(d-\ell))^i \prod_{j=1}^{i}\left[r - d\left(\frac{1+x(r-\ell)}{1+x(d-\ell)}\right)^j\right]},$$

which completes the proof. □

### 4.1.1 Number Levels

Theorem 4.10 for $d, r \to 1$ gives that the generating function for the number of set partitions of $\mathcal{P}_{n,k}$ according to the number of levels is given by

$$\frac{x^k}{\prod_{j=1}^{k}(1-x(\ell+j-1))} = \frac{x}{(1-x\ell)} \cdot \frac{\frac{x^{k-1}}{(1-x\ell)^{k-1}}}{\prod_{j=1}^{k-1}\left(1-\frac{jx}{1-x\ell}\right)},$$

which, by (2.9), implies

$$\frac{x^k}{\prod_{j=1}^{k}(1-x(\ell+j-1))} = \frac{x}{(1-x\ell)}\sum_{n\geq k-1}\text{Stir}(n,k-1)\frac{x^n}{(1-x\ell)^n}$$

$$= \sum_{n\geq k-1}\text{Stir}(n,k-1)\frac{x^{n+1}}{(1-x\ell)^{n+1}},$$

By using the fact that $\frac{1}{(1-x)^{k+1}} = \sum_{j\geq 0}\binom{k+j}{k}x^j$, see Exercise 2.10, we obtain

$$\frac{x^k}{\prod_{j=1}^{k}(1-x(\ell+j-1))} = \sum_{n\geq k-1}\sum_{j\geq 0}\binom{n+j}{n}\text{Stir}(n,k-1)x^{n+1+j}\ell^j.$$

This leads to the following result.

**Corollary 4.11** (Shattuck [320]) *The number of set partitions of $\mathcal{P}_{n,k}$ and $m$ levels is given by*

$$\binom{n-1}{m}\text{Stir}(n-1-m,k-1).$$

**Proof** (Direct Proof) We denote the set of all the set partitions $\pi \in \mathcal{P}_{n,k}$ such that $\text{lev}(\pi) = m$ by $A_n(k,m)$, and its cardinality by $a_n(k,m) = |A_n(k,m)|$.

Let us write a recurrence relation for the sequence $\{a_n(k,m)\}_{n,k,m}$ by considering whether either last letter in a set partition $\pi = \pi_1 \cdots \pi_n \in A_n(k,m)$ forms/does not form a level. (1) The set partition $\pi$ with $\pi_n \neq \pi_{n-1}$ can be decomposed as either $\pi = \pi'k$ with $\pi' \in A_{n-1}(k-1,m)$, or $\pi = \pi'c$ with $\pi' \in A_{n-1}(k,m)$ and $c \neq \pi'_{n-1}$. In this case there are

$$a_{n-1}(k-1,m) + (k-1)a_{n-1}(k,m)$$

set partitions. (2) The set partition $\pi$ with $\pi_n = \pi_{n-1}$ can be written as $\pi = \pi'\pi'_{n-1}$ with $\pi' \in A_{n-1}(k,m-1)$. In this case there are $a_{n-1}(k,m-1)$ set partitions.

Thus, the sequence $\{a_n(k,m)\}_{n,k,m}$ satisfies the following recurrence relation

$$a_n(k,m) = a_{n-1}(k-1,m) + (k-1)a_{n-1}(k,m) + a_{n-1}(k,m-1),$$

for $n \geq k > m > 0$ with the initial condition $a_n(k,0) = \text{Stir}(n-1,k-1)$ (see Corollary 4.13). Exercise 4.2 shows that the solution of this recurrence relation is given by $\binom{n-1}{m}\text{Stir}(n-1-m,k-1)$, which completes the proof. $\square$

**Definition 4.12** *A Carlitz set partition of $[n]$ is a set partition of $[n]$ without levels.*

The name Carliz is given to this set partition as that he introduced these "Carlitz compositions"; compositions without levels. For more details on Carlitz compositions, we refer the reader to [138]. Our next result tells how many Carlitz set partitions of $[n]$ there are?

**Corollary 4.13** (Shattuck [320]) *There is a bijection between the set $\mathcal{P}_{n,k}$ and the set of Carlitz set partitions of $\mathcal{P}_{n+1,k+1}$.*

**Proof** Let $B_{n+1,k+1} = \{\pi \in \mathcal{P}_{n+1,k+1} \mid \text{lev}(\pi) = 0\}$. Note that an $\ell$-level means a subword of $\ell$ identical letters, $\ell \geq 2$. We construct a bijection $f : \mathcal{P}_{n,k} \to B_{n+1,k+1}$ by mapping the set partition $\pi = \pi_1\pi_2\cdots\pi_n \in \mathcal{P}_{n,k}$ to another set partition $\pi' = f(\pi) \in B_{n+1,k+1}$ as follows. (1) if $\pi \in B_{n,k}$, define $\pi' = \pi(k+1)$. (2) Otherwise, (i) replace each member $\pi_j$ of an $\ell$-level with $c$ if $j$ is even (respectively, odd) and $\ell$ is odd (respectively, even), where $c = 1 + \max_{j \in [i]} \pi_j$ and $\pi_1\pi_2\cdots\pi_i$ is the subword immediately preceding the first $\pi_j$ to be replaced, $i = j-1$ ($c = 1$ if and only if the first $\ell$-level begins with $\pi_1 = 1$ and $\ell$ is even). It may be necessary to tag each designated $c$ for the next step; (ii) add 1 to all other letters $\geq c$ on the right of the first $c$; and (iii) insert $c$ at the end of the resulting word to obtaining $\pi'$.

Since the last letter $c$ of the image indicates the source, the map $f$ is reversible. $\square$

**Example 4.14** *For instance, the bijection in Corollary 4.13 can be illustrated as follows.*

(1) When $\pi = 12121312 \in B_{8,3}$, we have $\pi' = \pi 4 = 121213124 \in B_{9,4}$.
(2) The bijection for $112133321 \in \mathcal{P}_{9,3}$ gives

$$112133321 \to \bar{1}1213\bar{1}321 \to \bar{1}2324\bar{1}432 \to \pi' = 1232414321 \in B_{10,4}$$

(3) Similarly, the bijection for $121112233 \in \mathcal{P}_{9,3}$ gives

$$121112233 \to 121\bar{3}13\bar{2}33 \to 121\bar{3}13\bar{2}34 \to \pi' = 1213132343 \in B_{10,4}.$$

Another example, Theorem 4.10 with $\ell = 1$, $d = 1/v$, and $r = v$, gives the generating function for the number of set partitions of $\mathcal{P}_{n,k}$ blocks, according to the statistic ris $-$ des as

$$F_k(x,v) = \sum_{n \geq 0} \sum_{\pi \in \mathcal{P}_{n,k}} x^n v^{\mathrm{ris}(\pi) - \mathrm{des}(\pi)}$$

$$= \frac{x^k v^{k-1}}{(1 + x(1/v - 1))^k \prod_{j=1}^{k} \frac{v - \frac{1}{v}(\frac{1+x(v-1)}{1+x(1/v-1)})^j}{v - 1/v}}.$$

Hence, after simple algebraic operations (show the details!) we have

$$\frac{\partial}{\partial v} F_k(x,v)\,|_{v=1} = \frac{(k-1)(kx-2)x^k}{2 \prod_{j=1}^{k}(1-jx)},$$

which, by (2.9), implies that

$$\frac{\partial}{\partial v} F_k(x,v)\,|_{v=1} = \frac{1}{2}(k-1)(kx-2) \sum_{j \geq 1} \mathrm{Stir}(j,k) x^j. \qquad (4.9)$$

Thus, we can obtain the following result.

**Corollary 4.15** *We have*

$$\sum_{\pi \in \mathcal{P}_{n,k}} \mathrm{ris}(\pi) - \mathrm{des}(\pi) = (k-1)\mathrm{Stir}(n,k) - \binom{k}{2}\mathrm{Stir}(n-1,k).$$

*Moreover,*

$$\sum_{\pi \in \mathcal{P}_n} \mathrm{ris}(\pi) - \mathrm{des}(\pi) = \frac{1}{2}(\mathrm{Bell}_{n+1} - \mathrm{Bell}_n - \mathrm{Bell}_{n-1}).$$

**Proof** The first sum holds immediately from (4.9). Theorem 1.12 gives that

$$(k-1)\mathrm{Stir}(n,k) - \binom{k}{2}\mathrm{Stir}(n-1,k)$$
$$= \frac{1}{2}(\mathrm{Stir}(n+1,k) - \mathrm{Stir}(n,k-1) - \mathrm{Stir}(n-1,k-2))$$

and then

$$\sum_{\pi \in \mathcal{P}_n} \text{ris}(\pi) - \text{des}(\pi)$$
$$= \sum_{k \geq 1} \frac{1}{2}(\text{Stir}(n+1, k) - \text{Stir}(n, k-1) - \text{Stir}(n-1, k-2))$$
$$= \frac{1}{2}(\text{Bell}_{n+1} - \text{Bell}_n - \text{Bell}_{n-1}),$$

as claimed. □

Again, Theorem 4.10 with $d = 1$, $\ell = 1/v$, and $r = v$ gives that the generating function for the number of partitions of $\mathcal{P}_{n,k}$ according to the statistic of ris − lev is

$$P_{\text{ris}-\text{lev}}(x; v) = \sum_{n \geq 0} \sum_{\pi \in \mathcal{P}_n} x^n v^{\text{ris}(\pi) - \text{lev}(\pi)}$$
$$= \frac{x^k v^{k-1}}{(1 + x(1 - 1/v))^k \prod_{j=1}^{k} \frac{v - \left(\frac{1+x(v-1/v)}{1+x(1-1/v)}\right)^j}{v - 1/v}}.$$

Hence, after simple algebraic operations we have that

$$\frac{\partial}{\partial v} P_{\text{ris}-\text{lev}}(x; v) \big|_{v=1} = \frac{(k-1)x^k - kx^{k+1} + x^{k+2} \sum_{j=1}^{k} \frac{j(j-3)}{2(1-jx)}}{\prod_{j=1}^{k}(1 - jx)}.$$

Expanding at $x = 0$ with using (2.9) we get the following result. (Find the details!)

**Corollary 4.16** *Let* $1 \leq k \leq n$. *Then*

$$\sum_{\pi \in \mathcal{P}_{n,k}} \text{ris}(\pi) - \text{lev}(\pi) = (k-1)\text{Stir}(n, k) - k\text{Stir}(n-1, k)$$
$$+ \sum_{i=0}^{n-3} \left( \sum_{j=1}^{k} \frac{j^{i+1}(j-3)}{2} \right) \text{Stir}(n-2-i, k).$$

### 4.1.2 Nontrivial Rises and Descents

Let $\pi = \pi_1 \pi_2 \cdots \pi_n$ be any set partition. Recall that a rise (respectively, descent) is said to occur at $i$ if $\pi_i < \pi_{i+1}$ (respectively, $\pi_i > \pi_{i+1}$); see Definition 4.3.

**Definition 4.17** *Let* $\pi$ *be any set partition. We call a rise or a descent of* $\pi$ *at* $i$ *nontrivial if neither* $i$ *nor* $i+1$ *are the smallest elements of their respective blocks. Otherwise, the rise or the descent of* $\pi$ *is said to be trivial.*

**Example 4.18** *The set partition $\pi = 1123123 \in \mathcal{P}_7$ has four rises at positions $2, 3, 5, 6$ but it has two nontrivial rises at positions $5, 6$.*

The following lemma provides an explicit formula for the total number of nontrivial rises in all of the set partitions of $\mathcal{P}_{n,k}$.

**Lemma 4.19** *The number of nontrivial rises (descents) in all of the set partitions of $\mathcal{P}_{n,k}$ is given by*

$$\sum_{j=2}^{k} \binom{j}{2} \left( \sum_{i=2}^{n-k} j^{i-2} \mathrm{Stir}(n-i,k) \right).$$

**Proof** We leave the case of nontrivial descents as an interesting task to the reader and we focus only on the case of nontrivial rises. Fix $n$ and $k$. Let $i$ and $j$ be any two integers such that $2 \leq i \leq n-k$ and $2 \leq j \leq k$. Consider all the set partitions of $\mathcal{P}_{n,k}$, which may be decomposed uniquely as

$$\pi = \pi' j \alpha \beta, \tag{4.10}$$

where $\pi'$ is a set partition with $j-1$ blocks, $\alpha$ is a word in $[j]^i$ whose last two letters form a rise, and $\beta$ is possibly empty. The total number of nontrivial rises can be computed by finding the number of set partitions that may be expressed as in (4.10) for each $i$ and $j$ and then summing over all possible values of $i$ and $j$. Since there are $j^{i-2}$ choices for the first $i-2$ letters of $\alpha$, $\binom{j}{2}$ choices for the final two letters in $\alpha$, and $\mathrm{Stir}(n-i,k)$ choices for the remaining letters $\pi' j \beta$, which necessarily constitute a set partition of $\mathcal{P}_{n-i,k}$, we obtain that there are $\binom{j}{2} j^{i-2} \mathrm{Stir}(n-i,k)$ set partitions of $\mathcal{P}_{n,k}$ that may be given by (4.10). □

The definition of nontrivial rises and nontrivial descents motivates the following definition.

**Definition 4.20** *Let $\pi$ be any set partition that contains the subword pattern $\tau$. We say that $\tau$ occurs in $\pi$ as nontrivial occurrence if neither $i$ nor $i+1$ are the smallest elements of their respective blocks.*

We end this section by the following general remark. Since each set partition of $\mathcal{P}_{n,k}$ can be decomposed as $1\pi^{(1)} 2\pi^{(2)} \cdots k\pi^{(k)}$, then the generating function for the number of set partitions of $\mathcal{P}_{n,k}$ according to $\mathrm{nocc}_\tau$ the number nontrivial occurrences of the subword $\tau$ is given by

$$P_{\mathrm{nocc}}(x,y;q) = \sum_{n \geq 0} x^n \sum_{k=0}^{n} y^n \sum_{\pi \in \mathcal{P}_{n,k}} q^{\mathrm{nocc}_\tau} = 1 + \sum_{k \geq 1} x^k y^k \prod_{j=1}^{k} W_\tau(x;j),$$

where $W_\tau(x;j)$ is the generating function for the number of words of size $n$ over alphabet $[j]$ according to the number of occurrences of the subword pattern $\tau$ (see research Direction 4.7).

## 4.2 Peaks and Valleys

As discussed at the beginning of this chapter, rises, levels, and descents can be described as subword patterns of size two. Because there are only three such patterns, we were able to derive results for all patterns at once in Theorem 4.10. As the size of the pattern increases, the corresponding number of subword patterns increases rapidly, so generating function for one special type of subword patterns will be obtained at a time. Thus, we introduce the following notation to keep track the number of occurrences of a specific (subword) pattern.

**Definition 4.21** *We denote the number of occurrences of the (subword) pattern $\tau$ in a set partition $\pi$ by $\mathrm{occ}_\tau(\pi)$; in this context, $\mathrm{occ}_\tau$ is called a statistic on set partitions.*

We modify our notation for the generating functions, using only one indeterminate, $q$, to keep track of the number of occurrences of the pattern.

**Definition 4.22** *We denote the generating function $P_{\mathrm{occ}_\tau}(x, y; q)$ for the number of set partitions in $\mathcal{P}_{n,k}$ according to the statistic $\mathrm{occ}_\tau$ by $P_\tau(x, y; q)$, that is,*

$$P_\tau(x, y; q) = P_{\mathrm{occ}_\tau}(x, y; q) = \sum_{n \geq 0} \sum_{k=0}^{n} \sum_{\pi \in \mathcal{P}_{n,k}} x^n y^k q^{\mathrm{occ}_\tau(\pi)}.$$

*For simple notation, we set $P_\tau(x; q) = P_\tau(x, 1; q)$ to be the generating function for the number of set partitions of $[n]$ according to the statistic $\mathrm{occ}_\tau$. Moreover, we define $P_\tau(x, y; q | \theta_1 \cdots \theta_m)$ to be the generating function for the number of set partitions $\pi = \pi_1 \pi_2 \cdots \pi_n \in \mathcal{P}_{n,k}$ according to the statistic $\mathrm{occ}_\tau$ that start with $\theta_1 \cdots \theta_m$ (that is, $\pi_1 \cdot \pi_m = \theta_1 \cdots \theta_m$).*

Before we focus on longer patterns, let us summarize the results obtained for rises, levels, and drops in terms of subword patterns of size two. Setting $\ell, d = 1$ and $r = q$ in Theorem 4.10 for the subword pattern 12, taking $r, d = 1$ and $\ell = q$ in Theorem 4.10 for the subword pattern 11, and setting $r, \ell = 1$ and $d = q$ in Theorem 4.10 for the subword pattern 21, we obtain the results that are presented in Table 4.1.

Table 4.1: Generating functions for 2-letter patterns

| $\tau$ | $P_\tau(x, y; q)$ |
|---|---|
| 12 | $1 + \sum_{i \geq 1} \dfrac{x^i y^i q^{i-1}(q-1)^i}{\prod_{j=1}^{i}(q - (1 + x(q-1))^j)}$ |
| | Continued on next page |

| $\tau$ | $P_\tau(x,y;q)$ |
|---|---|
| 11 | $1 + \sum_{i\geq 1} \dfrac{x^i y^i}{\prod_{j=1}^{i}(1 - x(\ell + j - 1))}$ |
| 21 | $1 + \sum_{i\geq 1} \dfrac{x^i y^i (1-q)^i}{(1 + x(q-1))^i \prod_{j=1}^{i}\left(1 - \frac{q}{(1+x(q-1))^j}\right)}$ |

On several cases it is difficult to find the number of set partitions of $[n]$ according to the number of the blocks and number of occurrences of a pattern. Among these cases it is worth trying to enumerate the set of partitions of $\mathcal{P}_{n,k}$ according to number of occurrences of a pattern, where $k$ is arbitrary, which invites the following notation.

**Definition 4.23** *Fix $k$ to be any nonnegative integer number. We denote the generating function for the number of set partitions in $\mathcal{P}_{n,k}$ according to the statistic $\text{occ}_\tau$ by*

$$P_\tau(x;q;k) = \sum_{n\geq 0}\sum_{\pi \in \mathcal{P}_{n,k}} x^n q^{\text{occ}_\tau(\pi)}.$$

*Moreover, let $P_\tau(x;q;k|\theta_1\cdots\theta_m)$ be the generating function for the number of partitions in $\mathcal{P}_{n,k}$ according to the statistic $\text{occ}_\tau$ that starts $\theta_1\cdots\theta_m$ (that is, $\pi_1\cdot\pi_m = \theta_1\cdots\theta_m$).*

We now look for subword patterns of size three. There are a total of 13 subword patterns, namely, 111, 112, 121, 122, 123, 132, 211, 212, 213, 221, 231, 312, and 321. However, if we think of them in terms of rises, levels and descents, then there are only eight patterns, namely 123 = rise + rise, 122 = rise + level, peak = rise + descent, 112 = level + rise, 221 = level + descent, peak = descent + rise, 211 = descent + level, and 321 = descent + descent. Note that all the above patterns except the peak and valley patterns can be considered as a subfamily of general patterns. Thus, we first look for the peak and valley patterns, then we will be free to consider longer patterns.

Here, we find explicit formulas for the generating functions $P_{peak}(x;q;k)$ and $P_{valley}(x;q;k)$. Then we present several combinatorial proofs regrading our results. We start by counting these two statistics in words over alphabet $[k]$; see [138, Theorem 6.7].

### 4.2.1 Counting Peaks in Words

We start by discussing the pattern peak. To derive the generating function for the pattern peak in set partitions, we first derive results for the pattern peak in words over alphabet $[k]$. To do so, let $W_k(x,q)$ be the generating function for the number of words of size $n$ over the alphabet $[k]$ according to the number of peaks, that is,

$$W_k(x;q) = \sum_{n \geq 0} \left( x^n \sum_{\pi \in [k]^n} q^{peak(\pi)} \right).$$

To derive a recurrence relation for this generating function, we split the word into parts according to where the largest part occurs, as described in the proof of the next result.

**Lemma 4.24** *The generating function $W_k(x,q)$ satisfies the recurrence relation*

$$W_k(x;q) = \frac{x(q-1) + (1 - x(q-1))W_{k-1}(x;q)}{1 - x(1-q)(1-x) - x(x+q(1-x))W_{k-1}(x;q)},$$

*with the initial condition $W_0(x;q) = 1$.*

**Proof** Let $W_k^*(x;q)$ be the generating function for the number of words $\pi$ of size $n$ over the alphabet $[k]$ according to the number of peaks, such that $\pi$ contains the letter $k$. Let us write an equation for $W_k(x;q)$. Since each word in $[k]^n$ may or may not contain the letter $k$, we obtain

$$W_k(x;q) = W_{k-1}(x;q) + W_k^*(x;q). \quad (4.11)$$

Words $\pi$ that contain the letter $k$ can be decomposed as either $k$, $\pi'k$, $k\pi''$, $\pi'k\pi'''$, or $\pi'kk\pi''''$, where $\pi'$ is a nonempty word over the alphabet $[k-1]$, $\pi''$ is a nonempty word over the alphabet $[k]$, $\pi'''$ is a nonempty word over the alphabet $[k]$ which starts with a letter $a < k$, and $\pi''''$ is a word over the alphabet $[k]$. The corresponding generating functions for the number of words of these decompositions are given by

$$x,$$
$$x(W_{k-1}(x;q) - 1),$$
$$x(W_k(x;q) - 1),$$
$$x^2(W_{k-1}(x;q) - 1)W_k(x;q),$$
$$xq(W_{k-1}(x;q) - 1)(W_k(x;q) - xW_k(x;q) - 1),$$

respectively. Collecting these cases, we derive the following equation

$$W_k^*(x;q) = x(q-1) - x(q-1)W_{k-1}(x;q) + x((1-q)(1-x) + (x + q(1-x))W_{k-1}(x;q))W_k(x;q),$$

which, by (4.11), implies

$$W_k(x;q) = x(q-1) + (1 - x(q-1))W_{k-1}(x;q) + x((1-q)(1-x) + (x + q(1-x))W_{k-1}(x;q))W_k(x;q).$$

Solving for $W_k(x;q)$, we obtain

$$W_k(x;q) = \frac{x(q-1) + (1 - x(q-1))W_{k-1}(x;q)}{1 - x(1-q)(1-x) - x(x + q(1-x))W_{k-1}(x;q)},$$

as required. □

Lemma 4.24 gives a recurrence relation for the generating function $W_k(x;q)$ as presented in Theorem 6.7 in [138]. But here, we go further to find an explicit formula for the generating function $W_k(x;q)$ in terms of Chebyshev polynomials of the second kind (see Appendix D).

**Theorem 4.25** *The generating function $W_k(x;q)$ for the number of words of size $n$ over the alphabet $[k]$ according to the number of peaks is given by*

$$W_k(x;q) = \frac{x(q-1)(U_{k-1}(t) - U_{k-2}(t))}{U_k(t) - U_{k-1}(t) - (1 - x(q-1))(U_{k-1}(t) - U_{k-2}(t))},$$

*where $t = 1 + \frac{x^2}{2}(1-q)$ and $U_m$ is the $m$-th Chebyshev polynomial of the second kind.*

**Proof** The proof followed immediately by Lemma 4.24 and Proposition D.5. □

Note that the mapping $\pi_1 \cdots \pi_n \mapsto (k+1-\pi_1) \cdots (k+1-\pi_n)$ on $[k]^n$ gives that the generating function for the number of words of size $n$ over the alphabet $[k]$ according to the statistic $\text{occ}_{valley}$ is also given by $W_k(x;q)$.

Now, we consider the number of peaks (valleys) in all the words of size $n$ over the alphabet $[k]$. Theorem 4.25 yields

$$\lim_{q \to 1} W_k(x;q)$$

$$= \lim_{q \to 1} \frac{x(U_{k-1}(1) - U_{k-2}(1))}{\frac{d}{dq}(U_k(t) - 2U_{k-1}(t) + U_{k-2}(t))|_{q=1} + x(U_{k-1}(1) - U_{k-2}(1))}.$$

Fact D.4 gives $\lim_{q \to 1} W_k(x;q) = \lim_{q \to 1} \frac{x}{-kx^2+x} = \frac{1}{1-kx}$, which agrees with the generating function for the number of words of size $n$ over the alphabet $[k]$. In order to find $\frac{d}{dq} W_k(x;q)|_{q=1}$ using Theorem 4.25, first we define

$$f(q) = x(q-1)(U_{k-1}(t) - U_{k-2}(t))$$

and

$$g(q) = U_k(t) - U_{k-1}(t) - (1 - x(q-1))(U_{k-1}(t) - U_{k-2}(t)),$$

where $t = 1 + \frac{x^2}{2}(1-q)$. We leave to the reader to check that $f(1) = g(1) = 0$, $f'(1) = x$, $g'(1) = x - kx^2$, $f''(1) = -2x^3\binom{k}{2}$ and $g''(1) = 2x^4\binom{k+1}{3} - 2x^3\binom{k}{2}$,

where differentiation ia done with respect to $q$. Applying L'Hôpital's rule twice to
$$\lim_{q \to 1} \left( \frac{f'(q)g(q) - f(q)g'(q)}{g^2(q)} \right),$$
we obtain
$$\frac{d}{dq} W_k(x;q) \big|_{q=1} = \frac{\left(k\binom{k}{2} - \binom{k+1}{3}\right) x^3}{(1-kx)^2} = \frac{\left(2\binom{k}{3} + \binom{k}{2}\right) x^3}{(1-kx)^2},$$
which gives the following corollary. (For direct proof see Exercise 4.4.)

**Corollary 4.26** *The number of peaks (valleys) in all the words of size $n$ over the alphabet $[k]$ is given by*
$$(n-2)k^{n-3} \left( 2\binom{k}{3} + \binom{k}{2} \right).$$

### 4.2.2 Counting Peaks

Now our goal is to find an explicit formula for the generating function $P_{peak}(x;q;k)$ for the number of set partitions of $\mathcal{P}_{n,k}$ according to the number of peaks, that is,
$$P_{peak}(x;q;k) = \sum_{n \geq 0} x^n \sum_{\pi \in \mathcal{P}_{n,k}} q^{peak(\pi)}.$$

To derive the generating function $P_{peak}(x;q;k)$, we split the set partition into parts according to the leftmost occurrence of the largest letter in the set partition. At first, we will show that the derivation of this generating function can be expressed in terms the generating function $W_j(x;q)$, see Lemma 4.27. Then by our explicit formula for the generating function $W_j(x;q)$, we obtain explicit formula for $P_{peak}(x;q;k)$ (see Theorem 4.28).

**Lemma 4.27** *For all $k \geq 1$,*
$$P_{peak}(x;q;k) = x^k \prod_{j=1}^{k} (1 + x(1-q)W_j(x;q) + q(W_j(x;q) - 1)).$$

**Proof** Any set partition $\pi$ with exactly $k$ blocks can be decomposed as $\pi = \pi' k \pi''$, where $\pi'$ is a set partition with exactly $k-1$ blocks and $\pi''$ is a word over the alphabet $[k]$. Since $\pi''$ is either empty, starts with the letter $k$, or starts with a letter less than $k$, we obtain
$$P_{peak}(x;q;k) = xP_{peak}(x;q;k-1)(1+xW_k(x;q)+q(W_k(x;q)-xW_k(x;q)-1)).$$

Thus, by induction on $k$ together with the initial condition $P_{peak}(x;q;0) = 1$, we complete the proof.. □

Lemma 4.27 and Theorem 4.25 give the following result.

**Theorem 4.28** *For all $k \geq 1$, the generating function for the number of set partitions in $\mathcal{P}_{n,k}$ according to the number of peaks is given by*

$$P_{peak}(x;q;k) = x^k(1-q)^k \prod_{j=1}^{k} \frac{1+x+x^2(1-q) - \frac{U_j(t)-U_{j-1}(t)}{U_{j-1}(t)-U_{j-2}(t)}}{1+x(1-q) - \frac{U_j(t)-U_{j-1}(t)}{U_{j-1}(t)-U_{j-2}(t)}},$$

*where $t = 1 + \frac{x^2}{2}(1-q)$ and $U_m$ is the $m$-th Chebyshev polynomial of the second kind.*

The above theorem with $q = 0$ gives that the generating function for the number of set partitions in $\mathcal{P}_{n,k}$ without peaks is given by

$$P_{peak}(x;0;k) = x^k \prod_{j=1}^{k} \left(1 + \frac{x^2}{1+x - \frac{U_j(t)-U_{j-1}(t)}{U_{j-1}(t)-U_{j-2}(t)}}\right),$$

which, by (D.1), implies

$$P_{peak}(x;0;k) = x^k \prod_{j=1}^{k} \frac{U_{j-1}(t) - (1+x)U_{j-2}(t)}{(1-x)U_{j-1}(t) - U_{j-2}(t)},$$

where $t = 1 + \frac{x^2}{2}$ and $U_m$ is the $m$-th Chebyshev polynomial of the second kind. Thus, we can state the following result.

**Corollary 4.29** *The generating function for the number of set partitions of $\mathcal{P}_n$ without peaks is given by*

$$1 + \sum_{k \geq 1} P_{peak}(x;0;k) = \sum_{k \geq 0} x^k \prod_{j=1}^{k} \frac{U_{j-1}(t) - (1+x)U_{j-2}(t)}{(1-x)U_{j-1}(t) - U_{j-2}(t)},$$

*where $t = 1 + \frac{x^2}{2}$ and $U_m$ is the $m$-th Chebyshev polynomial of the second kind.*

Now, our aim is to find an explicit formula for the total number of peaks in all the set partitions of $[n]$ with exactly $k$ blocks. Theorem 4.28 gives

$$P_{peak}(x;1;k) = x^k \prod_{j=1}^{k} \lim_{q \to 1} \frac{(1-q)\left(1+x+x^2(1-q) - \frac{U_j(t)-U_{j-1}(t)}{U_{j-1}(t)-U_{j-2}(t)}\right)}{1+x(1-q) - \frac{U_j(t)-U_{j-1}(t)}{U_{j-1}(t)-U_{j-2}(t)}}$$

$$= x^k \prod_{j=1}^{k} \frac{1+x - \frac{U_j(1)-U_{j-1}(1)}{U_{j-1}(1)-U_{j-2}(1)}}{x + \frac{d}{dq}\left(\frac{U_j(t)-U_{j-1}(t)}{U_{j-1}(t)-U_{j-2}(t)}\right)|_{q=1}}.$$

Fact D.4 gives

$$P_{peak}(x;1;k) = x^k \prod_{j=1}^{k} \frac{-x}{-x+jx^2} = \frac{x^k}{\prod_{j=1}^{k}(1-jx)},$$

which agrees with the generating function for the number of set partitions of $\mathcal{P}_{n,k}$; see (2.9). Also, Theorem 4.28 gives

$$\frac{d}{dq}P_{peak}(x;q;k)\mid_{q=1} = P_{peak}(x;1;k) \sum_{j=1}^{k} \lim_{q \to 1} \left( \frac{\frac{d}{dq}h_j(q)}{h_j(q)} \right),$$

where $h_j(q) = \frac{(1-q)(1+x+x^2(1-q)-u_j)}{1+x(1-q)-u_j}$ and $u_j = \frac{U_j(t)-U_{j-1}(t)}{U_{j-1}(t)-U_{j-2}(t)}$ with $t = 1 + \frac{x^2}{2}(1-q)$. Let $u'_j(q) = \frac{d}{dq}u_j(q)$ and $u''_j(q) = \frac{d^2}{dq^2}u_j(q)$. Since $u_j(1) = 1$, $\lim_{q \to 1} h_j(q) = \frac{1}{1-jx}$ and $\lim_{q \to 1} u'_j(q) = -jx^2$, we have

$$\lim_{q \to 1} \frac{d}{dq}h_j(q) = \frac{x(j-1)}{1-jx} - x \lim_{q \to 1} \frac{1 - u_j(q) - (1-q)u'_j(q)}{(1+x(1-q)-u_j(q))^2}$$

$$= \frac{x(j-1)}{1-jx} - \frac{1}{2(1-jx)} \lim_{q \to 1} \frac{(1-q)u''_j(q)}{1+x(1-q)-u_j(q)}$$

$$= \frac{x(j-1)}{1-jx} - \frac{1}{2(1-jx)} \frac{u''_j(1)}{(x-jx^2)}.$$

It is easy to verify that $\lim_{q \to 1} u''_j(q) = -2\binom{j+1}{3}x^4 - 2\binom{j}{3}x^4$, which implies

$$\lim_{q \to 1} \frac{d}{dq}h_j(q) = \frac{x(j-1)}{1-jx} + \frac{x^3\binom{j+1}{3} + x^3\binom{j}{3}}{(1-jx)^2}.$$

Hence,

$$\frac{d}{dq}P_{peak}(x;q;k)\mid_{q=1} = P_{peak}(x;1;k) \sum_{j=1}^{k} \left( \frac{x^3\binom{j+1}{3} + x^3\binom{j}{3}}{(1-jx)} + x(j-1) \right).$$

Since $P_{peak}(x;1;k) = \sum_{n \geq 1} \text{Stir}(n,k)x^n$ and $\sum_{j=1}^{k} x(j-1) = x\binom{k}{2}$, we get the following result. (For direct proof see Exercise 4.6.)

**Corollary 4.30** *The total number of peaks in all the set partitions of $\mathcal{P}_{n,k}$ is given by*

$$\binom{k}{2}\text{Stir}(n-1,k) + \sum_{j=2}^{k}\left(j\binom{j}{2} - \binom{j+1}{3}\right)\left(\sum_{i=3}^{n-k} j^{i-3}\text{Stir}(n-i,k)\right).$$

### 4.2.3 Counting Valleys

In this section, our aim is to find an explicit formula for the generating function $P_{valley}(x; q; k)$ for the number of set partitions of $\mathcal{P}_{n,k}$ according to the number of valleys, that is,

$$P_{valley}(x; q; k) = \sum_{n \geq 0} \left( x^n \sum_{\pi \in \mathcal{P}_{n,k}} q^{valley(\pi)} \right).$$

To derive the generating function $P_{valley}(x; q; k)$, again, we split the set partitions into parts according to the leftmost occurrence of each letter in the set partition, as we describe in the proof of (4.14). Then, we define $W'_k(x; q)$ (respectively, $W''_k(x; q)$) to be the generating function for the number of words $\pi$ of size $n$ over the alphabet $[k]$ (respectively, $[k-1]$) according to the number of valleys in $k\pi(k+1)$ (respectively, $k\pi k$).

**Lemma 4.31** *For all $k \geq 2$,*

$$W'_k(x; q) = \frac{W''_k(x; q)}{1 - xW''_k(x; q)}$$

*and*

$$W''_k(x; q) = W''_{k-1}(x; q) + x(W''_{k-1}(x; q)W'_{k-1}(x; q) - 1) + xq,$$

*with $W'_1(x; q) = \frac{1}{1-x}$ and $W''_1(x; q) = 1$.*

**Proof** By the definitions we have that $W'_1(x; q) = \frac{1}{1-x}$. Now, let us write an equation for the generating function $W'_k(x; q)$. Each word $k\pi(k+1)$, where $\pi$ is a word over the alphabet $[k]$, can be decomposed as either

- $k\pi(k+1)$, where $\pi$ is a word over the alphabet $[k-1]$,
- or as $k\pi'k\pi''(k+1)$, where $\pi'$ is a word over the alphabet $[k-1]$ and $\pi''$ is a word over the alphabet $[k]$.

Note that the number of valleys in the word $k\pi(k+1)$ is the same as the number of valleys in $k\pi k$ for any word $\pi$ over the alphabet $[k-1]$. Therefore, for all $k \geq 2$,

$$W'_k(x; q) = W''_k(x; q) + xW''_k(x; q)W'_k(x; q),$$

which implies

$$W'_k(x; q) = \frac{W''_k(x; q)}{1 - xW''_k(x; q)}.$$

Similarly, each word $k\pi k$ such that $\pi$ is a word over the alphabet $[k-1]$ may be decomposed as either

- $k\pi k$, where $\pi$ is a word over the alphabet $[k-2]$,

- or as $k\pi'(k-1)\pi''k$, where $\pi'$ is a word over the alphabet $[k-2]$ and $\pi''$ is a word over the alphabet $[k-1]$.

Note that the number of valleys in the word $k\pi(k-1)$ is the same as the number of valleys in $(k-1)\pi(k-1)$ for any word $\pi$ over the alphabet $[k-2]$. Therefore, for all $k \geq 2$,

$$W_k''(x;q) = W_{k-1}''(x;q) + x(W_{k-1}''(x;q)W_{k-1}'(x;q) - 1) + xq,$$

where $xq$ accounts for the word $k(k-1)k$, which completes the proof. □

**Lemma 4.32** For all $k \geq 1$,

$$W_k''(x;q) = \frac{x(q-1)A_k}{A_{k+1} - (1-x^2(q-1))A_k},$$

where $A_k = (1 + x(1-x)(q-1))U_{k-2}(t) - U_{k-3}(t)$, $t = 1 + \frac{x^2}{2}(1-q)$, and $U_m$ is the $m$-th Chebyshev polynomial of the second kind.

**Proof** By Lemma 4.31 we have

$$W_k''(x;q) = x(q-1) + \frac{W_{k-1}''(x;q)}{1 - xW_{k-1}''(x;q)}, \tag{4.12}$$

with initial condition $W_1''(x;q) = 1$. Assume that $W_k''(x;q) = \frac{A_k'}{B_k'}$ for all $k \geq 1$. Thus,

$$\frac{A_k'}{B_k'} = x(q-1) + \frac{A_{k-1}'}{B_{k-1}' - xA_{k-1}'},$$

which implies $A_k' = x(q-1)B_{k-1}' + (1-x^2(q-1))A_{k-1}'$ and $B_k' = B_{k-1}' - xA_{k-1}'$, with $A_1' = B_1' = 1$, $A_2' = 1 + x(1-x)(q-1)$ and $B_2' = 1 - x$. Therefore,

$$A_k' = 2tA_{k-1}' - A_{k-2}', \quad A_1' = 1, \quad A_2' = 1 + x(1-x)(q-1). \tag{4.13}$$

Hence, by (D.1) and induction on $k$, we obtain $A_k' = A_k$. Using the recurrence $A_k' = x(q-1)B_{k-1}' + (1-x^2(q-1))A_{k-1}'$, we derive

$$B_k' = \frac{1}{x(q-1)}\left(A_{k+1}' - (1-x^2(q-1))A_k'\right),$$

which completes the proof. □

**Lemma 4.33** For all $k \geq 1$,

$$W_k'(x;q) = \frac{x(q-1)A_k}{A_{k+2} - (1-x^2(q-1))A_{k+1}},$$

where $A_k = (1 + x(1-x)(q-1))U_{k-2}(t) - U_{k-3}(t)$, $t = 1 + \frac{x^2}{2}(1-q)$, and $U_m$ is the $m$-th Chebyshev polynomial of the second kind.

**Proof** Lemma 4.31 and (4.12) imply $W'_k(x;q) = W''_{k+1}(x;q) - x(q-1)$. Thus, by Lemma 4.32, we obtain

$$W'_k(x;q) = \frac{x(q-1)(2tA_{k+1} - A_{k+2})}{A_{k+2} - (1 - x^2(q-1))A_{k+1}},$$

which, by (4.13), implies the result. □

Now, we are ready to count valleys in set partitions. From the fact that each set partition $\pi \in \mathcal{P}_{n,k}$ may be written uniquely as $\pi = 1\pi^{(1)} 2\pi^{(2)} \cdots k\pi^{(k)}$, where $\pi^{(j)}$ is a word over the alphabet $[j]$, it follows that the generating function $P_{valley}(x;q;k)$ is given by

$$P_{valley}(x;q;k) = x^k V_k(x;q) \prod_{j=1}^{k-1} W'_j(x;q), \tag{4.14}$$

where $V_k(x;q)$ is the generating function for the number of words $\pi$ over the alphabet $[k]$ according to the number of valleys in $k\pi$.

**Lemma 4.34** *For all $k \geq 1$,*

$$V_k(x;q) = \prod_{j=1}^{k} \frac{1}{1 - xW''_j(x;q)}.$$

**Proof** Since each word $k\pi$, where $\pi$ is a word over the alphabet $[k]$, has either the form $k\pi$, where $\pi$ is a word over the alphabet $[k-1]$, or the form $k\pi' k\pi''$, where $\pi'$ is a word over the alphabet $[k-1]$, and $\pi''$ is a word over the alphabet $[k]$, we obtain

$$V_k(x;q) = V'_k(x;q) + xW''_k(x;q)V_k(x;q), \tag{4.15}$$

where $V'_k(x;q)$ is the generating function for the number of words $\pi$ over the alphabet $[k-1]$ according to the number of valleys in $k\pi$. Similarly, each word $k\pi$, where $\pi$ is a word over the alphabet $[k-1]$, has either the form $k\pi$, where $\pi$ is a word over the alphabet $[k-2]$, or the from $k\pi'(k-1)\pi''$, where $\pi'$ is a word over the alphabet $[k-2]$ and $\pi''$ is a word over the alphabet $[k-1]$. Also, the number of valleys in $k\pi$ ($k\pi(k-1)$) is the same as the number of valleys in $(k-1)\pi$ $((k-1)\pi(k-1))$ for any word $\pi$ over the alphabet $[k-2]$. Therefore,

$$V'_k(x;q) = V'_{k-1}(x;q) + xW''_{k-1}(x;q)V_{k-1}(x;q). \tag{4.16}$$

Hence, by (4.15) and (4.16), we obtain

$$V_k(x;q) = \frac{V_{k-1}(x;q)}{1 - xW''_k(x;q)},$$

for all $k \geq 1$. Using this recurrence relation exactly $k$ times, we derive the requested result. □

Our next theorem provides an explicit formula for the generating function $P_{valley}(x; q; k)$.

**Theorem 4.35** *The generating function $P_{valley}(x; q; k)$ for the number of set partitions of $\mathcal{P}_{n,k}$ according to the number of valleys is given by*

$$P_{valley}(x; q; k)$$
$$= \frac{x^k}{(1-x)U_{k-1}(t) - U_{k-2}(t)} \prod_{j=1}^{k-1} \frac{U_{j-1}(t) - (1 - x(q-1))U_{j-2}(t)}{(1-x)U_{j-1}(t) - U_{j-2}(t)},$$

*where $t = 1 + \frac{x^2}{2}(1-q)$ and $U_m$ is the m-th Chebyshev polynomial of the second kind.*

**Proof** Lemma 4.34 and (4.14) give

$$P_{valley}(x; q; k) = x^k \prod_{j=1}^{k} \frac{1}{1 - xW_j''(x; q)} \prod_{j=1}^{k-1} W_j'(x; q).$$

By Lemmas 4.32 and 4.33 we obtain

$$P_{valley}(x; q; k) = \frac{x^k(A_{k+1} - (1 - x^2(q-1))A_k)}{A_{k+1} - A_k}$$
$$\cdot \prod_{j=1}^{k-1} \left( \frac{x(q-1)A_j}{A_{j+1} - A_j} \cdot \frac{A_{j+1} - (1 - x^2(q-1))A_j}{A_{j+2} - (1 - x^2(q-1))A_{j+1}} \right)$$
$$= \frac{x^k(A_2 - (1 - x^2(q-1))A_1)}{A_{k+1} - A_k} \prod_{j=1}^{k-1} \frac{x(q-1)A_j}{A_{j+1} - A_j},$$

where $A_k = (1 + x(1-x)(q-1))U_{k-2}(t) - U_{k-3}(t)$. Using the initial conditions $A_1 = 1$, $A_2 = 1 + x(1-x)(q-1)$ and the fact that

$$A_{j+1} - A_j = x(q-1)((1-x)U_{j-1}(t) - U_{j-2}(t)),$$

we derive

$$P_{valley}(x; q; k) = \frac{x^{k+1}(q-1)}{A_{k+1} - A_k} \prod_{j=1}^{k-1} \frac{x(q-1)A_j}{A_{j+1} - A_j}$$
$$= \frac{x^k}{(1-x)U_{k-1}(t) - U_{k-2}(t)} \prod_{j=1}^{k-1} \frac{(1 + x(1-x)(q-1))U_{j-2}(t) - U_{j-3}(t)}{(1-x)U_{j-1}(t) - U_{j-2}(t)},$$

which, by (D.1), gives the desired result. □

The above theorem for $q = 0$ gives the following result.

**Corollary 4.36** *The generating function for the number of set partitions of $\mathcal{P}_n$ without valleys is given by*

$$1 + \sum_{k \geq 1} P_{valley}(x;0;k) = \sum_{k \geq 0} \left( \frac{x^k \prod_{j=1}^{k-1}(U_{j-1}(t) - (1+x)U_{j-2}(t))}{\prod_{j=1}^{k}((1-x)U_{j-1}(t) - U_{j-2}(t))} \right),$$

*where $t = 1 + \frac{x^2}{2}$ and $U_m$ is the $m$-th Chebyshev polynomial of the second kind.*

Now, we turn our attention to counting valleys in all set partitions of $\mathcal{P}_{n,k}$. Theorem 4.35 together with Fact D.4 give

$$P_{valley}(x;1;k) = \frac{x^k}{(1-x)U_{k-1}(1) - U_{k-2}(1)} \prod_{j=1}^{k-1} \frac{U_{j-1}(1) - U_{j-2}(1)}{(1-x)U_{j-1}(1) - U_{j-2}(1)}$$

$$= \frac{x^k}{\prod_{j=1}^{k}(1-jx)},$$

which agrees with the generating function for the number of set partitions of $\mathcal{P}_{n,k}$; see (2.9). Also, Theorem 4.35 gives

$$\frac{\frac{d}{dq}P_{valley}(x;q;k)|_{q=1}}{P_{valley}(x;1;q)}$$

$$= \sum_{j=1}^{k-1} \frac{\lim_{q \to 1} \frac{d}{dq} \frac{U_{j-1}(t) - (1-x(q-1))U_{j-2}(t)}{(1-x)U_{j-1}(t) - U_{j-2}(t)}}{\frac{1}{1-jx}} - \frac{(1-x)U'_{k-1}(1) - U'_{k-2}(1)}{1-kx}$$

$$= \sum_{j=1}^{k-1} \left( U'_{j-1}(1) + xU_{j-2}(1) - U'_{j-2}(1) - \frac{(1-x)U'_{j-1}(1) - U'_{j-2}(1)}{1-jx} \right)$$

$$- \frac{(1-x)U'_{k-1}(1) - U'_{k-2}(1)}{1-kx},$$

which, by Fact D.4, implies that the generating function for the number of valleys in all the set partitions of $\mathcal{P}_{n,k}$ is given by

$$\frac{d}{dq}P_{valley}(x;q;k)|_{q=1}$$

$$= \frac{x^k}{\prod_{j=1}^{k}(1-jx)} \left( \sum_{j=1}^{k-1}\left(x(j-1) - x^2\binom{j}{2}\right) + \sum_{j=1}^{k} \frac{x^2\binom{j}{2} - x^3\binom{j+1}{3}}{1-jx} \right)$$

$$= \frac{x^k}{\prod_{j=1}^{k}(1-jx)} \left( x\binom{k-1}{2} - x^2\binom{k}{3} + \sum_{j=1}^{k} \frac{x^2\binom{j}{2} - x^3\binom{j+1}{3}}{1-jx} \right),$$

which leads to the following result (for another proof, see [230]).

**Corollary 4.37** *The total number of valleys in all the set partitions of $\mathcal{P}_{n,k}$ is given by*

$$\binom{k-1}{2}\mathrm{Stir}(n-1,k) - \binom{k}{3}\mathrm{Stir}(n-2,k)$$
$$+ \sum_{j=2}^{k}\binom{j}{2}\left(\sum_{i=2}^{n-k} j^{i-2}\mathrm{Stir}(n-i,k)\right) - \sum_{j=2}^{k}\binom{j+1}{3}\left(\sum_{i=3}^{n-k} j^{i-3}\mathrm{Stir}(n-i,k)\right).$$

---

## 4.3 Subword Patterns: $\ell$-Rises, $\ell$-Levels, and $\ell$-Descents

In this section, we will obtain generating function for three special types of subword patterns: long-rise or $\ell$-rises, long-level or $\ell$-levels, and long-descent or $\ell$-descents, that is, $12\cdots\ell$, $1^\ell = 11\cdots 1$ and $\ell\cdots 21$.

### 4.3.1 Long-Rise Pattern

Let $\tau = 12\cdots\ell$ be an increasing subword pattern. In this section, we consider the generating function $P_\tau(x;y;k)$ for the number of set partitions of $\mathcal{P}_{n,k}$ according to the statistic $\mathrm{occ}_\tau$. Note that each set partition $\pi$ with exactly $k$ blocks can be written as $\pi = \pi' k \sigma$, where $\pi'$ is a set partition with exactly $k-1$ blocks and $\sigma$ is a $k$-ary word. Thus

$$P_\tau(x;q;k) = xP_\tau^{(1)}(x;q;k-1)W_\tau(x;q;k), \tag{4.17}$$

where $W_\tau(x;q;k)$ is the generating function for the number of words of size $n$ over alphabet $[k]$ according to the statistic $\mathrm{occ}_\tau$, that is,

$$W_\tau(x;q;k) = \sum_{n\geq 0}\sum_{\pi\in[k]^n} x^n q^{\mathrm{occ}_\tau(\pi)}.$$

In order to find an explicit formula for the generating function $W_\tau(x;q;k)$, we consider the following general lemma.

**Lemma 4.38** *Let $\tau = 12\cdots\ell$. The generating function $W_\tau^{(s)}(x;q;k)$ for the number of words $\sigma(k+1)(k+2)\cdots(k+s)$ over alphabet $[k+s]$, $\sigma$ is word over alphabet $[k]$, of size $n+s$ according to the statistic $\mathrm{occ}_\tau$ satisfies*

$$W_\tau^{(s)}(x;q;k) = \frac{W_\tau^{(s)}(x;q;k-1) + xW_\tau^{(s+1)}(x;q;k-1) - xW_\tau^{(1)}(x;q;k-1)}{1 - xW_\tau^{(1)}(x;q;k-1)}$$

*for all $s = 0,1,\ldots,\ell-1$,*

$$W_\tau^{(\ell)}(x;q;k) = qW_\tau^{(\ell-1)}(x;q;k) \text{ and } W_\tau^{(0)}(x;q;k) = W_\tau(x;q;k),$$

*with the initial condition $W_\tau^{(s)}(x;q;k) = \frac{1}{1-kx}$ for all $s+k \leq \ell - 1$.*

**Proof** It is easy to verify that

$$W_\tau^{(\ell)}(x;q;k) = qW_\tau^{(\ell-1)}(x;q;k),$$
$$W_\tau^{(0)}(x;q;k) = W_\tau(x;q;k),$$
$$W_\tau^{(s)}(x;q;k) = \frac{1}{1-kx}$$

with $s+k \leq \ell - 1$. Now, assume $0 \leq s \leq \ell - 1$, and let us derive a recurrence relation for $W_\tau^{(s)}(x;q;k)$. Let $\pi$ be any word $\sigma$ over alphabet $[k]$ containing the letter $k$ exactly $m$ times, and let us consider the following two cases:

- The contribution for the case $m=0$ is $W_\tau^{(s)}(x;q;k-1)$.
- In the case $m > 0$, $\pi$ can be decomposed as

$$\pi = \pi^{(1)} k \pi^{(2)} k \cdots \pi^{(m)} k \pi^{(m+1)}.$$

By considering the two cases with $\pi^{(m+1)}$ being empty/nonempty, we obtain that the contribution of this case is given by

$$x^m (W_\tau^{(1)}(x;q;k-1))^{m-1} W_\tau^{(s+1)}(x;q;k-1)$$
$$+ x^m (W_\tau^{(1)}(x;q;k-1))^m (W_\tau^{(s)}(x;q;k-1) - 1).$$

Combining these two cases, we obtain

$$W_\tau^{(s)}(x;q;k)$$
$$= W_\tau^{(s)}(x;q;k-1) + \sum_{m \geq 1} x^m (W_\tau^{(1)}(x;q;k-1))^{m-1} W_\tau^{(s+1)}(x;q;k-1)$$
$$+ \sum_{m \geq 1} x^m (W_\tau^{(1)}(x;q;k-1))^m (W_\tau^{(s)}(x;q;k-1) - 1)$$

which implies

$$W_\tau^{(s)}(x;q;k)$$
$$= W_\tau^{(s)}(x;q;k-1)$$
$$+ \frac{xW_\tau^{(s+1)}(x;q;k-1) + xW_\tau^{(1)}(x;q;k-1)(W_\tau^{(s)}(x;q;k-1) - 1)}{1 - xW_\tau^{(1)}(x;q;k-1)}$$
$$= \frac{W_\tau^{(s)}(x;q;k-1) + xW_\tau^{(s+1)}(x;q;k-1) - xW_\tau^{(1)}(x;q;k-1)}{1 - xW_\tau^{(1)}(x;q;k-1)},$$

as required. □

Now we turn our attention to set partitions.

**Lemma 4.39** Let $\tau = 12\cdots\ell$. Then the generating function $P_\tau^{(s)}(x, y; k)$ for the number of set partitions $\pi(k+1)(k+2)\cdots(k+s)$, $\pi$ is a set partition of $[n]$ with exactly $k$ blocks, according to the statistic $\mathrm{occ}_\tau$ is given by

$$P_\tau^{(s)}(x;q;k) = xP_\tau^{(s+1)}(x;q;k-1) + xP_\tau^{(1)}(x;q;k-1)(W_\tau^{(s)}(x;q;k)-1),$$

for all $s = 1, 2, \ldots, \ell-1$,

$$P_\tau^{(\ell)}(x;q;k) = qP_\tau^{(\ell-1)}(x;q;k)$$

with the initial condition

$$P_\tau^{(s)}(x;q;k) = \frac{x^k}{(1-x)(1-2x)\cdots(1-kx)},$$

for all $s + k \leq \ell - 1$.

**Proof** From the definitions hold the following

$$P_\tau^{(\ell)}(x;q;k) = qP_\tau^{(\ell-1)}(x;q;k) \text{ and } P_\tau^{(s)}(x;q;k) = \frac{x^k}{\prod_{j=1}^k(1-jx)}$$

with $s + k \leq \ell - 1$. Now, let $0 \leq s \leq \ell - 1$ and let us write an equation for $P_\tau^{(s)}(x;q;k)$. Let $\pi$ be any set partition with $k$ blocks, then $\pi(k+1)(k+2)\cdots(k+s)$ can be decomposed as $\pi' k \pi''(k+1)(k+2)\cdots(k+s)$, where $\pi'$ is a set partition with exactly $k-1$ blocks and $\pi''$ is a word over alphabet $[k]$. If we consider either $\pi''$ is empty or nonempty word, we obtain

$$P_\tau^{(s)}(x;q;k) = xP_\tau^{(s+1)}(x;q;k-1) + xP_\tau^{(1)}(x;q;k-1)(W_\tau^{(s)}(x;q;k)-1),$$

as claimed. $\square$

Now, let us deal with the system of recurrence relations that are suggested by Lemma 4.38. By induction on $s$, there exists a solution of the following form

$$W_\tau^{(s)}(k) = W_\tau^{(s)}(x;q;k) = \frac{W_\tau^{'(s)}(k)}{W_\tau''(k)},$$

where

$$\begin{cases} W_\tau^{'(s)}(k) = W_\tau^{'(s)}(k-1) + xW_\tau^{'(s+1)}(k-1) - xW_\tau^{'(1)}(k-1), \\ W_\tau''(k) = W_\tau''(k-1) - xW_\tau^{'(1)}(k-1). \end{cases} \quad (4.18)$$

Hence, for all $k \geq 0$

$$W_\tau^{(s)}(k) = \frac{W_\tau^{'(s)}(k)}{1 - x\sum_{j=0}^{k-1} W_\tau^{'(1)}(j)}. \quad (4.19)$$

Now, let us rewrite the recurrence relations in Lemmas 4.38 and 4.39 and (4.18) in terms of matrices. Define

$$\mathbf{W}'_\tau(k) = \begin{pmatrix} W_\tau^{'(1)}(k) \\ W_\tau^{'(2)}(k) \\ \vdots \\ W_\tau^{'(\ell-1)}(k) \end{pmatrix} \text{ and } \mathbf{P}_\tau(k) = \begin{pmatrix} P_\tau^{(1)}(k) \\ P_\tau^{(2)}(k) \\ \vdots \\ P_\tau^{(\ell-1)}(k) \end{pmatrix}.$$

Then Lemmas 4.38-4.39 and (4.18) can be written as follows.

**Theorem 4.40** *Let $\tau = 12\cdots\ell$ be subword pattern. Then*

$$\mathbf{W}'_\tau(k) = \mathbf{A}^k \cdot \mathbf{1} \text{ and } \mathbf{P}_\tau(k) = x^k \left(\prod_{j=1}^k \mathbf{B}_j\right) \cdot \mathbf{1},$$

*where $\mathbf{1} = (1, 1, \cdots, 1)^T$ is a vector with $\ell - 1$ coordinates,*

$$\mathbf{A} = \mathbf{I} + x \begin{pmatrix} -1 & 1 & 0 & \cdots & 0 \\ -1 & 0 & 1 & \cdots & 0 \\ \vdots & & & & \\ -1 & 0 & 0 & & 1 \\ -1 & 0 & 0 & & q \end{pmatrix}, \mathbf{B}_j = \begin{pmatrix} W_\tau^{(1)}(j) - 1 & 1 & \cdots & 0 \\ W_\tau^{(2)}(j) - 1 & 0 & \cdots & 0 \\ \vdots & & & \\ W_\tau^{(\ell-2)}(j) - 1 & 0 & \cdots & 1 \\ W_\tau^{(\ell-1)}(j) - 1 & 0 & \cdots & q \end{pmatrix},$$

*and $\mathbf{I}$ the unit matrix of order $\ell - 1$ with*

$$W_\tau^{(s)}(j) = \frac{W_\tau^{'(s)}(j)}{1 - x\sum_{i=0}^{k-1} W_\tau^{'(1)}(i)}$$

*for all $j = 1, 2, \ldots, k$.*

**Proof** Lemma 4.38 together with (4.18) gives $\mathbf{W}'_\tau(k) = \mathbf{A} \cdot \mathbf{W}'_\tau(k-1)$, for all $k \geq 1$. Lemma 4.39 obtains $\mathbf{P}_\tau(k) = \mathbf{B}_k \cdot \mathbf{P}_\tau(k-1)$, for all $k \geq 1$. Thus, $\mathbf{P}_\tau(k) = x^k \left(\prod_{j=1}^k \mathbf{B}_j\right) \cdot \mathbf{P}_\tau(0)$. Using the initial condition $\mathbf{W}'_\tau(0) = \mathbf{P}_\tau(0) = \mathbf{1}$, we arrive at $\mathbf{W}'_\tau(k) = \mathbf{A}^k \cdot \mathbf{1}$ and $\mathbf{P}_\tau(k) = x^k \left(\prod_{j=1}^k \mathbf{B}_j\right) \cdot \mathbf{1}$, which complete the proof. □

For instance, the above theorem for $\tau = 12$ gives the generating function $P_{12}(x; q; k)$.

**Theorem 4.41** *The generating function for the number of set partitions of $\mathcal{P}_{n,k}$ according to the statistic $\mathrm{occ}_{12}$ is given by*

$$\frac{x^k q^{k-1}(1-q)^k}{\prod_{j=1}^k ((1-x+xq)^j - q)}.$$

**Proof** Theorem 4.40 for $\tau = 12$ gives $W_{12}^{'(1)}(x;q;k) = (1 - x + xq)^k$ for all $k \geq 0$. Thus, by (4.19) we have

$$W_{12}^{(1)}(x;q;k) = \frac{(1-x+xq)^k}{1 - x\sum_{j=0}^{k-1}(1-x+xq)^j} = \frac{(1-x+xq)^k}{1 - \frac{1-(1-x+xq)^k}{1-q}}$$

$$= \frac{(1-q)(1-x+xq)^k}{(1-x+xq)^k - q}. \tag{4.20}$$

Then Lemma 4.38 with $s = 0$ gives

$$W_{12}(x;q;k) = \frac{W_{12}(x;q;k-1)}{1 - \frac{x(1-q)(1-x+xq)^{k-1}}{(1-x+xq)^{k-1}-q}}$$

and $W_{12}(x;q;1) = \frac{1}{1-x}$. Thus, by induction on $k$, we obtain

$$W_{12}(x;q;k) = \frac{1}{\prod_{j=0}^{k-1}\left(1 - \frac{x(1-q)(1-x+xq)^j}{(1-x+xq)^j - q}\right)}$$

$$= \frac{\prod_{j=0}^{k-1}(1-x+xq)^j - q}{\prod_{j=1}^{k}(1-x+xq)^j - q} = \frac{1-q}{(1-x+xq)^k - q}. \tag{4.21}$$

On the other hand, Theorem 4.40 for $\tau = 12$ gives

$$P_{12}^{(1)}(x;q;k) = x^k \prod_{j=1}^{k}(W_{12}^{(1)}(x;q;j) - 1 + q),$$

and by (4.20) we get that

$$P_{12}^{(1)}(x;q;k) = \frac{x^k q^k (1-q)^k}{\prod_{j=1}^{k}(1-x+xq)^j - q}.$$

By combining (4.17) and (4.21) we complete the proof. □

As it is discussed in the previous section, see Corollaries 4.30 and 4.37, Theorem 4.41 leads to an explicit formula for the number rises in all set partitions of $\mathcal{P}_{n,k}$. (Try to do that!!) Here we give a direct proof for such formula.

**Corollary 4.42** (Shattuck [320]) *The total number rises in all set partitions of $\mathcal{P}_{n,k}$ is given by*

$$(k-1)\text{Stir}(n,k) + \sum_{j=2}^{k}\binom{j}{2}\sum_{i=2}^{n-k} j^{i-2}\text{Stir}(n-i,k).$$

**Proof** First consider the trivial rises (see Definition 4.20); there are $k-1$ trivial rises within each set partition of $\mathcal{P}_{n,k}$. Hence, their total is $(k-1)\mathrm{Stir}(n,k)$. To end the proof we show that the number of nontrivial rises (see Definition 4.20) is given by $\sum_{j=2}^{k} \binom{j}{2} \sum_{i=2}^{n-k} j^{i-2} \mathrm{Stir}(n-i,k)$.

Let $i,j$ be any integers such that $2 \leq i \leq n-k$ and $2 \leq j \leq k$. Consider all the set partitions of $\mathcal{P}_{n,k}$, which may be written as

$$\pi = \pi' j \pi'' \pi''', \tag{4.22}$$

where $\pi'$ is a set partition with exactly $j-1$ blocks, $\pi'' \in [j]^i$ whose last two letters form a rise, and $\pi'''$ is possibly empty. Then the total number of nontrivial rises can be derived by finding the number of set partitions which may be expressed as in (4.22) for fixed $i,j$ and then summing over all possible values of $i$ and $j$. As there are $j^{i-2}$ choices for the first $i-2$ letters of $\pi''$, $\binom{j}{2}$ choices for the final two letters in $\pi''$ (as the last letter must exceed its predecessor), and $\mathrm{Stir}(n-i,k)$ choices for the remaining letters $\pi' j \beta$ which necessarily constitute a set partition of $\mathcal{P}_{n-i,k}$, we obtain that there are $\binom{j}{2} j^{i-2} \mathrm{Stir}(n-i,k)$ nontrivial rises, which completes the proof. □

As we can see from Theorem 4.41, it is very hard to obtain an explicit formula in the case $\tau = 12\cdots\ell$ for $\ell \geq 3$. For instance, Mansour and Munagi [220] discussed the case $\tau = 123$, and they present a formula (in terms of matrices) for the generating function $P_{123}(x;q;k)$.

### 4.3.2 Long-Level Pattern

As we discussed at the beginning of this section, we are interested in studying the counting of the subword pattern $\tau = 1^\ell$, which consists of $\ell$ 1s. Note that for $\ell = 2$ we obtain the result for the statistic level and for $\ell = 3$ we obtain the result for the statistic level+level (3-letter pattern). In order to do that we need the following lemma, see Exercise 4.5.

**Lemma 4.43** *Let $1 \leq a \leq k$ and $\tau = 1^\ell$. Then the generating function $W_\tau(x;q;k,a)$ for the number of words $\sigma = \sigma_1 \sigma_2 \cdots \sigma_n$ over alphabet $[k]$ of size $n$ with $\sigma_1 = a$ according to the statistic $\mathrm{occ}_\tau$ is given by*

$$W_\tau(x;q;k,a) = \frac{x + x^2 + \cdots + x^{\ell-2} + x^{\ell-1}/(1-xq)}{1 - (k-1)(x + x^2 + \cdots + x^{\ell-2} + x^{\ell-1}/(1-xy))}.$$

**Theorem 4.44** *Let $\tau = 1^\ell$ and fix $k$. Then the generating function for the number of set partitions of $\mathcal{P}_{n,k}$ according to the statistic $\mathrm{occ}_\tau$ is given by*

$$P_\tau(x;q;k) = \frac{\left(x + x^2 + \cdots + x^{\ell-2} + \frac{x^{\ell-1}}{1-xq}\right)^k}{\prod_{j=0}^{k-1}\left(1 - j\left(x + x^2 + \cdots + x^{\ell-2} + \frac{x^{\ell-1}}{1-xq}\right)\right)}.$$

Moreover, the generating function for the number of set partitions of $[n]$ according to the statistic $\mathrm{occ}_\tau$ is given by

$$P_\tau(x,y;q) = 1 + \sum_{k\geq 1} \frac{y^k \left(x + x^2 + \cdots + x^{\ell-2} + \frac{x^{\ell-1}}{1-xq}\right)^k}{\prod_{j=0}^{k-1}\left(1 - j\left(x + x^2 + \cdots + x^{\ell-2} + \frac{x^{\ell-1}}{1-xq}\right)\right)}.$$

**Proof** We now obtain an explicit expression for $P_\tau(x;q;k)$ by means of Lemma 4.43. Note that each set partition $\pi$ with exactly $k$ blocks can be written as $\pi = \pi' k \pi''$, where $\pi'$ is a set partition with exactly $k-1$ blocks and $\pi''$ is a word over alphabet $[k]$. Clearly, the last letter in $\pi'$ does not equal $k$. Thus

$$P_\tau(x;q;k) = P_\tau(x;q;k-1)W_\tau(x;q;k,k),$$

for all $k \geq 1$ with initial condition $P_\tau(x;q;0) = 1$. Hence, by Lemma 4.43, for all $k \geq 1$,

$$P_\tau(x;q;k) = P_\tau(x,y;k-1) \frac{x + x^2 + \cdots + x^{\ell-2} + \frac{x^{\ell-1}}{1-xq}}{1 - (k-1)\left(x + x^2 + \cdots + x^{\ell-2} + \frac{x^{\ell-1}}{1-xq}\right)},$$

which implies that

$$P_\tau(x;q;k) = \frac{\left(x + x^2 + \cdots + x^{\ell-2} + \frac{x^{\ell-1}}{1-xq}\right)^k}{\prod_{j=0}^{k-1}\left(1 - j\left(x + x^2 + \cdots + x^{\ell-2} + \frac{x^{\ell-1}}{1-xq}\right)\right)}.$$

Thus, by $P_\tau(x;q) = 1 + \sum_{k\geq 1} P_\tau(x;q;k)y^k$ we complete the proof. $\square$

**Example 4.45** Theorem 4.44 for $\ell = 3$ gives

$$P_{111}(x,y;q) = 1 + \sum_{k\geq 1} \frac{y^k(x + x^2/(1-xq))^k}{\prod_{j=0}^{k-1}(1 - j(x + x^2/(1-xq)))}.$$

In particular, the generating function for the number of set partitions of $[n]$ that avoid the subword pattern $111$ is given by

$$P_{111}(x,y;0) = 1 + \sum_{k\geq 1} \frac{y^k x^k (1+x)^k}{\prod_{j=0}^{k-1}(1 - jx(1+x))}.$$

As a corollary of Theorem 4.44 we obtain the following result.

**Corollary 4.46** The total number of occurrences of the subword pattern $\tau = 1^\ell$ in all the set partitions of $\mathcal{P}_{n,k}$ is given by

$$k \sum_{i=0}^{n+1-\ell} \mathrm{Stir}(i,k) + \sum_{j=1}^{k}\sum_{i=0}^{n-\ell}(j^{n-\ell+1-i} - 1)\mathrm{Stir}(i,k). \qquad (4.23)$$

**Proof** Here we suggest two proofs: analytical proof and combinatorial proof. We start by the analytical proof. By Theorem 4.44 we obtain

$$\frac{d}{dq}P_\tau(x;q;k)|_{q=1} = \sum_{n\geq 0}\sum_{\pi\in\mathcal{P}_{n,k}} \mathrm{occ}_\tau(\pi)x^n$$

$$= \frac{kx^{k-1+\ell}}{(1-x)\prod_{j=1}^{k}(1-jx)} + \frac{x^{k+\ell}}{(1-x)\prod_{j=1}^{k}(1-jx)}\sum_{j=2}^{k}\frac{j-1}{1-jx},$$

which, by (2.9), is equivalent to

$$\sum_{n\geq 0}\sum_{\pi\in\mathcal{P}_{n,k}} \mathrm{occ}_\tau(\pi)x^n$$

$$= kx^{\ell-1}\sum_{n\geq 0}\mathrm{Stir}(n,k)x^n\sum_{n\geq 0}x^n + x^\ell \sum_{n\geq k}\mathrm{Stir}(n,k)x^n \sum_{j=1}^{k}\sum_{n\geq 0}(j^{n+1}-1)x^n.$$

Thus, by collecting the $x^n$ coefficient we obtain that the number of occurrences of the subword pattern $\tau = 1^\ell$ in all the set partitions of $[n]$ with exactly $k$ blocks is given by

$$k\sum_{i=0}^{n+1-\ell}\mathrm{Stir}(i,k) + \sum_{j=1}^{k}\sum_{i=0}^{n-\ell}(j^{n-\ell+1-i}-1)\mathrm{Stir}(i,k),$$

as required.

Now we give the details of the combinatorial proof (see [320]). Here, we show that the number of occurrences of the subword pattern $\tau = 1^\ell$ in all the set partitions of $\mathcal{P}_{n,k}$ is given by

$$(n-\ell+1)\mathrm{Stir}(n-\ell+1,k). \quad (4.24)$$

Decompose a set partition $\pi$ exactly as in (4.22) above except now $\alpha$ is a word over alphabet $[j]$ of size $i+\ell-1$ whose final $\ell$ letters are the same. To show that there are $k\mathrm{Stir}(n-\ell+1,k)$ $\ell$-levels that do start with the smallest element of a block, let $\pi' \in \mathcal{P}_{n-\ell+1,k}$ and suppose $b$ is the smallest element of block $a$, where $1 \leq a \leq k$. Increase all members of the set $\{b+1, b+2, \ldots, n-\ell+1\}$ by $\ell-1$ within $\pi'$ and add all members of the set $\{b+1, b+2, \ldots, b+\ell-1\}$ to block $a$ to obtain a $\ell$-level at $b$.

Also, counting the total number of $\ell$-levels in $\mathcal{P}_{n,k}$ can be derived by considering those $\ell$-levels caused by each set $\{i, i+1, \ldots, i+\ell-1\}$, $1 \leq i \leq n-\ell+1$. Note that there are $\mathrm{Stir}(n-\ell+1,k)$ $\ell$-levels in $\mathcal{P}_{n,k}$ caused by a set $\{i, i+1, \ldots, i+\ell-1\}$, upon regarding it as a single element. Hence there are $(n-\ell+1)\mathrm{Stir}(n-\ell+1,k)$ $\ell$-levels in all set partitions of $\mathcal{P}_{n,k}$. □

Comparing (4.23) and (4.24) yield the following identity

$$(m-k)\mathrm{Stir}(m,k) = \sum_{j=1}^{k}\sum_{i=1}^{m-k} j^i \mathrm{Stir}(m-i,k),$$

as has been shown in [320]. In [220], formulas were given which counted the members of $\mathcal{P}_{n,k}$ having a fixed number of levels. For instance, there are $\binom{n-1}{r}\mathrm{Stir}(n-r-1,k-1)$ set partitions in $\mathcal{P}_{n,k}$ with exactly $r$ occurrences of 2-levels. When $r = 0$, this reduces to the well-known fact that there are $\mathrm{Stir}(n-1, k-1)$ set partitions in $\mathcal{P}_{n,k}$ where no block contains two consecutive integers (see Corollary 4.13). The $r = 0$ case gives the general case as follows. First select $r$ members of $\{2, 3, \ldots, n\}$ to be the second numbers in $r$ 2-levels. Partition the remaining $n - r$ members of $[n]$ into $k$ blocks so that no two "consecutive" elements go in the same block, which can be done in $\mathrm{Stir}(n - r - 1, k - 1)$ ways. Then add the chosen $r$ elements of the set $[n]$ to the appropriate blocks so as to create $r$ 2-levels. On the other hand, there do not appear to be simple formulas for the number of set partitions of $\mathcal{P}_{n,k}$ having $r$ 2-rises (2-descents) for general $r$. The following result gives the number of set partitions of $\mathcal{P}_{n,k}$ having $r$ 3-levels and was considered by Theorem 4.44 analytically. Note that in [320] has been presented a combinatorial proof for the next result.

**Proposition 4.47** *The number of set partitions of $\mathcal{P}_{n,k}$ without 3-levels is given by*

$$\sum_{j=0}^{\lfloor \frac{n}{2} \rfloor} \binom{n-j}{j} \mathrm{Stir}(n-j-1, k-1),$$

*and the number of set partitions of $\mathcal{P}_{n,k}$ with $r$ occurrences of 3-levels, $r \geq 1$, is given by*

$$\sum_{j=1}^{r} \sum_{i=j}^{\lfloor \frac{n-r}{2} \rfloor} \binom{r-1}{j-1}\binom{i}{j}\binom{n-r-i}{i} \mathrm{Stir}(n-r-i-1, k-1). \qquad (4.25)$$

### 4.3.3 Long-Descent Pattern

In this section we obtain the generating function $P_{\ell\cdots 21}(x; q; k)$ for the number of set partitions of $\mathcal{P}_{n,k}$ according to the statistic $\mathrm{occ}_{\ell\cdots 21}$.

**Theorem 4.48** *Let $\ell \geq 1$ and $\tau = \ell \cdots 21$. Then the generating function for the number of set partitions of $\mathcal{P}_{n,k}$ according to the statistic $\mathrm{occ}_\tau$ is given by*

$$P_\tau(x, y; k) = x^k \prod_{j=0}^{k-1} \frac{W_\tau^{(1)}(x; q; j)}{1 - xW_\tau^{(1)}(x; q; j)}.$$

**Proof** Note that each set partition $\pi$ with exactly $k$ blocks can be written as $\pi = \pi' k \sigma$, where $\pi'$ is a set partition with exactly $k - 1$ blocks and $\sigma$ is a word over alphabet $[k]$. This leads to

$$P_\tau(x; q; k) = P_\tau(x; q; k-1) V_\tau(x; q; k),$$

where $V_\tau(x,y;k)$ is the generating function for the number $k$-ary words $k\pi$ of size $n$ according to the statistic $occ_\tau$. Each word $k\pi$ over alphabet $[k]$ can be written as $k\pi = k\pi^{(1)}k\pi^{(2)}\cdots k\pi^{(m)}$ with $m \geq 1$. Clearly, the occurrence of the subword $\tau$ are exactly the occurrences of the subword $12\cdots\ell$ in the reversal word of $k\pi$. So, by the definition of the generating function $W_\tau^{(s)}(x;q;k)$ (see Lemma 4.38), we derive

$$V_\tau(x;q;k) = \frac{xW_\tau^{(1)}(x;q;k-1)}{1 - xW_\tau^{(1)}(x;q;k-1)},$$

which completes the proof. □

For instance, Theorem 4.48 for $\tau = 21$ together with (4.20) give the following result.

**Corollary 4.49** *The generating function for the number of set partitions of $\mathcal{P}_{n,k}$ according to the statistic $occ_{21}$ is given by*

$$P_{21}(x;q;k) = x^k \prod_{j=0}^{k-1} \frac{(1-q)(1-x+xq)^j}{(1-x+xq)^{j+1} - q}.$$

Note that an explicit formula for generating function $P_{321}(x;q;k)$ can be found in [220].

As it is shown in the pervious section, see Corollaries 4.30 and 4.37, Corollary 4.49 obtains an explicit formula for the number descents in all set partitions of $[n]$ with exactly $k$ blocks, where we leave that to the interest reader. But, instead that we suggest a direct proof as follows.

**Corollary 4.50** (Shattuck [320]) *The total number descents in all set partitions of $\mathcal{P}_{n,k}$ is given by*

$$\binom{k}{2}\text{Stir}(n-1,k) + \sum_{j=2}^{k}\binom{j}{2}\sum_{i=k}^{n-2} j^{n-2-i}\text{Stir}(i,k).$$

**Proof** Similar arguments as in the proof of Corollary 4.42 show that the sum counts all descents where neither element is the smallest element of its block. In order to complete the proof, we need to show that $\binom{k}{2}\text{Stir}(n-1,k)$ counts all descents at $i$ for some $i$, where $i$ is the smallest member of its block in some set partition of $\mathcal{P}_{n,k}$. First we choose two numbers, $a, b$, in $[k]$ with $a < b$. Given $\pi \in \mathcal{P}_{n-1,k}$, let $m$ denote the smallest element of block $b$. Increase all elements of the set $\{m+1, m+2, \ldots, n-1\}$ in $\pi$ by one (leaving them within their blocks) and then add $m+1$ to block $a$. This produces a descent between the first element of block $b$ and an element of block $a$ within some set partition of $\mathcal{P}_{n,k}$. Thus, the total number descents at $i$, $i$ is the smallest element of its block in all set partitions of $\mathcal{P}_{n,k}$, is $\sum_{a<b}\text{Stir}(n-1,k) = \binom{k}{2}\text{Stir}(n-1,k)$, as required. □

## 4.4 Families of Subword Patterns

In this section we consider several families of $\ell$-letter subword patterns. For instance, the easiest generalization of the pattern 212 is the pattern $21^{\ell-2}2$, but more general version would be the patter $m\tau m$, where $\tau$ is a pattern of size $\ell - 2$ with letters from alphabet $[m-1]$. We divide our results to several subsections, where in each subsection we consider one of our families.

### 4.4.1 The Patterns $122\cdots 2, 11\cdots 12$

The first simplest family of subword patterns are the patterns $12^{\ell-1}$ and $1^{\ell-1}2$, which is the most obvious way to generalize subword patterns 122 and 112.

**Theorem 4.51** *For all $k \geq 0$,*

$$\mathcal{P}_{1^{\ell-1}2}(x;q;k) = \mathcal{P}_{12^{\ell-1}}(x;q;k).$$

**Proof** Each set partition $\pi = \pi_1\cdots\pi_n \in \mathcal{P}_{n,k}$ may be written uniquely as $\theta^{(1)}\cdots\theta^{(r)}$, where each $\theta^{(i)}$, $1 \leq i \leq r$, is a nonempty and nondecreasing word such that the rightmost letter of the word $\theta^{(i-1)}$ is larger than the leftmost letter of $\theta^{(i)}$ for all $i = 2, 3, \ldots, r$. For example, if $\pi = 1212233212444 \in \mathcal{P}_{13,4}$ then $\theta^{(1)} = 12$, $\theta^{(2)} = 12233$, $\theta^{(3)} = 2$ and $\theta^{(4)} = 12444$. Let the set of distinct letters in $\theta^{(i)}$ be $a_1 < a_2 < \cdots < a_t$, and write $\theta^{(i)} = a_{i_1}a_{i_2}\cdots a_{i_s}$ for some positive integers $s$ and $t$ with $1 \leq i_j \leq t$ for all $j$. Define $\theta'^{(i)} = a_{t+1-i_s}a_{t+1-i_{s-1}}\cdots a_{t+1-i_1}$ and let $\pi'$ be the set partition given by $\theta'^{(1)}\cdots\theta'^{(r)}$. For instance, in the above example we have that $\pi' = 1211223211124$. Thus, the map $\pi \mapsto \pi'$ is a bijection of $\mathcal{P}_{n,k}$ which changes each occurrence of $1^{\ell-1}2$ to an occurrence of $12^{\ell-1}$ and vice–versa. $\square$

Now, we give an explicit formula for the generating function $P_{12^{\ell-1}}(x,y,k)$.

**Theorem 4.52** *Let $\tau = 12\cdots 22 \in [2]^\ell$ be a subword pattern. Then*

$$P_\tau(x;q;k) = \frac{x^k(1 - x^{\ell-2}(1-q))^{k-1}}{\prod_{i=1}^{k}\left(1 - x\sum_{j=0}^{i-1}(1 - x^{\ell-1}(1-q))^j\right)}.$$

**Proof** Exercise 4.7 gives that the generating function for the number of words of size $n$ over the alphabet $[k]$ according to the statistic $occ_\tau$ is given by

$$Q(x;q;k) = \frac{1}{1 - x\sum_{i=0}^{k-1}(1 - x^{\ell-1}(1-q))^i}.$$

Now, let us consider the words $k\pi$, where $\pi$ is a word over the alphabet $[k]$.

By considering whether or not the first letter of $\pi$ is $k$, we obtain that the generating function for the number of words $k\pi$ of size $n$ over the alphabet $[k]$ according to the statistic occ$_\tau$ is given by

$$Q'(x;q;k) = (x^{\ell-1}q + x(1-x^{\ell-2}))Q(x;q;k).$$

Since each set partition $\pi$ of $\mathcal{P}_{n,k}$ may be written as $\pi = 1\pi^{(1)}2\pi^{(2)}\cdots k\pi^{(k)}$ such that each $\pi^{(i)}$ is a word over the alphabet $[i]$, we obtain that the generating function $P_\tau(x;q;k)$ is given by $\frac{x}{1-x}\prod_{j=2}^{k} Q'(x;q;j)$, as required. □

By differentiating the generating function $P_\tau(x;q;k)$ respect to $q$ and then substituting $q=1$, we obtain

$$\frac{d}{dq}P_\tau(x;q;k)\big|_{q=1}$$

$$= P_\tau(x,1,k)\left(x^{\ell-2}(k-1) - \sum_{i=1}^{k} \frac{\frac{d}{dq}\left(1 - x\sum_{j=0}^{i-1}(1-x^{\ell-1}(1-q))^j\right)\big|_{q=1}}{1-ix}\right)$$

$$= P_\tau(x,1,k)\left(x^{\ell-2}(k-1) + \sum_{i=1}^{k} \frac{x^\ell \binom{i}{2}}{1-ax}\right),$$

which implies the next result. First we need the following definition.

**Definition 4.53** *Given $\pi \in \mathcal{P}_n$ and $i \in [n]$. We will say that $i$ is* minimal *if $i$ is the smallest element of a block of $\pi$. Given $\pi \in \mathcal{P}_{n,k}$ and $\tau$ a subword pattern, we will call an occurrence of $\tau$ within $\pi$* primary *if no letter of $\tau$ corresponds to a minimal element of $\pi$.*

**Corollary 4.54** *The total number of occurrences of the pattern $12^{\ell-1}$ in all set partitions of $\mathcal{P}_{n,k}$ is given by*

$$(k-1)\mathrm{Stir}(n-\ell+2,k) + \sum_{j=2}^{k}\binom{j}{2}\left(\sum_{i=\ell}^{n-k} j^{i-\ell}\mathrm{Stir}(n-i,k)\right).$$

**Proof** (Combinatorial Proof) If the element $i$ starts block $B$ within a set partition $\pi \in \mathcal{P}_{n-\ell+2,k}$, then by increasing all elements of the set $\{i+1, i+2, \cdots, n-\ell+2\}$ by $\ell-2$ within $\pi$, leaving them within their current blocks, and adding the elements $i+1, i+2, \cdots, i+\ell-2$ to block $B$ to obtain an occurrence of the subword pattern $\tau = 12\cdots 22$ within a set partition of $\mathcal{P}_{n,k}$ in which the first 2 starts a block, say $B'$. There are clearly $(k-1)\mathrm{Stir}(n-\ell+2,k)$ occurrences of 12 in the block $B'$ within the set partitions of $\mathcal{P}_{n-\ell+2,k}$ since they occur each time a new block is started. Therefore, in order to complete the proof, we have to show that the total number of occurrences of $\tau$ in which the element corresponding to the 2 does not start a block is given by the sum. This can be obtained by Lemma 4.55 for $\tau = 12\cdots 22$, as required. □

**Lemma 4.55** Let $\tau$ be any subword pattern of size $\ell$ in the alphabet $[m]$. Then the total number of primary occurrences of $\tau$ in all of the set partitions of $\mathcal{P}_{n,k}$ is given by

$$\sum_{j=m}^{k} \binom{j}{m} \left( \sum_{i=\ell}^{n-k} j^{i-\ell} \operatorname{Stir}(n-i,k) \right).$$

**Proof** Let $i, j$ be any integers such that $\ell \leq i \leq n-k$ and $m \leq j \leq k$. Consider all the set partitions of $\mathcal{P}_{n,k}$ which can be written uniquely as

$$\pi = \pi' j \pi'' \pi''', \tag{4.26}$$

where $\pi'$ is a set partition with exactly $j-1$ blocks, $\pi'' \in [j]^i$ whose last $\ell$ letters constitute an occurrence of $\tau$, and $\pi'''$ is possibly empty. Clearly, there are $\binom{j}{m}$ choices for the final $\ell$ letters of $\pi''$ since these letters are to form an occurrence of $\tau$ (the smallest member of $[j]$ chosen will correspond to the 1 in $\tau$, the second smallest to 2, etc.). Since $\pi' j \pi''' \in \mathcal{P}_{n-i,k}$, we have $\operatorname{Stir}(n-i,k)$ such set partitions. Also, to insert $\pi''$ there are $j^{i-\ell} \binom{j}{m}$ possibilities ($j^{i-\ell}$ choices for the first $i-\ell$ letters of $\pi''$ and $\binom{j}{m}$ choices for the final $\ell$ letters). Thus, there are $\binom{j}{m} j^{i-\ell} \operatorname{Stir}(n-i,k)$ set partitions of $\mathcal{P}_{n,k}$ which may be written as in (4.26). By summing over all possible values of $i, j$, we complete the proof. □

### 4.4.2 The Patterns $22\cdots 21$, $211\cdots 1$

In this section, we find an explicit formula for the generating function $P_\tau(x; q; k)$ for $\tau = 2^{\ell-1}1$ and $\tau = 21^{\ell-1}$. Again, we need only investigate one case since the bijection used to establish Theorem 4.51 above also shows that the subword patterns $\tau = 2^{\ell-1}1$ and $\tau = 21^{\ell-1}$ are identically distributed on $\mathcal{P}_{n,k}$, where we decompose the canonical representations into nonincreasing words (that is, without rises) instead of decompose them into nondecreasing words as in the proof of Theorem 4.51.

**Theorem 4.56** For all $k \geq 0$,

$$P_{21^{\ell-1}}(x; q; k) = P_{2^{\ell-1}1}(x; q; k).$$

Now, we give an explicit formula for the generating function $P_{2^{\ell-1}1}(x; q; k)$.

**Theorem 4.57** Let $\tau = 2^{\ell-1}1$ be a subword pattern. Then

$$P_\tau(x; q; k) = x^k (q-1)^k \prod_{i=1}^{k} \frac{x^{\ell-2}(1 - x^{\ell-1}(1-q))^{i-1}}{1 - x^{\ell-2}(1-q) - (1 - x^{\ell-1}(1-q))^i}.$$

**Proof** Let $G_a(x; q)$ (respectively, $G'_a(x; q)$) denote the generating function for the number of words $\pi$ of size $n$ over the alphabet $[a]$ according to the number of occurrences of $\tau$ in $\pi$ (respectively, $a\pi$), where $a \geq 2$. Moreover, let $G_a'''(x;q)$ denote the generating function for the number of words of size $n$ over the alphabet $[a]$ which do not start with the letter $a$ (including the empty word) according to the statistic $\operatorname{occ}_\tau$. Then, we have $G_a'(x;q) = G_a(x;q) + (q-1)x^{\ell-2}(G_a'''(x;q) - 1)$ and $G_a(x;q) = G_a'''(x;q) + xG_a'(x;q)$. (Why?) Combining these relations give

$$G_a'(x;q) = \frac{(1 - x^{\ell-2}(1-q))G_a(x;q) + x^{\ell-2}(1-q)}{1 - x^{\ell-1}(1-q)}.$$

By Exercise 4.10 we have

$$G_a(x;q) = \frac{x^{\ell-2}(q-1)}{1 - x^{\ell-2}(1-q) - (1 - x^{\ell-1}(1-q))^a},$$

which implies

$$G_a'(x;q) = \frac{x^{\ell-2}(q-1)(1 - x^{\ell-1}(1-q))^{a-1}}{1 - x^{\ell-2}(1-q) - (1 - x^{\ell-1}(1-q))^a},$$

for all $a = 1, 2, \ldots, k$. By the fact that each set partition $\pi$ with exactly $k$ blocks may be written as $\pi = 1\pi^{(1)}2\pi^{(2)}\cdots k\pi^{(k)}$, where $\pi^{(i)}$ is a word over the alphabet $[i]$, we have that the generating function $P_\tau(x;q;k)$ can be expressed as $x^k \prod_{a=1}^k G_a'(x;q)$, which completes the proof. □

**Example 4.58** *Theorem 4.57 for $\ell = 3$ gives*

$$P_{221}(x;q;k) = x^{2k}(q-1)^k \prod_{i=1}^k \frac{(1 - x^2(1-q))^{i-1}}{1 - x(1-q) - (1 - x^2(1-q))^i}$$

*and substituting $q = 0$ in this implies that the generating function for the number of set partitions of $\mathcal{P}_{n,k}$ which avoid the subword pattern 221 is given by*

$$P_{221}(x;0;k) = x^{2k} \prod_{i=1}^k \frac{(1 - x^2)^{i-1}}{(1 - x^2)^i + x - 1}.$$

Now, we consider the total number of occurrences of the pattern $2^{\ell-1}1$ in all set partitions of $\mathcal{P}_{n,k}$. Theorem 4.57 gives

$$P_\tau(x;1;k) = \lim_{q \to 1} F_\tau(x;q;k) = x^k \prod_{j=1}^k \lim_{q \to 1} \left(\frac{f_j(q)}{g_j(q)}\right),$$

where

$$f_j(q) = x^{\ell-2}(q-1)(1 - x^{\ell-1}(1-q))^{j-1},$$
$$g_j(q) = 1 - x^{\ell-2}(1-q) - (1 - x^{\ell-1}(1-q))^j.$$

Since $f_j(1) = g_j(1) = 0$, $f_j'(1) = x^{\ell-2}$ and $g_j'(1) = x^{\ell-2} - jx^{\ell-1}$, where primes denote differentiation with respect to $q$, we have that

$$\lim_{q \to 1} \frac{f_j(q)}{g_j(q)} = \frac{1}{1-jx}$$

and hence $P_\tau(x; 1; k) = \frac{x^k}{(1-x)(1-2x)\cdots(1-kx)}$ which agrees with (2.9). Also, Theorem 4.57 implies

$$\frac{d}{dq} F_\tau(x; q; k) \Big|_{q=1} = F_\tau(x; 1; k) \sum_{j=1}^{k} \lim_{q \to 1} \frac{f_j'(q) g_j(q) - f_j(q) g_j'(q)}{f_j(q) g_j(q)}.$$

Since $f_j''(1) = 2(j-1)x^{2\ell-3}$ and $g_j''(1) = -j(j-1)x^{2\ell-2}$, two applications of L'Hôpital's rule implies

$$\lim_{q \to 1} \frac{f_j'(q) g_j(q) - f_j(q) g_j'(q)}{f_j(q) g_j(q)} = \lim_{q \to 1} \frac{f_j''(q) g_j'(q) - f_j'(q) g_j''(q)}{2 f_j'(q) g_j'(q)}$$

$$= \frac{f_j''(1)}{2 f_j'(1)} - \frac{g_j''(1)}{2 g_j'(1)}$$

$$= (j-1)x^{\ell-1} + \frac{\binom{j}{2} x^\ell}{1-jx},$$

which leads to

$$\frac{d}{dq} F_\tau(x; q; k) \Big|_{q=1} = F_\tau(x; 1; k) \left( x^{\ell-1} \binom{k}{2} + \sum_{j=1}^{k} \frac{x^\ell \binom{j}{2}}{1-jx} \right).$$

Therefore, we can state the following result.

**Corollary 4.59** *The total number of occurrences of the subword pattern $\tau = 2^{\ell-1}1$ in all of the set partitions of $\mathcal{P}_{n,k}$ is given by*

$$\binom{k}{2} \text{Stir}(n-\ell+1, k) + \sum_{j=2}^{k} \binom{j}{2} \left( \sum_{i=\ell}^{n-k} j^{i-\ell} \text{Stir}(n-i, k) \right).$$

**Proof** (Combinatorial Proof) By Lemma 4.55, it is sufficient to show that the total number of occurrences of the pattern $\tau$ within the set partitions of $\mathcal{P}_{n,k}$ in which the element of $[n]$ corresponding to the first 2 is minimal and equals $\binom{k}{2} \text{Stir}(n-\ell+1, k)$. Note that occurrences of $\tau$ within set partitions of $\mathcal{P}_{n,k}$ in which the first 2 is minimal are synonymous with occurrences of 21 within set partitions of $\mathcal{P}_{n-\ell+2,k}$ in which the 2 is minimal. So to end the proof, we show that there are $\binom{k}{2} \text{Stir}(r-1, k)$ occurrences of 21 in which the 2 is minimal within the set partitions of $\mathcal{P}_{r,k}$ as follows. Fix $1 \leq a < b \leq k$. Given a set partition in $\mathcal{P}_{r-1,k}$, let $m$ denote the smallest member of block that contains $b$. By increasing all members of $\{m+1, m+2, \ldots, r-1\}$ by one, leaving them within their blocks, and adding the element $m+1$ to block that contains $a$. This produces an occurrence of 21 at positions $m$ and $m+1$ within some set partition of $\mathcal{P}_{r,k}$. Thus, we have $\sum_{1 \leq a < b \leq k} \text{Stir}(n-\ell+1,k) = \binom{k}{2}\text{Stir}(n-\ell+1,k)$ such occurrences, which completes the proof. □

### 4.4.3 The Pattern $m\rho m$

In this section we find the generating function $P_{m\rho m}(x;q;k)$ in terms of the generating function $P_\rho(x;q;k)$.

**Theorem 4.60** *Let* $\tau = m\rho m \in [m]^\ell$ *be a subword pattern, where $\rho$ does not contain $m$. Then the generating function $P_\tau(x;q;k)$ is given by*

$$\frac{x^k}{(1-x)\cdots(1-(m-1)x)} \prod_{i=m}^{k} \frac{\frac{1}{1+\binom{i-1}{m-1}x^{\ell-1}(1-q)}}{1-(m-1)x - x\sum_{j=m-1}^{i-1} \frac{1}{1+\binom{j}{m-1}x^{\ell-1}(1-q)}}.$$

**Proof** Let $G_a(x;q)$ be the generating function for the number of words $\pi$ of size $n$ over the alphabet $[a]$ according to the number of occurrences of the subword pattern $\tau$ in $a\pi(a+1)$. Since $\pi$ either contains the letter $a$ (here we write $\pi = \pi' a \pi''$ where $\pi'$ is a word over the alphabet $[a-1]$) or does not, we derive

$$G_a(x;q) = G'_{a-1}(x;q) + x(G'_{a-1}(x;q) - x^{\ell-2}\binom{a-1}{m-1}(1-q))G_a(x;q),$$

which is equivalent to

$$G_a(x;q) = \frac{G'_{a-1}(x;q)}{1 - x(G'_{a-1}(x;q) - x^{\ell-2}\binom{a-1}{m-1}(1-q))}, \quad (4.27)$$

where $G'_a(x;q)$ is the generating function for the number of words of size $n$ over the alphabet $[a]$ according to the number occurrences of the subword pattern $\tau$. By [59, Theorem 2.7], we obtain

$$G'_a(x;q) = \frac{1}{1 - \min\{m-1,a\}x - x\sum_{j=m-1}^{a-1} \frac{1}{1+\binom{j}{m-1}x^{\ell-1}(1-q)}}, \quad (4.28)$$

for all $a \geq 1$. Thus, by the fact that each set partition $\pi \in \mathcal{P}_{n,k}$ can be written as $\pi = 1\pi^{(1)}2\pi^{(2)}\cdots k\pi^{(k)}$ such that $\pi^{(i)}$ is a word over the alphabet $[i]$, we obtain that the generating function $P_\tau(x;q;k)$ is given by $x^k \prod_{a=1}^{k} G_a(x;q)$. Hence, (4.27) and (4.28) give that the generating function $G_a(x;q)$ is given by

$$\frac{\frac{1}{1+\binom{a-1}{m-1}x^{\ell-1}(1-q)}}{1 - \min\{m-1,a-1\}x - x\sum_{j=m-1}^{a-2}\frac{1}{1+\binom{j}{m-1}x^{\ell-1}(1-q)} - \frac{x}{1+x^{\ell-1}\binom{a-1}{m-1}(1-q)}},$$

which completes the proof. □

**Example 4.61** *Theorem 4.60 for $\tau = 211\cdots 12 \in [2]^\ell$ gives*

$$P_\tau(x;q;k) = x^k \prod_{j=1}^{k} \frac{\frac{1}{1+(j-1)x^{\ell-1}(1-q)}}{1 - x - x\sum_{i=1}^{j-1}\frac{1}{1+ix^{\ell-1}(1-q)}}.$$

*In particular, when $\ell = 3$ and $q = 0$, we see that the generating function for the number of set partitions of $\mathcal{P}_{n,k}$ that avoid the pattern $\tau = 212$ is given by*

$$P_{212}(x;0;k) = x^k \prod_{j=0}^{k-1} \frac{1}{1 + jx^2 - x(1+jx^2)\sum_{i=0}^{j}\frac{1}{1+ix^2}}.$$

Differentiating the generating function in Theorem 4.60 gives

$$\frac{d}{dq}P_\tau(x;q;k)|_{q=1}$$

$$= P_\tau(x;1;k)\left(\sum_{a=m}^{k}\left[x^{\ell-1}\binom{a-1}{m-1} + \frac{1}{1-ax}\sum_{j=1}^{a-1}x^\ell\binom{j}{m-1}\right]\right)$$

$$= P_\tau(x;1;k)\left(x^{\ell-1}\binom{k}{m} + \sum_{j=1}^{k}\frac{x^\ell\binom{j}{m}}{1-jx}\right),$$

which leads to the following result.

**Corollary 4.62** *The total number of occurrences of the subword pattern $\tau = m\rho m \in [m]^\ell$, where $\rho$ does not contain $m$, in all of the set partitions of $\mathcal{P}_{n,k}$ is given by*

$$\binom{k}{m}\mathrm{Stir}(n-\ell+1,k) + \sum_{j=m}^{k}\binom{j}{m}\left(\sum_{i=\ell}^{n-k}j^{i-\ell}\mathrm{Stir}(n-i,k)\right).$$

**Proof** (Combinatorial Proof) By Lemma 4.55, the sum in the explicit formula counts all primary occurrences of the subword pattern $\tau$ within all set partitions of $\mathcal{P}_{n,k}$. Thus, to end the proof, we show that there are $\binom{k}{m}\mathrm{Stir}(n-\ell+1,k)$ total nonprimary occurrences of $\tau$. Note that the first letter of a nonprimary occurrence of $\tau$ in a set partition $\pi$ must correspond to a minimal element of $\pi$, with all the other letters comprising the occurrence nonminimal. Then, given $m$ numbers $a_1 < a_2 < \cdots < a_m$ in $[k]$ and $\pi \in \mathcal{P}_{n-\ell+1,k}$, let $j$ denote the smallest element of block that contains $a_m$. By increasing all elements of the set $\{j+1, j+2, \ldots, n-\ell+1\}$ by $\ell-1$ in $\pi$, leaving them within their current blocks, and then adding the element $j+i$ to the block that contains $a_r$, where $r$ denotes the $(i+1)$-st letter of the pattern $\tau$ for all $i$, $1 \le i \le \ell-1$. The obtaining set partition of $\mathcal{P}_{n,k}$ will have a nonprimary occurrence of $\tau$ starting at the $j$-th letter, as required. □

### 4.4.4 The Pattern $m\rho(m+1)$

In this section we find the generating function $P_{m\rho(m+1)}(x;q;k)$ in terms of the generating function $P_\rho(x;q;k)$.

**Theorem 4.63** *Let $\tau = m\rho(m+1) \in [m+1]^\ell$ be a subword pattern, where $\rho$ does not contain $m$ and $m+1$. Then, the generating function $P_\tau(x;q;k)$ with $k \geq m+1$ is given by*

$$\frac{x^k H_k(x;q)}{(1-x)\cdots(1-(m-1)x)} \prod_{a=m}^{k-1} \frac{G_a(x;q) - \binom{a-1}{m-1}x^{\ell-2}(1-q)}{1 - xG_a(x;q)},$$

*where*

$$H_a(x;q) = \frac{1}{1 - (m-1)x - x\sum_{i=m-2}^{a-2}\prod_{j=m-2}^{i}\left(1 - \binom{j}{m-1}x^{\ell-1}(1-q)\right)}$$

*with $a \geq m$,*

$$G_a(x;q) = \frac{\prod_{j=m-1}^{a-2}(1 - \binom{j}{m-1}x^{\ell-1}(1-q))}{1 - mx - x\sum_{i=m-1}^{a-3}\prod_{j=m-1}^{i}\left(1 - \binom{j}{m-1}x^{\ell-1}(1-q)\right)}$$

*with $a \geq m+1$, and $G_m(x;q) = \frac{1}{1-(m-1)x}$.*

**Proof** Let $G_a = G_a(x;q)$ (respectively, $G'_a = G'_a(x;q)$) be the generating function for the number of words $\pi$ of size $n$ over the alphabet $[a-1]$ according to the number occurrences of the pattern $\tau$ in $\pi a$ (respectively, $a\pi(a+1)$). Clearly, $G'_a = G_a - \binom{a-1}{m-1}x^{\ell-2}(1-q)$.

Note that each word $\pi$ over the alphabet $[a-1]$ either does not contain the letter $a-1$ or may be written as $\pi^{(1)}(a-1)\pi^{(2)}(a-1)\cdots\pi^{(s)}(a-1)\pi^{(s+1)}$ such that $\pi^{(j)}$ is a word over the alphabet $[a-2]$, for all $j$. Therefore,

$$G_a = G_{a-1} + \frac{xG_{a-1}G'_{a-1}}{1 - xG_{a-1}},$$

which implies

$$G_a = G_{a-1} + \frac{xG_{a-1}\left(G_{a-1} - \binom{a-2}{m-1}x^{\ell-2}(1-q)\right)}{1 - xG_{a-1}}$$

$$= \frac{G_{a-1}\left(1 - \binom{a-2}{m-1}x^{\ell-1}(1-q)\right)}{1 - xG_{a-1}}.$$

Hence, by induction on $a$, we have $G_a = \frac{1}{1-(a-1)x}$ for all $a = 1, 2, \ldots, m$, and

$$G_a = \frac{\prod_{j=m-1}^{a-2}(1 - \binom{j}{m-1}x^{\ell-1}(1-q))}{1 - mx - x\sum_{i=m-1}^{a-3}\prod_{j=m-1}^{i}\left(1 - \binom{j}{m-1}x^{\ell-1}(1-q)\right)},$$

for all $a \geq m+1$.

Let $H_a = H_a(x;q)$ be the generating function for the number of words $\pi$ of size $n$ over the alphabet $[a]$ according to the number of occurrences of the pattern $\tau$ in $\pi$. By [59, Theorem 2.11], we have

$$H_a = \frac{1}{1-(m-1)x - x\sum_{i=m-2}^{a-2} \prod_{j=m-2}^{i}\left(1 - \binom{j}{m-1}x^{\ell-1}(1-q)\right)},$$

for all $a \geq m$. Hence, the generating function for the number words $\pi$ of size $n$ over the alphabet $[a]$ according to the number of occurrences of the pattern $\tau$ in $a\pi(a+1)$ is given by (Explain?)

$$\frac{G'_a}{1-xH_a} = \frac{G_a - \binom{a-1}{m-1}x^{\ell-2}(1-q)}{1-xG_a},$$

for all $a \geq m$, and it is given by $\frac{1}{1-ax}$ for $a = 1, 2, \ldots, m-1$. Thus, by the fact that each set partition $\pi$ of $[n]$ with exactly $k$ blocks can be written as $\pi = 1\pi^{(1)} \cdots k\pi^{(k)}$ such that $\pi^{(i)}$ is a word over the alphabet $[i]$, we have that the generating function $P_\tau(x;q;k)$ is given by

$$\frac{x^k H_k}{(1-x)\cdots(1-(m-1)x)} \prod_{a=m}^{k-1} \frac{G_a - \binom{a-1}{m-1}x^{\ell-2}(1-q)}{1-xG_a},$$

as required. $\square$

**Example 4.64** Let $k \geq 2$. Then, Theorem 4.63 with $\tau = 213$ gives that the generating function for the number of set partitions of $\mathcal{P}_{n,k}$ according to the number occurrences of the subword pattern 213, which is given by

$$P_{213}(x;q;k) = \frac{x^k \prod_{a=2}^{k-1} \frac{G_a(x;q)-(a-1)x(1-q)}{1-xG_a(x;q)}}{(1-x)\left(1-x-x\sum_{i=0}^{k-2}\prod_{j=0}^{i}(1-jx^2(1-q))\right)},$$

where

$$G_a(x;q) = \frac{\prod_{j=1}^{a-2}(1-jx^2(1-q))}{1-2x - x\sum_{i=1}^{a-3}\prod_{j=1}^{i}(1-jx^2(1-q))}.$$

One may easily verify that

$$G_a(1) := G_a(x;1) = \frac{1}{1-(a-1)x} \qquad (4.29)$$

and

$$dG_a(1) := \frac{d}{dq}G_a(x;q)|_{q=1} = \frac{x^{\ell-1}\binom{a-1}{m}}{1-(a-1)x} + \frac{x^\ell \binom{a-1}{m+1}}{[1-(a-1)x]^2}. \qquad (4.30)$$

By combining these facts with Theorem 4.63 and the fact that

$$\frac{d}{dq}H_k(x;q)\,|_{q=1} = \frac{x^\ell}{1-kx}\sum_{i=m-2}^{k-2}\binom{i+1}{m} = \frac{x^\ell\binom{k}{m+1}}{1-kx},$$

we obtain

$$\frac{\frac{d}{dq}P_\tau(x;q;k)\,|_{q=1}}{P_\tau(x;1,k)} = \sum_{a=m}^{k-1}\left(dG_a(1) + \binom{a-1}{m-1}x^{\ell-2}\right)(1-(a-1)x)$$

$$+ \sum_{a=m}^{k-1}\frac{xdG_a(1)}{1-xG_a(1)} + \frac{x^\ell\binom{k}{m+1}}{1-kx}$$

$$= x^{\ell-2}\sum_{a=m}^{k-1}\left(\binom{a-1}{m-1} - (m-1)x\binom{a}{m}\right) + \sum_{a=m}^{k}\frac{x^\ell\binom{a}{m+1}}{1-ax}$$

$$= x^{\ell-2}\binom{k-1}{m} - x^{\ell-1}(m-1)\binom{k}{m+1} + \sum_{a=m}^{k}\frac{x^\ell\binom{a}{m+1}}{1-ax},$$

which yields the following result.

**Corollary 4.65** *The total number of occurrences of the subword pattern $\tau = m\rho(m+1) \in [m+1]^\ell$, where $\rho$ does not contain $m$ or $m+1$, in all of the partitions of $[n]$ with exactly $k$ blocks is given by*

$$\binom{k-1}{m}\mathrm{Stir}(n-\ell+2,k) - (m-1)\binom{k}{m+1}\mathrm{Stir}(n-\ell+1,k)$$

$$+ \sum_{j=m+1}^{k}\binom{j}{m+1}\left(\sum_{i=\ell}^{n-k}j^{i-\ell}\mathrm{Stir}(n-i,k)\right).$$

**Proof** (Combinatorial Proof) Again, by Lemma 4.55, it remains to show that the total number of nonprimary occurrences of $\tau$ is given by

$$\binom{k-1}{m}\mathrm{Stir}(n-\ell+2,k) - (m-1)\binom{k}{m+1}\mathrm{Stir}(n-\ell+1,k),$$

which is equivalent to

$$2\binom{k}{m+1}\mathrm{Stir}(n-\ell+1,k) + \binom{k-1}{m}\mathrm{Stir}(n-\ell+1,k-1),$$

by Theorem 1.12 and Fact 2.20(iii).

If an occurrence of $\tau$ is nonprimary within a set partition $\pi$, then the number corresponding to the block containing $m+1$ must be a minimal element of $\pi$. (why?) There are $\binom{k}{m+1}\mathrm{Stir}(n-\ell+1,k)$ nonprimary occurrences of $\tau$ in which the number corresponding to the block containing $m$ is not minimal. In order to show this, first set $m+1$ numbers $1 \leq a_1 < a_2 < \cdots < a_{m+1} \leq k$, and we denote the smallest element of block that contains $a_{m+1}$ in $\pi \in \mathcal{P}_{n-\ell+1,k}$ by $b$. Secondly, increase all members of $\{b, b+1, \ldots, n-\ell+1\}$ by $\ell-1$ within $\pi$, leaving all members within their current blocks, then adding the element $b+i$ to block that contains $a_r$, where $r$ denotes the $(i+1)$-st letter of the subword pattern $\tau$ for all $i$, $0 \leq i \leq \ell-2$. Since the minimal elements of each block that contains $a_i$ for $1 \leq i \leq m$ are all less than $b$, the obtaining set partition of $\mathcal{P}_{n,k}$ will have a nonprimary occurrence of $\tau$ starting at $b$ in which the number corresponding to $m$ is not minimal.

Similarly, there are $\binom{k}{m+1}\mathrm{Stir}(n-\ell+1,k)$ nonprimary occurrences of $\tau$ in which the number corresponding to the block containing $m$ is minimal but does not occur as a singleton block. In order to show this, set $m+1$ numbers $1 \leq a_1 < a_2 < \cdots < a_{m+1} \leq k$, and we denote the smallest element of block that contains $a_m$ of $\pi \in \mathcal{P}_{n-\ell+1,k}$ by $b$. Now, increase all members of $\{b, b+1, \ldots, n-\ell+1\}$ by $\ell-1$ within $\pi$, add $b$ to block that contains $a_{m+1}$, then add $b+i$ to block that contains $a_r$, where $r$ denotes the $(i+1)$-st letter of $\tau$, $1 \leq i \leq \ell-2$. Thus, the obtaining set partition of $\mathcal{P}_{n,k}$ contains a nonprimary occurrence of $\tau$, starting at $b$ in which the letter $b$ corresponding to $m$ is minimal but does not occur as a singleton block.

At the end, there are $\binom{k-1}{m}\mathrm{Stir}(n-\ell+1,k-1)$ nonprimary occurrences of $\tau$ in which the letter corresponding to $m$ occurs as a singleton block. In order to show that, again, set $1 \leq a_1 < a_2 < \cdots < a_m \leq k-1$, and we denote the smallest element of block that contains $a_m$ of $\pi \in \mathcal{P}_{n-\ell+1,k-1}$ by $b$. Increase all members of $\{b, b+1, \ldots, n-\ell+1\}$ by $\ell-1$ within $\pi$, add the singleton block $\{b\}$, and add $b+i$ to block that contains $a_r$, where $r$ denotes the $(i+1)$-st letter of $\tau$, $1 \leq i \leq \ell-2$. Thus, the obtaining set partition of $\mathcal{P}_{n,k}$ contains a nonprimary occurrence of $\tau$ starting at $b$ in which the singleton block $\{b\}$ occurs. □

### 4.4.5 The Pattern $(m+1)\rho m$

In this section we find the generating function $P_{(m+1)\rho m}(x;q;k)$ in terms of the generating function $P_\rho(x;q;k)$.

**Theorem 4.66** *Let $\tau = (m+1)\rho m \in [m+1]^\ell$ be a subword pattern, where $\rho$ does not contain $m$ and $m+1$. Then*

$$P_\tau(x;q;k) = x^k \prod_{a=1}^{k} \frac{G_a(x;q)}{1 - xG_a(x;q)},$$

*where*

$$G_a(x;q) = \frac{\prod_{j=m-1}^{a-2}(1 - \binom{j}{m-1})x^{\ell-1}(1-q))}{1 - mx - x\sum_{i=m-1}^{a-3}\prod_{j=m-1}^{i}\left(1 - \binom{j}{m-1}x^{\ell-1}(1-q)\right)}$$

*for $a \geq m+1$, and $G_a(x;q) = \frac{1}{1-(a-1)x}$ for $a \leq m$.*

**Proof** See Exercise 4.8. □

**Example 4.67** *Theorem 4.66 for $\tau = 312$ gives that the generating function $P_{312}(x;q;k)$ for the number of set partitions of $\mathcal{P}_{n,k}$ according to the number of occurrences of the pattern 312 is*

$$\frac{x^k \prod_{j=1}^{k-2}(1-jx^2(1-q))^{k-1-j}}{(1-x)(1-2x)\prod_{a=1}^{k-2}\left(1-2x-x\sum_{i=1}^{a}\prod_{j=1}^{i}(1-jx^2(1-q))\right)}.$$

Similar arguments as in the proof of Corollary 4.30, Theorem 4.66 shows the following result.

**Corollary 4.68** *The total number of occurrences of the subword pattern $\tau = (m+1)\rho m \in [m+1]^\ell$, where $\rho$ does not contain $m$ or $m+1$, in all of the partitions of $\mathcal{P}_{n,k}$ is given by*

$$\binom{k}{m+1}\mathrm{Stir}(n-\ell+1,k) + \sum_{j=m+1}^{k}\binom{j}{m+1}\left(\sum_{i=\ell}^{n-k}j^{i-\ell}\mathrm{Stir}(n-i,k)\right).$$

At the end we note that by using similar techniques as in the proof of the above theorems, ones can extend our study to include more general families. For instance, try to find an explicit formula for the generating function $P_\tau(x;q;k)$, where $\tau = m\rho m \rho' m$ is any word over alphabet $[m]$ such that $\rho\rho'$ is any word contains all the letters $1, 2, \ldots, m-1$ and does not contain the letter $m$.

---

## 4.5 Patterns of Size Three

In this section, we turn our attention to counting occurrences of subword patterns of size three within set partitions of $\mathcal{P}_{n,k}$. Theorem 4.44 implies

$$P_{111}(x;q;k) = \frac{(x+x^2+x^3/(1-xq))^k}{\prod_{j=0}^{k-1}(1-j(x+x^2+x^3/(1-xq)))}.$$

The expressions for $P_{123}(x;q;k)$ and $P_{321}(x;q;k)$ are more complicated and occur as Theorems 4.40 and 4.48. From our results in previous sections, we have the following:

$$P_{112}(x;q;k) = P_{122}(x;q;k) = \frac{x^k(1-x(1-q))^{k-1}}{\prod_{a=1}^{k}\left(1-x\sum_{j=0}^{a-1}(1-x^2(1-q))^j\right)},$$

$$P_{211}(x;q;k) = P_{221}(x;q;k) = \prod_{a=1}^{k} \frac{x^2(q-1)(1-x^2(1-q))^{a-1}}{1-x(1-q)-(1-x^2(1-q))^a},$$

$$P_{212}(x;q;k) = x^k \prod_{a=1}^{k} \frac{\frac{1}{1+(a-1)x^2(1-q)}}{1-x-x\sum_{j=1}^{a-1}\frac{1}{1+jx^2(1-q)}},$$

$$P_{213}(x;q;k) = \frac{x^k \prod_{a=2}^{k-1} \frac{G_a - (a-1)x(1-q)}{1-xG_a}}{(1-x)\left[1-x-x\sum_{i=0}^{k-2}\prod_{j=0}^{i}(1-jx^2(1-q))\right]},$$

$$P_{312}(x;q;k) = \frac{x^k \prod_{j=1}^{k-2}(1-jx^2(1-q))^{k-1-j}}{(1-x)(1-2x)\prod_{a=1}^{k-2}\left(1-2x-x\sum_{i=1}^{a}\prod_{j=1}^{i}(1-jx^2(1-q))\right)},$$

where

$$G_a = \frac{\prod_{j=1}^{a-2}(1-jx^2(1-q))}{1-2x-x\sum_{i=1}^{a-3}\prod_{j=1}^{i}(1-jx^2(1-q))}.$$

It remains the cases 121, 132, and 231. We will use more advanced algebraic techniques to derive both recurrence relations of the generating function for these patterns as well as find an explicit formula of the generating function for the total number of occurrences of these patterns.

**Theorem 4.69** *Let $G_k = P_{121}(x;q;k)$. Then the generating function $G_k$ satisfies the recurrence relation*

$$G_k = x \sum_{a=1}^{k-1} \sum_{j=0}^{k-1-a} \frac{x^{2j}(q-1)^j}{\prod_{i=0}^{j}(1-(k-a-i)x^2(q-1))} G_{k-j}$$

$$+ \sum_{a=1}^{k} \frac{x^{2(k-a)+1}(q-1)^{k-a}}{\prod_{i=0}^{k-1-a}(1-(k-a-i)x^2(q-1))} (G_a + G_{a-1})$$

*with the initial condition $G_1 = \frac{x}{1-x}$.*

**Proof** Let $G_k(a_s \cdots a_1) = G_k(x;q|a_s \cdots a_1)$ be the generating function for the number of set partitions $\pi = \pi_1 \cdots \pi_n \in \mathcal{P}_{n,k}$ such that $\pi_{n+1-j} = a_j$, $j = 1, 2, \ldots, s$, according to the number of occurrences of the pattern 121. Define $G_k = F_{121}(x;q;k)$. Let $1 \leq a \leq k-1$ and $a+1 \leq j \leq k-1$, from the

definitions we have
$$G_k(k) = xG_k + xG_{k-1},$$
$$G_k(a) = \sum_{j=1}^{k} G_k(ja) = x\sum_{j=1}^{a} G_k(j) + \sum_{j=a+1}^{k} G_k(ja)$$

and
$$G_k(ja) = \sum_{i=1, i\neq a}^{j} G_k(ija) + G_k(aja) + \sum_{i=j+1}^{k} G_k(ija)$$
$$= x^2 \sum_{i=1}^{j} G_k(i) + x^2(q-1)G_k(a) + x \sum_{i=j+1}^{k} G_k(ij),$$
$$G_k(ka) = x^2 G_k + x^2(q-1)G_k(a) + x^2 G_{k-1} + x^2(q-1)G_{k-1}(a).$$

Therefore,
$$G_k(k) = xG_k + xG_{k-1},$$
$$G_k(a) = \sum_{j=1}^{k} G_k(ja) = x\sum_{j=1}^{a} G_k(j) + \sum_{j=a+1}^{k} G_k(ja),$$
$$G_k(ja) = x^2 \sum_{i=1}^{j} G_k(i) + x^2(q-1)G_k(a) + x(G_k(j) - x\sum_{i=1}^{j} G_k(i))$$
$$= xG_k(j) + x^2(q-1)G_k(a),$$
$$G_k(ka) = x^2 G_k + x^2(q-1)G_k(a) + x^2 G_{k-1} + x^2(q-1)G_{k-1}(a).$$

Thus, for all $a = 1, 2, \ldots, k-1$,
$$G_k(a) = x\sum_{j=1}^{a} G_k(j) + \sum_{j=a+1}^{k} G_k(ja)$$
$$= x\sum_{j=1}^{a} G_k(j) + \sum_{j=a+1}^{k-1} (xG_k(j) + x^2(q-1)G_k(a))$$
$$+ x^2 G_k + x^2(q-1)G_k(a) + x^2 G_{k-1} + x^2(q-1)G_{k-1}(a)$$
$$= x\sum_{j=1}^{k} G_k(j) + (k-a)x^2(q-1)G_k(a) + x^2(q-1)G_{k-1}(a)$$
$$= xG_k + (k-a)x^2(q-1)G_k(a) + x^2(q-1)G_{k-1}(a),$$

with the initial condition $G_k(k) = xG_k + xG_{k-1}$. Then the generating function $G_k(a)$ satisfies the recurrence relation

$$G_k(a) = \frac{x}{1 - (k-a)x^2(q-1)} G_k + \frac{x^2(q-1)}{1 - (k-a)x^2(q-1)} G_{k-1}(a).$$

Iterating this recurrence relation, we have

$$G_k(a) = x \sum_{j=0}^{k-1-a} \frac{x^{2j}(q-1)^j}{\prod_{i=0}^{j}(1-(k-a-i)x^2(q-1))} G_{k-j}$$
$$+ \frac{x^{2(k-a)+1}(q-1)^{k-a}}{\prod_{i=0}^{k-1-a}(1-(k-a-i)x^2(q-1))}(G_a + G_{a-1}),$$

for all $a = 1, 2, \ldots, k$. By summing over all possible values of $a$, we obtain the requested result. □

**Corollary 4.70** *The generating function for the total number of occurrences of the subword pattern 121 in all the set partitions of $\mathcal{P}_{n,k}$ is given by*

$$\frac{x^{k+1}}{(1-x)(1-2x)\cdots(1-kx)} \sum_{j=2}^{k} \frac{\binom{j}{2}x^2 + 1 - jx}{1 - jx}.$$

**Proof** Theorem 4.69 for $q = 1$ gives $G_1(x;1) = \frac{x}{1-x}$ and

$$G_k(x;1) = (k-1)xG_k(x,1) + x(G_k(x,1) + G_{k-1}(x,1)),$$

which is equivalent to $G_k(x;1) = \frac{x^k}{\prod_{j=1}^{k}(1-jx)}$, as it is well known. Define $G'_k(x) = \frac{d}{dq}G_k(x;q)\big|_{q=1}$. Theorem 4.69 gives

$$G'_k(x) = x \sum_{a=1}^{k-1}((k-a)x^2 G_k(x,1) + G'_k(x)) + x \sum_{a=1}^{k-2} x^2 G_{k-1}(x,1)$$
$$+ x(G'_k(x) + G'_{k-1}(x)) + x^3(G_{k-1}(x,1) + G_{k-2}(x,1)),$$

which is equivalent to

$$G'_k(x) = \frac{x}{1-kx} G'_{k-1}(x)$$
$$+ \frac{x^3\binom{k}{2}G_k(x,1) + x^3(k-1)G_{k-1}(x,1) + x^3 G_{k-2}(x,1)}{1-kx}.$$

We complete the proof by substituting the formula of $G_k(x;1)$ together with using the initial condition $G'_1(x) = 0$, and iterating the above recurrence relation. □

The above theorem gives the following explicit formula for the total number of occurrences of 121 in all set partitions of $\mathcal{P}_{n,k}$.

**Corollary 4.71** *The total number of occurrences of the pattern 121 in all of the set partitions of $\mathcal{P}_{n,k}$ blocks is given by*

$$(k-1)\mathrm{Stir}(n-1,k) + \sum_{j=2}^{k} \binom{j}{2} f_{n,j} \left( \sum_{i=3}^{n-k} j^{i-3}\mathrm{Stir}(n-i,k) \right).$$

By using similar arguments as in the proof of Theorem 4.69, we obtain the following result.

**Theorem 4.72** *Let $G_k(a) = P_{132}(x; q; k|a)$ be the generating function for the number of set partitions $\pi = \pi_1 \cdots \pi_{n-1} a \in \mathcal{P}_{n,k}$ according to the number occurrences of the subword pattern 132, and let $G_k = \sum_{a=1}^{k} G_k(a) = P_{132}(x; q; k)$. Then the generating function $G_k(a)$ satisfies the recurrence relation*

$$G_k(a) = xG_k + x^2(q-1)(k-a)\sum_{j=1}^{a-1} G_k(j) + x^2(q-1)\sum_{j=1}^{a-1} G_{k-1}(j)$$

*with the initial conditions $G_k(k) = xG_k + xG_{k-1}$, $G_k(1) = xG_k$, $G_1 = G_1(1) = \frac{x}{1-x}$, $G_2(1) = \frac{x^3}{(1-x)(1-2x)}$, $G_2(2) = \frac{x^2}{1-2x}$ and $G_2 = \frac{x^2}{(1-x)(1-2x)}$.*

**Corollary 4.73** *The generating function for the total number of occurrences of the pattern 132 in all the set partitions of $\mathcal{P}_{n,k}$ is given by*

$$\frac{x^{k+2}}{(1-x)\cdots(1-kx)}\sum_{j=1}^{k-1}\frac{\binom{j}{2}\frac{3-2(j+1)x}{3}}{1-(j+1)x}.$$

**Proof** Define

$$G'_k(a) = \frac{d}{dq}G_k(x; q|a)\big|_{q=1} \text{ and } G'_k(x) = \frac{d}{dq}G_k(x; q)\big|_{q=1}.$$

Theorem 4.72 gives

$$G'_k(a) = xG'_k + x^2(k-a)\sum_{j=1}^{a-1} G_k(x; 1; j) + x^2\sum_{j=1}^{a-1} G_{k-1}(x; 1; j),$$

for all $a = 1, 2, \ldots, k-1$. By the facts that $G_k(x; 1|a) = xG_k(x; 1)$ and $G_k(x; 1) = \frac{x^k}{(1-x)\cdots(1-kx)}$, we have

$$G'_k(a) = xG'_k + x^3(k-a)(a-1)G_k(x; 1) + x^3(a-1)G_{k-1}(x; 1)$$

$$= xG'_k + \frac{x^{k+2}(a-1)(1-ax)}{(1-x)\cdots(1-kx)}.$$

By summing over all $a = 1, 2, \ldots, k-1$, we obtain

$$G'_k = kxG'_k + xG'_{k-1} + \frac{x^{k+2}}{(1-x)\cdots(1-kx)}\binom{k-1}{2}\frac{3-2kx}{3},$$

which is equivalent to

$$G'_k = \frac{x}{1-kx}G'_{k-1} + \frac{x^{k+2}}{(1-x)\cdots(1-kx)^2}\binom{k-1}{2}\frac{3-2kx}{3}.$$

Iterating the above recurrence relation using the initial condition $G'_1(x) = 0$, we obtain the requested result. □

By collecting the coefficients of the generating function in Corollary 4.73, we obtain the following explicit formula for the total number of occurrences of 132.

**Corollary 4.74** *The total number of occurrences of the subword pattern 132 in all of the set partitions of $\mathcal{P}_{n,k}$ is given by*

$$\binom{k}{3}\text{Stir}(n-2,k) + \sum_{j=3}^{k}\binom{j}{3}\left(\sum_{i=3}^{n-k}j^{i-3}\text{Stir}(n-i,k)\right).$$

Similar arguments also apply to the pattern 231.

**Theorem 4.75** *Let $G_k(a) = P_{231}(x;q;k|a)$ be the generating function for the number of set partitions $\pi = \pi_1 \cdots \pi_{n-1}a \in \mathcal{P}_{n,k}$ according to the number occurrences of the subword pattern 231, and let $G_k = \sum_{a=1}^{k} G_k(a) = P_{231}(x;q;k)$. Then the generating function $G_k(a)$ satisfies the recurrence relation*

$$G_k(a) = xG_k + x^2(q-1)\sum_{j=a+1}^{k}(k-j)G_k(j) + x^2(q-1)\sum_{j=a+1}^{k-1}G_{k-1}(j)$$

*with the initial conditions $G_k(k) = xG_k + xG_{k-1}$, $G_k(k-1) = xG_k$, $G_1 = G_1(1) = \frac{x}{1-x}$, $G_2(1) = \frac{x^3}{(1-x)(1-2x)}$, $G_2(2) = \frac{x^2}{1-2x}$ and $G_2 = \frac{x^2}{(1-x)(1-2x)}$.*

**Corollary 4.76** *The generating function for the total number of occurrences of the pattern 231 in all the set partitions of $\mathcal{P}_{n,k}$ is given by*

$$\frac{x^{k+1}}{(1-x)\cdots(1-kx)}\sum_{j=3}^{k}\frac{(j-2)(4j(j-1)x^2 + 3(1-3j)x + 6)}{6(1-jx)}.$$

**Proof** Define

$$G'_k(a) = \frac{d}{dq}G_k(x;q|a)\mid_{q=1} \text{ and } G'_k(x) = \frac{d}{dq}G_k(x;q)\mid_{q=1}.$$

Theorem 4.75 gives

$$G'_k(a) = xG'_k + x^2\sum_{j=a+1}^{k}(k-j)G_k(x;1;j) + x^2\sum_{j=a+1}^{k-1}G_{k-1}(x;1;j),$$

for all $a = 1, 2, \ldots, k-1$. By the facts that $G_k(x,1|a) = xG_k(x,1)$ and $G_k(x,1) = \frac{x^k}{(1-x)\cdots(1-kx)}$, we have

$$G'_k(a) = xG'_k + \frac{\binom{k-a}{2}x^{k+3}}{(1-x)\cdots(1-kx)}$$
$$+ \frac{(k-1-a)x^{k+2}}{(1-x)\cdots(1-(k-1)x)} + \frac{x^{k+1}}{(1-x)\cdots(1-(k-2)x)}.$$

By summing over all $a = 1, 2, \ldots, k-2$ and then by using the initial conditions $G'_k(k) = xG'_k + xG'_{k-1}$ and $G'_k(k-1) = xG'_k$, we have

$$G'_k = \frac{x}{1-kx}G'_{k-1}$$
$$+ \frac{x^{k+1}}{(1-x)\cdots(1-kx)}\frac{(k-2)(4k(k-1)x^2 + 3(1-3k)x + 6)}{6(1-kx)}.$$

Thus, iterating the above recurrence relation with using the initial condition $G'_2(x) = 0$, we obtain the requested result. $\square$

Equivalently, we also have the following explicit formula.

**Corollary 4.77** *The total number of occurrences of the subword pattern 231 in all of the set partitions of $[n]$ with exactly $k$ blocks is given by*

$$2\binom{k}{3}\text{Stir}(n-2,k) + \binom{k-1}{2}\text{Stir}(n-2,k-1)$$
$$+ \sum_{j=3}^{k}\binom{j}{3}\left(\sum_{i=3}^{n-k} j^{i-3}\text{Stir}(n-i,k)\right).$$

The combinatorial proof for Corollaries 4.71, 4.74, and 4.77 can be obtained in a similar way as discussed in the pervious sections.

---

## 4.6 Exercises

**Exercise 4.1** *Prove Lemma 4.8.*

**Exercise 4.2** *Show that $\binom{n-1}{m}\text{Stir}(n-1-m, k-1)$ is a solution for the recurrence relation*

$$a_n(k,m) = a_{n-1}(k-1,m) + (k-1)a_{n-1}(k,m) + a_{n-1}(k,m-1),$$

*for $n \geq k > m > 0$, and with the starting value $a_n(k,0) = \text{Stir}(n-1, k-1)$.*

**Exercise 4.3** *Show that the generating function $W_k(x,0)$ for the number of words of size $n$ over the alphabet $[k]$ without peaks (valleys) is given by*

$$W_k(x,0) = \frac{U_{k-1}(t) - U_{k-2}(t)}{(1-x)U_{k-1}(t) - U_{k-2}(t)},$$

*where $t = 1 + \frac{x^2}{2}$ and $U_m$ is the $m$-th Chebyshev polynomial of the second kind.*

**Exercise 4.4** *Give a direct proof for Corollary 4.26.*

**Exercise 4.5** *Prove Lemma 4.43.*

**Exercise 4.6** *Give a direct proof for Corollary 4.30.*

**Exercise 4.7** *Show that the generating function of words of size $n$ over the alphabet $[k]$ according to the statistic $occ_{122\cdots 2}$ is given by*

$$\frac{1}{1 - x \sum_{i=0}^{k-1}(1 - x^{\ell-1}(1-q))^i}.$$

**Exercise 4.8** *Prove Theorem 4.66.*

**Exercise 4.9** *Use Theorem 4.66 to derive the result of Corollary 4.37. Moreover, give a combinatorial proof for the result.*

**Exercise 4.10** *Show that the generating function for the number of words $\pi$ of size $n$ over the alphabet $[k]$ according to the number of occurrences of $2^{\ell-1}1$ in $\pi$ is given by*

$$\frac{x^{\ell-2}(q-1)}{1 - x^{\ell-2}(1-q) - (1 - x^{\ell-1}(1-q))^k}.$$

**Exercise 4.11** *Given integer $t > 1$, a $t$-succession is defined as the $t$ numbers $a, a+1, \ldots, a+t-1$, where $a > 0$. A set partition $\pi$ of $[n]$ is said to contain a $t$-succession if a block of $\pi$ does. Denote the set of set partitions of $[n]$ with exactly $k$ blocks and $r$ occurrences of $t$-successions by $C_t(n,k,r)$ and let $c_t(n,k,r) = |C_t(n,k,r)|$. Show that*
*(1) $c_2(n,k,r) = \binom{n-1}{r}\mathrm{Stir}(n-r-1, k-1)$.*
*(2) $c_2(n,r) = \sum_k c_2(n,k,r) = \binom{n-1}{r}\mathrm{Bell}(n-r-1)$.*

## 4.7 Research Directions and Open Problems

We now suggest several research directions, which are motivated both by the results and exercises of this and earlier chapter(s).

**Research Direction 4.1** *Theorem 4.40 suggests a formula for the generating function $P_{12\cdots\ell}(x; q; k)$. But this formula it is very complicated and not easy to deal with it even for $\ell = 3$ (see Theorem 4.41 for $\ell = 2$ and [220] for $\ell = 3$). Thus, an open question is to find an explicit formula for the generating function for the number of set partitions of $[n]$ (with exactly $k$ blocks) according to the number of occurrences of the subword pattern $12\cdots\ell$.*

160                    *Combinatorics of Set Partitions*

**Research Direction 4.2** *Theorems 4.69, 4.72, and 4.75 suggest the following question. Find an explicit formula for the generating function $P_\tau(x; q; k)$ when the subword pattern $\tau$ is either 121, 132, or 231 (for exact reference, see Mansour, Shattuck, and Yan [242].).*

**Research Direction 4.3** *The problem of counting subword patterns in set partitions has been extended to study of counting string patterns in set partitions. We say that the set partition $\pi$ of $[n[$ contains the string pattern $\tau$ if $\pi$ can be written as $\pi = \pi'\tau\pi''$. For instance, the problem of finding the number of set partitions of $[n]$ with exactly $r$ occurrences of the string pattern $\tau = 11\cdots 1 \in \mathcal{P}_t$ is equivalent to counting the number of t-successions in set partitions of $[n]$, see [258] (Exercise 4.11). Thus, it is naturally to ask the following: Fix a string pattern $\tau$, then find an explicit formula for the number of set partitions of $[n]$ with exactly $r$ occurrences of the string pattern $\tau$? For instance, the string pattern $12\cdots t$, or more generally the string pattern $11\cdots 122\cdots 2\cdots tt\cdots t$.*

**Research Direction 4.4** *Two subword pattern $\tau$ and $\nu$ are strongly tight Wilf-equivalence (respectively, tight Wilf-equivalence) if the number of set partitions of $\mathcal{P}_n$ that contain $\tau$ exactly $r$ times (respectively, does contain $\tau$) is the same as the number of set partitions of $\mathcal{P}_n$ that contain $\nu$ exactly $r$ times (respectively, does contain $\nu$), for all $n, r$ (respectively, for all $n$). We denote two patterns that are strongly tight Wilf-equivalence (respectively, tight Wilf-equivalence) by $\tau \stackrel{st}{\sim} \nu$ (respectively, by $\tau \stackrel{t}{\sim} \nu$). Table 4.1 shows that there are 3 tight Wilf-equivalence 2-letter subword patterns, namely, 11, 12, and 21. Section 4.5 gives (check the details!) that there are 11 tight Wilf-equivalence 3-letter subword patterns, namely, 111, 112 $\stackrel{t}{\sim}$ 122, 121, 211 $\stackrel{t}{\sim}$ 221, 212, 123, 132, 213, 231, 312, and 321.*

*(1) Classification of four letter subword patterns according to tight Wilf-equivalence: Table 4.2 contains the values of the sequences $\{\mathcal{P}_n(\tau)\}_{n=1}^{12}$ (obtained via an appropriate modification of the program given in Section H) for the four letter patterns $\tau$, which suggests that the tight Wilf-equivalence classes are given by*

$$1111, 1112 \stackrel{t}{\sim} 1222, 1121 \stackrel{t}{\sim} 1211 \stackrel{t}{\sim} 1221, 1122,$$

$$1123 \stackrel{t}{\sim} 1233, 1212, 1213 \stackrel{t}{\sim} 1223, 1231, 1232, 1234.$$

*Prove that these are indeed the tight Wilf-equivalence classes.*

**Table 4.2**: $\{\mathcal{P}_n(\tau)\}_{n=1}^{12}$ for four letter subword patterns $\tau$

| $\tau$ | $\{\mathcal{P}_n(\tau)\}_{n=1}^{12}$ for four letter subword patterns $\tau$ | | | | | |
|---|---|---|---|---|---|---|
| 1111 | 1 | 2 | 5 | 14 | 49 | 192 |
| | 832 | 3941 | 20197 | 111105 | 651899 | 4058287 |
| 1112, 1222 | 1 | 2 | 5 | 14 | 47 | 180 |
| | 770 | 3617 | 18434 | 101025 | 591230 | 3674212 |
| 1121, 1211 | 1 | 2 | 5 | 14 | 47 | 180 |
| 1221 | 771 | 3622 | 18458 | 101140 | 591820 | 3677432 |
| 1122 | 1 | 2 | 5 | 14 | 47 | 181 |
| | 776 | 3648 | 18596 | 101903 | 596221 | 3704076 |
| 1123, 1233 | 1 | 2 | 5 | 14 | 46 | 171 |
| | 707 | 3211 | 15851 | 84334 | 480244 | 2910237 |
| 1212 | 1 | 2 | 5 | 14 | 47 | 181 |
| | 775 | 3639 | 18535 | 101516 | 593769 | 3688179 |
| 1213, 1223 | 1 | 2 | 5 | 14 | 46 | 172 |
| | 714 | 3253 | 16102 | 85855 | 489742 | 2971786 |
| 1231 | 1 | 2 | 5 | 14 | 46 | 171 |
| | 710 | 3231 | 15972 | 85040 | 484529 | 2937229 |
| 1232 | 1 | 2 | 5 | 14 | 46 | 171 |
| | 708 | 3218 | 15894 | 84594 | 481844 | 2920406 |
| 1234 | 1 | 2 | 5 | 14 | 46 | 168 |
| | 672 | 2923 | 13676 | 68400 | 363730 | 2046611 |

(2) *Classification of five letter subword patterns according to tight Wilf-equivalence*: Table 4.3 contains the values of the sequences $\{\mathcal{P}_n(\tau)\}_{n=1}^{12}$ (obtained via an appropriate modification of the program given in Section H) for the five letter patterns $\tau$.

**Table 4.3**: $\{\mathcal{P}_n(\tau)\}_{n=1}^{12}$ for five letter subword patterns $\tau$

| $\tau$ | $\{\mathcal{P}_n(\tau)\}_{n=1}^{12}$ for five letter subword patterns $\tau$ | | | | | |
|---|---|---|---|---|---|---|
| 11111 | 1 | 2 | 5 | 15 | 51 | 200 |
| | 866 | 4095 | 20947 | 115018 | 673657 | 4186667 |
| 11112, 11212 | 1 | 2 | 5 | 15 | 51 | 198 |
| 12122, 12222 | 854 | 4032 | 20615 | 113198 | 663191 | 4123401 |
| 11121, 11221 | 1 | 2 | 5 | 15 | 51 | 198 |
| 12111, 12211 | 854 | 4032 | 20616 | 113203 | 663215 | 4123520 |
| 12221 | | | | | | |
| 11122, 11222 | 1 | 2 | 5 | 15 | 51 | 198 |
| | 854 | 4033 | 20621 | 113230 | 663361 | 4124337 |
| 11123, 11213 | 1 | 2 | 5 | 15 | 51 | 197 |
| 12133, 12333 | 845 | 3969 | 20203 | 110533 | 645676 | 4005002 |
| | | | | | *Continued on next page* | |

| $\tau$ | $\{\mathcal{P}_n(\tau)\}_{n=1}^{12}$ for five letter subword patterns $\tau$ | | | | | |
|---|---|---|---|---|---|---|
| 11211 | 1 | 2 | 5 | 15 | 51 | 198 |
|  | 854 | 4034 | 20627 | 113259 | 663500 | 4125047 |
| 11223, 12233 | 1 | 2 | 5 | 15 | 51 | 197 |
|  | 846 | 3976 | 20245 | 110786 | 647236 | 4014933 |
| 11231, 12231 | 1 | 2 | 5 | 15 | 51 | 197 |
| 12311, 12321 | 845 | 3969 | 20206 | 110553 | 645797 | 4005736 |
| 12331 | | | | | | |
| 11232, 12132 | 1 | 2 | 5 | 15 | 51 | 197 |
| 12232, 12322 | 845 | 3969 | 20204 | 110540 | 645719 | 4005265 |
| 12332 | | | | | | |
| 11233 | 1 | 2 | 5 | 15 | 51 | 197 |
|  | 845 | 3970 | 20211 | 110589 | 646060 | 4007670 |
| 11234, 12344 | 1 | 2 | 5 | 15 | 51 | 196 |
|  | 834 | 3879 | 19531 | 105653 | 610176 | 3742518 |
| 12112, 12212 | 1 | 2 | 5 | 15 | 51 | 198 |
|  | 854 | 4033 | 20620 | 113221 | 663298 | 4123927 |
| 12113, 12123 | 1 | 2 | 5 | 15 | 51 | 197 |
| 12213, 12223 | 845 | 3970 | 20210 | 110576 | 645936 | 4006609 |
| 12313, 12323 | | | | | | |
| 12121 | 1 | 2 | 5 | 15 | 51 | 198 |
|  | 855 | 4037 | 20638 | 113306 | 663722 | 4126174 |
| 12131 | 1 | 2 | 5 | 15 | 51 | 197 |
|  | 846 | 3976 | 20248 | 110806 | 647348 | 4015599 |
| 12134, 12234 | 1 | 2 | 5 | 15 | 51 | 196 |
| 12314, 12324 | 834 | 3880 | 19540 | 105724 | 610724 | 3746784 |
| 12334 | | | | | | |
| 12312 | 1 | 2 | 5 | 15 | 51 | 197 |
|  | 845 | 3971 | 20217 | 110615 | 646147 | 4007798 |
| 12341 | 1 | 2 | 5 | 15 | 51 | 196 |
|  | 834 | 3879 | 19538 | 105711 | 610607 | 3745688 |
| 12342 | 1 | 2 | 5 | 15 | 51 | 196 |
|  | 834 | 3879 | 19534 | 105679 | 610374 | 3744005 |
| 12343 | 1 | 2 | 5 | 15 | 51 | 196 |
|  | 834 | 3879 | 19532 | 105662 | 610247 | 3743067 |
| 12345 | 1 | 2 | 5 | 15 | 51 | 196 |
|  | 830 | 3826 | 19016 | 101181 | 572856 | 3434384 |

**Research Direction 4.5** *Theorem 4.10 presents the generating function for the number of set partitions of $\mathcal{P}_{n,k}$ according to the number rises, levels, and descents. Thus, its naturally to ask on generating function for the number of set partitions of $\mathcal{P}_{n,k}$ according to occurrences of two patterns or more. In particular, we can formulate the following research direction. Let $P_{\tau,\rho}(x)$ be the generating function for the number of set partitions of $\mathcal{P}_n$ that avoid both* the subwords pattern $\tau, \rho$. Find explicit formula for the generating function $P_{\tau,\rho}(x)$, where $\tau$ and $\rho$ any two subword patterns of size three.

**Research Direction 4.6** *Following the study of subword patterns on set partitions, the results of this chapter, and Exercises 2.12 and 2.13, we can define general type of subword patterns as follows. We say a sequence $a = a_1 a_2 \cdots a_n$ contains the $\phi$-subword pattern $\tau = \tau_1 \tau_2 \cdots \tau_k$ if the sequence $a$ can be written as $a' b_1 b_2 \cdots b_k a''$ such that the reduce form of $b_1 b_2 \cdots b_k$ equals $\tau$ and $b_j - b_i > \phi_j - \phi_i$ for all $k \geq j > i \geq 1$. Otherwise, we say that $a$ does avoids the $\phi$-subword pattern $\tau$. For instance, the set partition 1213 avoids 14-subword pattern 12, that is, there are no two adjacent letters with the right ones greater than the left ones by at least 3. The same set partition contain 13-subword pattern 12 (see third and fourth letters in the set partition). Indeed Exercise 2.13 says that the number of set partition of $[n]$ that avoid 12-subword pattern 12 is given by the n-th Catalan number. Our suggestion is to study the number of set partitions of $[n]$ that avoid a fixed $\phi$-subword pattern $\tau$.*

**Research Direction 4.7** *Find explicit formula for the generating function for the number of set partitions of $\mathcal{P}_{n,k}$ according to $\mathrm{nocc}_\tau$ the number nontrivial occurrences $\tau$, where $\tau$ is any subword pattern of size at most three or one of the general patterns that considered in this chapter.*

# Chapter 5

# Nonsubword Statistics on Set Partitions

To study statistics on a combinatorial structure it is very useful to obtain not just the enumeration of the number objects in the combinatorial structure, but to obtain results on the number of the objects in the structure that satisfy a ceratin set of conditions. For example, in Chapter 4 we considered subword-statistics on set partitions. More precisely, we studied the generating function for the number of set partitions of $\mathcal{P}_{n,k}$ according to the number occurrences of a fixed subword pattern $\tau$. In the current chapter, we will focus on other statistics, namely "nonsubword statistics", rather than subword statistics, where we will investigate various types of patterns that are not easy to express as subword patterns. These statistic motivated by the study of various types of statistics on the set of permutations, words and compositions, see the books of Bóna [48] and Heubach and Mansour [138].

As we discussed in Chapter 1, many authors have studied statistics on set partitions. Apart from the paper by Milne [252], who seems to have pioneered the study of set partitions statistics, distribution of which is given by the $q$-Stirling numbers, we mention Table 1.2. More precisely, he used Rota's [297] results to obtain a $q$-analog of Dobiński's formula (Theorem 3.2) for the number of set partitions of $[n]$. In the current chapter, starting from Milne's results, we will present several nonsubword statistics on set partitions. To define our statistics we need to fix a representation for our set partitions. So the question is whether the definition of a given statistic on set partitions depends on the representation that used for set partitions. Sometimes, the definition of a statistic in a representation of set partition can be simple/complicated compared to another representation. For instance, let $\pi$ be any canonical representation of a set partition: we define $f_i(\pi)$ to be the difference between the position of the rightmost occurrence of the letter $i$ and the leftmost occurrence of the letter $i$ in $\pi$. Note that if the letter $i$ does not occur in $\pi$, we define $f_i(\pi)$ to be 0. Then, we define the total statistic on $\pi$ to be $f(\pi) = \sum_{i\geq 1} f_i(\pi)$. As an example, if $\pi = 121341233212 \in \mathcal{P}_{12,4}$, then $f_1(\pi) = 11 - 1 = 10$, $f_2(\pi) = 12 - 2 = 10$, $f_3(\pi) = 9 - 4 = 5$ and $f_4(\pi) = 5 - 5 = 0$, which gives $f(\pi) = 10 + 10 + 5 + 0 = 25$. This statistic can be defined in different way if we use the block representation, namely $f(\pi) = \sum_{i\geq 1} \max B_i - \min B_i$ where $B_i$'s are the blocks of the set partition $\pi$. As an example, the block representation of the set partition

$\pi = 121341233212 \in \mathcal{P}_{12,4}$ is $\{1,3,6,11\}/\{2,7,9,12\}/\{4,8,9\}/\{5\}$, which leads to $f(\pi) = 11 - 1 + 12 - 2 + 9 - 4 + 5 - 5 = 25$. Hence, it is natural to divide our chapter into small sections, where each of these sections deal with a group of statistics which is defined by the same representation.

Our first section deals with the block representation of set partitions. Here, we will discuss several nonsubword statistics on set partitions, where each of them is defined on the block representations. For instance, in 2004, Wagner [350] studied the following three statistics on set partitions $\pi = B_1/B_2 \cdots /B_k$:

$$\sum_{i=1}^{k} i|B_i|, \qquad \sum_{i=1}^{k}(i-1)|B_i|, \qquad \sum_{i=1}^{k}(i-1)(|B_i|-1).$$

Then we use the rook representation of set partitions to study another set of statistics. More precisely, in Section 5.2 we define several statistics on the canonical and rook representations of set partitions. These statistics are based on the work of Wachs and White [349] in 1991. They presented several bijections between set partitions and rook placements on *stairstep Ferrers boards*, where they recovered the $q$-Stirling number of the second kind as given by Gould [118] (also, see [65]). Moreover, Wachs and White succeeded in extending the $q$-Stirling number of the second kind to the $p,q$-Stirling number of the second kind by studying distribution of pairs of statistics on set partitions.

In 1962, Rényi [288] studied the number of records (see Definition 5.14) in the set of permutations, see also [194, 116]. Only in 2008, Myers and Wilf [259] extended the study of the number of records to multiset permutations and words. In Section 5.3, based on the paper [190], we generalize the study of the number of records to set partitions. More precisely, we focus on the number of additional records (see Definition 5.16) and the statistic of the sum of the positions of these records. In both cases, we study the generating function for the number of set partitions of $[n]$ according to the number of blocks and number of (sum positions of the) *additional records*, see [190].

In Section 5.4, based on the paper [241], we will focus on the following statistics: the total number of positions between adjacent occurrences of a letter, and the total number of positions of the same letter lying between two letters which are strictly larger. More precisely, by using the graphical representation, see Definition 3.49, we will study the ordinary generating functions corresponding to the aforementioned statistics on $\mathcal{P}_{n,k}$. Among the results there are explicit formulas for the total value of statistics on $\mathcal{P}_{n,k}$ and $\mathcal{P}_n$, for which we provide both algebraic and combinatorial proofs.

In Section 5.6, we will introduce new type of patterns, called *generalized patterns*, see [237]. The name "generalized patterns" motivates from the research of generalized pattern on set of permutations, words, and compositions as discussed in the books of Bóna [48] and Heubach and Mansour [138]. The study of generalized patterns in set partition was initiated by work of Goyt [120], where he studied the number of set partitions that avoid a set of generalized patterns and characterized several known statistics (such as *inversion number* and *major index*) in terms of the generalized patterns.

In Section 5.7, we will present the analog of the *major index* statistic on set permutations and words for the case of set partitions. The main goal of this section is to present results of Chen, Gessel, Yan, and Yang [71], where they introduced and studied a new statistic, called the *p*-major index, on the set partitions of $\mathcal{P}_n$. In particular, they showed that the two statistics two-crossing number and *p*-major index have the same distribution over $\mathcal{P}_n$.

## 5.1 Statistics and Block Representation

In this section we will focus on several statistics that are defined on the block representation of set partitions. Actually, a statistic on the block representation of set partition can be described as a function of the cardinalities of the blocks $B_1, B_2, \ldots, B_k$ where the elements of the set $[n]$ are regarded as labels of $n$ unit masses, where the masses with labels in block $B_i$ are placed at $x = i$.

**Definition 5.1** *Let $\pi = B_1/B_2/\cdots/B_k$ be any block representation of a set partition in $\mathcal{P}_{n,k}$. We define the following statistics*

$$w^*(\pi) = \sum_{i=1}^{k} i|B_i|, \quad w(\pi) = \sum_{i=1}^{k}(i-1)|B_i|, \quad \tilde{w}(\pi) = \sum_{i=1}^{k}(i-1)(|B_i|-1).$$

*Clearly, $w(\pi) = w^*(\pi) - n$ and $\tilde{w}(\pi) = w^*(\pi) - n - \binom{k}{2}$. In other words, if the elements of the set $[n]$ are regarded as labels of $n$ unit masses, then $w^*(\pi)$ is the moment about $x = 0$ of the mass configuration. The statistics $w(\pi)$ and $\tilde{w}(\pi)$ admit of similar interpretations. Note that, in Definition 5.74, the statistic $w^*(\pi)$ will be called* dual major index.

Moreover, we associate the following three polynomials to our three statistics:

$$S_q^*(n,k) = \sum_{\pi \in \mathcal{P}_{n,k}} q^{w^*(\pi)}, \qquad S_q(n,k) = \sum_{\pi \in \mathcal{P}_{n,k}} q^{w(\pi)},$$

and

$$\tilde{S}_q(n,k) = \sum_{\pi \in \mathcal{P}_{n,k}} q^{\tilde{w}(\pi)}.$$

For example, since there is only one set partition on [1], namely $\pi = \{1\}$, we have $S_q^*(1,1) = q$, $S_q(1,1) = 1$ and $\tilde{S}_q(1,1) = 1$. Our next result presents explicit formulas for the generating functions for the sequences $\{S_q^*(n,k)\}_{n\geq 0}$, $\{S_q(n,k)\}_{n\geq 0}$ and $\{\tilde{S}_q(n,k)\}_{n\geq 0}$. The statistic $w$ introduced by Milne [252] to answer a question of Garsia. Also, the same statistic has been defined by using the rook placement presentation as we will see. Indeed, the statistics as given in the above definitions are the most common $q$-analog of Stirling numbers of the second kind in the literature as they defined in [65, 118]. More recent, the same statistics has been reconsidered by Wagner [350].

**Theorem 5.2** (Wagner [350]) *We have*

$$\sum_{n\geq 0} \tilde{S}_q(n,k) x^n = \frac{x^k}{(1-[1]_q x)(1-[2]_q x)\cdots(1-[k]_q x)}$$

$$\sum_{n\geq 0} S_q^*(n,k) x^n = \frac{q^{\binom{k+1}{2}} x^k}{(1-q[1]_q x)(1-q[2]_q x)\cdots(1-q[k]_q x)}.$$

**Proof** Let $a_{n,k,j}$ be the number of set partitions $\pi = B_1/B_2/\cdots/B_k$ of $\mathcal{P}_{n,k}$ such that $w^*(\pi) = j$. The element $n$ of the set $[n]$ satisfies either

- $B_k = \{n\}$; there are clearly $a_{n-1,k-1,j-k}$ such set partitions $\pi$,

- $n \in B_i$ such that $|B_i| > 1$, where $1 \leq i \leq k$; there are clearly $a_{n-1,k,j-i}$ set partitions.

Combining these two cases, we obtain

$$S_q^*(n,k) = \sum_{j\geq 0} a_{n,k,j} q^j = \sum_{j\geq 0} a_{n-1,k-1,j} q^{j+k} + \sum_{i=1}^{k} q^i \sum_{j\geq 0} a_{n-1,k,j} q^j$$
$$= q^k S_q^*(n-1,k-1) + q[k]_q S_q^*(n-1,k).$$

By the definitions we have that $S_q^*(0,0) = 1$ and $S_q^*(n,0) = S_q^*(0,k) = 0$ for all $n, k \geq 1$. By using similar arguments as above, we can derive

$$S_q(n,k) = q^{k-1} S_q(n-1,k-1) + [k]_q S_q(n-1,k), \qquad (5.1)$$
$$\tilde{S}_q(n,k) = \tilde{S}_q(n-1,k-1) + [k]_q \tilde{S}_q(n-1,k). \qquad (5.2)$$

Theorem 3.20 gives

$$\tilde{S}_q(n,k) = \sum_{d_1+\cdots+d_k=n-k,\, d_i\geq 0} \prod_{j=1}^{k} [j]_q^{d_j},$$

$$\sum_{n\geq 0} \tilde{S}_q(n,k) x^n = \frac{x^k}{(1-[1]_q x)(1-[2]_q x)\cdots(1-[k]_q x)}.$$

By using the relations $S_q(n,k) = q^{\binom{k}{2}} \tilde{S}_q(n,k)$ and $S_q^*(n,k) = q^n S_q(n,k)$ which hold immediately from Definition 5.1, we derive

$$\sum_{n\geq 0} S_q^*(n,k) x^n = \frac{q^{\binom{k+1}{2}} x^k}{(1-q[1]_q x)(1-q[2]_q x)\cdots(1-q[k]_q x)},$$

which completes the proof. □

Note that the $q$-Stirling numbers $\tilde{S}_q(n,k)$ has been found by Johnson [151], which are different from $q$-Stirling numbers that are given by Milne as described in Theorem 3.15. Indeed, the work of Johnson was motivated by the results of Gessel [113] on studying the *celebrated exponential formula*.

## 5.2 Statistics and Canonical and Rook Representations

In this section we will focus on several statistics that are defined via canonical representation and rook representation of set partitions. In order to state our interested statistics we need to fix the leftmost and rightmost occurrences of each letter in a word.

**Definition 5.3** *Let $\pi = \pi_1 \pi_2 \cdots \pi_n$ be any word in $[k]^n$. A letter $\pi_i$ is said to be* leftmost *(respectively,* rightmost*) letter of $\pi$ is $\pi_j \neq \pi_i$ for all $j = 1, 2, \ldots, i-1$ (respectively, $j = i+1, i+2, \ldots, n$). We denote the set of leftmost and rightmost of letters of $\pi$ by*

$$L(\pi) = \{i \in [n] \mid \pi_i \text{ is the leftmost letter of } \pi\},$$
$$R(\pi) = \{i \in [n] \mid \pi_i \text{ is the rightmost letter of } \pi\},$$

*respectively.*

**Example 5.4** *If $\pi = 121431234 \in [4]^9$, then $L(\pi) = \{1, 2, 4, 5\}$ and $R(\pi) = \{6, 7, 8, 9\}$.*

Now we are ready to define our statistics.

**Definition 5.5** *Let $\pi = \pi_1 \pi_2 \cdots \pi_n$ be any word in $[k]^n$ and let $i = 1, 2, \ldots, n$. We define*

$$\mathrm{lbig}_i(\pi) = |\{j \in L(\pi) \mid j < i \text{ and } \pi_j > \pi_i\}|,$$
$$\mathrm{lsmall}_i(\pi) = |\{j \in L(\pi) \mid j < i \text{ and } \pi_j < \pi_i\}|,$$
$$\mathrm{rbig}_i(\pi) = |\{j \in R(\pi) \mid j > i \text{ and } \pi_j > \pi_i\}|,$$
$$\mathrm{rsmall}_i(\pi) = |\{j \in R(\pi) \mid j > i \text{ and } \pi_j < \pi_i\}|.$$

*The total of these statistics over all $i$ are given by*

$$\mathrm{lbig}(\pi) = \sum_{i=1}^{n} \mathrm{lbig}_i(\pi), \qquad \mathrm{lsmall}(\pi) = \sum_{i=1}^{n} \mathrm{lsmall}_i(\pi),$$
$$\mathrm{rbig}(\pi) = \sum_{i=1}^{n} \mathrm{rbig}_i(\pi), \qquad \mathrm{rsmall}(\pi) = \sum_{i=1}^{n} \mathrm{rsmall}_i(\pi).$$

**Example 5.6** *Continuing the pervious example, we have that* $\mathrm{lbig}_1(\pi) = 0$, $\mathrm{lbig}_2(\pi) = 0$, $\mathrm{lbig}_3(\pi) = 1$, $\mathrm{lbig}_4(\pi) = 0$, $\mathrm{lbig}_5(\pi) = 1$, $\mathrm{lbig}_6(\pi) = 3$, $\mathrm{lbig}_7(\pi) = 2$, $\mathrm{lbig}_8(\pi) = 1$, *and* $\mathrm{lbig}_9(\pi) = 0$, *which implies that* $\mathrm{lbig}(\pi) = 8$. *Similarly, it can be shown that* $\mathrm{lsmall}(\pi) = 11$, $\mathrm{rbig}(\pi) = 15$, *and* $\mathrm{rsmall}(\pi) = 6$.

Note that the statistics lsmall has a very simple interpretation:

$$\mathrm{lsmall}(\pi) = \sum_{i=1}^{n}(\pi_i - 1),$$

for any set partition $\pi = \pi_1\pi_2\cdots\pi_n \in \mathcal{P}_n$. In order to study the statistics lbig, lsmall, rbig and rsmall on set partitions, we use the terminology of the rook placement of set partitions as described in Section 3.2.4.4. We start by defining several statistics on rook placements.

**Definition 5.7** *For any n-th triangular board, we denote the set of all cells in column i by* $\mathrm{Col}_i$ *and the set of all cells in row j by* $\mathrm{Row}_j$, *where the columns numbered by increasing order from right to left, and the rows numbered by increasing order from top to bottom. Let r be any rook in* $\mathrm{R}_{n,k}$, *we define*

$$\mathrm{EC}(r) = \{i \mid \mathrm{Col}_i \text{ of } r \text{ has no rook}\}$$

*and*

$$\mathrm{ER}(r) = \{j \mid \mathrm{Row}_j \text{ of } r \text{ has no rook}\}.$$

*We define inversion vectors on* $r \in \mathrm{R}_{n,k}$:

1. *Delete all cells to the left of and below each rook ("south" and "west" of each rook) from the board, including the cell containing the rook, and delete all columns with no rooks. Then the* column-southwest inversion vector *is defined by*

$$\mathrm{csw}_i(r) = |\{a \in \mathrm{Col}_i \mid a \text{ has not been deleted}\}|.$$

2. *Delete all cells to the left and above each rook ("north" and "west" of each rook), including the cell containing the rook (but not empty columns). Then the column* northwest inversion vector *is defined by*

$$\mathrm{cnw}_i(r) = |\{a \in Cal_i \mid a \text{ has not been deleted}\}|.$$

3. *Delete all cells to the left and above each rook, including the cell containing the rook, and all rows with no rooks. Then the* row-northwest inversion vector *is defined by*

$$\mathrm{rnw}_i(r) = |\{a \in \mathrm{Row}_i \mid a \text{ has not been deleted}\}|.$$

*Let r be any rook, by the inversion vectors we define the following three statistics:*

$$\mathrm{csw}(r) = \sum_{i=1}^{n} \mathrm{csw}_i(r), \ \mathrm{cnw}(r) = \sum_{i=1}^{n} \mathrm{cnw}_i(r), \ \mathrm{rnw}(r) = \sum_{i=1}^{n} \mathrm{rnw}_i(r).$$

**Example 5.8** *Let $r \in R_{9,5}$ as described in Figure 5.1. Then the column-*

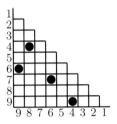

**FIGURE 5.1**: A rook placement

*southwest inversion, northwest inversion, and row-northwest inversion are given by Figure 5.2.*

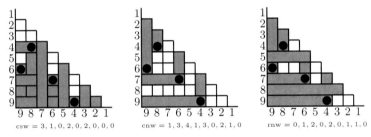

**FIGURE 5.2**: The vectors $\mathrm{csw}(r)$, $\mathrm{cnw}(r)$, and $\mathrm{rnw}(r)$

*Thus,* $\mathrm{csw}(r) = 8$, $\mathrm{cnw}(r) = 15$ *and* $\mathrm{rnw}(r) = 7$.

Note that the statistic csw was discovered by Garsia and Remmel [111]. Also, other "inversion numbers" are possible, including another described in [111], but all of them are trivially equivalent to these three statistics as described in Definition 5.7.

Firstly, we study the generating functions for the joint distribution for certain pair of the above statistics. This leads to a strong relation between our statistics lbig and lsmall and our new statistics in rook placements in Definition 5.7.

**Theorem 5.9** (Wachs and White [349, Corollary 4.6]) *For all $1 \leq k \leq n$,*

$$\sum_{\pi \in \mathcal{P}_{n,k}} q^{\mathrm{lbig}(\pi)} p^{\mathrm{lsmall}(\pi)} = \mathrm{Stir}_{p,q}(n,k) = \sum_{r \in R_{n,k}} q^{\mathrm{csw}(r)} p^{\mathrm{cnw}(r)}, \quad (5.3)$$

172                Combinatorics of Set Partitions

where $\mathrm{Stir}_{p,q}$ is the $p,q$-Stirling number of the second kind, see Appendix C.5.

**Proof** Firstly, we establish a bijection $f$ showing the following identity

$$\sum_{\pi \in \mathcal{P}_{n,k}} q^{\mathrm{lbig}(\pi)} p^{\mathrm{lsmall}(\pi)} = \sum_{r \in \mathrm{R}_{n,k}} q^{\mathrm{csw}(r)} p^{\mathrm{cnw}(r)}.$$

Given $\pi = \pi_1 \pi_2 \cdots \pi_n \in \mathcal{P}_{n,k}$, we define $f(\pi) = r \in \mathrm{R}_{n,k}$ as follows. Place rooks successively in columns from right to left. If $i \in L(\pi)$ then no rook is placed in Col$_i$. Otherwise, place a rook in the with available $\pi_i$ cell from the bottom in Col$_i$ such that cells attacked by rooks already placed are not available. Clearly, $f$ is a bijection and that it sends $L(\pi)$ to $\mathrm{EC}(\mathrm{f}(\pi))$. Therefore, (i) $\mathrm{csw}(\mathrm{f}(\pi)) = \mathrm{lbig}(\pi)$, (ii) $\mathrm{cnw}(\mathrm{f}(\pi)) = \mathrm{lsmall}(\pi)$, and (iii) $\mathrm{EC}(\mathrm{f}(\pi)) = L(\pi)$. The equality $\sum_{\pi \in \mathcal{P}_{n,k}} q^{\mathrm{lbig}(\pi)} p^{\mathrm{lsmall}(\pi)} = \mathrm{Stir}_{p,q}(n,k)$ follows immediately from the obvious recursion on set partitions, see the first lines of the proof of Theorem 5.2.  □

**Example 5.10** Let $\pi = 112132242 \in \mathcal{P}_{9,4}$, then $L(\pi) = \{1,3,5,8\}$ and $f(\pi)$ is the rook placement in Figure 5.3. Thus $\mathrm{lbig}(\pi) = \mathrm{csw}(\mathrm{f}(\pi)) = 5$ and

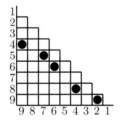

**FIGURE 5.3**: The rook placement $f(112132242)$

$\mathrm{lsmall}(\pi) = \mathrm{cnw}(\mathrm{f}(\pi)) = 9$.

In particular, if we set either $p = 1$ or $q = 1$ on the right-hand side of (5.3), then we obtain the result of Garsia and Remmel [111].

Now let us consider the statistics rbig, rsmall, and rnw. As we will see the consideration of these statistics is more complicated comparing to the consideration of the statistics lbig, lsmall, csw, and cnw. However, the aim of the next step is to present a bijection that sends rsmall to csw and rbig to cnw which obtains a result analogous to Theorem 5.9. Composing this bijection with the inverse of the bijection $f$ in the proof of Theorem 5.9 gives a bijection that sends rsmall to lbig and rbig to lsmall. Before we present our bijection, we must establish a relationship between the statistics rsmall and rbig that holds immediately from the definitions.

**Lemma 5.11** (Wachs and White [349, Proposition 5.1]) *For any* $\pi \in \mathcal{P}_{n,k}$,

$$\mathrm{rbig}_i(\pi) + \mathrm{rsmall}_i(\pi) = |\{j \in R(\pi) \mid j \geq i\}| - 1.$$

## Nonsubword Statistics on Set Partitions

**Theorem 5.12** (Wachs and White [349, Corollary 5.3]) *For all $1 \leq k \leq n$,*

$$\sum_{\pi \in \mathcal{P}_{n,k}} q^{\text{rsmall}(\pi)} p^{\text{rbig}(\pi)} = \sum_{r \in \text{R}_{n,k}} q^{\text{csw}(r)} p^{\text{cnw}(r)} \tag{5.4}$$

$$= \sum_{\pi \in \mathcal{P}_{n,k}} q^{\text{lbig}(\pi)} p^{\text{lsmall}(\pi)} = \text{Stir}_{p,q}(n,k), \tag{5.5}$$

*where* $\text{Stir}_{p,q}$ *is the $p,q$-Stirling number of the second kind, see Appendix C.5.*

**Proof** For any subset $B$ of $[n]$, define $\overline{B}$ to be the set $\{n+1-i \mid i \in B\}$. First, we establish a bijection $g : \mathcal{P}_{n,k} \to \text{R}_{n,k}$. Let $\pi \in \mathcal{P}_{n,k}$ and let us define $g(\pi) = r \in \text{R}_{n,k}$ as follows. Let $\text{EC}(r) = \overline{R(\pi)}$ and let $E$ be the set of cells in these columns. Now, place rooks successively in rows from top to bottom, starting with first row. In row Row$_i$, let $U_i$ be the set of cells that are not attacked by previously placed rooks. Place a rook in a cell of $U_i$ so that exactly rsmall$_i(\pi)$ cells in $U_i$ are to the right of the rook. In the case lsmall$_i(\pi) \geq |U_i|$, we leave Row$_i$ empty. For instance, when

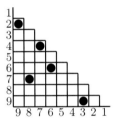

**FIGURE 5.4**: The rook placement $g(11213114233)$

$\pi = 11213114233 \in \mathcal{P}_{11,4}$ then $L(\pi) = \{1,3,5,8\}$, $R(\pi) = \{5,6,8,9\}$, $\overline{R(\pi)} = \{1,2,4,5\}$, (rsmall$_1(\pi), \ldots,$ rsmall$_9(\pi)$) = $(0,0,1,0,2,0,0,1,0)$ and $g(\pi)$ is the rook placement in Figure 5.4.

Next, we verify that the map $g$ is well-defined. If $i \notin L(\pi)$, then $\pi_i$ belongs to the set $\{\pi_j | j \in L(\pi) \text{ and } j < i\}$, which implies that

$$\text{rsmall}_i(\pi) \leq |\{j \in L(\pi) \mid j < i\}| - |\{j \in R(\pi) \mid j < i\}|.$$

Otherwise, if $i \in L(\pi)$, then $\pi_i$ is larger than the letters $\{\pi_j \mid j \in L(\pi)$ and $j < i\}$. Therefore,

$$\text{rsmall}_i(\pi) = |\{j \in L(\pi) \mid j < i\}| - |\{j \in R(\pi) \mid j < i\}|.$$

On the other hand, $|U_i| = i - 1 - v_i - |\{j \in R(\pi) \mid j < i\}|$, where $v_i$ is the number of rooks placed in the top $i-1$ rows. But by induction we can assume that the empty rows among the first $i-1$ rows correspond to $\{j \in L(\pi) \mid j < i\}$, which gives $v_i = (i-1) - |\{j \in L(\pi) \mid j < i\}|$, and then

$$|U_i| = |\{j \in L(\pi) \mid j < i\}| - |\{j \in R(\pi) \mid j < i\}|.$$

Hence, if $i \in L(\pi)$, then $g$ will assign a rook to some cell in Row$_i$, otherwise $g$ leaves Row$_i$ empty. Note that the empty rows correspond exactly to $L(\pi)$. Since the columns corresponding to $R(\pi)$ are empty and there are as many empty rows as there are empty columns, these are the only empty columns.

Is quite clear how to reverse the map $g$. The empty columns give $R(\pi)$ and rsmall$(\pi)$ can be easily reconstructed. It is then easy to reconstruct $\pi$ from rsmall$(\pi)$ and $R(\pi)$. Thus, $g$ is one-to-one. Hence, by Theorem 3.56, $g$ is a bijection.

Finally, we show that $g$ is a bijection from $\mathcal{P}_{n,k}$ to $\mathcal{R}_{n,k}$ with the following properties: (i) ER$(g(\pi)) = L(\pi)$, (ii) EC$(g(\pi)) = \overline{R(\pi)}$, (iii) csw$(g(\pi)) = $ rsmall$(\pi)$, and (iv) cnw$(g(\pi)) = $ rbig$(\pi)$. It is obvious from the construction of $g$ that (i)-(iii) hold. We now show (iv). If $i \notin L(\pi)$, then Row$_i$ of $g(\pi)$ has a rook in some column, say Col$_j$. Let $C$ be the set of all the cells in the columns Col$_{n-i+1}, \ldots,$ Col$_n$ and let $R$ be all the columns in the first $i-1$ rows in the columns Col$_{j+1}, \ldots,$ Col$_{n-i}$. Thus, by Lemma 5.11 we have

$$\text{rbig}_i(\pi) = |\{a \in R(\pi) \mid a > i\}| - 1 - \text{rsmall}_i(\pi)$$
$$= \text{number of empty columns in } C - \text{rsmall}_i(\pi)$$
$$= \text{number of empty rows in } C - \text{number rooks in } R$$
$$= \text{number of empty rows in } C - \text{number non-empty rows of } R$$
$$= \text{number of empty rows of } C \cup R$$
$$= \text{number of cells in Col}_j \text{ below Row}_i \text{ with no rook to their left}$$
$$= \text{cnw}_j(g(\pi)).$$

Therefore, rbig$(\pi) = \sum_{i \notin L(\pi)} \text{rbig}_i(\pi) + \binom{k}{2} = \text{cnw}(g(\pi))$, as required. □

We note that the map $f^{-1} \circ g$ is a bijection from $\mathcal{P}_{n,k}$ to $\mathcal{P}_{n,k}$ that maps rsmall to lbig, rbig to lsmall, and also sends $\overline{R}$ to $L$. Also, the map $f^{-1} \circ $ transpose $\circ \, g$ is a bijection from $\mathcal{P}_{n,k}$ to $\mathcal{P}_{n,k}$ which maps rsmall to lbig and $L$ to $L$.

Additionally, Wachs and White [349] constructed other bijections similar to $f, g$ in Theorems 5.9 and 5.12 to compare distributions of our statistics on set partitions and rook placements. In particular, they showed the following result (see Exercise 5.9).

**Theorem 5.13** (Wachs and White [349, Corollary 6.2]) *For all $1 \leq k \leq n$,*

$$\sum_{r \in \mathrm{R}_{n,k}} q^{\mathrm{csw}(r)} = \sum_{r \in \mathrm{R}_{n,k}} q^{\mathrm{rnw}(r)} = \widetilde{\mathrm{Stir}}_q(n,k), \tag{5.6}$$

*where $\widetilde{\mathrm{Stir}}_q(n,k)$ is the $q$-Stirling number of the second kind, see C.3.*

## 5.3 Records and Weak Records

The main goal of this section is to study the statistic of the "number of records" (various types of records: strong records, weak records, and additional records) on set partitions. Actually, the statistic of the number of records was initiated by Rényi [288] (1962) in the set of permutations: see also [194, 116]. Only in 2008, Myers and Wilf [259] extended the study of the number of records to multiset permutations and words. Another result, Kortchemski [197] studied the statistic srec, namely the "sum of the positions of all the records".

**Definition 5.14** *Let $\pi = \pi_1\pi_2\cdots\pi_n$ be any permutation of size n. An element $\pi_i$ in $\pi$ is a record if $\pi_i > \pi_j$ for all $j \in [i-1]$. Furthermore, the position of this record is i. In the literature records are also referred to as left–to–right maxima or outstanding elements. We define the statistic $\mathsf{srec}(\pi)$ to be the sum over the positions of all records in $\pi$.*

**Example 5.15** *For example, let $\pi = 312465$ be a permutation in $S_6$, then $\pi_1 = 3$, $\pi_4 = 4$, and $\pi_5 = 6$ are the records of $\pi$ at positions 1, 4 and 53, respectively. In this case, $\mathsf{srec}(\pi) = 1 + 4 + 5 = 10$.*

Note that the number of records in a set partition $\pi$ is exactly the number of the blocks in $\pi$. Hence, the number of set partitions of $\mathcal{P}_n$ with exactly $k$ records is given by $\mathrm{Stir}(n,k)$ (see Definition 1.11). Thus, the research interests not on number records on $\mathcal{P}_n$ but on various variation of records on $\mathcal{P}_n$, as we will see in the next lines.

In [281], Prodinger studied the statistic srec for words over the alphabet $\mathbb{N}$ equipped with geometric probabilities $\mathbb{P}(X = j) = pq^{j-1}$ with $p + q = 1$. Moreover, he found the expected value of the "sum of the positions of strong records" in random geometrically distributed words of size $n$. Previously, Prodinger [280] studied the number of "strong and weak records" in samples of geometrically distributed random variables, see Section 8.4. The number of records have recently also been investigated for the set of compositions by Knopfmacher and Mansour in [189].

**Definition 5.16** *A strong record in a word $\pi = \pi_1\pi_2\cdots\pi_n$ is an element $\pi_i$ such that $\pi_i > \pi_j$ for all $j \in [i - 1]$ (that is, it must be strictly larger than elements to the left) and a weak record is an element $\pi_i$ such that $\pi_i \geq \pi_j$ for all $j \in [i - 1]$ (must be only larger or equal to elements to the left). An additional weak record is an weak record that are not also strong record. We denote the number of strong, weak and additional records in $\pi$ by* strongrec, weakrec, addrec, *respectively. Furthermore, the position $i$ is called the position of the strong record (weak record, additional weak record). Let $\mathsf{sumrec}(\pi)$ (respectively, $\mathsf{sumwrec}(\pi)$) be the sum of the positions of the strong records (respectively, additional weak records) in $\pi$.*

**Example 5.17** Let $\pi = 1313142211331$ be a word in $[4]^{13}$. The strong records in $\pi$ are given by $\pi_1 = 3$, $\pi_2 = 3$, and $\pi_6 = 4$, the weak records in $\pi$ are given by $\pi_1 = 1$, $\pi_2 = 3$, $\pi_4 = 3$, and $\pi_6 = 4$, and the additional weak records in $\pi$ is given by $\pi_4 = 3$. Moreover, the sum of the positions of the strong records, weak records, and additional weak records in $\pi$ is given by 9, 13, and 4, respectively.

It is natural with respect to such words to consider records of a set partition. The statistic strong records on set partitions corresponds to the well studied statistic *number of blocks*, namely blo, in the set partitions: the number of set partitions of $n$ with exactly $k$ strong records is given by Stir(n, k) the Stirling number of the second kind (see Definition 1.11 and Theorem 1.12). In [190], the statistic number of the weak records has been studied. In addition, considered the statistic *sum of positions of records* in a set partition. For the number of strong records in set partitions, we refer to [107].

**Theorem 5.18** (Flajolet and Sedgewick [107, Example III.11]) *The mean number of strong records over all set partitions of $[n]$ is $\text{Bell}_{n+1}/\text{Bell}_n - 1$. Asymptotically, the mean and variance of the number of strong records over all set partitions of $[n]$ are given by $\frac{n}{\log n}$ and $\frac{n}{\log^2 n}$, respectively.*

**Proof** The proof of the first part of this theorem obtains from Corollary 3.17 while the proof of the second part follows from Theorem 8.60. □

In the next three subsections we study the number of additional weak records, the sum of positions of records, and sum of positions of additional weak records in set partitions. Our method is based on finding the corresponding generating functions and then obtaining the mean and variance of each statistic.

### 5.3.1 Weak Records

Our first aim is to study the number of weak records in set partitions of $[n]$. It turns out that the number of weak records over the number of strong records is comparatively small. Thus, it is more interesting to investigate *additional weak records*, see Definition 5.16. To do so, we need the following generating functions.

**Definition 5.19** *Let $P_{\text{strongrec,addrec}}(x, y; q)$ (see Definition 4.1) be the generating function for the number of set partitions of $\mathcal{P}_n$ according to the statistics of the number of strong records and the number of additional weak records. Moreover, let $P_{\text{strongrec,addrec}}(x, y; q; k)$ be the generating function for the number of set partitions of $\mathcal{P}_{n,k}$ according to the statistics of the number of strong records and the number of additional weak records, that is,*

$$P_{\text{strongrec,addrec}}(x, y; q; k) = \sum_{n \geq 0} x^n \sum_{\pi \in \mathcal{P}_{n,k}} y^{\text{strongrec}(\pi)} q^{\text{addrec}(\pi)}.$$

Since any set partitions can be decomposed as

$$\pi = 1\pi^{(1)} 2\pi^{(2)} \cdots k\pi^{(k)} \tag{5.7}$$

for some $k$, where $\pi^{(j)}$ denotes an arbitrary word over an alphabet $[j]$, including the empty word, we have

$$P_{\text{strongrec,addrec}}(x, y; q; k) = x^k y^k \prod_{j=1}^{k} f_j(x, q),$$

where $f_j(x, q)$ is the generating function for the number of words over alphabet $[j]$ according to the number occurrences of the letter $j$. Considering the first letter in each word shows that $f_j(x, q) = 1 + (j - 1 + q) x f_j(x, q)$, where 1 counts the empty word, which implies that $f_j(x, q) = \frac{1}{1-(j-1+q)x}$. Hence,

$$P_{\text{strongrec,addrec}}(x, y; q) = \sum_{k \geq 1} \frac{y^k x^k}{\prod_{j=1}^{k}(1 - (j + q - 1)x)}.$$

Now, by similar techniques used in Example 2.59, we find the exponential generating function for the number of set partitions of $[n]$ according to the statistics of the number of strong records and the number of additional weak records. At first we expand the generating function $P_{\text{strongrec,addrec}}(x, y; q)$ into partial fractions:

$$P_{\text{strongrec,addrec}}(x, y; q) = \sum_{k \geq 1} \frac{y^k}{\prod_{j=1}^{k}(1/x - (j + q - 1))}$$

$$= \sum_{m \geq 1} \frac{a_m}{1/x - (m + q - 1)}.$$

The coefficient $a_m$ can be found by multiplying by $\frac{1}{x} - (m + q - 1)$ and setting $x = \frac{1}{m+q-1}$:

$$a_m = \sum_{k \geq 1} \frac{y^k (1/x - (m + q - 1))}{\prod_{j=1}^{k}(1/x - (j + q - 1))} \bigg|_{x=1/(m+q-1)}$$

$$= \sum_{k \geq m} \frac{y^k}{\prod_{j=1, j \neq m}^{k}(1/x - (j + q - 1))} \bigg|_{x=1/(m+q-1)}$$

$$= \sum_{k \geq m} y^k \prod_{j=1, j \neq m}^{k} (m - j)^{-1} = \frac{1}{(m-1)!} \sum_{k \geq m} \frac{(-1)^{k-m} y^k}{(k-m)!} = \frac{y^m e^{-y}}{(m-1)!}.$$

Therefore,

$$\sum_{k \geq 1} \frac{y^k x^k}{\prod_{j=1}^{k}(1 - (j + q - 1)x)} = \sum_{m \geq 1} \frac{y^m e^{-y} x}{(m-1)!(1 - (m + q - 1)x)}.$$

Now we expand $\frac{x}{1-(m+q-1)x}$ into a geometric series:

$$\sum_{k\geq 1} \frac{y^k x^k}{\prod_{j=1}^{k}(1-(j+q-1)x)} = \sum_{m\geq 1} \frac{y^m e^{-y}}{(m+q-1)(m-1)!} \sum_{n\geq 1}((m+q-1)x)^n.$$

Since we would like to work with the exponential generating function rather than the ordinary generating function, we introduce a factor $\frac{1}{n!}$:

$$\sum_{m\geq 1} \frac{y^m e^{-y}}{(m+q-1)(m-1)!} \sum_{n\geq 1} \frac{((m+q-1)x)^n}{n!}$$

$$= \sum_{m\geq 1} \frac{y^m e^{-y}}{(m+q-1)(m-1)!} \left(e^{(m+q-1)x} - 1\right).$$

In order to simplify this, we differentiate with respect to $x$ to derive

$$\sum_{m\geq 1} \frac{y^m e^{-y} e^{(m+q-1)x}}{(m-1)!} = y e^{y e^x + qx - y}. \tag{5.8}$$

Note that by substituting $y = q = 1$ we obtain that $e^{e^x + x - 1}$, which is indeed the derivative of the exponential generating function $e^{e^x - 1}$ of the Bell numbers, see Example 2.58. Also, Equation (5.8) can be interpreted in another way: as we defined, $y$ marks the number of blocks in a set partition, while $q$ marks the number of elements (other than 1) in the first block. Indeed, $ye^{qx}$ generates a single block (to which the element 1 is added), while $e^{y(e^x-1)}$ generates an arbitrary number of additional blocks. In addition, there is a simple bijection that shows this identity: in a set partition, replace every 1 between the first occurrence of $r$ and the first occurrence of $r+1$ (if any) by $r$ and vice versa. Then the additional weak records are exactly mapped to elements of the first block.

By finding the coefficient of $x^n$ in the generating function $ye^{ye^x + qx - y}$ with $y = 1$, and by finding the coefficient of $x^n y^k$ in the generating function $ye^{ye^x + qx - y}$, we obtain an explicit formula for the number of set partitions with a prescribed number of strong and number of additional weak records.

**Theorem 5.20** *The number of set partitions of $\mathcal{P}_{n,k}$ with exactly $\ell$ additional weak records is given by*

$$\binom{n-1}{\ell} \mathrm{Stir}(n-1-\ell, k-1).$$

*The number of set partitions of $\mathcal{P}_n$ with exactly $\ell$ additional weak records is*

$$\binom{n-1}{\ell} \mathrm{Bell}_{n-1-\ell}.$$

From (5.8), we see that the exponential generating function for the total number of additional weak records over all set partition of $\mathcal{P}_{n,k}$ is given by

$$\frac{d}{dq} y e^{ye^x + qx - y}\bigg|_{q=1} = yxe^x e^{y(e^x - 1)}. \tag{5.9}$$

This generating function also arises in another interesting context, see Exercise 3.7. Comparing the exponential generating functions in (5.9) and in Exercise 3.7(iii), we notice that these exponential generating functions differ only by a factor of $y$. This can be shown as follows: the total number of elements that are in the same block as the element 1 in all set partitions in $\mathcal{P}_{n,k}$ is exactly the total number of singletons, excluding those of the form $\{1\}$, in all set partitions in $\mathcal{P}_{n,k+1}$ (Explain that combinatorially? (See Exercise 5.3.)

By the above theorem, can be derived asymptotically the number of additional weak records over all set partitions of $[n]$, as will be discussed in Section 8.5.3.

### 5.3.2 Sum of Positions of Records

Let $p_{\text{sumrec}}(n, k)$ be the generating function for the number of set partitions in $\mathcal{P}_{n,k}$ according to the statistic sumrec, that is, $p_{\text{sumrec}}(n, k) = \sum_{\pi \in \mathcal{P}_{n,k}} q^{\text{sumrec}(\pi)}$. The aim of this section is to study the generating function $p_{\text{sumrec}}(n, k)$ (for another approach, we refer the reader to [190]). Notice that a set partition of $[n]$ is obtained by adding $n$ to a set partition of $[n-1]$. Thus, if $n$ is added as a singleton, then sumrec changes by $n$; otherwise, it remains unchanged. Therefore,

$$p_{\text{sumrec}}(n, k) = k p_{\text{sumrec}}(n-1, k) + q^n p_{\text{sumrec}}(n-1, k-1).$$

For $q = 1$, this recurrence reduces to the recursion for the Stirling numbers of the second kind (see Theorem 1.12). By Exercise 5.5, we obtain

$$\text{sr}(n, k) = \frac{d}{dq} p_{\text{sumrec}}(n, k)\bigg|_{q=1} = \sum_{\pi \in \mathcal{P}_{n,k}} \text{sumrec}(\pi)$$

$$= k \text{Stir}(n+1, k) - (n+1)\text{Stir}(n, k-1).$$

Let $P_{\text{sumrec}}(x, y; q)$ be the exponential generating function

$$P_{\text{sumrec}}(x, y; q) = \sum_{n \geq 0} \sum_{k \geq 0} \frac{y^k x^n}{n!} p_{\text{sumrec}}(n, k).$$

Then the recurrence relation for $p_{\text{sumrec}}(n, k)$ can be written in terms of generating functions as follows.

**Lemma 5.21** *The generating function $P_{\text{sumrec}}(x, y; q)$ satisfies*

$$\frac{\partial}{\partial x} P_{\text{sumrec}}(x, y; q) = y \frac{\partial}{\partial y} P_{\text{sumrec}}(x, y; q) + qy P_{\text{sumrec}}(qx, y; q). \tag{5.10}$$

As we can guess, (5.10) can not be solved explicitly. However, one can determine explicit expressions for the derivatives with respect to $q$ at $q = 1$, since this amounts to solving linear partial differential equations of the form $S_x(u,x) = uS_u(u,x) + uS(u,x) + f(u,x)$, which can be done by standard techniques (see [273]). By using computer algebra (such as Maple), we obtain

$$\frac{\partial}{\partial q}P_{\mathsf{sumrec}}(x,y;q)\Big|_{q=1} = ye^{y(e^x-1)}\left(ye^x(e^x - x - 1) + e^x - 1\right) \quad (5.11)$$

and

$$\frac{\partial^2}{\partial q^2}P_{\mathsf{sumrec}}(x,y;q)\Big|_{q=1} + \frac{\partial}{\partial q}P_{\mathsf{sumrec}}(x,y;q)\Big|_{q=1}$$
$$= \frac{y}{2}e^{y(e^x-1)}\Big(2y^3e^{4x} - y^2e^{3x}(4y(x+1) - 13)$$
$$+ 2ye^{2x}\left(y^2(x+1)^2 - y(x^2 + 8x + 10) + 8\right)$$
$$+ e^x\left(y^2(2x^2 + 10x + 7) - 2y(x^2 + 7x + 9) + 2\right) + 2(y-1)\Big).$$

Hence, we can state the following result.

**Theorem 5.22** *The mean and variance of* sumrec, *taken over all set partitions of $[n]$, are given by*

$$\mathrm{mean} = \frac{\mathrm{Bell}_{n+2} - \mathrm{Bell}_{n+1} - (n+1)\mathrm{Bell}_n}{\mathrm{Bell}_n},$$

$$\mathrm{variance} = \frac{\mathrm{Bell}_{n+4} - \frac{3}{2}\mathrm{Bell}_{n+3} - (2n + \frac{7}{2})\mathrm{Bell}_{n+2} - \frac{1}{2}\mathrm{Bell}_{n+1}}{\mathrm{Bell}_n} - \mathrm{mean}^2.$$

**Proof** In order to find the mean, we need to find an explicit formula for the coefficient of $x^n$ in the generating function $\frac{\partial}{\partial q}P_{\mathsf{sumrec}}(x,y;q)\Big|_{y=q=1}$. By (5.11) together with $y = 1$ gives

$$\frac{\partial}{\partial q}P_{\mathsf{sumrec}}(x,y;q)\Big|_{y=q=1} = e^{e^x-1}\left(e^{2x} - xe^x - 1\right).$$

By Example 2.58, we obtain that $e^{e^x-1} = \sum_{n\geq 0} \mathrm{Bell}_n \frac{x^n}{n!}$, then by differentiating twice we get

$$e^x e^{e^x-1} = \sum_{n\geq 0} \mathrm{Bell}_{n+1}\frac{x^n}{n!},$$

$$e^{2x}e^{e^x-1} + e^x e^{e^x-1} = \sum_{n\geq 0} \mathrm{Bell}_{n+2}\frac{x^n}{n!}.$$

All this together leads to

$$\frac{\partial}{\partial q}P_{\mathsf{sumrec}}(x,y;q)\Big|_{y=q=1} = \sum_{n\geq 0}(\mathrm{Bell}_{n+2} - \mathrm{Bell}_{n+1} - n\mathrm{Bell}_n - \mathrm{Bell}_n)\frac{x^n}{n!},$$

which is equivalent to

$$\left.\frac{\partial}{\partial q}P_{\mathsf{sumrec}}(x,y;q)\right|_{y=q=1} = \sum_{n\geq 0}(\mathrm{Bell}_{n+2} - \mathrm{Bell}_{n+1} - (n+1)\mathrm{Bell}_n)\frac{x^n}{n!},$$

Hence, the mean is given by $\frac{\mathrm{Bell}_{n+2}-\mathrm{Bell}_{n+1}-(n+1)\mathrm{Bell}_n}{\mathrm{Bell}_n}$. By similar techniques one obtain an explicit formula for variance. (Check!!) □

In Section 8.5.3 we will see that the mean is asymptotically $\sim \frac{n^2}{\log^2 n}$, while the variance is $\sim \frac{n^3}{2\log^3 n}$.

### 5.3.3 Sum of Positions of Additional Weak Records

In this section, we will discuss the statistic sumwrec, the sum of positions of all additional weak records. Following the methods that used in the previous section, let $p_{\mathsf{sumwrec}}(n,k) = \sum_{\pi \in \mathcal{P}_{n,k}} q^{\mathsf{sumwrec}(\pi)}$, when $n$ is added to a set partition of $[n-1]$, it becomes a new weak record if and only if it is added to the last block, which gives

$$p_{\mathsf{sumwrec}}(n,k) = (k-1+q^n)p_{\mathsf{sumwrec}}(n-1,k) + p_{\mathsf{sumwrec}}(n-1,k-1),$$

which, by induction, implies

$$\mathsf{sumwrec}(n,k) = \left.\frac{d}{dq}p_{\mathsf{sumwrec}}(n,k)\right|_{q=1} = \sum_{\pi \in \mathcal{P}_{n,k}} \mathsf{sumwrec}(\pi)$$

$$= \frac{(n-1)(n+2)}{2}\mathrm{Stir}(n-1,k).$$

Moreover, this implies that the mean of sumwrec over all set partitions of $[n]$ is given by

$$\frac{(n-1)(n+2)\mathrm{Bell}_{n-1}}{2\mathrm{Bell}_n}.$$

Again, we can introduce the exponential generating function

$$P_{\mathsf{sumwrec}}(x,y;q) = \sum_{n\geq 0}\sum_{k\geq 0} \frac{y^k x^n}{n!} p_{\mathsf{sumwrec}}(n,k;q).$$

Then the recurrence relation for $p_{\mathsf{sumwrec}}(n,k;q)$ can be written as

$$\frac{\partial}{\partial x}P_{\mathsf{sumwrec}}(x,y;q)$$
$$= y\frac{\partial}{\partial y}P_{\mathsf{sumwrec}}(x,y;q) + qP_{\mathsf{sumwrec}}(qx,y;q) + (y-1)P_{\mathsf{sumwrec}}(x,y;q).$$

By applying the same ideas as in the previous section (we leave the details for the interest reader), we derive explicit expressions for derivatives with respect to $q$ at $q=1$, so that we also find the variance.

**Theorem 5.23** *The* mean *and* variance *of the statistic* sumwrec, *taken over all set partitions of* $[n]$, *are given by*

$$\text{mean} = \frac{(n+2)(n-1)\text{Bell}_{n-1}}{2\text{Bell}_n},$$

$$\text{variance} = \frac{(n-1)(2n^2+5n+6)\text{Bell}_{n-1}}{6\text{Bell}_n}$$

$$+ \frac{(n-1)(3n^3+5n^2-10n-24)\text{Bell}_{n-2}}{12\text{Bell}_n} - \text{mean}^2.$$

*Asymptotically,* mean *and* variance *are* $\sim \frac{n}{2}\log n$ *and* $\sim \frac{n^2}{3}\log n$.

## 5.4 Number of Positions Between Adjacent Occurrences of a Letter

In this section, according to [241], we study the following statistics on $\mathcal{P}_{n,k}$: the total number of positions between adjacent occurrences of a letter, and the total number of positions of the same letter lying between two letters which are strictly larger. To define our statistics, we use the graphical representation, see Definition 3.49.

**Definition 5.24** *For given a graphical representation of a set partition $\pi$ of $[n]$, we say that the two points $(j, i)$ and $(j, i')$ lying on the vertical line $x = j$ have $j$-distance $m$ if there are $m$ points in the interior of the subset of the first quadrant of $\mathbb{Z}^2$ bounded by the line segment between $(j, i)$ and $(j, i')$ and the horizontal lines emanating in the positive direction from these points.*

*We denote the total sum of the $j$-distances between any two adjacent points lying on the line $x = j$ in the graphical representation of the set partition $\pi$ by $\text{dis}_j(\pi)$. Define $\text{dis}(\pi) = \sum_{j\geq 1} \text{dis}_j(\pi)$ for any set partition $\pi$ of $[n]$.*

**Example 5.25** *For example, if $\pi = 123124222 \in \mathcal{P}_9$, then $\text{dis}_1(\pi) = \text{dis}_2(\pi) = 2$ and $\text{dis}_3(\pi) = \text{dis}_4(\pi) = 0$. Moreover, $\text{dis}(\pi) = 4$.*

Clearly, the statistics $\text{dis}_j$ (and $\text{dis}$) can be defined directly by the canonical representation of the set partition as follows. Let $\pi = \pi_1 \pi_2 \cdots \pi_n \in \mathcal{P}_{n,k}$, two elements $\pi_i, \pi_{i'}$ have $j$-distance $m$ if

$$\pi_i = \pi_{i'} = j \text{ and } |\{\pi_s \mid \pi_s > j, i < s < i'\}| = m.$$

By using the pattern terminology, the statistic $\text{dis}_j$ counts the number of occurrences of the pattern 1-2-1 in which the 1's correspond to adjacent occurrences of the letter $j$, which implies that the dis statistic counts the total number of occurrences of the pattern 1-2-1 in which the 1's correspond to adjacent occurrences of the same letter.

**Remark 5.26** *The statistics* $\text{dis}_j$ *and* $\text{dis}$ *also have an interpretation directly in terms of blocks. Let* $\pi = B_1/B_2/\cdots/B_k \in \mathcal{P}_{n,k}$, *fix* $j$ *such that* $1 \leq j \leq n$, *and let* $B_j = \{b_1, b_2, \ldots b_r\}$ *with* $b_1 < b_2 < \cdots < b_r$. *For each consecutive pair* $(b_i, b_{i+1})$, $1 \leq i \leq r-1$, *consider the number of elements* $c \in [n]$ *occurring in blocks of* $\pi$ *to the right of* $B_j$ *such that* $b_i < c < b_{i+1}$. *Doing this for each pair* $b_i$, $b_{i+1}$, *and then summing the resulting numbers gives* $\text{dis}_j$ *and by summing the results for all blocks gives* $\text{dis}$.

From these discussions, we see the same statistic can be defined in different representations of the set partitions, where here we interested on the graphical representation because our statistics can be easily described and defined, as we discussed before.

**Definition 5.27** *Let* $F_n(y; q_1, q_2, \ldots)$ *be the generating function for the number of set partitions in* $\mathcal{P}_{n,k}$ *according to the statistics* $\text{dis}_1, \text{dis}_2, \ldots$, *that is,*

$$F_n(y; q_1, q_2, \ldots) = \sum_{k \geq 0} \sum_{\pi \in \mathcal{P}_{n,k}} y^k \prod_{j \geq 1} q_j^{\text{dis}_j(\pi)}.$$

Now, let us write a recurrence relation for the sequence $\{F_n(y; q_1, \ldots)\}_{n \geq 0}$. Assume that there are $j+1$ occurrences of the letter 1 within a set partition in $\mathcal{P}_{n,k}$ at positions $1, i_1, i_2, \ldots, i_j$. From the definitions, the contribution of the case $j = 0$ is given by $y F_{n-1}(y; q_2, q_3, \ldots)$ and the contribution of the case $j > 0$ is given by

$$F_{n-1-j}(y; q_2, q_3, \ldots) \sum_{2 \leq i_1 < i_2 < \cdots < i_j \leq n} q_1^{i_j - j - 1}.$$

Hence, by combining these contributions, we obtain the following recurrence relation:

$$F_n(y; q_1, q_2, \ldots) = y F_{n-1}(y; q_2, q_3, \ldots) \\ + y \sum_{j=1}^{n-1} \left( F_{n-1-j}(y; q_2, q_3, \ldots) \sum_{2 \leq i_1 < i_2 < \cdots < i_j \leq n} q_1^{i_j - j - 1} \right) \quad (5.12)$$

with the initial condition $F_0(y; q_1, q_2, \ldots) = 1$. In order to simplify the rightmost sum in (5.12) we use the following lemma, where the proof is left as Exercise 5.7.

**Lemma 5.28** *Let*

$$a_{n,j} = \sum_{2 \leq i_1 < i_2 < \cdots < i_j \leq n} x^{i_j - 1}$$

*for all* $0 < j < n$. *Then*

$$a_{n,j} = \frac{x^j}{(1-x)^j} - \frac{x^n}{(1-x)^j} \sum_{i=0}^{j-1} \binom{n-1-j+i}{i} (1-x)^i.$$

Thus, Lemma 5.28 and (5.12) yield

$$F_n(y; q_1, q_2, \ldots) = yF_{n-1}(y; q_2, q_3, \ldots)$$
$$+ y \sum_{j=1}^{n-1} F_{n-1-j}(y; q_2, q_3, \ldots) \left( \frac{1}{(1-q_1)^j} - \sum_{i=0}^{j-1} \binom{n-1-j+i}{i} \frac{q_1^{n-j}}{(1-q_1)^{j-i}} \right) \quad (5.13)$$

with the initial condition $F_0(y; q_1, q_2, \ldots) = 1$. By multiplying (5.13) by $x^n$ and then summing over all $n \geq 1$, we obtain

$$P_{\text{dis}_1, \text{dis}_2, \ldots}(x, y; q_1, q_2, \ldots)$$
$$= 1 + \frac{xy}{1 - x/(1-q_1)} P_{\text{dis}_1, \text{dis}_2, \ldots}(x, y; q_2, q_3, \ldots)$$
$$- \frac{yx^2 q_1}{1 - q_1} \sum_{n \geq 0} x^n \left[ \sum_{j=0}^{n} F_{n-j}(y; q_2, q_3, \ldots) \sum_{i=0}^{j} \binom{n-j+i}{i} \frac{q_1^{n-j}}{(1-q_1)^{j-i}} \right]$$
$$= 1 + \frac{xy}{1 - x/(1-q_1)} P_{\text{dis}_1, \text{dis}_2, \ldots}(x, y; q_2, q_3, \ldots)$$
$$- \frac{yx^2 q_1}{1 - q_1} \sum_{n \geq 0} x^n \left[ \sum_{m=0}^{n} F_m(y; q_2, q_3, \ldots) \sum_{i=0}^{n-m} \binom{m+i}{i} \frac{q_1^m}{(1-q_1)^{n-m-i}} \right]$$
$$= 1 + \frac{xy}{1 - x/(1-q_1)} P_{\text{dis}_1, \text{dis}_2, \ldots}(x, y; q_2, q_3, \ldots)$$
$$- \frac{yx^2 q_1}{1 - q_1} \sum_{k \geq 0} \left[ x^k q_1^k F_k(y; q_2, q_3, \ldots) \sum_{n \geq k} \frac{x^{n-k}}{(1-q_1)^{n-k}} \sum_{i=0}^{n-k} \binom{k+i}{i} (1-q_1)^i \right]$$
$$= 1 + \frac{xy}{1 - x/(1-q_1)} P_{\text{dis}_1, \text{dis}_2, \ldots}(x, y; q_2, q_3, \ldots)$$
$$- \frac{yx^2 q_1}{1 - q_1} \sum_{k \geq 0} \left[ x^k q_1^k F_k(y; q_2, q_3, \ldots) \sum_{n \geq 0} \frac{x^n}{(1-q_1)^n} \sum_{i=0}^{n} \binom{k+i}{i} (1-q_1)^i \right].$$

Using the fact that

$$\frac{1}{(1-x)^{m+1}(1-\frac{x}{1-t})} = \sum_{i \geq 0} \frac{x^i}{(1-t)^i(1-x)^{m+1}} = \sum_{i \geq 0} \sum_{j \geq 0} \binom{m+j}{j} \frac{x^{i+j}}{(1-t)^i}$$
$$= \sum_{n \geq 0} \sum_{j=0}^{n} \binom{m+j}{j} \frac{x^n}{(1-t)^{n-j}}$$
$$= \sum_{n \geq 0} \frac{x^n}{(1-t)^n} \sum_{j=0}^{n} \binom{m+j}{j} (1-t)^j$$

we obtain the following general result.

**Theorem 5.29** *The generating function* $F(x, y; q_1, q_2, \ldots)$ *satisfies*

$$P_{\mathrm{dis}_1,\mathrm{dis}_2,\ldots}(x, y; q_1, q_2, \ldots)$$
$$= 1 + \frac{xy(1-q_1)}{1-q_1-x} P_{\mathrm{dis}_1,\mathrm{dis}_2,\ldots}(x, y; q_2, q_3, \ldots)$$
$$- \frac{yx^2 q_1}{(1-x)(1-q_1-x)} P_{\mathrm{dis}_1,\mathrm{dis}_2,\ldots}\left(\frac{xq_1}{1-x}, y; q_2, q_3, \ldots\right).$$

In the next section we will use the above theorem to study several particular cases.

### 5.4.1 The Statistic dis

Define
$$P_{\mathrm{dis}}(x, y; q) = P_{\mathrm{dis}_1,\mathrm{dis}_2,\ldots}(x, y; q, q, \ldots).$$

Theorem 5.29 gives

$$P_{\mathrm{dis}}(x, y; q)$$
$$= 1 + \frac{xy(1-q)}{1-q-x} P_{\mathrm{dis}}(x, y; q) - \frac{yx^2 q}{(1-x)(1-q-x)} P_{\mathrm{dis}}\left(\frac{xq}{1-x}, y; q\right).$$

Equivalently,

$$P_{\mathrm{dis}}(x, y; q) = \frac{1-q-x}{1-q-x-xy(1-q)} - \frac{yx^2 q P_{\mathrm{dis}}\left(\frac{xq}{1-x}, y; q\right)}{(1-x)(1-q-x-xy(1-q))}.$$

Iterating the above recurrence relation infinite number of times leads to the following result.

**Theorem 5.30** *The generating function* $P_{\mathrm{dis}}(x, y; q)$ *for the number of set partitions in* $\mathcal{P}_n$ *according to the statistic* dis *is given by*

$$\sum_{j \geq 0} \frac{(-1)^j y^j x^{2j} q^{j^2} (1-q)^j (1-q-x)}{\prod_{i=1}^{j}(1-q-x(1-q^i)) \prod_{i=0}^{j}(1-q-x(1+yq^i - yq^{i+1}))}.$$

Note that Theorem 5.30 for $q = 1$ gives

$$P_{\mathrm{dis}}(x, y; 1)$$
$$= \sum_{j \geq 0} \lim_{q \to 1} \left( \frac{(-1)^j y^j x^{2j} q^{j^2} (1-q)^j (1-q-x)}{\prod_{i=1}^{j}(1-q-x(1-q^i)) \prod_{i=0}^{j}(1-q-x(1+yq^i - yq^{i+1}))} \right)$$
$$= \sum_{j \geq 0} y^j x^j \lim_{q \to 1} \frac{1}{\prod_{i=1}^{j} 1 - x \lim_{q \to 1} \frac{1-q^i}{1-q}} = \sum_{j \geq 0} \frac{y^j x^j}{\prod_{i=1}^{j}(1-ix)}, \qquad (5.14)$$

which agrees with that the generating function for the number set partitions in $\mathcal{P}_{n,k}$ is given by $\frac{x^k}{\prod_{i=1}^{k}(1-ix)}$, see Example 2.56.

**Corollary 5.31** *The generating function for* $\mathrm{total}_{\mathcal{P}_n}(\mathrm{dis}) = \sum_{\pi \in \mathcal{P}_n} \mathrm{dis}(\pi)$ *is given by*

$$x \sum_{j \geq 0} \frac{y^j x^j}{\prod_{i=0}^{j}(1-ix)} \left( \sum_{i=2}^{j} \frac{\binom{i}{2}}{1-ix} \right).$$

**Proof** Let

$$P'_{\mathrm{dis}}(x,y;1) = \frac{d}{dq} P_{\mathrm{dis}}(x,y;q) \mid_{q=1}.$$

Theorem 5.30 gives

$P'_{\mathrm{dis}}(x,y;1)$
$$= \sum_{j \geq 0} \lim_{q \to 1} \frac{(-1)^j y^j x^{2j} q^{j^2} (1-q)^j (1-q-x) f_j(x,y;q)}{\prod_{i=1}^{j}(1-q-x(1-q^i)) \prod_{i=0}^{j}(1-q-x(1+yq^i-yq^{i+1}))},$$

where

$f_j(x,y;q)$
$$= \frac{j^2}{q} - \frac{1}{1-q-x} - \frac{j - \sum_{i=1}^{j} \frac{1-iq^{i-1}x}{1-x\frac{1-q^i}{1-q}}}{1-q} + \sum_{i=0}^{j} \frac{1+yq^{i-1}x(i-(i+1)q)}{1-q-x(1+yq^i-yq^{i+1})}.$$

By (5.14), we obtain

$$P'_{\mathrm{dis}}(x,y;1) = \sum_{j \geq 0} \frac{x^j y^j}{\prod_{i=0}^{j} 1 - ix} \lim_{q \to 1} f_j(x,y;q). \tag{5.15}$$

On the other hand, by l'Hôpital rule we have

$$\lim_{q \to 1} f_j(x,y;q) = j^2 + y(j+1) - \frac{j}{x} - \lim_{q \to 1} \frac{j - \sum_{i=1}^{j} \frac{1-iq^{i-1}x}{1-x\frac{1-q^i}{1-q}}}{1-q}$$

$$= j^2 + y(j+1) - \frac{j}{x} - \sum_{i=1}^{j} \lim_{q \to 1} \frac{d}{dq}\left( \frac{1-iq^{i-1}x}{1-x\frac{1-q^i}{1-q}} \right)$$

$$= j^2 + y(j+1) - \frac{j}{x} + \sum_{i=1}^{j} \frac{\binom{i}{2}y}{1-ix}.$$

Hence, by (5.15),

$$P'_{\mathrm{dis}}(x,y;1) = \sum_{j \geq 0} \frac{y^j x^j}{\prod_{i=0}^{j} 1 - ix} \left( j^2 + y(j+1) - \frac{j}{x} + x \sum_{i=1}^{j} \frac{\binom{i}{2}}{1-ix} \right)$$

$$= \sum_{j \geq 0} \frac{y^j x^j}{\prod_{i=0}^{j} 1 - ix} \left( j^2 + y(j+1) - \frac{y(j+1)}{1-(j+1)x} + x \sum_{i=1}^{j} \frac{\binom{i}{2}}{1-ix} \right),$$

which is equivalent to

$$P'_{\text{dis}}(x, y; 1) = \sum_{j \geq 0} \frac{y^j x^j}{\prod_{i=0}^{j} 1 - ix} \left( j^2 - \frac{y(j+1)^2 x}{1 - (j+1)x} + x \sum_{i=1}^{j} \frac{\binom{i}{2}}{1 - ix} \right)$$

$$= x \sum_{j \geq 0} \frac{y^j x^j}{\prod_{i=0}^{j} 1 - ix} \left( \sum_{i=1}^{j} \frac{\binom{i}{2}}{1 - ix} \right),$$

as required. $\square$

By the above corollary and Example 2.56, we obtain the following result.

**Corollary 5.32** *Let $1 \leq k \leq n$. Then*

$$\text{total}_{\mathcal{P}_{n,k}}(\text{dis}) = \sum_{i=2}^{k} \left( \binom{i}{2} \sum_{j=k}^{n-1} i^{n-1-j} \text{Stir}(j, k) \right).$$

The above formula can be simplified. To do so, we convert the ordinary generating function to an exponential generating function as describe in Example 2.59:

$$P'_{\text{dis}}(x, y; 1) = \sum_{j \geq 0} \frac{y^j}{\prod_{i=1}^{j}(\frac{1}{x} - i)} \left( \sum_{i=1}^{j} \frac{\binom{i}{2}}{\frac{1}{x} - i} \right) = \sum_{j \geq 1} \left( \frac{a_j(r)}{(\frac{1}{x} - j)^2} + \frac{b_j(r)}{\frac{1}{x} - j} \right)$$

for certain $a_j(y)$ and $b_j(y)$ that depend only on $y$. To determine these values, we expand around $z = x^{-1} = m$ for fixed $m$. For $j \geq m$, one has

$$\frac{y^j}{\prod_{i=1}^{j}(x^{-1} - i)} \left( \sum_{i=1}^{j} \frac{\binom{i}{2}}{x^{-1} - i} \right)$$

$$= \frac{y^j}{z - m} \prod_{\substack{i=1 \\ i \neq m}}^{j} (z - i)^{-1} \left( \frac{\binom{m}{2}}{z - m} + \sum_{\substack{i=1 \\ i \neq m}}^{j} \frac{\binom{i}{2}}{m - i} + O(z - m) \right)$$

$$= \frac{y^j}{z - m} \prod_{\substack{i=1 \\ i \neq m}}^{j} (m - i)^{-1} \cdot \prod_{\substack{i=1 \\ i \neq m}}^{j} \left( 1 + \frac{z - m}{m - i} \right)^{-1}$$

$$\cdot \left( \frac{\binom{m}{2}}{z - m} + \sum_{\substack{i=1 \\ i \neq m}}^{j} \frac{\binom{i}{2}}{m - i} + O(z - m) \right),$$

which implies

$$\frac{y^j}{\prod_{i=1}^{j}(x^{-1}-i)}\left(\sum_{i=1}^{j}\frac{\binom{i}{2}}{x^{-1}-i}\right)$$

$$=\frac{y^j(-1)^{j-m}}{(m-1)!(j-m)!(z-m)}\cdot\left(1-\sum_{\substack{i=1\\i\neq m}}^{j}\frac{z-m}{m-i}+O((z-m)^2)\right)$$

$$\cdot\left(\frac{\binom{m}{2}}{z-m}+\sum_{\substack{i=1\\i\neq m}}^{j}\frac{\binom{i}{2}}{m-i}+O(z-m)\right)$$

$$=\frac{y^j(-1)^{j-m}}{(m-1)!(j-m)!(z-m)}\cdot\left(\frac{\binom{m}{2}}{z-m}+\sum_{\substack{i=1\\i\neq m}}^{j}\frac{\binom{i}{2}-\binom{m}{2}}{m-i}+O(z-m)\right)$$

$$=\frac{y^j(-1)^{j-m}}{(m-1)!(j-m)!(z-m)}$$

$$\cdot\left(\frac{\binom{m}{2}}{z-m}-\frac{j^2-j+2+2mj-4m}{4}+O(z-m)\right)$$

$$=\frac{y^j(-1)^{j-m}\binom{m}{2}}{(m-1)!(j-m)!(z-m)^2}-\frac{y^j(-1)^{j-m}(j^2-j+2+2mj-4m)}{4(m-1)!(j-m)!(z-m)}+O(1).$$

Hence, we have

$$a_m(y)=\sum_{j\geq m}\frac{y^j(-1)^{j-m}\binom{m}{2}}{(m-1)!(j-m)!}=\frac{r^m\binom{m}{2}}{(m-1)!}\sum_{k\geq 0}\frac{(-y)^k}{k!}$$

$$=\frac{y^m\binom{m}{2}e^{-y}}{(m-1)!},$$

and, similarly,

$$b_m(y)=-\frac{y^m e^{-y}(y^2-4my+3m^2-5m+2)}{4(m-1)!}.$$

Now we pass the following ordinary generating function

$$P'_{\text{dis}}(x,y;1)=\sum_{j\geq 1}a_j(y)\sum_{k\geq 0}(k+1)j^k x^{k+2}+\sum_{j\geq 1}b_j(y)\sum_{k\geq 0}j^k x^{k+1}$$

to the exponential generating function (see Example 2.59)

$$\begin{aligned}
E(x,y) &= \sum_{j\geq 1} a_j(y) \sum_{k\geq 0} \frac{(k+1)j^k x^{k+2}}{(k+2)!} + \sum_{j\geq 1} b_j(y) \sum_{k\geq 0} \frac{j^k x^{k+1}}{(k+1)!} \\
&= \sum_{j\geq 1} a_j(y) \frac{e^{jx}(jx-1)+1}{j^2} + \sum_{j\geq 1} b_j(y) \frac{e^{jx}-1}{j} \\
&= \sum_{j\geq 1} \frac{y^j \binom{j}{2} e^{-y}}{(j-1)!} \cdot \frac{e^{jx}(jx-1)+1}{j^2} \\
&\quad - \sum_{j\geq 1} \frac{y^j e^{-y}(y^2 - 4jy + 3j^2 - 5j + 2)}{4(j-1)!} \cdot \frac{e^{jx}-1}{j} \\
&= x \sum_{j\geq 2} \frac{y^j e^{-y} e^{jx}}{2(j-2)!} - \sum_{j\geq 1} \frac{(j-1)y^j e^{-y}(e^{jx}-1)}{2j!} \\
&\quad - \sum_{j\geq 1} \frac{(y^2 - 4jy + 3j^2 - 5j + 2)y^j e^{-y}(e^{jx}-1)}{4j!}.
\end{aligned}$$

Therefore,

$$\begin{aligned}
E(x,y) &= \frac{xy^2 e^{2x}}{2} e^{y(e^x-1)} - \sum_{j\geq 1} \frac{(y^2 - 4jy + 3j^2 - 3j)y^j e^{-y}(e^{jx}-1)}{4j!} \\
&= \frac{xy^2 e^{2x}}{2} e^{y(e^x-1)} - \sum_{j\geq 0} \frac{y^{j+2} e^{-y}(e^{jx}-1)}{4j!} \\
&\quad + \sum_{j\geq 1} \frac{y^{j+1} e^{-y}(e^{jx}-1)}{(j-1)!} - \sum_{j\geq 2} \frac{3y^j e^{-y}(e^{jx}-1)}{4(j-2)!} \\
&= \frac{xy^2 e^{2x}}{2} e^{y(e^x-1)} - \frac{y^2}{4}\left(e^{y(e^x-1)} - 1\right) \\
&\quad + y^2\left(e^x e^{y(e^x-1)} - 1\right) - \frac{3y^2}{4}\left(e^{2x} e^{y(e^x-1)} - 1\right) \\
&= \frac{y^2}{4} e^{y(e^x-1)} \left((2x-3)e^{2x} + 4e^x - 1\right).
\end{aligned}$$

By extracting coefficients, we obtain (do that!)

$$\begin{aligned}
\text{total}_{\mathcal{P}_{n,k}}(\text{dis}) &= \frac{k(k-1)(2n-3k)}{4} \text{Stir}(n-1,k) \\
&\quad + \frac{n-k+1}{2} \text{Stir}(n-1,k-2) \\
&\quad + \frac{(k-1)(4n-5k+2)}{4} \text{Stir}(n-1,k-1).
\end{aligned}$$

By using Theorem 1.12 several times, we can state the following result.

**Theorem 5.33** *Let $1 \leq k \leq n$. Then*

$$\text{total}_{\mathcal{P}_{n,k}}(\text{dis}) = \frac{(n+1-k)\text{Stir}(n+1,k) - \left(\binom{k}{2}+n\right)\text{Stir}(n,k)}{2}.$$

It is well known that $\text{Stir}(n,k) = \frac{k^n}{k!} + O((k-1)^n)$ for fixed $k$, see Chapter 8; hence we have the following immediate corollary.

**Corollary 5.34** *For fixed $k$,*

$$\text{total}_{\mathcal{P}_{n,k}}(\text{dis}) \sim \frac{(k-1)k^n}{4k!}(2n-3k) + O(n(k-1)^n),$$

*and the mean of* $\text{total}_{\mathcal{P}_{n,k}}(\text{dis})$ *is asymptotically*

$$\frac{(k-1)(2n-3k)}{4} + O\left(n\left(\frac{k-1}{k}\right)^n\right).$$

For $\text{total}_{\mathcal{P}_n}(\text{dis})$, we set $y=1$ in the exponential generating function to find that the exponential generating function for the total of statistic dis over all set partitions of $\mathcal{P}_n$ is given by

$$\frac{1}{4}e^{e^x-1}\left((2x-3)e^{2x} + 4e^x - 1\right),$$

and again extracting coefficients gives the following result.

**Theorem 5.35** *For all $n$,*

$$\text{total}_{\mathcal{P}_n}(\text{dis}) = \frac{2n+7}{4}\text{Bell}_{n+1} - \frac{3}{4}\text{Bell}_{n+2} - \frac{2n+1}{4}\text{Bell}_n.$$

Since $\frac{\text{Bell}_{n+1}}{\text{Bell}_n} = \frac{n}{\log n} + O\left(\frac{n \log \log n}{\log^2 n}\right)$ (see Theorem 8.54), we obtain that the mean of $\text{total}_{\mathcal{P}_n}(\text{dis})$ is asymptotically

$$\frac{n^2}{2\log n} + O\left(\frac{n^2 \log \log n}{\log^2 n}\right).$$

### 5.4.2 The Statistic $m$-Distance

In this section, we investigate the generating function $P_{\text{dis}_1}(x,y;q)$, which is the generating function for the number of set partitions in $\mathcal{P}_{n,k}$ according to the statistic $\text{dis}_1$. Theorem 5.29 gives

$$P_{\text{dis}_1}(x,y;q) = 1 + \frac{xy(1-q)P_{\text{dis}_1}(x,y;1)}{1-q-x} - \frac{yx^2qP_{\text{dis}_1}\left(\frac{xq}{1-x},y;1\right)}{(1-x)(1-q-x)}.$$

By substituting $q = 1$ at first and then solving, we obtain

$$P_{\text{dis}_1}(x,y;q) = 1 + \frac{xy(1-q)}{1-q-x} \sum_{j\geq 0} \frac{y^j x^j}{\prod_{i=1}^{j} 1-ix}$$
$$- \frac{yx^2 q}{1-q-x} \sum_{j\geq 0} \frac{y^j x^j q^j}{\prod_{i=0}^{j} 1-x(1+iq)},$$

which implies the following result.

**Theorem 5.36** *The generating function for the number of set partitions in $\mathcal{P}_{n,k}$ according to the statistic $\text{dis}_1$ is given by*

$$P_{\text{dis}_1}(x,y;q) = 1 + \frac{xy}{1-q-x} \sum_{j\geq 0} y^j x^j \left[ \frac{1-q}{\prod_{i=1}^{j} 1-ix} - \frac{xq^{j+1}}{\prod_{i=0}^{j} 1-x(1+iq)} \right].$$

For instance, the generating function for the number of partitions $\pi \in \mathcal{P}_{n,k}$ with $d_1(\pi) = 0$ is given by

$$P_{\text{dis}_1}(x,y;q) = 1 + \frac{xy}{1-x} \sum_{j\geq 0} \frac{y^j x^j}{\prod_{i=1}^{j}(1-ix)},$$

which implies that the number of set partitions $\pi \in \mathcal{P}_{n,k}$ having $\text{dis}_1(\pi) = 0$ is given by $\sum_{j=k-1}^{n-1} \text{Stir}(j, k-1)$ if $n \geq k \geq 1$. Thus, the number of set partitions $\pi \in \mathcal{P}_n$ with $\text{dis}_1(\pi) = 0$ is given by $\sum_{j=0}^{n-1} \text{Bell}_j$ for all $n \geq 1$.

Now, by differentiating the generating function $P_{\text{dis}_1}(x,y;q)$ with respect to $q$ at $q = 1$, and extracting the coefficient of $y^k$, we have

$$[y^k]\left(\frac{d}{dq}P_{\text{dis}_1}(x,y;q)\,|_{q=1}\right) = \frac{y^k}{\prod_{i=0}^{k} 1-ix} \sum_{i=1}^{k} \frac{(i-1)x}{1-ix}, \qquad (5.16)$$

which leads to the following result.

**Corollary 5.37** *Let $1 \leq k \leq n$. Then*

$$\text{total}_{\mathcal{P}_{n,k}}(\text{dis}_1) = \sum_{i=2}^{k}\left((i-1)\sum_{j=k}^{n-1} i^{n-1-j}\text{Stir}(j,k)\right).$$

Corollary 5.37 may be generalized to find the $\text{total}_{\mathcal{P}_{n,k}}(\text{dis}_m)$ as follows. At first, delete all the letters $1, 2, \ldots, m-1$ from the set partition $\pi$ and denote the resulting set partition by $\pi'$. Then counting $\text{dis}_1(\pi')$ over all possible set partitions $\pi'$ obtains the following result.

**Corollary 5.38** For all $n > k \geq 1$, $\text{total}_{\mathcal{P}_{n,k}}(\text{dis}_m)$ is given by

$$\sum \left[ \prod_{s=1}^{m-1} \binom{n-j_s}{i_s} \sum_{i=2}^{k+1-m} (i-1) \sum_{j=k+1-m}^{n-j_m} i^{n-j_m-j} \text{Stir}(j, k+1-m) \right],$$

where $j_s = i_1 + \cdots + i_{s-1} + s$ and the external sum is over all $(i_1, \ldots, i_{m-1})$ such that $0 \leq i_s \leq n - j_s$ for all $s = 1, 2, \ldots, m-1$.

Similar to the case of the statistic dis, one derives a simpler formula for $\text{dis}_1$ as well as its asymptotic behavior by means of exponential generating functions. At first, by using the partial fraction decomposition, we have

$$\sum_{k \geq 1} \frac{y^k x^k}{\prod_{i=1}^{k}(1-ix)} \sum_{i=1}^{k} \frac{(i-1)x}{1-ix}$$
$$= \sum_{j \geq 2} \frac{y^j e^{-y}}{(j-2)!(x^{-1}-j)^2} - \sum_{j \geq 1} \frac{y^j e^{-y}(j-y-1)}{(j-1)!(x^{-1}-j)},$$

and as an immediate consequence the exponential generating function

$$\sum_{j \geq 2} \frac{y^j e^{-y}}{(j-2)!} \cdot \frac{e^{jx}(jx-1)+1}{j^2} - \sum_{j \geq 1} \frac{y^j e^{-y}(j-y-1)}{(j-1)!} \cdot \frac{e^{jx}-1}{j}.$$

Note that the sums do not evolve directly to elementary functions. Thus, we differentiate with respect to $x$ to obtain, after some simplifications,

$$\sum_{j \geq 2} \frac{y^j e^{-y}}{(j-2)!} \cdot xe^{jx} - \sum_{j \geq 1} \frac{y^j e^{-y}(j-y-1)}{(j-1)!} \cdot e^{jx} = y^2 e^x (xe^x - e^x + 1)e^{y(e^x-1)}.$$

By extracting coefficients implies that $\text{total}_{\mathcal{P}_{n,k}}(\text{dis}_1)$, we derive

$$(n-k)\text{Stir}(n-2, k-2) + k(k-1)(n-k-1)\text{Stir}(n-2, k)$$
$$+ (k-1)(2n-2k-1)\text{Stir}(n-2, k-1).$$

Simplifying this, using Theorem 1.12, gives the following result.

**Theorem 5.39** For all $1 \leq k \leq n$,

$$\text{total}_{\mathcal{P}_{n,k}}(\text{dis}_1) = (n-k)\text{Stir}(n, k) - (n-1)\text{Stir}(n-1, k).$$

Moreover, we can derive the following asymptotic result which parallels Corollary 5.34.

**Corollary 5.40** For fixed $k$,

$$\text{total}_{\mathcal{P}_{n,k}}(\text{dis}_1) \sim \frac{(k-1)(n-k-1)k^{n-1}}{k!} + O(n(k-1)^n),$$

and the mean of $\text{total}_{\mathcal{P}_{n,k}}(\text{dis}_1)$ is asymptotically

$$\frac{(k-1)(n-k-1)}{k} + O\left(n\left(\frac{k-1}{k}\right)^n\right).$$

By setting $y = 1$ in the exponential generating function, we obtain $e^x(xe^x - e^x + 1)e^{e^x-1}$, which implies the following theorem for $\text{total}_{\mathcal{P}_{n,k}}(\text{dis}_1)$ which parallels Theorem 5.35.

**Theorem 5.41** *For all $n$,*

$$\text{total}_{\mathcal{P}_n}(\text{dis}_1) = (n+1)\text{Bell}_n - \text{Bell}_{n+1} - (n-1)\text{Bell}_{n-1}.$$

*The mean of $\text{total}_{\mathcal{P}_n}(\text{dis}_1)$ is asymptotically $n + O(n/\log n)$.*

### 5.4.3 Combinatorial Proofs

Now, we provide direct combinatorial proofs of Corollaries 5.32 and 5.37 and Theorems 5.39 and 5.41. In order to do that, we first consider some additional statistics on $\mathcal{P}_{n,k}$.

**Definition 5.42** *Let $\pi = \pi_1\pi_2\cdots\pi_n \in \mathcal{P}_{n,k}$, we define the statistic prm by $\text{prm}(\pi) = j - 1$, where $\pi_j = 1 < \pi_i$ for all $i = j+1, j+2, \ldots, n$, that is, $j$ denotes the position of the rightmost of the letter 1 in $\pi$.*

The total of the statistic prm on set partitions $\mathcal{P}_{n,k}$ is given by the following lemma.

**Lemma 5.43** *Let $1 \leq k \leq n$, $\text{total}_{\mathcal{P}_{n,k}}(\text{prm}) = (n-k)\text{Stir}(n,k)$.*

**Proof** In the case $k = 1$ there is only one set partition, namely $\pi = 11\cdots 1 \in \mathcal{P}_{n,1}$, thus $\text{prm}(\pi) = n - 1$. Also, in the case $k = n$ there is only one set partition, namely $\pi = 12\cdots n \in \mathcal{P}_{n,n}$, thus $\text{prm}(\pi) = 0$. Therefore, we can assume that $2 \leq k \leq n-1$ and let $\pi = \pi_1\pi_2\cdots\pi_n \in \mathcal{P}_{n,k}$. By considering either (i) $\pi_n = 1$, (ii) $\pi_n > 1$ with $\pi' = \pi_1\pi_2\cdots\pi_{n-1} \in \mathcal{P}_{n-1,k}$, or (iii) $\pi_n = k$ with $\pi' \in \mathcal{P}_{n-1,k-1}$, we obtain

$$\begin{aligned}\text{total}_{\mathcal{P}_{n,k}}&(\text{prm}) \\ &= (n-1)\text{Stir}(n-1,k) + (k-1)(n-1-k)\text{Stir}(n-1,k) \\ &\quad + (n-k)\text{Stir}(n-1,k-1) \\ &= (n-k)(\text{Stir}(n-1,k-1) + k\text{Stir}(n-1,k)) \\ &= (n-k)\text{Stir}(n,k)\end{aligned}$$

(use Theorem 1.12), which completes the induction. □

Now, we define two more statistics as follows on set partitions.

**Definition 5.44** Let $\pi = \pi_1\pi_2\cdots\pi_n \in \mathcal{P}_{n,k}$, we denote the number of distinct elements of $\{2,3,\ldots,k\}$ which occur to the left of the rightmost 1 in $\pi$ by $\alpha(\pi)$ and we denote the number of positions to the right of the rightmost 1 in $\pi$ and not corresponding to the initial occurrence of a letter by $\beta(\pi)$.

**Example 5.45** For instance, if $\pi = 12321342523 \in \mathcal{P}_{11,5}$, then $\alpha(\pi) = 2$ and $\beta(\pi) = 4$.

We start by stating the following relation between the statistics $\alpha$ and $\beta$ on the set partitions.

**Lemma 5.46** For all $1 \leq k \leq n$, $\text{total}_{\mathcal{P}_{n,k}}(\alpha) = \text{total}_{\mathcal{P}_{n,k}}(\beta)$.

**Proof** Let $\pi \in \mathcal{P}_{n,k}$. By the definitions we have that

$$\beta(\pi) = (n-k) - \text{prm}(\pi) + \alpha(\pi).$$

Thus, our claim is equivalent to

$$\sum_{\pi \in \mathcal{P}_{n,k}} \alpha(\pi) = \sum_{\pi \in \mathcal{P}_{n,k}} (n - k - \text{prm}(\pi) + \alpha(\pi)),$$

which reduces to

$$\sum_{\pi \in \mathcal{P}_{n,k}} \text{prm}(\pi) = (n-k) \sum_{\pi \in \mathcal{P}_{n,k}} 1 = (n-k)\text{Stir}(n,k).$$

Hence, by Lemma 5.43 we complete the proof. □

**Direct proof of Corollary 5.37**: Now, let us provide a direct proof of Corollary 5.37. First we state the following lemma where we leave the proof as an interesting exercise (see [241]).

**Lemma 5.47** Let $\gamma(\pi)$ to be the number of positions in $\pi \in \mathcal{P}_{n,k}$ corresponding to letters greater than 1 and not corresponding to an initial occurrence of a letter. Then the total of the statistic $\gamma$ over all set partitions in $\mathcal{P}_{n,k}$ is given by

$$\sum_{i=2}^{k} \left( (i-1) \sum_{j=k}^{n-1} i^{n-1-j} \text{Stir}(j,k) \right).$$

Thus, from the definitions and Lemma 5.46, we may write

$$\text{total}_{\mathcal{P}_{n,k}}(\gamma) = \text{total}_{\mathcal{P}_{n,k}}(d_1) - \text{total}_{\mathcal{P}_{n,k}}(\alpha) + \text{total}_{\mathcal{P}_{n,k}}(\beta) = \text{total}_{\mathcal{P}_{n,k}}(d_1),$$

which implies the formula of $\text{total}_{\mathcal{P}_{n,k}}(\text{dis}_1)$ as given in Corollary 5.37. □

**Direct proof of Corollary 5.32**: The above proof can be extended to explain Corollary 5.32 as follows. We first extend our statistics.

**Definition 5.48** For each $i \in [k-1]$, let $\alpha_i(\pi)$ denote the number of distinct elements of the set $\{i+1, i+2, \ldots, k\}$ which occur to the left of the rightmost $i$ in $\pi \in \mathcal{P}_{n,k}$, let $\beta_i(\pi)$ denote the number of positions to the right of the rightmost $i$ in $\pi$ that do not correspond to an initial occurrence of a letter but are larger than $i$, and let $\gamma_i(\pi)$ to be the number of positions of $\pi$ greater than $i$ and not corresponding to an initial occurrence of a letter.

Clearly, $\alpha = \alpha_1$ and $\beta = \beta_1$. Also, for each $i$,

$$\text{total}_{\mathcal{P}_{n,k}}(\alpha_i) = \text{total}_{\mathcal{P}_{n,k}}(\beta_i),$$

by Lemma 5.46, deleting all occurrence of the letters $1, 2, \ldots, i-1$ within set partitions of $\mathcal{P}_{n,k}$ and then considering $\alpha_1$ and $\beta_1$ on the obtaining set partitions, as in the proof of Corollary 5.38. Also, we can state that

$$\sum_{i=1}^{k-1} \text{total}_{\mathcal{P}_{n,k}}(\gamma_i) = \sum_{i=2}^{k} \left( \binom{i}{2} \sum_{j=k}^{n-1} i^{n-1-j} \text{Stir}(j,k) \right). \quad (5.17)$$

In order to show this, we fix $i$ and $j$ such that $2 \leq i \leq k \leq j \leq n-1$, and we consider the set of all ordered pairs $(\pi, a)$, where

$$\pi = \pi' i \pi'' \pi''' \in \mathcal{P}_{n,k}, \quad (5.18)$$

$a$ is a positive integer strictly less than the last letter of $\pi''$, $\pi'$ is a set partition with exactly $i-1$ blocks, $\pi''$ is nonempty word in $[i]^{n-j}$ whose last letter is not 1 and $\pi'''$ is any word. Thus, the right side of (5.17) gives the total number of ordered pairs as $i$ and $j$ vary over all possible values. Similarly, $\text{total}_{\mathcal{P}_{n,k}}(\gamma_i)$ equals the total number of ordered pairs $(\pi, a)$ with $a = i$ for each $i \in [k-1]$, which gives (5.17).

By the definitions and the fact that $\text{total}_{\mathcal{P}_{n,k}}(\alpha_i) = \text{total}_{\mathcal{P}_{n,k}}(\beta_i)$, we have for each $i \in [k-1]$,

$$\text{total}_{\mathcal{P}_{n,k}}(\gamma_i) = \text{total}_{\mathcal{P}_{n,k}}(\text{dis}_i) - \text{total}_{\mathcal{P}_{n,k}}(\alpha_i) + \text{total}_{\mathcal{P}_{n,k}}(\beta_i)$$
$$= \text{total}_{\mathcal{P}_{n,k}}(\text{dis}_i),$$

which gives

$$\text{total}_{\mathcal{P}_{n,k}}(\text{dis}) = \sum_{i=1}^{k-1} \text{total}_{\mathcal{P}_{n,k}}(\text{dis}_i) = \sum_{i=1}^{k-1} \text{total}_{\mathcal{P}_{n,k}}(\gamma_i),$$

from which Corollary 5.32 follows from (5.17). $\square$

**Direct proof of Theorems 5.39 and 5.41**: Again, by using similar arguments in the above proof we can explain Theorems 5.39 and 5.41. At first we state the following lemma.

**Lemma 5.49** *Let $\pi = \pi_1\pi_2\cdots\pi_n \in \mathcal{P}_{n,k}$ and define $\delta(\pi)$ to be the number of positions $i > 1$ such that $\pi_i = 1$. Then*

$$\text{total}_{\mathcal{P}_{n,k}}(\delta) = (n-1)\text{Stir}(n-1,k).$$

**Proof** Follows immediately from the fact that there are $\text{Stir}(n-1,k)$ occurrences of 1 in the $i$-th position, taken over $\mathcal{P}_{n,k}$, for each $i$, $2 \leq i \leq n$. □

Since there are $(n-k)\text{Stir}(n,k)$ positions in all of the set partitions of $\mathcal{P}_{n,k}$ which do not correspond to initial occurrences of letters, we see that

$$\text{total}_{\mathcal{P}_{n,k}}(\gamma) = (n-k)\text{Stir}(n,k) - (n-1)\text{Stir}(n-1,k),$$

upon subtracting all noninitial 1's. However, by Corollary 5.37 we have

$$\text{total}_{\mathcal{P}_{n,k}}(\gamma) = \text{total}_{\mathcal{P}_{n,k}}(\text{dis}_1).$$

By summing the expression in Theorem 5.39 over all $k$ and using the fact that $\sum_{k=1}^{n} k\text{Stir}(n,k) = \text{Bell}_{n+1} - \text{Bell}_n$, we obtain Theorem 5.41. □

Note that ones can derive another expression for $\text{total}_{\mathcal{P}_n}(\text{dis}_1)$ as

$$\text{total}_{\mathcal{P}_{n,k}}(\text{dis}_1) = \sum_{i=1}^{n-k} i S_{n-1-i,k-1}\binom{n-1}{i+1},$$

see Exercise 5.10.

## 5.5 The Internal Statistic

At first, we define internal points in the graphical representation of a set partition, see Definition 3.49.

**Definition 5.50** *Suppose the set partition $\pi = \pi_1\pi_2\cdots\pi_n$ is represented graphically as in Definition 3.49. We will call a point $P$ in the graph of $\pi$ internal if there exist points in the graph both above and below it lying to its right. In other words, a point $P = (\pi_i, i)$ in the graph of $\pi$ is internal if and only if there exist $j$ and $k$ with $j < i < k$ such that $\pi_i < \min\{\pi_j, \pi_k\}$. We call the corresponding letter $\pi_i$ in $\pi$ an internal letter.*

For instance, in the set partition as described in Figure 3.2, only the points $(1,4)$ and $(2,5)$ are internals.

**Definition 5.51** *Given $\pi \in \mathcal{P}_{n,k}$ and $m \in [k-1]$, let $\text{int}_m(\pi)$ be the number of internal points of $\pi$ whose x-coordinate is $m$. We define $\text{int}(\pi) = \sum_{m=1}^{k-1} \text{int}_m(\pi)$ to be the total number of internal points of $\pi$.*

**Example 5.52** *For instance, if $\pi = 123214321431 \in \mathcal{P}_{12,4}$ then $\text{int}_1(\pi) = 2$, $\text{int}_2(\pi) = 2$, $\text{int}_3(\pi) = 1$ and $\text{int}(\pi) = 5$.*

**Definition 5.53** *Let $F_n(y; q_1, q_2, \ldots)$ be the generating function for the number of set partitions in $\mathcal{P}_{n,k}$ according to the statistics $\text{int}_1$, $\text{int}_2$, $\ldots$, that is,*

$$F_n(y; q_1, q_2, \ldots) = \sum_{k \geq 0} \sum_{\pi \in \mathcal{P}_{n,k}} y^k \prod_{j \geq 1} q_j^{\text{int}_j(\pi)}.$$

Let $n \geq 2$, assume that in a set partition of $[n]$ the letter 1 occurs exactly $j+1$ times, where $1 \leq j \leq n-1$. If $j = n-1$ then there is only one possible set partition, so let $j \leq n-2$. Let be $j-k+1$ of the 1's appear before the first occurrence of the letter 2 and $i$ of the 1's are internal (so the set partition ends with $11\cdots 1 = 1^{k-i}$). Then the $i$ 1's are to be distributed in $n-j-2$ positions in $\binom{n-j-3+i}{i}$ possible ways. Therefore, the total $\text{int}_1$ weight contribution is

$$\sum_{k=0}^{j} \sum_{i=0}^{k} q_1^i \binom{n-j-3+i}{i} = \sum_{i=0}^{j} q_1^i \binom{n-j-3+i}{i}(j-i+1).$$

Hence, for $n \geq 2$,

$$\begin{aligned}F_n(y; q_1, q_2, \ldots) &= y + yF_{n-1}(y; q_2, q_3, \ldots) \\ &+ y \sum_{j=1}^{n-2} F_{n-1-j}(y; q_2, q_3, \ldots) \sum_{i=0}^{j} q_1^i \binom{n-j-3+i}{i}(j-i+1).\end{aligned} \quad (5.19)$$

Define $H(x, y; q_1, q_2, \ldots) = P_{\text{int}_1, \text{int}_2, \ldots}(x, y; q_1, q_2, \ldots) - 1$. By multiplying both sides of (5.19) by $x^n$ and summing over $n \geq 2$, we obtain

$$P_{\text{int}_1, \text{int}_2, \ldots}(x, y; q_1, q_2, \ldots) = 1 + \frac{yx^2}{1-t} + xyP_{\text{int}_1, \text{int}_2, \ldots}(x, y; q_2, q_3, \ldots) \quad (5.20)$$

$$+ y \sum_{n \geq 3} x^n \left( \sum_{j=1}^{n-2} F_{n-1-j}(y; q_2, q_3, \ldots) \sum_{i=0}^{j} q_1^i \binom{n-j-3+i}{i}(j-i+1) \right).$$

Observe that

$$y \sum_{n \geq 3} x^n \left( \sum_{j=1}^{n-2} F_{n-1-j}(y; q_2, q_3, \ldots) \sum_{i=0}^{j} q_1^i \binom{n-3-j+i}{i}(j+1-i) \right)$$

$$= yx^3 \sum_{n \geq 0} x^n \left( \sum_{j=0}^{n} F_{n+1-j}(y; q_2, q_3, \ldots) \sum_{i=0}^{j+1} q_1^i \binom{n-1-j+i}{i}(j+2-i) \right)$$

$$= yx^3 \sum_{n \geq 0} x^n \left( \sum_{j=1}^{n+1} F_j(y; q_2, q_3, \ldots) \sum_{i=0}^{n-j+2} q_1^i \binom{j-2+i}{i}(n+3-j-i) \right)$$

$$= yx^2 \sum_{j \geq 1} x^j F_j(y; q_2, q_3, \ldots) \left( \sum_{n \geq 0} x^n \sum_{i=0}^{n+1} q_1^i \binom{j-2+i}{i}(n+2-i) \right) \quad (5.21)$$

and

$$\sum_{n\geq 0} x^n \sum_{i=0}^{n+1} q_1^i \binom{j-2+i}{i}(n+2-i)$$

$$= \sum_{n\geq 0} x^n q_1^{n+1}\binom{j-1+n}{n+1} + \sum_{i\geq 0} x^i q_1^i \binom{j-2+i}{i}\sum_{n\geq 0} x^n(n+2)$$

$$= \frac{1}{x}\left(\frac{1}{(1-xq_1)^{j-1}} - 1\right) + \left(\frac{1}{(1-xq_1)^{j-1}}\right)\left(\frac{2-x}{(1-x)^2}\right),$$

so that (5.20) and (5.21) imply

$$P_{\text{int}_1,\text{int}_2,\ldots}(x,y;q_1,q_2,\ldots)$$
$$= 1 + \frac{yx^2}{1-x} + xy P_{\text{int}_1,\text{int}_2,\ldots}(x,y;q_2,q_3,\ldots)$$
$$- xy(P_{\text{int}_1,\text{int}_2,\ldots}(x,y;q_2,q_3,\ldots) - 1)$$
$$+ \frac{xy(1-xq_1)}{(1-x)^2}\left(P_{\text{int}_1,\text{int}_2,\ldots}\left(\frac{x}{1-xq_1},y;q_2,q_3,\ldots\right) - 1\right),$$

which obtains the following recurrence relation for the generating function $H(x,y;q_1,q_2,\ldots)$.

**Theorem 5.54** *The generating function $H(t,r;q_1,q_2,\ldots)$ satisfies*

$$H(x,y;q_1,q_2,\ldots) = \frac{xy}{1-x} + \frac{xy(1-xq_1)}{(1-x)^2} H\left(\frac{x}{1-xq_1},y;q_2,q_3,\ldots\right).$$

In the next sections we use the above theorem to study several particular cases.

### 5.5.1 The Statistic int

Let $H(x,y;q) := H(x,y;q,q,\ldots)$, so $P_{\text{int}}(x,y;q) = H(x,y;q)+1$. Theorem 5.54 gives

$$H(x,y;q) = \frac{xy}{1-x} + \frac{xy(1-xq)}{(1-x)^2} H\left(\frac{x}{1-xq},y;q\right).$$

By iterating this recurrence relation, we obtain the following result.

**Theorem 5.55** *The generating function $P_{\text{int}}(x,y;q)$ for the number of set partitions in $\mathcal{P}_{n,k}$ according to the statistic int is given by*

$$1 + \sum_{j\geq 1} \frac{y^j x^j}{1-x-(j-1)qx}\prod_{i=1}^{j-1}\frac{1-ixq}{(1-x-(i-1)qx)^2}.$$

By differentiating the generating function $P_{\text{int}}(x, y; q)$ with respect to $q$ at $q = 1$, and then finding the coefficient of $y^k$, we obtain

$$[y^k]\left(\frac{d}{dq}P_{\text{int}}(x, y; q)\big|_{q=1}\right) = \frac{y^{k+1}}{\prod_{i=1}^{k}(1-ix)}\left(\frac{1}{1-kx} + \sum_{i=1}^{k}\frac{i-2}{1-ix}\right),$$

which implies the following result.

**Corollary 5.56** *For all $1 \leq k \leq n$,*

$$\text{total}_{\mathcal{P}_{n,k}}(\text{int})$$
$$= \sum_{j=k}^{n-1} k^{n-1-j}\text{Stir}(j,k) + \sum_{i=1}^{k}\left((i-2)\sum_{j=k}^{n-1} i^{n-1-j}\text{Stir}(j,k)\right).$$

It seems that there is no nice formula for $\text{total}_{\mathcal{P}_{n,k}}(\text{int})$ analogous to Theorems 5.33 and 5.35 (the partial fraction method gives very complicated exponential generating function), one can at least again determine the asymptotic behavior. For fixed $k$, note that the generating function is rational, with a dominant double pole at $\frac{1}{k}$. By means of the classical singularity analysis (see Chapter 8), we get the expansion

$$\frac{t^{k+1}}{\prod_{i=1}^{k}(1-it)}\left(\frac{1}{1-kt} + \sum_{i=1}^{k}\frac{i-2}{1-it}\right)$$
$$= \frac{k-1}{k \cdot k!(1-tk)^2} + \frac{3-k-k^2-kH_k}{k \cdot k!(1-tk)} + O(1),$$

where $H_k = \sum_{j=1}^{k}\frac{1}{j}$ denotes the $k$-th harmonic number, which leads to the following result.

**Corollary 5.57** *For fixed $k$,*

$$\text{total}_{\mathcal{P}_n}(\text{int}) \sim \frac{k^{n-1}}{k!}((k-1)n + 2 - k^2 - kH_k) + O(n(k-1)^n),$$

*and the* **mean** *of int over all such partitions is therefore asymptotically*

$$\frac{(k-1)n}{k} + \frac{2}{k} - k - H_k + O\left(n\left(\frac{k-1}{k}\right)^n\right).$$

For the total over all set partitions one has to proceed differently: by Exercise 5.13, the total int is at most the total $\text{dis}_1$, which is $(n+1)\text{Bell}_n - \text{Bell}_{n+1} - (n-1)\text{Bell}_{n-1}$, by Theorem 5.41. On the other hand, it is greater than

$$\sum_{k=1}^{n}\sum_{i=1}^{k}\left((i-2)\sum_{j=k}^{n-1} i^{n-1-j}\text{Stir}(j,k)\right),$$

by Corollary 5.56. The partial fractions technique can then be applied to this expression to yield the exponential generating function $(x-1)(e^x-1)e^{e^x+x-1}$, as in the proof of Theorem 5.41 above. Extracting coefficients yields the following theorem. (Check the details!!)

**Theorem 5.58** *For all $n$,*

$$(n+1)\text{Bell}_n - \text{Bell}_{n+1} - 2(n-1)\text{Bell}_{n-1}$$
$$\leq \text{total}_{\mathcal{P}_n}(\text{int})$$
$$\leq (n+1)\text{Bell}_n - \text{Bell}_{n+1} - (n-1)\text{Bell}_{n-1}$$

*and so the average* int *is* $\frac{\text{total}_{\mathcal{P}_n}(\text{int})}{\text{Bell}_n} = n + O(n/\log n)$.

**Proof** Simply note that the difference between the two sides of the inequality is $(n-1)\text{Bell}_{n-1} = O(\text{Bell}_n \log n)$. Therefore,

$$\frac{\text{total}_{\mathcal{P}_n}(\text{int})}{\text{Bell}_n} = \frac{(n+1)\text{Bell}_n - \text{Bell}_{n+1} - (n-1)\text{Bell}_{n-1}}{\text{Bell}_n} + O(\log n)$$
$$= n + O(n/\log n),$$

as required. □

### 5.5.2 The Statistic $\text{int}_1$

By fixing $q_2 = q_3 = \cdots = 1$ in Theorem 5.54 and using the $q = 1$ case, we obtain an explicit formula for the generating function $P_{\text{int}_1}(x, y; q)$ for the number of set partitions in $\mathcal{P}_{n,k}$ according to the statistic $\text{int}_1$ as follows.

**Theorem 5.59** *The generating function $P_{\text{int}_1}(x, y; q)$ is given by*

$$P_{\text{int}_1}(x, y; q) = 1 + \frac{xy}{1-x} + \frac{xy(1-xq)}{(1-x)^2} \sum_{j\geq 1} y^j x^j \prod_{i=1}^{j} \frac{1}{1-(q+i)x}.$$

Note that Theorem 5.59 for $q = 0$ gives that the number of set partitions $\pi$ in $\mathcal{P}_{n,k}$ such that $\text{int}_1(\pi) = 0$ is given by

$$4 \sum_{j=k-1}^{n-1} (n-j)\text{Stir}(j, k-1)$$

if $2 \leq k \leq n$, and therefore the number of set partitions $\pi \in \mathcal{P}_n$ with $\text{int}_1(\pi) = 0$ is given by

$$1 + \sum_{j=1}^{n-1} (n-j)\text{Bell}_j$$

for all $n \geq 2$. This also follows directly from the definitions since within a set partition $\pi \in \mathcal{P}_n$ with $\mathrm{int}_1(\pi) = 0$, the 1's can only occur as runs at the very beginning or at the very end.

By differentiating the generating function $P_{\mathrm{int}_1}(x, y; q)$ with respect to $q$ at $q = 1$, and then finding the coefficient of $y^k$, $k \geq 2$, we obtain

$$[y^k]\left(\frac{d}{dq}G(x,y;q)|_{q=1}\right) = \frac{x^{k+1}}{\prod_{i=1}^{k}(1-ix)}\left(-\frac{1}{1-x} + \sum_{i=1}^{k-1}\frac{1}{1-(i+1)x}\right),$$

which implies the following result.

**Corollary 5.60** *For all* $1 \leq k \leq n$,

$$\sum_{i=2}^{k}\left(\sum_{j=k}^{n-1} i^{n-1-j}\mathrm{Stir}(j,k)\right) - \sum_{j=k}^{n-1}\mathrm{Stir}(j,k).$$

Note that the combinatorial proof for Corollary 5.56 applies to Corollary 5.60 as well. There are $i^{n-1-j}$ choices for the word $\pi'$, which now must end in a 1. Thus, the first sum counts the total number of secondary 1's over $\mathcal{P}_{n,k}$. From this, we subtract the total number of secondary 1's which aren't internal, which is given by $\sum_{j=k}^{n-1}\mathrm{Stir}(j,k)$ (to show this, let $i=1$ in (A.5)).

Finally, let us determine the asymptotic behavior for $\mathrm{total}_{\mathcal{P}_{n,k}}(\mathrm{int}_1)$. For fixed $k$, we can again expand around the dominant pole $t = \frac{1}{k}$:

$$\frac{x^{k+1}}{\prod_{i=1}^{k}(1-ix)}\left(-\frac{1}{1-x} + \sum_{i=1}^{k-1}\frac{1}{1-(i+1)x}\right)$$

$$= \frac{1}{k \cdot k!(1-kx)^2} - \frac{2(2k-1)}{(k-1)k \cdot k!(1-kx)} + O(1),$$

which leads to the following result.

**Corollary 5.61** *For fixed* $k$,

$$\mathrm{total}_{\mathcal{P}_{n,k}}(\mathrm{int}_1) \sim \frac{k^{n-1}}{k!}\left(n - \frac{3k-1}{k-1}\right) + O(n(k-1)^n),$$

*and the* **mean** *of* $\mathrm{int}_1$ *over all such set partitions is asymptotically*

$$\frac{n}{k} - \frac{3k-1}{k(k-1)} + O\left(n\left(\frac{k-1}{k}\right)^n\right).$$

For $\mathrm{total}_{\mathcal{P}_n}(\mathrm{int}_1)$ we can state the following result.

**Theorem 5.62** *For all* $n$,

$$\mathrm{total}_{\mathcal{P}_n}(\mathrm{int}_1) = (n-1)\mathrm{Bell}_{n-1} - 2\mathrm{Bell}_n \leq \mathrm{total}_{\mathcal{P}_n}(\mathrm{int}_1) \leq (n-1)\mathrm{Bell}_{n-1},$$

*and so the* **mean** *of* $\text{int}_1$ *is*

$$\frac{\text{total}_{\mathcal{P}_n}(\text{int}_1)}{\text{Bell}_n} = \log n + O(\log\log n).$$

**Proof** First observe

$$\sum_{i=2}^{k}\left(\sum_{j=k}^{n-1} i^{n-1-j}\text{Stir}(j,k)\right) - \sum_{j=k}^{n-1}\text{Stir}(j,k)$$
$$= \sum_{i=1}^{k}\left(\sum_{j=k}^{n-1} i^{n-1-j}\text{Stir}(j,k)\right) - 2\sum_{j=k}^{n-1}\text{Stir}(j,k)$$

and

$$\sum_{k=1}^{n}\sum_{j=k}^{n-1}\text{Stir}(j,k) = \sum_{j=1}^{n-1}\sum_{k=1}^{j}\text{Stir}(j,k)$$
$$= \sum_{j=1}^{n-1}\text{Bell}_j < \sum_{j=0}^{n-1}\binom{n-1}{j}\text{Bell}_j = \text{Bell}_n.$$

Furthermore, one obtains (check!!), as in the proofs of Theorems 5.35 and 5.41,

$$\sum_{k=1}^{n}\sum_{i=1}^{k}\left(\sum_{j=k}^{n-1} i^{n-1-j}\text{Stir}(j,k)\right) = (n-1)\text{Bell}_{n-1},$$

which completes the proof. □

---

## 5.6 Statistics and Generalized Patterns

Following Section 3.2.1, we start this section by introducing new type patterns, namely *generalized patterns*, on the block representation of set partitions. The name "generalized patterns" motivates from the research of generalized pattern on set of permutations, words, and compositions as discussed in the books of Bóna [48] and Heubach and Mansour [138].

A set partition $\pi = B_1/B_2/\cdots/B_k$ defines a notion of adjacency of blocks, where we consider $B_i$ as being adjacent to both $B_{i-1}$ and $B_{i+1}$. Instead to present a formal "complicated" definition for our new patterns, we will focus on an example. Let us consider the set partition $\pi = 157/26/348$ in $\mathcal{P}_8$ and the pattern $\tau = 13/2$. As we can see $\pi$ has 17 occurrences of the pattern $\tau$, namely $15/2, 15/3, 15/4, 17/2, 17/6, 17/3, 17/4, 57/6, 5/26, 26/3, 26/4, 5/48,$

5/38, 7/48, 7/48, 6/48, and 6/38. Now, suppose that an occurrence of $\tau$ must appear in $\pi$ as adjacent blocks, then we have only 8 occurrences of the pattern $\tau$, namely, 15/2, 17/2, 17/6, 57/6, 5/26, 26/3, 26/4, and 6/48. We define $\tau$ with the adjacency blocks by the *generalized pattern* 13|2. In general, we will denote block adjacency using the vertical line.

**Example 5.63** *An occurrence of the generalized pattern* 14|2/3/5 *in set partition* $B_1/B_2/\cdots/B_k$ *is a set subpartition* $ab/c/d/e$ *such that* $a < c < d < b < e$ *and there exist three different numbers* $1 \leq i, i', i'', i''' \leq k$ *such that* $a, b \in B_i$, $c \in B_{i'}$, $d \in B_{i''}$ *and* $e \in B_{i'''}$, *where* $|i - i'| = 1$.

Another restriction can be required that the elements that represent 1 and 3 in the occurrence of the pattern $\tau = 13/2$ are adjacent. For example, the set partition $\pi = 157/26/34 \in \mathcal{P}_7$ has seven occurrences of the pattern $\tau$ such that the letters 1 and 3 are adjacent, namely, 15/2, 15/3, 15/4, 57/6, 5/26, 26/3, and 26/4. We denote this pattern with the requirement that 1 and 3 are adjacent by $\widehat{13}/2$. In general, we denote element adjacency by placing an arc over the elements, which must be adjacent.

**Example 5.64** *An occurrence of the generalized pattern* $\widehat{14}|2/\widehat{35}$ *in set partition* $B_1/B_2/\cdots/B_k$ *is a set subpartition* $ab/c/de$ *such that* $a < c < d < b < e$ *and there exist three different numbers* $1 \leq i, i', i'' \leq k$ *such that* $a, b \in B_i$, $c \in B_{i'}$ *and* $d, e \in B_{i''}$, *where* $|i - i'| = 1$, $b$ *followed immediately* $a$ *in the block* $B_i$ *and* $e$ *followed immediately* $d$ *in the block* $B_{i''}$.

Goyt [120] studied the number of set partitions that avoid a set of generalized patterns. In this context, we follow Definitions 3.23 and 3.32, and we denote the set of all set partitions of $\mathcal{P}_n$ that avoid a generalized pattern $\tau$ (respectively, a set of generalized patterns $T$) by $\mathcal{P}_n(\tau)$ (respectively, $\mathcal{P}_n(T)$). In particular, Goyt showed the following result.

**Proposition 5.65** (Goyt [120, Lemma 5.1]) *For all* $n \geq 0$,

(i) $\mathcal{P}_n(1/2/3) = \mathcal{P}_n(1|2/3) = \mathcal{P}_n(1/2|3) = \mathcal{P}_n(1|2|3)$,

(ii) $\mathcal{P}_n(1/23) = \mathcal{P}_n(1|23) = \mathcal{P}_n(1/\widehat{23}) = \mathcal{P}_n(1|\widehat{23})$,

(iii) $\mathcal{P}_n(13/2) = \mathcal{P}_n(\widehat{13}/2) = \mathcal{P}_n(13|2) = \mathcal{P}_n(\widehat{13}|2)$,

(iv) $\mathcal{P}_n(123) = \mathcal{P}_n(\widehat{123}) = \mathcal{P}_n(1\widehat{23}) = \mathcal{P}_n(\widehat{12}3)$,

(v) $\mathcal{P}_n(12/3) = \mathcal{P}_n(\widehat{12}/3)$,

(vi) $\mathcal{P}_n(12|3) = \mathcal{P}_n(\widehat{12}|3)$.

**Proof** The proof of all these cases are very similar. Thus, we only present the details of the proof of the fact that $\mathcal{P}_n(12|3) = \mathcal{P}_n(\widehat{12}|3)$. It is obvious if a set partition of $[n]$ contains the generalized pattern $\widehat{12}|3$, then it contains the generalized pattern 12|3. Now let $\pi = B_1/B_2/\cdots/B_k$ be any set partitions of $[n]$ contains an occurrence $ab/c$ of the generalized pattern 12|3. Assume $a, b \in B_i$ and $c \in B_j$ such that $|i - j| = 1$. If there is an element $b' \in B_i$ such that $a < b' < b$, then $ab'/c$ is an occurrence of 12|3. Hence, we can assume that there is no element $b' \in B_i$ such that $a < b' < b$. Thus, $\pi$ contains $\widehat{12}|3$, which completes the proof. □

In Section 3.2.1 (see Table 3.1) are presented explicit formulas for the number of set partitions that avoid a pattern of size three. Thus, in order to complete the enumeration of a generalized pattern of size three, we have to study the cardinality of the set $\mathcal{P}_n(12|3)$.

**Theorem 5.66** (Goyt [120, Lemma 5.1]) *We have*

$$|\mathcal{P}_n(12|3)| = |\mathcal{P}_{n-1}(12|3)| + 1 + \sum_{k=0}^{n-3} \binom{n-2}{k} |\mathcal{P}_k(12|3)|$$

*with* $|\mathcal{P}_0(12|3)| = |\mathcal{P}_1(12|3)| = 1$.

**Proof** Let $\pi = B_1/B_2/\cdots/B_k$ be any set partition in $\mathcal{P}_n(12|3)$. It is not hard to see that if $|B_i| \geq 1$ then $|B_{i+1}| = |B_{i-1}| = 1$, and if $B_{i+1} = \{a\}$, then $a$ is less than each element, not the first element, in the block $B_i$. Now let us consider the number elements in the first block:

- If $|B_1| = 1$, then $\pi$ avoids 12|3 if and only if Reduce($B_2/\cdots/B_k$) avoids 12|3.

- If $2 \leq |B_1| = k < n$, then $B_2 = 2$ and thus $\pi$ avoids 12|3 if and only if Reduce($B_3/\cdots/B_k$) avoids 12|3.

- If $|B_1| = n$ then $\pi = 12\cdots n$ and it avoids 12|3.

Thus,

$$|\mathcal{P}_n(12|3)| = |\mathcal{P}_{n-1}(12|3)| + 1 + \sum_{k=0}^{n-3} \binom{n-2}{k} |\mathcal{P}_{n-2-k}(12|3)|$$

with $|\mathcal{P}_0(12|3)| = |\mathcal{P}_1(12|3)| = 1$, as required. □

We refer the interested reader to [120], where the above results have been extended to study the number of even (odd) set partitions avoiding a three letter generalized pattern.

Now we will see that the language of generalized patterns is another language for writing several known statistics on set partitions. We start by recalling the inversion statistic on set partitions.

**Definition 5.67** Let $\pi = B_1/B_2/\cdots/B_k$ be any set partition and $b \in B_i$. We will say that $(b, B_j)$ is an inversion if $b > \min B_j$ and $i < j$. We define the inversion number of $\pi$, written $\mathrm{inv}(\pi)$, to be the number of inversions in $\pi$.

**Example 5.68** For instance, if $\pi = 137/26/45$ is a set partition, then $(3, B_2)$, $(7, B_2)$, $(7, B_3)$ and $(6, B_3)$ are the inversions in $\pi$. Thus $\mathrm{inv}(\pi) = 4$.

The inversion statistic was defined by Johnson [150, 151] and, independently, by Deodhar (1998 Ph.D. thesis). Indeed, the original definitions were formulated differently and not in terms of inversions. This statistic on set partitions has several applications. Johnson [150, 151] used it to present a combinatorial interpretation of the $q$-analog of the exponential formula (see Gessel [113]) and a $q$-analog of Faá di Bruno's formula for the $n$-th derivative of a composite function.

**Theorem 5.69** The generating function $S_q(n,k) = \sum_{\pi \mathcal{P}_{n,k}} q^{\mathrm{inv}(\pi)}$ satisfies

$$S_q(n+1, k) = S_q(n, k-1) + [k]_q S_q(n, k),$$

for all $1 \leq k \leq n+1$.

**Proof** Let us write an equation for $S_q(n, k)$. For any set partition of $\mathcal{P}_{n+1,k}$ there are two possibilities: either $n+1$ forms a single block, or the block containing $n+1$ has more than one element. In the first case, the contribution is $S_q(n, k-1)$, while in the second case, the element $n+1$ can be placed into one of the $k$ blocks of a set partition of $\mathcal{P}_{n,k}$, which leads to contribution $(1 + q + q^2 + \cdots + q^{k-1})S_q(n,k)$. The proof is followed by adding these two cases. □

As a remark, the above theorem and Theorem 5.2 (see the recurrence relation for $\tilde{S}_q(n,k)$ in the proof) suggested that the statistics $\tilde{w}$ and inv have the same distribution on the set partitions of $\mathcal{P}_{n,k}$. (Why!!)

The inversion number of a set partition can be characterized as number occurrences of a generalized pattern as follows.

**Proposition 5.70** (Goyt [120, Proposition 6.1]) For any $\pi \in \mathcal{P}_n$, the inversion number of $\pi$ equals number of occurrences of the generalized pattern $\widehat{1}3/\widehat{2}$, that is, number occurrences of the set subpartition $ac/b$ in $\pi$ such that $a < b < c$, where $a = \min B_i$, $b = \min B_j$ and $i < j$.

**Proof** Let $\pi = B_1/B_2/\cdots/B_k$ be any set partition. Let $c \in B_i$ and $(c, B_j)$ be an inversion, if $a = \min B_i$ and $b = \min B_j$ then $(c, B_j)$ corresponds to an occurrence of $\widehat{1}3/\widehat{2}$. On the other hand, if $ac/b$ is an occurrence of $\widehat{1}3/\widehat{2}$, then $a = \min B_i$ and $b = \min B_j$ where $i < j$ since $a < b$. Also, $b > c = \min B_j$. Thus, the occurrence $ac/b$ gives the inversion $(c, B_j)$. □

As a corollary from the above two results we state that the generating function $\sum_{\pi \mathcal{P}_{n,k}} q^{\widehat{13/\widehat{2}}(\pi)}$ is given by $S_q(n,k)$ (see Theorem 5.69).

**Definition 5.71** *Let $\pi = B_1/B_2/\cdots/B_k$ be any set partition. We will say that $b$ is a descent in block $B_i$ if $b > \min B_{i+1}$ and $b \in B_i$. We denote the number descents in the block $B_i$ by $\mathrm{des}_i$. We define the major index of $\pi$ to be $\mathrm{maj}(\pi) = \sum_{i=1}^{k-1} \mathrm{ides}_i(\pi)$.*

**Example 5.72** *Let $\pi = 137/26/45 \in \mathcal{P}_7$. So $\mathrm{des}_1(\pi) = 2$ and $\mathrm{des}_2(\pi) = 1$. Thus $\mathrm{maj}(\pi) = 2 + 2 = 4$.*

**Proposition 5.73** (Goyt [120, Proposition 6.2]) *For any $\pi \in \mathcal{P}_n$,*

$$\mathrm{maj}(\pi) = \mathrm{occ}_{\widehat{13}/\widehat{2}}(\pi) + \mathrm{occ}_{\widehat{1}/\widehat{24}|\widehat{3}}(\pi). \tag{5.22}$$

**Proof** Let $\pi = B_1/B_2/\cdots/B_k$ be any set partition. Let $\beta_1 = \widehat{13}/\widehat{2}$ and $\beta_2 = \widehat{1}/\widehat{24}|\widehat{3}$. We proceed the proof into two steps, where we show that

1. $b$ is a descent in block $B_i$ if and only if $b$ represents the 3 in an occurrence of $\beta_1$, or, for $i \geq 2$, the 4 in an occurrence of $\beta_2$, and

2. each descent $b$ in block $B_i$ contributes $i$ to the right-hand side of (5.22). To see (1), let $b$ be a descent in block $B_i$. If $a = \min B_i$ and $c = \min B_{i+1}$ then $ab/c$ is a occurrence of $\beta_1$ where $b$ represents the 3. Also, if $i \geq 2$ and we let $d = \min B_j$ with $j < i$ then $d/ab/c$ is an occurrence of $\beta_2$, where $b$ represents the 4. On the other side, let $ab/c$ be an occurrence of $\beta_1$, then $c = \min B_{i+1}$ for some $i$, and $b$ is a descent of block $B_i$. Also, an occurrence $d/ab/c$ of $\beta_2$ with $c = \min B_{i+1}$ for some $i \geq 2$ gives the descent $b$ in block $B_i$.

Now, let us prove (2). If $b$ is a descent in block $B_i$, then exists exactly one occurrence of $\beta_2$ with $b$ representing 3, since the 1 and 2 in $\beta_1$ must be represented by $a = \min B_i$ and $c = \min B_{i+1}$, respectively. Now, if $b$ represents the 4 in a occurrence of $\beta_2$ then the 2 and 3 must be represented by $a = \min B_i$ and $c = \min B_{i+1}$, respectively. But now the 1 may be represented by the minimum of any block $B_j$ with $j < i$. So the total contribution of the two patterns is $i$, as required. □

**Definition 5.74** *Let $\pi = B_1/B_2/\cdots/B_k$ be any set partition. We will say that $b$ is an ascent in block $B_i$ if $b > \min B_{i-1}$ and $b \in B_i$. We denote the number of ascents in the block $B_i$ by $\mathrm{asc}_i$. We define the dual major index of $\pi$ to be $\widehat{\mathrm{maj}}(\pi) = \sum_{i=1}^{k}(i-1)|B_i|$. Also, we define the dual inversion number of $\pi$, written $\widehat{\mathrm{inv}}(\pi)$, to be the number of pairs $(b, B_j)$ such that $b \in B_i$, $b > \min B_j$, and $i > j$. We will call these pairs dual inversions. Clearly, $\widehat{\mathrm{inv}}(\sigma) = \widehat{\mathrm{maj}}(\sigma)$ for any set partition $\sigma$, since every ascent causes $i - 1$ dual inversions.*

**Example 5.75** *Let $\pi = 137/26/45 \in \mathcal{P}_7$. The ascents of $\pi$ are $2, 6, 4, 5$ which gives $\mathrm{asc}_2(\pi) = \mathrm{asc}_3(\pi) = 2$. Also $\widehat{\mathrm{maj}}(\pi) = \widehat{\mathrm{inv}}(\pi) = 2 + 4 = 6$.*

Note that the statistics $\widehat{\mathrm{inv}}$ and $\widehat{\mathrm{maj}}$ are equivalent to the statistic $w$ which is defined in Definition 5.1. For enumeration the number of set partitions according to the statistic $w$ we refer to Theorem 5.2. Again, Goyt [120] characterized the statistic $\widehat{\mathrm{maj}}$ in terms of generalized patterns. Note that Theorem 5.2 provides an explicit formula for the generating function for the number of set partitions of $\mathcal{P}_{n,k}$ according to the statistic $\widehat{\mathrm{maj}}$.

**Proposition 5.76** (Goyt [120, Proposition 6.3]) *For any $\pi \in \mathcal{P}_n$,*

$$\widehat{\mathrm{inv}}(\pi) = \mathrm{occ}_{\widehat{1/2}}(\pi) + \mathrm{occ}_{\widehat{1/23}}(\pi).$$

**Proof** See Exercise 5.8. □

Moreover, Goyt [120] defined four other statistics (see Definition 5.5) on set partitions and characterized them in terms of generalized patterns. He proved the following result, where we omit the proof and leave it to the interested reader.

**Proposition 5.77** (Goyt [120, Propositions 6.4–6.5]) *Let $\pi = B_1/\cdots/B_k$ be any set partition and $b \in B_i$. Then for any set partition $\sigma$ we have*

- $\mathrm{lbig}(\sigma) = \mathrm{occ}_{\widehat{13/2}}(\sigma)$.
- $\mathrm{lsmall}(\sigma) = \mathrm{occ}_{\widehat{1/2}}(\sigma) + \mathrm{occ}_{\widehat{1/23}}(\sigma)$.
- $\mathrm{rbig}(\sigma) = \mathrm{occ}_{\widehat{1/23}}(\sigma) + \mathrm{occ}_{\widehat{13/24}}(\sigma) + \mathrm{occ}_{\widehat{1/2}}(\sigma) + \mathrm{occ}_{\widehat{12/3}}(\sigma)$.
- $\mathrm{rsmall}(\sigma) = \mathrm{occ}_{\widehat{13}/2}(\sigma) + \mathrm{occ}_{\widehat{14}/23}(\sigma)$.

See Section 5.2 to find more results on the statistics that mentioned in the above proposition.

**Definition 5.78** Let $\pi$ be any set partition given by its standard representation, see Definition 3.50. We say that the two edges $(i_1, j_1)$ and $(i_2, j_2)$ are

- *crossing if $i_1 < i_2 < j_1 < j_2$,*
- *nesting if $i_1 < i_2 < j_2 < j_1$,*
- *alignment if $i_1 < j_1 \leq i_2 < j_2$,*

Let $\mathrm{cr}(\pi)$, $\mathrm{ne}(\pi)$ and $\mathrm{al}(\pi)$ be the number of crossings, nestings and alignments in $\pi$, respectively. More generally, we say that $k$ edges $(i_1, j_1), \ldots, (i_k, j_k)$ are $k$-crossing if $i_1 < i_2 < \cdots < i_k < j_1 < j_2 < \cdots < j_k$ and are $k$-nesting if $i_1 < i_2 < \cdots < i_k < j_k < \cdots < j_2 < j_1$. Let $\mathrm{cr}_k(\pi)$ and $\mathrm{ne}_k(\pi)$ be the number of $k$-crossings and $k$-nestings of $\pi$. Clearly, $\mathrm{cr} = \mathrm{cr}_2$ and $\mathrm{ne} = \mathrm{ne}_2$.

By the definition, these statistics (for exact enumeration on these statistics, see Section 5.8) can be characterized as generalized pattern statistics.

**Proposition 5.79** (Goyt [120, Propositions 6.6–6.7]) *For any set partition $\pi$ we have*

- $\operatorname{cr}(\pi) = \operatorname{occ}_{\widehat{13}/\widehat{24}}(\pi).$
- $\operatorname{ne}(\pi) = \operatorname{occ}_{\widehat{14}/\widehat{23}}(\pi).$
- $\operatorname{al}(\pi) = \operatorname{occ}_{\widehat{12}/\widehat{34}}(\pi) + \operatorname{occ}_{\widehat{1234}}(\pi) + \operatorname{occ}_{\widehat{123}}(\pi),$

*and more generally,*

- $\operatorname{cr}_k(\pi) = \operatorname{occ}_{\widehat{1(k+1)}/\widehat{2(k+2)}/\cdots/\widehat{k(2k)}}(\pi).$
- $\operatorname{ne}_k(\pi) = \operatorname{occ}_{\widehat{1(2k)}/\widehat{2(2k-1)}/\cdots/\widehat{k(k+1)}}(\pi).$

After a very nice characterization of several known statistics on set partitions, Goyt and Sagan [122] considered the sets $\mathcal{P}_n(13/2)$ and $\mathcal{P}_n(13/2, 123)$. In particular, they proved that the statistics lsmall and rbig (see Definition 5.5) are equidistributed over the set partitions in either $\mathcal{P}_n(13/2)$ or $\mathcal{P}_n(13/2, 123)$, where the proof is followed by showing that the complement map, see Definition 3.26, is bijective map and it exchanges the statistics lsmall and rbig.

**Theorem 5.80** (Goyt and Sagan [122, Theorem 1.1]) *For all $n$,*

$$\sum_{\pi \in \mathcal{P}_n(13/2)} q^{\operatorname{lsmall}(\pi)} = \sum_{\pi \in \mathcal{P}_n(13/2)} q^{\operatorname{rbig}(\pi)}$$

*and*

$$\sum_{\pi \in \mathcal{P}_n(13/2,123)} q^{\operatorname{lsmall}(\pi)} = \sum_{\pi \in \mathcal{P}_n(13/2,123)} q^{\operatorname{rbig}(\pi)}$$

In the next result we find an explicit formula for the generating function $\sum_{\pi \in \mathcal{P}_n(13/2)} q^{\operatorname{rbig}(\pi)}$.

**Theorem 5.81** (Goyt and Sagan [122, Theorem 2.1]) *The map $\phi$ is a bijection such that $\operatorname{rbig}(\pi) = |\phi(\pi)|$ for any set partition $\pi \in \mathcal{P}_n(13/2)$. Moreover, for all $n \geq 1$,*

$$\sum_{\pi \in \mathcal{P}_n(13/2)} q^{\operatorname{rbig}(\pi)} = \sum_{\pi \in \mathcal{P}_n(13/2)} q^{\operatorname{rbig}(\pi)} = \prod_{j=1}^{n-1}(1+q^j).$$

**Proof** We denote the set of sequences $a_1 a_2 \cdots a_k$ with $n \geq a_1 > a_2 > \cdots > a_k \geq 1$ by $DP_n$. Define $\phi : \mathcal{P}_n \to DP_n$ by

$$\phi(B_1/B_2/\cdots/B_k) = \left(\sum_{i=1}^{k}|B_i|\right)\left(\sum_{i=1}^{k-1}|B_i|\right)\cdots\left(\sum_{i=1}^{1}|B_i|\right).$$

Clearly, $\phi(\pi) \in DP_n$ and for any $\lambda \in DP_n$ there exists a set partition $\pi =$

$B_1/B_2/\cdots/B_k$ such that $\phi(\pi) = \lambda$ with $B_i = (\lambda_{i-1}+1)(\lambda_{i-1}+2)\cdots\lambda_i$ and $\lambda_0 = 0$. Since $\lambda_j = \sum_{i=1}^{j}|B_i|$ is the contribution of $B_{j+1}$ to rbig (and $B_1$ makes no contribution) we have $\text{rbig}(\pi) = |\lambda| = \sum_{i=1}^{k}\lambda_i = |\phi(\pi)|$, as required. □

Now we turn our focus to the distribution of rbig over $\mathcal{P}_n(13/2, 123)$.

**Theorem 5.82** (Goyt and Sagan [122]) *For all $n \geq 0$,*

$$F_{\text{rbig}}(n) = \sum_{\pi \in \mathcal{P}_n(13/2, 123)} q^{\text{rbig}(\pi)} = \sum_{j \geq 0} q^{\binom{j}{2} + \binom{n-j}{2}} \begin{bmatrix} j \\ n-j \end{bmatrix}_q.$$

**Proof** Let us write a recurrence relation for $F_{\text{rbig}}(n)$. Since any set partition in $\mathcal{P}_n(13/2, 123)$ is a layered and each of its blocks has size either one or two, the last block has form $n$ or $(n-1)n$. The first case contributes $n-1$ to rbig and the latter case contributes $n-2$ to rbig. Thus, $F_{\text{rbig}}(n)$ satisfies the recurrence relation

$$F_{\text{rbig}}(n) = q^{n-1} F_{\text{rbig}}(n-1) + q^{n-2} F_{\text{rbig}}(n-2)$$

with the initial condition $F_{\text{rbig}}(0) = F_{\text{rbig}}(1) = 1$. Note that $F_{\text{rbig}}(n)$ when $q = 1$ equals $\text{Fib}_{n+1}$ as we shown in Table 3.2.

Let $F_{\text{rbig}}(x)$ be the generating function for the sequence $\{F_{\text{rbig}}(n)\}_{n \geq 0}$, that is, $F_{\text{rbig}}(x) = \sum_{n \geq 0} F_{\text{rbig}}(n) x^n$. Rewriting the recurrence relation in terms of generating function we obtain $F_{\text{rbig}}(x) = 1 + x(1+x) F_{\text{rbig}}(xq)$. By using this relation infinity number of times, we derive

$$F_{\text{rbig}}(x) = \sum_{j \geq 0} x^j q^{\binom{j}{2}} \prod_{i=0}^{j-1} (1 + xq^i).$$

Using the well known identity $\prod_{i=0}^{j-1}(1+xq^i) = \sum_{m \geq 0} q^{\binom{m}{2}} \begin{bmatrix} j \\ m \end{bmatrix}_q$ and finding the coefficient of $x^n$, we get the following result. □

For another proof and for applications of the polynomial $F_{\text{rbig}}(n)$ we refer to [122].

## 5.7 Major Index

Two of the classical statistics on set of permutations are the inversion number statistic (occurrences of the subsequence pattern 21) and the major index statistic (the sum of the positions $i$ such that $\pi_i > \pi_{i+1}$). MacMahon

[212] showed that these two statistics are equidistributed on the rearrangement class of any word. A statistic equidistributed with inv is called *Mahonian*. The main goal of this section is to present result of Chen, Gessel, Yan, and Yang [71], where they introduced a new statistic, called the *p-major index* and denoted pmaj, on the set partitions of $[n]$.

**Definition 5.83** *A set partition $\pi \in \mathcal{P}_n$ is a matching if and only if the degree of each vertex in the standard representation of $\pi$ ($\mathcal{G}_\pi$) is exactly one. In particular, a permutation $\pi$ of size $m$ can be represented as a matching $M_\pi$ of $[2m]$ with arcs connecting $m + 1 - \pi_i$ and $i + m$ for $i = 1, 2, \ldots, m$.*

Clearly, under the map $\pi \mapsto M_\pi$ we obtain $\mathrm{cr}_2(M_\pi) = \mathrm{inv}(\pi)$, where inv is the inversion number statistic on the set of permutations and $\mathrm{cr}_2$ is the crossing number statistic on the set of Matchings.

**Definition 5.84** *For any set partition $\pi = B_1/B_2/\cdots/B_k \in \mathcal{P}_{n,k}$, define*

$$\min(\pi) = \{\min(B_i) \mid i = 1, 2, \ldots, k\}, \quad \max(\pi) = \{\max(B_i) \mid i = 1, 2, \ldots, k\}.$$

*We denote the set of partitions $\pi$ of $[n]$ with $\min(\pi) = S$ and $\max(\pi) = T$ by $\mathcal{P}_n(S, T)$.*

For instance, for $\pi = 1567/248/3$, $\min(\pi) = \{1, 2, 3\}$ and $\max(\pi) = \{3, 7, 8\}$. Note that each point (vertex) $v$ in $\mathcal{G}_\pi$ of a set partition $\pi$ is either

- a left-hand endpoint if $v \in \min(\pi) \setminus \max(\pi)$, or

- a right-hand endpoint if $v \in \max(\pi) \setminus \min(\pi)$, or

- an isolated point if $v \in \min(\pi) \cap \max(\pi)$, or

- a left-hand and right-hand endpoint if $v \notin \min(\pi) \cup \max(\pi)$.

In particular, the set $[n] \setminus \max(\pi)$ (respectively, $[n] \setminus \min(\pi)$) contains all the points which are the left-hand (respectively, right-hand) endpoints of some arcs. Setting $\min(\pi) = S$ and $\max(\pi) = T$ equivalents to fixing the type of each vertex in $[n]$. Since $\mathcal{G}_\pi$ uniquely determines $\pi$, we can identify a set partition $\pi \in \mathcal{P}_n$ with the set of arcs of $\mathcal{G}_\pi$. Hence, the set $\mathcal{P}_n(S, T)$ is in one-to-one correspondence with the set of matchings between the sets $[n] \setminus T$ and $[n] \setminus S$ such that $i < j$ whenever $i \in [n] \setminus T$ and $j \in [n] \setminus S$ and $i$ is matched to $j$. In this context a matching is referred as a *good matching*. Denote the set of all good matchings from $[n] \setminus T$ to $[n] \setminus S$ by $\mathcal{M}_n(S, T)$.

Motivating by Definitions 5.71 and 5.74 we define a new statistic on the set partitions.

**Definition 5.85** *Let $\mathcal{G}_\pi$ be the standard representation of the set partition $\pi \in \mathcal{P}_n$. We define $\mathrm{pmaj}(\pi)$ by the following procedure*

- Label the arc $(i_d, j_d)$ by $d$, where the arcs of $\pi$ are $(i_1, j_1), \ldots, (i_k, j_k)$ with $i_1 > \cdots > i_k$.

- Associate a sequence $\sigma(r)$ to each right-hand end point $r$ as follows. Suppose that the right-hand endpoints are $r_1 > \cdots > r_k$. Let us define $\sigma(r_i)$ recursively: $\sigma(r_1) = a$, where $a$ label of the arc that has right-hand endpoint $r_1$. After we define $\sigma(r_i)$, assume that the left-hand endpoints of the arcs labeled $a_1, a_2, \ldots, a_t$ lie between $r_{i+1}$ and $r_i$, and we process the sequence $\sigma(r_{i+1})$. Then $\sigma(r_{i+1})$ is obtained from $\sigma(r_i)$ by deleting entries $a_1, \ldots, a_t$ and adding $b$ before the leftmost letter, where $b$ is the label for the arc whose right-hand endpoint is $r_{i+1}$.

- Finally, let $\mathrm{pmaj}(\pi) = \sum_{i=1}^{k} \mathrm{des}(\sigma(r_i))$.

**Example 5.86** Let $\pi = 15/27/34/68$, then $r_1 = 8$, $r_2 = 7$, $r_3 = 5$ and $r_4 = 4$, see Figure 5.5.

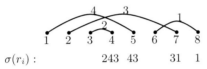

**FIGURE 5.5**: The standard representation of the partition $15/27/34/68$.

Then $\sigma(r_1) = 1$, $\sigma(r_2) = 31$, $\sigma(r_3) = 43$ and $\sigma(r_4) = 243$, which implies that $\mathrm{des}(\sigma(r_1)) = 0$, $\mathrm{des}(\sigma(r_2)) = 1$, $\mathrm{des}(\sigma(r_3)) = 1$ and $\mathrm{des}(\sigma(r_4)) = 1$. Therefore, the p-major index of $\pi$ is $\mathrm{pmaj}(\pi) = 3$.

**Theorem 5.87** (Chen, Gessel, Yan, and Yang [71, Theorem 1]) Let $S, T \subseteq [n]$ such that $|S| = |T|$. Then

$$\sum_{\pi \in \mathcal{P}_n(S,T)} q^{\mathrm{cr}_2(\pi)} = \prod_{i \notin T} (1 + q + \cdots + q^{h(i)-1}),$$

where $h(i) = |T \cap [i+1, n]| - |S \cap [i+1, n]|$.

**Proof** Let $\overline{S} = [n] \backslash S$ and $\overline{T} = [n] \backslash T = \{i_1, \ldots, i_k\}$ such that $i_1 < \cdots < i_k$. Let $M$ be the set of sequences $a_1 a_2 \cdots a_k$ with $1 \leq a_j \leq h(i_j)$ for each $j = 1, 2, \ldots, k$. We proceed the proof by presenting a bijection $M \to \mathcal{M}_n(S, T)$ (see Sainte-Catherine [307]). Let $a = a_1 a_2 \cdots a_k \in M$, and let us define a matching $a' \in \mathcal{M}_n(S, T)$ as follows:

- There are exactly $h(i_k)$ elements in $\overline{S}$ that are greater than $i_k$, so we list them in increasing order $1, 2, \ldots, h(i_k)$. Then we match $i_k$ to the $a_k$-th element, and mark it by star.

- Assume that the elements $i_{j+1}, \ldots, i_k$ are matched to some elements in $\overline{S}$, and let us process the element $i_j$. So, there are exactly $h(i_j)$ elements in $\overline{S}$ that are greater than $i_j$ and not marked by star, list these elements in increasing order, match $i_j$ to the $a_j$-th of them, and mark it by star.

At the end when $j = 1$, we obtain a good matching $a'$ in $\mathcal{M}_n(S, T)$. The map $a \to a'$ gives the required bijection.

Now, let $\pi \in \mathcal{P}_n$ such that the arc set of $\mathcal{G}_\pi$ is $a' \in \mathcal{M}_n(S, T)$. By our bijection, the number of 2-crossings formed by arcs $(i_j, v)$ and $(u, w)$ with $u < i_j < w < v$ equals $a_j - 1$, which gives $\mathrm{cr}_2(\pi) = \sum_{i=1}^{k}(a_j - 1)$ and then

$$\sum_{\pi \in \mathcal{P}_n(S,T)} q^{\mathrm{cr}_2(\pi)} = \sum_{a_1 \cdots a_k \in M} q^{\sum_{j=1}^{k}(a_j - 1)} = \prod_{i \notin T}(1 + q + \cdots + q^{h(i)-1}),$$

as required. □

**Example 5.88** *For example let* $n = 6$, $S = \{1, 2\}$ *and* $T = \{4, 6\}$. *Then* $\overline{T} = \{1, 2, 3, 5\}$, $\overline{S} = \{3, 4, 5, 6\}$, *and* $h(1) = 1$, $h(2) = h(3) = 2$ *and* $h(5) = 2$. *Figure 5.6 presents the bijection from $M$ and $\mathcal{P}_n(S, T)$.*

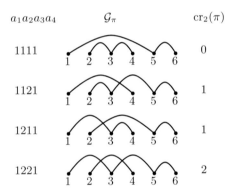

**FIGURE 5.6**: The $\mathrm{cr}_2(\pi)$ for $\pi \in \mathcal{P}_n(\{1, 2\}, \{4, 6\})$

In order to state our next result of finding explicit formula for the generating function for the number of set partitions in $\mathcal{P}_n(S, T)$ according to the statistic pmaj, we need the following lemma on permutations (for another version see [129]).

**Lemma 5.89** (Chen, Gessel, Yan, and Yang [71, Lemma 5]) *Let* $\sigma = \sigma_1 \cdots \sigma_n$ *be any permutation of size $n$. Define*

$$\sigma^{(j)} = (\sigma_1 + 1) \cdots (\sigma_j + 1) 1 (\sigma_{j+1} + 1) \cdots (\sigma_n + 1),$$

for all $j = 0, 1, \ldots, n$. Then the major indices of the permutations
$$\sigma^{(0)}, \ldots, \sigma^{(n)}$$
are all distinct and run from $\mathrm{maj}(\sigma^{(0)})$ to $\mathrm{maj}(\sigma^{(0)}) + n$ in some order.

**Proof** Clearly, $\mathrm{maj}(\sigma^{(0)}) = \mathrm{maj}(\sigma) + \mathrm{des}(\sigma)$. Let $d_i$ be the number of descents of $\sigma$ that are greater than $i$. Then
$$\mathrm{maj}(\sigma^{(i)}) = \begin{cases} \mathrm{maj}(\sigma) + d_i, & \text{if } \sigma_i > \sigma_{i+1}, \\ \mathrm{maj}(\sigma) + i + d_i, & \text{otherwise.} \end{cases}$$

It is easy to verify that the major indices of the permutations $\sigma^{(0)}, \ldots, \sigma^{(n)}$ are all distinct and run from $\mathrm{maj}(\sigma^{(0)})$ to $\mathrm{maj}(\sigma^{(0)}) + n$ in some order. □

**Theorem 5.90** (Chen, Gessel, Yan and Yang [71, Theorem 2]) *Let $S, T \subseteq [n]$ such that $|S| = |T|$. Then,*
$$\sum_{\pi \in \mathcal{P}_n(S,T)} q^{\mathrm{pmaj}(\pi)} = \prod_{i \notin T} (1 + q + \cdots + q^{h(i)-1}),$$
*where $h(i) = |T \cap [i+1, n]| - |S \cap [i+1, n]|$.*

**Proof** Let us find the contribution of the arc with label 1 to the generating function $\sum_{\pi \in \mathcal{P}_n(S,T)} q^{\mathrm{pmaj}(\pi)}$. Here, we identify the set $\mathcal{P}_n(S,T)$ with the set $\mathcal{M}_n(S,T)$ of good matchings. For $\pi \in \mathcal{P}_n(S,T)$, let $i_k$ be the maximal element in the set $\overline{T} = [n] \setminus T$, which is the left-hand endpoint of the arc labeled 1 in Definition 5.85. Set
$$A = \overline{T \cup \{i_k\}} \text{ and } B = \overline{S \cup \{j_{h(i_k)}\}}.$$

For any good matching $a$ between $A$ and $B$ let $a_s$, $1 \leq s \leq h(i_k)$, be the good matching obtained from $a$ by joining the edge $(i_k, j_s)$ and replacing each edge $(i', j_r)$ by $(i', j_{r+1})$ for all $r > s$. Thus, the arc labeling of $a_s$ can be derived from $a$ by labeling the arc $(i_k, j_s)$ by 1 and adding 1 to the label of each arc of $a$. Now assume that $\sigma^{(j_1)} = b_1 b_2 \cdots b_{h(i_k)-1}$ for the matching $a$. Thus, Definition 5.85 gives
$$\mathrm{pmaj}(a_s) = \mathrm{pmaj}(a)$$
$$+ \mathrm{maj}((b_1 + 1) \cdots (b_{s-1} + 1) 1 (b_s + 1) \cdots (b_{h(i_k)-1} + 1))$$
$$- \mathrm{maj}(b_1 \cdots b_{h(i_k)-1}).$$

On the other hand, Lemma 5.89 gives that the values
$$\mathrm{maj}((b_1 + 1) \cdots (b_{s-1} + 1) 1 (b_s + 1) \cdots (b_{h(i_k)-1} + 1)) - \mathrm{maj}(b_1 \cdots b_{h(i_k)-1})$$
are all distinct and run from $0, 1, \ldots, h(i_k) - 1$ in some order. Hence,
$$\sum_{\pi \in \mathcal{P}_n(S,T)} q^{\mathrm{pmaj}(\pi)} = (1 + q + \cdots + q^{h(i_k)-1}) \sum_{\pi \in \mathcal{P}_{n-1}(A,B)} q^{\mathrm{pmaj}(\pi)}.$$

By induction on $k$ we complete the proof. □

**Example 5.91** *Continuing the pervious example, Figure 5.7 presents the p-major indices for the partitions in Figure 5.6.*

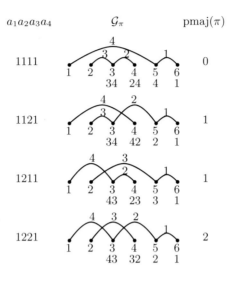

**FIGURE 5.7**: The pmaj($\pi$) for $\pi \in \mathcal{P}_n(\{1,2\},\{4,6\})$

By combining Theorem 5.87 and Theorem 5.90 we obtain the following result.

**Corollary 5.92** (Chen, Gessel, Yan, and Yang [71, Corollary 3]) *Let $S, T \subseteq [n]$ such that $|S| = |T|$. Then the two statistics $\mathrm{cr}_2$ and pmaj have the same distribution over the set $\mathcal{P}_n(S,T)$, that is,*

$$\sum_{\pi \in \mathcal{P}_n(S,T)} q^{\mathrm{cr}_2(\pi)} = \sum_{\pi \in \mathcal{P}_n(S,T)} q^{\mathrm{pmaj}(\pi)}.$$

In particular, when $n = 2m$, $S = [m]$ and $T = [m+1, 2m]$, the set $\mathcal{P}_n(S,T)$ is in one-to-one correspondence with the set $\mathcal{M}_n(S,T)$, which is one-to one with the set of permutations of size $m$ (the $i$-th letter of the permutation is given by $j_i - m$, where $j_i$ is the right-hand endpoint of the arc that has left-hand endpoint $i$). Thus, in this case the above corollary tells us that the statistics inv and maj have the same distribution over the set of permutations of size $n$, as showed by Foata [109]. Chen et al [71] did not stop with stating the above corollary, but presented a bijection

$$\phi : \mathcal{P}_n(S,T) \to \mathcal{P}_n(S,T)$$

such that $\mathrm{pmaj}(\pi) = \mathrm{cr}_2(\phi(\pi))$. This bijection provides a generalization for Foata's second fundamental transformation which is used to show the equidistribution of the permutation statistics inv and maj. The construction details for the bijection $\phi$ are given in [71, Theorem 8].

## 5.8 Number of Crossings, Nestings and Alignments

Following [159], we construct an involution on set partitions which keeps track of the numbers of crossings, nestings and alignments of two edges (see Definition 5.78). We derive then the symmetric distribution of the numbers of crossings and nestings in set partitions, which generalizes Klazar's [184] result on Matchings. We consider the numbers of crossings, nestings, and alignments of two edges in $\pi$, respectively.

**Definition 5.93** *Let $\pi$ be any set partition. Recall that a block of $\pi$ has exactly one element is said to be singleton. Let $B$ be any block of a set partition whose at least two elements. An element $b \in B$ is*

- *an opener if $b = \min B$;*
- *a closer if $b = \max B$;*
- *a transient if it is neither opener or closer of $B$.*

The set of openers, closers, singletons and transients of $\pi$ will be denoted by $\mathcal{O}(\pi)$, $\mathcal{C}(\pi)$, $\mathcal{S}(\pi)$ and $\mathcal{T}(\pi)$, respectively. The 4-tuple $(\mathcal{O}(\pi), \mathcal{C}(\pi), \mathcal{S}(\pi), \mathcal{T}(\pi))$ is called the type of $\pi$.

Note that the type in the above definition is not the same type of a set partition which defined in Definition 5.84.

**Example 5.94** *For the set partition $\pi = \{1, 5, 8\}, \{2\}, \{3, 4, 7\}, \{6\}$, we have $\mathcal{O}(\pi) = \{1, 3\}$, $\mathcal{C}(\pi) = \{7, 8\}$, $\mathcal{S}(\pi) = \{2, 6\}$ and $\mathcal{T}(\pi) = \{4, 5\}$.*

**Definition 5.95** *A 4-tuple $\lambda = (\mathcal{O}, \mathcal{C}, \mathcal{S}, \mathcal{T})$ of subsets of $[n]$ is a set partition type of $[n]$ if there exists a set partition of $[n]$ whose type is $\lambda$. Denote by $\mathcal{P}_n(\lambda)$ the set of partitions of type $\lambda$. In particular, a set partition type $\lambda$ is a matching type if $\lambda = (\mathcal{O}, \mathcal{C}) := (\mathcal{O}, \mathcal{C}, \emptyset, \emptyset)$. Denote by $M_{2n}(\gamma)$ the set of matchings of type $\gamma$.*

Klazar [184] proved the symmetric distribution of the numbers of crossings and nestings of two edges in perfect matchings. Note that Chen et al. [69] have found other interesting results on the crossings and nestings on matchings and set partitions. Moreover, we refer the reader to Krattenthaler's paper [201] for a more general context of related problems and to Chapter 6. Later, Kasraoui and Zeng [159] showed the following result, which we leave as an exercise.

**Theorem 5.96** *(Kasraoui and Zeng [159]) For each set partition type $\lambda$ of $[n]$ there is an involution $\rho : \mathcal{P}_n(\lambda) \mapsto \mathcal{P}_n(\lambda)$ preserving the number of alignments, and exchanging the numbers of crossings and nestings. In other words, for each $\pi \in \mathcal{P}_n(\lambda)$, we have $\mathrm{cr}(\rho(\pi)) = \mathrm{ne}(\pi)$, $\mathrm{ne}(\rho(\pi)) = \mathrm{cr}(\pi)$ and $\mathrm{al}(\rho(\pi)) = \mathrm{ali}(\pi)$.*

As a corollary of the above theorem we obtain that

**Corollary 5.97** *We have*

$$\sum_{\pi \in \mathcal{P}_n} p^{cr(\pi)} q^{ne(\pi)} t^{al(\pi)} = \sum_{\pi \in \mathcal{P}_n} p^{ne(\pi)} q^{cr(\pi)} t^{al(\pi)}.$$

*In particular, the number of set partitions of $[n]$ with no crossing is equal to the number of set partitions of $[n]$ with no nesting.*

---

## 5.9 Exercises

**Exercise 5.1** *Show that Theorem 5.2 gives*

$$S_q(n+1, k) = \sum_{j=0}^{n} \binom{n}{j} q^j S_q(j, k-1),$$

$$\tilde{S}_q(n+1, k) = \sum_{j=0}^{n} \binom{n}{j} q^{j-k+1} \tilde{S}_q(j, k-1),$$

$$S_q^*(n+1, k) = q^{n+1} \sum_{j=0}^{n} \binom{n}{j} S_q^*(j, k-1),$$

*for all $n \geq 0$ and $k \geq 1$.*

**Exercise 5.2** *According to the notation of Theorem 5.2, prove that*

$$\tilde{S}_{-1}(n, k) = \binom{n - \lfloor k/2 \rfloor - 1}{n - k},$$

*and give a combinatorial proof.*

**Exercise 5.3** *Show that the total number of elements that are in the first block (excluding 1 itself) in all set partitions in $\mathcal{P}_{n,k}$ is exactly the total number of singletons, excluding those of the form $\{1\}$, in all set partitions in $\mathcal{P}_{n,k+1}$.*

**Exercise 5.4** *Let $f_{n,k}$ be the number of set partitions of $[n+1]$ with the largest singleton $k+1$. Show that*

$$f(x, y) = \sum_{n \geq 0} \sum_{k \geq 0} f_{n+k,k} \frac{x^n}{n!} \frac{y^k}{k!} = e^{e^{x+y} - x - 1}$$

*and derive $f_{n,k} = \frac{1}{e} \sum_{j \geq 0} \frac{j^k (j-1)^n}{j!}$. For extensions, see [337, 338].*

**Exercise 5.5** *Define the polynomials $a(n,k;q)$ by the recurrence relation*

$$a(n,k;q) = ka(n-1,k;q) + q^n a(n-1,k-1;q)$$

*for all $1 \leq k \leq n$ with the initial conditions $a(0,0;q) = 1$ and $a(n,0;q) = 0$ for all $n \geq 1$. Define $b(n,k) = \frac{d}{dq}a(n,k;q)\big|_{q=1}$. Show that*

$$b(n,k) = k\mathrm{Stir}(n+1,k) - (n+1)\mathrm{Stir}(n,k-1).$$

**Exercise 5.6** *Show that the exponential generating function for the sum of the statistic* sumrec *over all the set partitions of $[n]$ according to number of $k$ blocks ($p$ marks the number of blocks) is given by*

$$pe^{p(e^x-1)}(pe^x(e^x-x-1)+e^x-1) = \frac{d}{dx}pe^{p(e^x-1)}(e^x-x-1).$$

**Exercise 5.7** *Prove Lemma 5.28.*

**Exercise 5.8** *Prove Proposition 5.76.*

**Exercise 5.9** *Prove Theorem 5.13.*

**Exercise 5.10** *Show that* $\mathrm{total}_{\mathcal{P}_{n,k}}(\mathrm{dis}_1) = \sum_{i=1}^{n-k} i\mathrm{S}_{n-1-i,k-1}\binom{n-1}{i+1}$.

**Exercise 5.11** *Show that the number of set partitions $\pi$ of $[n]$ having $\mathrm{int}(\pi) = 0$ is given by $\mathrm{Fib}_{2n-1}$.*

**Exercise 5.12** *Show that the number of set partitions $\pi \in \mathcal{P}_{n,k}$ having $\mathrm{int}(\pi) = 0$ is given by $\binom{n+k-2}{n-k}$.*

**Exercise 5.13** *Prove that* $\mathrm{total}_{\mathcal{P}_{n,k}}(\mathrm{int}) \leq \mathrm{total}_{\mathcal{P}_{n,k}}(\mathrm{dis}_1)$, *for all $1 \leq k \leq n$.*

**Exercise 5.14** *Provide a combinatorial explanation for the expression in Corollary 5.56 for* $\mathrm{total}_{\mathcal{P}_{n,k}}(\mathrm{int})$ *over $\mathcal{P}_{n,k}$.*

**Exercise 5.15** *Define $S_q(n,k) = \sum_{\pi \in \mathcal{P}_{n,k}} q^{\mathrm{inv}(\pi)}$ as in Theorem 5.69. Prove*

$$S_q(n+1,k) = \sum_{j=0}^{n} S_q(n,j)S_q(n-j,k-1),$$

*for all $n \geq 0$ and $k \geq 1$.*

**Exercise 5.16** In [336] has been extended the study of the statistic patterns from the set partitions to the set of ordered set partitions. An ordered set partition of $[n]$ is a set partition of $[n]$ where the blocks of the set partition are ordered arbitrarily. We denote the set of ordered set partitions of $[n]$ with exactly $k$ blocks by $\mathcal{OP}_{n,k}$. Note that most known statistics on $\mathcal{P}_{n,k}$ can be defined in terms of inversions between the letters (integers) in a set partition $\pi$ and the openers and closers of the blocks of $\pi$. The opener of a block is its least element and the closer is its greatest element. For example, the (ordered) set partition $\{1,4,6\},\{2,3,7\},\{5\}$ of $[7]$ has openers $1,2,5$ and closers $5,6,7$. In [336] studied several statistics in terms of openers and closers on the set of ordered set partitions (for other research papers see [142, 143, 160]).

Now, we present a such example. For any ordered set partition $\pi \in \mathcal{OP}_{n,k}$, we define $ros_i(\pi)$ to be the number all elements $j$ such that $j < i$, $j$ is an opener of $\pi$ and the number elements of the block that contains $j$ is greater that the number elements of the block that contains $i$. Let $ros(\pi) = \sum_{i=1}^{n} ros_i(\pi)$ and $ROS(\pi) = ros(\pi) + \binom{k}{2}$. Prove that the generating function for the number of ordered set partitions $\pi \in \mathcal{OP}_{n,k}$ according to the statistic $ROS$ is given by $[k]_q! S_q(n,k)$ (see Theorem 5.2 and (5.1)).

**Exercise 5.17** In this exercise, we consider set partitions of $[n]$ in which the position of the individual player counts, where the $i$-th player presented by $i$-th letter. This extension has been introduced in [372]. For instance in the set partition $\{1,2\},\{3,4\}$ is a coalition of players 1 and 2 is playing against a coalition of players 3 and 4.

(i) Show that when the position (signified by the order in the partition) of the individual player counts, then there are 24 set partitions for the case of a coalition of players 1 and 2 is playing against a coalition of player 3 and 4.

(ii) Find a recursive formula that count the number of set partitions of $[n]$ such that the position of the individual player counts.

**Exercise 5.18** Set partitions has been extended to a general combinatorial structure, namely $k$-covers. The set $B$ is called a $k$-covers of order $r$ of $[n]$ if $B$ is a collection of $r$ nonempty, not necessarily distinct subsets (blocks) of $[n]$ with the property that each element belongs to exactly $k$ members of $B$. This notion was introduced by the Comtet [80] in 1968. The $k$-covers $B$ is called proper if the members of $B$ are distinct and it is called restricted if each element of $B$ belongs to a distinct collection of elements of $B$. Clearly, proper 1-covers are set partitions of $[n]$. Enumeration of $k$-covering and proper $k$-covering has been considered by Baroti [19], Comtet [80], Bender [26], and Devitt and Jackson [91].

(1) Let $a_k(n,r)$ be the number of $k$-covers of order $r$ of $[n]$, show that generating function $\sum_{n \geq 0} \sum_{r \geq 0} a_k(n,r) \frac{x^r y^n}{n!}$ is given by

$$A_k(x,y) = e^{-\sum_{j=1}^{k} \frac{x^j}{j}} \sum_{i_1,\ldots,i_k \geq 0} \frac{x^{1 \cdot i_1 + \cdots + k \cdot i_k}}{i_1! \cdots i_k! 2^{i_2} 3^{i_3} \cdots k^{i_k}} e^{y f_k(i_1,\ldots,i_k)},$$

where $f_k(i_1,\ldots,i_k) = [z^k]\left(\prod_{j=1}^k (1+u^j)^{i_j}\right)$.

(2) Let $b_k(n,r)$ be the number of proper k-covers of order r of $[n]$, show that generating function $B_k(x,y) = \sum_{n\geq 0}\sum_{r\geq 0} b_k(n,r)\frac{x^r y^n}{n!}$ is given by

$$e^{\sum_{j=1}^k \frac{(-x)^j}{j}}\sum_{i_1,\ldots,i_k\geq 0}\frac{(-1)^{3i_2+\cdots+(k+1)i_k}x^{1\cdot i_1+\cdots+k\cdot i_k}}{i_1!\cdots i_k! 2^{i_2} 3^{i_3}\cdots k^{i_k}}e^{yf_k(i_1,\ldots,i_k)},$$

where $f_k(i_1,\ldots,i_k) = [z^k]\left(\prod_{j=1}^k (1+u^j)^{i_j}\right)$.

(3) Let $c_k(n,r)$ and $d_k(n,r)$ be the number restricted k-covers and restricted proper k-covers of order r of $[n]$, show that the generating functions $\sum_{n\geq 0}\sum_{r\geq 0} c_k(n,r)\frac{x^r y^n}{n!}$ and $\sum_{n\geq 0}\sum_{r\geq 0} d_k(n,r)\frac{x^r y^n}{n!}$ are given by $A_k(x,\log(1+y))$ and $B_k(x,\log(1+y))$, respectively.

**Exercise 5.19** Let $\pi$ be any set partition with blocks $B_1, B_2,\ldots, B_m$. Define $Nin(\pi)$ to be the set of all the vectors $(x,y)$ where $x = \min B_i$, $y \in B_j$, $y \neq B_j$ and $i < j$. Define $S_{n,k} = \sum_{\pi \in \mathcal{P}_{n,k}} q^{|Nin(\pi)|}$, see [30]. Show that

$$S_{n,k} = S_{n-1,k-1} - [n-1]_q S_{n-1,k}.$$

**Exercise 5.20** As we mentioned, Milne [252] and Wachs and White [349] studied several statistics on set partitions. Later, Simion [322] (see also [374]) extended the study of statistics on set partitions to statistics on noncrossing set partitions. In this exercise, we give an example for her results. For a set partition $\pi = \pi_1\pi_2\cdots\pi_n \in \mathcal{P}_n$, define $ls(\pi) = \sum_{j=1}^n |\{\pi_j \mid \pi_j < \pi_i, j < i\}|$, $lb(\pi) = \sum_{j=1}^n |\{\pi_j \mid \pi_j > \pi_i, j < i\}|$, $rs(\pi) = \sum_{j=1}^n |\{\pi_j \mid \pi_j < \pi_i, j > i\}|$ and $rb(\pi) = \sum_{j=1}^n |\{\pi_j \mid \pi_j > \pi_i, j > i\}|$. Show

$$\sum_{\pi\in\mathcal{NC}(n,k)} q^{rb(\pi)} = \sum_{\pi\in\mathcal{NC}(n,k)} q^{ls(\pi)}$$

and

$$\sum_{\pi\in\mathcal{NC}(n,k)} q^{rs(\pi)} = \sum_{\pi\in\mathcal{NC}(n,k)} q^{lb(\pi)}.$$

**Exercise 5.21** Show that the number of set partitions of $\mathcal{P}_n(1212)$ without singletons is equal to the number of set partitions of $\mathcal{P}_n(1221)$ without singletons.

---

## 5.10 Research Directions and Open Problems

We now suggest several research directions, which are motivated both by the results and exercises of this and earlier chapter(s).

**Research Direction 5.1** *Following Exercise 5.18, ones can ask the following question. Denote the set of all k-covers (proper k-covers, restricted k-covers, restricted proper k-covers) of order r of $[n]$ by $\mathcal{COV}_k(n,r)$ ($\mathcal{PCOV}_k(n,r)$, $\mathcal{RCOV}_k(n,r)$, $\mathcal{RPCOV}_k(n,r)$). Let $\alpha$ be any statistic on the set $\mathcal{COV}_k(n,r)$ (or $\mathcal{PCOV}_k(n,r)$, $\mathcal{RCOV}_k(n,r)$, $\mathcal{RPCOV}_k(n,r)$). Study the generating functions*

$$A_\alpha(x,y) = \sum_{n\geq 0}\sum_{r\geq 0}\sum_{\pi\in\mathcal{COV}_k(n,r)} q^{\alpha(\pi)}\frac{x^r y^n}{n!},$$

$$B_\alpha(x,y) = \sum_{n\geq 0}\sum_{r\geq 0}\sum_{\pi\in\mathcal{PCOV}_k(n,r)} q^{\alpha(\pi)}\frac{x^r y^n}{n!},$$

$$C_\alpha(x,y) = \sum_{n\geq 0}\sum_{r\geq 0}\sum_{\pi\in\mathcal{RCOV}_k(n,r)} q^{\alpha(\pi)}\frac{x^r y^n}{n!},$$

$$D_\alpha(x,y) = \sum_{n\geq 0}\sum_{r\geq 0}\sum_{\pi\in\mathcal{RPCOV}_k(n,r)} q^{\alpha(\pi)}\frac{x^r y^n}{n!}.$$

*Our statistic $\alpha$ can be defined as any statistic in this chapter or in the book. For instance, largest block, the largest minimal element in the blocks, number elements in the first block, number occurrences of two consecutive elements in the same block, and number singleton blocks.*

**Research Direction 5.2** *Let $S, T \subseteq [n]$ such that $|S| = |T|$. Then Theorems 5.87 and 5.90 suggested two statistics $\alpha = \mathrm{cr}_2$ and $\alpha = \mathrm{pmaj}$, respectively, such that*

$$\sum_{\pi\in\mathcal{P}_n(S,T)} q^{\alpha(\pi)} = \prod_{i\notin T}(1+q+\cdots+q^{|T\cap[i+1,n]|-|S\cap[i+1,n]|-1}). \qquad (5.23)$$

*The question is to formulate other statistics $\alpha$ on the set partition $\mathcal{P}_n(S,T)$ satisfy (5.23).*

**Research Direction 5.3** *Theorems 5.9 and 5.12 studied the statistics lbig, rbig, lsmall, and rsmall on $\mathcal{P}_n$. Then Theorem 5.80 extends these statistics to $\mathcal{P}_n(13/2)$ and $\mathcal{P}_n(13/2, 123)$. So our research suggestion can be formulated as follows. Assume $\alpha$ is a statistic on $\mathcal{P}_n$, $\alpha$ may or may not defined in this chapter. The question is to find and explicit formula for $P_{\tau;\alpha}(n) = \sum_{\pi\in\mathcal{P}_n(\tau)}q^{\alpha(\pi)}$, where $\tau$ is fixed pattern. For instance, find $P_{\tau;\mathrm{dis_m}}(n)$, $P_{\tau;\mathrm{int_m}}(n)$, $P_{\tau;\mathrm{cr}_2}(n)$, and $P_{\tau;\mathrm{pmaj}}(n)$, where $\tau$ is any pattern of size three.*

**Research Direction 5.4** *As consequence of Definition 1.21 one can ask to study the number of set partitions of $[n]$ according to the number occurrences of a fixed substring pattern $\tau$. For instance, the number of set partitions of*

[n] with exactly m occurrences of the substring 1 equals the number of set partitions of [n] with exactly m elements in the first block which is given by $\text{Bell}_{n-m}$. Note that containing occurrences of a subword pattern is the same as containing occurrences of a set of substring patterns. For example, counting occurrences of the subword pattern 12 equivalents counting occurrences of the substrings ab such that $a < b$.

**Research Direction 5.5** *As consequence of Remark 5.26 we can define a new concept of a pattern. Let $\tau'$ be any word over alphabet $\{2, 3, \ldots, k\}$. We say that the set partition $\pi$ contains the distance pattern $\tau = 1\tau'1$ if $\pi$ can be decomposed as $\pi = \pi' a \pi'' a \pi'''$ such that $\pi''$ has a subsequence on the alphabet $\{a+1, a+2, \ldots\}$ of type $\tau'$ and $\pi''$ does not contain the letter a. Otherwise, we say that $\pi$ avoids the distance pattern $\tau$. For instance, the set partition 1213123441312 contains the distance pattern 12341 but it avoids the distance pattern 12431. Our research question is to count the number of set partitions of $\mathcal{P}_n$ according to occurrences of a fixed distance pattern.*

# Chapter 6

# Avoidance of Patterns in Set Partitions

## 6.1 History and Connections

In Chapter 4, we considered the enumeration of subword patterns. In the current chapter, we will focus on pattern avoidance rather than enumeration. Additionally, we will investigate various types of patterns, namely subsequence patterns, generalized patterns, and partially ordered patterns.

Permutation patterns or avoiding patterns research has becomes an important interest in enumerative combinatorics, as evidenced by posting each year a special volume on permutation patterns in an international journal. For example, a special volume in *Annals of Combinatorics*, in *Electronic Journal of Combinatorics*, in *London Mathematical Society Lecture Note Series*, in *Pure Mathematics and Applications*, and hundreds of articles elsewhere in journals (see [48], [138] and references therein for an overview). The subject permutation patterns has proved to be useful language in mathematics in general and in enumerative combinatorics in particular, where this subject has applications to computer science, algebraic geometry, theory of Kazhdan–Luszting polynomials, singularities of Schubert varieties [206], various sorting algorithms [194, Chapter 2.2.1], and sortable permutations [353].

Pattern avoidance was first studied for $S_n$ that avoid a subsequence pattern in $S_3$. The first known explicit solution seems to date back to Hammersley [132], where he found the number of permutations in $S_n$ that avoid the subsequence pattern of 321. In [194, Chapter 2.2.1] and [193, Chapter 5.1.4], Knuth shows that for any $\tau \in S_3$, we have $|S_n(\tau)| = \frac{1}{n+1}\binom{2n}{n}$ (see [327, Sequence A000108]). Other researchers considered restricted permutations in the 1970s and early 1980s (see, for example, [295], [299], and [300]) but the first systematic study was not undertaken until 1985, when Simion and Schmidt [325] found the number of permutations in $S_n$ that avoid any subset of subsequence patterns in $S_3$.

Burstein [55] extended the study of pattern avoidance in permutations to the study of pattern avoidance in words, where he determined the number of words over alphabet $[k]$ of size $n$ that avoid any subset of subsequence patterns in $S_3$. Later, Burstein and Manosur [58] considered subsequence patterns with repeated letters. Recently, subsequence pattern avoidance has been studied for compositions (see [138] and references therein).

In 1996, Klazar [176] (see also [178, 185]) extended the study of pattern avoidance in permutations, words, and compositions to subsequence pattern avoidance in set partitions, where he determined the number of set partitions of $[n]$ that avoid either the subsequence pattern 1212 or the subsequence pattern 1221 and showed that the number of such set partitions in both cases is given by the $n$-th Catalan number $\frac{1}{n+1}\binom{2n}{n}$. Later, Sagan [306] determined the number of set partitions of $[n]$ that avoid any subsequence pattern $\tau$ of size three, namely $\tau = 111, 112, 121, 122, 123$. This result followed by several articles on noncrossing and nonnesting set partitions as we discussed in Theorem 3.52 (see also Section 7.2). The determination all the equivalence classes of set partitions has been done only for patterns of size three and dealing with $k$-noncrossing and $k$-nonnesting set partitions until Jelínek and Mansour [146] found all the equivalence classes of set partitions of size at most seven.

This chapter is divided into three main sections which correspond to three type of patterns: subsequence patterns, generalized patterns, and partially ordered patterns. In Section 6.2, based on [146] we determine all the equivalence classes of subsequence patterns of size at most seven, where our classification is largely based on several new infinite families of pairs of equivalent patterns. For example, we prove that there is a bijection between $k$-noncrossing $(12\cdots k12\cdots k$-avoiding$)$ and $k$-nonnesting $(12\cdots kk\cdots 21$-avoiding$)$ set partitions. In Section 6.3, based on [237] we present several known and new results on generalized 3-letter patterns, those that have the some adjacency requirements. We derive results for permutation and multi-permutation patterns of types $(1,2)$ and $(2,1)$, which are the only generalized patterns not investigated in pervious section. Finally, in Section 6.4, we discus results on partially ordered patterns. Kitaev [171] introduced these patterns in the context of permutations, extending the generalized patterns defined by Babson and Steingrímsson [13]. Later, Kitaev and Mansour [173] investigated avoidance of partially ordered patterns in words, and Heubach, Kitaev, and Mansour [135] extended to the case of compositions.

Since in this chapter we will deal almost exclusively with pattern avoidance, we introduce a simple notation for the set of set partitions, avoiding a certain pattern, the number of such set partitions and the corresponding (ordinary) generating function. In order to highlight the fact that we consider avoidance, we use the notation $AP$ instead of $P$ - think $A$ for avoidance - for the relevant sets, numbers, and generating functions.

**Definition 6.1** *We denote the set of all set partitions of $\mathcal{P}_n$ (respectively, set partitions of $\mathcal{P}_{n,k}$) that avoid the pattern $\tau$ by $\mathcal{P}_n(\tau)$ (respectively, $\mathcal{P}_{n,k}(\tau)$. We denote the number of set partitions in this set by $P_n(\tau)$ (respectively, $P_{n,k}(\tau)$). The corresponding ordinary generating functions are given by*

$P_\tau(x) = \sum_{n\geq 0} P_n(\tau)x^n$ and

$$P_\tau(x,y) = \sum_{n,k\geq 0} P_{n,k}(\tau)x^n y^m = \sum_{k\geq 0} P_\tau(x;k)y^k.$$

Clearly, $P_\tau(x,y) = P_\tau(x,y;0)$ and $P_\tau(x) = P_\tau(x;0)$ (see Definition 4.1). As before, we are interest in determining which nonsubword patterns are equivalent in the sense that they are avoided by the same number (thus the same generating functions) of set partitions. We follow the terminology in the research literature and refer to this equivalence as Wilf-equivalence.

**Definition 6.2** *Two (nonsubword) patterns $\tau$ and $\nu$ are called Wilf-equivalent if the number of set partitions of $[n]$ that avoid $\tau$ is the same as the number of set partitions of $[n]$ that avoid $\nu$, for all $n$. We denote two nonsubword patterns $\tau$ and $\nu$ that are Wilf-equivalent by $\tau \sim \nu$. Strongly, they are called strong Wilf-equivalent if the number of set partitions of $[n]$ with exactly $k$ blocks that avoid $\tau$ is the same as the number of set partitions of $[n]$ with exactly $k$ blocks that avoid $\nu$, for all $n$ and $k$. We denote two patterns $\tau$ and $\nu$ that are strong Wilf-equivalent by $\tau \stackrel{s}{\sim} \nu$.*

Clearly, strong Wilf-equivalent implies Wilf-equivalent; that is, if $\tau$ and $\nu$ are strong Wilf-equivalent then $\tau$ and $\nu$ are Wilf-equivalent. But the opposite direction does not necessary hold, as we explain in the following example.

**Example 6.3** *Since there is only one set partition of $[n]$, namely $11\cdots 1$, that avoids the subsequence pattern 12, we obtain that $P_n(12) = 1$ and $P_{n,k}(12) = \delta_{k=1}$. Since there is only one set partition of $[n]$, namely $12\cdots n$, that avoids the subsequence pattern 11, we obtain that $P_n(11) = 1$ and $P_{n,k}(11) = \delta_{n=k}$. Thus, we can state that*

$$P_{11}(x) = P_{12}(x) = \frac{1}{1-x}$$

*and*

$$P_{12}(x,y) = 1 + \frac{x}{1-x}y, \quad P_{11}(x,y) = \frac{1}{1-xy}.$$

*Hence, 11 and 12 are Wilf-equivalent but are not strong Wilf-equivalent, that is, $11 \sim 12$ and $11 \not\stackrel{s}{\sim} 12$.*

---

## 6.2 Avoidance of Subsequence Patterns

We now look at subsequence patterns that, unlike the subword patterns, do not have any adjacency requirements. This type of pattern was originally studied in the context of permutations, generalized to the case of words and compositions (see [47, 138, 172]), and is referred to in the literature as patterns - classical patterns.

**Definition 6.4** *A sequence (permutation, word, composition, or set partition) $\pi = \pi_1\pi_2\cdots\pi_n$ contains a subsequence pattern $\tau = \tau_1\tau_2\cdots\tau_m$ if there exists a m-term subsequence of $\pi$ such that its reduced form (see Definition 1.19) equals $\tau$. Otherwise we say that $\pi$ avoids the subsequence pattern $\tau$ or is $\tau$-avoiding.*

**Example 6.5** *The set partition $\pi = 12131113$ of $[8]$ contains the subsequence pattern $122$ three times (as the subsequences $\pi_1\pi_4\pi_8 = 133$, $\pi_2\pi_4\pi_8 = 233$ and $\pi_3\pi_4\pi_8 = 133$) and avoids the subsequence pattern $1221$.*

In subsequent sections we introduce the notion of dashes into patterns to indicate the positions in the pattern where there is no adjacency requirement. With this notion, a subsequence pattern $\tau_1\tau_2\cdots\tau_k$ could be written as $\tau_1\text{-}\tau_2\text{-}\cdots\text{-}\tau_k$, as there is no adjacency requirements at all. Since we deal exclusively with subsequence patterns in this section, we omit the dashes and we write as mentioned in the literature (see Section 5.6).

Now, let us establish some notional conventions that will be used throughout this section.

**Definition 6.6** *For m-term sequence $\theta = \theta_1\theta_2\cdots\theta_m$ and an integer $a$, we define $\theta + a$ to be the sequence $(\theta_1 + a)(\theta_2 + a)\cdots(\theta_m + a)$.*

In order to arrive to our goal, determination of all equivalence classes of set partitions of size at most seven, we start with general theory on pattern avoidance in fillings of restricted diagrams.

### 6.2.1 Pattern-Avoiding Fillings of Diagrams

We start by presenting the tools that will be useful in our study of pattern–avoidance and proving our key results, where we introduce a general relationship between pattern–avoidance in set partitions and pattern-avoidance in fillings of restricted shapes.

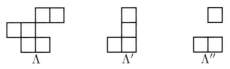

**FIGURE 6.1**: Three diagrams

**Definition 6.7** *A* diagram *is a finite set of cells of the two–dimensional square grid. The $ij$-th cell or $(i,j)$ cell of a diagram $\Lambda$ is the cell in $i$-th row and $j$-th column of $\Lambda$.*

**Example 6.8** *Figure 6.1 presents three diagrams $\Lambda, \Lambda', \Lambda''$.*

**Definition 6.9** *A* filling *of a diagram is to write a nonnegative integer into each cell. We number the rows of diagrams from bottom to top so that the "first row" of a diagram is its bottom row, and we number the columns from left to right. For any diagram, or any matrix, or any filling of a diagram $\Lambda$, we denote the number of rows and columns of $\Lambda$ by $row(\Lambda)$ and $col(\Lambda)$, respectively.*

Note that we are interest in applying the same convention, also, to matrices and to fillings. We always assume that each row and each column of a diagram is nonempty. For instance, when we refer to a diagram with $r$ rows (columns), it is assumed that each of the $r$ rows (columns) contains at least one cell of the diagram. Also, we assume that there is a unique empty diagram with no rows and no columns.

**Example 6.10** *Figure 6.2 presents a diagram $\Lambda$ and its filling with $row(\Lambda) = 4$ and $col(\Lambda) = 5$.*

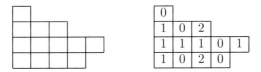

**FIGURE 6.2**: A diagram and its filling

In this section, we mostly use very special type of shape diagrams, namely, Ferrers diagrams and stack polyominoes.

**Definition 6.11** *A* Ferrers diagram *or* Ferrers shape *is a diagram whose cells are arranged into contiguous rows and columns such that*

- *the length of any row is at least the length of any row above it,*

- *the rows are* right-justified, *that is, the rightmost cells of the rows appear in the same column.*

**Example 6.12** *Figure 6.3 presents a Ferrers diagram, but Figure 6.2 does not.*

**FIGURE 6.3**: A Ferrers diagram

Note that the main reason of drawing Ferrers diagrams as right-justified rather than left-justified shapes, which is different from standard practice, is our definition will be more intuitive in the context of our results.

**Definition 6.13** *A stack polyomino $\Pi$ is a diagram such that*

- *its cells are arranged into contiguous rows and columns;*
- *if a column intersects the $i$-th row, then it intersects the $j$-th row with $1 \leq j < i$, for all $i = 1, 2, \ldots, row(\Pi)$.*

**Example 6.14** *Figure 6.4 presents a stack polyomino, but Figure 6.2 does not.*

**FIGURE 6.4**: A stack polyomino

Clearly, every Ferrers shape is also a stack polyomino. But a stack polyomino can be regarded as a union of a Ferrers shape and a vertically reflected copy of another Ferrers shape (see Figure 6.4).

**Definition 6.15** *A 0-1 matrix is a matrix that all its entries equal to 0 or 1. A 0-1 filling is a filling that only uses values 0 and 1. In 0-1 filling, a 0-cell (respectively, 1-cell) of a filling is a cell that is filled with value 0 (respectively, 1). A* semi-standard *is a 0-1 filling such that each of its columns contains exactly one 1-cell. A 0-1 filling is called* sparse *if every column has at most one 1-cell. A column (respectively, row) of a 0-1 filling is called* zero column *(respectively,* zero row*) if it contains no 1-cell.*

**Example 6.16** *The 0-1 fillings of the middle diagram $\lambda'$ in Figure 6.1 are given by*

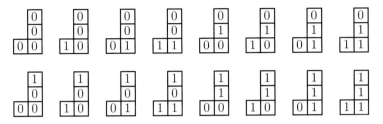

*Among these 0-1 fillings there are three semi-standard fillings and eight sparse fillings.*

In the literature, there are several possibilities to define the concept avoiding a pattern in fillings, but the following definition seems to be the most useful (at least in our book) and most common.

**Definition 6.17** *Let $M = (m_{ij})_{i\in[r], j\in[c]}$ be a 0-1 matrix with $r$ rows and $c$ columns. Let $\Lambda$ be a filling of a diagram. We say that $\Lambda$ contains $M$ if $\Lambda$ contains $r$ distinct rows $i_1 < \cdots < i_r$ and $c$ distinct columns $j_1 < \cdots < j_c$ such that*

- *each of the rows $i_1, \ldots, i_r$ intersects all columns $j_1, \ldots, j_c$ in a cell that belongs to the underlying diagram of $\Lambda$,*

- *if $m_{k\ell} = 1$ for some $k, \ell$, then the $(i_k, j_\ell)$ cell of $\Lambda$ has a nonzero value.*

*Otherwise, if $\Lambda$ does not contain $M$, we say that $\Lambda$ avoids $M$. We denote the number of semi-standard fillings of $\Lambda$ that avoid $M$ by $ss_\Lambda(M)$.*

**Example 6.18** *Let*

$$M = \begin{pmatrix} 1 & 1 \\ 0 & 0 \end{pmatrix}, \quad M' = \begin{pmatrix} 1 & 0 \\ 1 & 1 \end{pmatrix} \quad \text{and} \quad \Lambda = \begin{array}{|c|c|c|c|c|c|} \hline & & 1 & 0 & 0 & 0 \\ \hline & 1 & 0 & 1 & 0 & 0 \\ \hline 1 & 0 & 0 & 0 & 1 & 0 \\ \hline 0 & 0 & 0 & 0 & 0 & 1 \\ \hline \end{array}.$$

*Then the filling diagram $\Lambda$ contains the matrix $M$ (for example see rows $1, 3$ and columns $3, 5$), but it avoids the matrix $M'$.*

**Definition 6.19** *We say that two matrices $M$ and $M'$ are Ferrers-equivalent, denoted by $M \stackrel{F}{\sim} M'$, if for every Ferrers shape $\Omega$, $ss_\Omega(M) = ss_\Omega(M')$. We say that $M$ and $M'$ are stack-equivalent, denoted by $M \stackrel{s}{\sim} M'$, if the equality holds even for semi-standard fillings of an arbitrary stack polyomino.*

In last decade, pattern–avoidance in the fillings of diagrams has received a lot of attention. Apart from semi-standard fillings, several authors have investigated *standard* fillings with exactly one 1-cell in each row and each column (see [14, 331], as well as general fillings with nonnegative integers (see [87, 201]). In addition, nontrivial results were obtained for fillings of more general shapes (see [301]). These results often study the cases when the forbidden pattern $M$ is the *identity matrix* (the $r \times r$ matrix, $I_r$, with $m_{ij} = \delta_{i=j}$) or the *anti-identity matrix* (the $r \times r$ matrix, $J_r$, with $m_{ij} = \delta_{i=j=r+1}$).

In our next steps, mostly we will deal with semi-standard fillings (or just fillings).

**Remark 6.20** *Let $M$ and $M'$ be two Ferrers-equivalent 0-1 matrices with a 1-cell in every column. Let $f$ be a bijection between $M$-avoiding and $M'$-avoiding semi-standard fillings of Ferrers shapes. Then $f$ can be extended into a bijection between $M$-avoiding and $M'$-avoiding sparse fillings of Ferrers shapes. To see this, let $\Lambda$ be any sparse $M$-avoiding filling of a Ferrers shape $\Omega$, so the nonzero columns of $\Lambda$ form a semi-standard filling of a (not necessarily contiguous) subdiagram of $\Omega$, and then we apply $f$ to this subfilling to transform $\Lambda$ into a sparse $M'$-avoiding filling of $\Omega$. Similarly, this fact holds for stack polyominoes instead of Ferrers shapes.*

In order to translate the language of set partitions to the language of fillings, we introduce the following notation.

**Definition 6.21** *Let $\theta = \theta_1 \theta_2 \cdots \theta_m$ be any sequence of positive integers, and let $d \geq \max_{i \in [m]} \theta_i$ be an integer. We denote the 0-1 matrix with $d$ rows and $m$ columns which has a $(i, j)$ 1-cell if and only if $\theta_j = i$ by $M(\theta, d)$.*

**Example 6.22** *Let $\pi = 12345412$ be a set partition of $[8]$ with exactly 5 blocks and $\tau = 121$ be a set partition of $[3]$ with exactly 2 blocks. Then*

$$M(\pi, 5) = \begin{pmatrix} 0 & 0 & 0 & 0 & 1 & 0 & 0 & 0 \\ 0 & 0 & 0 & 1 & 0 & 1 & 0 & 0 \\ 0 & 0 & 1 & 0 & 0 & 0 & 0 & 0 \\ 0 & 1 & 0 & 0 & 0 & 0 & 0 & 1 \\ 1 & 0 & 0 & 0 & 0 & 0 & 1 & 0 \end{pmatrix} \text{ and } M(\tau, 2) = \begin{pmatrix} 0 & 1 & 0 \\ 1 & 0 & 1 \end{pmatrix}.$$

Now we are ready to give a correspondence between set partitions and fillings of Ferrers diagrams.

**Lemma 6.23** *Let $\theta$ and $\theta'$ be two nonempty sequences over the alphabet $[k]$, and let $\tau$ be any set partition. If $M(\theta, k)$ is Ferrers-equivalent to $M(\theta', k)$ then*

$$\sigma = 12 \cdots k(\tau + k)\theta \sim \sigma' = 12 \cdots k(\tau + k)\theta'.$$

**Proof** Let $\pi \in \mathcal{P}_{n,m}$, $M = M(\pi, m)$, and let $\tau$ be a set partition with $m'$ blocks and $T = M(\tau, m')$. Now let us color the cells of $M$ black and gray as follows:

- If $\tau$ is empty, then the $(i,j)$ cell is colored gray if and only if row $i$ has at least one 1-cell strictly to the left of column $j$.

- If $\tau$ is nonempty, then the $(i,j)$ cell is colored gray if and only if the submatrix of $M$ induced by the rows $i+1,\ldots,m$ and columns $1,\ldots,j-1$ contains $T$.

- Otherwise, the cell is colored black.

For instance, if $\pi = 12345412$ and $\tau = 121$ as in Example 6.22, then $M$ with black and gray colors has the following form:

$$M = \begin{pmatrix} 0 & 0 & 0 & 0 & 1 & 0 & 0 & 0 \\ 0 & 0 & 0 & 1 & 0 & 1 & 0 & 0 \\ 0 & 0 & 1 & 0 & 0 & 0 & 0 & 0 \\ 0 & 1 & 0 & 0 & 0 & 0 & 0 & 1 \\ 1 & 0 & 0 & 0 & 0 & 0 & 1 & 0 \end{pmatrix},$$

where the gray (black) cell is presented as gray (black) entry in the matrix $M$. Clearly, the gray cells form a Ferrers diagram (see the above example), and the entries of the matrix $M$ form a sparse filling $\Lambda$ of this diagram. Also, the leftmost 1-cell of each row is always black, and any 0-cell of the same row to the left of the leftmost 1-cell is black too.

It is not hard to see that the set partition $\pi$ avoids $\sigma$ if and only if the filling $\Lambda$ of the gray diagram avoids $M(\theta,k)$, and $\pi$ avoids $\sigma'$ if and only if $\Lambda$ avoids $M(\theta',k)$. By the assumption that $M(\theta,k) \stackrel{F}{\sim} M(\theta',k)$, we get that there is a bijection $f$ that maps $M(\theta,k)$-avoiding fillings of Ferrers shapes onto $M(\theta',k)$-avoiding fillings of the same shape. Thus, by Remark 6.20, $f$ can be extended to bijection $\tilde{f}$ of sparse fillings. By $\tilde{f}$, we construct a bijection between the sets $\mathcal{P}_n(\sigma)$ and $\mathcal{P}_n(\sigma')$ as follows. For a set partition $\pi \in \mathcal{P}_{n,m}(\sigma)$, we take $M$ and $\Lambda$ as above. Note that, $\Lambda$ is $M(\theta,k)$-avoiding. By the bijection $\tilde{f}$, we map $\Lambda$ into an $M(\theta',k)$-avoiding sparse filling $\Lambda' = \tilde{f}(\Lambda)$, while the filling of the black cells of $M$ remains the same. We thus obtain a new matrix $M'$.

By coloring the cells of $M'$ black and gray, we see that each cell of $M'$ receives the same color as the corresponding cell of $M$, even though the occurrences of $T$ in $M'$ need not correspond exactly to the occurrences of $T$ in $M$. Indeed, if $\tau$ is nonempty (similarly, $\tau$ is empty), then for each gray cell $c$ of $M$, there is an occurrence of $T$ to the left and above $c$ consisting entirely of black cells. This occurrence is contained in $M'$ as well, which shows that the color of the cell $c$ remains gray in $M'$.

From our construction, $M'$ has exactly one 1-cell in each column; hence there is a sequence $\pi'$ over the alphabet $[m]$ such that $M' = M(\pi',m)$. So, $\pi'$ is a canonical sequence of a set partition. To show this, note that for every $i \in [m]$, the leftmost 1-cell of $M$ in $i$-th row is black, and the preceding 0-cells in $i$-th row are black too. Thus, the leftmost 1-cell of $i$-th row in $M$ is also the leftmost 1-cell of $i$-th row in $M'$, which implies that the first occurrence of the symbol $i$ in $\pi$ appears at the same place as the first occurrence of $i$ in $\pi'$. Therefore, $\pi'$ is a set partition. Moreover, the gray cells of $M'$ avoid $M(\theta', k)$, which implies that $\pi'$ avoids $\sigma'$. Hence, the map $\pi \mapsto \pi'$ is invertible and describes a bijection between the sets $\mathcal{P}_n(\sigma)$ and $P_n(\sigma')$, which completes the proof. □

Generally, we will see that the relation $12 \ldots k\theta \sim 12 \ldots k\theta'$ does not give that $M(\theta, k)$ and $M(\theta', k)$ are Ferrers equivalent. For instance, we will show that $12112 \sim 12212$, even though $M(112, 2)$ is not Ferrers equivalent to $M(212, 2)$ (see Section 6.2.12.2). On the other hand, by the following lemma, the relation $12 \ldots k\theta \sim 12 \ldots k\theta'$ establishes a weaker equivalence between pattern-avoiding fillings.

**Lemma 6.24** *Let $\theta$ be a nonempty sequence over the alphabet $[k]$ and let $\sigma = 12 \cdots k\theta$. For every $n$ and $m$, there is a bijection $f$ that maps the set $\mathcal{P}_{n,m}(\sigma)$ onto the set of all the $M(\theta, k)$-avoiding fillings $\Lambda$ of Ferrers shapes with $\mathrm{col}(\Lambda) = n - m$ and $\mathrm{row}(\Lambda) \leq m$.*

**Proof** Let $\pi \in \mathcal{P}_{n,m}(\sigma)$ and let $M = M(\pi, m)$. We consider the same black and gray coloring of $M$ as in the proof of Lemma 6.23 (see the case $\tau =$). For instance, if $\pi = 123454144$, then $M$ with black and gray colors has the following form

$$M = \begin{pmatrix} 0 & 0 & 0 & 0 & 1 & 0 & 0 & 0 & 0 \\ 0 & 0 & 0 & 1 & 0 & 1 & 0 & 1 & 1 \\ 0 & 0 & 1 & 0 & 0 & 0 & 0 & 0 & 0 \\ 0 & 1 & 0 & 0 & 0 & 0 & 0 & 0 & 0 \\ 1 & 0 & 0 & 0 & 0 & 0 & 1 & 0 & 0 \end{pmatrix}$$

where the gray (black) cell is presented as gray (black) entry in the matrix $M$. Clearly,

- $M$ has exactly $m$ black 1-cells;

- each 1-cell is black if and only if it is the leftmost 1-cell of its row;

- if $c_i$ is the column containing the black 1-cell in row $i$, then either $c_i$ is the rightmost column of $M$, or column $c_i + 1$ is the leftmost column of $M$ with exactly $i$ gray cells.

Now, let $\Lambda$ be the filling constructed by the gray cells. In our above example,

$$\Lambda = \begin{array}{|c|c|c|c|c|c|c|} \hline & & & & 0 & 0 & 0 & 0 \\ \hline & & & 0 & 1 & 0 & 1 & 1 \\ \hline & & 0 & 0 & 0 & 0 & 0 & 0 \\ \hline & 0 & 0 & 0 & 0 & 0 & 0 & 0 \\ \hline 0 & 0 & 0 & 0 & 0 & 1 & 0 & 0 \\ \hline \end{array}.$$

As was found in the proof of Lemma 6.23, the filling $\Lambda$ is a sparse $M(\theta, k)$-avoiding filling of a Ferrers shape. Note that the filling $\Lambda$ has exactly one zero column of height $i$, and this column, which corresponds to $c_{i+1}$, is the rightmost of all the columns of $\Lambda$ with height at most $i$, for all $i = 1, 2, \cdots, m-1$.

Let $\Lambda^-$ be the subfilling of $\Lambda$ induced by all the nonzero columns of $\Lambda$. In our example,

$$\Lambda^- = \begin{array}{|c|c|c|c|} \hline 0 & 0 & 0 & 0 \\ \hline 1 & 0 & 1 & 1 \\ \hline 0 & 0 & 0 & 0 \\ \hline 0 & 0 & 0 & 0 \\ \hline 0 & 1 & 0 & 0 \\ \hline \end{array}.$$

Observe that $\Lambda^-$ is a semi-standard $M(\theta, k)$-avoiding filling of a Ferrers shape with exactly $n-m$ columns and at most $m$ rows, and let us define $f(\pi) = \Lambda^-$. We claim that the map $f$ is invertible. To show this, let $\Pi$ be a filling of a Ferrers shape with $n - m$ columns and at most $m$ rows. We insert $m-1$ zero columns $c_2, c_3, \ldots, c_m$ into the filling $\Pi$ as follows: each column $c_i$ has height $i - 1$, and it is inserted immediately after the rightmost column of $\Pi \cup \{c_2, \ldots, c_{i-1}\}$ that has height at most $i - 1$. Note that the filling obtained by this operation corresponds to the gray cells of the original matrix $M$. Let us call this sparse filling $\Lambda$. By adding a new 1-cell on top of each zero column of $\Lambda$ and adding a new 1-cell in front of the bottom row, we yield a semi-standard filling of a diagram with $n$ columns and $m$ rows such that the diagram can be completed into a matrix $M = M(\pi, m)$, where $\pi$ is a $\sigma$-avoiding set partition, as required. $\square$

Lemma 6.23 presents a way of dealing with set partition patterns of the form $12 \cdots k(\tau + k)\theta$, where $\theta$ is a word over $[k]$ and $\tau$ is a set partition. To consider other forms such as $12 \cdots k\theta(\tau + k)$, we need a correspondence between set partitions and fillings of stack polyominoes.

**Lemma 6.25** *Let $\tau$ be any set partition, and let $\theta$ and $\theta'$ be any two nonempty sequences over the alphabet $[k]$ such that $M(\theta, k) \stackrel{s}{\sim} M(\theta', k)$. Then*

$$\sigma = 12 \cdots k\theta(\tau + k) \sim \sigma' = 12 \cdots k\theta'(\tau + k).$$

**Proof** Let $\pi \in \mathcal{P}_{n,m}$, $\tau$ be any set partition with exactly $m'$ blocks and set $M = M(\pi, m)$. Similarly as in the proof of Lemma 6.23, we color the cells of $M$ black and gray as follows: A $(i, j)$ cell of $M$ is colored gray, if

- the submatrix of $M$ formed by the intersection of the top $m - i$ rows and the rightmost $n - j$ columns contains $M(\tau, m')$, and

- the matrix $M$ has at least one 1-cell in row $i$ appearing strictly to the left of column $j$.

Otherwise, the cell is colored black. Clearly, the gray cells form a stack polyomino and the matrix $M$ induces a sparse filling $\Lambda$ of this polyomino.

As in the proof of Lemma 6.23, it is not hard to check that the set partition $\pi$ above avoids $\sigma$ ($\sigma'$) if and only if the filling $\Lambda$ avoids $M(\theta, k)$ ($M(\theta', k)$). The rest of the proof can be obtained by using similar arguments as in the proof of Lemma 6.23. Let us assume that $M(\theta, k)$ and $M(\theta', k)$ are stack-equivalent via a bijection $f$. So, Remark 6.20 extends $f$ to a bijection $\tilde{f}$ between $M(\theta, k)$-avoiding and $M(\theta', k)$-avoiding sparse fillings of a given stack polyomino.

Now, let $\pi \in \mathcal{P}_{n,m}(\sigma)$ and define $M$ and $\Lambda$ as above. By applying $\tilde{f}$ on the filling $\Lambda$, we obtain an $M(\theta', k)$-avoiding filling $\Lambda'$; the filling of the black cells of $M$ remains the same, which yields a matrix $M'$ and a sequence $\pi'$ such that $M' = M(\pi', k)$. Also, it can be verified that the gray cells of $M'$ are the same as the gray cells of $M$. By (ii) the leftmost 1-cell of each row of $M$ is unaffected by this transform. This implies that the first occurrence of $i$ in $\pi'$ is at the same place as the first occurrence of $i$ in $\pi$, which implies that $\pi'$ is a set partition. Hence, $\pi'$ avoids $\sigma'$, and the transform $\pi \mapsto \pi'$ is a bijection from $\mathcal{P}_n(\sigma)$ to $\mathcal{P}_n(\sigma')$, which completes the proof. □

The following simple proposition about pattern-avoidance in fillings will turn out to be very helpful in the classification of pattern avoidance in set partitions.

**Proposition 6.26** *Let $\theta$ be any nonempty sequence over the alphabet $[k-1]$. Then*
$$M(\theta, k) \stackrel{s}{\sim} M(\theta+1, k).$$
*Moreover, if $\theta$ and $\theta'$ are two sequences over $[k-1]$ and*

- *if $M(\theta, k-1) \stackrel{F}{\sim} M(\theta', k-1)$ then $M(\theta, k) \stackrel{F}{\sim} M(\theta', k)$;*
- *if $M(\theta, k-1) \stackrel{s}{\sim} M(\theta', k-1)$ then $M(\theta, k) \stackrel{s}{\sim} M(\theta', k)$.*

**Proof** Define $M = M(\theta, k)$, $M^- = M(\theta, k-1)$ and $M^+ = M(\theta+1, k)$. Note that a filling $\Lambda$ of a stack polyomino avoids $M$ if and only if the filling obtained by erasing the topmost cell of every column of $\lambda$ avoids $M^-$, and $\Lambda$ avoids $M'$ if and only if the filling obtained by erasing the bottom row of $\Lambda$ avoids $M^-$. Therefore, it remains to describe a bijection between $M$-avoiding and $M^+$-avoiding fillings. Set an $M$-avoiding filling $\Lambda$. In every column of this filling, move the topmost element into the bottom row, and move every other element into the row directly above it. This gives an $M^+$-avoiding filling. Similarly, we can prove the second part of the theorem. □

We remark that if $\theta$ is a sequence over $[k-2]$, Proposition 6.26 from
$$M(\theta, k) \stackrel{s}{\sim} M(\theta+1, k) \stackrel{s}{\sim} M(\theta+2, k).$$

Lemmas 6.23 and 6.25 suggest that the first part of Proposition 6.26 can be expressed in terms of pattern-avoiding set partitions; we leave the proof to the interested reader (see [146]).

**Corollary 6.27** *Let $\theta$ be any nonempty sequence over $[k-1]$ and $\tau$ be any set partition, then*

$$12\cdots k(\tau+k)\theta \sim 12\cdots k(\tau+k)(\theta+1)$$

*and*

$$12\cdots k\theta(\tau+k) \sim 12\cdots k(\theta+1)(\tau+k).$$

To derive more results on classification of pattern-avoidance on set partitions, we state another result related to pattern-avoidance in Ferrers diagrams. First, let us present the following notation.

**Definition 6.28** *For any two matrices $A$ and $B$, we define $\begin{pmatrix} A & 0 \\ 0 & B \end{pmatrix}$ to be the matrix with a copy of $A$ in the top left corner and a copy of $B$ in the bottom right corner.*

The idea of the next result is not new! Actually, it has already been used by Backelin, West, and Xin [14] on standard fillings of Ferrers diagrams, and later extended by de Mier [87] for fillings with arbitrary integers. Here, we apply it to semi-standard fillings (see Exercise 6.3).

**Lemma 6.29** *Let $A$ and $A'$ be any two Ferrers equivalent matrices and let $B$ be any another matrix. Then*

$$\begin{pmatrix} B & 0 \\ 0 & A \end{pmatrix} \stackrel{F}{\sim} \begin{pmatrix} B & 0 \\ 0 & A' \end{pmatrix}.$$

Indeed, Lemma 6.29 does not directly obtain new pairs of equivalent set partition patterns. But it allows us to show the following result.

**Proposition 6.30** *Let $\theta_1 > \theta_2 > \cdots > \theta_m$ and $\nu_1 > \nu_2 > \cdots > \nu_m$ be two strictly decreasing sequences over the alphabet $[k]$, and let $r_1, \ldots, r_m$ be positive integers. Define weakly decreasing sequences $\theta = \theta_1^{r_1}\theta_2^{r_2}\cdots\theta_m^{r_m}$ and $\nu = \nu_1^{r_1}\nu_2^{r_2}\cdots\nu_m^{r_m}$. Then*

$$M(\theta, k) \stackrel{F}{\sim} M(\nu, k).$$

*Moreover, for any set partition $\tau$,*

$$12\cdots k(\tau+k)\theta \sim 12\cdots k(\tau+k)\nu.$$

**Proof** Let $j$ be a minimal element such that $\theta_i = \nu_i$ for all $i = 1, 2, \ldots, m-j$. We proceed with the proof by induction over $j$. If $j = 0$, then $\theta = \nu$ and the claim holds. Assume $j > 0$ and without loss of generality assume that $\theta_{m-j+1} - \nu_{m-j+1} = d > 0$. Consider the sequence $\nu_1' > \nu_2' > \cdots > \nu_m'$ such that $\nu_i' = \nu_i$ for every $i = 1, 2, \ldots, m-j$ and $\nu_i' = \nu_i + d$ for every $i > m-j$. The sequence $\{\nu_i'\}_{i=1}^m$ is strictly decreasing, and its first $m-j+1$ terms are equal to $\theta_i$. Define $\nu' = (\nu'_1)^{r_1}(\nu'_2)^{r_2}\cdots(\nu'_m)^{r_m}$. Thus, by the induction hypothesis we have that $M(\theta, k) \stackrel{F}{\sim} M(\nu', k)$.

In order to obtain that $M(\nu, k) \stackrel{F}{\sim} M(\nu', k)$, we write $\nu = \nu^{(0)}\nu^{(1)}$, where $\nu^{(0)}$ is the prefix of $\nu$ containing all the symbols of $\nu$ greater than $\nu_{m-j+1}$ and $\nu^{(1)}$ is the suffix of the remaining symbols. Note that $\nu' = \nu^{(0)}(\nu^{(1)}+d)$. We can write $M(\nu, k) = \begin{pmatrix} B & 0 \\ 0 & A \end{pmatrix}$ and $M(\nu', k) = \begin{pmatrix} B & 0 \\ 0 & A' \end{pmatrix}$, where $A = M(\nu^{(1)}, \nu_{m-j} - 1)$ and $A' = M(\nu^{(1)} + d, \nu_{m-j} - 1)$. Hence, by Proposition 6.26, the assumption $A \stackrel{F}{\sim} A'$ and Lemma 6.29, we get $M(\nu, k) \stackrel{F}{\sim} M(\nu', k)$, which completes the induction step and implies the first claim of the proposition. The last part of the proposition follows directly from Lemma 6.23. □

Now, as consequence of the above work on filling diagrams, we obtain our results on classification subsequence patterns in set partitions. We present our results in several separate subsections, where each one deals with a very special type of subsequence pattern.

### 6.2.2 Basic Facts and Patterns of Size Three

We start by enumeration two general cases $1^m$ and $12\cdots m$ and present the enumeration of three-letter patterns. Note that Example 6.3 gives an explicit formula for the number of set partitions of $[n]$ that avoid a subsequence pattern of size two. In particular, this example shows $11 \sim 12$ and $11 \stackrel{s}{\not\sim} 12$.

**Fact 6.31** *The exponential generating function for the number set partitions of $[n]$ that avoid the subsequence pattern $1^m$ is given by*

$$\sum_{n \geq 0} \frac{P_n(1^m)}{n!} x^n = e^{\sum_{j=1}^{m-1} \frac{x^j}{j!}}.$$

**Proof** See Exercise 6.5. □

**Fact 6.32** *A set partition avoids the subsequence pattern $12\cdots m$ if and only if it has fewer than $m$ blocks. The corresponding exponential generating function is equal to*

$$\sum_{j=0}^{m-1} \frac{(e^x - 1)^j}{j!}.$$

*and the corresponding ordinary generating function is given by*

$$\sum_{j=0}^{m-1} \frac{x^j}{(1-x)(1-2x)\cdots(1-jx)}.$$

**Proof** See Exercise 6.6. □

As we discussed in the introduction, Sagan [306] has determined the Wilf-equivalence classes $\mathcal{P}_n(\tau)$ for the five subsequence patterns $\tau$ of size three, namely $\tau = 111, 112, 121, 122, 123$. We summarize the relevant results in Table 6.1.

**Table 6.1**: Number of partitions in $\mathcal{P}_n(\tau)$, where $\tau \in \mathcal{P}_3$

| $\tau$ | $P_n(\tau)$ |
|---|---|
| 111 | [327, Sequence A000085] |
| 112, 121, 122, 123 | $2^{n-1}$ |

The cases 111 and 123 can be obtained from Facts 6.31 and 6.32, respectively. The rest of the cases, namely $112, 121, 122$, we exercise (see Exercise 6.1). Note that the number of set partitions of $\mathcal{P}_n(111)$ is the same as the number of involutions on $n$ letters (Show that!), where an *involution* is a permutation $\pi = \pi_1 \pi_2 \cdots \pi_n$ such that $\pi_{\pi_i} = i$ for all $i = 1, 2, \ldots, n$.

### 6.2.3 Noncrossing and Nonnesting Set Partitions

As we will discuss in Section 7.2, there are several different definitions of crossings and nestings in a set partition. In this chapter, we are interested in considering a set partition as a word, that is, in defining our restrictions on the canonical form of the set partition (see Definition 1.3). More precisely, we define noncrossing and nonnesting set partitions as follows.

**Definition 6.33** *A set partition is $k$-noncrossing if it avoids the subsequence pattern $12 \cdots k 12 \cdots k$, and it is $k$-nonnesting if it avoids the subsequence pattern $12 \cdots k k \cdots 21$.*

Lemma 6.23 shows that a bijection between $k$-noncrossing set partitions and $k$-nonnesting set partitions can be constructed from a bijection between $I_k$-avoiding and $J_k$-avoiding semi-standard fillings of Ferrers diagrams. We refer the reader to [201], where Krattenthaler has presented an excellent comprehensive summary of the connections between $I_r$-avoiding and $J_r$-avoiding fillings of a fixed Ferrers diagram under additional restrictions for row-sums and column-sums.

Here, we state the result about the correspondence between $I_k$-avoiding and $J_k$-avoiding fillings of diagrams, which can be used as a weaker version of Theorem 13 in [201]. Note that in [201], the bijection between $I_k$-avoiding and $J_k$-avoiding fillings with fixed each row sum and each column sum is not explicitly constructed. Additionally, Theorem 13 in [201] was presented for any fillings with nonnegative integers; however, the previous remark shows that the result holds even when restricted to semi-standard fillings.

**Theorem 6.34** (Krattenthaler [201]) *For every Ferrers diagram $\Pi$ and every nonnegative integer $k$, there is a bijection between the $I_k$-avoiding semi-*

standard fillings of $\Pi$ and the $J_k$-avoiding semi-standard fillings of $\Pi$. Moreover, the bijection preserves the number of 1-cells in every row.

**Definition 6.35** *Let $A, B$ be any two subsets of a set $C$, and let $\alpha$ be any statistic defined on $C$, that is, $\alpha(c)$ is a nonnegative integer number for any $c \in C$. We say that a bijection $f$ between sets $A$ and $B$ preserves the statistic $\alpha$ if and only if $\alpha(b) = \alpha(a)$ with $b = f(a)$, $a \in A$ and $b \in B$.*

Theorem 6.34 and Lemma 6.23 give there is a bijection between $k$-noncrossing set partitions of $[n]$ and $k$-nonnesting set partitions of $[n]$. Actually, we even obtain the following refinement.

**Corollary 6.36** *For every $n, k \geq 0$, there is a bijection between $k$-noncrossing set partitions of $[n]$ and $k$-nonnesting partitions of $[n]$. Moreover, the bijection preserves the number of blocks, the size of each block, and the smallest element of every block.*

Translating Lemma 6.23 for $\theta = 12 \cdots k$ and $\theta' = k(k-1) \cdots 1$ into the terminology of pattern-avoiding set partitions gives the following result.

**Corollary 6.37** *Let $\tau$ be any set partition and let $k$ be any positive integer. Then*
$$12 \cdots k(\tau + k)12 \cdots k \sim 12 \cdots k(\tau + k)k(k-1) \cdots 1.$$
*Moreover,*
$$12 \cdots k(\tau + k)12 \cdots k \overset{s}{\sim} 12 \cdots k(\tau + k)k(k-1) \cdots 1.$$

Our next result needs a general concept of diagrams and fillings, which is given by the next definition.

**Definition 6.38** *A diagram is convex, if for any two cells in a either column or row, the elements between are also cells of the diagram. It is intersection-free, if any two columns are comparable, that is, the set of row coordinates of cells in one column is contained in the set of row coordinates of cells in the other. A moon polyomino is a convex and intersection-free diagram.*

**Example 6.39** *Figure 6.5 presents a convex diagram but not intersection-free, since the first and the last columns are incomparable.*

We stated the moon polynomino definition to extend Lemma 6.25 as we extended Lemma 6.23 to obtain Corollary 6.40. This extension can be achieved by using Proposition 5.3 in [301], Theorem 6.42, due to Rubey, where he has proved that the matrices $I_k$ and $J_k$ are stack-equivalent, rather than just Ferrers-equivalent. More precisely, Rubey's result deals with fillings of *moon polyominoes* with prescribed row-sums. However, since a transposed copy of a stack polyomino is a special case of a moon polyomino, Rubey's result applies to fillings of stack polyominoes with prescribed column sums as well. Combining this result with Lemma 6.25, we obtain the following result.

**FIGURE 6.5**: A moon polyomino

**Corollary 6.40** *For any set partition $\tau$ and any positive integer $k$,*
$$12\cdots k12\cdots k(\tau+k) \sim 12\cdots kk(k-1)\cdots 1(\tau+k).$$

At end of the section, we note that several general enumeration results has been found. For example, in Section 6.2.6 has been considered counting of set partitions of $\mathcal{P}_n(12\cdots k12)$. In [254] has been studied the generating function for the number of nonnesting set partitions of $[n]$, where the authors used the algebraic kernel method together with a linear operator to describe a coefficient extraction process of the generating function $P_{12\cdots kk\cdots 21}(x)$.

### 6.2.4 The Patterns $12\cdots(k+1)12\cdots k$, $12\cdots k12\cdots(k+1)$

The main goal of this section is to prove that
$$12\cdots(k+1)12\cdots k \sim 12\cdots k12\cdots(k+1).$$

Again, this result is a consequence of the fillings of polyominoes (see Section 6.2.1).

**Definition 6.41** *Let $\Pi$ be a stack polyomino. The content of $\Pi$ is the sequence of the column heights of $\Pi$, listed in nondecreasing order.*

For instance, the content of the stack polyomino in Figure 6.4 is given by the sequence $1, 1, 2, 3$. The critical point of our proof can be stated as follows.

**Theorem 6.42** *(Rubey [301, Proposition 5.3]) Let $\Pi$ and $\Pi'$ be two stack polyominoes with the same content, and let $k \geq 1$ be any positive integer. Then there is a bijection between the $I_k$-avoiding semi-standard fillings of $\Pi$ and the $I_k$-avoiding semi-standard fillings of $\Pi'$.*

Indeed, this theorem is a special case of Proposition 5.3 in [301], where it deals with arbitrary nonnegative integer fillings, rather than semi-standard fillings. However, see the last paragraph of Section 4 in [301], it is easy to see that Rubey's bijection maps semi-standard fillings to semi-standard fillings. Clearly, Theorem 6.42 implies that $I_k$ and $J_k$ are stack-equivalent. The number of $J_k$-avoiding fillings of a stack polyomino $\Pi$ equals the number of $I_k$-avoiding fillings of the mirror image of $\Pi$, which, by Theorem 6.42, is the same as the number of $I_k$-avoiding fillings of $\Pi$.

### 6.2.4.1 The Pattern $12\cdots(k+1)12\cdots k$

Let us now study in more detail the set partitions that avoid the subsequence pattern $12\cdots k(k+1)12\cdots k$. In order to do that, we need the following definitions.

**Definition 6.43** *Let $\pi = \pi_1 \cdots \pi_n$ be a set partition. We say that an element $\pi_i$ is left-dominating if $\pi_i \geq \pi_j$ for each $j < i$. We say that a left-dominating element $\pi_i$ left-dominates an element $\pi_j$, if $\pi_i > \pi_j$, $i < j$ and $\pi_i$ is the rightmost left-dominating element with these two properties. Note that, if $\pi_j$ not left-dominating, then it is left-dominated by a unique left-dominating element. But, a left-dominating element is not left-dominated by any other element. If an element is not left-dominating, we call it simply left-dominated.*

*The left shadow of $\pi$ is the sequence $\overline{\pi}$ obtained by replacing each left-dominated element by the symbol '$*$'. We say that a non-star symbol $i$ left-dominates an occurrence of a star, if $i$ is the rightmost non-star to the left of the star.*

**Example 6.44** *Let $\pi = 1232321445$ be a set partition of $[10]$. The left-dominating elements of $\pi$ are $\pi_1 = 1$, $\pi_2 = 2$, $\pi_3 = 3$, $\pi_5 = 3$, $\pi_8 = 4$, $\pi_9 = 4$ and $\pi_{10} = 5$. The left-dominated elements in $\pi$ are $\pi_4 = 2$, $\pi_6 = 2$ and $\pi_7 = 1$. Thus, the left shadow of $\pi$ is the sequence $\overline{\pi} = 123*3**445$. In $\overline{\pi}$, the leftmost occurrence of $3$ left-dominates a single star, while the second occurrence of $3$ left-dominates two stars.*

It is not hard to see that a sequence $\overline{\pi}$ over the alphabet $[m] \cup \{*\} = \{1, 2, \ldots, m, *\}$ is a left shadow of a set partition with $m$ blocks if and only if it satisfies the following:

- each symbol in $[m]$ appears at least once,
- the non-star symbols of $\overline{\pi}$ form a nondecreasing sequence, and
- no occurrence of the symbol 1 may left-dominate an occurrence of $*$. Any other non-star symbol may left-dominate any number of stars, and each star is dominated by a non-star.

**Definition 6.45** *A sequence that satisfies the above three conditions is called a left-shadow sequence. Note that a left-shadow sequence is uniquely determined by the multiplicities of its non-star symbols and by the number of stars dominated by each non-star.*

**Definition 6.46** *Let $\pi$ be a set partition. We define $\Lambda = \Lambda(\pi)$ to be the semi-standard filling of a Ferrers diagram such that*

1. *the columns of $\Lambda$ correspond to the left-dominated elements of $\pi$. The $i$-th column of $\Lambda$ has height $j$ if the $i$-th left-dominated element of $\pi$ is dominated by an occurrence of $j + 1$, and*

2. the $i$-th column of $\Lambda$ has a 1-cell in row $j$ if the $i$-th left-dominated element of $\pi$ is equal to $j$.

**Example 6.47** *The semi-standard filling $\Lambda(123232144)$ (see Example 6.44) is given by*

| 0 | 1 | 1 |
|---|---|---|
| 1 | 0 | 0 |

Note that the shape of the underlying diagram of $\Lambda(\pi)$ is characterized by the left shadow of $\pi$; that is, the number of columns of height $h$ in $\Lambda$ equals the number of stars in the left shadow which are dominated by an occurrence of $h+1$. It is easy to see that the left shadow $\overline{\pi}$ and the filling $\Lambda(\pi)$ together uniquely determine the set partition $\pi$. In fact, for every semi-standard filling $\Lambda'$ with the same shape as $\Lambda(\pi)$, there is a (unique) set partition $\pi'$ with the same left-shadow as $\pi$ and $\Lambda(\pi') = \Lambda'$. As a straightforward application of the above definitions, we state the following observation.

**Fact 6.48** *A set partition $\pi$ avoids the subsequence pattern*

$$12\cdots k(k+1)12\cdots k$$

*if and only if the filling $\Lambda(\pi)$ avoids $I_k$.*

### 6.2.4.2 The Pattern $12\cdots k12\cdots(k+1)$

Let us now focus on the set partitions that avoid the subsequence pattern $12\cdots k12\cdots(k+1)$. As in the previous case, we need the following definitions.

**Definition 6.49** *Let $\pi = \pi_1\cdots\pi_n$ be a set partition. We say that an element $\pi_i$ is* right-dominating *if either $\pi_i \geq \pi_j$ for each $j > i$ or $\pi_i > \pi_j$ for each $j < i$. If $\pi_i$ is not right-dominating, we say that it is* right-dominated. *We say that $\pi_i$* right-dominates *$\pi_j$ if $\pi_i$ is the leftmost right-dominating element appearing to the right of $\pi_j$, and $\pi_j$ itself is not right-dominating.*

*The* right shadow *$\widetilde{\pi}$ of $\pi$ is obtained by replacing each right-dominated element of $\pi$ by a star.*

**Example 6.50** *Let $\pi = 122134233122 \in \mathcal{P}_{12}$. The right-dominating elements of $\pi$ are $\pi_1 = 1$, $\pi_2 = 2$, $\pi_5 = 3$, $\pi_6 = 4$, $\pi_8 = 3$, $\pi_9 = 3$, $\pi_{11} = 2$ and $\pi_{12} = 2$. The right-dominated elements of $\pi$ are $\pi_3 = 2$, $\pi_4 = 1$, $\pi_7 = 2$ and $\pi_{10} = 1$. Then the right shadow of the partition $\pi = 122134233122$ is the sequence $12**34*33*22$.*

A sequence $\widetilde{\pi}$ over the alphabet $[m] \cup \{*\}$ is the right shadow of a set partition with $m$ blocks if and only if it satisfies the following

- the non-star symbols of $\widetilde{\pi}$ form a subsequence $(1, 2, \ldots, m, s_1, s_2, \ldots, s_p)$ where the sequence $s_1 s_2 \cdots s_p$ is nonincreasing, and

- no occurrence of the symbol 1 may right-dominate an occurrence of $*$. Any other non-star symbol may right-dominate any number of stars, and each star is right-dominated by a non-star.

**Definition 6.51** *A right-shadow sequence is a sequence that satisfies the above two conditions. Note that a right-shadow sequence is uniquely determined by the multiplicities of its non-star symbols and by the number of stars right-dominated by each non-star.*

**Definition 6.52** *Let $\pi$ be a set partition. We define $\Pi = \Pi(\pi)$ to be the semi-standard filling of a stack polyomino such that*

1. *the columns of $\Pi$ correspond to the right-dominated elements of $\pi$. The $i$-th column of $\Pi$ has height $j$ if the $i$-th right-dominated element of $\pi$ is dominated by an occurrence of $j+1$, and*

2. *the $i$-th column of $\Pi$ has a 1-cell in row $j$ if the $i$-th right-dominated element of $\pi$ equals $j$.*

Let $\Pi'$ be the underlying diagram of $\Pi(\pi)$. Note that $\Pi'$ is uniquely characterized by the right shadow $\widetilde{\pi}$ of the set partition $\pi$, although there may be different right shadows corresponding to the same shape $\Pi'$. The sequence $\widetilde{\pi}$ and the filling $\Pi(\pi)$ together characterize the set partition $\pi$. For a fixed $\widetilde{\pi}$, the map $\pi \mapsto \Pi(\pi)$ describes a bijection between set partitions with right shadow $\widetilde{\pi}$ and fillings of $\Pi'$. As a straightforward application of the above definitions, we state the following observation.

**Fact 6.53** *A set partition $\pi$ avoids the subsequence pattern*

$$12\cdots k12\cdots(k+1)$$

*if and only if the filling $\Pi(\pi)$ avoids $I_k$.*

### 6.2.4.3 The Equivalence

Now, we ready to state the main result of the current subsection and prove it.

**Theorem 6.54** *For any $k \geq 1$,*

$$\tau = 12\cdots k(k+1)12\cdots k \sim \tau' = 12\cdots k12\cdots k(k+1).$$

**Proof** Let $\pi$ be a set partition with exactly $m$ blocks that avoids $\tau$, let $\overline{\pi}$ be its left shadow, and let $\Lambda(\pi)$ be its filling as described in Definition 6.46. Let $\Lambda'$ denote the underlying shape of $\Lambda(\pi)$. Fact 6.48 implies that $\Lambda(\pi)$ avoids $I_k$. Let $\widetilde{\sigma}$ be the right-shadow sequence determined as follows:

1. For each symbol $i \in [m]$, the number of occurrences of $i$ in $\overline{\pi}$ equals the number of its occurrences in $\widetilde{\sigma}$.

2. For any $i$ and $j$, the number of stars left-dominated by the $j$-th occurrence of $i$ in $\overline{\pi}$ equals the number of stars right-dominated by the $j$-th occurrence of $i$ in $\tilde{\sigma}$.

Clearly, $\tilde{\sigma}$ is unique. Now, let $\Sigma$ be the stack polyomino whose columns correspond to the stars of $\tilde{\sigma}$, where the $i$-th column has height $h$ if the $i$-th star of $\tilde{\sigma}$ is right-dominated by $h+1$. By Theorem 6.42, there is a bijection $f$ between the $I_k$-avoiding fillings of $\Lambda'$ and the $I_k$-avoiding fillings of $\Sigma$. This bijection transforms $\Lambda(\pi)$ into a filling $\Pi$ of $\Sigma$. Define a set partition $\sigma$ by replacing the $i$-th star in $\tilde{\sigma}$ by the row-index of the 1-cell in the $i$-th column of $\Pi$. By construction, $\sigma$ is a set partition with right shadow $\tilde{\sigma}$, and $\Pi(\sigma) = \Pi$. By Fact 6.53, $\sigma$ avoids $\tau'$. It is easy to check that this transformation is invertible, which completes the proof. □

### 6.2.5 Patterns of the Form $1(\tau + 1)$

Here, we find an explicit formula for the exponential generating function for the number of set partitions of $\mathcal{P}_n(1(\tau + 1))$ in terms of the exponential generating function for the number of set partitions of $\mathcal{P}_n(\tau)$ (see Exercise 6.7).

**Theorem 6.55** *Let $\tau$ be any pattern, and let $F_\tau(x)$ be the exponential generating function for the number of set partitions of $\mathcal{P}_n(\tau)$. Then for every $n \geq 1$,*

$$\mathcal{P}_n(1(\tau+1)) = \sum_{i=0}^{n-1} \binom{n-1}{i} P_i(\tau).$$

*Moreover,*

$$F_{1(\tau+1)}(x) = 1 + \int_0^x F_\tau(t) e^t dt.$$

More generally, we can state the following result.

**Theorem 6.56** *Let $P_n(\rho; a_1, a_2, \ldots, a_m)$ be the number of set partitions that avoid $\rho$ such that the $i$-th block has $a_i$ elements. Then for any set partition $\tau$,*

$$P_n(1(\tau+1); a_1, a_2, \ldots, a_m) = \binom{n-1}{a_1-1} P_{n-a_1}(\tau; a_2, a_3, \ldots, a_m).$$

As a consequence of Theorem 6.55 we obtain the following result, where we leave the proof as an exercise (see Exercise 6.8).

**Corollary 6.57** *If $\tau \sim \tau'$ (respectively, $\tau \stackrel{s}{\sim} \tau'$) then $1(\tau+1) \sim 1(\tau'+1)$ (respectively, $1(\tau+1) \stackrel{s}{\sim} 1(\tau'+1)$), and more generally,*

$$12 \cdots k(\tau+k) \sim 12 \cdots k(\tau'+k) \ (\textit{respectively,} \ 12 \cdots k(\tau+k) \stackrel{s}{\sim} 12 \cdots k(\tau'+k)).$$

In particular, since $123 \sim 122 \sim 112 \sim 121$, we see that for every $m \geq 2$

$$12 \cdots (m-1)m(m+1) \sim 12 \cdots (m-1)mm$$
$$\sim 12 \cdots (m-1)(m-1)m$$
$$\sim 12 \cdots (m-1)m(m-1).$$

Conversely, if $1(\tau+1) \sim 1(\tau'+1)$ (respectively, $1(\tau+1) \overset{s}{\sim} 1(\tau'+1)$), then $\tau \sim \tau'$ (respectively, $\tau \overset{s}{\sim} \tau'$).

### 6.2.6 The Patterns $12 \cdots k1$, $12 \cdots k12$

Here, we will investigate finding an explicit formulas for the generating functions for the number of set partitions of either $\mathcal{P}_n(12 \cdots k1)$ or $\mathcal{P}_n(12 \cdots k12)$, see [168, 229]

**Table 6.2:** The cardinality $|\mathcal{P}_n(12 \cdots k12)|$ for $2 \leq k \leq 5$ and $1 \leq n \leq 10$

| $k \backslash n$ | 1 | 2 | 3 | 4 | 5 | 6 | 7 | 8 | 9 | 10 |
|---|---|---|---|---|---|---|---|---|---|---|
| 2 | 1 | 2 | 5 | 14 | 42 | 132 | 429 | 1430 | 4862 | 16796 |
| 3 | 1 | 2 | 5 | 15 | 51 | 188 | 731 | 2950 | 12235 | 51822 |
| 4 | 1 | 2 | 5 | 15 | 52 | 202 | 856 | 3868 | 18313 | 89711 |
| 5 | 1 | 2 | 5 | 15 | 52 | 203 | 876 | 4112 | 20679 | 109853 |

Table 6.2 presents the cardinality of the set $\mathcal{P}_n(12 \cdots k12)$, where $k = 2, 3, 4, 5$. We will show that the ordinary generating function $F_k(x) = \sum_{n \geq 0} |\mathcal{P}_n(12 \cdots k12)| x^n$ for the number of set partitions of $\mathcal{P}_n$ is rational in $x$ and $\sqrt{(1-kx)^2 - 4x^2}$. Here, we deal with recurrence relations in terms of ordinary generating functions with a single index $\ell$. Let us denote by $F_{k,\ell}(x)$ the generating function for the number of set partitions in

$$\mathcal{P}_n(12 \cdots k12 \cdots d; \ell)$$
$$= \{\pi = \pi_1 \pi_2 \cdots \pi_n \in \mathcal{P}_n(12 \cdots k12 \cdots d) \mid \pi_1 < \pi_2 < \cdots \pi_\ell \geq \pi_{\ell+1}\},$$

that is,

$$F_{k,\ell}(x) = \sum_{n \geq 0} |\mathcal{P}_n(12 \cdots k12; \ell)| x^n.$$

Note that we define $F_{k,0}(x)$ to be 1 (for the empty set partition). Clearly,

$$F_k(x) = \sum_{i \geq 0} F_{k,i}(x).$$

Our calculations are based on finding a system of linear recurrence relations for the generating functions $\{F_{k,\ell}(x)\}_{\ell \geq 0}$. As we will see later, the recurrences contain the expression $\sum_{i\geq\ell} F_{k,i}(x)$, so we define

$$\overline{F}_{k,\ell}(x) = \sum_{i\geq\ell} F_{k,i}(x).$$

Obviously, the generating function $\overline{F}_{k,\ell}(x)$ is well defined, since

$$\overline{F}_{k,\ell}(x) = F_k(x) - \sum_{i=0}^{\ell-1} F_{k,i}(x).$$

Our first lemma finds a recurrence relation for $F_{k,\ell}(x)$, where $1 \leq \ell \leq k-1$.

**Lemma 6.58** *For all $1 \leq \ell \leq k-1$,*

$$F_{k,\ell}(x) = \ell x \overline{F}_{k,\ell}(x) + x^\ell.$$

**Proof** Let $\pi = \pi_1 \cdots \pi_n \in \mathcal{P}_n(12\cdots k12; \ell)$. If $n > \ell$ then $\pi_{\ell+1} \leq \ell$. Thus $\pi \in \mathcal{P}_n(12\cdots k12)$ if and only if the reduced form of $\pi' = \pi_1 \cdots \pi_\ell \pi_{\ell+2} \cdots \pi_n$ is a set partition in $\cup_{j\geq\ell} \mathcal{P}_{n-1}(12\cdots k12; j)$. Therefore,

$$F_{k,\ell}(x) = \ell x(F_{k,\ell}(x) + F_{k,\ell+1}(x) + \cdots) + x^\ell = \ell x \overline{F}_{k,\ell}(x) + x^\ell,$$

where $x^\ell$ counts the unique set partition $12\cdots\ell \in \mathcal{P}_\ell(12\cdots k12; \ell)$, as claimed. □

The above lemma with the definition of $\overline{F}_{k,\ell}(x)$ gives the following system:

$$\begin{cases} F_{k,0}(x) & = 1 \\ x\sum_{i=0}^{0} F_{k,i}(x) + F_{k,1}(x) & = xF_k(x) + x \\ 2x\sum_{i=0}^{1} F_{k,i}(x) + F_{k,2}(x) & = 2xF_k(x) + x^2 \\ \vdots & \\ (k-1)x\sum_{i=0}^{k-2} F_{k,i}(x) + F_{k,k-1}(x) & = (k-1)xF_k(x) + x^{k-1}. \end{cases} \quad (6.1)$$

Now, we find an explicit formula for $F_{k,\ell}(x)$ in terms of $F_k(x)$.

**Lemma 6.59** *For all $1 \leq \ell \leq k-1$,*

$$F_{k,\ell}(x) = \sum_{i=0}^{\ell} (-1)^{i+\ell}(ixF_k(x) + x^i)\beta_{i,\ell},$$

*where $\beta_{\ell,\ell} = 1$ and $\beta_{i,\ell} = \ell x \prod_{j=i+1}^{\ell-1}(jx-1)$ for $i = 0, 1, \ldots, \ell-1$.*

**Proof** By using Cramer's Rule on (6.1), we obtain

$$F_{k,\ell}(x) = \sum_{i=0}^{\ell} (-1)^{i+\ell}(ixF_k(x) + x^i)|\mathbf{A}_i|,$$

where

$$\mathbf{A}_i = \begin{pmatrix} (i+1)x & 1 & 0 & \cdots & 0 & 0 \\ (i+2)x & (i+2)x & 1 & \cdots & 0 & 0 \\ \vdots & & & & & \\ (\ell-1)x & (\ell-1)x & (\ell-1)x & \cdots & (\ell-1)x & 1 \\ \ell x & \ell x & \ell x & \cdots & \ell x & \ell x \end{pmatrix}.$$

Thus, by the formula

$$\begin{vmatrix} ax & 1 & 0 & \cdots & 0 & 0 \\ (a+1)x & (a+1)x & 1 & \cdots & 0 & 0 \\ \vdots & & & & & \\ (b-1)x & (b-1)x & (b-1)x & \cdots & (b-1)x & 1 \\ bx & bx & bx & \cdots & bx & bx \end{vmatrix} = bx \prod_{j=a}^{b-1}(jx-1),$$

which holds by induction on $b \geq a$ (Prove it!), we derive

$$F_{k,\ell}(x) = \sum_{i=0}^{\ell} (-1)^{i+\ell}(ix F_k(x) + x^i)\beta_{i,\ell},$$

as required. □

The above lemma with the definitions give the following result (see Exercise 6.21).

**Corollary 6.60** *We denote the set of all set partitions $\pi_1 \pi_2 \cdots \pi_n \in \mathcal{P}_n$ such that $\pi_1 < \cdots < \pi_j \geq \pi_{j+1}$ and $j \leq \ell$ by $\mathcal{Q}_\ell(n)$. Then the generating function $Q_\ell(x)$ for the number of set partitions in $\mathcal{Q}_\ell(n)$ is given by*

$$\sum_{n \geq 0} \#\mathcal{Q}_\ell(n) x^n = \frac{1 + \sum_{j=1}^{\ell} \sum_{i=0}^{j}(-1)^{i+j} x^i \beta_{i,j}}{1 - \sum_{j=1}^{\ell} \sum_{i=0}^{j}(-1)^{i+j} ix\beta_{i,j}},$$

*where $\beta_{\ell,\ell} = 1$ and $\beta_{i,\ell} = \ell x \prod_{j=i+1}^{\ell-1}(jx-1)$ for $i = 0, 1, \ldots, \ell-1$.*

**Theorem 6.61** *Let $k \geq 2$. Then the ordinary generating function $J_k(x)$ for the number of $12 \cdots k1$-avoiding set partitions of $[n]$ is given by*

$$\frac{1 - x + (1-x)\sum_{j=1}^{k-2}\sum_{i=0}^{j}(-1)^{i+j}x^i\beta_{i,j} + \sum_{i=0}^{k-1}(-1)^{i+k-1}x^i\beta_{i,k-1}}{1 - x - x(1-x)\sum_{j=1}^{k-2}\sum_{i=0}^{j}(-1)^{i+j}i\beta_{i,j} - x\sum_{i=0}^{k-1}(-1)^{i+k-1}i\beta_{i,k-1}},$$

*where $\beta_{\ell,\ell} = 1$ and $\beta_{i,\ell} = \ell x \prod_{j=i+1}^{\ell-1}(jx-1)$ for $i = 0, 1, \ldots, \ell-1$.*

**Proof** Let $J_{k,\ell}(x)$ be the ordinary generating function for the number of $12\cdots k1$-avoiding set partitions $\pi = \pi_1\pi_2\cdots\pi_n \in \mathcal{P}_n$ such that $\pi_1 < \pi_2 < \cdots < \pi_\ell \geq \pi_{\ell+1}$. Then, a similar argument as in the proof of Lemma 6.59 gives that

$$J_{k,\ell}(x) = \sum_{i=0}^{\ell}(-1)^{i+\ell}(ixJ_k(x) + x^i)\beta_{i,\ell}, \quad \ell = 1, 2, \ldots, k-1, \tag{6.2}$$

with $J_{k,0}(x) = 1$. On the other hand,

$$J_{k,\ell}(x) = x^{\ell+1-k}J_{k,k-1}(x), \quad \ell = k, k+1, k+2, \ldots. \tag{6.3}$$

(Why?) Therefore, using the fact that $J_k(x) = \sum_{\ell\geq 0} J_{k,\ell}(x)$, (6.2) and (6.3), we can write

$$J_k(x) = 1 + \sum_{\ell=0}^{k-1}\sum_{i=0}^{\ell}(-1)^{i+\ell}(ixJ_k(x) + x^i)\beta_{i,\ell} + \sum_{\ell\geq k} x^{\ell+1-k}J_{k,k-1}(x).$$

Again, by (6.2), we have

$$J_k(x) = 1 + \sum_{\ell=0}^{k-1}\sum_{i=0}^{\ell}(-1)^{i+\ell}(ixJ_k(x) + x^i)\beta_{i,\ell}$$

$$+ \frac{x}{1-x}\sum_{i=0}^{k-1}(-1)^{i+k-1}(ixJ_k(x) + x^i)\beta_{i,k-1}.$$

The solution of this equation completes the proof. $\square$

**Example 6.62** *Theorem 6.61 for $k = 2, 3$ gives $J_2(x) = \frac{1-x}{1-2x}$ and $J_3(x) = \frac{1-2x}{(1-x)(1-3x)}$.*

In order to enumerate the set $\mathcal{P}_n(12\cdots k12)$, we need more notation. Let

$$H_k(x,y) = \sum_{\ell=0}^{k-2} F_{k,\ell}(x)y^\ell, \quad F_k(x,y) = \sum_{\ell\geq 0} F_{k,\ell}(x)y^\ell.$$

From Lemma 6.59, we can observe

$$H_k(x,y) = \sum_{\ell=0}^{k-2} y^\ell\left(\sum_{i=0}^{\ell}(-1)^{i+\ell}(ixF_k(x,1) + x^i)\beta_{i,\ell}\right)$$

and

$$F_k(x,1) = F_k(x),$$

where $\beta_{\ell,\ell} = 1$ and $\beta_{i,\ell} = \ell x\prod_{j=i+1}^{\ell-1}(jx-1)$, for $i = 0, 1, \ldots, \ell-1$. The generating function $F_{k,\ell}(x)$ with $\ell \geq k-1$ satisfies the following recurrence relation, and we leave the proof to the interested reader (for full details, see [229]).

**Lemma 6.63** *For all $j \geq 0$,*

$$F_{k,k-1+j}(x) = x^{k-1+j} + \sum_{i=0}^{j-1} x^{j+1-i}\overline{F}_{k,k-1+i}(x) + (k-1)x\overline{F}_{k,k-1+j}(x).$$

Now we state an explicit formula for the generating function $F_k(x,y)$.

**Theorem 6.64** *We have*

$$\left(1 + \frac{x^2y^2}{(1-y)(1-xy)} + \frac{(k-1)xy}{1-y}\right)(F_k(x,y) - H_k(x,y))$$
$$= \frac{(xy)^{k-1}}{1-xy} + \frac{y^{k-1}}{1-y}\left(\frac{x^2y}{1-xy} + (k-1)x\right)(F_k(x,1) - H_k(x,1)).$$

**Proof** Lemmas 6.58 and 6.63 give

$$F_k(x,y) - \frac{1}{1-xy}$$
$$= \sum_{j=1}^{k-2} jx\overline{F}_{k,j}(x)y^j + (k-1)x\sum_{j\geq k-1}\overline{F}_{k,j}(x)y^j + \frac{x^2y}{1-xy}\sum_{j\geq k-1}\overline{F}_{k,j}(x)y^j$$
$$= \sum_{j=1}^{k-2}(F_{k,j}(x) - x^j)y^j + \frac{xy^{k-1}}{1-y}\left(\frac{xy}{1-xy} + k - 1\right)(F_k(x,1) - H_k(x,1))$$
$$- \frac{xy}{1-y}\left(\frac{xy}{1-xy} + k - 1\right)(F_k(x,y) - H_k(x,y)),$$

which is equivalent to

$$\left(1 + \frac{x^2y^2}{(1-y)(1-xy)} + \frac{(k-1)xy}{1-y}\right)(F_k(x,y) - H_k(x,y))$$
$$= \frac{(xy)^{k-1}}{1-xy} + \frac{y^{k-1}}{1-y}\left(\frac{x^2y}{1-xy} + (k-1)x\right)(F_k(x,1) - H_k(x,1)),$$

as claimed. □

**Theorem 6.65** *The generating function for the number of set partitions of $\mathcal{P}_n(12\cdots k12)$ is given by*

$$\frac{\frac{x^{k-1}y_k}{1-xy_k} + \sum_{j=0}^{k-2}\sum_{i=0}^{j}(-1)^{i+j}x^i\beta_{i,j}}{1 - \sum_{j=0}^{k-2}\sum_{i=0}^{j}(-1)^{i+j}ix\beta_{i,j}},$$

*where*

$$y_k = \frac{1 - (k-2)x - \sqrt{(1-kx)^2 - 4x^2}}{2x(1-(k-2)x)},$$

$\beta_{j,j} = 1$ and $\beta_{i,j} = jx\prod_{s=i+1}^{j-1}(sx-1)$, *for $i = 0, 1, \ldots, j-1$.*

**Proof** The functional equation in the statement of Proposition 6.64 can be solved systematically using the *kernel method* technique (see [15]): Let

$$y = y_k = \frac{1 - (k-2)x - \sqrt{(1-kx)^2 - 4x^2}}{2x(1 - (k-2)x)}$$

be one of the roots of the equation $1 + \frac{x^2 y^2}{(1-y)(1-xy)} + \frac{(k-1)xy}{1-y}$. Then Proposition 6.64 gives

$$F_k(x, 1) - H_k(x, 1) = \frac{x^{k-1} y_k}{1 - x y_k}.$$

Therefore, Lemma 6.59 gives

$$F_k(x, 1) - \sum_{j=0}^{k-2} \sum_{i=0}^{j} (-1)^{i+j} (ix F_k(x, 1) + x^i) \beta_{i,j} = \frac{x^{k-1} y_k}{1 - x y_k},$$

which completes the proof. □

The case of 12312-avoiding has been received a lot of attention from other researchers. Note that Theorem 6.65 shows that the generating function for the number of 12312-avoiding set partitions of $[n]$ is given by

$$\frac{3 - 3x - \sqrt{1 - 6x + 5x^2}}{2(1-x)}.$$

This generating function counts other combinational structures. So the question was to find a bijections between 12312-avoiding set partitions and these combinatorial structures. For instance, Kim [168] found bijections between the sets 2-*distant noncrossing set partitions*, 12312-avoiding set partitions, 3-*Motzkin paths*, *UH-free Schröder paths*, and Schröder paths without *peaks at even height* (see also [159], [318], and [366]).

**Definition 6.66** A $k$-*distant crossing* of a set partition $\pi$ is a set of two edges $(i_1, j_1)$ and $(i_2, j_2)$ of the standard representation of $\pi$ satisfying $i_1 < i_2 \leq j_1 < j_2$ and $j_1 - i_2 \geq k$.

A *Schröder path* of size $2n$ is a lattice path of size $2n$ consisting of steps $U = (1, 1)$, $D = (1, -1)$ and $H = (2, 0)$ and does not go below the $x$-axis. A $UH$-*free Schröder path* is a Schröder path such that there are no two consecutive steps $UH$, $U$ step followed by $H$ step. A *peak* is a pair of consecutive steps $UD$. The height of the peak is the height of the $D$ step.

A $k$-*Motzkin path* is a lattice path of size $n$ consisting of steps $U = (1, 1)$, $D = (1, -1)$ and $k$ coloured steps of type $H = (1, 0)$.

More precisely, Kim [168] described a direct bijection between 2-distant noncrossing set partitions and 12312-avoiding set partitions. A direct bijection between the set $\mathcal{P}_n(12312)$ and the set of $UH$-free Schröder paths of size $2(n-1)$ is obtained by Yan [366].

### 6.2.7 Patterns Equivalent to $12\cdots m$

Now we turn back our attention to continue our classifications of subsequence patterns. Fact 6.32 (also, see Exercise 6.6) gives that the number of set partitions of $\mathcal{P}_n(12\cdots m)$ is given by

$$P_n(12\cdots m) = \sum_{i=0}^{m-1} \mathrm{Stir}(n,i).$$

As an application of the previous results, we present two classes of subsequence patterns that are equivalent to the subsequence pattern $12\cdots m$. This result gives an alternative combinatorial interpretation of the Stirling numbers $\mathrm{Stir}(n,i)$.

**Theorem 6.67** *For every $k \geq 2$, the following subsequence patterns are equivalent:*

(i) $12\cdots(k+1)$,

(ii) $12\cdots(k-1)kd$ *with* $d \in [k]$,

(iii) $12\cdots(k-1)dk$ *with* $d \in [k-1]$.

**Proof** Corollary 6.57 shows

$$12\cdots(k+1) \sim 12\cdots(k-1)kk \sim 12\cdots(k-1)(k-1)k.$$

But by Corollary 6.27, we obtain the following equivalences

$$12\cdots(k-1)kk \sim 12\cdots(k-1)kd \text{ and } 12\cdots(k-1)(k-1)k \sim 12\cdots(k-1)dk,$$

as claimed. □

### 6.2.8 Binary Patterns

In this subsection, we focus on the avoidance of subsequence *binary* patterns. At first we consider the subsequence patterns of the form $1^k 21^\ell$ (We have already seen that $112 \sim 121$ (see Section 6.2.2). In order to do that, we need some preparation.

**Definition 6.68** *Let $\pi$ be a set partition, so $\pi$ can be uniquely expressed as $1\pi^{(1)}\cdots 1\pi^{(t-1)}1\pi^{(t)}$, where $\pi^{(i)}$ is a (possibly empty) maximal subword of $\pi$ that do not contain the letter 1. In this context, this representation will be called the* chunk representation *of $\pi$, and the sequence $\pi^{(i)}$ will be referred to as the $i$-th chunk of $\pi$. By concatenating the chunks into a sequence $\pi^{(1)}\cdots\pi^{(t)}$ and then taking the reduced form (subtracting 1 from every letter), we obtain a canonical form of a set partition; we denote this set partition by $\pi^-$.*

**Example 6.69** If $\pi = 121321142313 \in \mathcal{P}_{12,4}$ then $\pi^{(1)} = 2$, $\pi^{(2)} = 32$, $\pi^{(3)} = \emptyset$, $\pi^{(4)} = 423$ and $\pi^{(5)} = 3$. So $\pi^- = 1213122$.

We start by characterizing the set $\mathcal{P}_n(1^j 21^k)$.

**Lemma 6.70** *Let $\pi$ be a set partition that has $t$ occurrences of the letter $1$ and let $\pi^{(i)}$ be the $i$-th chunk of $\pi$. Let $j \geq 1$ and $k \geq 0$ be two integers. Then the set partition $\pi$ avoids $1^j 21^k$ if and only if*

*(i) the set partition $\pi^-$ avoids $1^j 21^k$.*

*(ii) for every $i$ such that $j \leq i \leq t-k$, $\pi^{(i)}$ is empty.*

**Proof** Clearly, (i) and (ii) are necessary. To verify that they are sufficient, we argue by contradiction. Let $\pi$ be a set partition that satisfies (i) and (ii), and assume that $\pi$ has a subsequence $a^j b a^k$ for two symbols $a < b$. If $a = 1$ we have a contradiction with (ii), and if $a > 1$, then $\pi^-$ contains the sequence $(a-1)^j (b-1)(a-1)^k$, which leads to contradiction with (i). □

**Theorem 6.71** *For any three integers $j, k, m$ with $1 \leq j, k \leq m$,*

$$1^j 21^{m-j} \sim 1^k 21^{m-k}.$$

*Moreover, $1^j 21^k \stackrel{s}{\sim} 1^{j+k} 2$.*

**Proof** We proceed with the proof by showing that there is a bijection

$$f : \mathcal{P}_n(1^k 21^{m-k}) \to \mathcal{P}_n(1^m 2),$$

for every $m > k \geq 1$. We construct the map $f$ by induction on the number of blocks of $\pi$. If $\pi$ has one block, namely $\pi = 1^n$, then we define $f(\pi) = \pi$. Assume that $f$ has been defined for all set partitions with at most $b-1$ blocks, let $\pi \in \mathcal{P}_{n,b}(1^k 21^{m-k})$ and let $t$ be the size of the first block of $\pi$. Let $\pi^{(1)}, \ldots, \pi^{(t)}$ be the chunks of $\pi$ and let $\pi^-$ be given as above, and let us define $\sigma = f(\pi^-)$. This is well defined, since $\pi^- \in \mathcal{P}_{n-t,b-1}(1^k 21^{m-k})$. Now, let $\nu = \sigma + 1$. We express $\nu$ as a concatenation of the form $\nu = \nu^{(1)} \cdots \nu^{(t)}$, where the size of $\nu^{(i)}$ equals the size of $\pi^{(i)}$. Lemma 6.70 gives that the chunk $\theta^{(i)}$ (and hence also $\nu^{(i)}$) is empty whenever $k \leq i \leq t - m + k$. Now, we set

- $f(\pi) = 1\nu^{(1)} 1\nu^{(2)} 1 \cdots 1\nu^{(t-1)} 1\nu^{(t)}$ when $t < m$, and

- $f(\pi) = 1\nu^{(1)} 1\nu^{(2)} 1 \cdots 1\nu^{(k-1)} 1\nu^{(t-m+k+1)} \cdots 1\nu^{(t-1)} 1\nu^{(t)} 1^{t-m+1}$ when $t \geq m$.

By Lemma 6.70 it is easy to see that $f(\pi)$ avoids $1^m 2$, and then $f$ is a bijection between the sets $\mathcal{P}_n(1^k 21^{m-k})$ and $\mathcal{P}_n(1^m 2)$ (moreover, $f$ preserves the number of blocks and the size of each block), which completes the proof. □

Using our results on fillings, we can add another subsequence pattern to the equivalence class covered by Theorem 6.71, which is a consequence of Corollary 6.27 with $k = 2$ and $\theta = 1^{m-1}$.

**Theorem 6.72** *For every* $m \geq 1$, $12^m \sim 121^{m-1}$.

Using Fact 6.31, and Theorems 6.71, 6.72 and 6.55, we obtain the following result (we leave the details of the proof as an exercise (see Exercise 6.9).

**Corollary 6.73** *Let $m$ be a positive integer, let $\tau$ be any subsequence pattern from the set*
$$T = \{1^k 21^{m-k} \mid 1 \leq k \leq m\} \cup \{12^m\}.$$
*The exponential generating function $F(x)$ for the number of set partitions of $\mathcal{P}_n(\tau)$, $\tau \in T$, is given by*
$$F(x) = 1 + \int_0^x \exp\left(t + \sum_{i=1}^{m-1} \frac{t^i}{i!}\right) dt.$$

Now we consider another type of binary subsequence patterns, namely, the subsequence pattern of the form $12^k 12^{m-k}$ with $1 \leq k \leq m$. For a fixed $m$, these subsequence patterns are all equivalent. In order to see this, it suffices to show that the matrices $M(2^{k-1} 12^{m-k}, 2)$ are all Ferrers-equivalent, and then we apply Lemma 6.23. Thus, we will define a bijection between pattern-avoiding fillings which gives the Ferrers-equivalence of these matrices. In addition, we will see that this bijection has additional properties that will be useful in proving more complicated criteria for set partition-equivalence that cannot be obtained only from Lemma 6.23.

**Definition 6.74** *Let $\Lambda$ be a sparse filling of a stack polyomino $\Pi$ and let $t \geq 1$ be an integer. A sequence $\theta_1 \theta_2 \cdots \theta_t$ of 1-cells in $\Lambda$ is called a decreasing chain if for every $i \in [t-1]$ the column containing $\theta_i$ is to left of the column containing $\theta_{i+1}$ and the row containing $\theta_i$ is above the row of $\theta_{i+1}$. An increasing chain is defined analogously.*

*A filling is $t$-falling if it has at least $t$ rows, and in its bottom $t$ rows, the leftmost 1-cells of the nonzero rows form a decreasing chain. Clearly, a $t$-falling semi-standard filling of a stack polyomino $\Pi$ only exists if the leftmost column of $\Pi$ has height at least $t$.*

**Example 6.75** *Let $\Lambda$ be the following sparse filling of a stack polyomino*

|   |   | 1 |   |   |
|---|---|---|---|---|
|   | 1 | 1 | 0 |   |
| 1 | 0 | 0 | 0 | 1 |

*Then $(1,1)$ 1-cell, $(2,2)$ 1-cell, and $(4,3)$ 1-cell are an increasing chain, and $(4,3)$ 1-cell and $(5,1)$ 1-cell are a decreasing chain.*

Avoidance of Patterns in Set Partitions            253

In the rest of this subsection, we use the following notation.

**Definition 6.76** Let $\theta_{p,q}$ denote the sequence $2^p 12^q$ and $\bar{\theta}_{p,q}$ denote the sequence $1^p 21^q$, where $p, q$ are nonnegative integers.

**Lemma 6.77** For every $p, q \geq 0$, the matrix $M(\theta_{p,q}, 2)$ is stack-equivalent to the matrix $M(\theta_{p+q,0}, 2)$. In addition, if $p \geq 1$ then for every stack polyomino $\Pi$, there is a bijection $f$ between the $M(\theta_{p,q}, 2)$-avoiding and $M(\theta_{p+q,0}, 2)$-avoiding semi-standard fillings of $\Pi$ satisfies the following properties:

- the bijection $f$ preserves the number of 1-cells in every row, and
- both $f$ and $f^{-1}$ map t-falling fillings to t-falling fillings, for every $t \geq 1$.

**Proof** Fix $p, q \geq 0$, let $M = M(\theta_{p,q}, 2)$, $M' = M(\theta_{p+q,0}, 2)$, and $\Pi$ be a stack polyomino. We proceed with the proof by induction over the number of rows of $\Pi$. If $\Pi$ has only one row, then the bijection is given by the identity map. Now, let us assume that $\Pi$ has $r \geq 2$ rows and we are presented with a semi-standard filling $\Lambda$ of $\Pi$. Let $\Pi^-$ be the diagram obtained from $\Pi$ by deleting the $r$-th row as well as every column that contains a 1-cell of $\Lambda$ in the $r$-th row. The filling $\Lambda$ induces on $\Pi^-$ a semi-standard filling $\Lambda^-$. We claim that the filling $\Lambda$ avoids $M$ if and only if

(i) the filling $\Lambda^-$ avoids $M$, and

(ii) if the $r$-th row of $\Lambda$ contains $m$ 1-cells in columns $c_1 < c_2 < \cdots < c_m$ and if $m \geq p + q$, then for every $i$ with $p \leq i \leq m - q$, the column $c_i$ is either the rightmost column of the $r$-th row of $\Pi$, or $c_i + 1 = c_{i+1}$.

Clearly, (i) and (ii) are necessary. We prove that they are sufficient. (i) guarantees that $\Lambda$ does not contain any copy of $M$ that would be confined to the first $r - 1$ rows and (ii) guarantees that $\Lambda$ has no copy of $M$ that would intersect the $r$-th row.

Now, we define recursively our bijection between $M$-avoiding and $M'$-avoiding fillings. Let $\Lambda$ be an $M$-avoiding filling of $\Pi$, and let $\Lambda^-$ and $c_1, \ldots, c_m$ as defined above. By the induction hypothesis, we have a bijection between $M$-avoiding and $M'$-avoiding fillings of the shape $\Pi^-$, which maps $\Lambda^-$ to a filling $\tilde{\Lambda}^-$ of $\Pi^-$. Let $\tilde{\Lambda}$ be the filling of $\Pi$ that has the same value as $\Lambda$ in the $r$-th row, and the columns not containing a 1-cell in the $r$-th row are filled according to $\tilde{\Lambda}^-$. Note that $\tilde{\Lambda}$ contains no copy of $M'$ in its first $r - 1$ rows, and it contains no copy of $M$ that would intersect the $r$-th row.

If $\tilde{\Lambda}$ has fewer than $p + q$ 1-cells in the $r$-th row, we define $f(\Lambda) = \tilde{\Lambda}$, otherwise we modify $\tilde{\Lambda}$ as follows: For every $i = 1, 2, \ldots, q$,

- consider the columns with indices strictly between $c_{m-q+i}$ and $c_{m-q+i+1}$ (if $i = q$, we take all columns to the right of $c_m$ that intersect the last row), and

- remove these columns from $\tilde{\Lambda}$ and reinsert them between the columns $c_{p+i-1}$ and $c_{p+i}$ (which used to be adjacent by (ii) above).

Note that these transformations preserve the relative left-to-right order of all the columns that do not contain a 1-cell in their $r$-th row. So, the obtained filling still has no copy of $M'$ in the first $r-1$ rows. By construction, the filling satisfies (ii) for the values $p' = p+q$ and $q' = 0$ used instead of the original $p$ and $q$. Hence, it is a $M'$-avoiding filling. Thus, this construction gives a bijection $f$ between $M$-avoiding and $M'$-avoiding fillings.

Clearly, the bijection $f$ preserves the number of 1-cells in each row. So, it remains to check that if $p \geq 1$, then $f$ preserves the $t$-falling property. Fix $t$, let $r = row(\Pi)$ and let us consider the following three cases:

- $r < t$: Then no filling of $\Pi$ is $t$-falling.
- $r = t$: Here $\Lambda$ is $t$-falling if and only if $\Lambda^-$ is $(t-1)$-falling and the $r$-th row is either empty or has a 1-cell in the leftmost column of $\Pi$. These conditions are preserved by $f$ and $f^{-1}$, provided $p \geq 1$.
- $r > t$: The filling $\Lambda$ is $t$-falling if and only if $\Lambda^-$ is $t$-falling.

Now, we obtain the required result from the induction hypothesis and from the fact that the relative position of the 1-cells of the first $r-1$ rows does not change when we transform $\tilde{\Lambda}$ into $f(\Lambda)$. □

Lemma 6.77 implies several results about pattern avoidance in set partitions.

**Corollary 6.78** *For any set partition $\tau$ and for any $k \geq 2$ and $p, q \geq 0$,*

$$12\cdots k(\tau+k)\theta_{p,q} \sim 12\cdots k(\tau+k)\theta_{p+q,0},$$

*and*

$$12\cdots k\theta_{p,q}(\tau+k) \sim 12\cdots k\theta_{p+q,0}(\tau+k).$$

**Proof** Lemma 6.77 shows that the two matrices $M(\theta_{p,q}, 2)$ and $M(\theta_{p+q,0}, 2)$ are Ferrers-equivalent. Thus, Proposition 6.26 states

$$M(\theta_{p,q}, k) \stackrel{F}{\sim} M(\theta_{p+q,0}, k)$$

for any $k \geq 2$. Hence, Lemma 6.23 implies that the first equivalence holds. Similarly, Lemma 6.25 shows that the second equivalence holds. □

Corollaries 6.73 and 6.78 give the following result.

**Theorem 6.79** *For every sequence $\theta$ over the alphabet $[m]$, for every $p \geq 1$ and $q \geq 0$,*

$$12\cdots m(m+1)^p(m+2)(m+1)^q\theta \stackrel{s}{\sim} 12\cdots m(m+1)^{p+q}(m+2)\theta.$$

Next, we present two theorems that make use of the $t$-falling property.

**Theorem 6.80** *Let $\tau$ be any set partition with $k$ blocks, let $p \geq 1$ and $q \geq 0$. Then*
$$\sigma = \tau(\overline{\theta}_{p,q} + k) \sim \sigma' = \tau(\overline{\theta}_{p+q,0} + k).$$

**Proof** Let $\pi \in \mathcal{P}_{n,m}$, and let $M = M(\pi, m)$. We color the cells of $M$ black and gray, where a $(i,j)$ cell is gray if and only if the submatrix of $M$ formed by the intersection of the first $i-1$ rows and $j-1$ columns of $M$ contains $M(\tau, k)$. It is not hard to verify that for each gray $(i,j)$ cell there is an occurrence of $M(\tau, k)$ that appears in the first $i-1$ rows and the first $j-1$ columns and which consists entirely of black cells. Therefore, for any matrix $M'$ obtained from $M$ by modifying the filling of $M$'s gray cells, the gray cells of $M'$ appear exactly at the same positions as the gray cells of $M$.

Let $\Lambda$ be the diagram formed by the gray cells of $M$, and let $\Pi$ be the filling of $\Lambda$ by the values from $M$. Clearly, $\Lambda$ is an upside-down copy of a Ferrers shape, and the set partition $\pi$ avoids $\sigma$ (respectively $\sigma'$) if and only if $\Pi$ avoids $M(\overline{\theta}_{p,q}, 2)$ (respectively, $M(\overline{\theta}_{p+q,0}, 2)$). Now, assume that the set partition $\pi$ is $\sigma$-avoiding, and let us describe a procedure to transform $\pi$ into a set partition $\pi'$ that avoids $\sigma'$.

We first turn the filling $\Pi$ and the diagram $\Lambda$ upside down, which transforms $\Lambda$ into a Ferrers shape $\Lambda'$, and transforms the $M(\overline{\theta}_{p,q}, 2)$-avoiding filling $\Pi$ into an $M(\theta_q^p, 2)$-avoiding filling $\overline{\Pi}$ of $\Lambda'$. By applying the bijection $f$ of Lemma 6.77 on $\overline{\Pi}$, ignoring the zero columns of $\overline{\Pi}$, we obtain a filling $\overline{\Pi}' = f(\overline{\Pi})$ which avoids $M(\overline{\theta}_{p+q,0}, 2)$. Again, we turn this filling upside down, obtaining a $M(\overline{\theta}_{p+q,0}, 2)$-avoiding filling $\Pi'$ of $\Lambda$. Then, by filling the gray cells of $M$ with the values of $\Pi'$ while the filling of the black cells remains the same, we obtain a matrix $M'$. The matrix $M'$ has exactly one 1-cell in each column, so there is a sequence $\pi'$ over the alphabet $[m]$ such that $M' = M(\pi', m)$.

Obviously, the sequence $\pi'$ avoids $\sigma'$. Thus, it remains to show that $\pi'$ is a set partition. To do so, we use the preservation of the $t$-falling property. Let $c_i$ be the leftmost 1-cell of the $i$-th row of $M$, let $c'_i$ be the leftmost 1-cell of the $i$-th row of $M'$. We know that the cells $c_1, \ldots, c_m$ form an increasing chain, because $\pi$ was a set partition. We want to show that the cells $c'_1, \ldots, c'_m$ form an increasing chain as well. Let $s$ be the largest index such that the cell $c_s$ is black in $M$ (we define $s = 0$ if no such cell exists). Note that the cells $c_1, \ldots, c_s$ are black and the cells $c_{s+1}, \ldots, c_m$ are gray in $M$. We have $c_i = c'_i$ for every $i = 1, 2, \ldots, s$. If $s > 0$, we also see that all the gray 1-cells of $M$ are in the columns to the right of $c_s$. This means that even in the matrix $M'$ all the gray 1-cells are to the right of $c_s$, because the empty columns of $\Pi$ must remain empty in $\Pi'$. In particular, all the cells $c'_{s+1}, \ldots, c'_m$ appear to the right of $c'_s$. Thus, it remains to show that $c'_{s+1}, \ldots, c'_m$ form an increasing chain. We know that the cells $c_{s+1}, \ldots, c_m$ form an increasing chain in $M$ and in $\Pi$. When $\Pi$ is turned upside down, this chain becomes a decreasing chain $\overline{c}_{s+1}, \ldots, \overline{c}_m$ in

$\overline{\Pi}$. This chain shows that $\overline{\Pi}$ is $(m-s)$-falling. Lemma 6.77 shows that $\overline{\Pi}'$ is a $(m-s)$-falling, therefore it contains a decreasing chain $\overline{c'_{s+1}}, \ldots, \overline{c'_m}$ in its bottom $m-s$ rows. This decreasing chain corresponds to an increasing chain $c'_{s+1}, \ldots, c'_m$ in $M'$, showing that $\pi'$ is a set partition, as required.

Hence, the above construction can be reversed, which constructs a bijection between the sets $\mathcal{P}_n(\sigma)$ and $\mathcal{P}_n(\sigma')$. □

Our next result can be proved by a similar technique, but the argument is slightly more technical. Thus, we leave the proof as an exercise for the interested reader (see Exercise 6.10).

**Theorem 6.81** *Let $\theta$ be an arbitrary sequence over the alphabet $[k]$, let $p \geq 1$ and $q \geq 0$. Then*

$$12 \cdots k(\overline{\theta}_{p,q} + k)\theta \sim 12 \cdots k(\overline{\theta}_{p+q,0} + k)\theta.$$

### 6.2.9 Patterns Equivalent to $12^k 13$

In this section, we consider the following sets of subsequence patterns

$$\mathcal{L}_t^+ = \{12^{p+1} 12^q 32^r : p, q, r \geq 0, p + q + r = t\}$$
$$\mathcal{L}_t^- = \{12^{p+1} 32^q 12^r : p, q, r \geq 0, p + q + r = t\}$$
$$\mathcal{L}_t = \mathcal{L}_t^+ \cup \mathcal{L}_t^-,$$

where $t$ is a nonnegative integer. The main goal of this section is to show that all the subsequence patterns in $\mathcal{L}_t$ are equivalent. Throughout this section, we assume that $t$ is arbitrary but fixed. For simple notation, we write $\mathcal{L}^+, \mathcal{L}^-$ and $\mathcal{L}$ instead of $\mathcal{L}_t^+, \mathcal{L}_t^-$ and $\mathcal{L}_t$, if there is no risk of confusion. We start eith the following definition.

**Definition 6.82** *Let $\sigma$ be a set pattern over the alphabet $\{1, 2, 3\}$, let $\pi$ be a set partition with $m$ blocks, and let $k \leq m$ be an integer. We say that $\pi$ contains $\sigma$ at level $k$, if there are letters $\ell, h \in [m]$ such that $\ell < k < h$, and the set partition $\pi$ contains a subsequence $\theta$ made of the letters $\{\ell, k, h\}$ which is order-isomorphic to $\sigma$.*

**Example 6.83** *The set partition $\pi = 123132311422211$ contains $\sigma = 121223$ at level 3, because $\pi$ contains the subsequence 131334, but $\pi$ avoids $\sigma$ at level 2, because $\pi$ has no subsequence of the form $\ell 2\ell 22h$ with $\ell < 2 < h$.*

The plan is to show that for any $\sigma, \sigma' \in \mathcal{L}$ and $k$, there is a bijection $f_k$ that maps the set partitions avoiding $\sigma$ at level $k$ to the set partitions avoiding $\sigma'$ at level $k$, while preserving $\sigma'$-avoidance at all levels $j < k$ and preserving $\sigma$-avoidance at all levels $j > k+1$. Then by composing the maps $f_k$

for $k = 2, \ldots, n-1$, we obtain a bijection between the sets $\mathcal{P}_n(\sigma)$ and $\mathcal{P}_n(\sigma')$. Indeed, this idea, to the best of my knowledge is due to Jelínek and Mansour [146]. In order to discuss the plan details, we have to create a new type of words, namely "landscape words".

**Definition 6.84** *Let $\pi$ be any set partition and fix a level $k \geq 2$. A letter of $\pi$ is called $k$-low ($k$-high!) if it is smaller (greater) than $k$. A $k$-low cluster ($k$-high cluster) is a maximal consecutive sequence of $k$-low letters ($k$-high letters) in $\pi$. The $k$-landscape of $\pi$ is a word over the alphabet $\{L, k, H\}$ obtained from $\pi$ by replacing each $k$-low cluster with a single letter $L$ and each $k$-high cluster with a single letter $H$.*

*A word $w$ over the alphabet $\{L, k, H\}$ is called a $k$-landscape word if*

- *the first letter of $w$ is $L$, the second letter of $w$ is $k$.*

- *no two letters $L$ are consecutive in $w$, no two letters $H$ are consecutive in $w$.*

*Clearly, the landscape of a set partition is a landscape word.*

*Two $k$-landscape words $w$ and $w'$ are said to be compatible, if each of the three letters $\{L, k, H\}$ has the same number of occurrences in $w$ as in $w'$. For simplicity, we drop the prefix $k$ from these terms, if the value of $k$ is clear from the context.*

**Example 6.85** *Consider the set partition $\pi = 12313231422211$. It has five 3-low clusters, namely $12, 1, 2, 1$ and $22211$, it has one 3-high cluster $4$, and its 3-landscape is $L3L3L3LHL$.*

Note that for two compatible $k$-landscape words $\vartheta, \vartheta'$, we have a natural bijection between set partitions with landscape $\vartheta$ and set partitions with landscape $\vartheta'$. If $\pi$ has landscape $\vartheta$, then we map $\pi$ to the set partition $\pi'$ of landscape $\vartheta'$ which has the same $k$-low clusters and $k$-high clusters as $\pi$. Moreover, the $k$-low and $k$-high clusters appear in the same order in $\pi$ as in $\pi'$. It is easy to show that these rules define a unique set partition $\pi'$, which provides a bijection between set partitions of landscape $\vartheta$ and set partitions of landscape $\vartheta'$ which will be called *the $k$-shuffle from $\vartheta$ to $\vartheta'$*. On the next lemma we establish the key property of shuffles.

**Lemma 6.86** *Let $\vartheta$ and $\vartheta'$ be two compatible $k$-landscape words. Let $\pi$ be a set partition with $k$-landscape $\vartheta$ and let $\pi'$ be the set partition obtained from $\pi$ by the shuffle from $\vartheta$ to $\vartheta'$. Let $\sigma$ be a subsequence pattern from $\mathcal{L}$, and let $j$ be an integer.*

(i) *If $\sigma$ does not end with the letter $1$ and $j > k$, then $\pi'$ contains $\sigma$ at level $j$ if and only if $\pi$ contains $\sigma$ at level $j$.*

(ii) *If $\sigma$ does not end with the letter $3$ and $j < k$, then $\pi'$ contains $\sigma$ at level $j$ if and only if $\pi$ contains $\sigma$ at level $j$.*

**Proof** (i) Let $\sigma = 12^{p+1}32^q 12^r \in \mathcal{L}^-$ (similarly, $\sigma \in \mathcal{L}^+$). By assumption, we have $r > 0$. Assume that $\pi$ contains $\sigma$ at a level $j > k$. In particular, $\pi$ has a subsequence $\theta = \ell j^{p+1} h j^q \ell j^r$, with $\ell < j < h$.

- If $k < \ell$, then all the letters of $\theta$ are $k$-high. By the fact that the shuffle preserves the relative order of high letters, we obtain that the set partition $\pi'$ contains the subsequence $\theta$.

- If $k \geq \ell$, the shuffle preserves the relative order of the letters $j, h$, which are all high. Let $x, y$ be the two letters of $\theta$ directly adjacent to the second occurrence of $\ell$ in $\theta$ (if $q > 0$, both these letters are equal to $j$, otherwise one of them is $h$ and the other $j$). The two letters are both high, but they have to appear in different $k$-high clusters. After the shuffle, the two letters $x, y$ will be in different clusters and separated by a nonhigh symbol $\ell' \leq k$. Since the first occurrence of $\ell'$ in $\pi'$ precedes any occurrence of $j$, the set partition $\pi'$ contains a subsequence $\ell' j^{p+1} h j^q \ell' j^r$, which has reduce form $\sigma$.

Now, we show that the shuffle preserves the occurrence of $\sigma$ at level $j$. Since the inverse of the shuffle from $\vartheta$ to $\vartheta'$ is the shuffle from $\vartheta'$ to $\vartheta$, we obtain that the inverse of a shuffle preserves the occurrence of $\sigma$ at level $j$ as well.

(ii) We prove this by using a similar argument as in the proof of (i). Assume that $\pi$ contains $\sigma$ at a level $j < k$. Thus, $\pi$ contains a subsequence over the alphabet $\{\ell < j < h\}$, which had reduce form $\sigma$.

- If $h < k$, the letters of $\theta$ are low and hence preserved by the shuffle.

- Let $h \geq k$ and let $x, y$ be the two letters of $\theta$ adjacent to the letter of $h$. Recall that $\sigma$ does not end with the letter 3; thus, $x, y$ are both well defined. So, the letters $x, y$ have to appear in two distinct low clusters. After the shuffle is performed there will be a nonlow symbol $h'$ between $x, y$. Hence, the set partition $\pi'$ contains $\sigma$.

This completes the proof. □

**Definition 6.87** *For a given $k$, a set partition $\pi$ of $[n]$ is called a $k$-hybrid if $\pi$ avoids $\sigma'$ at every level $j < k$ and $\pi$ avoids $\sigma$ at every level $j \geq k$.*

Now, we use shuffles as basic building blocks for our bijection. The first example is the following result.

**Lemma 6.88** *For all $p, q, r \geq 0$,*

$$\sigma = 12^{p+1} 12^q 32^r \sim \sigma' = 12^{p+1} 32^q 12^r.$$

**Proof** Fix $p, q, r \geq 0$, and define $t = p + q + r$. We prove that for every $k = 2, 3, \ldots, n-1$ there is a bijection $f_k$ between $k$-hybrids and $(k+1)$-hybrids. Since 2-hybrids are precisely the set partitions of $\mathcal{P}_n(\sigma)$ and $n$-hybrids are precisely the set partitions of $\mathcal{P}_n(\sigma')$, we obtain the requested result.

Fix $k$. Note that a set partition $\pi$ contains $\sigma$ at level $k$ if and only if its $k$-landscape $\vartheta$ contains a subsequence $k^{p+1} L k^q H k^r$. Similarly, $\pi$ contains $\sigma'$ at level $k$ if and only if $\vartheta$ contains a subsequence $k^{p+1} H k^q L k^r$.

Let $\pi$ be a $k$-hybrid with landscape $\vartheta$. If $\pi$ has at most $t$ occurrences of $k$, then it is also a $(k+1)$-hybrid and we set $f_k(\pi) = \pi$. Otherwise, let $\vartheta = xyz$ such that $x$ is the shortest prefix of $\vartheta$ that has $p + 1$ letters $k$ and $z$ is the shortest suffix of $\vartheta$ that has $r$ letters $k$. By assumption, $x, z$ do not overlap. Let $\overline{y}$ be the word obtained by reversing the order of the letters of $y$, and define $\vartheta' = x\overline{y}z$. Note that $\vartheta'$ is a landscape word compatible with $\vartheta$, and that $\vartheta$ avoids $k^{p+1} L k^q H k^r$ if and only if $\vartheta'$ avoids $k^{p+1} H k^q L k^r$. By applying the shuffle from $\vartheta$ to $\vartheta'$ to $\pi$ that transforms it into a set partition $\pi' = f_k(\pi)$, Lemma 6.86 shows that $\pi'$ is a $(k+1)$-hybrid. Hence, $f_k$ is the requested bijection. □

Another result in the same spirit is the following lemma, where the proof is left as an exercise (see Exercise 6.11).

**Lemma 6.89** *For every $p, q, r \geq 0$, $12^{p+2} 12^q 32^r \sim 12^{p+1} 12^q 32^{r+1}$.*

By Corollary 6.78, we already know that for any $p, q \geq 0$, $12^{p+1} 12^q 3 \sim 12^{p+q+1} 13$, so with Lemmas 6.88 and 6.89, we can state the main result of this section.

**Theorem 6.90** *For every $t$, the subsequence patterns in $\mathcal{L}_t$ are (strong) Wilf-equivalent.*

### 6.2.10 Landscape Patterns

We show that with a little bit work on landscape words using the previous argument as described in the previous subsection, new Wilf-equivalences can be achieved. Throughout this section, we use the following definition.

**Definition 6.91** *We say that $\tau$ is a 1-2-4 pattern if $\tau$ has the form $123\theta$ where $\theta$ is a sequence that*

- *has exactly one occurrence of the symbol 1 and of the symbol 4,*

- *all its remaining symbols are equal to 2, and*

- *the symbol 4 is neither the first nor the last symbol of $\theta$.*

*Similarly, a 1-3-4 pattern is a pattern of the form $123\nu$ where $\nu$ is a sequence that*

260                    Combinatorics of Set Partitions

- has one occurrence of the symbol 1 and of the symbol 4,
- all its other symbols are equal to 3, and
- the symbol 1 is not the last symbol of $\nu$.

Note that we decided to avoid the patterns $1232^p 12^q 4$, $12342^p 12^q$, and $1233^p 43^q 1$ from the set of 1-2-4 and 1-3-4 patterns defined above, because some of the arguments we will need in the following discussion (namely in Lemma 6.97) would become more complicated if these special types of patterns were allowed. We need not be too concerned about this constraint, because we have already dealt with the patterns of the three excluded types in Corollary 6.78 and Theorem 6.81. More precisely, Corollary 6.78 gives $1232^p 12^q 4 \sim 1232^{p+q} 14$ and $12342^p 12^q \sim 12342^{p+q} 1$, while Theorem 6.81 gives $1233^p 43^q 1 \sim 1233^{p+q} 41$.

For our arguments, we extend some of the terminology of Section 6.2.9.

**Definition 6.92** Let $\tau$ be a 1-2-4 pattern, $k$ be a natural number, and $\pi$ be a set partition. We say that $\pi$ contains $\tau$ at level $k$, if $\pi$ has a subsequence $\theta$ such that Reduce$(\theta) = \tau$ and the occurrences of the letter 2 in $\tau$ correspond to the occurrences of the letter $k$ in $\theta$. Similarly, if $\tau$ is a 1-3-4 pattern, we say that a partition $\pi$ contains $\tau$ at level $k$ if $\pi$ has a subsequence $\theta$ such that Reduce$(\theta) = \tau$ and the occurrences of the letter $k$ in $\theta$ corresponding to the letter 3 in $\tau$.

Now, the plan is to present an analogue of Lemma 6.86 for 1-2-4 and 1-3-4 patterns. In order to do that, we will define very special types of $k$-shuffles.

**Definition 6.93** Let $\vartheta$ be a $k$-landscape word. We say that two occurrences of the letter $H$ in $\vartheta$ are separated if there is at least one occurrence of $L$ between them. Similarly, two letters $L$ are separated if there is at least one $H$ between them. We also say that two clusters of a set partition are separated if the corresponding letters of the landscape word are separated.

**Example 6.94** Let $\vartheta = LkLkHkkHLkH$ a $k$-landscape word. In $\vartheta$, neither the first two occurrences of $L$ nor the first two occurrences of $H$ are separated, while the second and third occurrence of $H$, as well as the second and third occurrence of $L$ are separated.

**Definition 6.95** Let $\vartheta$ and $\vartheta'$ be two $k$-landscape words. We say that $\vartheta$ and $\vartheta'$ are H-compatible if they are compatible, and if for any $i, j$, the $i$-th and $j$-th occurrence of $H$ in $\vartheta$ are separated if and only if the $i$-th and $j$-th occurrence of $H$ in $\vartheta'$ are separated. Similarly, an L-compatible pair of words can be defined.

**Example 6.96** The two compatible words

$$\vartheta = LkHkkkHL \text{ and } \vartheta' = LkHkkLHk$$

are L-compatible (since the two occurrences of L are separated in both words) but they are not H-compatible (the two letters H are not separated in $\vartheta$ but they are separated in $\vartheta'$).

The next lemma explains the relevance of the above definitions; in its proof is used the same argument as in the proof of the first part of Lemma 6.86, which stated that the occurrence of $\tau$ is preserved by the shuffle as long as $\tau$ does not end with a 1. For full details, we refer the reader to [146].

**Lemma 6.97** *Let $k$ be an integer.*

(1) *Let $\vartheta, \vartheta'$ be two L-compatible $k$-landscape words, and let $\tau$ be a 1-2-4 pattern. Let $\pi$ be an any set partition, and let $\pi'$ be the set partition obtained from $\pi$ by the $k$-shuffle from $\vartheta, \vartheta'$. For every $j < k$, $\pi$ contains $\tau$ at level $j$ if and only if $\pi'$ contains $\tau$ at level $j$. Moreover, if the last letter of $\tau$ equals 2, then the previous equivalence also holds for every $j > k$.*

(2) *Let $\vartheta, \vartheta'$ be two H-compatible $k$-landscape words, and let $\tau$ be a 1-3-4 pattern. Let $\pi$ be an arbitrary set partition, and let $\pi'$ be the set partition obtained from $\pi$ by the $k$-shuffle from $\vartheta, \vartheta'$. For every $j > k$, $\pi$ contains $\tau$ at level $j$ if and only if $\pi'$ contains $\tau$ at level $j$. Moreover, if the last letter of $\tau$ equals 3, then the previous equivalence also holds for every $j < k$.*

By Lemma 6.97, we may show all the following equivalence relations

$$1232^p 412^q \sim 1232^p 42^q 1, \qquad 1232^p 142^q \sim 12312^p 42^q,$$
$$123^{p+1} 143^q \sim 123^{p+1} 13^q 4, \qquad 123^{p+1} 413^q \sim 12343^p 13^q,$$

for every $p, q \geq 0$, see the next four lemmas.

**Lemma 6.98** *For all $p, q \geq 1$, $\tau = 1232^p 412^q \sim \tau' = 1232^p 42^q 1$.*

**Proof** To prove our lemma, it is sufficient to establish a bijection $f_k$ between $k$-hybrids and $(k+1)$-hybrids. We say that a $k$-high cluster of $\pi$ is *extra-high* if it contains a letter greater than $k + 1$. We claim that $\pi$ contains $\tau$ at level $k$ if and only if by scanning the $k$-landscape $\vartheta$ of $\pi$ from left to right we may find the following: the leftmost high cluster, $p$ occurrences of the letter $k$, an extra-high cluster, a low cluster, and followed by $q$ occurrences of $k$. To show this, it suffices to notice that the leftmost high cluster contains the letter $k + 1$, and to the left of this cluster we may always find all the letters of $[k]$ in increasing order. Similarly, $\pi$ contains $\tau'$ at level $k$ if and only if it contains, left-to-right, the leftmost high-cluster, $p$ occurrences of $k$, an extra-high cluster, $q$ occurrences of $k$ and a low cluster.

Now suppose that $\pi$ is a $k$-hybrid set partition. Let H' be the leftmost extra-high cluster of $\pi$ such that between H' and the leftmost high cluster of

$\pi$ there are at least $p$ occurrences of $k$. If this cluster does not exist, or if $\pi$ has at most $q-1$ of the letter $k$ to the right of H′, then $\pi$ avoids both $\tau$ and $\tau'$ at level $k$, and in this case we define $f_k(\pi) = \pi$. Otherwise, let $\vartheta$ be the $k$-landscape of $\pi$, and let us write $\vartheta$ as

$$\vartheta = x\mathrm{H}'yk_q\vartheta^{(1)}k_{q-1}\vartheta^{(2)}\cdots k_1\vartheta^{(q)},$$

where H′ represents the extra-high cluster defined above, and $k_i$ represents the $i$-th letter $k$ in $\pi$, counted from the right. The letters $x, y$ and $\vartheta^{(1)}, \ldots, \vartheta^{(q)}$ above refer to the corresponding subwords of $\vartheta$ appearing between these symbols.

By this construction, none of the $\vartheta^{(i)}$'s contains the letter $k$, so each of them is an alternating sequence over the alphabet $\{\mathrm{L}, \mathrm{H}\}$, possibly empty. Since $\pi$ avoids $\tau$ at level $k$, the subword $y$ does not contain the letter L. We write $\vartheta^{(1)} = H^*\vartheta^{(1)-}$ as follows:

- If the first letter of $\vartheta^{(1)}$ is H, then we put $H^* = \mathrm{H}$ and $\vartheta^{(1)-}$ is equal to $\vartheta^{(1)}$ with the first letter removed.

- If $\vartheta^{(1)}$ does not start with H, then $H^*$ is the empty string and $\vartheta^{(1)-} = \vartheta^{(1)}$.

Define the word $\vartheta'$ by

$$\vartheta' = x\mathrm{H}'\vartheta^{(1)-}k_1\vartheta^{(2)}k_2\vartheta^{(3)}k_3\cdots k_{q-1}\vartheta^{(q)}k_qH^*y.$$

It is easy to verify that $\vartheta'$ is a landscape word (note that neither $y$ nor $\vartheta^{(1)-}$ can start with the letter H), and that $\vartheta'$ is L-compatible with $\vartheta$ (recall that $y$ contains no L).

Let $\pi'$ be the set partition obtained from $\pi$ by the shuffle from $\vartheta$ to $\vartheta'$. Since the words $\vartheta$ and $\vartheta'$ share the same prefix up to the letter H′, we conclude that the prefix of $\pi$ through the cluster H′ is not affected by the shuffle. So, the shuffle preserves the property that H′ is the leftmost extra-high cluster with at least $p$ letters $k$ between H′ and the leftmost high cluster of $\pi'$. It is routine to verify that $\pi'$ avoids $\tau'$ at level $k$. Thus, Lemma 6.97 shows $\pi'$ is a $(k+1)$-hybrid set partition. It is easy to see that for any given $(k+1)$-hybrid set partition $\pi'$, we may uniquely invert the procedure above and obtain a $k$-hybrid set partition $\pi$. By defining $f_k(\pi) = \pi'$, we obtain the bijection between $k$-hybrids and $(k+1)$-hybrids, as required. □

The proofs of the next three lemmas follow the same basic arguments as the proof of Lemma 6.98. The only difference is in the decompositions of the corresponding landscape words $\vartheta, \vartheta'$. Thus, we only focus on the differences.

**Lemma 6.99** *For all $p, q \geq 1$, $\tau = 1232^p142^q \sim \tau' = 12312^p42^q$.*

**Proof** A set partition $\pi$ contains $\tau$ at level $k$ if and only if it contains, from left to right, the leftmost high cluster, $p$ copies of $k$, a low cluster, an extra-high cluster, and $q$ copies of $k$. Similarly, we characterize $\tau'$.

We denote the leftmost high cluster of $\pi$ by $H_1$. Let $H'$ be the rightmost extra-high cluster of $\pi$ that has the property that there are at least $q$ occurrences of $k$ to the right of $H'$. If $H'$ does not exist, or if there are fewer than $p$ occurrences of $k$ between $H_1$ and $H'$, then $\pi$ contains neither $\tau$ nor $\tau'$ at level $k$ and we put $f_k(\pi) = \pi$. Otherwise, let $\vartheta$ be the landscape of $\pi$, and let us decompose

$$\vartheta = x H_1 \vartheta^{(1)} k_1 \vartheta^{(2)} k_2 \cdots \vartheta^{(p)} k_p y H' z$$

where none of the $\vartheta^{(i)}$ contains $k$, and $y$ avoids L. Write $\vartheta^{(p)} = \vartheta^{(p)-} H^*$ such that $\vartheta^{(p)-}$ does not end with the letter H and $H^*$ is equal either to H or to the empty string, depending on whether $\vartheta^{(p)}$ ends with H or not. Then define $\vartheta' = x H_1 \overline{y} k_1 H^* \vartheta^{(1)} k_2 \vartheta^{(2)} \cdots k_p \vartheta^{(p)-} H' z$, where $\overline{y}$ is the reversal of $y$. The rest of the proof is analogous to Lemma 6.98. □

Now we apply the same arguments to 1-3-4 patterns.

**Lemma 6.100** *For any $p \geq 0$ and $q \geq 1$, $\tau = 123^{p+1}13^q 4 \sim \tau' = 123^{p+1}143^q$.*

**Proof** Note that a $k$-hybrid is a set partition that avoids $\tau$ at every level $j \geq k$ and that avoids $\tau'$ at every level below $k$. Let us say that a $k$-cluster of a set partition $\pi$ is *extra-low* if it contains a letter smaller than $k-1$. A set partition contains $\tau$ at level $k$ if and only if it has $p+1$ occurrences of $k$ followed by an extra-low cluster, followed $q$ occurrences of $k$, followed by a high cluster. Similarly, a set partition contains $\tau'$ at level $k$ if and only if it has $p+1$ copies of $k$, followed by an extra-low cluster, followed by a high cluster, followed by $q$ copies of $k$.

Let $\pi$ be a $k$-hybrid set partition. Let $L'$ be the leftmost extra-low cluster of $\pi$ that has at least $p+1$ copies of $k$ to its left. If $L'$ does not exist, or if it has fewer than $q$ copies of $k$ to its right, then we set $f_k(\pi) = \pi$. Otherwise, we decompose the landscape word $\vartheta$ of $\pi$ as

$$\vartheta = x L' \vartheta^{(1)} k_1 \cdots \vartheta^{(q-1)} k_{q-1} \vartheta^{(q)} k_q y,$$

where the $\vartheta^{(i)}$ do not contain the letter $k$. By assumption, $y$ avoids H. Now, we write $y = L^* y^-$ such that $L^*$ is either an empty word or a single letter L, and $y^-$ does not start with L. Then we define $\vartheta'$ by

$$\vartheta' = x L' y^- k_1 L^* \vartheta^{(1)} k_2 \cdots \vartheta^{(q-1)} k_q \vartheta^{(q)}.$$

The words $\vartheta, \vartheta'$ are H-compatible. Now, it remains to define the bijection between $k$-hybrids and $(k+1)$-hybrids as we described in the proofs of the two previous lemma. □

**Lemma 6.101** *For every $p, q \geq 0$, $\tau = 123^{p+1}413^{q+1} \sim \tau' = 12343^p 13^{q+1}$.*

**Proof** As in the above proofs, we set $\pi$ to be a $k$-hybrid set partition. Let $L'$ be the rightmost extra-low cluster that has at least $q+1$ copies of $k$ to its right. If $L'$ has at least $p+1$ copies of $k$ to its left, then the landscape $\vartheta$ of $\pi$ can be written as

$$\vartheta = \mathrm{L}k_1\vartheta^{(1)} \cdots k_p\vartheta^{(p)}k_{p+1}y\mathrm{L}'z.$$

Next, we set $\vartheta^{(p)} = \vartheta^{(p)-}L^*$ as usual and we define

$$\vartheta' = \mathrm{L}k_1 L^* \overline{y} k_2 \vartheta^{(1)} k_3 \vartheta^{(2)} \cdots \vartheta^{(p-1)} k_{p+1} \vartheta^{(p)-}\mathrm{L}'z.$$

The rest is the same as in the proofs of the previous lemmas. □

All the above four lemmas can be summarized as follows.

**Theorem 6.102** *For every $p, q \geq 0$,*

(1) $1232^p 412^q \sim 1232^p 42^q 1$

(2) $1232^p 142^q \sim 12312^p 42^q$

(3) $123^{p+1} 143^q \sim 123^{p+1} 13^q 4$

(4) $123^{p+1} 413^q \sim 12343^p 13^q$

**Proof** If $p$ and $q$ are both positive, the results follow directly from the four preceding lemmas. If $p = 0$, (2) and (4) are trivial, (1) is a special case of Corollary 6.78, and (3) is covered by Lemma 6.100. If $q = 0$, (1) and (3) are trivial, (2) is a special case of Corollary 6.78, and (4) follows from Theorem 6.81. □

### 6.2.11 Patterns of Size Four

Now we ready to use our general theorems to consider the classification of the equivalence classes of the patterns of size at most seven. We start by the classification of the equivalence classes of the patterns of size four. Here we need to prove a very special equivalence, namely, $1212 \sim 1123$.

#### 6.2.11.1 The Pattern 1123

Unlike in the previous arguments, we do not establish a bijection between pattern-avoiding classes, but rather we show that $P_n(1123) = \frac{1}{n+1}\binom{2n}{n}$. It is well known that the number of noncrossing partitions of $\mathcal{P}_n(1212)$ is given by the $n$-th Catalan number (for example, see [204]),, which implies that $P_n(1123) = P_n(1212) = \mathrm{Cat}_n$ for all $n \geq 0$. We obtain this by proving that $P_n(1123)$ equals the number of Dyck paths of size $2n$ (see Example 2.61). Let $\mathcal{D}_{n,k}$ be the set of Dyck paths of size $2n$ whose last up-step is followed by exactly $k$ down-steps. Define $d_{n,k} = \mathcal{D}_{n,k}$.

**Lemma 6.103** *The sequence* $\{d_{n,k}\}_{n,k}$ *is given by*

$$d(1,1) = 1 \tag{6.4}$$

$$d(n,k) = 0 \quad if \quad k < 1 \quad or \quad k > n \tag{6.5}$$

$$d(n,k) = \sum_{j=k-1}^{n-1} d(n-1,j) \quad for \quad n \geq 2, n \geq k \geq 1. \tag{6.6}$$

**Proof** See Exercise 6.12. □

We now focus on the set $\mathcal{P}_n(1123)$. At first, we will give a correspondence between 1123-avoiding set partitions and 123-avoiding sequences. We start with the following definition.

**Definition 6.104** *A 123-avoiding sequence is a sequence* $\theta_1 \cdots \theta_\ell$ *of positive integers that avoids 123. We define the* rank *of a sequence to be equal to* $\ell + m - 1$, *where* $\ell$ *is the size of the sequence and* $m = \max_{i \in [\ell]} \theta_i$ *is the largest element of the sequence. We denote the set of 123-avoiding sequences of rank* $n$ *with last element equal to* $k$ *by* $\mathcal{T}_{n,k}$, *and we define* $t_{n,k} = |\mathcal{T}_{n,k}|$

**Example 6.105** *There are five 123-avoiding sequences of rank 3:* $(1,1,1)$, $(1,2), (2,1), (2,2)$ *and* $(3)$. *There are fourteen 123-avoiding sequences of rank 4:* $(1,1,1,1), (1,1,2), (1,2,1), (1,2,2), (2,1,1), (2,1,2), (2,2,1), (2,2,2),$ $(1,3), (2,3), (3,1), (3,2), (3,3),$ *and* $(4)$.

**Proposition 6.106** *A 1123-avoiding set partition* $\pi \in \mathcal{P}_{n,m}$ *has the following form*

$$\pi = 123 \cdots (m-2)(m-1)\theta \tag{6.7}$$

*where* $\theta \in \mathcal{T}_{n,m}$. *Conversely, if* $\theta \in \mathcal{T}_{n,m}$ *then* $\pi$ *given by (6.7) is a canonical representation of a 1123-avoiding set partition of* $[n]$.

*In particular, the number of set partitions of* $\mathcal{P}_n(1123)$ *with last element* $k$ *is given by* $t_{n,k}$.

**Proof** Let $\pi = \pi_1 \cdots \pi_n \in \mathcal{P}_{n,m}(1123)$. Clearly, $\pi_i = i$ for every $i \in [m-1]$. It follows that $\pi$ can be written as $\pi = 123 \cdots (m-2)(m-1)\theta$, where the sequence $\theta$ has size $l = n - m + 1$ and maximum element equal to $m$. In particular, $\theta \in \mathcal{T}_{n,k}$. Now, we show that $\theta$ is a 123-avoiding sequence. If $\theta$ contained a subsequence $xyz$ for $x < y < z$, then the original set partition would contain a subsequence $xxyz$, which is forbidden. It follows that $\theta$ obtained from a 1123-avoiding set partition $\pi$ has all the required properties. Conversely, if $\theta$ is a 123-avoiding sequence of rank $n$ and maximum element $m$, then $\pi = 12 \cdots (m-1)\theta$ is a set partition of $\mathcal{P}_{n,m}(1123)$, and the last element of $\pi$ equals the last element of $\theta$. □

By the previous proposition, $t_{n,k}$ equals the number of set partitions of $\mathcal{P}_n(1123)$ with last element equal to $k$. In order to prove that set partitions of $\mathcal{P}_n(1123)$ have the same enumeration as Dyck paths of size $2n$, it suffices to show that $d_{n,k} = t_{n,k}$ for each $n, k$. The next result presents a recurrence relation for $t_{n,k}$; for the proof, we refer the reader to [146].

**Proposition 6.107** *The numbers $t_{n,k}$ satisfy*

$$t(1,1) = 1 \tag{6.8}$$
$$t(n,k) = 0 \quad \text{if} \quad k < 1 \quad \text{or} \quad k > n \tag{6.9}$$
$$t(n,k) = \sum_{i=k-1}^{n-1} t(n-1, i) \quad \text{for} \quad n \geq 2, n \geq k \geq 1 \tag{6.10}$$

The following results are direct consequences of Propositions 6.106 and 6.107, where we leave the proofs to the reader.

**Theorem 6.108** *The number of $1123$-avoiding matchings of size $n$ with last element equal to $k$ equals the number of Dyck paths of size $2n$ whose last up-step is followed by $k$ down-steps. Moreover, the number of $1123$-avoiding matchings of size $n$ is $\mathrm{Cat_n}$.*

**Remark 6.109** *A matching of size $2n$ is a partition of the set $[2n]$ into $n$ disjoint pairs. We denote the set of all matchings of size $2n$ by $\mathcal{M}_n$. Note that a matching can be represented via the canonical form (see Definition 1.3) of the corresponding set partition, which is a sequence of integers in which each integer $i \in [n]$ occurs exactly twice, and the first occurrence of $i$ precedes the first occurrence of $i+1$. It is not hard to see that the cardinality of the set $\mathcal{M}_n$ is given by the double factorial $(2n-1)$, for all $n \geq 1$. The research on matching does not received as much attention as the research on set partitions. But in the last decade several papers have been published, where each has focused on our statistics or pattern avoidance problem on the set of matchings. At first let us explain the meaning of a pattern in a set of matchings.*

*A partial pattern of size $k$ is a sequence of integers from the set $[k]$ in which each $i \in [k]$ appears at least once and at most twice, and the first occurrence of $i$ always precedes the first occurrence of $i+1$. Given a partial pattern $\sigma$ and a matching $\mu$, we say that $\mu$ avoids $\sigma$ if $\mu$ avoids $\sigma$ as set partitions. We denote the set of all matchings of size $2n$ that avoid the partial pattern $\sigma$ by $\mathcal{M}_n(\sigma)$. Two partial patterns $\tau$ and $\sigma$ are equivalent if $|\mathcal{M}_n(\tau)| = |\mathcal{M}_n(\sigma)|$.*

*On pattern equivalences of matchings, Jelínek and Mansour [148] described several families of equivalent pairs of patterns in matchings, where many of these equivalences are based on the study of filling diagrams and set partitions. As in set partitions, they verified by computer enumeration that these families contain all the equivalences among patterns of length at most six.*

*A lot of research has been done on the enumeration problem on matchings and a book can be writtem as a sequel to the current book. In any case, here we* give some example of this research. For example, the case of $k$-noncrossing and $k$-nonnesting on matchings have been considered by Chen, Deng, Du, Stanley, and Yan [69]. In particular, they showed that the generating function for the number of $k$-noncrossing ($k$-nonnesting) matchings of size $2n$ is given by

$$\det(J_{i-j}(x) - J_{i+j}(x))_{1 \leq i,j \leq k-1},$$

where $J_m(x) = \sum_{j \geq 0} \frac{x^{m+2j}}{j!(m+j)!}$. Another example; [145] presented a bijection between 1123-avoiding matchings of size $2n$ and lattice paths in the first quarter from $(0,2)$ to $(2n,0)$ with steps $(1,1)$ and $(1,-1)$, which shows that the number of 1123-avoiding matchings of size $2n$ is given by $\frac{3}{n+2}\binom{2n}{n-1}$. Also, they combinatorially showed that the number of 1132-avoiding matchings of size $2n$ is given by $\frac{1}{2}\binom{2n}{n}$. For the interested reader, we give here some references as a starting point to the subject: [70, 72, 73, 94, 130, 182].

Theorem 6.108 derives the closed-form expression for $t_{n,k}$. Since the number of Dyck paths that end with an up-step followed by $k$ down-steps equals the number of nonnegative lattice paths from $(0,0)$ to $(2n-k-1, k-1)$, we may apply standard arguments for the enumeration of nonnegative lattice paths to obtain the formula $t_{n,k} = \frac{k}{n}\binom{2n-k-1}{n-1}$ (see Exercise 6.13).

### 6.2.11.2 Classification of Patterns of Size Four

Theorem 6.108 and the general results presented in the previous sections allow us to fully classify patterns of size four by their equivalence classes as presented in Table 6.3.

**Table 6.3**: Number of set partitions of $\mathcal{P}_n(\tau)$, where $\tau \in \mathcal{P}_4$

| $\tau$ | $P_n(\tau)$ | Reference |
|---|---|---|
| 1111 | [327, Sequence A001680] | Fact 6.31 |
| 1122 | (6.11) | (6.11) |
| 1123, 1212, 1221 | Cat$_n$ | Theorem 6.108 |
| 1112, 1121, 1211, 1222 | [327, Sequence A005425] | Corollary 6.73 |
| 1213, 1223, 1231, 1232, 1233, 1234 | [327, Sequence A007051] | Fact 6.32 |

Note that Klazar [178] has shown that the number of set partitions of $\mathcal{P}_n(1122)$ is given by

$$P_n(1122) = \sum_{j \geq 0, i \geq 3, i+2j \leq n} (j+1)^2 \binom{n}{i+2j} j! + \sum_{j=0}^{\lfloor n/2 \rfloor} \binom{n}{2j} j!. \quad (6.11)$$

Also, he enumerated the cases of 1231-avoiding set partitions and 12341-avoiding set partitions. Later, Mansour and Severini [229] enumerated the general case $12\cdots k1$-avoiding set partitions as discussed in Theorem 6.61.

268    Combinatorics of Set Partitions

The case of 1212-avoiding is known as noncrossing set partitions, where at first they were investigated by Kreweras [204] and Poupard [276]; for more details, see Section 7.2. Later, Mansour and Severini [229] enumerated the general case of $12\cdots k12$-avoiding set partitions as we showed in Theorem 6.65.

## 6.2.12  Patterns of Size Five

To complete the characterization of the equivalence of subsequence patterns of size five, we need to consider one more isolated case, namely, the pattern 12112. First, our goal is to show that this subsequence pattern is equivalent to the three subsequence patterns 12221, 12212, and 12122. The latter three patterns are all equivalent by Corollary 6.78. Thus, it is sufficient to prove that $12112 \sim 12212$.

Note that the proof involving the pattern 12112 does not use the notion of Ferrers equivalence. In fact, the matrix $M(2,112)$ is not Ferrers-equivalent to the three Ferrers-equivalent matrices $M(2,221)$, $M(2,212)$, and $M(2,122)$.

### 6.2.12.1  The Equivalence $12112 \sim 12212$

At first, we introduce more terminology and notation that we need throughout the proof of the equivalence $12112 \sim 12212$.

**Definition 6.110** *Let $\theta = \theta_1\theta_2\cdots\theta_n \in [m]^n$ such that every element of $[m]$ appears in $\theta$ at least once. For $i \in [m]$ let $f_i$ and $\ell_i$ denote the index of the first and the last occurrence of the letter $i$ in $\theta$.*

**Example 6.111** *If $\theta = 213112332$ then $f_1 = 2$, $f_2 = 1$, $f_3 = 3$, $\ell_1 = 5$, $\ell_2 = 9$ and $\ell_3 = 3$.*

**Definition 6.112** *Let $k \in [m]$. We say that the sequence $\theta$ is a $k$-semicanonical sequence ($k$-sequence for short), if*

- *for every $i,j$ with $1 \leq i < k$ and $i < j$, we have $f_i < f_j$,*
- *for every $i,j$ with $k \leq i < j \leq m$, we have $\ell_i < \ell_j$.*

Note that $m$-semicanonical sequences are precisely the canonical representation of set partitions of $\mathcal{P}_{n,m}$, while the 1-canonical sequences are precisely the sequences satisfying $\ell_i < \ell_{i+1}$ for $i \in [m-1]$. Also, for every $k \in [m]$ and set partition $\pi = \pi_1\cdots\pi_n \in \mathcal{P}_{n,m}$, there is exactly one $k$-sequence $\theta = \theta_1\cdots\theta_n$ with the property $\theta_i = \theta_j$ if and only if $\pi_i = \pi_j$. In particular, fixing $n,m$, the number of $k$-sequences is independent of $k$, and each set partition of $\mathcal{P}_{n,m}$ is represented by a unique $k$-sequence.

The following lemma plays the critical role in the proof of our equivalence.

**Lemma 6.113** *Fix $n, m$. Then the number of 12112-avoiding $k$-sequences is independent of $k$. Thus, for every $k \in [m]$, the number of 12112-avoiding $k$-sequences of size $n$ with $m$ letters equals the number of 12112-avoiding set partitions of $n$ with $m$ blocks.*

At first, we show how this lemma helps to obtain our equivalence.

**Theorem 6.114** *We have $12112 \sim 12212$. Actually, for every $n, k$, there is a bijection between $\mathcal{P}_{n,k}(12112)$ and $\mathcal{P}_{n,k}(12212)$.*

**Proof** Fix $m, n$. We know that the set $\mathcal{P}_{n,m}(12112)$ corresponds to the set $m$-semicanonical words of $[m]^n$, and Lemma 6.113 gives that these words are in bijection with 1-semicanonical 12112-avoiding words of $[m]^n$. It remains to give a bijection between the 12112-avoiding 1-sequences and the 12212-avoiding set partitions.

Let $\theta$ be a 1-semicanonical 12112-avoiding sequence with $m$ letters and size $n$, reverse the order of letters in $\theta$, and then take the complement, that is, replace each letter $i$ by $m - i + 1$. So, this transform is an involution which maps 12112-avoiding 1-sequences onto 12212-avoiding $m$-sequences, which are exactly the set partitions of $\mathcal{P}_{n,m}(12212)$. $\square$

For the proof, we refer the reader to [146]. Here, we only present a brief sketch of the main idea of the proof of Lemma 6.113. Fix $m, n$, and assume that each sequence we consider has size $n$ and $m$ distinct letters. In the following arguments, it is often convenient to represent a sequence $\theta = \theta_1 \cdots \theta_n$ by the matrix $M(\theta, m)$; see Definition 6.21. A matrix representing a $k$-sequence will be called a *$k$-matrix*, and a matrix representing a 12112-avoiding sequence will be simply called a 12112-*avoiding matrix*. According to our notation and terminology, we use the term *sparse matrix* for a 0-1 matrix with at most one 1-cell in each column, and the term *semi-standard matrix* for a 0-1 matrix with exactly one 1-cell in each column (see Definition 6.15). For a 0-1 matrix $M$, we set $f_i = f_i(M)$ and $\ell_i = \ell_i(M)$ to be the column-index of the first and the last 1-cell in the $i$-th row of $M$.

The idea is to construct a bijection that transforms a $(k+1)$-matrix $M$ into a $k$-matrix, which ignores 12112-avoidance for a while. Let the last 1-cell in row $k$ of $M$ be in column $c$; in this context the row $k$ is called *the key row of $M$*. If the last 1-cell in row $k + 1$ appears to the right of column $c$, then $M$ is already a $k$-matrix and as required. Otherwise, if row $k + 1$ has no 1-cell to the right of $c$, then we swap the key row $k$ with the row $k + 1$ to obtain a new matrix $M'$ whose key row is $k+1$. Repeating this invertible procedure until the key row is either the topmost row of the matrix, or the row above the key row has a 1-cell to the right of column $c$. This procedure transforms the original $(k+1)$-matrix into a $k$-matrix. Unfortunately, this simplistic method does not preserve 12112-avoidance. However, the idea is to present an algorithm which follows the same basic structure as the procedure above, where we swap the key row with the row above it. For a full description of this procedure, we refer the reader to [146].

### 6.2.12.2 Classification of Patterns of Size Five

Table 6.4 presents the equivalence classes of subsequence patterns of size five, together with the reference to the appropriate result. By Corollary 6.57, if $\tau$ and $\sigma$ are two equivalent subsequence patterns of size four, then $1(\tau+1)$ and $1(\sigma+1)$ are equivalent subsequence patterns of size five, and vice versa. Thus, the classification of the subsequence patterns of size four given in Table 6.3 describes the equivalences among subsequence patterns of size four with only one occurrence of 1. Also, to save space, we omit the singleton equivalence classes and the enumeration data. For this reason, Table 6.4 only presents the references for the subsequence patterns with at least two occurrences of 1. Note that in the following tables we give the reference of the result in the book without giving the name of the reference such as Theorem, Corollary, etc. The full listing of all the classes, together with enumeration data that shows the distinction between the classes, is available in Appendix I.

Table **6.4**: Nonsingleton equivalence classes of subsequence patterns of $\mathcal{P}_5$

| $\tau$ | Reference | $\tau$ | Reference |
|---|---|---|---|
| 12133, 12233 | 6.27 | 11223, 11232 | 6.80 |
| 12112, 12122, 12212, 12221 | 12112 $\sim$ 12122 by 6.114, the rest by 6.78 | 11112, 11121, 11211, 12111, 12222 | 6.71, 6.72 |
| 12113, 12223, 12232, 12311, 12322, 12333 | 6.27 | 12314, 12324, 12334, 12341, 12342, 12343, 12344, 12345 | 6.67 |
| 12123, 12132, 12134, 12213, 12231, 12234, 12312, 12321, 12323, 12331, 12332 | 12332 $\sim$ 12331 by 6.30, 12234 $\sim$ 12134 by 6.27, the rest by 6.90 | | |

### 6.2.13 Patterns of Size Six

Table 6.5 lists the equivalence classes of subsequence patterns of size six. Again, in order to save space, we omit the singleton equivalence classes and the enumeration data. The full listing of all the classes, together with enumeration data that shows the distinction between the classes, is available in Appendix I. As before, we provide references for the subsequence patterns that contain at least two occurrences of 1 and we give the reference of the result in the book without giving the name of the reference such as Theorem, Corollary, etc.

**Table 6.5**: Nonsingleton equivalence classes of subsequence patterns of $\mathcal{P}_6$

| Pattern | Reference | Pattern | Reference |
|---|---|---|---|
| 123413, 123424 | 6.27 | 123134, 123143 | 6.102 |
| 121345, 122345 | 6.27 | 121344, 122344 | 6.27 |
| 121234, 122134 | 6.78 | 122133, 121233 | 6.78 |
| 112334, 112343 | 6.80 | 121134, 122234 | 6.27 |
| 123121, 123232 | 6.27 | 121333, 122333 | 6.27 |
| 121133, 122233 | 6.27 | 111223, 111232 | 6.80 |
| 112223, 112232, 112322 | 6.80 | 123123, 123312, 123321 | 6.37 |

| Pattern | Reference |
|---|---|
| 123144, 123244, 123344 | 6.27 |
| 121334, 122334, 121343, 122343 | 121334 $\sim$ 122334 and 121343 $\sim$ 122343 by 6.27 |
| 122311, 123311, 123211, 123322 | 122311 $\sim$ 123211 by 6.81, the rest by 6.30 |
| 122221, 121222, 122122, 122212 | 6.78 |
| 111112, 111121, 111211, 112111, 121111, 122222 | 6.71, 6.72 |
| 121113, 122232, 122322, 123111, 123222, 123333, 122223 | 121113 $\sim$ 122223 and 123111 $\sim$ 123222 by 6.27 |
| 123114, 123224, 123334, 123343, 123411, 123422, 123433, 123444 | 123114 $\sim$ 123224 and 123411 $\sim$ 123422 by 6.27 |
| 123415, 123425, 123435, 123445, 123451, 123452, 123453, 123454, 123455, 123456 | 6.67 |
| 121223, 121232, 121322, 122123, 122132, 122213, 122231, 122312, 122321, 123112, 123122, 123212, 123221, 123223, 123233, 123323, 123331, 123332 | 123112 $\sim$ 123223 by 6.27, 123331 $\sim$ 123332 $\sim$ 123221 by 6.30, the rest by 6.90 |
| 123124, 123145, 123214, 123234, 123243, 123245, 123324, 123341, 123342, 123345, 123412, 123421, 123423, 123431, 123432, 123434, 123441, 123442, 123443 | 123421 $\sim$ 123431 $\sim$ 123441 $\sim$ 123432 by 6.30, 123124 $\sim$ 123234, 123145 $\sim$ 123245, 123214 $\sim$ 123324, 123341 $\sim$ 123342, 123412 $\sim$ 123423 by 6.27 |

## 6.2.14 Patterns of Size Seven

As before, Table 6.6 lists the nonsingleton classes of subsequence patterns of size seven. The full listing of all the classes, together with enumeration data that shows the distinction between the classes, is available in the home page of the author.

**Table 6.6**: Nonsingleton equivalence classes of subsequence patterns of $\mathcal{P}_7$

| Pattern | Reference | Pattern | Reference |
| --- | --- | --- | --- |
| 1234514, 1234525 | 6.27 | 1234351, 1234352 | 6.27 |
| 1234315, 1234425 | 6.27 | 1233451, 1233452 | 6.27 |
| 1234132, 1234243 | 6.27 | 1234213, 1234324 | 6.27 |
| 1233441, 1233442 | 6.27 | 1213453, 1223453 | 6.27 |
| 1213435, 1223435 | 6.27 | 1213456, 1223456 | 6.27 |
| 1213455, 1223455 | 6.27 | 1213345, 1223345 | 6.27 |
| 1213443, 1223443 | 6.27 | 1213434, 1223434 | 6.27 |
| 1234131, 1234242 | 6.27 | 1233134, 1233143 | 6.102 |
| 1234133, 1234244 | 6.27 | 1231334, 1231433 | 6.102 |
| 1234113, 1234224 | 6.27 | 1231242, 1232142 | 6.102 |
| 1232412, 1232421 | 6.102 | 1231124, 1232234 | 6.27 |
| 1231214, 1232324 | 6.27 | 1213344, 1223344 | 6.27 |
| 1123445, 1123454 | 6.80 | 1232114, 1233224 | 6.27 |
| 1212345, 1221345 | 6.78 | 1212344, 1221344 | 6.78 |
| 1233312, 1233321 | 6.78 | 1212333, 1221333 | 6.78 |
| 1121334, 1121343 | 6.80 | 1211134, 1222234 | 6.27 |
| 1112334, 1112343 | 6.80 | 1231212, 1232323 | 6.27 |
| 1232121, 1233232 | 6.27 | 1231221, 1232332 | 6.27 |
| 1232112, 1233223 | 6.27 | 1231122, 1232233 | 6.27 |
| 1122334, 1122343 | 6.80 | 1211345, 1222345 | 6.27 |
| 1211344, 1222344 | 6.27 | 1213444, 1223444 | 6.27 |
| 1211333, 1222333 | 6.27 | 1231112, 1232223 | 6.27 |
| 1231121, 1232232 | 6.27 | 1231211, 1232322 | 6.27 |
| 1213333, 1223333 | 6.27 | 1211133, 1222233 | 6.27 |
| 1111223, 1111232 | 6.80 | 1123334, 1123343, 1123433 | 6.80 |
| Pattern | | Reference | |
| 1212234, 1221234, 1222134 | | 6.78 | |
| 1212234, 1221234, 1222134 | | 6.78 | |
| 1233122, 1233212, 1233221 | | 6.78 | |
| 1212233, 1221233, 1222133 | | 6.78 | |
| 1234513, 1234524, 1234535 | | 6.27 | |
| | | Continued on next page | |

| Pattern | Reference |
|---|---|
| 1234135, 1234245, 1234254 | 1234135 $\sim$ 1234245 by 6.27 |
| 1231456, 1232456, 1233456 | 6.27 |
| 1231455, 1232455, 1233455 | 6.27 |
| 1234231, 1234341, 1234342 | 6.27 |
| 1231145, 1232245, 1233345 | 6.27 |
| 1234313, 1234424, 1233413 | 1233413 $\sim$ 1234313 by 6.102 and 1234313 $\sim$ 1234424 by 6.27 |
| 1234121, 1234232, 1234343 | 6.27 |
| 1231444, 1232444, 1233444 | 6.27 |
| 1231144, 1232244, 1233344 | 6.27 |
| 1112223, 1112232, 1112322 | 6.80 |
| 1231244, 1232144, 1232344, 1233244 | 6.27 |
| 1213445, 1213454, 1223445, 1223454 | 6.27 |
| 1212334, 1212343, 1221334, 1221343 | 1212334 $\sim$ 1221334 and 1212343 $\sim$ 1221343 by 6.78, 1212334 $\sim$ 1212343 by 6.80 |
| 1234434, 1234441, 1234442, 1234443 | 6.27 |
| 1234155, 1234255, 1234355, 1234455 | 6.27 |
| 1231245, 1232145, 1232345, 1233245 | 1231245 $\sim$ 1232345 by 6.27, 1231245 $\sim$ 1232145 by 6.78 |
| 1211334, 1211343, 1222334, 1222343 | 6.27 |
| 1223111, 1232111, 1233111, 1233222 | 1223111 $\sim$ 1232111 by 6.81, 1232111 $\sim$ 1233111 $\sim$ 1233222 by 6.30 |
| 1122223, 1122232, 1122322, 1123222 | 6.80 |
| 1222311, 1223211, 1232211, 1233311, 1233322 | 1222311 $\sim$ 1223211 $\sim$ 1232211 by 6.81, 1232211 $\sim$ 1233311 $\sim$ 1233322 by 6.27 |
| 1212222, 1221222, 1222122, 1222212, 1222221 | 6.78 |
| 1231445, 1231454, 1232445, 1232454, 1233445, 1233454 | 6.27 |
| 1213334, 1213343, 1213433, 1223334, 1223343, 1223433 | 1213334 $\sim$ 1213343 $\sim$ 1213433 by 6.80, 1213334 $\sim$ 1223334 by 6.27 |
| | Continued on next page |

| Pattern | Reference |
|---|---|
| 1111112, 1111121, 1111211, 1112111, 1121111, 1211111, 1222222 | 6.71, 6.72 |
| 1211113, 1222223, 1222232, 1222322, 1223222, 1231111, 1232222, 1233333 | 6.27 |
| 1231114, 1232224, 1233334, 1233343, 1233433, 1234111, 1234222, 1234333, 1234444 | 6.27 |
| 1234115, 1234225, 1234335, 1234445, 1234454, 1234511, 1234522, 1234533, 1234544, 1234555 | 6.27 |
| 1234516, 1234526, 1234536, 1234546, 1234556, 1234561, 1234562, 1234563, 1234564, 1234565, 1234566, 1234567 | 6.67 |
| 1231234, 1233214, 1233412, 1233421, 1234123, 1234234, 1234312, 1234321, 1234412, 1234421, 1234423, 1234431, 1234432 | $1233412 \sim 1233421$, $1234312 \sim 1234321$ and $1234412 \sim 1234421$ by 6.78, $1231234 \sim 1233214$ by 6.40, $1231234 \sim 1234123$ by 6.54, $1234123 \sim 1234321$ by 6.37, $1233421 \sim 1234321$ by 6.81, $1234321 \sim 1234421 \sim 1234431 \sim 1234432$ by 6.30 |
| 1212223, 1212232, 1212322, 1213222, 1221223, 1221232, 1221322, 1222123, 1222132, 1222213, 1222231, 1222312, 1222321, 1223122, 1223212, 1223221, 1231222, 1232122, 1232212, 1232221, 1232333, 1233233, 1233323, 1233331, 1233332 | $1232221 \sim 1233331 \sim 1233332$ by 6.30, the rest by 6.90 |
| | Continued on next page |

| Pattern | Reference |
|---|---|
| 1234125, 1234156, 1234215, 1234235, 1234256, 1234325, 1234345, 1234354, 1234356, 1234435, 1234451, 1234452, 1234453, 1234456, 1234512, 1234521, 1234523, 1234531, 1234532, 1234534, 1234541, 1234542, 1234543, 1234545, 1234551, 1234552, 1234553, 1234554 | $1234521 \sim 1234531 \sim 1234541 \sim 1234551 \sim 1234532$ by 6.30, $1234125 \sim 1234235$, $1234156 \sim 1234256$, $1234215 \sim 1234325$, $1234451 \sim 1234452$ and $1234512 \sim 1234523$ by 6.27 |
| 1231224, 1232124, 1232214, 1232334, 1232343, 1232433, 1233234, 1233243, 1233324, 1233341, 1233342, 1233423, 1233431, 1233432, 1234112, 1234122, 1234212, 1234221, 1234223, 1234233, 1234323, 1234331, 1234332, 1234334, 1234344, 1233411, 1233422, 1234211, 1234311, 1234322, 1234411, 1234422, 1234433 | 6.27 |

Finally, we note that the raw enumeration data for subsequence patterns of size six and seven are available in Appendix I. The data were obtained with a computer program, whose source code is given in Appendix H.

## 6.3 Generalized Patterns

In this section we will discuss enumeration and avoidance of other types (not of subsequence, subword) of patterns in set partitions. The subsequence and subword patterns are very special type of patterns of the so-called generalized patterns, which were introduced by Babson and Steingrímsson [13] in the context of permutations to study *Mahonian statistics*.

**Definition 6.115** *A generalized pattern of size $k$ is a word consisting of $k$ letters in which two adjacent letters may or may not be separated by a dash. The absence of a dash between two adjacent letters in the pattern indicates that the corresponding letters in the set partition must be adjacent. If the pattern $\tau$ is of the form $\tau = \tau^{(1)}\text{-}\tau^{(2)}\text{-}\cdots\text{-}\tau^{(\ell)}$ where the $\tau^{(i)}$ are all subword patterns of sizes $j_i$, then $\tau$ is of* type $(j_1, j_2, \ldots, j_\ell)$.

Clearly, a subword pattern is a generalized pattern that has no dashes, and a subsequence pattern is a generalized pattern that has dashes between all pairs of adjacent letters.

**Definition 6.116** *A sequence (permutation, word, composition or set partition) $\sigma$ contains a generalized pattern $\tau$ of size $k$ if $\tau$ equals the reduced form of any $k$-term subsequence of $\sigma$ that follows the adjacency requirements given by $\tau$. Otherwise, we say that $\sigma$ avoids the generalized pattern $\tau$ or is $\tau$-avoiding.*

**Example 6.117** *Let $\tau = 123\text{-}2\text{-}2$. Then $\tau$ has the type $(3,1,1)$, since $\tau^{(1)} = 123$, $\tau^{(2)} = 2$ and $\tau^{(3)} = 2$. Furthermore, $\tau$ occurs in $\sigma$ if there is a subsequence $\sigma_i \sigma_{i+1} \sigma_{i+2} \sigma_j \sigma_\ell$ with $i+2 < j < \ell$ and $\sigma_i < \sigma_{i+1} = \sigma_j = \sigma_\ell < \sigma_{i+2}$. Thus, the set partition $1234153452443$ contains two occurrences of $123\text{-}2\text{-}1$, namely, $\sigma_2 \sigma_3 \sigma_4 \sigma_7 \sigma_{13} = 23433$ and $\sigma_7 \sigma_8 \sigma_9 \sigma_{11} \sigma_{12} = 34544$.*

In this section, we study generalized patterns of size three for set partition. The only 3-letter generalized patterns that we have not yet discussed are those with one dash, that is, the patterns of type either $(2,1)$ or $(1,2)$; see the next subsections. These patterns were studied by Claesson [78] for permutations, Burstein and Mansour [58] for words, and by Heubach and Mansour [138] for compositions.

### 6.3.1 Patterns of Type $(1,2)$

In this section' we focus on generalized patterns of type $(1,2)$, where our results are summarized in Table 6.7.

**Table 6.7**: Three-letter generalized patterns of type $(1,2)$

| $\tau$ | Reference | $\tau$ | Reference |
|---|---|---|---|
| 1-11 | 6.122 | 1-32 | 6.127 |
| 2-21, 1-12, 1-21, 2-12 | 6.125 | 1-23 | 6.127 |
| 2-11, 3-21, 3-12 | 6.120 | 2-31 | 2.17-2.18 |
| 1-22 | 6.124 | 2-13 | 2.13-2.16 |

Most our results in this section related to finding an explicit formula for the generating function for the number of set partitions of $\mathcal{P}_{n,k}(\tau)$, namely, $P_\tau(x;k)$ (see Definition 6.1). In order to express our formulas, we need the following notation.

**Definition 6.118** *Define $W_\tau(x;k)$ to be the generating function for the number of words of size $n$ over the alphabet $[k]$ that avoid the pattern $\tau$.*

We start with a theorem considering a general class of patterns.

**Theorem 6.119** Let $\tau = \ell\text{-}\tau'$ be a generalized pattern with one dash such that $\tau'$ is a subword pattern over the alphabet $[\ell - 1]$. Then

$$P_\tau(x; k) = \frac{x^k}{\prod_{j=1}^{\ell-1}(1 - jx)} \prod_{j=\ell}^{k} \frac{W_{\tau'}(x; j - 1)}{1 - xW_{\tau'}(x; j - 1)}.$$

**Proof** Note that each member $\pi \in \mathcal{P}_{n,k}$ can be written as

$$\pi = 1\pi^{(1)} 2\pi^{(2)} \cdots k\pi^{(k)},$$

where $\pi^{(j)}$ is a word over the alphabet $[j]$ which we decompose further as $\pi^{(j,1)} j \cdots \pi^{(j,s)} j \pi^{(j,s+1)}$, where each $\pi^{(j,i)}$ is a word over the alphabet $[j - 1]$. Thus, in terms of generating functions, we may write

$$P_\tau(x; k) = x^k \prod_{j=1}^{k} \frac{W_{\tau'}(x; j - 1)}{1 - xW_{\tau'}(x; j - 1)}.$$

By the definitions, we have $W_{\tau'}(x; j) = \frac{1}{1-jx}$ for all $j = 0, 1, \ldots, \ell - 2$, which completes the proof. $\square$

**Example 6.120** Fix $k \geq 1$. From Theorem 6.119 for $\tau = 3\text{-}12$ and the fact that $W_{12}(x; j) = \frac{1}{(1-x)^j}$ we obtain

$$P_{3\text{-}12}(x; k) = \frac{x^k}{(1-x)(1-2x)} \prod_{j=3}^{k} \frac{1/(1-x)^{j-1}}{1 - x/(1-x)^{j-1}}$$

$$= x^k \prod_{j=1}^{k} \frac{1}{(1-x)^{j-1} - x}.$$

Another example; using Theorem 6.119 for $\tau = 3\text{-}21$ together with the fact that $W_{21}(x; j) = \frac{1}{(1-x)^j}$, we obtain

$$P_{3\text{-}21}(x; k) = x^k \prod_{j=1}^{k} \frac{1}{(1-x)^{j-1} - x}.$$

Also, Theorem 6.119 for $\tau = 2\text{-}11$ together with $W_{11}(x; j) = \frac{1+x}{1-(j-1)x}$ (see [59, Section 2]) gives

$$P_{2\text{-}11}(x; k) = x^k \prod_{j=1}^{k} \frac{1+x}{1 - (j-1)x - x^2}.$$

**Theorem 6.121** Let $\tau = 1\text{-}1^{m-1}$ be a generalized pattern with one dash having size $m$, where $m \geq 3$. Then

$$P_\tau(x; k) = x^k \prod_{j=1}^{k} \frac{1 - x^{m-1}}{1 - jx + (j-1)x^{m-1}}.$$

**Proof** For all $n \geq j \geq 2$,

$$P_{n,j}(\tau) = \sum_{i=1}^{m-1} P_{n-i,j-1}(\tau) + (j-1) \sum_{i=1}^{m-2} P_{n-i,j}(\tau),$$

where the first term counts all set partitions of $\mathcal{P}_{n,j}(\tau)$ such that the letter $j$ can be followed by no letter other than $j$, and second term counts all set partitions of $\mathcal{P}_{n,j}(\tau)$ that end in a run of exactly $i$ letters of the same kind for some $i$, $1 \leq i \leq m-2$, and the letter $j$ is followed by a letter other than $j$ on at least one occasion. Therefore, by multiplying the above recurrence by $x^n$ and summing over all $n \geq j$, we obtain

$$P_\tau(x;j) = \frac{P_\tau(x;j-1) \sum_{i=1}^{m-1} x^i}{1-(j-1)\sum_{i=1}^{m-2} x^i} = \frac{x(1-x^{m-1})P_\tau(x;j-1)}{1-jx+(j-1)x^{m-1}}.$$

By iterating this with using the initial condition $P_\tau(x;1) = \frac{x(1-x^{m-1})}{1-x}$, we complete the proof. $\square$

**Example 6.122** Theorem 6.121 for $m=3$ gives

$$P_{1\text{-}11}(x;k) = x(1+x)^k \left( x^{k-1} \prod_{j=1}^{k-1} \frac{1}{1-jx} \right).$$

Thus, by (2.9), the number of set partitions in $\mathcal{P}_{n,k}(1\text{-}11)$ is given by

$$\sum_{j=0}^{\min\{k,n-k\}} \text{Stir}(n-j-1, k-1) \binom{k}{j}.$$

See Exercise 6.15 for a combinatorial explanation.

Using similar arguments as in the proof of Theorem 6.121, we obtain the following result (see [231]).

**Theorem 6.123** Let $\tau = 1\text{-}2^{m-1}$ be a generalized pattern with one dash having size $m$, where $m \geq 3$. Then

$$P_\tau(x;k) = \frac{x^k}{1-x} \prod_{j=2}^{k} \frac{1-x^{m-2}}{1-jx+(j-2)x^{m-1}+x^m}.$$

**Example 6.124** Theorem 6.123 for $m=3$ gives

$$P_{1\text{-}22}(x;k) = \frac{x^k}{1-x} \prod_{j=2}^{k} \frac{1}{1-(j-1)x-x^2}.$$

Taking $k = 2$ in this implies that $P_{n,2}(1\text{-}22)$ has cardinality $\text{Fib}_{n-1}$, where $\text{Fib}_n$ denotes the $n$-th Fibonacci number with $\text{Fib}_0 = \text{Fib}_1 = 1$. Try to verify directly using the interpretation for $\text{Fib}_n$ in terms of square-and-domino tilings of length $n$!

**Proposition 6.125** *The number of set partitions of either*
*(1) $P_{n,k}(1\text{-}21)$ or $P_{n,k}(1\text{-}12)$ is given by $\binom{n-1}{k-1}$.*
*(2) $P_{n,k}(2\text{-}21)$ or $P_{n,k}(2\text{-}12)$ is given by $\binom{n+\binom{k}{2}-1}{n-k}$.*

**Proof** (1) Note that a set partition in $\mathcal{P}_{n,k}(1\text{-}21)$ cannot have a descent; hence, it must be increasing. So we have $\binom{n-1}{k-1}$ set partitions in $P_{n,k}(1\text{-}21)$. A set partition in $\mathcal{P}_{n,k}(1\text{-}12)$ must be of the form $12\cdots k\alpha$, where $\alpha$ is a word of size $n-k$ in the alphabet $[k]$ and decreasing, which means that there are $\binom{n-1}{k-1}$ set partitions in $\mathcal{P}_{n,k}(1\text{-}12)$ as well.

(2) Let $\pi = 1\pi^{(1)}2\pi^{(2)}\ldots k\pi^{(k)} \in \mathcal{P}_{n,k}(2\text{-}21)$, where each $\pi^{(i)} \in [i]^{a_i}$ and $a_1 + a_2 + \cdots + a_k = n - k$. So each $\pi^{(i)}$ must be increasing in order to avoid an occurrence of 2-21. Hence, there are $\sum \prod_{i=1}^{k} \binom{a_i + i - 1}{i-1}$ set partitions in $\mathcal{P}_{n,k}(2\text{-}21)$, where the sum is over all $k$-tuples $(a_1, a_2, \ldots, a_k)$ of nonnegative integers having sum $n-k$. On the other hand, this quantity is exactly the coefficient of $x^{n-k}$ in the generating function $\prod_{i=1}^{k} \frac{1}{(1-x)^i} = \frac{1}{(1-x)^{\binom{k+1}{2}}}$ (Check?), which is $\binom{n-k+\binom{k+1}{2}-1}{n-k} = \binom{n+\binom{k}{2}-1}{n-k}$. We leave the case 2-12 to the interested reader. □

Burstein and Mansour [59] studied the generating function $W_\tau(x; k)$ for several families of subword patterns $\tau$. From these results we can find explicit formulas for the generating functions $P_{1\text{-}322\cdots 2}(x; k)$ and $P_{1\text{-}233\cdots 3}(x; k)$ as follows; for proofs, see Exercise 6.16.

**Theorem 6.126** *Let $\tau = 1\text{-}32^{m-2}$ and $\tau' = 1\text{-}23^{m-2}$ be two generalized patterns of size $m \geq 3$. Then*

$$P_\tau(x; k) = x^k(1 - x^{m-2})^{\binom{k-1}{2}} \prod_{j=1}^{k} \frac{x^{m-3}}{x^{m-3} - x^{m-2} - 1 + (1 - x^{m-2})^{j-1}},$$

*and*

$$P_{\tau'}(x; k) = x^k(1 - x^{m-3} + x^{m-2})^{k-2} \prod_{j=1}^{k} \frac{x^{m-3}}{x^{m-3} - x^{m-2} - 1 + (1 - x^{m-2})^{j-1}},$$

*with $P_{\tau'}(x; 1) = \frac{x}{1-x}$.*

**Example 6.127** *Letting $m = 3$ in Theorem 6.126 gives*

$$P_{1\text{-}32}(x; k) = x^k(1 - x)^{\binom{k-1}{2}} \prod_{j=1}^{k} \frac{1}{(1-x)^{j-1} - x}.$$

and
$$P_{1\text{-}23}(x;k) = x^{2k-2} \prod_{j=1}^{k} \frac{1}{(1-x)^{j-1} - x},$$

with $P_{1\text{-}23}(x;1) = \frac{x}{1-x}$.

### 6.3.2 Patterns of Type $(2,1)$

We start with a general theorem that follows immediately from the fact that each set partition $\pi$ with exactly $k$ blocks may be uniquely written as $\pi = \pi' k \sigma$, where $\pi'$ is a set partition with exactly $k-1$ blocks and $\sigma$ is a word over the alphabet $[k]$.

**Theorem 6.128** *Let $\tau = \tau'\text{-}\ell$ be a generalized pattern with one dash such that $\tau'$ is a subword pattern over the alphabet $[\ell - 1]$. Then*

$$P_\tau(x;k) = xW_\tau(x;k)P_{\tau'}(x;k-1).$$

**Example 6.129** *The formula for $P_{12\text{-}3}(x;k)$ follows from Theorem 6.128 with $\tau = 12\text{-}3$ and is also obvious from the definitions: $P_{12\text{-}3}(x;1) = \frac{x}{1-x}$, $P_{12\text{-}3}(x;2) = \frac{x^2}{(1-x)(1-2x)}$ and $P_{12\text{-}3}(x;k) = 0$ for all $k \geq 3$.*

*Using the facts that $W_{21\text{-}3}(x;k) = \prod_{j=1}^{k} \frac{(1-x)^{j-1}}{(1-x)^{j-1} - x}$ (see Theorem 3.6 of [59]) and $P_{21}(x;k-1) = \frac{x^{k-1}}{(1-x)^{k-1}}$, Theorem 6.128 with $\tau = 21\text{-}3$ gives*

$$P_{21\text{-}3}(x;k) = \frac{x^k}{1-x} \prod_{j=2}^{k} \frac{(1-x)^{j-2}}{(1-x)^{j-1} - x}.$$

*By Theorem 3.2 of [59] we have $W_{11\text{-}2}(x;k) = \prod_{j=0}^{k-1} \frac{1-(j-1)x}{1-jx-x^2}$ and by Theorem 4.10 we can state*

$$P_{11}(x;k-1) = \frac{x^{k-1}}{\prod_{j=1}^{k-2}(1-jx)}.$$

*Thus, Theorem 6.128 with $\tau = 11\text{-}2$ gives*

$$P_{11\text{-}2}(x;k) = x^k(1+x) \prod_{j=0}^{k-1} \frac{1}{1-jx-x^2}.$$

The next result shows that $1^{m-1}\text{-}1 \sim 1\text{-}1^{m-1}$.

**Theorem 6.130** *For each positive integer $k$,*

$$P_{1^{m-1}\text{-}1}(x;k) = P_{1\text{-}1^{m-1}}(x;k).$$

**Proof** Let us write a set partition $\pi \in \mathcal{P}_{n,k}(1\text{-}1^{m-1})$ as

$$\pi = \alpha^{(1)}\beta^{(1)} \cdots \alpha^{(r)}\beta^{(r)}$$

such that each $\alpha^{(i)}$ is a (maximal) sequence of consecutive 1's and each $\beta^{(i)}$ is a word in the alphabet $\{2, 3, \ldots, k\}$ (with $\alpha^{(r)}$ possibly empty). Define $\pi^{(1)} = \alpha^{(r)}\beta^{(1)} \cdots \alpha^{(1)}\beta^{(r)}$ to be the set partition obtained by reversing the order of the runs of consecutive 1's. Now reverse the order of the runs of consecutive 2's in $\pi^{(1)}$ and denote the obtaining set partition by $\pi^{(2)}$, and, likewise, reverse the order of the runs for each subsequent letter of $[k]$. The resulting set partition $\pi' = \pi^{(k)}$ will belong to $\mathcal{P}_{n,k}(1^{m-1}\text{-}1)$ since all runs (except possibly the last) of a given letter will have a size at most $m - 2$, and it is easy to see that the map $\pi \mapsto \pi'$ is a bijection. $\square$

Note that the proof of Theorem 6.130 shows that the statistics recording the number of occurrences of $1^{m-1}\text{-}1$ and $1\text{-}1^{m-1}$ are identically distributed on $\mathcal{P}_{n,k}$. Next, we look at the patterns 12-1, 12-2, 21-1, and 21-2, where we leave the details of the proofs as an exercise (see Exercise 6.17).

**Proposition 6.131** *The number of set partitions of $\mathcal{P}_{n,k}$ avoiding*
  (1) *either 12-1 or 12-2 is given by $\binom{n-1}{k-1}$.*
  (2) *either 21-1 or 21-2 is given by $\binom{n+\binom{k}{2}-1}{n-k}$.*

In order to present our next results, we extend the definition of the generating function $P_\tau(x; k)$ as follows.

**Definition 6.132** *Let $P_\tau(x; k | a_s \cdots a_1)$ denote the generating function for the number of set partitions $\pi = \pi_1\pi_2 \cdots \pi_n$ of $\mathcal{P}_{n,k}(\tau)$ such that $\pi_{n+1-j} = a_j$ for all $j = 1, 2, \ldots, s$.*

The next proposition gives a recursive way of finding the generating function $P_{23\text{-}1}(x; k | a)$ for all positive integers $a$ and $k$.

**Proposition 6.133** *We have $P_{23\text{-}1}(x; k) = \sum_{a=1}^{k} P_{23\text{-}1}(x; k | a)$, where the generating functions $P_{23\text{-}1}(x; k | a)$ satisfy the recurrence relation*

$$P_{23\text{-}1}(x; k | a)$$
$$= \sum_{i=1}^{a-1} \left( \frac{x^3 \sum_{j=1}^{i} P_{23\text{-}1}(x; k-1 | j)}{((1-x)^{k-a} - x)((1-x)^{k-i} - x)} \prod_{j=i+1}^{a-1} \left(1 + \frac{x^2}{(1-x)^{k-j} - x}\right) \right)$$
$$+ \frac{x^2}{(1-x)^{k-a} - x} \sum_{j=1}^{a} P_{23\text{-}1}(x; k-1 | j),$$

*for $1 \leq a \leq k - 2$, with the initial conditions $P_{23\text{-}1}(x; k | k) = xP_{23\text{-}1}(x; k) + xP_{23\text{-}1}(x; k-1)$ and $P_{23\text{-}1}(x; k | k-1) = xP_{23\text{-}1}(x; k)$ if $k \geq 3$, with $P_{23\text{-}1}(x; 2 | 1) = \frac{x^3}{(1-x)(1-2x)}$ and $P_{23\text{-}1}(x; 2 | 2) = \frac{x^2}{1-2x}$.*

**Proof** It is not hard to verify the initial conditions. From definitions, we have that $P_{23\text{-}1}(x;k) = \sum_{a=1}^{k} P_{23\text{-}1}(x;k|a)$. Also, for all $1 \le a \le k-2$,

$$\begin{aligned} P_{23\text{-}1}(x;k|a) &= \sum_{b=1}^{a} P_{23\text{-}1}(x;k|b,a) \\ &\quad + P_{23\text{-}1}(x;k|a+1,a) + \sum_{b=a+2}^{k} P_{23\text{-}1}(x;k|b,a) \\ &= x\sum_{j=1}^{a} P_{23\text{-}1}(x;k|j) \\ &\quad + xP_{23\text{-}1}(x;k|a) + \sum_{j=a+2}^{k} P_{23\text{-}1}(x;k|j,a). \end{aligned} \quad (6.12)$$

We now find a formula for $P_{23\text{-}1}(x;k|b,a)$, where $k > b > a+1 \ge 2$:

$$\begin{aligned} P_{23\text{-}1}(x;k|b,a) &= \sum_{j=1}^{k} P_{23\text{-}1}(x;k|j,b,a) \\ &= \sum_{j=1}^{a} P_{23\text{-}1}(x;k|j,b,a) + \sum_{j=a+1}^{b-1} P_{23\text{-}1}(x;k|j,b,a) \\ &\quad + \sum_{j=b}^{k} P_{23\text{-}1}(x;k|j,b,a) \\ &= x^2 \sum_{j=1}^{a} P_{23\text{-}1}(x;k|j) + x\sum_{j=b}^{k} P_{23\text{-}1}(x;k|j,a), \end{aligned} \quad (6.13)$$

with

$$\begin{aligned} P_{23\text{-}1}&(x;k|k,a) \\ &= \sum_{j=1}^{a} P_{23\text{-}1}(x;k|j,k,a) + P_{23\text{-}1}(x;k|k,k,a) \\ &= x^2 \sum_{j=1}^{a} P_{23\text{-}1}(x;k|j) + x^2 \sum_{j=1}^{a} P_{23\text{-}1}(x;k-1|j) \\ &\quad + xP_{23\text{-}1}(x;k|k,a). \end{aligned} \quad (6.14)$$

By (6.12)-(6.13), we have

$$\begin{aligned} P_{23\text{-}1}(x;k|b,a) &- xP_{23\text{-}1}(x;k|a) \\ &= -x^2 P_{23\text{-}1}(x;k|a) - x\sum_{j=a+2}^{b-1} P_{23\text{-}1}(x;k|j,a), \end{aligned}$$

with the initial condition $P_{23\text{-}1}(x;k|a+1,a) = xP_{23\text{-}1}(x;k|a)$ (which is obvious from the definitions). Thus, by induction on $b$, we obtain

$$P_{23\text{-}1}(x;k|b,a) = x(1-x)^{b-a-1} P_{23\text{-}1}(x;k|a). \quad (6.15)$$

for all $b = a+1, a+2, \ldots, k-1$. Therefore, (6.12), (6.14), and (6.15) imply

$$P_{\text{23-1}}(x;k|a) = x\sum_{j=1}^{a} P_{\text{23-1}}(x;k|j) + xP_{\text{23-1}}(x;k|a)\sum_{b=a+1}^{k-1}(1-x)^{b-a-1}$$
$$+ \frac{x^2}{1-x}\sum_{j=1}^{a}(P_{\text{23-1}}(x;k|j) + P_{\text{23-1}}(x;k-1|j)),$$

which is equivalent to

$$P_{\text{23-1}}(x;k|a)$$
$$= \frac{x}{(1-x)^{k-a}}\left(\sum_{j=1}^{a} P_{\text{23-1}}(x;k|j) + x\sum_{j=1}^{a} P_{\text{23-1}}(x;k-1|j)\right).$$

Hence, by induction on $a$, we obtain

$$P_{\text{23-1}}(x;k|a)$$
$$= \sum_{i=1}^{a-1}\left(\frac{x^3 \sum_{j=1}^{i} P_{\text{23-1}}(x;k-1|j)}{((1-x)^{k-a}-x)((1-x)^{k-i}-x)}\prod_{j=i+1}^{a-1}\left(1 + \frac{x^2}{(1-x)^{k-j}-x}\right)\right)$$
$$+ \frac{x^2}{(1-x)^{k-a}-x}\sum_{j=1}^{a} P_{\text{23-1}}(x;k-1|j),$$

for all $1 \leq a \leq k-2$, as required. □

**Example 6.134** *The above proposition can be used to find an explicit formula for the generating function $P_{\text{23-1}}(x;k)$ for any given $k$. For $k=2$, we have $P_{\text{23-1}}(x;2|1) = \frac{x^3}{(1-x)(1-2x)}$, $P_{\text{23-1}}(x;2|2) = \frac{x^2}{1-2x}$ and $P_{\text{23-1}}(x;2) = \frac{x^2}{(1-x)(1-2x)}$. For $k=3$, we have*

$$\begin{aligned}P_{\text{23-1}}(x;3|1) &= \frac{x^2}{1-3x+x^2}P_{\text{23-1}}(x;2|1) = \frac{x^5}{(1-x)(1-2x)(1-3x+x^2)},\\ P_{\text{23-1}}(x;3|2) &= xP_{\text{23-1}}(x;3),\\ P_{\text{23-1}}(x;3|3) &= xP_{\text{23-1}}(x;3) + \frac{x^3}{(1-x)(1-2x)}.\end{aligned}$$

*By summing the above equations, we obtain $(1-2x)P_{\text{23-1}}(x;3) = \frac{x^3}{1-3x+x^2}$, which implies*

$$P_{\text{23-1}}(x;3) = \frac{x^3}{(1-3x+x^2)(1-2x)}.$$

*Similarly, one can also find*

$$P_{\text{23-1}}(x;4) = \frac{x^4(1-5x+8x^2-4x^3+x^4)}{(1-x)(1-2x)^2(1-3x+x^2)(1-4x+3x^2-x^3)}.$$

There are comparable recurrences satisfied by the generating functions $PP_{\text{22-1}}(x;k|a)$, the proof of which we leave as an exercise (see Exercise 6.18).

**Proposition 6.135** We have $P_{22\text{-}1}(x;k) = \sum_{a=1}^{k} P_{22\text{-}1}(x;k|a)$, where the generating functions $P_{22\text{-}1}(x;k|a)$ satisfy the recurrence relation

$$P_{22\text{-}1}(x;k|a)$$
$$= (1+x)x^2 \sum_{i=1}^{a-1} \left( \frac{\prod_{j=i+1}^{a-1}(1-(k-1-j)x)}{\prod_{j=i}^{a}(1-(k-j)x-x^2)} P_{22\text{-}1}(x;k-1|i) \right)$$
$$+ \frac{x}{1-(k-a)x-x^2} P_{22\text{-}1}(x;k-1|a),$$

for all $1 \leq a \leq k-2$, with the initial conditions $P_{22\text{-}1}(x;k|k) = xP_{22\text{-}1}(x;k) + xP_{22\text{-}1}(x;k-1)$ and $P_{22\text{-}1}(x;k|k-1) = x(P_{22\text{-}1}(x;k) - x^2 P_{22\text{-}1}(x;k) - x^2 P_{22\text{-}1}(x;k-1))$ if $k \geq 2$, with $P_{22\text{-}1}(x;1|1) = P_{22\text{-}1}(x;1) = \frac{x}{1-x}$.

**Example 6.136** The above proposition can be used to find an explicit formula for the generating function $P_{22\text{-}1}(x;k)$ for any given $k$. For instance, the above proposition for $k=2$ yields

$$P_{22\text{-}1}(x;2|1) = (x-x^3)P_{22\text{-}1}(x;2) - \frac{x^4}{1-x},$$
$$P_{22\text{-}1}(x;2|2) = xP_{22\text{-}1}(x;2) + \frac{x^2}{1-x}.$$

By summing the above equations, we get $P_{22\text{-}1}(x;2) = (2x-x^3)P_{22\text{-}1}(x;2) + x^2(1+x)$, which gives

$$P_{22\text{-}1}(x;2) = \frac{x^2(1+x)}{(1-x)(1-x-x^2)}.$$

Similarly, we have

$$P_{22\text{-}1}(x;3) = \frac{x^3(1+x-2x^2-x^3)}{(1-x)^2(1-x-x^2)(1-2x-x^2)}.$$

Finally, there are similar relations involving the generating functions $P_{32\text{-}1}(x;k|a)$; we leave the proof to the interested reader.

**Proposition 6.137** We have $P_{32\text{-}1}(x;k) = \sum_{a=1}^{k} P_{32\text{-}1}(x;k|a)$, where the generating functions $P_{32\text{-}1}(x;k|a)$ satisfy the recurrence relation

$$P_{32\text{-}1}(x;k|a)$$
$$= x^2 \sum_{i=1}^{a-1} \left( \frac{(1-x)^{k-1-i} \prod_{j=i+1}^{a-1} \frac{(1-x)^{k-j}}{(1-x)^{k-j}-x}}{((1-x)^{k-a}-x)((1-x)^{k-i}-x)} \right) P_{32\text{-}1}(x;k-1|i)$$
$$+ x \frac{(1-x)^{k-1-a}}{(1-x)^{k-a}-x} P_{32\text{-}1}(x;k-1|a),$$

for all $1 \leq a \leq k-2$, with the initial conditions

$$P_{32\text{-}1}(x;k|k) = xP_{32\text{-}1}(x;k) + xP_{32\text{-}1}(x;k-1),$$
$$P_{32\text{-}1}(x;k|k-1) = xP_{32\text{-}1}(x;k),$$

if $k \geq 3$, $P_{32\text{-}1}(x;2|1) = \frac{x^3}{(1-x)(1-2x)}$ and $P_{32\text{-}1}(x;2|2) = \frac{x^2}{1-2x}$.

**Example 6.138** *The above proposition can be used to find an explicit formula for the generating function $P_{32\text{-}1}(x;k)$ for any given $k$. For $k = 2$, we have*

$$P_{32\text{-}1}(x;2|1) = \frac{x^3}{(1-x)(1-2x)}, \quad P_{32\text{-}1}(x;2|2) = \frac{x^2}{1-2x},$$

*and then* $P_{32\text{-}1}(x;2) = \frac{x^2}{(1-x)(1-2x)}$. *For $k = 3$, we have*

$$P_{32\text{-}1}(x;3|1) = \frac{x(1-x)}{1-3x+x^2} P_{32\text{-}1}(x;2|1) = \frac{x^4}{(1-2x)(1-3x+x^2)},$$
$$P_{32\text{-}1}(x;3|2) = xP_{32\text{-}1}(x;3),$$
$$P_{32\text{-}1}(x;3|3) = xP_{32\text{-}1}(x;3) + \frac{x^3}{(1-x)(1-2x)}.$$

*By summing the above equations, we obtain*

$$(1-2x)P_{32\text{-}1}(x;3) = \frac{x^3}{(1-x)(1-3x+x^2)},$$

*which implies*

$$P_{32\text{-}1}(x;3) = \frac{x^3}{(1-3x+x^2)(1-2x)(1-x)}.$$

Our results are summarized in Table 6.8.

**Table 6.8**: Three-letter generalized patterns of type $(2,1)$

| $\tau$ | Reference |
|---|---|
| 11-1 | 6.130 |
| 11-2, 12-3, 21-3 | 6.129 |
| 12-2, 21-1, 12-1, 21-1 | 6.131 |
| 22-1, 23-1, 32-1 | No explicit formula |
| 31-2, 13-2 | Open |

Now, we use the generating trees tools to study the number of set partitions of $[n]$ avoiding either 23-1, 22-1, or 32-1. From Example 2.82 we obtain the following result.

**Theorem 6.139** *The generating tree $T(23\text{-}1)$ for $\mathcal{P}_n(23\text{-}1)$ is given by*

**Root:** $(1, 1, 1)$

**Rules:** $(a, b, c) \rightsquigarrow (a, b, a) \cdots (a, b, c)(c, b, c+1) \cdots (c, b, b)(c, b+1, b+1)$.

Let $H_{23\text{-}1}(t; a, b, c)$ be the generating function for the number of set partitions of level $n$ that are labeled $(a, b, c)$ in the generating tree $T(23\text{-}1)$, as described in Theorem 6.139 above. Define $H_{23\text{-}1}(t, u, v, w) = \sum_{a,b,c} H_{23\text{-}1}(t; a, b, c) u^a v^b w^c$. Then Theorem 6.139 implies

$$H_{23\text{-}1}(t, u, v, w)$$
$$= tuvw + t \sum_{a,b,c} H_{23\text{-}1}(t; a, b, c) u^a v^b (w^a + \cdots + w^c)$$
$$+ t \sum_{a,b,c} H_{23\text{-}1}(t; a, b, c) u^c v^b (w^{c+1} + \cdots + w^b)$$
$$+ t \sum_{a,b,c} H_{23\text{-}1}(t; a, b, c) u^c v^{b+1} w^{b+1}$$
$$= tuvw + t \sum_{a,b,c} H_{23\text{-}1}(t; a, b, c) u^a v^b \frac{w^a - w^{c+1}}{1 - w}$$
$$+ t \sum_{a,b,c} H_{23\text{-}1}(t; a, b, c) u^c v^b \frac{w^{c+1} - w^{b+1}}{1 - w}$$
$$+ t \sum_{a,b,c} H_{23\text{-}1}(t; a, b, c) u^c v^{b+1} w^{b+1}$$
$$= tuvw + \frac{t}{1-w}(H_{23\text{-}1}(t, uw, v, 1) - wH_{23\text{-}1}(t, u, v, w))$$
$$+ \frac{tw}{1-w}(H_{23\text{-}1}(t, 1, v, uw) - H_{23\text{-}1}(t, 1, vw, u)) + tvwH_{23\text{-}1}(t, 1, vw, u),$$

which gives the following result.

**Theorem 6.140** *The generating function $H_{23\text{-}1}(t, u, v, w)$ satisfies*

$$H_{23\text{-}1}(t, u, v, w)$$
$$= tuvw + \frac{t}{1-w}(H_{23\text{-}1}(t, uw, v, 1) - wH_{23\text{-}1}(t, u, v, w))$$
$$+ \frac{tw}{1-w}(H_{23\text{-}1}(t, 1, v, uw) - H_{23\text{-}1}(t, 1, vw, u)) + tvwH_{23\text{-}1}(t, 1, vw, u).$$

Using the above theorem, we see that the first fifteen values of the sequence recording the number of partitions of $\mathcal{P}_n(23\text{-}1)$, $n = 1, 2, \ldots, 15$, are 1, 2, 5, 14, 42, 132, 430, 1444, 4983, 17634, 63906, 236940, 898123, 3478623, and 13761820.

**Theorem 6.141** *The generating tree* $T(22\text{-}1)$ *for* $\mathcal{P}_n(22\text{-}1)$ *is given by*

**Root:** $(1,1,1)$

**Rules:** $(a,b,c) \rightsquigarrow \begin{matrix}(a,b,a)\cdots(a,b,c-1)(c,b,c)\\(a,b,c+1)\cdots(c,b,b)(c,b+1,b+1).\end{matrix}$

**Proof** We label each set partition $\pi = \pi_1\cdots\pi_n$ of $[n]$ by $(a,b,c)$, where $c = \pi_n$, $b = \max_{1\le i \le n} \pi_i$, and

$$a = \begin{cases} 1, & \text{if there is no } i \text{ such that } \pi_i = \pi_{i+1}; \\ \max\{\pi_i : \pi_i = \pi_{i+1}\}, & \text{otherwise.} \end{cases}$$

Clearly, the set partition of $[1]$ is labeled by $(1,1,1)$ and $c \ge a$ (otherwise, $\pi$ would contain the pattern 22-1). If we have a set partition $\pi$ associated with a label $(a,b,c)$, then each child of $\pi$ is a set partition of the form $\pi' = \pi c'$, where $c' = a, a+1, \ldots, b+1$, for if $c' < a$, then $\pi'$ would contain 22-1, which is not allowed. Thus, we have the four cases:

- if $c > c' \ge a$, then $\pi'$ is labeled by $(a,b,c')$,
- if $c' = c$, then $\pi'$ is labeled by $(c,b,c)$,
- if $b \ge c' > c$, then $\pi'$ is labeled by $(a,b,c')$,
- if $c' = b+1$, then $\pi'$ is labeled by $(a,b+1,b+1)$.

Combining the above cases yields our generating tree. □

Let $H_{22\text{-}1}(t;a,b,c)$ be the generating function for the number of partitions of level $n$ that are labeled by $(a,b,c)$ in the generating tree $T(22\text{-}1)$, as described in Theorem 6.141. Define

$$H_{22\text{-}1}(t,u,v,w) = \sum_{a,b,c} H_{22\text{-}1}(t;a,b,c) u^a v^b w^c.$$

Then, using similar arguments as in the proof of Theorem 6.140 above (the details are leaved to the reader), we obtain the following result.

**Theorem 6.142** *The generating function* $H_{22\text{-}1}(t,u,v,w)$ *satisfies*

$$H_{22\text{-}1}(t,u,v,w)$$
$$= tuvw + \frac{t}{1-w}(H_{22\text{-}1}(t,uw,v,1) - H_{22\text{-}1}(t,u,v,w)) + tH_{22\text{-}1}(t,1,v,uw)$$
$$+ \frac{tw}{1-w}(H_{22\text{-}1}(t,u,v,w) - H_{22\text{-}1}(t,u,vw,1)) + tvwH_{22\text{-}1}(t,u,vw,1).$$

Using the above theorem, we see that the first fifteen values of the sequence recording the number of partitions of $\mathcal{P}_n(22\text{-}1)$, $n = 1, 2, \ldots, 15$), are 1, 2, 5, 14, 44, 153, 585, 2445, 11109, 54570, 288235, 1628429, 9792196, 623191991, and 419527536.

**Theorem 6.143** *The generating tree $T(32\text{-}1)$ for $\mathcal{P}_n(32\text{-}1)$ is given by*

**Root:** $(1,1,1)$

**Rules:** $\begin{aligned}(a,b,c) &\rightsquigarrow (a,b,a)\cdots(c-1,b,c-1)\\ (a,b,c) &\cdots (a,b,b)(a,b+1,b+1).\end{aligned}$

**Proof** We label each set partition $\pi = \pi_1\cdots\pi_n \in \mathcal{P}_n$ by $(a,b,c)$, where $c = \pi_n$, $b = \max_{1 \leq i \leq n} \pi_i$, and

$$a = \begin{cases} 1, & \text{if there is no } i \text{ such that } \pi_i > \pi_{i+1}; \\ \max\{\pi_{i+1} : \pi_i > \pi_{i+1}\}, & \text{otherwise.} \end{cases}$$

We label the set partition $\{1\}$ of $[1]$ by $(1,1,1)$, and clearly $c \geq a$ (otherwise, $\pi$ would contain $32-1$). If we have a set partition $\pi$ associated with a label $(a,b,c)$, then each child of $\pi$ is a set partition of the form $\pi' = \pi c'$, where $c' = a, a+1,\ldots, b+1$, for if $c' < a$, then $\pi'$ would contain $32-1$, which is not allowed. Thus, we have the four cases:

- if $c > c' \geq a$, then $\pi'$ is labeled by $(c', b, c')$,
- if $c' = c$, then $\pi'$ is labeled by $(a, b, c)$,
- if $b \geq c' \geq c+1$, then $\pi'$ is labeled by $(a, b, c')$,
- if $c' = b+1$, then $\pi'$ is labeled by $(a, b+1, b+1)$.

Combining the above cases yields our generating tree. □

Let $H_{32\text{-}1}(t; a, b, c)$ be the generating function for the number of partitions of level $n$ that are labeled by $(a,b,c)$ in the generating tree $T(32\text{-}1)$, as described in Theorem 6.143. Define

$$H_{32\text{-}1}(t,u,v,w) = \sum_{a,b,c} H_{32\text{-}1}(t; a, b, c) u^a v^b w^c.$$

Then, using similar arguments as in the proof of Theorem 6.140 above, we obtain

**Theorem 6.144** *The generating function $H_{32\text{-}1}(t, u, v, w)$ satisfies*

$$H_{32-1}(t,u,v,w)$$
$$= tuvw + \frac{t}{1-uw}(H_{32\text{-}1}(t, uw, v, 1) - H_{32\text{-}1}(t, 1, v, uw))$$
$$+ \frac{t}{1-w}(H_{32\text{-}1}(t, u, v, w) - wH_{32\text{-}1}(t, u, vw, 1)) + tvw H_{32\text{-}1}(t, u, vw, 1).$$

Using the above theorem, we see that the first fifteen values of the sequence recording the number of partitions of $\mathcal{P}_n(32\text{-}1)$, $n = 1, 2, \ldots, 15$, are 1, 2, 5, 15, 51, 189, 747, 3110, 13532, 61198, 286493, 1383969, 6881634, 35150498, and 184127828.

We end this section with the following notes. The enumeration problem of finding the generating function $P_\tau(x; k)$, where $\tau$ is a pattern of type $(1, 2)$, was completed in the first subsection above. However, the problem of finding the generating function $P_\tau(x; k)$, where $\tau$ is a pattern of type $(2, 1)$, is not complete; see Research Direction 6.1.

## 6.4 Partially Ordered Patterns

Like the other patterns, partially ordered patterns POPs were originally studied on the set of permutations [170], then on the set of $k$-ary words [173], and finally on the set of compositions [135].

**Definition 6.145** *A partially ordered generalized pattern (or just POGP) $\xi$ is a word in reduced form, where if the letters a and b are incomparable in a POP $\xi$, then the relative size of the letters corresponding to a and b in a set partition $\pi$ is unimportant in an occurrence of $\xi$ in the set partition $\pi$. Letters shown with the same number of primes are comparable to each other (for example, $1'$ and $3'$ are comparable), while letters shown without primes are comparable to all letters of the alphabet.*

Note that we always use $\xi$ to refer to a pattern from a partially ordered set, and $\tau$ for a pattern from an (ordered) alphabet. Furthermore, in this section any pattern that does not contain dashes is a subword pattern.

**Definition 6.146** *For a given POP $\xi = \xi_1\xi_2 \cdots \xi_k$, we say that a subword pattern $\tau = \tau_1\tau_2 \cdots \tau_k$ is a linear extension of $\xi$ if $\xi_i < \xi_j$ implies that $\tau_i < \tau_j$. A generalized pattern is a linear extension of a generalized POP if it has the same adjacency requirements as the POP and obeys the rules for linear extension of subwords. A POP $\xi$ occurs in a sequence (permutation, word, composition or set partition) if any of its linear extensions occurs in the sequence. Otherwise, the sequence avoids the POP.*

**Example 6.147** *The simplest nontrivial example of a POP, that differs from the ordinary generalized patterns is $\tau = 1'\text{-}2\text{-}1''$, where the second letter is the greatest one and the first and the last letters are incomparable to each other. The set partition $\pi = 12132$ has four occurrences of $\tau$, namely, 121, 132, 232, and 132. Clearly, counting peaks in a set partitions equivalent to counting the pattern $\tau$.*

**Example 6.148** In $\tau = 2'2''1$, the three letters must be adjacent, the last letter is the smallest, and the first two letters are incomparable. The set partition $\pi = 123321241$ has three occurrences of $\tau$, namely, 332, 321, and 241.

In $\tau = 1'\text{-}1''2$, the last letter is the greatest, the last two letters must be adjacent, and the first two letters are incomparable. The set partition $\pi = 121132$ has three occurrences of $\tau$, namely, 113 (twice) and 213.

Recall that $P_{n,k}(\tau)$ denotes the number of partitions of $\mathcal{P}_{n,k}$ avoiding the POP $\tau$ and $P_\tau(x;k) = \sum_{n \geq 0} P_{n,k}(\tau) x^n$ denotes the generating function for the sequence $\{P_{n,k}(\tau)\}_{n \geq 0}$. Also, recall that $AW_{n,k}(\tau)$ denotes the number of $k$-ary words of $[k]^n(\tau)$ and $AW_\tau(x;k) = \sum_{n \geq 0} AW_{n,k}(\tau) x^n$ denotes the corresponding generating function.

**Example 6.149** For example, if $\tau = 1'\text{-}2\text{-}1''$, then each member of $P_{n,k}(\tau)$ has the form $11\cdots 122\cdots 2\cdots kk\cdots k$ such that each letter from $[k]$ occurs at least once, which implies

$$P_{1'\text{-}2\text{-}1''}(x;k) = \left(\frac{x}{1-x}\right)^k. \tag{6.16}$$

Now, we consider the following two classes of POGP's.

**Definition 6.150** Let $\{\tau_0, \tau_1, \ldots, \tau_s\}$ be a set of subword patterns. A multi-pattern is of the form $\tau = \tau_0\text{-}\tau_1\text{-}\cdots\text{-}\tau_s$ and a shuffle pattern is of the form $\tau = \tau_0\text{-}a_1\text{-}\tau_1\text{-}a_2\text{-}\cdots\text{-}\tau_{s-1}\text{-}a_s\text{-}\tau_s$, where each letter of $\tau_i$ is incomparable with any letter of $\tau_j$ whenever $i \neq j$. In addition, the letters $a_i$ are either all greater or all smaller than any letter of $\tau_j$ for any $i$ and $j$.

**Example 6.151** For example, $\tau = 1'\text{-}2\text{-}1''$ is a shuffle pattern, while $\tau = 1'\text{-}1''\text{-}1'''$ is a multi-pattern. From the definitions, we see that one can obtain a multi-pattern from a shuffle pattern simply by removing all of the letters $a_i$.

### 6.4.1 Patterns of Size Three

There is a connection between multi-avoidance of generalized patterns and POPs; see [170]. That is, to avoid a single POGP is equivalent to avoiding a collection of generalized patterns. For example, to avoid $\tau = 1'2'\text{-}3\text{-}1''$ is to simultaneously avoid the 5 patterns: 12-3-1, 12-3-2, 12-4-3, 13-4-2, and 23-4-1. Another example, a set partition $\pi = \pi_1 \pi_2 \cdots \pi_n$ avoids the pattern $\tau = 2'12''$ if there exists no index $i \leq n-2$ such that $\pi_i > \pi_{i+1} < \pi_{i+2}$, which is equivalent to simultaneously avoiding 212, 312, and 213. Occurrences of $\tau = 2'12''$ are called valleys, and Theorem 4.35 presents the generating function for the number of set partitions in $\mathcal{P}_{n,k}$ according to the number of valleys. Similarly, a peak corresponds to an occurrence of $\tau = 1'21''$, and a similar formula for $P_{1'21''}(x;k)$ is given in Theorem 4.28. In this subsection, we find explicit formulas and/or recurrences for $P_\tau(x;k)$ in the cases when

$\tau = 12'2''$, $21'1''$, $2'2''1$, or $1'1''2$ as well as in the comparable cases when $\tau$ has the form $x$-$yz$, such as $\tau = 2'$-$2''1$. In some instances, results may be generalized to longer patterns.

**Theorem 6.152** Let $\tau = 12'2''$. For all $k \geq 2$,

$$P_\tau(x;k) = \frac{x^{2k-2}}{\prod_{j=1}^{k}(1 - x - (j-1)x^2)},$$

with $P_\tau(x;1) = \frac{x}{1-x}$.

**Proof** Fix $n \geq k \geq 2$. Set partitions $\pi = \pi_1\pi_2\cdots\pi_n \in \mathcal{P}_{n,k}(\tau)$ with exactly $n - j$ 1's may be created from set partitions $\pi' = \pi'_1\pi'_2\cdots\pi'_j \in \mathcal{P}_{j,k-1}$ on the letters $\{2, 3, \ldots, k\}$ by placing at least one 1 before $\pi'_1$ at the beginning of $\pi'$ as well as at least one 1 between $\pi'_{i-1}$ and $\pi'_i$ for $2 \leq i \leq j$. Thus, from each $\pi' \in \mathcal{P}_{j,k-1}$, there are exactly $\binom{n-2j-1+(j+1)-1}{(j+1)-1} = \binom{n-j-1}{j}$ set partitions of $\mathcal{P}_{n,k}(\tau)$ that may be formed as described and ending in 1 and exactly $\binom{n-j-1}{j-1}$ set partitions of $\mathcal{P}_{n,k}(\tau)$ that may be formed which do not. Hence, there are $\binom{n-j-1}{j} + \binom{n-j-1}{j-1} = \binom{n-j}{j}$ set partitions of $\mathcal{P}_{n,k}(\tau)$ which may be formed altogether. Then we have

$$P_{n,k}(\tau) = \sum_{j=k-1}^{\lfloor n/2 \rfloor} \binom{n-j}{j} \mathrm{Stir}(j;k-1), \qquad k \geq 2, \quad (6.17)$$

with $P_{n,k}(\tau) = 1$ for all $n$. Multiplying (6.17) by $x^n$ and summing over all $n \geq k$ implies

$$\sum_{n \geq k} p_\tau(n;k)x^n = \sum_{j \geq k-1} \mathrm{Stir}(j;k-1)x^j \sum_{n \geq 2j} \binom{n-j}{j} x^{n-j}$$

$$= \sum_{j \geq k-1} \mathrm{Stir}(j;k-1)x^j \cdot \frac{x^j}{(1-x)^{j+1}}$$

$$= \frac{1}{1-x} \sum_{j \geq k-1} \mathrm{Stir}(j;k-1) \left(\frac{x^2}{1-x}\right)^j.$$

Hence, the desired result now follows from (2.9). □

Next, we consider a pattern which generalizes $\tau = 21'1''$.

**Theorem 6.153** Let $m \geq 3$, $\tau = ba_1a_2\cdots a_{m-1}$ with $b > \max\{a_1, \ldots, a_{m-1}\}$ and the $a_i$ are incomparable. Then for all $k \geq 1$,

$$P_\tau(x;k) = \prod_{j=1}^{k} \frac{x - (j-1)x^m}{1 - jx + (j-1)^{m-1}x^m}.$$

**Proof** Let $\pi \in \mathcal{P}_{n,k}(\tau)$, so $\pi$ can be written as $\pi = \pi^{(1)}\pi^{(2)}\cdots\pi^{(k)}$, where $\pi^{(1)}$ is a non-empty string of 1's and each $\pi^{(j)}$, $j \geq 2$, is of the form

$$\pi^{(j)} = j\pi^{(j,1)}j\pi^{(j,2)}\cdots j\pi^{(j,r)}, \tag{6.18}$$

where $r \geq 1$ and each $\pi^{(j,s)}$ is a possibly empty $(j-1)$-ary word having a size of at most $m-2$. To complete the proof, note that words of the form (6.18) are enumerated by the generating function

$$A_j(x) = \frac{x\sum_{i=0}^{m-2}(j-1)^i x^i}{1-x\sum_{i=0}^{m-2}(j-1)^i x^i} = \frac{x - (j-1)^{m-1}x^m}{1 - jx + (j-1)^{m-1}x^m},$$

with $P_\tau(x;k) = \frac{x}{1-x}\prod_{j=2}^{k} A_j(x)$. □

**Example 6.154** *Taking $m = 3$ in Theorem 6.153 yields*

$$P_{2'1''}(x;k) = x^k \prod_{j=1}^{k} \frac{1 + (j-1)x}{1 - x - (j-1)x^2},$$

*for all $k \geq 1$.*

We next consider a pattern that generalizes $\tau = 2'\text{-}2''1$.

**Theorem 6.155** *Let $m \geq 3$ and $\tau = 2'\text{-}2''1\cdots 1$ has size $m$. For all $k \geq 2$,*

$$P_\tau(x;k) = \frac{x^k(1-x^{m-2})^{\binom{k-1}{2}}}{(1-x)(1-2x+x^{m-1})} \prod_{j=3}^{k} \frac{x^{m-3}}{x^{m-3} - 1 + (1-x^{m-2})^j},$$

*with $P_\tau(x;1) = \frac{x}{1-x}$.*

**Proof** Let $\pi = 1\pi^{(1)}2\pi^{(2)}\cdots k\pi^{(k)} \in \mathcal{P}_{n,k}(\tau)$, where each $\pi^{(s)}$ is an $s$-ary word. If $3 \leq j \leq k$, then it must be the case that the word $(j-1)\pi^{(j)}$ avoids $\sigma = 21\cdots 1 \in [2]^{m-1}$, and $\pi^{(2)}$ must avoid $\sigma$. Given $i \geq 1$, let $AW_\sigma^{(i)}(x;j)$ denote the generating function for the number of $j$-ary words $\theta$ of size $n$ such that $j\theta$ avoids $\sigma$. By [59] we have

$$AW_\sigma(x;j) = \frac{1}{1 - x^{3-m}(1 - (1-x^{m-2})^j)}$$

and

$$AW_\sigma^{(j+1)}(x;j) = \frac{(1-x^{m-2})^j}{1 - x^{3-m}(1 - (1-x^{m-2})^j)},$$

for all $j \geq 1$. From the definitions, we have

$$AW_\sigma^{(i)}(x;j) = AW_\sigma^{(i+1)}(x;j) + x^{m-2}AW_\sigma^{(i)}(x;j),$$

for all $i = 1, 2, \ldots, j$, and two applications of this yields

$$AW_\sigma^{(j-1)}(x;j) = \frac{1}{(1-x^{m-2})^2} AW_\sigma^{(j+1)}(x;j)$$

$$= \frac{(1-x^{m-2})^{j-2}}{1-x^{3-m}(1-(1-x^{m-2})^j)}.$$

Thus, we have

$$P_\tau(x;k) = \frac{x^k}{1-x} AW_\sigma(x;2) \prod_{j=3}^{k} AW_\sigma^{(j-1)}(x;j)$$

$$= \frac{x^k}{(1-x)(1-2x+x^{m-1})} \prod_{j=3}^{k} \frac{x^{m-3}(1-x^{m-2})^{j-2}}{x^{m-3}-1+(1-x^{m-2})^j},$$

which completes the proof. □

By similar arguments as in the proof of Theorem 6.155, we obtain the generating function $P_{2'\text{-}1\cdots12''}$ that implies the following result.

**Corollary 6.156** *Let $2'\text{-}1 \cdots 12''$ and $2'\text{-}2''1 \cdots 1$ be POP's of size $m \geq 3$. For all $k \geq 1$,*

$$P_{2'\text{-}1\cdots12''}(x;k) = P_{2'\text{-}2''1\cdots1}(x;k).$$

**Example 6.157** *Taking $m = 3$ in Theorem 6.155 yields*

$$P_{2'\text{-}2''1}(x;k) = \frac{x^k}{(1-x)^{2k-1}},$$

which implies

$$Ap_{2'\text{-}2''1}(n;k) = \binom{n+k-2}{2k-2}, \tag{6.19}$$

for all $n \geq k \geq 1$.

We conclude with the pattern $\tau = 1'\text{-}1''2$. Note that clearly $P_{n,k}(\tau) = 0$ for all $n$ if $k \geq 3$, with $P_{n,2}(\tau) = n-1$ for all $n \geq 2$ since set partitions of $\mathcal{P}_{n,2}(\tau)$ must be of the form $12\alpha_2\alpha_1$, where $\alpha_i$ denotes a possibly empty string of the letter $i$. This observation may be generalized as follows.

**Theorem 6.158** *Let $m \geq 3$ and $\tau = 1'\text{-}1''2 \cdots 2$ that size $m$. For all $k \geq 2$,*

$$P_\tau(x;k) = \frac{x^k(1-x^{m-2})(1-x^{m-3})^{k-2}}{1-x} \prod_{j=2}^{k} \frac{x^{m-3}}{x^{m-3}-1+(1-x^{m-2})^j}$$

with $P_\tau(x;1) = \frac{x}{1-x}$.

**Proof** Let $k \geq 2$ and let us write $\pi \in \mathcal{P}_{n,k}(\tau)$ as

$$\pi = \alpha^{(1)} \alpha^{(2)} \beta^{(2)} \alpha^{(3)} \beta^{(3)} \cdots \alpha^{(k)} \beta^{(k)},$$

where each $\alpha^{(j)}$ is a nonempty string of the letter $j$ and each $\beta^{(j)}$, $j \geq 2$, is a (possibly empty) $j$-ary word not starting with the letter $j$. If $\alpha^{(1)}$ has a size of one, then $\alpha^{(2)}$ can have any size, whereas if $\alpha^{(1)}$ has a size of two or more, then $\alpha^{(2)}$ must have a size of at most $m - 3$. In terms of generating functions, this case contributes

$$x \left( \frac{x}{1-x} \right) + \left( \frac{x^2}{1-x} \right) \left( \frac{x - x^{m-2}}{1-x} \right) = \frac{x^2(1 - x^{m-2})}{(1-x)^2}.$$

Furthermore, in order to avoid an occurrence of $\tau$, note that each string $\alpha^{(j)}$, $j \geq 3$, must have a size of at most $m - 3$ and that each word $\beta^{(j)}$, $j \geq 2$, must avoid the pattern $\sigma = 12 \cdots 2 \in [2]^{m-1}$. Now the generating function for $j$-ary words which avoid $\sigma$ and do not start with the letter $j$ is $AW_\sigma(x;j) - xAW_\sigma(x;j)$, where $AW_\sigma(x;j)$ is obtained by substituting $y = 0$ Theorem 2.2 of [59] and is thus given by

$$AW_\sigma(x;j) = \frac{x^{m-3}}{x^{m-3} - 1 + (1 - x^{m-2})^j}.$$

Therefore, $P_\tau(x;k)$, $k \geq 2$, is given by

$$P_\tau(x;k) = \frac{x^2(1 - x^{m-2})}{(1-x)^2} \left( \frac{x - x^{m-2}}{1-x} \right)^{k-2} \prod_{j=2}^{k} \frac{x^{m-3}(1-x)}{x^{m-3} - 1 + (1 - x^{m-2})^j}$$

$$= \frac{x^k(1 - x^{m-2})(1 - x^{m-3})^{k-2}}{1-x} \prod_{j=2}^{k} \frac{x^{m-3}}{x^{m-3} - 1 + (1 - x^{m-2})^j},$$

as required. $\square$

### 6.4.2 Shuffle Patterns

Let us consider the shuffle pattern $\phi = \tau\text{-}\ell\text{-}\tau$ (see Definition 6.150), where $\ell$ is the greatest letter in $\phi$ and each letter in the left $\tau$ is incomparable with every letter in the right $\tau$.

**Theorem 6.159** *Let $\phi$ be the shuffle pattern $\tau\text{-}\ell\text{-}\tau$. Then*

$$P_\phi(x;k) = \frac{x^k}{\prod_{j=1}^{\ell-1}(1 - jx)} g_{k-\ell}$$

$$+ \sum_{j=0}^{k-\ell} \frac{x^{j+1} P_\tau(x; k-1-j) AW_{\phi;\tau}(x; k-1-j)}{(1 - xAW_\tau(x; k-1-j))^2} g_{j-1},$$

*where $g_j = \prod_{i=0}^{j} \frac{AW_\tau(x;k-1-i)}{1-xAW_\tau(x;k-1-i)}$ and $AW_{\phi;\tau}(x;k) = AW_\phi(x;k) - AW_\tau(x;k)$.*

**Proof** Let $P_{\phi;\tau}(x;k) = P_\phi(x;k) - P_\tau(x;k)$. We show how to obtain a recurrence relation on $k$ for $P_\phi(x;k)$. Suppose $\pi \in \mathcal{P}_{n,k}(\phi)$ is such that it contains the letter $k$ exactly $d \geq 1$ times. Clearly, $\pi$ can be written in the following form:
$$\pi = \pi^{(0)} k \pi^{(1)} k \cdots k \pi^{(d)},$$
where $\pi^{(0)}$ is a set partition with exactly $k-1$ blocks avoiding $\phi$ and $\pi^{(j)}$ is a $\phi$-avoiding $(k-1)$-ary word, for $j = 1, 2, \ldots, d$. There are two possibilities: either $\pi^{(j)}$ avoids $\tau$ for all $j$, or there exists $j_0$ such that $\pi^{(j_0)}$ contains $\tau$ and for any $j \neq j_0$, the word $\pi^{(j)}$ avoids $\tau$. In the first case, the generating function for the number of such words is given by $x^d P_\tau(x; k-1) AW_\tau^d(x; k-1)$, while in the second case, it is given by

$$x^d P_{\phi;\tau}(x; k-1) AW_\tau^d(x; k-1)$$
$$+ dx^d P_\tau(x; k-1) AW_\tau^{d-1}(x; k-1) AW_{\phi;\tau}(x; k-1).$$

In the last expression, the quantity
$$x^d P_{\phi;\tau}(x; k-1) AW_\tau^d(x; k-1)$$
counts set partitions in the second case above with $j_0 = 0$, and the expression
$$dx^d P_\tau(x; k-1) AW_\tau^{d-1}(x; k-1) AW_{\phi;\tau}(x; k-1)$$
counts such partitions with $j_0 > 0$. Note that the difference $P_{\phi;\tau}(x; k-1)$ (resp., $AW_{\phi;\tau}(x; k-1)$) is the generating function for the number of set partitions (resp., words) avoiding $\phi$ and containing $\tau$. Therefore,

$$P_\phi(x;k)$$
$$= \sum_{d \geq 1} x^d P_\tau(x; k-1) AW_\tau^d(x; k-1) + x^d P_{\phi;\tau}(x; k-1) AW_\tau^d(x; k-1)$$
$$+ \sum_{d \geq 1} dx^d P_\tau(x; k-1) AW_\tau^{d-1}(x; k-1) AW_{\phi;\tau}(x; k-1),$$

or, equivalently, for $k \geq \ell$,

$$P_\phi(x;k)$$
$$= \frac{xP_\tau(x; k-1) AW_\tau(x; k-1)}{1 - xAW_\tau(x; k-1)} + \frac{xP_{\phi;\tau}(x; k-1) AW_\tau(x; k-1)}{1 - xAW_\tau(x; k-1)}$$
$$+ \frac{xP_\tau(x; k-1) AW_{\phi;\tau}(x; k-1)}{(1 - xAW_\tau(x; k-1))^2}$$
$$= \frac{xP_\phi(x; k-1) AW_\tau(x; k-1)}{1 - xAW_\tau(x; k-1)} + \frac{xP_\tau(x; k-1) AW_{\phi;\tau}(x; k-1)}{(1 - xAW_\tau(x; k-1))^2}.$$

Iterating the above recurrence relation and noting the initial condition

$P_\phi(x;k) = \frac{x^k}{\prod_{j=1}^k (1-jx)}$ for all $k \leq \ell - 1$, we obtain

$$P_\phi(x;k) = \frac{x^k}{\prod_{j=1}^{\ell-1}(1-jx)} g_{k-\ell}$$
$$+ \sum_{j=0}^{k-\ell} \frac{x^{j+1} P_\tau(x;k-1-j) AW_{\phi;\tau}(x;k-1-j)}{(1-xAW_\tau(x;k-1-j))^2} g_{j-1},$$

as claimed. □

**Example 6.160** Let $\phi = 1'$-$2$-$1''$. Here $\tau = 1$, so $AW_\tau(x;k) = 1$ and $P_\tau(x;k) = 0$ for all $k \geq 1$, since only the empty word avoids $\tau$. Hence, by Theorem 6.159, we have $P_\phi(x;k) = \frac{x^k}{(1-x)^k}$, which gives (6.16).

More generally, consider a shuffle pattern of the form $\phi = \tau$-$\ell$-$\nu$, where $\ell$ is the greatest letter of the pattern and every letter in $\tau$ is incomparable with every letter in $\nu$.

**Theorem 6.161** Let $\phi$ be the shuffle pattern $\tau$-$\ell$-$\nu$. Then

$$P_\phi(x;k) = \frac{x^k}{\prod_{j=1}^{\ell-1}(1-jx)} g_{k-\ell}$$
$$+ \sum_{j=0}^{k-\ell} \frac{x^{j+1} P_\tau(x;k-1-j)(AW_\phi(x;k-1-j) - AW_\nu(x;k-1-j))}{(1-xAW_\tau(x;k-1-j))(1-xAW_\nu(x;k-1-j))} g_{j-1},$$

where

$$g_j = \prod_{i=0}^j \frac{AW_\nu(x;k-1-i)}{1-xAW_\nu(x;k-1-i)}$$

and $AW_{\phi;\tau}(x;k) = AW_\phi(x;k) - AW_\tau(x;k)$.

**Proof** As in the previous proof, define $P_{\phi;\tau}(x;k) = P_\phi(x;k) - P_\tau(x;k)$. We proceed as in the proof of Theorem 6.159. Suppose $\pi \in \mathcal{P}_{n,k}(\phi)$ is such that it contains the letter $k$ exactly $d \geq 1$ times. Clearly, $\pi$ can be written as $\pi = \pi^{(0)} k \pi^{(1)} k \cdots k \pi^{(d)}$, where $\pi^{(0)}$ is a set partition with exactly $k-1$ blocks and $\pi^{(j)}$ is a $\phi$-avoiding word on $k-1$ letters, for $j = 1, 2, \ldots, d$. There are two possibilities: either $\pi^{(j)}$ avoids $\tau$ for all $j$, or there exists $j_0$ such that $\pi^{(j_0)}$ contains $\tau$, $\pi^{(j)}$ avoids $\tau$ for all $j = 0, 1, \ldots, j_0 - 1$ and $\pi^{(j)}$ avoids $\nu$ for any $j = j_0 + 1, \ldots, d$. In the first case, the generating function for the number of such set partitions is given by $x^d P_\tau(x;k-1) AW_\tau^d(x;k-1)$. In the second case, we have

$$x^d P_{\phi;\tau}(x;k-1) AW_\nu^d(x;k-1)$$
$$+ x^d P_\tau(x;k-1) \sum_{j=1}^d AW_\tau^{j-1}(x;k-1) AW_\nu^{d-j}(x;k-1) AW_{\phi;\tau}(x;k-1).$$

Therefore, we get

$$P_\phi(x;k)$$
$$= \sum_{d\geq 1} x^d P_\tau(x;k-1)AW_\tau^d(x;k-1) + x^d P_{\phi;\tau}(x;k-1)AW_\nu^d(x;k-1)$$
$$+ \sum_{d\geq 1} x^d P_\tau(x;k-1) \sum_{j=1}^{d} AW_\tau^{j-1}(x;k-1)AW_\nu^{d-j}(x;k-1)AW_{\phi;\tau}(x;k-1)$$
$$= \frac{xP_\tau(x;k-1)AW_\tau(x;k-1)}{1-xAW_\tau(x;k-1)} + \frac{xP_{\phi;\tau}(x;k-1)AW_\nu(x;k-1)}{1-xAW_\nu(x;k-1)}$$
$$+ xP_\tau(x;k-1)AW_{\phi;\tau}(x;k-1) \sum_{d\geq 0} x^d \sum_{j=0}^{d} AW_\tau^j(x;k-1)AW_\nu^{d-j}(x;k-1).$$

By the identity $\sum_{n\geq 0} x^n \sum_{j=0}^{n} p^j q^{n-j} = \frac{1}{(1-xp)(1-xq)}$, we then have

$$P_\phi(x;k) = \frac{xP_\tau(x;k-1)AW_\tau(x;k-1)}{1-xAW_\tau(x;k-1)} + \frac{xP_{\phi;\tau}(x;k-1)AW_\nu(x;k-1)}{1-xAW_\nu(x;k-1)}$$
$$+ \frac{xP_\tau(x;k-1)AW_{\phi;\tau}(x;k-1)}{(1-xAW_\tau(x;k-1))(1-xAW_\nu(x;k-1))}$$
$$= \frac{xP_\phi(x;k-1)AW_\nu(x;k-1)}{1-xAW_\nu(x;k-1)}$$
$$+ \frac{xP_\tau(x;k-1)AW_{\phi;\nu}(x;k-1)}{(1-xAW_\tau(x;k-1))(1-xAW_\nu(x;k-1))}.$$

Iterating the above recurrence relation and noting the initial conditions $P_\phi(x;k) = \frac{x^k}{\prod_{j=1}^{k}(1-jx)}$ for all $k \leq \ell - 1$, we obtain

$$P_\phi(x;k) = \frac{x^k}{\prod_{j=1}^{\ell-1}(1-jx)} g_{k-\ell}$$
$$+ \sum_{j=0}^{k-\ell} \frac{x^{j+1}P_\tau(x;k-1-j)AW_{\phi;\nu}(x;k-1-j)}{(1-xAW_\tau(x;k-1-j))(1-xAW_\nu(x;k-1-j))} g_{j-1},$$

as claimed. □

**Example 6.162** Let $\phi = 1'\text{-}3\text{-}1'2''$. Here, $\tau = 1$ and $\nu = 12$, so $AW_\tau(x;k) = 0$ and $AW_\nu(x;k) = \frac{1}{(1-x)^k}$ for all $k \geq 1$. Hence, by Theorem 6.161, we have

$$P_\phi(x;k) = x^k \prod_{j=0}^{k-1} \frac{1}{(1-x)^j - x}.$$

## 6.5 Exercises

**Exercise 6.1** *Find an explicit formula for the number of set partitions of $\mathcal{P}_n(\tau)$, where $\tau = 112, 121, 122$.*

**Exercise 6.2** *Prove the second claim of Proposition 6.26.*

**Exercise 6.3** *Prove Lemma 6.29.*

**Exercise 6.4** *Prove or disprove that Lemma 6.29 holds if the matrices $\begin{pmatrix} B & 0 \\ 0 & A \end{pmatrix}$ and $\begin{pmatrix} B & 0 \\ 0 & A' \end{pmatrix}$ are replaced with $\begin{pmatrix} A & 0 \\ 0 & B \end{pmatrix}$ and $\begin{pmatrix} A' & 0 \\ 0 & B \end{pmatrix}$ respectively.*

**Exercise 6.5** *Prove Fact 6.31.*

**Exercise 6.6** *Prove Fact 6.32.*

**Exercise 6.7** *Prove Theorem 6.55.*

**Exercise 6.8** *Prove Corollary 6.57.*

**Exercise 6.9** *Prove Corollary 6.73.*

**Exercise 6.10** *Prove Theorem 6.81.*

**Exercise 6.11** *Prove Lemma 6.11.*

**Exercise 6.12** *Let $\mathcal{D}_{n,k}$ be the set of Dyck paths of size $2n$ whose last up-step is followed by exactly $k$ down-steps. Define $d_{n,k} = \mathcal{D}_{n,k}$. Then the numbers $d_{n,k}$ are determined by the following set of recurrences:*

$$d(1,1) = 1,$$
$$d(n,k) = 0 \quad \text{if} \quad k < 1 \quad \text{or} \quad k > n,$$
$$d(n,k) = \sum_{j=k-1}^{n-1} d(n-1,j) \quad \text{for} \quad n \geq 2, n \geq k \geq 1.$$

**Exercise 6.13** *Show that the number of Dyck paths of size $2n$ that ends with an up-step followed by $k$ down-steps is given by $\frac{k}{n}\binom{2n-k-1}{n-1}$.*

**Exercise 6.14** *Let $spe_n$ be the number of Schröder paths of size $2n$ without peaks at even level. Prove that*

1. *the number of set partitions of $[n+1]$ that avoid the subsequence pattern 12312 is given by $spe_n$.*

2. the number of set partitions of $[n+1]$ that avoid the subsequence pattern 12321 is given by $spe_n$.

**Exercise 6.15** The number of partitions in $P_{n,k}(1\text{-}11)$ is given by
$$\sum_{j=0}^{\min\{k,n-k\}} \operatorname{Stir}(n-j-1, k-1)\binom{k}{j}.$$

**Exercise 6.16** Show that
$$P_{1\text{-}32\cdots 2}(x;k) = \frac{x^{(m-2)k}(1-x^{m-2})^{\binom{k-1}{2}}}{\prod_{j=1}^{k} x^{m-3}(1-x) - 1 + (1-x^{m-2})^{j-1}},$$
$$P_{1\text{-}23\cdots 3}(x;k) = \frac{x^{(m-2)k}(1-x^{m-3}(1-x))^{k-2}}{\prod_{j=1}^{k} x^{m-3}(1-x) - 1 + (1-x^{m-2})^{j-1}}$$
with $P_{1\text{-}23\cdots 3}(x;1) = \frac{x}{1-x}$.

**Exercise 6.17** Let $n \geq k \geq 1$. Prove
(1) $P_{n,k}(12\text{-}1) = P_{n,k}(12\text{-}2) = \binom{n-1}{k-1}$.
(2) $P_{n,k}(21\text{-}1) = P_{n,k}(21\text{-}2) = \binom{n+\binom{k}{2}-1}{n-k}$.

**Exercise 6.18** Show that the generating function $P_{22\text{-}1}(x;k)$ is given by $\sum_{a=1}^{k} P_{22\text{-}1}(x;k|a)$, where the generating functions $P_{22\text{-}1}(x;k|a)$ satisfy the recurrence relation
$$P_{22\text{-}1}(x;k|a)$$
$$= (1+x)x^2 \sum_{i=1}^{a-1} \left( \frac{\prod_{j=i+1}^{a-1}(1-(k-1-j)x)}{\prod_{j=i}^{a}(1-(k-j)x-x^2)} P_{22\text{-}1}(x;k-1|i) \right)$$
$$+ \frac{x}{1-(k-a)x-x^2} P_{22\text{-}1}(x;k-1|a),$$
for all $1 \leq a \leq k-2$, with the initial conditions $P_{22\text{-}1}(x;k|k) = xP_{22\text{-}1}(x;k) + xP_{22\text{-}1}(x;k-1)$ and $P_{22\text{-}1}(x;k|k-1) = x(P_{22\text{-}1}(x;k) - x^2 P_{22\text{-}1}(x;k) - x^2 P_{22\text{-}1}(x;k-1))$ if $k \geq 2$, with $P_{22\text{-}1}(x;1|1) = P_{22\text{-}1}(x;1) = \frac{x}{1-x}$.

**Exercise 6.19** Let $k \geq 1$. Then $P_{2'\text{-}1\text{-}2'',2'\text{-}2''\text{-}1}(x;k) = \frac{x^k}{(1-x)^{2k-1}}$.

**Exercise 6.20** If $\tau = 2\text{-}1'\text{-}1''$, then $Q_\tau(x,y) = \frac{(1-x)(1-x-xy)-x^2y}{(1-x-xy)^2-x^2y}$.

**Exercise 6.21** . Prove Corollary 6.60.

**Exercise 6.22** Find a bijection between the set of 2-distant noncrossing set partitions of $[n]$ and set $\mathcal{P}_n(12312)$.

**Exercise 6.23** *As a consequence of the classification equivalence classes of subsequence patterns in the set partitions of* $\mathcal{P}_n$, *find all the equivalence classes of subsequence patterns of* $\mathcal{P}_3$ *and* $\mathcal{P}_4$ *in the set of prefect matchings of* $[2n]$; *see [147, 148]. A prefect matching on* $[2n]$ *is a set partition of* $[2n]$ *such that each block has exactly size two.*

---

## 6.6 Research Directions and Open Problems

We now suggest several research directions which are motivated both by the results and exercises of this chapter.

**Research Direction 6.1** *Table 6.4 presents complete classification equivalence classes of subsequence patterns of size five. So the next question is to find the cardinality of each class, that is, to find an explicit formula for the number of set partitions of* $[n]$ *that avoid a subsequence pattern of size five. Fact 6.32 gives a formula for cardinality of the set* $\mathcal{P}_n(12345)$ *and Fact 6.31 gives a formula for cardinality of the set* $\mathcal{P}_n(11111)$. *Theorem 6.55 together with Table 6.3 show how to find an explicit formula for the cardinality of the set* $\mathcal{P}_n(\tau)$, *where* $\tau = 12332, 12233, 12333, 12222$. *Thus, we know the formula for the first, third, fourth, ninth, 20th, and 21st class in table 6.4. These suggest the following question: Derive an explicit formula for (the generating function for) the number of set partitions avoiding a subsequence pattern of size five. Actually the same question can be asked about Table 6.5 and Table 6.6 for patterns of size six and seven, respectively.*

*Sometimes it is good to ask finding bijections between our sets* $\mathcal{P}_n(\tau)$ *and other combinatorial structures; such bijections can be used to find an explicit formula for the cardinality of the sets* $\mathcal{P}_n(\tau)$. *For instance, Yan [366] showed that both* $\mathcal{P}_{n+1}(12312)$ *and* $\mathcal{P}_{n+1}(12321)$ *are in one-to-one correspondence with Schröder paths of size* $2n$ *without peaks at an even level; see Exercise 6.14.*

**Research Direction 6.2** *As we showed in Section 6.3, by using generating functions, we obtained recurrence relations satisfied by* $P_\tau(x; k)$ *in the cases when* $\tau$ *equals 23-1, 22-1 or 32-1, and by using generating trees, we obtained functional equations satisfied by* $P_\tau(x; k)$ *in these cases. However, we were unable to find explicit formulas for* $P_\tau(x; k)$. *We also failed to find explicit formulas for the generating function* $P_\tau(x; k)$ *in the cases when* $\tau$ *equals either 13-2 or 31-2; see Research Direction 6.1. On the other hand, we can write recurrence relations for these cases that are analogous to the case 23-1 above, for example, but the recurrence relations here are more complicated and require two indices. See Table 6.8. Hence, we suggest the following questions:*

*(1) Derive an explicit expression for the generating function for the number of*

set partitions avoiding the generalized pattern 23-1, for which a recursion was given in Proposition 6.133.

(2) Derive an explicit expression for the generating function for the number of set partitions avoiding the generalized pattern 22-1, for which a recursion was given in Proposition 6.135.

(3) Derive an explicit expression for the generating function for the number of set partitions avoiding the generalized pattern 32-1, for which a recursion was given in Proposition 6.137.

(4) Derive an explicit expression for the generating function for the number of set partitions avoiding the generalized pattern 13-2.

(5) Derive an explicit expression for the generating function for the number of set partitions avoiding the generalized pattern 31-2.

**Research Direction 6.3** *Section 6.2.6 suggests the following research question. Find an explicit formula for the generating function for the number of $12\cdots k12\cdots d$-avoiding set partitions of $[n]$. Note that the case $d = 0$ it is trivial, the case $d = 1$, and the case $d = 2$ have been described in Theorems 6.61 and 6.65*

**Research Direction 6.4** *Following Exercise 5.18, one can write each $k$-covers of order $r$ of $[n]$ as a sequence, called the linear representation, on $r$ letters that each letter can be colored by one of the colors of the set $[k]$. Thus, if we impose the definition of avoiding subsequence (or avoiding subword) patterns in the linear representation of $k$-covers, then we derive an extension of our results. More precisely, denote the set of all the linear representation of $k$-covers (proper $k$-covers, restricted $k$-covers, restricted proper $k$-covers) of order $r$ of $[n]$ that avoid a pattern $\tau$ by $\mathcal{COV}_{\tau;k}(n,r)$ ($\mathcal{PCOV}_{\tau;k}(n,r)$, $\mathcal{RCOV}_{\tau;k}(n,r)$, $\mathcal{RPCOV}_{\tau;k}(n,r)$). Then the research question if to find the cardinalities of these sets for a fixed pattern.*

**Research Direction 6.5** *Note that enumeration of $k$-noncrossing set partitions, set partitions of $[n]$ that avoid the pattern $12\cdots k12\cdots k$, has received a lot of attention in last few years. Denote such set of set partitions by $NC_k(n)$. Up to now, it is well known that the number of 2-noncrossing set partitions of $[n]$ is given by the n-th Catalan number $NC_2(n) = \text{Cat}_n$. Only on 2006, Bousquet-Mélou and Xin [50] (see also [69]) counted the 3-noncrossing set partitions of $[n]$, where they showed that $NC_3(0) = NC_3(1) = 1$ and for $n \geq 0$,*

$$9n(n+3)NC_3(n) - 2(5n^2+32n+42)NC_3(n+1) + (n+7)(n+6)NC_3(n+2) = 0,$$

*and they conjectured that the number of $k$-noncrossing set partitions of $[n]$ are not P-recursive, for $k \geq 4$. Prove/disprove this conjecture.*

For a set partition $\pi \in \mathcal{P}_n$ in standard representation, let $\pi^m$ denote the set partition obtained from $\pi$ by reflecting in the vertical line $x = (n+1)/2$. We denote the set of all set partitions $\pi$ of $[n]$ such that $\pi = \pi^m$ by $\mathcal{P}_n^m$.

Later, Xin and Zhang [363] enumerated bilaterally symmetric 3-noncrossing set partitions of $\mathcal{P}_n^m$. Thus, one can ask prove/disprove that the number of $k$-noncrossing set partitions of $\mathcal{P}_n^m$ is $P$-recursive for $k \geq 4$.

**Research Direction 6.6** *Our results on pattern avoidance in set partitions can be extended to cover the case of counting occurrences of a pattern in set partitions. More precisely, we can formulate our suggestion as follows. We say that the two patterns $\tau, \tau'$ are $r$-Wilf-equivalent, denoted $\tau' \sim_r \tau'$ if the number of set partitions of $\mathcal{P}_n$ that contain the pattern $\tau$ exactly once is the same number of set partitions of $\mathcal{P}_n$ that contain the pattern $\tau'$ exactly once. Also, let $P_{\tau;r}(x)$ be the generating function for the number $P_n(\tau;r)$ of set partitions of $[n]$ that contain the pattern $\tau$ exactly once. Thus, one can ask to find all the $r$-Wilf-equivalent classes for patterns of size at most seven for set partitions and find explicit formulas for $P_{\tau;r}(x)$, when $\tau$ is a fixed pattern of size of at most five.*

*In the case $r = 1$, with help of computer, we can create the number of set partitions of $[n]$ that contain the pattern $\tau$ exactly once, where $\tau$ has size at most seven; see Tables 6.9–6.11. In the case of subsequence patterns of size three, it is not hard to see that $P_n(123;1) = 2^{n-2}-1$, $P_n(121;1) = (n-1)2^{n-4}$, and $P_n(112;1) = P_n(122;1) = (n-2)2^{n-3}$, which implies there is only one nontrivial 1-Wilf-equivalent class, namely, $112 \sim_1 122$; see Table 6.9.*

**Table 6.9**: *The number $a_\tau(n)$ of set partitions of $[n]$ that contain exactly once a subsequence pattern of size three, where $n = 3, 4, \ldots, 10$*

| $\tau\backslash n$ | 3 | 4 | 5 | 6 | 7 | 8 | 9 | 10 |
|---|---|---|---|---|---|---|---|---|
| 111 | 1 | 4 | 20 | 80 | 350 | 1456 | 6384 | 27840 |
| 112 | 1 | 4 | 12 | 32 | 80 | 192 | 448 | 1024 |
| 122 | 1 | 4 | 12 | 32 | 80 | 192 | 448 | 1024 |
| 121 | 1 | 3 | 8 | 20 | 48 | 112 | 256 | 576 |
| 123 | 1 | 3 | 7 | 15 | 31 | 63 | 127 | 255 |

*The next table suggests there are exactly two nontrivial 1-Wilf-equivalence class for subsequence patterns of size four on set partitions, namely, $1121 \sim_1 1211$ and $1223 \sim_1 1234$. Prove or disprove that?*

**Table 6.10**: *The number $a_\tau(n)$ of set partitions of $[n]$ that contain a subsequence pattern of size four, where $n = 4, 5, \ldots, 10$*

| $\tau\backslash n$ | 4 | 5 | 6 | 7 | 8 | 9 | 10 |
|---|---|---|---|---|---|---|---|
| 1111 | 1 | 5 | 30 | 175 | 980 | 5796 | 34860 |

| | | | | | | |
|---|---|---|---|---|---|---|
| 1112 | 1 | 6 | 31 | 148 | 697 | 3282 | 15633 |
| 1121 | 1 | 5 | 25 | 116 | 541 | 2531 | 12031 |
| 1122 | 1 | 8 | 43 | 201 | 889 | 3858 | 16710 |
| 1123 | 1 | 6 | 28 | 120 | 493 | 1975 | 7785 |
| 1211 | 1 | 5 | 25 | 116 | 541 | 2531 | 12031 |
| 1212 | 1 | 6 | 28 | 120 | 495 | 2002 | 8008 |
| 1213 | 1 | 5 | 19 | 66 | 221 | 727 | 2367 |
| 1221 | 1 | 7 | 36 | 167 | 740 | 3197 | 13585 |
| 1222 | 1 | 6 | 30 | 140 | 645 | 2982 | 13972 |
| 1223 | 1 | 6 | 25 | 90 | 301 | 966 | 3025 |
| 1231 | 1 | 6 | 26 | 99 | 353 | 1213 | 4076 |
| 1232 | 1 | 5 | 18 | 58 | 179 | 543 | 1636 |
| 1233 | 1 | 7 | 34 | 142 | 547 | 2005 | 7108 |
| 1234 | 1 | 6 | 25 | 90 | 301 | 966 | 3025 |

The next table suggests there are exactly four nontrivial 1-Wilf-equivalence class for subsequence patterns of size four on set partitions, namely

$$11121 \sim_1 11211 \sim_1 12111, \quad 12123 \sim_1 12323, \quad 12342 \sim_1 12343$$

and

$$12334 \sim_1 12345.$$

*Prove or disprove that?*

**Table 6.11**: *The number $a_\tau(n)$ of set partitions of $[n]$ that contain a subsequence pattern of size five, where $n = 5, 6, \ldots, 11$*

| $\tau \backslash n$ | 5 | 6 | 7 | 8 | 9 | 10 | 11 |
|---|---|---|---|---|---|---|---|
| 11111 | 1 | 6 | 42 | 280 | 1890 | 12852 | 90552 |
| 11112 | 1 | 8 | 51 | 315 | 1920 | 11820 | 74352 |
| 11121 | 1 | 7 | 43 | 263 | 1589 | 9753 | 61335 |
| 11122 | 1 | 10 | 74 | 485 | 3043 | 18823 | 116766 |
| 11123 | 1 | 8 | 55 | 344 | 2082 | 12423 | 74075 |
| 11211 | 1 | 7 | 43 | 263 | 1589 | 9753 | 61335 |
| 11212 | 1 | 8 | 55 | 346 | 2124 | 12956 | 79443 |
| 11213 | 1 | 7 | 42 | 237 | 1320 | 7368 | 41560 |
| 11221 | 1 | 9 | 63 | 401 | 2477 | 15228 | 94359 |
| 11222 | 1 | 10 | 74 | 487 | 3066 | 19021 | 118189 |
| 11223 | 1 | 10 | 69 | 412 | 2301 | 12455 | 66558 |
| 11231 | 1 | 8 | 52 | 313 | 1827 | 10572 | 61321 |
| 11232 | 1 | 9 | 57 | 319 | 1699 | 8884 | 46286 |
| 11233 | 1 | 10 | 69 | 417 | 2391 | 13457 | 75529 |
| *Continued on next page* | | | | | | | |

| $\tau\backslash n$ | 5 | 6 | 7 | 8 | 9 | 10 | 11 |
|---|---|---|---|---|---|---|---|
| 11234 | 1 | 9 | 57 | 322 | 1749 | 9391 | 50351 |
| 12111 | 1 | 7 | 43 | 263 | 1589 | 9753 | 61335 |
| 12112 | 1 | 8 | 55 | 348 | 2148 | 13185 | 81304 |
| 12113 | 1 | 8 | 48 | 259 | 1357 | 7086 | 37331 |
| 12121 | 1 | 7 | 46 | 280 | 1686 | 10161 | 61848 |
| 12122 | 1 | 8 | 55 | 345 | 2108 | 12782 | 77805 |
| 12123 | 1 | 9 | 55 | 290 | 1430 | 6827 | 32083 |
| 12131 | 1 | 7 | 39 | 204 | 1045 | 5345 | 27575 |
| 12132 | 1 | 9 | 54 | 279 | 1357 | 6452 | 30482 |
| 12133 | 1 | 10 | 65 | 358 | 1836 | 9154 | 45339 |
| 12134 | 1 | 8 | 46 | 235 | 1140 | 5397 | 25256 |
| 12211 | 1 | 9 | 63 | 405 | 2531 | 15755 | 98718 |
| 12212 | 1 | 8 | 55 | 346 | 2122 | 12926 | 79072 |
| 12213 | 1 | 10 | 65 | 358 | 1837 | 9148 | 45011 |
| 12221 | 1 | 9 | 65 | 424 | 2668 | 16577 | 103051 |
| 12222 | 1 | 8 | 50 | 300 | 1785 | 10752 | 66276 |
| 12223 | 1 | 9 | 55 | 291 | 1453 | 7125 | 35017 |
| 12231 | 1 | 10 | 67 | 387 | 2095 | 10990 | 56675 |
| 12232 | 1 | 8 | 45 | 226 | 1097 | 5316 | 26083 |
| 12233 | 1 | 11 | 78 | 458 | 2440 | 12336 | 60721 |
| 12234 | 1 | 9 | 55 | 290 | 1428 | 6787 | 31618 |
| 12311 | 1 | 9 | 60 | 360 | 2070 | 11724 | 66327 |
| 12312 | 1 | 9 | 56 | 303 | 1540 | 7602 | 37026 |
| 12313 | 1 | 9 | 58 | 322 | 1647 | 8036 | 38166 |
| 12314 | 1 | 9 | 54 | 273 | 1262 | 5545 | 23652 |
| 12321 | 1 | 9 | 56 | 302 | 1523 | 7430 | 35655 |
| 12322 | 1 | 8 | 45 | 226 | 1097 | 5316 | 26083 |
| 12323 | 1 | 9 | 55 | 290 | 1430 | 6827 | 32083 |
| 12324 | 1 | 8 | 42 | 185 | 747 | 2886 | 10924 |
| 12331 | 1 | 10 | 69 | 413 | 2311 | 12482 | 66077 |
| 12332 | 1 | 10 | 66 | 367 | 1874 | 9145 | 43546 |
| 12333 | 1 | 10 | 70 | 430 | 2500 | 14218 | 80410 |
| 12334 | 1 | 10 | 65 | 350 | 1701 | 7770 | 34105 |
| 12341 | 1 | 10 | 66 | 364 | 1821 | 8588 | 38991 |
| 12342 | 1 | 9 | 53 | 261 | 1173 | 5013 | 20821 |
| 12343 | 1 | 9 | 53 | 261 | 1173 | 5013 | 20821 |
| 12344 | 1 | 11 | 79 | 471 | 2535 | 12807 | 62023 |
| 12345 | 1 | 10 | 65 | 350 | 1701 | 7770 | 34105 |

**Research Direction 6.7** *Fix $m \geq 1$. Let $\mathcal{P}_n^m$ be the set of all words $\pi = \pi_1\pi_2\cdots\pi_n$ such that $\pi_1 = 1$ and $\pi_i \leq \pi_{i-1} + m$ for all $i = 2, 3, \ldots, n$. Clearly, $\mathcal{P}_n^1 = \mathcal{P}_n$. According to our results, one can ask to extend our classification and* enumeration (and maybe the statistics that presented in the previous chapters) on $\mathcal{P}_n$ to the set $\mathcal{P}_n^m$. For instance, if $f(x)$ is the generating function for the number of words in $\mathcal{P}_n^m$ that avoid the pattern 1212, then it is easy to verify that $f(x)$ satisfies the relation $f(x) = \frac{1}{1-mxf(x)}$, which implies that the number of such words of size $n$ is given by $\mathrm{Cat}_n m^n$.

**Research Direction 6.8** Let $\mathcal{P}'_n$ be the set partitions of $n$ without singletons. For instance, $\mathcal{P}'_1$ is an empty set, $\mathcal{P}'_2 = \{11\}$, $\mathcal{P}'_3 = \{111\}$ and $\mathcal{P}'_4 = \{1111, 1122, 1212, 1221\}$. As consequence of our results, one can ask to restrict the combinatorial problems that we discussed in previous chapters and the current chapter to the set $\mathcal{P}'_n$. For instance, if $f(x)$ is the generating function $\sum_{n \geq 0} |\mathcal{P}'_n(1212)| x^n$, then it is not hard to see that $f(x)$ satisfies $f(x) = 1 + x^2 f^2(x)/(1 - xf(x))$, which implies that the number of set partitions of $\mathcal{P}'_n(1212)$ is given by the n-th Riordan number; see [327, Sequence A005043].

# Chapter 7

## Multi Restrictions on Set Partitions

In this chapter, we focus on two goals: multiple pattern avoidance and restrictions that cannot be clearly characterized by patterns. Let us start our discussion with the first goal. While the case of set partitions avoiding a single pattern has attracted much attention (see Chapter 6), the case of multiple pattern avoidance remains less investigated. In particular, it is natural, as the next step, to consider set partitions avoiding either pairs or triples (or even set) of patterns. The study of multiple pattern avoidance on set partitions was initiated by Goyt [120] when he extended the results of Sagan [306] to cover the cardinalities of the set partitions in $\mathcal{P}_n$ that avoid a subset of $\mathcal{P}_3$; see Table 3.2. Following Goyt, several research papers on multiple pattern avoidance were published with the focus on the connections between set partitions that avoid a set of patterns and combinatorial structures. For instance, Mansour and Shattuck [236, 232, 233, 235, 239] presented connections between set partitions that avoid three patterns, left Mozkin numbers (see Section 7.5), Sequence A054391 (see Section 7.6), Generalized Catalan numbers (see Section 7.7), Catalan numbers (see Section 7.7) and Pell numbers (see Section 7.8). Actually, each of these connections focused on avoiding very special sets of patterns. This changed when Jelínek, Mansour, and Shattuck [149] classified the equivalence classes among pair of patterns of several general types as we will discus see in the next sections. First, they classified pairs of patterns $(\sigma, \tau)$ where $\sigma \in \mathcal{P}_3$ is a pattern with at least two distinct letters and $\tau \in \mathcal{P}_k(\sigma)$. They provided an upper bound for the number of equivalence classes, and provided an explicit formula for the generating function of all such avoidance classes, showing that in all cases this generating function is rational; see Section 7.1. Next, they focused on the set of noncrossing set partitions that avoid a pattern, that is, avoiding a pair of patterns of the form $(1212, \tau)$, where $\tau \in \mathcal{P}_k(1212)$. Here also, they provided several general equivalence criteria for pattern pairs of this type, and showed that these criteria account for all the equivalences observed when $\tau$ has a size of at most six; see Sections 7.2 and 7.3. The most popular subset of set partitions is the subset of noncrossing set partitions (see the introduction of the book), which has received a lot of attention from different branches of mathematics such as enumerative combinatorics and algebraic combinatorics (see also Chapters 9 and 10 for applications in computer science and physics). Maybe the main reason for the wide applications of noncrossing set partitions is this subset enumerated by Catalan numbers; see Section 7.2. At the end, Jelínek, Mansour, and Shattuck

[149] performed a full classification of the equivalence classes of all the pairs $(\sigma, \tau)$, where $\sigma, \tau \in \mathcal{P}_4$; see Section 7.4.

Now we discuss the second goal of this chapter, namely restrictions that cannot be clearly characterized by patterns. As we classified our statistics intotwo groups the subword and the nonsubword statistics (see Chapters 4 and 5), we also classify our restrictions. The main reason for distinguishing our restrictions is that sometimes it is easy to define our restrictions without using our patterns. For instance, for a given block representation of set partition $\pi = B_1/B_2/\cdots/B_k \in \mathcal{P}_{n,k}$ we define the *variation* of $\pi$ to be $\sum_{i=1}^{k-1} |a \min B_i - \min B_{i+1}|$ with constant $a$. Here the question is to find the number of set partitions of $\mathcal{P}_n$ that have no variation. As we can see, the variation condition can be formulated by using the canonical representation, but it is complicated. As we have shown in Chapter 6 and the first part of this chapter, we considered restrictions which are defined by pattern avoidance in the canonical representation of set partitions. So, the second goal of this chapter is to give several interesting examples of restrictions on set partitions that interested combinatorialists around the world. The first example, the study of $d$-regular set partitions (Section 7.9), has received a lot of attention. It seems that the first consideration of $d$-regular set partitions goes back to Prodinger [278], who called them $d$-*Fibonacci partitions*. Since then several authors have given different proofs and extensions for the result of Prodinger; for example see [68, 77, 156]. The study of $d$-regular set partitions followed by several research papers such as set partitions satisfy certain set of conditions on the block representations. More precisely, Chu and Wei [77] introduced the restriction $|i - j| \geq \ell$, where $x_i, x_j$ are any two elements in the same block of the set partition of $A = \{x_1, x_2, \ldots, x_n\}$, as discussed in Section 7.10. In Sections 7.11 and 7.12, we give two other examples of restrictions where the first deals with singletons and largest singletons in the set partitions, and the second deals with introducing a new subset of set partitions that has nice combinatorial properties, namely, *block connected set partitions*.

## 7.1 Avoiding a Pattern of Size Three and Another Pattern

We start with the study of classes of set partitions that avoid a pair of subsequence patterns $(\sigma, \tau)$, where $\sigma$ is a subsequence pattern in $\mathcal{P}_3$. We may assume that $\tau$ does not contain $\sigma$, otherwise $\mathcal{P}_n(\sigma, \tau) = \mathcal{P}_n(\sigma)$. Note that

- In Section 3.2.1 it has been shown that $P_n(112) = P_n(121) = P_n(122) = P_n(123) = 2^{n-1}$.

- A set partition avoids 111 if and only if each of its blocks has a size of at most two. Such a set partition is known as a *partial matching*. Pattern avoidance in partial matching has been considered by Jelínek and Mansour [148].

We therefore focus on the subsequence patterns 112, 121, 122, and 123.

**Definition 7.1** *We say that a pair of subsequence patterns* $(\sigma, \tau)$ *is a* $(3, k)$-*pair if* $\sigma \in \{112, 121, 122, 123\}$ *and* $\tau$ *is a subsequence pattern of size* $k$ *that avoids* $\sigma$.

Our main results are devoted to prove a general criteria for equivalences among $(3, k)$-pairs, and then finding an explicit formula for the generating functions of set partitions avoiding an arbitrary given $(3, k)$-pair; see Corollaries 7.22 and 7.23. The results of this section are presented below according to the pattern of size three that we avoid.

### 7.1.1 The Patterns 112, 121

First, let us consider the $(3, k)$-pairs $(112, \tau)$ and $(121, \tau)$, where we show that the two avoidance classes $\mathcal{P}_n(121)$ and $\mathcal{P}_n(112)$ are closely related. Recalling Definition 1.21, is obvious the canonical representation of each set partition that avoids either 121 or 112.

**Fact 7.2** *A set partition* $\tau$ *avoids* 121 *(*112*) if and only if* $\tau$ *has the form* $1^{a_1} 2^{a_2} \cdots m^{a_m}$ ($12 \cdots m m^{a_m - 1} (m-1)^{a_{m-1} - 1} \cdots 1^{a_1 - 1}$), *for some* $m \geq 1$ *and some sequence* $\mathbf{a} = (a_1, \ldots, a_m)$ *of positive integers.*

Note that the number of set partitions of $\mathcal{P}_n(112)$ ($\mathcal{P}_n(121)$) is given by $2^{n-1}$, which is the same as the number of compositions of $[n]$. This motivates the following definitions.

**Definition 7.3** *For a composition* $\mathbf{a} = (a_1, \cdots, a_m)$, *let* $\tau_{121}(\mathbf{a})$ *denote the 121-avoiding pattern* $1^{a_1} 2^{a_2} \cdots m^{a_m}$ *and let* $\tau_{112}(\mathbf{a})$ *denote the 112-avoiding pattern* $12 \cdots m m^{a_m - 1} (m-1)^{a_{m-1} - 1} \cdots 1^{a_1 - 1}$.

Note that $\tau_{112}(\mathbf{a})$ is the unique 112-avoiding set partition with $m$ blocks whose $i$-th block has size $a_i$, and similarly for $\tau_{121}(\mathbf{a})$.

**Definition 7.4** *Let* $\mathbf{a} = (a_1, \cdots, a_m)$ *and* $\mathbf{b} = (b_1, \cdots, b_k)$ *be two compositions. We say that* $\mathbf{b}$ *dominates* $\mathbf{a}$, *if there is an $m$-tuple of indices* $i_1, i_2, \ldots, i_m$ *such that* $1 \leq i_1 < i_2 < \cdots < i_m \leq k$, *and* $a_j \leq b_{i_j}$ *for each* $j \in [m]$. *In other words,* $\mathbf{b}$ *dominates* $\mathbf{a}$ *if* $\mathbf{b}$ *contains a $m$-term subsequence whose every component is greater than or equal to the corresponding component of* $\mathbf{a}$.

**Example 7.5** *The composition* $\mathbf{b} = 213214$ *dominates the composition* $\mathbf{a} = 123$ *since* $a_1 < b_1$, $a_2 < b_3$ *and* $a_3 < b_6$.

310                 *Combinatorics of Set Partitions*

We present the following simple fact without proof.

**Fact 7.6** *For any two compositions* **a** *and* **b**, *the following are equivalent:*

- **b** *dominates* **a**,
- $\tau_{112}(\mathbf{b})$ *contains* $\tau_{112}(\mathbf{a})$,
- $\tau_{121}(\mathbf{b})$ *contains* $\tau_{121}(\mathbf{a})$.

Note that the above fact shows that the sets $\mathcal{P}_n(112)$ and $\mathcal{P}_n(121)$ are ordered by containment and the set of all integer compositions ordered by domination consists of three isomorphic *posets*, with size-preserving isomorphisms identifying a composition **a** with $\tau_{112}(\mathbf{a})$ and $\tau_{121}(\mathbf{a})$. We thus have the following result.

**Corollary 7.7** *For any integer composition* **a**, *the* $(3,k)$-*pairs* $(112, \tau_{112}(\mathbf{a}))$ *and* $(121, \tau_{121}(\mathbf{a}))$ *are equivalent.*

**Definition 7.8** *For two compositions* **a** *and* **a**′, *let us write* $\mathbf{a} \stackrel{d}{\sim} \mathbf{a}'$ *if for every* $n$, *the number of compositions of size* $n$ *dominating* **a** *is equal to the number of compositions of size* $n$ *dominating* **a**′. *For a composition* $\mathbf{a} = (a_1, \ldots, a_m)$, *let* $M(\mathbf{a})$ *denote the multiset* $\{a_1, \ldots, a_m\}$.

Fact 7.6 states that $\mathbf{a} \stackrel{d}{\sim} \mathbf{a}'$ if and only if $(112, \tau_{112}(\mathbf{a})) \sim (112, \tau_{112}(\mathbf{a}'))$; that is if and only if $(121, \tau_{121}(\mathbf{a})) \sim (121, \tau_{121}(\mathbf{a}'))$.

**Lemma 7.9** *Assume either (1)* **a** *and* **a**′ *are any two compositions with* $M(\mathbf{a}) = M(\mathbf{a}')$, *or (2)* $\mathbf{a} = (a_1, \ldots, a_{m-1}, 2)$ *and* $\mathbf{a}' = (a_1, \ldots, a_{m-1}, 1, 1)$ *two compositions. Then* $\mathbf{a} \stackrel{d}{\sim} \mathbf{a}'$.

**Proof** The proofs of (1) and (2) are very similar. Thus, we give only the details of the proof of (1). In this case, it is enough to show (1) when **a**′ is obtained from **a** by exchanging the order of two consecutive elements. Suppose $\mathbf{a} = (a_1, \ldots, a_m)$ and $\mathbf{a}' = (a_1, \ldots, a_{j-1}, a_{j+1}, a_j, a_{j+2}, \ldots, a_m)$. Let $\mathbf{b} = (b_1, \ldots, b_k)$ be a composition of size $n$ that dominates **a** with (i) $i \in [k]$ the smallest index such that $(b_1, \ldots, b_i)$ dominates $(a_1, \ldots, a_{j-1})$ and (ii) $s \in [k]$ the largest index such that $(b_{s+1}, \ldots, b_k)$ dominates $(a_{j+2}, \ldots, a_m)$. Since **b** dominates **a**, $i+2 \leq s$ and $(b_{i+1}, \ldots, b_s)$ dominates $(a_j, a_{j+1})$. Define $\mathbf{b}' = (b_1, \ldots, b_i, b_s, b_{s-1}, \ldots, b_{i+1}, b_{s+1}, \ldots, b_k)$. Clearly, **b**′ dominates **a**′, and the mapping $\mathbf{b} \mapsto \mathbf{b}'$ is a size-preserving bijection between compositions that dominate **a** and those that dominate **a**′, which completes the proof. □

Lemma 7.9 shows that every composition **a** is $\stackrel{d}{\sim}$-equivalent to a composition **a**′ that has the property that its components are weakly decreasing and none of them is equal to 2. This invites the following definition.

**Definition 7.10** *A 2-free integer partition is a composition of size $n$ such that its components are weakly decreasing and none of them is equal to 2.* Let $\xi_n$ be the number of 2-free integer partitions of size $n$.

Note that the sequence $\{\xi_n\}_{n\geq 0}$ is listed as Sequence A027336 in [327]. Basic estimates on the number of integer partitions (see, for instance, [4]) imply the bound $\xi_k = 2^{O(\sqrt{k})}$.

**Definition 7.11** *Let $\mathbf{a} = (a_1, a_2, \ldots, a_m)$ be any composition, and define $P_{112,\tau_{112}(\mathbf{a})}(x,y)$ to be the generating function for the number of set partitions of $\mathcal{P}_{n,k}(112, \tau_{112}(\mathbf{a}))$, that is,*

$$P_{112,\tau_{112}(\mathbf{a})}(x,y) = \sum_{n,k\geq 0} P_{n,k}(112, \tau_{112}(\mathbf{a})) x^n y^k.$$

The next theorem presents an explicit formula for the generating function $P_{112,\tau_{112}(\mathbf{a})}(x,y)$ for any composition $\mathbf{a}$.

**Theorem 7.12** *For any composition $\mathbf{a} = (a_1, a_2, \ldots, a_m)$,*

$$P_{112,\tau_{112}(\mathbf{a})}(x,y) = \sum_{j=0}^{m-1} \frac{x^{a_1+\cdots+a_j} y^j (1-x)}{\prod_{i=1}^{j+1}(1-x(1+y)+x^{a_i}y)}.$$

**Proof** Let $n \geq 1$, $\pi \in \mathcal{P}_{n,k}(112, \tau_{112}(\mathbf{a}))$, and let $r$ be the size of the first block of $\pi$. Let us write an equation for the generating function $P_{112,\tau_{112}(\mathbf{a})}(x,y)$. We consider the following two cases: (1) $1 \leq r \leq a_1 - 1$, (2) $r \geq a_1$. In the first case, $\pi$ must be of the form $1\pi' 1^{r-1}$, where $\pi'$ is some set partition on the letters $\{2, 3, \ldots\}$ avoiding $\{112, \tau_{112}(\mathbf{a})\}$, which contributes to our equation

$$xy P_{112,\tau_{112}(\mathbf{a})}(x,y) + x^2 y P_{112,\tau_{112}(\mathbf{a})}(x,y) + \cdots + x^{a_1-1} y P_{112,\tau_{112}(\mathbf{a})}(x,y)$$
$$= \frac{x - x^{a_1}}{1 - x} y P_{112,\tau_{112}(\mathbf{a})}(x,y).$$

In the second case, $\pi$ must be of the form $1\pi' 1^{r-1}$, where $\pi'$ is a set partition on the letters $\{2, 3, \ldots\}$ avoiding $\{112, \tau_{112}(\mathbf{a}')\}$ since $r \geq a_1$, where $\mathbf{a}' = (a_2, \ldots, a_m)$. Thus, this case contributes to our equation

$$x^{a_1} y P_{112,\tau_{112}(\mathbf{a}')}(x,y) + x^{a_1+1} y P_{112,\tau_{112}(\mathbf{a}')}(x,y) + \cdots$$
$$= \frac{x^{a_1} y}{1-x} P_{112,\tau_{112}(\mathbf{a}')}(x,y),$$

where we define $P_{112,\tau_{112}(\mathbf{a}')}(x,y)$ to be 0 when $m=1$. By adding these two contributions, we obtain

$$P_{112,\tau_{112}(\mathbf{a})}(x,y) = 1 + \frac{x-x^{a_1}}{1-x} y P_{112,\tau_{112}(\mathbf{a})}(x,y) + \frac{x^{a_1}y}{1-x} P_{112,\tau_{112}(\mathbf{a}')}(x,y),$$

which is equivalent to

$$P_{112,\tau_{112}(\mathbf{a})}(x,y)$$
$$= \frac{1-x}{1-x(1+y)+x^{a_1}y} + \frac{x^{a_1}y}{1-x(1+y)+x^{a_1}y} P_{112,\tau_{112}(\mathbf{a'})}(x,y).$$

Iterating this recurrence completes the proof. □

By applying Theorem 7.12 for compositions of size at most three, we derive Table 7.1.

**Table 7.1**: Generating functions $F_\mathbf{a}(x,y)$, where **a** is a nonempty composition of size at most three

| **a** | $F_\mathbf{a}(x,y)$ | **a** | $F_\mathbf{a}(x,y)$ | **a** | $F_\mathbf{a}(x,y)$ |
|---|---|---|---|---|---|
| 1 | 1 | 11 | $1+\frac{xy}{1-x}$ | 2 | $\frac{1-x}{1-x(1+y)+x^2y}$ |
| 111 | $1+\frac{xy}{1-x}+\frac{x^2y}{(1-x)^2}$ | 12 | $1+\frac{xy}{1-x(1+y)+x^2y}$ | 21 | $\frac{1-x+x^2y}{1-x(1+y)+x^2y}$ |
| 3 | $\frac{1-x}{1-x(1+y)x^2y}$ | | | | |

## 7.1.2 The Pattern 123

Now, we consider the case $(3,k)$-pair $(123,\tau)$. Clearly, a set partition avoids 123 if and only if it has at most two blocks. We will distinguish two cases, depending on whether $\tau$ has a single block or whether it has two blocks. We start with the trivial case that $\tau$ has one block, where we leave the proof to the reader.

**Fact 7.13** *A set partition avoids* $(123,1^m)$ *if and only it has at most two blocks and each block has size at most* $m-1$. *Moreover, the generating function for the number set partitions of* $P_n(123,1^m)$ *is given by*

$$\sum_{n\geq 0} P_n(123,1^m)x^n = 1 + \sum_{a=1}^{m-1}\sum_{b=0}^{m-1}\binom{a+b-1}{b}x^{a+b}.$$

**Example 7.14** *The generating function of the class* $P_n(123,111)$ *is given by* $1+x+2x^2+3x^3+3x^4$.

To investigate the pair $(123,\tau)$ with $\tau$ has two blocks, we state the following general result.

**Theorem 7.15** *Let* $m\geq 2$ *be an integer. Let* $\tau=\tau_1\tau_2\cdots\tau_k$ *be a set partition with exactly* $m$ *blocks such that* $\tau_i=i$ *for each* $i\in[m-1]$. *Then the generating* function $\sum_{n\geq 0} P_n(12\cdots(m+1),\tau)x^n$ is given by

$$\left(\sum_{n\geq 0} P_n(12\cdots(m+1))x^n\right) - \left(\frac{x}{1-(m-1)x}\right)^{k-m} \prod_{j=1}^{m} \frac{x}{1-jx}.$$

**Proof** Let $A_n$ be the set of partitions of $\mathcal{P}_n(12\cdots(m+1))$ but contain $\tau$, that is, $Q_n = \mathcal{P}_n(12\cdots(m+1)) \setminus \mathcal{P}_n(12\cdots(m+1),\tau)$. Define $A(x)$ to be the generating function for $|A_n|$, that is, $A(x) = \sum_{n\geq 0} |A_n| x^n$. It is enough to show

$$A(x) = \left(\frac{x}{1-(m-1)x}\right)^{k-m} \frac{x}{1-mx} \prod_{j=1}^{m-1} \frac{x}{1-jx}. \tag{7.1}$$

To verify that, let $\pi \in A_n$, so $\pi$ has exactly $m$ blocks, and let us write $\tau = 12\cdots(m-1)\tau_m\cdots\tau_k$. Since $\pi$ contains $\tau$ as a pattern and $\tau$ has $m$ blocks, the set partition $\pi$ can be decomposed as

$$\pi = 1\pi^{(1)} 2\pi^{(2)} \cdots (m-1)\pi^{(m-1)} \tau_m \pi^{(m)} \tau_{m+1} \pi^{(m+1)} \cdots \tau_k \pi^{(k)}$$

such that

- for $1 \leq j < m-1$, $\pi^{(j)}$ is an arbitrary word over the alphabet $[j]$,
- for $m-1 \leq j < k$, $\pi^{(j)}$ is an arbitrary word over $[m] \setminus \{\tau_{j+1}\}$, and
- $\pi^{(k)}$ is an arbitrary word over $[m]$.

Conversely, any sequence with such a decomposition is a set partition of $A_n$. Hence, 7.1 holds. $\square$

Theorem 7.15 for $m = 2$ and $\sum_{n\geq 0} P_n(123)x^n = (1-x)/(1-2x)$ gives our next result.

**Corollary 7.16** *For every $k$, the $(3,k)$-pairs of the form $(123,\tau)$ where $\tau$ has two blocks are all equivalent, and the generating function of any such pair is*

$$\sum_{n\geq 0} P_n(123,\tau)x^n = \sum_{i=0}^{k-1} \left(\frac{x}{1-x}\right)^i.$$

Comparing the generating function of the previous corollary with the formula of Theorem 7.12, we can say even more.

**Corollary 7.17** *For every $k$ and every set partition $\tau \in \mathcal{P}_k$ with two blocks, the $(3,k)$-pair $(123,\tau)$ is equivalent to the $(3,k)$-pair $(112, 12\cdots k)$.*

### 7.1.3 The Pattern 122

Now, we study the case $(3,k)$-pair $(122, \tau)$. A set partition $\tau$ avoids 122 if and only if each block of $\tau$ except possibly the first one has size one, or equivalently, any number greater than 1 appears at most once in (the canonical representation of) $\tau$. For every $k$, all the $(3,k)$-pairs of the form $(122, \tau)$ are equivalent to $(112, 1^k)$. We first describe a bijection between 122-avoiding and 123-avoiding partitions which, under suitable assumptions, preserves containment. Let $\tau = \tau_1 \tau_2 \cdots \tau_k$ be a 122-avoiding set partition. Define a set partition $f(\tau) = \tau_1' \tau_2' \cdots \tau_k'$ by putting $\tau_i' = 1$ if $\tau_i = 1$ and $\tau_i' = 2$ if $\tau_i > 1$. For instance, if $\pi = 1123145$, then $f(\pi) = 1122122$. Note that the mapping $f$ defined by these properties is a bijection between the sets $\mathcal{P}_n(122)$ and $\mathcal{P}_n(123)$. From this bijection, with a little work (see [149]), we can state the following classification.

**Proposition 7.18** *For any set partition $\tau \in \mathcal{P}_k(122)$ with at least two blocks, the $(3,k)$-pair $(122, \tau)$ is equivalent to the $(3,k)$-pair $(123, f(\tau))$.*

Using Proposition 7.18 together with Corollary 7.17, one obtains the following result.

**Corollary 7.19** *For any $k$ and any set partition $\tau \in \mathcal{P}_k(122)$ with at least two blocks, the $(3,k)$-pair $(122, \tau)$ is equivalent to the $(3,k)$-pair $(112, 12 \cdots k)$.*

It remains to deal with $(3,k)$-pairs of the form $(122, 1^k)$. It turns out that these pairs are also equivalent to all the other $(3,k)$-pairs of the form $(122, \tau)$, and the proof is left to the interested reader.

**Proposition 7.20** *The $(3,k)$-pair $(122, 1^k)$ is equivalent to the $(3,k)$-pair $(122, 12 \cdots k)$.*

Using Corollary 7.19 and Proposition 7.20, one achieves the main goal of this subsection.

**Corollary 7.21** *For any $k$, the $(3,k)$-pairs of the form $(122, \tau)$ are all equivalent, and they are equivalent to the pair $(112, 12 \cdots k)$.*

Note that the results of this section imply the following (see [149]).

**Corollary 7.22** *For each $k$ and each $(3,k)$-pair $(\sigma, \tau)$, the generating function for the sequence $\{P_n(\sigma, \tau)\}_{n \geq 0}$ is rational.*

**Corollary 7.23** *For each $k \geq 3$, the $(3,k)$-pairs form at most $1 + \xi_k$ equivalence classes, where $\xi_k$ is the number of 2-free integer partitions (See Sequence A027336 in [327]).*

We end this section by presenting all the equivalence classes of $(3, 4)$-pairs; see Table 7.2.

Table 7.2: The equivalence classes of $(3,4)$-pairs

| $(\tau, \tau')$ | $\sum_{n>0} P_n(\tau, \tau') x^n$ |
|---|---|
| $(123, 1111)$ | $\sum_{i=1}^{3} \sum_{j=0}^{3} \binom{i+j-1}{j} x^{i+j}$ |
| $(121, 1111) \sim (112, 1111)$ | $\frac{1}{1-x-x^2-x^3}$ |
| $(121, 1222) \sim (121, 1112) \sim (112, 1211)$ $\sim (112, 1222)$ | $\frac{1-x+x^3}{(1-x)(1-x-x^2)}$ |
| $(112, 1231) \sim (112, 1232) \sim (112, 1233)$ $\sim (121, 1122) \sim (121, 1123) \sim (121, 1223)$ $\sim (121, 1233) \sim (121, 1234) \sim (122, 1111)$ $\sim (122, 1112) \sim (122, 1121) \sim (122, 1123)$ $\sim (122, 1211) \sim (122, 1213) \sim (122, 1231)$ $\sim (122, 1234) \sim (112, 1221) \sim (123, 1112)$ $\sim (123, 1121) \sim (123, 1122) \sim (123, 1211)$ $\sim (123, 1212) \sim (123, 1221) \sim (123, 1222)$ $\sim (112, 1234)$ | $\sum_{i=0}^{3} \frac{x^i}{(1-x)^i}$ |

## 7.2 Pattern Avoidance in Noncrossing Set Partitions

In our next step, we study classes of set partitions that avoid a pair of subsequence patterns $(1212, \tau)$. We may assume that $\tau$ does not contain 1212, otherwise $\mathcal{P}_n(1212, \tau) = \mathcal{P}_n(1212)$. Note that Theorem 3.52 shows that $P_n(1212) = \text{Cat}_n$.

Recall that a set partition $\pi$ is $k$-*noncrossing* if it avoids the subsequence pattern $12 \cdots k 12 \cdots k$, and $\pi$ is $k$-*nonnesting* if it avoids the subsequence pattern $12 \cdots kk \cdots 21$; see Definition 6.33. Let us point out that there are several different concepts of "crossings" and "nestings" used in the literature. Klazar [176] has considered two blocks $B, B'$ of a set partition to be crossing (respectively, nesting) if there are four elements $x_1 < y_1 < x_2 < y_2$ (respectively, $x_1 < y_1 < y_2 < x_2$) such that $x_1, x_2 \in B$ and $y_1, y_2 \in B'$, and similarly for $k$-crossings and $k$-nestings. Note that Klazar's definition makes no assumption about the relative order of the minimal elements of $B$ and $B'$, which allows more general configurations to be considered as crossing or nesting. Thus, Klazar's $k$-noncrossing and $k$-nonnesting set partitions are a proper subset of our $k$-noncrossing and $k$-nonnesting set partitions, except for 2-noncrossing set partitions where the two concepts coincide. (Why?)

Another approach to crossings in set partitions was introduced by Chen et al. [68, 69]. They use the so-called *standard representation* (see Definition 3.50), where a set partition of $[n]$ with blocks $B_1, B_2, \ldots, B_k$ is represented by a graph on the vertex set $[n]$, with $a, b \in [n]$ connected by an edge if they belong to the same block and there is no other element of this block between them. In this terminology, a set partition is $k$-crossing (or $k$-nesting) if the representing graph contains $k$ edges that are pairwise crossing (or nesting), where two edges $e_1 = \{a < b\}$ and $e_2 = \{a' < b'\}$ are crossing (or nesting) if $a < a' < b < b'$ (or $a < a' < b' < b$, respectively). Let us call such set partitions *graph-k-crossing* and *graph-k-nesting*, to avoid confusion with our own terminology of Definition 6.33. It is not difficult to see that a set partition is graph-2-noncrossing if and only if it is 2-noncrossing, but for nestings and for $k$-crossings with $k > 2$, the two concepts are incomparable. For instance the set partition 12121 is graph-2-nonnesting but it contains 1221, while 12112 is graph-2-nesting and avoids 1221. Similarly, 1213123 has no graph-3-crossing and contains 123123, while 1232132 has a graph-3-crossing and avoids 123123. Chen et al. [69] showed that the number of graph-$k$-noncrossing and graph-$k$-nonnesting partitions of $[n]$ is equal. This is true also for $k$-noncrossing and $k$-nonnesting set partitions as has been described in Corollary 6.36. Actually, we showed the following general result (see Corollary 6.37): If $\tau$ is a set partition and $k$ is a positive integer, then the subsequence patterns $12 \cdots k(\tau + k) 12 \cdots k$ and $12 \cdots k(\tau + k) k(k-1) \cdots 1$ are Wilf-equivalent. Moreover, the number of set partitions in $\mathcal{P}_{n,m}(12 \cdots k(\tau + k) 12 \cdots k)$ is the same as the number of set partitions in $\mathcal{P}_{n,m}(12 \cdots k(\tau + k)k(k - 1) \cdots 1)$, for all $m = 1, 2, \ldots, n$. It is interesting to note that the proofs of both these results are based on a reduction to theorems on pattern avoidance in the fillings of Ferrers diagrams (this is only implicit in [69]; a direct construction is given by Krattenthaler [201]) as described in Chapter 6, although the constructions employed in the proofs of these results are quite different.

A general result on distribution of the numbers of crossings and nestings in set partitions has been found by Kasraoui and Zeng [159] as described in Theorem 5.96, which shows that there is an involution $\rho : \mathcal{P}_n(\lambda) \mapsto \mathcal{P}_n(\lambda)$ (see Definition 5.95) preserving the number of alignments, and exchanging the numbers of crossings and nestings. Another general result is due to Poznanovik and Yan [277]. In order to present such a result we need the following definitions and notation.

**Definition 7.24** *Let $\mathcal{T}(\pi)$ be the subtree rooted by $\pi$ of the generating tree $\mathcal{T}$ of the set partitions as described in Figure 2.79. We denote the children of a set partition $\pi$ with exactly $k$ blocks by $\pi^{(0)}, \pi^{(1)}, \ldots, \pi^{(k)}$, where*

$$\pi^{(0)} = \{1\}, B_1 + 1, \ldots, B_k + 1,$$

*and*

$$\pi^{(i)} = \{1\} \cup (B_i + 1), B_2 + 1, \ldots, B_{i-1} + 1, B_{i+1} + 1, \ldots, B_k + 1,$$

*when $B + 1$ denotes the set $\{a + 1 \mid a \in B\}$. Define $\mathcal{T}(\pi, \ell)$ to be the set of all set partitions in $\mathcal{T}(\pi)$ at level $\ell$ and $\mathcal{T}(\pi, \ell, m)$ to be the set of all set partitions in $\mathcal{T}(\pi, \ell)$ with exactly $m$ blocks.*

**Example 7.25** *The first levels of the generating tree $\mathcal{T}$ of the set partitions are presented in the following figure:*

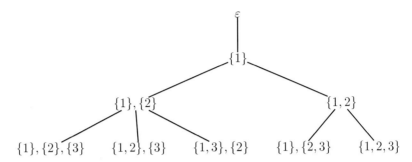

**FIGURE 7.1**: Generating tree for set partitions

In particular the children of the set partition $\pi = \{1\}, \{2\}$ are $\pi^{(0)} = \{1\}, \{2\}, \{3\}$, $\pi^{(1)} = \{1, 2\}, \{3\}$ and $\pi^{(2)} = \{1, 3\}, \{2\}$.

**Definition 7.26** *For each $b_i = \min B_i$ of a set partition $\pi$, define $\alpha_i(\pi)$ to be the number of edges $(p, q)$ in the standard representation of $\pi$ such that $p < b_i < q$ and $\beta_i(\pi)$ to be the number of edges $(p, q)$ in the standard representation of $\pi$ such that $p < q < b_i$.*

**Example 7.27** *Let $\pi = \{1, 3, 6\}, \{2, 4, 5\}, \{7\}$ be a set partition of $[7]$. Then $\alpha_1(\pi) = 0, \alpha_2(\pi) = 1, \alpha_3(\pi) = 0$ and $\beta_1(\pi) = \beta_2(\pi) = 0, \beta_3(\pi) = 4$.*

It is obvious that the numbers $\alpha_i(\pi), \beta_i(\pi)$ satisfy $\alpha_0(\pi^{(j)}) = \beta_0(\pi^{(j)}) = 0$ where $j = 0, 1, \ldots, k$, and otherwise satisfy $\alpha_i(\pi^{(0)}) = \alpha_{i-1}(\pi)$, $\beta_i(\pi^{(0)}) = \beta_{i-1}(\pi)$,

$$\alpha_i(\pi^{(j)}) = \begin{cases} \alpha_{i-1}(\pi) + 1, & i = 2, 3, \ldots, j \\ \alpha_i(\pi), & j+1 \leq i \leq k. \end{cases}$$

$$\beta_i(\pi^{(j)}) = \begin{cases} \beta_{i-1}(\pi), & i = 2, 3, \ldots, j \\ \beta_i(\pi) + 1, & j+1 \leq i \leq k. \end{cases}$$

for all $j = 1, 2, \ldots, k$.

**Definition 7.28** *Let $G$ be an abelian group (see Appendix E) and $a, b \in G$. Define the statistic $s_{a,b} : \cup_{n \geq 0} \mathcal{P}_n \mapsto G$ as $s_{a,b}(\pi) = \mathrm{cr}(\pi)\mathrm{a} + \mathrm{ne}(\pi)\mathrm{b}$ (See Definition 5.78 for the statistics $\mathrm{cr}$ and $\mathrm{ne}$). We define $s_{a,b}(A)$ to be the multiset $\{s_{a,b}(\pi) \mid \pi \in A\}$.*

By the relations for the numbers $\alpha_i(\pi), \beta_i(\pi)$, we obtain $s_{a,b}(\pi^{(0)}) = s_{a,b}(\pi^{(1)}) = s_{a,b}(\pi)$ and $s_{a,b}(\pi^{(j)}) = s_{a,b}(\pi) + \alpha_j(\pi)a + \beta_j(\pi)b$ for $j \geq 1$. More generally, we can state the following result.

**Lemma 7.29** *(Poznanović and Yan [277]) Let $\pi$ be any nonempty set partition. Then*

$$s_{a,b}((\pi^{(0)})^{(j)}) = \begin{cases} s_{a,b}(\pi), & j = 0, 1 \\ s_{a,b}(\pi^{(j-1)}), & j \geq 2 \end{cases}$$

*and for $i \geq 1$,*

$$s_{a,b}((\pi^{(i)})^{(j)}) = \begin{cases} s_{a,b}(\pi^{(i)}), & j = 0, 1 \\ s_{a,b}(\pi^{(i)}) + s_{a,b}(\pi^{(j-1)}) - s_{a,b}(\pi^{(1)}) + a, & j = 2, 3, \ldots, i \\ s_{a,b}(\pi^{(i)}) + s_{a,b}(\pi^{(j)}) - s_{a,b}(\pi^{(1)}) + b, & j \geq i+1 \end{cases}$$

The above lemma is critical for proving the next result which obtained by Poznanovik and Yan [277]; we leave the proof to the interest reader.

**Theorem 7.30** *(Poznanović and Yan [277]) Let $\pi$ be any set partition. We have*

- *if $s_{a,b}(T(\pi, \ell)) = s_{a,b}(T(\pi', \ell))$ for $\ell = 0, 1$, then $s_{a,b}(T(\pi, \ell, m)) = s_{a,b}(T(\pi', \ell, m))$ for all $\ell, m \geq 0$.*

- *if $s_{a,b}(T(\pi, \ell)) = s_{b,a}(T(\pi', \ell))$ for $\ell = 0, 1$, then $s_{a,b}(T(\pi, \ell, m)) = s_{b,a}(T(\pi', \ell, m))$ for all $\ell, m \geq 0$.*

*In other words, if the statistic $s_{a,b}$ coincides on the first two levels of the trees $T(\pi)$ and $T(\pi')$ then it coincides on $T(\pi, \ell, m)$ and $T(\pi', \ell, m)$ for all $\ell, m \geq 0$, and similarly for the pair of statistics $s_{a,b}, s_{b,a}$.*

Note that the conditions of the above theorem imply that $\pi$ and $\pi'$ have the same number of blocks. But they are not necessarily set partitions of the same $[n]$. As a direct corollary, we obtain a result of Kasraoui and Zeng [160]; see Corollary 5.97.

**Corollary 7.31** *We have*

$$\sum_{\pi \in \mathcal{P}_n} p^{\mathrm{cr}(\pi)} q^{\mathrm{ne}(\pi)} t^{\mathrm{al}(\pi)} = \sum_{\pi \in \mathcal{P}_n} p^{\mathrm{ne}(\pi)} q^{\mathrm{cr}(\pi)} t^{\mathrm{al}(\pi)}.$$

**Proof** We use $G = (\mathbb{Z} \oplus \mathbb{Z}, +)$, $a = (1, 0)$, $b = (0, 1)$ and $\pi = \pi' = \{1\}$. Any set partition $\mu \in \mathcal{P}_{n,k}$ has exactly $n-k$ edges. Hence, $\mathrm{cr}(\pi) + \mathrm{ne}(\pi) + \mathrm{al}(\pi) = \binom{n-k}{2}$. Thus, the result follows from Theorem 7.30. □

This result has been extended (see [277]) to present an explicit formula for the generating function for the number of set partitions of $[n]$ according to the statistics $cr$ and $ne$ (see Exercise 7.9).

**Corollary 7.32** *The generating function $\sum_{n \geq 0} \sum_{\pi \in \mathcal{P}_n} p^{\mathrm{cr}(\pi)} q^{\mathrm{ne}(\pi)} z^n$ for the* number of set partitions of $[n]$ according to the number of crossings and nestings is given by

$$1 + \cfrac{z}{1 - ([1]_{p,q}+1)z - \cfrac{[1]_{p,q}z^2}{1 - ([2]_{p,q}+1)z - \cfrac{[2]_{p,q}z^2}{\ddots}}},$$

where $[r]_{p,q} = \frac{p^r - q^r}{p-q}$.

A different formula for the generating function $\sum_{n\geq 0} \sum_{\pi \in \mathcal{P}_n} p^{\mathrm{cr}(\pi)} q^{\mathrm{ne}(\pi)} z^n$ was given in [160] (see Exercise 7.9).

On the other hand, combinatorial structures and noncrossing/nonnesting set partitions have been connected by several results (see Theorem 3.52). This is because the number of set partitions of $[n]$ is given by the $n$-th Catalan number, and Catalan numbers enumerated lot of combinatorial structures as described in [332] and Chapter 2.

**Definition 7.33** *We denote the set of noncrossing set partitions of $\mathcal{P}_n$ by $\mathrm{NC}_n$, that is, $\mathrm{NC}_n = \mathcal{P}_n(1212)$, and the set of noncrossing set partitions of $\mathcal{P}_{n,k}$ by $\mathrm{NC}_{n,k}$, that is, $\mathrm{NC}_{n,k} = \mathcal{P}_{n,k}(1212)$.*

Here we give a beautifully simple bijection between the set of noncrossing set partitions and the set of ordered trees, as described in Section 3.2.4.3.

**Theorem 7.34** *(Dershowitz and Zaks [90]) There is a bijection between the set $\mathrm{NC}_n$ and the set of ordered trees of $n+1$ vertices.*

**Proof** Given a tree $T$ with $n+1$ vertices and $k$ internal nodes, traverse it in preorder (visit the root, then recursively visit its subtrees from left to right) and label each vertex when first visited, where the root of the tree is not labeled. The sets of labels of the children of the internal vertices form the blocks of the set partition. It is an easy matter to check that this is a noncrossing set partition of $\mathrm{NC}_{n,k}$. $\square$

As we said earlier, the research on noncrossing set partitions does not end with this simple bijection or the bijection in Theorem 3.52. For instance, Natarajan [260] presented a bijection between noncrossing set partitions of $[2n+1]$ into $n+1$ blocks such that no block contains two consecutive elements, and the set of sequences $\{s_i\}_1^n$ such that $1 \leq s_i \leq i$, and if $s_i = j$, then $s_{i-r} \leq j - r$ for $r \in [j-1]$.

Finally, the set of noncrossing set partitions has greatly interested researchers in algebraic combinatorics, which a topic for another book! As we said in the introduction of the book (see Table 1.6), we will not focus of the algebraic interpretations of the noncrossing set partitions, but here we will give a few references that will give the reader a starting point for further reading. Formally, the study of noncrossing set partitions dates back to the 1970s, when Kreweras [204] showed that the noncrossing set partitions of $A$ form a lattice, ordered by refinement, and then obtained several interesting properties of this lattice (see also [298, 360]). Enumeration of several objects on this lattice have been considered, for instance in [35, 97, 98, 99, 101, 255, 322, 323, 326, 328, 334, 341] and in [10, 11, 12, 33, 34, 213, 263, 287, 302] and [7, 9] and references therein. Note that the first uniform bijection between nonnesting and noncrossing set partitions was constructed in [9], and the proof involves new and interesting combinatorics in the classical types and consequently showed some conjectural properties of the Panyushev map and two cyclic sieving phenomena by Bessis and Reiner [34]. For a more general study of algebraic constructions on set partitions, see [131, 289].

Now we are ready to investigate our main goal, which is to study set partitions that avoid the subsequence pattern 1212 and another subsequence pattern.

### 7.2.1 CC-Equivalences

Recall Definition 2.63. We start by explaining the title of this subsection, that is, creating an equivalence relation between the pairs $(1212, \sigma)$.

**Definition 7.35** *We write $\sigma \stackrel{nc}{\sim} \tau$ if $(1212, \sigma)$ is equivalent to $(1212, \tau)$. If $\sigma \stackrel{nc}{\sim} \tau$, then we say that $\sigma$ and $\tau$ are nc-equivalent ("nc" stands for "noncrossing").*

*We say that a set partition $\pi$ is connected, if it cannot be written as $\pi = \sigma[\tau]$, where $\sigma$ and $\tau$ are nonempty set partitions. Note that a noncrossing set partition is connected if and only if its last element belongs to the first block. For any set partition $\pi$, there is a unique sequence of nonempty connected partitions $\sigma_1, \ldots, \sigma_m$ such that $\pi = \sigma_1[\sigma_2][\sigma_3]\cdots[\sigma_m]$. In this context, we call the set partitions $\sigma_i$ the components of $\pi$.*

*We say that two noncrossing set partition patterns $\sigma$ and $\tau$ are cc-equivalent, denoted by $\sigma \stackrel{cc}{\sim} \tau$, if there is a bijection $f$ from the set $\mathcal{P}_n(1212, \sigma)$ to the set $\mathcal{P}_n(1212, \tau)$ such that for every $\pi \in \mathcal{P}_n(1212, \sigma)$ the set partition $f(\pi)$ has the same size and the same number of components as $\pi$. In particular, cc-equivalence is a refinement of nc-equivalence.*

The cc-equivalence is introduced in [149]. This helps to present a general cc-equivalences between the pairs $(1212, \sigma)$, where the proofs are based of finding bijections between the two classes in the question. For example, the next theorem presents two such cc-equivalences; we leave the proof as a noneasy exercise (for full details, see [149]).

**Theorem 7.36** *Let $\sigma$, $\rho$ and $\tau$ (possibly empty) be any three noncrossing set partitions.*

*(1) If $\rho \stackrel{cc}{\sim} \rho'$, then $\sigma[\rho][\tau] \stackrel{cc}{\sim} \sigma[\rho'][\tau]$.*

(2) If $\sigma$ is connected, then $\sigma[\rho] \stackrel{cc}{\sim} \rho[\sigma]$.

The above theorem can be extended to multiple patterns (more than two patterns).

**Theorem 7.37** *Let $\sigma^{(1)}, \ldots, \sigma^{(k)}$ be a k-tuple of noncrossing set partitions, and let $\nu = \nu_1 \cdots \nu_k$ be a permutation of size $k$. Then the set partition $\sigma^{(1)}[\sigma^{(2)}] \cdots [\sigma^{(k)}]$ is cc-equivalent to the set partition $\sigma_{\nu_1}[\sigma^{(\nu_2)}] \cdots [\sigma^{(\nu_k)}]$.*

**Proof** Without loss of generality, we assume that all the $\sigma^{(i)}$ are connected and that $\nu$ is a transposition of adjacent elements. Suppose that for some $i < k$ we have $\nu = 12 \cdots (i-1)(i+1)i(i+2) \cdots k$. Theorem 7.36(1) shows that $\sigma^{(i)}[\sigma^{(i+1)}] \stackrel{cc}{\sim} \sigma^{(i+1)}[\sigma^{(i)}]$, and then from Theorem 7.36(2) we obtain the requested result, by setting $\sigma = \sigma^{(1)}[\sigma^{(2)}] \cdots [\sigma^{(i-1)}]$, $\rho = \sigma^{(i)}[\sigma^{(i+1)}]$, $\rho' = \sigma^{(i+1)}[\sigma^{(i)}]$, and $\tau = \sigma^{(i+2)}[\sigma^{(i+3)}] \cdots [\sigma^{(k)}]$. □

**Theorem 7.38** *Let $\sigma^{(1)}, \ldots, \sigma^{(k)}$ be a k-tuple of noncrossing set partitions, let $s < k$ be an index such that the set partition $\sigma_s$ is empty, or connected, or contains only singleton blocks. Then the set partition $\sigma = 1[\sigma^{(1)}]1 \cdots [\sigma^{(k)}]1$ is cc-equivalent to $\sigma' = 1[\sigma^{(1)}]1 \cdots 1[\sigma^{(s-1)}]1[\sigma^{(s+1)}]1[\sigma^{(s)}]1[\sigma^{(s+2)}]1 \cdots 1[\sigma^{(k)}]1$.*

**Proof** Let $NC_{n,p}$ be the set of noncrossing set partitions of $NC_n$ with exactly $p$ components. We proceed by induction. Fix an integer $n$, and suppose that for every $n' < n$ and for every $p$, $|NC_{n',p}(\sigma)| = |NC_{n',p}(\sigma')|$. Let $g$ be a bijection from $NC_{n'}(\sigma)$ to $NC_{n'}(\sigma')$ preserving the number of components. Without loss of generality, we assume that $g$ has the property that $g(\pi) = \pi$ for any set partition $\pi$ that avoids both $\sigma$ and $\sigma'$. So to complete the induction step, we need to find a bijection $f$ mapping $NC_n(\sigma)$ to $NC_n(\sigma')$ such that it preserves the number of components.

Let $\pi \in NC_n(\sigma)$. If $\pi$ is disconnected, it can be written as $\pi = \pi^{(1)}[\pi^{(2)}] \cdots [\pi^{(m)}]$ with $m > 1$ and $\pi^{(i)}$ connected. Thus, we define $f(\pi)$ to be $g(\pi^{(1)})[g(\pi^{(2)})] \cdots [g(\pi^{(m)})]$. Clearly, $f$ satisfies all the claimed properties. Now, assume that $\pi$ is connected, which implies that $\pi$ can be uniquely written as $\pi = 1[\pi^{(1)}]1[\pi^{(2)}]1 \cdots 1[\pi^{(m)}]1$ for some $\sigma$-avoiding noncrossing set partitions $\pi^{(i)}$.

To simplify our proof, we use the following terminology. For a set partition $\rho$, an occurrence $I = (i_1, \ldots, i_\ell)$ of $\rho$ in $\pi$ is a *top-level occurrence* if it maps the elements of the first block of $\rho$ to the elements of the first block of $\pi$. Otherwise, we say that $I$ is a *deep occurrence*. Note that if $\rho$ is connected, then any deep occurrence of $\rho$ in $\pi$ must correspond to an occurrence of $\rho$ in one of the set partitions $\pi^{(1)}, \ldots, \pi^{(m)}$. For integers $1 \leq i \leq j \leq m+1$, define $\pi(i,j) = 1[\pi^{(i)}]1 \cdots 1[\pi^{(j-1)}]1$. For an integer $i \leq m+1$, define $\pi(\leq i) = \pi(1,i)$ and $\pi(\geq i) = \pi(i, m+1)$. We apply analogous notation for other connected partitions as well. Let $\overline{\pi}^{(i)} = g(\pi^{(i)})$ and $\overline{\pi} = 1[\overline{\pi}^{(1)}]1 \cdots 1[\overline{\pi}^{(m)}]1$.

Now we are ready to prove our claim. By induction hypothesis, for any $i \in [m]$, $\overline{\pi}^{(i)}$ is $\sigma'$-avoiding and $\overline{\pi}^{(i)}$ contains $\sigma$ if and only if $\pi^{(i)}$ contains $\sigma'$.

Consequently, $\overline{\pi}$ has no deep occurrence of $\sigma'$ and $\overline{\pi}$ has a deep occurrence of $\sigma$ if and only if $\pi$ has a deep occurrence of $\sigma'$. Using the fact that $\overline{\pi}^{(i)} = \pi^{(i)}$ whenever $\pi^{(i)}$ avoids both $\sigma$ and $\sigma'$, we see that for any $h, i \in [m+1]$, $\pi(h,i)$ has a top-level occurrence of $1[\sigma^{(j)}]1$ if and only if $\overline{\pi}(h,i)$ does. Thus, $\overline{\pi}$ has no top-level occurrence of $\sigma$, and $\overline{\pi}$ has a top-level occurrence of $\sigma'$ if and only if $\pi$ does.

Fix $a \in [m+1]$ be the smallest index such that $\overline{\pi}(\leq a)$ has a top-level occurrence of $\sigma(\leq s)$, and fix $b \in [m+1]$ be the largest index such that $\overline{\pi}(\geq b)$ has a top-level occurrence of $\sigma(\geq s+2)$. If such $a$ or $b$ do not exist, or if $a+2 > b$, then $\overline{\pi}$ has no top-level occurrence of either $\sigma$ or $\sigma'$, and we define $f(\pi) = \overline{\pi}$. Thus, we assume that $a + 2 \leq b$.

Let $c$ be the smallest integer from $\{a+1, a+2, \ldots, b\}$ such that $\overline{\pi}(a,c)$ has a top-level occurrence of $1[\sigma_s]1$. If no such $c$ exists, we again define $f(\pi) = \overline{\pi}$. Otherwise, define a set partition $\widehat{\pi} = 1[\widehat{\pi}^{(1)}]1 \cdots [\widehat{\pi}^{(m)}]1$, by setting

$$\widehat{\pi}(\leq a) = \overline{\pi}(\leq a), \quad \widehat{\pi}(\geq b) = \overline{\pi}(\geq b), \quad \widehat{\pi}(a, a+b-c) = \overline{\pi}(c, b),$$
$$\widehat{\pi}(a+b-c, b) = 1[\overline{\pi}^{(c-1)}]1[\overline{\pi}^{(c-2)}]1 \ldots 1[\overline{\pi}^{(a+1)}]1[\overline{\pi}^{(a)}]1.$$

Note that $\widehat{\pi}(a+b-c, b)$ has a top-level occurrence of $1[\sigma_s]1$, while $\widehat{\pi}(a+b-c+1, b)$ does not. (Why?) Since $\overline{\pi}$ has no top-level occurrence of $\sigma$, we know that $\overline{\pi}(c,b)$ has no top-level occurrence of $1[\sigma_{s+1}]1$. This shows that $\widehat{\pi}$ has no top-level occurrence of $\sigma'$, and therefore $\widehat{\pi}$ is a $\sigma'$-avoiding set partition. We then define $f(\pi) = \widehat{\pi}$. It is not difficult to see that $f$ is a bijection between the sets $\mathrm{NC}_n(\sigma)$ and $\mathrm{NC}_n(\sigma')$ that preserves the number of components. □

By using the main arguments in the proof of Theorem 7.38, namely the induction and the terminology of top-level occurrences, we obtain another result (for full details, we refer the reader to [149]).

**Theorem 7.39** *Let $\sigma^{(1)}, \ldots, \sigma^{(k)}$ be a $k$-tuple of noncrossing set partitions, let $s < k$ be an index, and let $\sigma'$ be a set partition cc-equivalent to $\sigma^{(s)}$. Then the pattern $\sigma = 1[\sigma^{(1)}]1 \cdots 1[\sigma^{(k)}]1$ is cc-equivalent to*

$$\sigma' = 1[\sigma^{(1)}]1 \cdots 1[\sigma^{(s-1)}]1[\sigma']1[\sigma^{(s+1)}]1 \cdots 1[\sigma^{(k)}]1.$$

### 7.2.2 Generating Functions

Now, we will use generating function techniques as tools in our proofs. Let us therefore introduce the following notation.

**Definition 7.40** *For a set partition $\tau$, we denote the generating function for the number of set partitions of $\mathrm{NC}_n(\tau)$ by $NCP_\tau(x) = P_{1212,\tau}(x)$, and we denote the generating function of the set of nonempty connected set partitions of $\mathrm{NC}_n(\tau)$ by $CP_\tau(x)$.*

**Theorem 7.41** *Let $\sigma$ and $\tau$ be two possibly empty connected set partitions. Then $1[\sigma]1[\tau] \stackrel{cc}{\sim} 1[\tau]1[\sigma]$.*

**Proof** Here we only present the basic idea of the proof; for full details, see [149]. The first step in the proof presents a formulation the statement of the theorem in terms of the generating function, as follows. First consider the case where $\sigma, \tau \neq \emptyset$. Let $G_{\sigma,\tau}(x)$ denote the generating function of noncrossing set partitions that avoid $1[\sigma]1[\tau]$ but contain $\sigma[\tau]$, that is,

$$G_{\sigma,\tau}(x) = NCP_{1[\sigma]1[\tau]}(x) - NCP_{\sigma[\tau]}(x).$$

Theorem 7.36 shows that $\sigma[\tau]$ and $\tau[\sigma]$ are nc-equivalent, that is,

$$NCP_{\sigma[\tau]}(x) = NCP_{\tau[\sigma]}(x).$$

Thus, to show that $1[\sigma]1[\tau]$ is nc-equivalent to $1[\tau]1[\sigma]$, it is enough to prove that $G_{\sigma,\tau}(x) = G_{\tau,\sigma}(x)$. We will derive a formula for $G_{\sigma,\tau}(x)$ from which the previous identity will easily follow. To do so, we introduce a special type of set partition. We say that a set partition $\pi$ is $\rho$-*minimal* if it is connected, noncrossing, contains $\rho$, but avoids $1[\rho]1$. Let $M_\rho(x)$ be the generating function of the set of $\rho$-minimal set partitions. If $\rho$ is a connected set partition, a noncrossing set partition $\pi$ avoids $\rho$ if and only if each component of $\pi$ avoids $\rho$. In particular, we have the identity

$$NCP_\rho(x) = \frac{1}{1 - CP_\rho(x)}. \tag{7.2}$$

With some work it can be shown that the generating function $G_{\sigma,\tau}(x)$ is given by

$$G_{\sigma,\tau}(x) = \frac{M_\sigma(x)Z(x)}{1 - CP_\sigma(x)Z(x)} \frac{1}{1 - CP_{1[\sigma]1}(x)Z(x)} \frac{M_\tau(x)}{1 - CP_\tau(x)}, \tag{7.3}$$

where

$$Z(x) = \frac{1}{1 - xNCP_\tau(x)} = \frac{1}{1 - \frac{x}{1 - CP_\tau(x)}}.$$

Using the identity

$$CP_{1[\rho]1}(x) = x + \frac{x^2}{1 - x - CP_\rho(x)} = \frac{x}{1 - \frac{x}{1 - CP_\rho(x)}},$$

which is valid for any connected noncrossing partition $\rho$, (7.3) simplifies into

$$G_{\sigma,\tau}(x) = \frac{M_\sigma(x)M_\tau(x)}{1 - \frac{x(1-CP_\sigma(x))(1-CP_\tau(x))}{(1-x-CP_\sigma(x))(1-x-CP_\tau(x))}}$$

$$\cdot \frac{1}{1 - x - CP_\sigma(x) - CP_\tau(x) + CP_\sigma(x)CP_\tau(x)},$$

with the first fraction in the above expression being equal to

$$(1 - CP_{1[\sigma]1}(x)Z(x))^{-1}$$

and the second being equal to

$$Z(x)(1 - CP_\sigma(x)Z(x))^{-1}(1 - CP_\tau(x))^{-1}.$$

This implies that $G_{\sigma,\tau}(x) = G_{\tau,\sigma}(x)$, which completes the proof for the case when both $\sigma$ and $\tau$ are nonempty.

Similar arguments have been used to show that the theorem holds when $\sigma$ or $\tau$ is empty. □

Note that in the previous theorem, the assumption that $\sigma$ and $\tau$ are connected is necessary, as shown, for instance, by the two patterns $1[1]1[12] = 12134$ and $1[12]1[1] = 12314$, which are not nc-equivalent. Also, nc-equivalence in the conclusion cannot in general be replaced with cc-equivalence. For example, taking $\sigma$ empty and $\tau = 1$, we see that $1[\sigma]1[\tau] = 112$, while $1[\tau]1[\sigma] = 121$. Since 112 and 121 do not have the same number of components, it is easy to see that they cannot be cc-equivalent.

**Theorem 7.42** *Let $\sigma$ be a connected set partition, and let $\tau = 1[\rho]$ for some set partition $\rho$. If $\sigma \stackrel{nc}{\sim} \tau$ then $1[\sigma] \stackrel{nc}{\sim} \tau 1 = 1[\rho]1$.*

**Proof** By assumption, we have $NCP_\sigma(x) = NCP_\tau(x)$. Note that a set partition $\pi$ avoids $1[\sigma]$ if and only if it can be written as $1[\pi^{(1)}]1[\pi^{(2)}]1 \cdots 1[\pi^{(k)}]$ for some $k$, where each $\pi^{(i)}$ is a $\sigma$-avoiding set partition ($\sigma$ is connected). Therefore, we have the identity

$$NCP_{1[\sigma]}(x) = \frac{1}{1 - xNCP_\sigma(x)}.$$

Consider now the set partition $\tau 1 = 1[\rho]1$. Since this set partition is connected, we see that $\pi$ avoids $\tau 1$ if and only if each component of $\pi$ avoids $\tau 1$. Moreover, a connected set partition $\pi = \pi_1 \cdots \pi_n$ avoids $\tau 1$ if and only if $\pi_1 \cdots \pi_{n-1}$ avoids $\tau$. This implies

$$NCP_{\tau 1}(x) = \frac{1}{1 - xNCP_\tau(x)}.$$

Since $NCP_\sigma(x) = NCP_\tau(x)$, we get that $NCP_{1[\sigma]}(x) = NCP_{\tau 1}(x)$. □

**Example 7.43** *Theorem 7.42 for $\sigma = 11$ and $\tau = 12$ with the fact that $\sigma \stackrel{nc}{\sim} \tau$, we imply that $1[\sigma] = 122 \stackrel{nc}{\sim} \tau 1 = 121$. Applying the theorem again to this new pair of patterns, we obtain that $1221 \stackrel{nc}{\sim} 1232$, and a third application reveals that $12332 \stackrel{nc}{\sim} 12321$.*

Generalizing this example into a straightforward induction argument, we derive the next result.

**Corollary 7.44** *For any $k$,*
$$12\cdots(k-1)kk(k-1)\cdots32 \overset{nc}{\sim} 12\cdots(k-1)k(k-1)\cdots21,$$
$$12\cdots(k-1)kk(k-1)\cdots21 \overset{nc}{\sim} 12\cdots k(k+1)k\cdots32.$$

Note that the set partitions avoiding $12\cdots(k-1)k(k-1)\cdots21$ are precisely those that do not have a $k$-tuple of pairwise nested blocks.

**Theorem 7.45** *We have $12333 \overset{nc}{\sim} 12321$.*

**Proof** By (7.2), for a connected partition $\pi$, $NC_\pi(x) = \frac{1}{1-CP_\pi(x)}$, and for arbitrary $\tau$, we have
$$CP_{1[\tau]1}(x) = \frac{x}{1-xNCP_\tau(x)}.$$

Combining these two identities and simplifying, we deduce that
$$NCP_{12321}(x) = \frac{1-3x+x^2}{(1-x)(1-3x)} = 1 + \sum_{n\geq 1} \frac{3^{n-1}+1}{2} x^n.$$

Let us consider the pattern $\tau = 12333$. The generating function for the empty set partition together with those that have a single block is, of course, $1/(1-x)$. Also, a noncrossing set partition with at least two blocks avoids $\tau$ if and only if it has a decomposition of the form
$$11\cdots 12[\rho^{(1)}]2[\rho^{(2)}]2\cdots 2[\rho^{(k)}]1[\sigma^{(1)}]1[\sigma^{(2)}]1\cdots 1[\sigma^{(\ell)}]$$
for some $k \geq 1$ and $\ell \geq 0$, where the $\rho^{(i)}$ and $\sigma^{(j)}$ are 111-avoiding noncrossing set partitions. By Example 2.65 we have
$$NCP_{111}(x) = 1 + xNCP_{111}(x) + (xNCP_{111}(x))^2.$$

Therefore,
$$NCP_{12333}(x) = \frac{1}{1-x} + \frac{x^2 NCP_{111}(x)}{(1-x)(1-xNCP_{111}(x))^2},$$
from which the result easily follows. □

We remark that the counting of 12333-avoiding noncrossing set partitions (and therefore also 12321-avoiding noncrossing set partitions) has been encountered before in different contexts (see [31, 296] and [327, Sequence A124302]).

Again, by Theorem 7.42 we can extend the equivalence $12333 \overset{nc}{\sim} 12321$ to an infinite sequence of equivalences.

**Corollary 7.46** *For every $k \geq 3$,*
$$12\cdots k(k+1)k(k-1)\cdots 2 \overset{nc}{\sim} 12\cdots(k-1)kkk(k-2)\cdots 1,$$
$$12\cdots(k-1)k(k-1)\cdots 1 \overset{nc}{\sim} 12\cdots kkk(k-2)\cdots 2.$$

### 7.2.3 CC-Equivalences of Patterns of Size Four

As an application of the above theorems, one may completely identify the Wilf-equivalence classes corresponding to the pairs $(1212, \tau)$, where $\tau$ is a noncrossing set partitions of size four.

**Table 7.3**: Nonsingleton cc-equivalences of patterns of size four

| CC-Equivalence | Reference |
|---|---|
| $1112 \stackrel{nc}{\sim} 1222$ | Theorem 7.37, [232] |
| $1211 \stackrel{nc}{\sim} 1121$ | Theorem 7.38, [235] |
| $1123 \stackrel{nc}{\sim} 1223 \stackrel{nc}{\sim} 1233$ | Theorem 7.37 |
| $1232 \stackrel{nc}{\sim} 1213 \stackrel{nc}{\sim} 1221 \stackrel{nc}{\sim} 1122$ | Theorem 7.37, Theorem 7.41 Corollary 7.44, [238] |

Some results of this table have been investigated by different research papers:

- The cc-equivalence $1122 \stackrel{nc}{\sim} 1213$ has been proved by Mansour and Shattuck [238]. In particular, they showed that the number of 1122-avoiding (1213-avoiding) noncrossing set partitions of $[n]$ is given by

$$\mathrm{Fib}_{2n-2}$$

for all $n \geq 0$ (see Exercise 7.10). Moreover, they counted the set partitions of $\mathcal{P}_n(1212, 1221)$ and set partitions of $\mathcal{P}_n(1212, 1232)$ according to several statistics.

- The cc-equivalence $1211 \stackrel{nc}{\sim} 1121$ has shown by Mansour and Shattuck [235]. In particular, they found that the generating function for the number of 1211-avoiding (1121-avoiding) noncrossing set partitions of $[n]$ is given by

$$\frac{(1-x)^2 - \sqrt{1 - 4x + 2x^2 + x^4}}{2x^2}.$$

In addition, they found another three classes, namely $\mathcal{P}_n(1121, 1221)$, $\mathcal{P}_n(1112, 1123)$, and $\mathcal{P}_n(1122, 1123)$, enumerated by the same generating function.

- The cc-equivalence $1112 \stackrel{nc}{\sim} 1222$ has been found by Mansour and Shattuck [232], as has been described in Section 7.5.

### 7.2.4 CC-Equivalences of Patterns of Size Five

As an application of the above theorems, one may completely identify the Wilf-equivalence classes corresponding to $(1212, \tau)$, where $\tau$ is a noncrossing set partition of size five; see Table 7.4.

Table 7.4: Nonsingleton cc-equivalences of patterns of size five

| CC-Equivalence | References |
|---|---|
| $12222 \overset{nc}{\sim} 11112$ | Theorem 7.37 |
| $12211 \overset{nc}{\sim} 11221$ | Theorem 7.38 |
| $11231 \overset{nc}{\sim} 12311$ | Theorem 7.38 |
| $12342 \overset{nc}{\sim} 12314$ | Theorem 7.37 |
| $12331 \overset{nc}{\sim} 12231$ | Theorem 7.39 and Theorem 7.37 |
| $12331 \overset{nc}{\sim} 12231$ | Theorem 7.39 and Theorem 7.37 |
| $11121 \overset{nc}{\sim} 11211 \overset{nc}{\sim} 12111$ | Theorem 7.38 |
| $11122 \overset{nc}{\sim} 12221 \overset{nc}{\sim} 11222$ | Theorem 7.37 and Theorem 7.41 |
| $12343 \overset{nc}{\sim} 12134 \overset{nc}{\sim} 12324$ | Theorem 7.37 |
| $12233 \overset{nc}{\sim} 11233 \overset{nc}{\sim} 11223$ | Theorem 7.37 |
| $12113 \overset{nc}{\sim} 12322 \overset{nc}{\sim} 12232 \overset{nc}{\sim} 11213$ | Theorem 7.37 and Theorem 7.38 |
| $12234 \overset{nc}{\sim} 12334 \overset{nc}{\sim} 12344 \overset{nc}{\sim} 11234$ | Theorem 7.37 |
| $12332 \overset{nc}{\sim} 12333 \overset{nc}{\sim} 12133 \overset{nc}{\sim} 11123$ $\overset{nc}{\sim} 12213 \overset{nc}{\sim} 12321 \overset{nc}{\sim} 12223 \overset{nc}{\sim} 11232$ | Theorem 7.37, Theorem 7.41, Corollary 7.44 and Corollary 7.46 |

Moreover, the number of noncrossing set partitions of $[n]$ that avoid a pattern of size four, five and six is given in Tables I.2–I.4, where the results of this section can show all the Wilf cc-equivalences as described in this table.

## 7.3 General Equivalences

In the study of multi-avoidance, the concept of strong equivalence (see Definition 6.2) becomes relevant through the following simple result.

**Theorem 7.47** *Let $\rho$ be a pattern of the form $12\cdots(k-1)k^a$ for some $a,k \geq 1$. Suppose that the two patterns $\sigma \sim_s \tau$, then $(\rho,\sigma) \sim (\rho,\tau)$.*

**Proof** Note that a set partition avoids $\rho$ if and only if it either has fewer than $k$ blocks, or for every $i \geq k$, its $i$-th block has size less than $a$. In other words, avoidance of $\rho$ can be characterized as a property of block sizes. Now assume that $\sigma \sim_s \tau$ via a bijection $f$. Since $f$ preserves the sizes of each block, we know that a set partition $\pi$ avoids $\rho$ if and only if $f(\pi)$ avoids $\rho$. In particular, $f$ maps the set of $(\rho,\sigma)$-avoiding set partitions bijectively to the set of $(\rho,\tau)$-avoiding set partitions. □

We can use a similar argument for bijections that preserve the number of blocks.

**Definition 7.48** *Let us say that two patterns $\sigma$ and $\tau$ are* **block-count equivalent** *if there is a bijection $f$ between $\sigma$-avoiding and $\tau$-avoiding set partitions, with the property that for any $\sigma$-avoiding set partition $\pi$, $\pi$ has the same size and the same number of blocks as $f(\pi)$.*

Obviously, two set partitions that are strong equivalent are also block-count equivalent. But there are examples of patterns that are block-count equivalent but not strong equivalent; Try to find a such example in Chapter 6! Corollary 6.27 and Theorems 6.55 and 6.114 give the following result.

**Theorem 7.49** *We have*
  *(1) If $k$ is an integer, $\tau$ is a set partition, and $\theta$ is any sequence of numbers from the set $[k-1]$, then the following pairs of set partitions are block-count equivalent: $12 \cdots k[\tau]\theta$ and $12 \cdots k[\tau](\theta+1)$; $12 \cdots k\theta[\tau]$ and $12 \cdots k(\theta+1)[\tau]$.*
  *(2) If two set partitions $\sigma$ and $\tau$ are block-count equivalent, then $1[\sigma]$ and $1[\tau+]$ are block-count equivalent as well.*
  *(3) The two patterns $12112$ and $12212$ are block-count equivalent.*

Here is how we may use block-count equivalence for our purposes.

**Theorem 7.50** *Let $k \geq 1$ be an integer. If $\sigma$ and $\tau$ are block-count equivalent, then the pairs of patterns $(12 \cdots k, \sigma)$ and $(12 \cdots k, \tau)$ are equivalent.*

**Proof** Clearly, a set partition avoids $12 \cdots k$ if and only if it has at most $k-1$ blocks. If $\sigma$ and $\tau$ are block-count equivalent via a bijection $f$, then $f$ is also a bijection between the sets $\mathcal{P}_n(12 \cdots k, \sigma)$ and $\mathcal{P}_n(12 \cdots k, \tau)$. □

Now we provide several general results applicable to infinite families of pattern-avoiding classes. Our first argument involves patterns $\sigma = 1^a 2 1^b$ with $a \geq 1$ and $b \geq 0$. Let $k = a + b + 1$ be the size of $\sigma$.

**Definition 7.51** *For a set of patterns $T$, let $T'$ denote the set $\{1[\tau], \tau \in T\}$. For a set of patterns $R$, let $\mathcal{P}_n(\sigma, R)$ denote the set of set partitions of size $n$ that avoid the pattern $\sigma$ as well as all the patterns in $R$, and let $\mathcal{P}_n(\sigma, R; i)$ denote the set of set partitions in $\mathcal{P}_n(\sigma, R)$ whose first block has size $i$. Let $P_n(\sigma, R)$ and $P_n(\sigma, R; i)$ denote the cardinality of $\mathcal{P}_n(\sigma, R)$ and $\mathcal{P}_n(\sigma, R; i)$, respectively.*

Then by considering the number elements in the first block, we obtain the following result (see [149]).

**Lemma 7.52** *Let $\sigma = 1^a 2 1^b$ with $a \geq 1$ and $b \geq 0$. For any set of patterns $T$,*

$$P_n(\sigma, T') = \sum_{i=1}^{k-2} P_{n-i}(\sigma, T) \binom{n-1}{i-1} + \sum_{i=k-1}^{n} P_{n-i}(\sigma, T) \binom{n-i+k-3}{k-3}.$$

Note that the formula in the previous lemma does not depend on $a$ and $b$. Thus, we can state the following result.

**Corollary 7.53** *Let $\sigma = 1^a 21^b, \rho = 1^c 21^d \in \mathcal{P}_k$. Let $T$ and $U$ be two sets of patterns. If $\{\sigma\} \cup T \sim \{\rho\} \cup U$ then $\{\sigma\} \cup T' \sim \{\rho\} \cup U'$.*

**Proposition 7.54** *Let $T$ be any set of patterns. Then for every $n \geq 1$,*

$$P_n(T') = \sum_{k=0}^{n-1} \binom{n-1}{k} P_k(T).$$

*Consequently, if $T$ and $R$ are sets of patterns such that $T \sim R$, then $T' \sim R'$.*

**Proof** It is enough to observe that a partition $\pi$ avoids $T'$ if and only if the subpartition of $\pi$ obtained by removing the first block of $\pi$ avoids $T$. Therefore, there are exactly $\binom{n-1}{k} P_k(T)$ partitions in $\mathcal{P}_n(T')$ whose first block has size $n - k$. □

**Corollary 7.55** *We have $(1222, 1234) \sim (1211, 1234)$.*

**Proof** By Proposition 7.54 and Lemma 7.52, we have

$$P_n(1222, 1234) = \sum_{k=0}^{n-1} \binom{n-1}{k} P_k(111, 123)$$

$$= 1 + (n-1) + 2\binom{n-1}{2} + 3\binom{n-1}{3} + 3\binom{n-1}{4},$$

$$P_n(1211, 1234) = P_{n-1}(123, 1211) + (n-1)P_{n-2}(123, 1211)$$

$$+ 1 + \sum_{k=1}^{n-3} P_k(123, 1211)(k+1).$$

From Theorem 7.15, we deduce that $P_n(123, 1211) = n + \binom{n-1}{2}$. Substituting into the above expression and simplifying shows that both $P_n(1211, 1234)$ and $P_n(1222, 1234)$ are equal to $(n^4 - 6n^3 + 19n^2 - 22n + 16)/8$ for $n \geq 1$. □

Reasoning similar to that used in the proof of Lemma 7.52 above yields the following result, whose proof we omit.

**Proposition 7.56** *Let $T$ be any set of patterns. Let $a_n(R)$ (respectively, $b_n(R)$) denote the number of set partitions of $[n]$ that avoid all the patterns in $R$ as well as both 1112 and 1121 (respectively, 1121 and 1211). Then, for all $n \geq 4$,*

$$a_n(T') = a_{n-1}(T) + (n-1)a_{n-2}(T) + a_{n-3}(T) + a_{n-4}(T) + \cdots + a_0(T),$$
$$b_n(T') = b_{n-1}(T) + (n-1)b_{n-2}(T) + b_{n-3}(T) + b_{n-4}(T) + \cdots + b_0(T).$$

## 7.4  Two Patterns of Size Four

In this section we focus on pair of patterns of size four, where sometimes we obtain the results by considering one pattern of size four and another general pattern of a specific form. For instance, in the next theorem, we consider the case of avoiding 1213 and another pattern of the form $1[\tau]$.

**Theorem 7.57** *Let $\rho = 1[\sigma]$ be any pattern of size at least two. Then the generating function $P_{1213,\rho}(x)$ is given by*

$$1 + \frac{x}{1-x} P_{1213,\sigma}(x) + \frac{x^2}{(1-x)(1-2x)}(P_{1213,\sigma}(x) - 1).$$

*Moreover, if $(1213, \tau) \sim (1213, \tau')$, then $(1213, 1[\tau]) \sim (1213, 1[\tau'])$.*

**Proof** Let us write an equation for the generating function $H_\rho(x)$. For each nonempty set partition $\pi \in \mathcal{P}_n(1213, \rho)$, either the first block of $\pi$ contains only 1 or it contains 1 and 2 or it contains 1 and at least one other element, not 2. The contributions from the first two cases are $xP_{1213,\sigma}(x)$ and $x(P_{1213,\rho}(x) - 1)$, respectively. Each set partition $\pi$ in the last case must have the form $12\pi'1\alpha$, where $\pi'$ does not contain 1 and $\alpha$ is a (possibly empty) word on $\{1, 2\}$, which implies a contribution of $\frac{x^2}{1-2x}(P_{1213,\sigma}(x) - 1)$. Hence,

$$P_{1213,\rho}(x) = 1 - x + xP_{1213,\sigma}(x) + xP_{1213,\rho}(x) + \frac{x^2}{1-2x}(P_{1213,\sigma}(x) - 1),$$

which completes the proof. □

**Example 7.58** *We consider some specific examples. Theorem 7.57 together with the fact that $P_{1213,112}(x) = \frac{1-x}{1-2x}$ give*

$$P_{1213,1223}(x) = 1 + \frac{x(1 - 3x + 3x^2)}{(1-2x)^2(1-x)}.$$

*Since $112 \sim 123$, we then have $(1213, 1223) \sim (1213, 1234)$, by Theorem 7.57. Using the same reasoning as in the proof of Theorem 7.57 gives*

- $P_{1213,1231}(x) = 1 + xP_{1213,1231}(x) + x(P_{1213,1231}(x) - 1) + \frac{x^3}{(1-x)(1-2x)}$, *which implies $P_{1213,1231}(x) = P_{1213,1223}(x)$, whence $(1213, 1223) \sim (1213, 1231)$;*

- $P_{1213,1221}(x) = 1 + xP_{1213,1221}(x) + x(P_{1213,1221}(x) - 1) + \frac{x^3}{(1-x)^3}$, *which implies*

$$P_{1213,1221}(x) = \frac{1 - 4x + 6x^2 - 3x^3 + x^4}{(1-x)^3(1-2x)};$$

- $P_{1213,1232}(x) = 1 + xP_{1213,121}(x) + x(P_{1213,1232}(x) - 1) + \frac{x^3}{(1-x)^2(1-2x)}$, whence $P_{1213,1232}(x) = P_{1213,1221}(x)$.

*Similarly, we obtain*

$$P_{1213,1233}(x) = P_{1213,1221}(x) \text{ and } P_{1213,1121}(x) = \frac{(1-x-x^2)(1-x)^2}{1+4x+4x^2-2x^4}.$$

**Example 7.59** Let $L_\sigma(x) = P_{1231,\sigma}(x)$. In this example, we study the generating function $L_\sigma(x)$ for a couple of cases.

First, we consider the case $\sigma = 1121$. Note that each nonempty set partition $\pi$ that avoids 1231 and 1121 can be expressed as $\pi = 11\cdots 1[\pi']$, $\pi = 122\cdots 211\cdots 1\pi''$, or $\pi = 122\cdots 211\cdots 122\cdots 2[\pi''']$, where $\pi''$ starts with 3 if nonempty. Thus, the generating function $L_{1121}(x)$ satisfies

$$L_{1121}(x) = 1 + \frac{x}{1-x} L_{1121}(x) + \frac{x^2}{1-x}(L_{1121}(x) - 1) + \frac{x^4}{(1-x)^3} L_{1121}(x),$$

*whence*

$$L_{1121}(x) = \frac{(1-x-x^2)(1-x)^2}{1+4x+4x^2-2x^4}.$$

Next, we consider the case $\sigma = 1232$. From the structure, each nonempty set partition $\pi$ that avoids 1231 and 1232 can be written as $\pi = 1[\pi']$ or as $\pi = 1\pi''$, where $\pi''$ is nonempty, or as $\pi = 122\cdots 21\alpha[\pi''']$, where $\alpha$ is a (possibly empty) word in $\{1,2\}$. Thus the generating function $L_{1232}(x)$ satisfies

$$L_{1232}(x) = 1 + xL_{121}(x) + x(L_{1232}(x) - 1) + \frac{x^3}{(1-x)(1-2x)} L_{121}(x),$$

*whence*

$$L_{1232}(x) = 1 + \frac{x(1 - 3x + 3x^2)}{(1-2x)^2(1-x)}.$$

The remaining results in this section are of a more specific nature and cover most of the equivalences in the table below left to be shown concerning the avoidance of two patterns of size four. We start with the following result, where we omit the proof (it is based on direct enumerations either by finding a recurrence relation, or by using generating function techniques) and leave it to the interested reader.

**Proposition 7.60** *Let* $n \geq 2$. *Then*

$$P_n(1221, 1232) = P_n(1221, 1223) = 1 + (n-1)2^{n-2},$$
$$P_n(1212, 1123) = P_n(1212, 1233)$$
$$= P_n(1212, 1223)$$
$$= P_n(1122, 1221) = 1 + (n-1)2^{n-2},$$
$$P_n(1123, 1234) = P_n(1122, 1233) = 2^{n-5}(n^2 - n + 14).$$

We now consider the case of avoiding the pair $(1122, 1223)$.

**Proposition 7.61** *We have*

$$P_{1122,1223}(x) = \frac{1 - 4x + 5x^2 - x^3}{(1-x)(1-2x)^2}.$$

**Proof** Let $f_{a_1 a_2 \cdots a_m} = f_{a_1 a_2 \cdots a_m}(x)$ denote the generating function for the number of members $\pi = \pi_1 \pi_2 \cdots \pi_n \in \mathcal{P}_n(1122, 1223)$, where $n \geq m$ such that $\pi_1 \pi_2 \cdots \pi_m = a_1 a_2 \cdots a_m$, and let $f^*_{a_1 a_2 \cdots a_m} = f^*_{a_1 a_2 \cdots a_m}(x)$ denote the generating function counting the same set partitions with the further restriction that $\max_{1 \leq i \leq n}(\pi_i) = \max_{1 \leq i \leq m}(a_i)$. By direct enumerations, one can obtain

$$F^*(x) := 1 + \sum_{k \geq 1} f^*_{12 \cdots k}(x) = \frac{1 - 4x + 5x^2 - 2x^3 + x^4}{(1-x)(1-2x)^2}. \quad (7.4)$$

Now, we consider the generating functions $\tilde{f}_{12\cdots k} = \tilde{f}_{12\cdots k}(x)$, $k \geq 1$, for the number of set partitions $\pi = \pi_1 \pi_2 \cdots \pi_n \in \mathcal{P}_n(1122, 1223)$ having length at least $k$ such that $\pi_1 \pi_2 \cdots \pi_k = 12 \cdots k$ and $\pi_{k+1} \leq k$ (if it exists). From the definitions, we have

$$\tilde{f}_{12\cdots k} = x^k + f_{12\cdots k1} + \sum_{j=2}^{k} f_{12\cdots kj}$$

$$= x^k + f_{12\cdots k1} + \sum_{j=2}^{k} x^{k-j+1} f^*_{12\cdots j}, \quad k \geq 1, \quad (7.5)$$

for if $2 \leq j \leq k$, then $f_{12\cdots kj} = x^{k-j+1} f^*_{12\cdots j}$, since the letters $j+1, j+2, \ldots, k$ can only appear once in a set partition enumerated by $f_{12\cdots kj}$, with no letters greater than $k$ occurring. Furthermore, we have

$$f_{12\cdots k1} = \frac{x^{k+1}}{(1-x)^{k-1}(1-2x)}, \quad k \geq 1, \quad (7.6)$$

the enumerated set partition $\pi$ having the form

$$\pi = 1\alpha_0 1 \alpha_1 \alpha_k \alpha_{k-1} \cdots \alpha_2,$$

where $\alpha_0 = 23 \cdots k$, $\alpha_1$ is a (possibly empty) word obtained by replacing the 2's occurring in a word in $\{1, 2\}$ successively with the letters $k+1, k+2, \ldots$, and $\alpha_i$, $2 \leq i \leq k$, is a sequence which is either empty or is nonempty and of the form $i1^a$ for some $a \geq 0$.

Summing (7.5) over $k \geq 1$, and using (7.6) then gives

$$P_{1122,1223}(x) = 1 + \sum_{k \geq 1} \widetilde{f}_{12\cdots k}(x)$$

$$= \frac{1}{1-x} + \sum_{k \geq 1} \frac{x^{k+1}}{(1-x)^{k-1}(1-2x)} + \sum_{j \geq 2} f^*_{12\cdots j}(x) \sum_{k \geq j} x^{k-j+1}$$

$$== \frac{1}{1-x} + \frac{x^2(1-x)}{(1-2x)^2} + \frac{x}{1-x}\left(F^*(x) - \frac{x}{1-x} - 1\right),$$

which yields the requested result, by (7.4). □

**Proposition 7.62** *If* $n \geq 0$, *then* $P_n(1123, 1233) = P_n(1231, 1233) = P_n(1123, 1232)$. *In addition, we have*

$$P_{1122,1232}(x) = \frac{1 - 4x + 6x^2 - 3x^3 + x^4}{(1-x)^3(1-2x)}.$$

**Proof** See Exercise 7.17. □

**Proposition 7.63** *The generating functions* $P_{1123,1211}(x)$ *and* $P_{1123,1222}(x)$ *are given by*

$$\frac{(1-x^2)\sqrt{(1-x)(1-x-4x^2)}}{2x^2(1-3x+x^2)} - \frac{1 - 3x - 2x^2 + 14x^3 - 15x^4 + 3x^5}{2x^2(1-x)^2(1-3x+x^2)}$$

**Proof** See Exercise 7.18. □

Combining the results of previous sections yields a complete solution to the problem of identifying all of the equivalence classes of $(4,4)$-pairs. It should be observed that any pattern pair not represented in the table below belongs to a Wilf class of size one, such classes being determined by numerical evidence (note that there are $\binom{15}{2} - 84 = 21$ singleton classes in all).

**Table 7.5**: Nonsingleton equivalences of pair patterns of size four

| |
|---|
| $(1121,1232) \stackrel{7.52}{\sim} (1112,1223) \stackrel{7.52}{\sim} (1121,1223) \stackrel{7.52}{\sim} (1211,1232)$ $\stackrel{7.58,7.52}{\sim} (1213,1223) \stackrel{7.58}{\sim} (1213,1234) \stackrel{7.58}{\sim} (1213,1231) \stackrel{7.47}{\sim} (1231,1234)$ $\stackrel{7.15}{\sim} (1232,1234) \stackrel{7.15}{\sim} (1223,1234) \stackrel{7.15}{\sim} (1233,1234) \stackrel{7.54}{\sim} (1222,1233)$ $\stackrel{7.54}{\sim} (1223,1232) \stackrel{7.54}{\sim} (1223,1233) \stackrel{7.54}{\sim} (1232,1233) \stackrel{7.59,7.54}{\sim} (1231,1232)$ $\stackrel{7.60}{\sim} (1123,1212) \stackrel{7.60}{\sim} (1122,1221) \stackrel{7.60}{\sim} (1212,1223) \stackrel{7.61}{\sim} (1122,1223)$ $\stackrel{7.60}{\sim} (1221,1223) \stackrel{7.60}{\sim} (1221,1232) \stackrel{7.60}{\sim} (1212,1233) \stackrel{7.47}{\sim} (1221,1233)$ |
| $(1212,1232) \sim (1112,1213) \sim (1123,1223) \sim (1221,1231)$ $\sim (1123,1213) \sim (1212,1213) \sim (1212,1221) \sim (1222,1223)$ $\sim (1122,1212) \sim (1222,1232) \sim (1211,1231)$, see [238] |
| Continued on next page |

| |
|---|
| $(1112,1233) \stackrel{7.47}{\sim} (1211,1233) \stackrel{7.47}{\sim} (1121,1233) \stackrel{7.52}{\sim} (1112,1234)$ $\stackrel{7.47}{\sim} (1121,1234) \stackrel{7.47}{\sim} (1211,1234) \stackrel{7.55}{\sim} (1222,1234)$ |
| $(1213,1221) \stackrel{7.58}{\sim} (1213,1232) \stackrel{7.58}{\sim} (1213,1233) \stackrel{7.47}{\sim} (1231,1233)$ $\stackrel{7.62}{\sim} (1123,1232) \stackrel{7.62}{\sim} (1123, 1233) \stackrel{7.62}{\sim} (1122,1232)$ |
| $(1112,1123) \stackrel{7.80}{\sim} (1211,1212) \stackrel{7.80}{\sim} (1122,1123) \stackrel{7.80}{\sim} (1121,1221)$ $\stackrel{7.80}{\sim} (1121,1212)$ |
| $(1211,1221) \stackrel{7.65}{\sim} (1112,1212) \stackrel{7.65}{\sim} (1212,1222) \stackrel{7.65}{\sim} (1221,1222)$ |
| $(1211,1222) \stackrel{7.47}{\sim} (1121,1222) \stackrel{7.47}{\sim} (1112,1222)$ |
| $(1111,1121) \stackrel{7.47}{\sim} (1111,1211) \stackrel{7.47}{\sim} (1111,1112)$ |
| $(1212,1234) \stackrel{7.47}{\sim} (1221,1234)$ |
| $(1112,1232) \stackrel{7.52}{\sim} (1211,1223)$ |
| $(1222,1231) \stackrel{7.47}{\sim} (1213,1222)$ |
| $(1111,1223) \stackrel{7.47}{\sim} (1111,1232)$ |
| $(1123,1211) \stackrel{7.63,7.63}{\sim} (1123,1222)$ |
| $(1123,1234) \stackrel{7.60}{\sim} (1122, 1233)$ |
| $(1121,1231) \stackrel{7.58,7.59}{\sim} (1121,1213)$ |
| $(1111,1213) \stackrel{7.47}{\sim} (1111,1231)$ |
| $(1111,1212) \stackrel{7.47}{\sim} (1111,1221)$ |
| $(1121,1211) \stackrel{7.56}{\sim} (1112,1121)$ |

## 7.5 Left Motzkin Numbers

Define $\mathcal{R}_n$ to be the set $\mathcal{P}_n(111, 1212)$ and $\mathcal{M}_n$ to be the set of all Motzkin paths of size $n$; see Example 2.62. This example shows that $|\mathcal{R}_n|$ is given by the $n$-th Motzkin number. In this section, we will be interested in lattice paths that are extension of Motzkin paths, namely "Motzkin left factors".

**Definition 7.64** *Let $\mathcal{L}_n$ denote the set of all paths of length $n$ using $u$, $d$, and $\ell$ steps starting from the origin and not dipping below the x-axis. Such paths are called* Motzkin left factors; *for example, see [2, Page 111] or [155, Page 9]. Let $L_n = |\mathcal{L}_{n-1}|$ if $n \geq 1$, with $L_0 = 1$.*

It is well known that the generating function $\sum_{n\geq 0} L_n x^n$ is given by

$$\frac{1 - 3x + \sqrt{1 - 2x - 3x^2}}{2(1 - 3x)} \qquad (7.7)$$

and the sequence $\{L_n\}_{n\geq 0}$ satisfies the relation

$$L_{n+1} = M_n + \sum_{k=0}^{n-1} M_k L_{n-k}, \qquad n \geq 1, \tag{7.8}$$

with $L_0 = L_1 = 1$.

In this section, following Mansour and Shattuck [232], we identify four classes of set partitions each avoiding a pair of classical patterns of length four and each enumerated by Motzkin left factors. The main result of this section can be formulated as follows.

**Theorem 7.65** *If $n \geq 0$, then $P_n(\pi, \pi') = L_n$ for the following sets $(\pi, \pi')$:*

*(i)* $(1222, 1212)$  *(ii)* $(1112, 1212)$  *(iii)* $(1211, 1221)$  *(iv)* $(1222, 1221)$.

This theorem has been proved in [232] as follows. The proofs of the first two cases are more or less combinatorial, while, in the last two, they are algebraic, based on solving functional equation by using the kernel method [15].

### 7.5.1  The Pairs $(1222, 1212)$ and $(1112, 1212)$

In this section, we count the set partitions either in $\mathcal{P}_n(1222, 1212)$ or $\mathcal{P}_n(1112, 1212)$.

**Proposition 7.66**  *Let $\mathcal{A}_n = \mathcal{P}_n(1222, 1212)$. Then for $n \geq 0$, $|\mathcal{A}_n| = L_n$.*

**Proof** Fix $n \geq 1$. We proceed with the proof by defining a bijection $g$ between the sets $\mathcal{A}_n$ and $\mathcal{L}_{n-1}$. Fix $f$ to be the bijection between the sets $\mathcal{R}_m$ and $\mathcal{M}_m$ as described in Example 2.65. Now, we define the bijection $g$. First observe that any set partition $\pi \in \mathcal{A}_n$ may be written as $\pi = 1[\pi^{(1)}]1[\pi^{(2)}]\cdots 1[\pi^{(r)}]$ for some $r \geq 1$ such that $\pi^{(i)} \in \mathcal{P}_{n_i}(111, 1212)$ for some $n_i \geq 0$ for all $i$. Then let $g(\pi) = f(\pi^{(1)})Uf(\pi^{(2)})\cdots Uf(\pi^{(r)})$. To reverse $g$, suppose $\alpha \in \mathcal{L}_{n-1}$ terminates at the point $(n-1, r-1)$ for some $r \geq 1$. Given $0 \leq i \leq r-1$, let $s_i$ denote the right-most step of $\alpha$ which either lies along the line $y = i$ as an $\ell$ or touches it from above as a $D$ or touches it from below as a $U$ (in the case when $i = 0$, only the first two conditions would apply). Write $\alpha = \alpha_0 \alpha_1 \cdots \alpha_{r-1}$, where $\alpha_0$ counts all steps of $\alpha$ up to and including $s_0$ and $\alpha_i$, $1 \leq i \leq r-1$, is the sequence of steps starting with the $U$ directly following step $s_{i-1}$ and ending at step $s_i$. So, $\alpha_i = U\alpha'_i$ if $i \geq 1$, with $\alpha'_i$ and $\alpha_0$ possibly empty Motzkin paths. Then define $g^{-1}(\alpha)$ by $1[f^{-1}(\alpha_0)]1[f^{-1}(\alpha_1)]\cdots 1[f^{-1}(\alpha_{r-1})]$.  □

The next result follows immediately from the above proposition and Theorem 7.36. In [232] has been suggested another proof that it is very similar to the proof of the above proposition.

**Proposition 7.67** *Let $\mathcal{B}_n = \mathcal{P}_n(1112, 1212)$. Then for $n \geq 0$, $|\mathcal{B}_n| = L_n$.*

## 7.5.2 The Pair $(1211, 1221)$

To deal with this case, we divide up the set partitions in question according to the size of the maximal increasing initial run. Actually, this idea has been used before in this book, and we will see that this idea can be used on other cases. Let $f_k(x)$ denote the generating function for the number of set partitions $\pi = \pi_1 \pi_2 \cdots \pi_n \in \mathcal{P}_n(1211, 1221)$, where $n \geq k$, such that $\pi_1 \pi_2 \cdots \pi_k = 12 \cdots k$ with $\pi_{k+1} \leq k$ (if there is a $(k+1)$-st letter).

**Lemma 7.68** *For all $k \geq 1$,*

$$f_k(x) = x^k + x^k \overline{f}_1(x) + \sum_{j=1}^{k-1} x^{j+1} \overline{f}_{k-j}(x),$$

*with initial condition $f_0(x) = 1$, where $\overline{f}_k(x) = \sum_{i \geq k} f_i(x)$.*

**Proof** Since a set partition with the maximal increasing run has size one just have either one letter, or start $11$, we have $f_1(x) = x + x\overline{f}_1(x)$. Thus the lemma holds when $k = 1$. Let $k \geq 2$; a set partition $\pi$ enumerated by $f_k(x)$ can be written as either (i) $12 \cdots k$, (ii) $12 \cdots k j \pi'$ with $1 \leq j \leq k-1$, or (iii) $12 \cdots kk\pi''$. The first case contributes $x^k$. In Case (ii), the word $\pi'$ contains no letters in $[j]$, for otherwise if it contained a letter in $[j-1]$, then there would be an occurrence of $1221$ and if it contained the letter $j$, then there would be an occurrence of $1211$. Thus, the letters $(j+1)(j+2) \cdots k\pi'$, taken together, reduce a set partition of the form enumerated by $\overline{f}_{k-j}(x)$. Thus, the contribution in this case toward the generating function $f_k(x)$ is $x^{j+1} \overline{f}_{k-j}(x)$. Similar arguments give a contribution of $x^k \overline{f}_1(x)$ for Case (iii). By combining these three cases, we complete the proof. □

**Proposition 7.69** *The number of set partitions of $\mathcal{P}_n(1211, 12221)$ is given by $L_n$.*

**Proof** Define the generating function $f(x,y) = \sum_{k \geq 0} f_k(x) y^k$. Multiplying the recurrence in statement of Lemma 7.68 by $y^k$ and summing over $k \geq 1$ gives

$$f(x,y) = 1 + \frac{xy}{1-xy} + \frac{xy}{1-xy} \overline{f}_1(x) + \sum_{k \geq 2} \left( \sum_{j=1}^{k-1} x^{j+1} \overline{f}_{k-j}(x) \right) y^k$$

$$= 1 + \frac{xy}{1-xy} + \frac{xy}{1-xy}(f(x,1) - 1) + \sum_{j \geq 1} x^{j+1} \sum_{k \geq j+1} \overline{f}_{k-j}(x) y^k$$

$$= 1 + \frac{xy}{1-xy} f(x,1) + \sum_{j \geq 1} x^{j+1} y^j \sum_{k \geq 1} \overline{f}_k(x) y^k.$$

By $\overline{f}_k(x) = \sum_{i \geq k} f_i(x)$, we obtain

$$f(x,y) = 1 + \frac{xy}{1-xy}f(x,1) + \frac{x^2y}{1-xy}\sum_{k \geq 1} y^k \sum_{i \geq k} f_i(x)$$

$$= 1 + \frac{xy}{1-xy}f(x,1) + \frac{x^2y}{1-xy}\sum_{i \geq 1} f_i(x) \sum_{k=1}^{i} y^k.$$

So

$$f(x,y) = 1 + \frac{xy}{1-xy}f(x,1) + \frac{x^2y^2}{(1-xy)(1-y)}\sum_{i \geq 1} f_i(x)(1-y^i)$$

$$= 1 + \frac{xy}{1-xy}f(x,1) + \frac{x^2y^2}{(1-xy)(1-y)}(f(x,1) - f(x,y)),$$

which implies

$$\left(1 + \frac{x^2y^2}{(1-xy)(1-y)}\right)f(x,y) = 1 + \frac{xy}{1-xy}\left(1 + \frac{xy}{1-y}\right)f(x,1). \quad (7.9)$$

By the *kernel method* (see [15]) techniques, if we assume that $y = y_0$ in (7.9), where $y_0$ satisfies $1 + \frac{x^2y_0^2}{(1-xy_0)(1-y_0)} = 0$, that is, $y_0 = \frac{1+x-\sqrt{1-2x-3x^2}}{2x(1+x)}$, then

$$\sum_{n \geq 0} P_n(1211, 1221)x^n = f(x,1) = \frac{1}{1 - \frac{xy_0}{1-xy_0}} = \frac{1 - 3x + \sqrt{1-2x-3x^2}}{2(1-3x)},$$

which implies the required result. □

### 7.5.3 The Pair $(1222, 1221)$

Once again, we divide up the set of set partitions in question according to the statistic which records the size of the maximal increasing initial run. Let $k \geq 1$, define $f_k(x)$ (respectively, $g_k(x)$) to be the generating function for the number of partitions $\pi \in \mathcal{P}_n$ having at least $k$ letters and avoiding the patterns 1222 (respectively, 111) and 1221) such that $\pi_1\pi_2\cdots\pi_k = 12\cdots k$ with $\pi_{k+1} \leq k$ (if there is a $(k+1)$-st letter). With similar arguments as in the proof of Lemma 7.68, one can obtain the following relations involving the generating functions $g_k(x)$ and $f_k(x)$ (see [232]).

**Lemma 7.70** *For $k \geq 1$,*

$$g_k(x) = x^k + \sum_{j=1}^{k} x^{j+1}\overline{g}_{k-j}(x),$$

$$f_k(x) = x^k + x\overline{f}_k(x) + \sum_{j=2}^{k} x^{j+1}\overline{g}_{k-j}(x)$$

with initial condition $g_0(x) = f_0(x) = 1$, where $\bar{g}_k(x) = \sum_{i \geq k} g_i(x)$ and $\bar{f}_k(x) = \sum_{i \geq k} f_i(x)$.

**Proposition 7.71** *The number of set partitions of $\mathcal{P}_n(1222, 12221)$ is given by $L_n$.*

**Proof** Define the generating functions $g(x,y) = \sum_{k \geq 0} g_k(x) y^k$ and $f(x,y) = \sum_{k \geq 0} f_k(x) y^k$. Lemma 7.70 gives

$$g(x,y) = \frac{1}{1-xy} + \sum_{k \geq 1} y^k \sum_{j=1}^{k} x^{j+1} \bar{g}_{k-j}(x)$$

$$= \frac{1}{1-xy} + \frac{x^2 y}{1-xy} \sum_{k \geq 0} y^k \bar{g}_k(x)$$

$$= \frac{1}{1-xy} + \frac{x^2 y}{(1-xy)(1-y)} (g(x,1) - yg(x,y)),$$

$$f(x,y) = \frac{1}{1-xy} + x \sum_{k \geq 1} y^k \bar{f}_k(x) + \sum_{k \geq 2} y^k \sum_{j=2}^{k} x^{j+1} \bar{g}_{k-j}(x)$$

$$= \frac{1}{1-xy} + \frac{xy}{1-y} \sum_{i \geq 1} (1-y^i) f_i(x) + \frac{x^3 y^2}{(1-xy)(1-y)} \sum_{i \geq 0} (1-y^{i+1}) g_i(x)$$

$$= \frac{1}{1-xy} + \frac{xy}{1-y} (f(x,1) - f(x,y)) + \frac{x^3 y^2}{(1-xy)(1-y)} (g(x,1) - yg(x,y)).$$

This implies

$$\left(1 + \frac{x^2 y^2}{(1-xy)(1-y)}\right) g(x,y) = \frac{1}{1-xy} + \frac{x^2 y}{(1-xy)(1-y)} g(x,1), \quad (7.10)$$

$$\left(1 + \frac{xy}{1-y}\right) f(x,y)$$

$$= \frac{1}{1-xy} + \frac{xy}{1-y} f(x,1) + \frac{x^3 y^2}{(1-xy)(1-y)} (g(x,1) - yg(x,y)). \quad (7.11)$$

To solve these functional equations, we use the kernel method technique. In this case, if we assume that $y = y_0$ in (7.10), where $y_0$ satisfies $1 + \frac{x^2 y_0^2}{(1-xy_0)(1-y_0)} = 0$, that is, $y_0 = \frac{1+x-\sqrt{1-2x-3x^2}}{2x(1+x)}$, then

$$\sum_{n \geq 0} \mathcal{P}_n(111, 1221) x^n = g(x,1) = \frac{y_0}{1-xy_0} = \frac{1-x-\sqrt{1-2x-3x^2}}{2x^2}.$$

Moreover, (7.10) gives

$$g(x, 1/(1-x)) = \frac{1-x}{1-3x} (1 - xg(x,1)). \quad (7.12)$$

Now, if we assume that $y = y_1 = \frac{1}{1-x}$ in (7.11), then

$$f(x,1) = \frac{1-x}{1-2x} - \frac{x^2}{1-2x}\left(g(x,1) - \frac{1}{1-x}g(x,1/(1-x))\right),$$

which, by (7.12), implies

$$f(x,1) = \frac{1 - 3x + \sqrt{1-2x-3x^2}}{2(1-3x)},$$

as required. □

Note that several of the above results has been refined either by adding a parameter which records the number of blocks of a set partition, or counting the set partitions in the question according to some statistics such number descents or inversions. For exact details, we refer the reader to [232].

## 7.6 Sequence A054391

In this section we link between some classes of set partitions and Sequence A054391 in [327]. This sequence, $a_n$, has the generating function given by

$$\sum_{n \geq 0} a_n x^n = 1 - \frac{2x^2}{2x^2 - 3x + 1 - \sqrt{1-2x-3x^2}}. \quad (7.13)$$

The first few $a_n$ values, starting with $n = 0$, are given by 1, 1, 2, 5, 14, 41, 123, 374, …. Mansour and Shattuck [233] identified six classes of the set partitions of $[n]$ each avoiding a classical pattern of size four and another of size five and each enumerated by the number $a_n$. More precisely, they proved the following result.

**Theorem 7.72** *If $n \geq 0$, then $P_n(\pi, \pi') = a_n$ for the following sets $(\pi, \pi')$:*

(1) $(1212, 12221)$  (2) $(1212, 11222)$  (3) $(1212, 11122)$
(4) $(1221, 12311)$  (5) $(1221, 12112)$  (6) $(1221, 12122)$.

Note that Theorem 7.36 shows that $(1212, 11122) \sim (1212, 11222)$ and Theorem 7.37 gives $(1212, 12221) \sim (1212, 11222)$. In [233] is presented a direct bijection showing the equivalence $12221 \overset{nc}{\sim} 11222$. Now, it remains to consider the cases (1), (4), (5), and (6). In order to show this, we give algebraic proofs for cases (1), (4), and (6) and find one-to-one correspondences between cases (5) and (6).

### 7.6.1 The Pairs $(1212, 12221), (1212, 11222), (1212, 11122)$

We start by counting set partitions of $NC_n(12221)$.

**Proposition 7.73** *The generating functions $NCP_{12221}(x)$ is given by*

$$1 - \frac{2x^2}{2x^2 - 3x + 1 - \sqrt{1 - 2x - 3x^2}}.$$

**Proof** Note that each nonempty set partition $\pi \in \mathcal{P}_n(1212, 12221)$ may be decomposed as either $\pi = 1[\pi']$ or as $\pi = 1[\pi^{(1)}]1[\pi^{(2)}]\cdots 1[\pi^{(r)}]$, for some $r \geq 2$, where $\pi^{(i)}$ avoids $\{1212, 111\}$ if $1 \leq i \leq r-1$ and $\pi^{(r)}$ avoiding $\{1212, 12221\}$. Thus

$$NCP_{12221}(x) = 1 + xNCP_{12221}(x) + \frac{x^2 NCP_{12221}(x) NCP_{111}(x)}{1 - xNCP_{111}(x)},$$

which is equivalent to

$$NCP_{12221}(x) = \frac{1 - xNCP_{111}(x)}{1 - x - xNCP_{111}(x)}.$$

Thus, by Example 2.65 we obtain the first case. □

### 7.6.2 The Pair $(1221, 12311)$

Again, to establish this case, we divide up the set of set partitions in question according to the statistic of the size of the maximal increasing initial run. To do so, for $k \geq 1$, let $f_k(x)$ denote the generating function for the number of partitions $\pi$ of $[n]$ having at least $k$ letters and avoiding the patterns 1221 and 12311 such that $\pi_1\pi_2\cdots\pi_k = 12\cdots k$ with $\pi_{k+1} \leq k$ (if there is a $(k+1)$-st letter).

**Lemma 7.74** *For $k \geq 2$,*

$$f_k(x) = x^k + x^k \overline{f}_1(x) + x^{k-1} \overline{f}_2(x) + \sum_{j=2}^{k-1} x^{k+1-j} \overline{f}_j(x),$$

*with initial conditions $f_0(x) = 1, f_1(x) = x + x\overline{f}_1(x)$, where $\overline{f}_k(x) = \sum_{i \geq k} f_i(x)$.*

**Proof** Since a set partition in the case $k = 1$ may just have one letter or start with 11, we have that $f_1(x) = x + x\overline{f}_1(x)$. If $k = 2$, then a partition $\pi$ enumerated by $f_2(x)$ must either be 12 or start with 121 or 122, which implies

$$f_2(x) = x^2 + x\overline{f}_2(x) + x^2\overline{f}_1(x).$$

If $k \geq 3$, then we consider the following cases concerning the set partitions enumerated by $f_k(x)$: (i) $12\cdots k$, (ii) $12\cdots kj\pi'$ with $1 \leq j \leq k-2$, (iii) $12\cdots k(k-1)\pi'$, or (iv) $12\cdots kk\pi'$. The contribution of the first case is $x^k$. In the second case, the word $\pi'$ contains no letters in $[j]$, for otherwise there would be an occurrence of 1221 if it contained a letter in $[j-1]$ or an occurrence of 12311 if it contained the letter $j$. Thus, the letters $(j+1)(j+2)\cdots k\pi'$, taken together, reduce a set partition of the form enumerated by $\overline{f}_{k-j}(x)$, which implies that the contribution in this case is $x^{j+1}\overline{f}_{k-j}(x)$. A similar argument gives a contribution of $x^{k-1}\overline{f}_2(x)$ for the third case. Finally, in Case (iv), no letter in $[k-1]$ occurs in $\pi'$, with the second $k$ extraneous, which implies a contribution of $x^k\overline{f}_1(x)$. We complete the proof by combining all of the cases. □

**Proposition 7.75** *The generating function $\sum_{n\geq 0} P_n(1221, 12311)$ is given by*

$$1 - \frac{2x^2}{2x^2 - 3x + 1 - \sqrt{1 - 2x - 3x^2}}.$$

**Proof** Define the generating function $F(x,y) = \sum_{k\geq 0} f_k(x) y^k$. Clearly, $\overline{f}_1(x) = f_1(x) + f_2(x) + \cdots = F(x,1) - 1$, $f_1(x) = x + x\overline{f}_1(x) = xF(x,1)$, and $\overline{f}_2(x) = F(x,1) - f_1(x) - 1 = (1-x)F_1(x,1) - 1$. By multiplying the recurrence relation in statement of Lemma 7.74 by $y^k$ and summing over $k \geq 2$, we obtain

$F(x,y)$

$$= 1 + f_1(x)y + \sum_{k\geq 2}\left(x^k + x^k\overline{f}_1(x) + x^{k-1}\overline{f}_2(x) + \sum_{j=2}^{k-1} x^{k+1-j}\overline{f}_j(x)\right) y^k$$

$$= 1 + xyF(x,1) + \frac{x^2y^2}{1-xy} + \frac{x^2y^2(F(x,1)-1)}{1-xy}$$

$$+ \frac{xy^2((1-x)F(x,1)-1)}{1-xy} + \sum_{k\geq 2}\left(\sum_{j=2}^{k-1} x^{k+1-j}\overline{f}_j(x)\right) y^k$$

$$= 1 + xyF(x,1) + \frac{xy^2}{1-xy}[F(x,1)-1] + \frac{x^2y}{1-xy}\sum_{j\geq 2}\overline{f}_j(x)y^j$$

$$= 1 + xyF(x,1) + \frac{xy^2}{1-xy}[F(x,1)-1] + \frac{x^2y}{1-xy}\sum_{i\geq 2} f_i(x)\sum_{j=2}^{i} y^j.$$

Thus,

$$F(x,y) = 1 + xyF(x,1) + \frac{xy^2}{1-xy}[F(x,1)-1]$$

$$+ \frac{x^2y}{1-xy}\left[\frac{y^2}{1-y}(F(x,1) - f_1(x) - 1) - \frac{y}{1-y}(F(x,y) - f_1(x)y - 1)\right],$$

which implies

$$\left(1 + \frac{x^2y^2}{(1-xy)(1-y)}\right) F(x,y)$$
$$= \frac{1-xy-xy^2+x^2y^2}{1-xy} + \frac{xy(1-xy-y^2+2xy^2)}{(1-xy)(1-y)} F(x,1). \quad (7.14)$$

By the kernel method technique, if we assume that $y = y_0$ in (7.14), where $y_0$ satisfies $1 + \frac{x^2 y_0^2}{(1-xy_0)(1-y_0)} = 0$, that is, $y_0 = \frac{1+x-\sqrt{1-2x-3x^2}}{2x(1+x)}$, then

$$\sum_{n\geq 0} P_n(12311,1221)x^n = F(x,1) = -\frac{(1-y_0)(1-xy_0-xy_0^2+x^2y_0^2)}{xy_0(1-xy_0-y_0^2+2xy_0^2)}$$

$$= 1 - \frac{2x^2}{2x^2 - 3x + 1 - \sqrt{1-2x-3x^2}},$$

as claimed. $\square$

### 7.6.3 The Pairs $(1221, 12112), (1221, 12122)$

First, we construct a bijection $\mathcal{P}_n(1221, 12122) \to \mathcal{P}_n(1221, 12112)$ and then show that the generating function for $P_n(1221, 12122)$ is given by (7.13). To present the bijection, we need the following lemma, where we omit the proof and leave it to the interested reader (see [233]).

**Lemma 7.76**

*(1) Suppose $\pi \in \mathcal{P}_n(1221, 12122)$ has at least one occurrence of the pattern 12112. Let $b \geq 2$ be the smallest letter such that there exists $a \in [b-1]$ for which there is a subsequence in $\pi$ given by abaab. Then (i) the element $b$ occurs exactly twice in $\pi$ and (ii) the element $a$ is uniquely determined.*

*(2) Suppose $\pi = \pi_1 \pi_2 \cdots \pi_n \in \mathcal{P}_n(1221, 12122)$ contains at least one occurrence of the pattern 12112 and let $a$ and $b$ be as defined in (1). Write $\pi = \pi^{(1)} b \pi^{(2)} b \pi^{(3)}$, where $\pi^{(3)}$ is possibly empty. Then we have the following: (i) Only letters greater than $b$ can occur in $\pi^{(2)}$, with the exception of $a$, and no letter other than $a$ can occur more than once in $\pi^{(2)}$; (ii) Any letter occurring in $\pi^{(2)}$ can occur at most once in $\pi^{(3)}$, all of whose letters are greater than $b$.*

**Proposition 7.77** *For all $n \geq 0$,*

$$P_n(1221, 12122) = P_n(1221, 12112).$$

**Proof** Let $A = \mathcal{P}_n(1221, 12122)$ and $B = \mathcal{P}_n(1221, 12112)$; we proceed with the proof by describing a bijection $f : A \to B$. Suppose $\pi \in A$. If $\pi \in B$, then let $f(\pi) = \pi$; otherwise, $\pi' = \pi$ contains at least one occurrence of the pattern 12112. Let $b_0$ denote the smallest letter $b$ in $\pi'$ for which there exists a subsequence *abaab* for some $a < b$, and let $a_0$ denote the corresponding unique letter $a$ (see Lemma 7.76(1)). Set $\pi' = \pi'^{(1)} b_0 \pi'^{(2)} b_0 \pi'^{(3)}$ as in Lemma 7.76(2), where $\pi^{(2)} = \alpha^{(1)} a_0 \cdots \alpha^{(r-1)} a_0 \alpha^{(r)}$. Let $\pi'' = \pi^{(1)} b_0 \overline{\pi}^{(2)} b_0 \pi^{(3)}$, where $\overline{\pi}^{(2)} = \alpha^{(1)} a_0 \alpha^{(2)} b_0 \cdots \alpha^{(r-1)} b_0 \alpha^{(r)}$ in which we have changed all but the first $a_0$ occurring in $\pi^{(2)}$ to $b_0$. Clearly, this replaces all of the occurrences of 12112 involving $a_0$ and $b_0$ with ones of 12122. Using Lemma 7.76(2), one can check that no occurrences of 1221 is introduced.

If $\pi'$ has no occurrences of 12112, then let $f(\pi) = \pi'$. Otherwise, let $b_1$ denote the smallest letter $b$ in $\pi'$ for which there exists a subsequence *abaab* for some $a < b$. Thus, $b_1 > b_0$. By similar argument as in the proof of Lemma 7.76(1), we can show that a letter $a < b_1$ for which the subsequence $ab_1 aab_1$ occurs in $\pi'$ is uniquely determined, which we will denote by $a_1$. Let $\pi''$ denote the set partition obtained by changing all of the letters $a_1$, except the first, coming after the leftmost $b_1$ to $b_1$. So, no occurrence of 1221 are introduced (from the minimality of $b_1$). Now repeat the above process, considering $\pi''$.

Since $b_0 < b_1 < b_2 < \cdots$, the process must end in a finite number, say $m$, of steps, with the resulting set partition $\tilde{\pi} \in \mathcal{P}_n(1221, 12112)$. Define $f(\pi) = \tilde{\pi}$. Clearly, the largest $b$ for which there exists $a < b$ such that *ababb* occurs in $\tilde{\pi}$ is $b = b_{m-1}$ whenever $m \geq 1$. This follows from the fact one can check that no occurrences of 12122 in which the letter 2 corresponds to a letter greater than $b_i$ are introduced in the mapping $\pi$ to $\pi'$, $\pi'$ to $\pi''$, $\cdots$. If $m \geq 1$, then one can also see that the largest $a < b_{m-1}$ for which there is a subsequence of the form $ab_{m-1}ab_{m-1}b_{m-1}$ in $\tilde{\pi}$ is $a = a_{m-1}$. So to reverse the process, we first consider the largest letter $b$ (if it exists) for which *ababb* occurs in $\tilde{\pi}$ for some $a < b$ and then consider the largest such $a$ corresponding to this $b$. One can then change the letters accordingly to reverse the final step of the process describing $f$ and the other steps can be similarly reversed, going from last to first. $\square$

Now, we find an explicit formula for the generating function for the number of set partitions of $\mathcal{P}_n(1221, 12122)$. In order to achieve this, we will consider the following three types of generating functions:

- For all $k \geq 1$, let $f_k(x)$ be the generating function for the number of set partitions $\pi = \pi_1 \pi_2 \cdots \pi_n \in \mathcal{P}_n(1221, 12122)$ such that $\pi_1 \pi_2 \cdots \pi_k = 12 \cdots k$ and $\pi_{k+1} \leq k$. We define $f_0(x) = 1$.

- For all $k \geq 2$, let $h_k(x)$ be the generating function for the number of set partitions $\pi = \pi_1 \pi_2 \cdots \pi_n \in \mathcal{P}_n(1221, 12122)$ such that $\pi_1 \pi_2 \cdots \pi_k = 12 \cdots k$ and $\pi_{k+1} = 1$.

- For all $k \geq 2$, let $g_k(x)$ be the generating function for the number of set partitions $\pi = \pi_1 \pi_2 \cdots \pi_n \in \mathcal{P}_n(1221, 12122)$ such that $\pi_1 \pi_2 \cdots \pi_k = 12 \cdots k$, $\pi_{k+1} = 1$ and $\pi_j \neq 1$ for all $j = k+2, \ldots, n$.

We define the following two variable generating functions:

$$F(x,y) = \sum_{k \geq 0} h_k(x) y^k, \quad H(x,y) = \sum_{k \geq 2} h_k(x) y^k, \quad G(x,y) = \sum_{k \geq 2} g_k(x) y^k.$$

Our aim is to find an explicit formula for $F(x,1)$, which is the generating function for the sequence $\{P_n(1221, 12122)\}_{n\geq 0}$. The next lemma provides relations which we will need between these generating functions.

**Lemma 7.78** *We have*

$$F(x,y) = 1 + \frac{xy}{1-xy}F(x,1) + \frac{1}{1-xy}H(x,y),$$

$$H(x,y) = G(x,y) + \frac{xy}{1-y}(yH(x,1) - H(x,y)),$$

*and*

$$\left(1 + \frac{x^2y^2}{(1-y)(1-xy)}\right)G(x,y) = \frac{x^3y^2}{1-xy} + \frac{x^2y^2}{(1-y)(1-xy)}G(x,1)$$
$$+ \frac{x^4y^2(2-x)}{(1-x)(1-xy)}F(x,1) + \frac{x^3y^2}{(1-x)(1-xy)}H(x,1).$$

**Proof** Here we only present the proof of the last relation, where the first two are similar (see [233]). Let $a_\tau(x)$ (respectively, $b_\tau(x)$) be the generating function for the number of set partitions $\pi = \pi_1\pi_2\cdots\pi_n \in \mathcal{P}_n(1221, 12122)$ such that $\pi_1\pi_2\cdots\pi_k = \tau$ (respectively, $\pi_1\pi_2\cdots\pi_k = \tau$ and $\pi_j \neq 1$ for all $j \geq k+1$). By the definitions, we have

$$g_k(x) = x^{k+1} + \sum_{j=2}^{k-1} b_{12\cdots k1j}(x) + b_{12\cdots k1k}(x) + b_{12\cdots k1(k+1)}(x)$$
$$= x^{k+1} + \sum_{j=2}^{k-1} x^j g_{k+1-j}(x) + x^{k+2}F(x,1) + b_{12\cdots k1(k+1)}(x),$$

and for all $\ell \geq k+1$,

$$b_{12\cdots k1(k+1)\cdots \ell}(x) = x^{\ell+1} + \sum_{j=2}^{k} b_{12\cdots k1(k+1)\cdots \ell j}(x)$$
$$+ \sum_{j=k+1}^{\ell} b_{12\cdots k1(k+1)\cdots \ell j}(x) + b_{12\cdots k1(k+1)\cdots (\ell+1)}(x)$$
$$= x^{\ell+1} + \sum_{j=2}^{k} x^j g_{\ell+1-j}(x) + b_{12\cdots k1(k+1)\cdots (\ell+1)}(x)$$
$$+ \sum_{j=k+1}^{\ell-1} x^j h_{\ell+1-j}(x) + x^{\ell+1}(F(x,1) - 1).$$

Hence, by summing over all $\ell \geq k+1$, we obtain

$$g_k(x)$$
$$= \frac{x^{k+1}}{1-x} + x^{k+2}F(x,1) + \frac{x^{k+2}}{1-x}(F(x,1)-1) + \frac{x^{k+1}}{1-x}H(x,1)$$
$$+ \sum_{j=2}^{k-1} x^j g_{k+1-j}(x) + \sum_{j=3}^{k} \frac{x^j - x^{k+1}}{1-x} g_{k+2-j}(x) + \frac{x^2 - x^{k+1}}{1-x} \sum_{\ell \geq k} g_\ell(x).$$

Multiplying the above recurrence by $y^k$, and summing over all $k \geq 2$, yields

$$G(x,y)$$
$$= \frac{x^3 y^2}{(1-x)(1-xy)} + \frac{x^4 y^2}{1-xy}F(x,1) + \frac{x^4 y^2}{(1-x)(1-xy)}(F(x,1)-1)$$
$$+ \frac{x^3 y^2}{(1-x)(1-xy)}H(x,1) + \frac{x^2 y}{1-xy}G(x,y) + \frac{x^3 y}{1-xy}G(x,y)$$
$$+ \frac{x^2 y}{(1-x)(1-y)}(yG(x,1) - G(x,y)) - \frac{x^3 y^2}{(1-x)(1-xy)}G(x,1),$$

which implies the required result. □

Now we are ready to find an explicit formula for the generating function $F(x,1)$. Lemma 7.78 gives a system of functional equations which we will solve using the kernel method (see [139]). Lemma 7.78 implies

$$F(x,1) = \frac{1-x}{1-2x} + \frac{1}{1-2x}H(x,1), \qquad (7.15)$$

and replacing $y$ by $\frac{1}{1-x}$ gives

$$H(x,1) = (1-x)G\left(x, \frac{1}{1-x}\right). \qquad (7.16)$$

Replacing $y$ first by $y_1 = \frac{1+x-\sqrt{1-2x-3x^2}}{2x(1+x)}$ and then by $y_2 = \frac{1}{1-x}$ gives

$$G(x,1)$$
$$= \frac{x^3 y_1^2}{1-xy_1} + \frac{x^4 y_1^2(2-x)}{(1-x)(1-xy_1)}F(x,1) + \frac{x^3 y_1^2}{(1-x)(1-xy_1)}H(x,1), \quad (7.17)$$

$$G\left(x, \frac{1}{1-x}\right)$$
$$= \frac{x^3 y_2}{1-3x} + \frac{x^4 y_2^2(2-x)}{1-3x}F(x,1) + \frac{x^3 y_2^2}{1-3x}H(x,1) - \frac{x}{1-3x}G(x,1). \quad (7.18)$$

By solving the system of equations (7.15)–(7.18), we arrive at the following result.

**Proposition 7.79** *The generating function for $\sum_{n\geq 0} P_n(1221, 12122)x^n$ is given by*

$$1 - \frac{2x^2}{2x^2 - 3x + 1 - \sqrt{1 - 2x - 3x^2}}.$$

---

## 7.7 Catalan and Generalized Catalan Numbers

In this section, let $w_n$ denote the *generalized Catalan number* defined by the recurrence

$$w_n = 2w_{n-1} + \sum_{i=1}^{n-2} w_i w_{n-2-i}, \quad n \geq 2, \tag{7.19}$$

with $w_0 = w_1 = 1$. The first few $w_n$ values for $n \geq 0$ are 1, 1, 2, 5, 13, 35, 97, 275, .... Using (7.19), one can show that the $w_n$ has the generating function given by

$$\sum_{n\geq 0} w_n x^n = \frac{(1-x)^2 - \sqrt{1 - 4x + 2x^2 + x^4}}{2x^2}. \tag{7.20}$$

The numbers $w_n$ count, among other things, the Catalan paths of size $n$ having no occurrences of $DDUU$ (see [310]), or, equivalently, the Catalan paths of size $n+1$ having no $UUDD$. They also count the subset of the permutations of size $n$ avoiding the two generalized patterns 1-3-2 and 12-34 (see Example 2.10 of [214]). For further information on these numbers, see also Sequence A025242 of [327]. Here, we provide new combinatorial interpretations for the $w_n$ in terms of set partitions, showing that they enumerate certain two-pattern avoidance classes. Mansour and Shattuck [235] proved the following result (see Exercise 7.12).

**Theorem 7.80** *For all $n \geq 0$, $P_n(1212, 1211) = w_n$. Moreover,*

$$(1211, 1212) \sim (1121, 1212) \sim (1121, 1221) \sim (1112, 1123) \sim (1122, 1123).$$

In [239], Mansour and Shattuck found a new relation between set partitions and Catalan numbers. In particular, they proved the following result.

**Theorem 7.81** *If $n \geq 0$, then $P_n(u, v) = \text{Cat}_n$ for the following sets $(u, v)$:*

(1) $(1222, 12323)$   (2) $(1222, 12332)$   (3) $(1211, 12321)$
(4) $(1211, 12312)$   (5) $(1121, 12231)$   (6) $(1121, 12132)$
(7) $(1112, 12123)$   (8) $(1112, 12213)$.

They proved this theorem (we omit the proof and leave it as an exercise; see Exercise 7.13) by giving algebraic proofs for cases (1), (3), (5), (6), and (7) and finding one-to-one correspondences between cases (1) and (2), (3) and (4), and

(7) and (8). They established (3), (5), (6), and (7) by using the *kernel method* (see [15]) to solve the functional equations that arise once certain parameters have been introduced. Our bijection between cases (3) and (4) and between (7) and (8) are of an algorithmic nature and systematically replace occurrences of 12312 (respectively, 12213) with ones of 12321 (respectively, 12123) without introducing 1211 (respectively, 1112).

## 7.8 Pell Numbers

In this subsection we link our set partitions to the Pell numbers. The $n$-th Pell number, denoted by $\text{Pell}_n$, is defined via the following recurrence relation (see [327, Sequence A000129])

$$\text{Pell}_n = 2\text{Pell}_{n-1} + \text{Pell}_{n-2}, \quad \text{Pell}_1 = 1 \text{ and } \text{Pell}_2 = 2. \quad (7.21)$$

In 2011, Mansour and Shattuck [236] found six different classes of set partitions where each of these classes is characterized by avoidance of three subsequence patterns and enumerated by the $n$-th Pell number. Moreover, they presented several $q$-analog of the Pell numbers by studying different statistics on these classes. We start with the following theorem.

**Theorem 7.82** *Let $n \geq 1$. Then $P_n(\tau, \tau', \tau'') = \text{Pell}_n$ for the following three subsequence patterns $(\tau, \tau', \tau'')$:*

(1) $(1112, 1212, 1213)$  (2) $(1121, 1212, 1213)$  (3) $(1121, 1221, 1231)$
(4) $(1211, 1212, 1213)$  (5) $(1211, 1221, 1231)$  (6) $(1212, 1222, 1232)$.

**Proof** Given $\pi \in \mathcal{P}_n$, we denote the set partition of $\mathcal{P}_{n+1}$ obtained by adding 1 to each letter of $\pi$ and then writing a 1 in front by $inc(\pi)$. For instance, if $\pi = 12123 \in \mathcal{P}_5$, then $inc(\pi) = 123234 \in \mathcal{P}_6$.

We prove the first and second cases, the third and fifth cases are given as Exercise 7.1, while the fourth and sixth cases will be considered later in more details; see the next two subsections.

(1) Set $\mathcal{A}_n = \mathcal{P}_n(1112, 1212, 1213)$. Note that any set partition $\pi \in \mathcal{A}_n$ with $n \geq 3$ is characterized by one of the following forms: $\pi = 1^n$, $\pi = 1[\pi']1^a$ or $\pi = 11[\pi']1^b$, where $0 \leq a \leq n-2$, $0 \leq b \leq n-3$, and $\pi'$ is any nonempty set partition avoiding the three patterns. This implies that the sequence $a_n = |\mathcal{A}_n|$ satisfies the recurrence relation $a_n = 2a_{n-1} + a_{n-2}$ with $a_1 = 1$ and $a_2 = 2$. Hence, by (7.21) we have that $a_n = \text{Pell}_n$ for all $n \geq 1$.

(2) Set $\mathcal{B}_n = \mathcal{P}_n(1121, 1212, 1213)$. It is easy to verify that $\pi$ can be decomposed as either (i) $\pi = 1^a[\pi']$ or (ii) $\pi 1[\pi']2^b$, where $2 \leq a \leq n$, $0 \leq b \leq n-2$, and $\pi'$, $\pi''$ denote, respectively, possibly empty and nonempty set partitions

which avoid all three patterns. This implies that the sequence $b_n = |\mathcal{B}_n|$ satisfies the recurrence relation $b_n = 2b_{n-1} + b_{n-2}$ with $b_1 = 1$ and $b_2 = 2$, which completes the proof. □

### 7.8.1 Counting $\mathcal{P}_n(1211, 1212, 1213)$ by inv

Now, we count the set partitions of $\mathcal{A}_n = \mathcal{P}_n(1211, 1212, 1213)$ according to the number of inversions.

**Lemma 7.83** *Let* $a_n(y,q) = \sum_{\pi \in \mathcal{A}_n} y^{\text{blo}(\pi)} q^{\text{inv}(\pi)}$ *with* $a_0(y,q) = 1$. *We have*

$$a_n(y,q) = (y+1)a_{n-1}(y,q) + yq^{n-2}a_{n-2}(y,q) \qquad (7.22)$$

*with* $a_1(y,q) = x$ *and* $a_2(y,q) = y + y^2$.

**Proof** By the definitions we have that $a_1(y,q) = y$ and $a_2(y,q) = y + y^2$. So from now we assume that $n \geq 3$. First note that each set partition $\pi \in \mathcal{A}_n$ can be written as either $\pi = 1[\pi']$ with $\pi' \in \mathcal{A}_{n-1}$, $\pi = 1\pi'$ with $\pi' \in \mathcal{A}_{n-1}$, or $\pi = 1[\pi'']1$ with $\pi'' \in \mathcal{A}_{n-2}$. (Why?) Then the total weights of the set partitions of $\mathcal{A}_n$ formed in these three steps are given by $ya_{n-1}(y,q)$, $a_{n-1}(y,q)$ and $yq^{n-2}a_{n-2}(y,q)$, respectively. Thus, $a_n(y,q) = (y+1)a_{n-1}(y,q) + yq^{n-2}a_{n-2}(y,q)$, which completes the proof. □

Note that taking $y = q = 1$ in (7.22) gives $|\mathcal{A}_n| = a_n(1,1) = \text{Pell}_n$, for all $n \geq 1$. Taking $y = -q = 1$ in (7.22) gives

$$a_n(1,-1) = 2a_{n-1}(1,-1) + (-1)^n a_{n-2}(1,-1)$$

with $a_1(1,-1) = 1$ and $a_2(1,-1) = 2$. Considering even and odd cases for $n$, we obtain from this recurrence by induction, $a_n(1,-1) = \text{Fib}_{\lfloor \frac{3n}{2} \rfloor}$, $n \geq 1$.

**Proposition 7.84** *The generating function* $A(x, y; q) = \sum_{n \geq 0} a_n(y,q) x^n$ *is given by*

$$A(x, y; q) = \sum_{j \geq 0} \frac{y^j (1 - q^j x - yq^{2j} x^2) q^{j(j-1)} x^{2j}}{(1 - (1+y)x) \cdots (1 - (1+y)q^j x)}. \qquad (7.23)$$

**Proof** Set $A(x) = A(x, y; q)$. Multiplying both sides of (7.22) by $x^n$ and summing over $n \geq 3$ gives

$$A(x) - 1 - yx - (y + y^2)x^2 = (y+1)x(A(x) - 1 - xy) + yx^2(A(qx) - 1),$$

equivalently,

$$A(x) = \frac{1 - x - yx^2}{1 - (y+1)x} + \frac{yx^2}{1 - (y+1)x} A(qx).$$

Iterating this recurrence an infinite number of times yields the required result. □

Note that $A(x; 1, 1) = \frac{1-x-x^2}{1-2x-x^2}$, which is the generating function for the sequence $\{\delta_{n,0} + \text{Pell}_n\}_{n \geq 0}$.

**Theorem 7.85** *For all $n \geq 1$,*

$$a_n(y, q) = \sum_{j=0}^{\lfloor \frac{n-1}{2} \rfloor} y^{j+1} q^{j^2} \binom{n-1-j}{j}_q (1+x)^{n-1-2j}, \qquad (7.24)$$

*with $a_0(y, q) = 1$.*

**Proof** Using the fact that $\frac{t^n}{(1-t)(1-tq)\cdots(1-tq^n)} = \sum_{m \geq 0} \binom{m}{n}_q t^m$ (see Appendix C), one may expand the denominators in the expression above for $A(x)$ to obtain

$$A(x) = \sum_{i,j \geq 0} \frac{y^j(1 - q^j x - yq^{2j}x^2)q^{j(j-1)}x^{j+i}}{(1+y)^{j-i}} \binom{i}{j}_q.$$

This implies that the coefficient of $x^n$ in $A(x)$ is given by

$$\sum_{j \geq 0} y^j q^{j(j-1)} \binom{n-j}{j}_q (1+y)^{n-2j} - \sum_{j \geq 0} y^j q^{j^2} \binom{n-1-j}{j}_q (1+y)^{n-1-2j}$$

$$- \sum_{j \geq 0} y^{j+1} q^{j(j+1)} \binom{n-2-j}{j}_q (1+y)^{n-2-2j}$$

$$= \sum_{j \geq 0} y^j q^{j(j-1)} \binom{n-j}{j}_q (1+y)^{n-2j} - \sum_{j \geq 0} y^j q^{j^2} \binom{n-1-j}{j}_q (1+y)^{n-1-2j}$$

$$- \sum_{j \geq 1} y^j q^{j(j-1)} \binom{n-1-j}{j-1}_q (1+y)^{n-2j}$$

$$= \sum_{j \geq 0} y^j q^{j(j-1)} (1+y)^{n-2j} \left[ \binom{n-j}{j}_q - \binom{n-1-j}{j-1}_q - \frac{q^j}{1+y} \binom{n-1-j}{j}_q \right]$$

$$= \sum_{j \geq 0} y^j q^{j(j-1)} (1+y)^{n-2j} \left[ q^j \binom{n-1-j}{j}_q - \frac{q^j}{1+y} \binom{n-1-j}{j}_q \right]$$

$$= \sum_{j \geq 0} y^{j+1} q^{j^2} \binom{n-1-j}{j}_q (1+y)^{n-1-2j},$$

where we used the fact that $\binom{n-j}{j}_q - \binom{n-1-j}{j-1}_q = q^j \binom{n-1-j}{j}_q$ (see Appendix C), which completes the proof. □

Note that (7.24) for $x = q = 1$ reduces to the well-known formula, Exercise 2.9(ii),

$$\text{Pell}_n = \sum_{j=0}^{\lfloor \frac{n-1}{2} \rfloor} \binom{n-1-j}{j} 2^{n-1-2j}$$

Recall the familiar result that the total number of inversions occurring within all of the permutations of length $n$ is given by the formula $\frac{1}{2}\binom{n}{2}n!$ if $n \geq 1$ (see, for example, [333]). We end this section by studying the total number of inversions over all of the set partitions of $\mathcal{A}_n$.

**Theorem 7.86** *The total number of inversions occurring within all the members of $\mathcal{A}_n$ is given by*

$$\frac{n-1}{8}\left(n(\mathrm{Pell}_{n-1}+\mathrm{Pell}_{n-2})-\mathrm{Pell}_n\right). \tag{7.25}$$

*In particular, the average number of inversions is approximately*

$$\frac{2-\sqrt{2}}{8}n^2 - \frac{3-\sqrt{2}}{8}n + \frac{1}{8}$$

*as $n$ grows large.*

**Proof** From Appendix C we can state that $f_{n,j} = \frac{d}{dq}\binom{n}{j}_q\big|_{q=1} = \binom{n}{2}\binom{n-2}{j-1}$ for all $j = 0, 1, \ldots, n$. Now we can use Theorem 7.85 to find the total number of inversions occurring within all of the set partitions of $\mathcal{A}_n$,

$$\frac{d}{dq}a_n(y,q)\big|_{y=q=1}$$
$$= \sum_{j=0}^{\lfloor\frac{n-1}{2}\rfloor}\left(j^2\binom{n-1-j}{j}+\binom{n-1-j}{2}\binom{n-3-j}{j-1}\right)2^{n-1-2j},$$

which is equivalent to

$$\frac{d}{dq}a_n(y,q)\big|_{y=q=1}$$
$$= \sum_{j=0}^{\lfloor\frac{n-1}{2}\rfloor}\left(j^2\binom{n-1-j}{j}+\frac{j(n-1-2j)}{2}\binom{n-1-j}{j}\right)2^{n-1-2j}$$
$$= \frac{n-1}{2}\sum_{j=0}^{\lfloor\frac{n-1}{2}\rfloor}j\binom{n-1-j}{j}2^{n-1-2j}.$$

To complete the proof of the first statement, we need to prove

$$b_n = \sum_{j=0}^{\lfloor\frac{n-1}{2}\rfloor}j\binom{n-1-j}{j}2^{n-1-2j} = \frac{n(\mathrm{Pell}_{n-1}+\mathrm{Pell}_{n-2})-\mathrm{Pell}_n}{4}, \tag{7.26}$$

In order to do that, observe that for $n \geq 3$,

$$b_n = \sum_{j=1}^{\lfloor \frac{n-1}{2} \rfloor} (n-1-j)\binom{n-2-j}{j-1} 2^{n-1-2j}$$

$$= \sum_{j=0}^{\lfloor \frac{n-3}{2} \rfloor} (n-2-j)\binom{n-3-j}{j} 2^{n-3-2j}$$

$$= (n-2)\sum_{j=0}^{\lfloor \frac{n-3}{2} \rfloor} \binom{n-3-j}{j} 2^{n-3-2j} - b_{n-2},$$

so that $b_n = (n-2)\text{Pell}_{n-2} - b_{n-2}$. Thus (7.26) follows by induction.

The second statement is an immediate consequence of the first upon dividing through by $\text{Pell}_n$ and noting $\lim_{n\to\infty} \left(\frac{\text{Pell}_{n-1}}{\text{Pell}_n}\right) = \sqrt{2} - 1$, which follows from Exercise 2.9(iii). □

### 7.8.2 Counting $\mathcal{P}_n(1212, 1222, 1232)$ by Comaj

Here, we count the set partitions of $\mathcal{B}_n = \mathcal{P}_n(1212, 1222, 1232)$ according to the comajor index, $\text{Comaj}(\pi_1\pi_2\cdots\pi_n) := \sum_{\pi_i < \pi_{i+1}} i$. First, we state the following lemma.

**Lemma 7.87** *Let $b_n(y,q) = \sum_{\pi \in \mathcal{B}_n} y^{\text{blo}(\pi)} q^{\text{Comaj}(\pi)}$ with $b_0(y,q) = 1$. Then*

$$b_n(y,q) = (yq^{n-1} + 1)b_{n-1}(y,q) + yq^{n-2}b_{n-2}(y,q) \tag{7.27}$$

*with $b_1(y,q) = y$ and $b_2(y,q) = y + y^2 q$.*

**Proof** Let $\pi = 1\pi^{(1)} 2\pi^{(2)} \cdots k\pi^{(k)} \in \mathcal{B}_n$ such that $\pi^{(i)}$ is a word over alphabet $[i]$. Since $\pi$ avoids 1212, the $\pi^{(i)}$ cannot increase. Since $\pi$ avoids 1232, no $\pi^{(i)}$ can contain a letter $j$ with $1 < j < i$, which implies $\pi^{(i)} = i^a 1^b$ if $i \geq 2$ for some $a, b \geq 0$. Since $\pi$ avoids 1222, we must have $a = 0$ or $a = 1$, that is, $\pi^{(i)} = i1^b$ or $\pi^{(i)} = 1^b$. Now, assume that $\pi = 1\pi^{(1)} 2\pi^{(2)} \cdots k\pi^{(k)}$ such that $\pi^{(i)}$ is a word over alphabet $[i]$. If $\pi \in \mathcal{B}_{n-1}$, then either write $k+1$ just after $\pi$ or write 1 just after $\pi$ to obtain two set partitions of $\mathcal{B}_n$ and if $\pi \in \mathcal{B}_{n-2}$, then write the letter $k+1$ twice just after $\pi$ to obtain a set partition of $\mathcal{B}_n$. Note that all set partitions of $\mathcal{B}_n$, $n \geq 3$, arise exactly once in this way.

Observe that the total weights of the set partitions of $\mathcal{B}_n$ formed by the above three steps are given by $yq^{n-1}b_{n-1}(y,q)$, $b_{n-1}(y,q)$, and $yq^{n-2}b_{n-2}(y,q)$, respectively, which implies the desired result. □

Note that if we set $y = q$ in (7.27), then the recurrence is satisfied by the Pell polynomials $\text{Pell}_n(q)$ introduced in [309] but with the initial conditions of $q$ and $q + q^3$ replacing those of $P_n(q)$ of 1 and $1 + q$.

**Proposition 7.88** Let $B(x) = \sum_{n\geq 0} b_n(y,q)x^n$. Then

$$B(x) = 1 + \sum_{n\geq 1} y^n q^{\binom{n}{2}} x^n \frac{(1+x)\cdots(1+q^{n-2}x)}{(1-x)\cdots(1-q^{n-1}x)}.$$

**Proof** Let $B'(x) = B(x) - 1$. Multiplying both sides of (7.27) by $x^n$ and summing over $n \geq 3$ gives

$$B'(x) = \frac{xy}{1-x} + \frac{xy(1+x)}{1-x} B'(qx).$$

Iterating this relation an infinite number of times yields the following result. $\square$

**Theorem 7.89** For all $n \geq 1$,

$$b_n(y,q) = \sum_{j=0}^{\lfloor \frac{n-1}{2} \rfloor} y^{j+1} q^{j^2} \binom{n-1-j}{j}_q \prod_{i=j+1}^{n-j-1} (1+yq^i) \qquad (7.28)$$

with $b_0(y,q) = 1$.

**Proof** By using the facts (see Appendix C)

$$\frac{t^n}{(1-t)(1-tq)\cdots(1-tq^n)} = \sum_{m\geq 0} \binom{m}{n}_q t^m$$

and

$$(1+t)(1+qt)\cdots(1+q^{n-1}t) = \sum_{j=0}^{n} q^{\binom{j}{2}} \binom{n}{j}_q t^j,$$

the expression $B'(x) = B(x) - 1$ can be written as

$$\sum_{\substack{m\geq 0, \\ n\geq 1}} y^n q^{\binom{n}{2}} x^{m+1} \binom{m}{n-1}_q (1+x)\cdots(1+q^{n-2}x)$$

$$= \sum_{\substack{k,m\geq 0, \\ n\geq 1}} y^n q^{\binom{n}{2}+\binom{k}{2}} x^{m+k+1} \binom{m}{n-1}_q \binom{n-1}{k}_q.$$

Then the coefficient of $x^r$, $r \geq 1$, is given by

$$\sum_{m=0}^{r-1} \sum_{n=r-m}^{m+1} y^n q^{\binom{n}{2}+\binom{r-1-m}{2}} \binom{m}{n-1}_q \binom{n-1}{r-1-m}_q$$

$$= \sum_{m=0}^{r-1} q^{\binom{r-1-m}{2}} \binom{m}{r-1-m}_q \sum_{n=r-m}^{m+1} y^n q^{\binom{n}{2}} \binom{2m+1-r}{n+m-r}_q,$$

where we used the following identity $\binom{a}{b}_q \binom{b}{c}_q = \binom{a}{c}_q \binom{a-c}{b-c}_q$. By replacing $m$ by $r-1-m$, and then replacing $n$ by $n+m+1$, we derive that the coefficient is

$$\sum_{m=0}^{r-1} q^{\binom{m}{2}} \binom{r-1-m}{m}_q \sum_{n=m+1}^{r-m} y^n q^{\binom{n}{2}} \binom{r-2m-1}{n-m-1}_q$$

$$= \sum_{m=0}^{r-1} q^{\binom{m}{2}} \binom{r-1-m}{m}_q \sum_{n=0}^{r-2m-1} y^{n+m+1} q^{\binom{n+m+1}{2}} \binom{r-2m-1}{n}_q$$

$$= \sum_{m=0}^{r-1} y^{m+1} q^{m^2} \binom{r-1-m}{m}_q \sum_{n=0}^{r-2m-1} (yq^{m+1})^n \binom{r-2m-1}{n}_q q^{\binom{n}{2}}$$

$$= \sum_{m=0}^{r-1} y^{m+1} q^{m^2} \binom{r-1-m}{m}_q \prod_{i=0}^{r-2m-2} (1+yq^{i+m+1}),$$

by the $q$-binomial theorem (see Appendix C) and the fact

$$\binom{n+m+1}{2} = \binom{n}{2} + \binom{m+1}{2} + n(m+1).$$

This completes the proof. □

Using this result, one can obtain a simple closed-form expression for the total comajor index of all the set partitions of $\mathcal{B}_n$.

**Theorem 7.90** *The total comajor index of all the set partitions of $\mathcal{B}_n$ is given by*

$$\frac{2n^2 \text{Pell}_n - n(3\text{Pell}_n - \text{Pell}_{n-1}) + \text{Pell}_n}{8}.$$

*In particular, the average comajor index value is approximately*

$$\frac{1}{4}n^2 - \frac{4-\sqrt{2}}{8}n + \frac{1}{8}$$

*as $n$ grows large.*

**Proof** By Theorem 7.89 and the proof of Theorem 7.86, we obtain

$$\frac{d}{dq} b_n(y,q) \Big|_{y=q=1}$$

$$= \sum_{j=0}^{\lfloor \frac{n-1}{2} \rfloor} \frac{d}{dq} \left( q^{j^2} \binom{n-1-j}{j} \prod_{i=j+1}^{n-j-1} (1+q^i) \right)_{q=1}$$

$$= \sum_{j=0}^{\lfloor \frac{n-1}{2} \rfloor} j^2 \binom{n-1-j}{j} 2^{n-1-2j} + \sum_{j=0}^{\lfloor \frac{n-1}{2} \rfloor} \binom{n-1-j}{2} \binom{n-3-j}{j-1} 2^{n-1-2j}$$

$$+ \sum_{j=0}^{\lfloor \frac{n-1}{2} \rfloor} \binom{n-1-j}{j} 2^{n-2-2j} \left( \binom{n-j}{2} - \binom{j+1}{2} \right)$$

$$= \sum_{j=0}^{\lfloor \frac{n-1}{2} \rfloor} \left( 2j^2 + j(n-1-2j) + \binom{n-j}{2} - \binom{j+1}{2} \right) \binom{n-1-j}{j} 2^{n-2-2j}$$

$$= \sum_{j=0}^{\lfloor \frac{n-1}{2} \rfloor} \left( \binom{n}{2} - j \right) \binom{n-1-j}{j} 2^{n-2-2j}$$

$$= \frac{1}{2} \binom{n}{2} \sum_{j=0}^{\lfloor \frac{n-1}{2} \rfloor} \binom{n-1-j}{j} 2^{n-1-2j} - \frac{1}{2} \sum_{j=0}^{\lfloor \frac{n-1}{2} \rfloor} j \binom{n-1-j}{j} 2^{n-1-2j}$$

$$= \frac{1}{2} \binom{n}{2} \text{Pell}_n - \frac{1}{8} \left( n(\text{Pell}_{n-1} + \text{Pell}_{n-2}) - \text{Pell}_n \right)$$

$$= \frac{1}{8} \left( 2n^2 \text{Pell}_n - n(3\text{Pell}_n - \text{Pell}_{n-1}) + \text{Pell}_n \right).$$

The second statement is an immediate consequence of the first upon dividing through by $\text{Pell}_n$ and noting $\lim_{n\to\infty} \left( \frac{\text{Pell}_{n-1}}{\text{Pell}_n} \right) = \sqrt{2} - 1$, which follows from Exercise 2.9(iii). □

---

## 7.9 Regular Set Partitions

The study of $d$-regular set partitions has received a lot of attention. It seems that the first consideration of $d$-regular set partitions goes back to Prodinger [278], who called them $d$-*Fibonacci partitions*. Regular set partitions have been used as a formal language in the enumeration of RNA secondary structures (An RNA sequence can be presented as a sequence over the alphabet $\{A, C, G, U\}$, where molecule $A$ for adenine, $C$ for cytosine, $G$ for guanine and $U$ for uracil; these single-stranded molecules fold onto themselves by the so-called *Watson–Crick rules*: A forms base pairs with $U$ and $C$ forms base pairs with $G$. A helical structure can be formed based on the sequence of molecules and the rules). The RNA secondary structure has been presented, in mathematical modeling, as a sequence of numbers $1, 2, \ldots, n$ along with some base pairs, where we have the restriction that all base pairs are allowed except for any two adjacent numbers $i$ and $i+1$, and there do not exist two base pairs $(i, j)$ and $(k, \ell)$ with $i < k < j < \ell$. In this setting, the set $R(n, k)$ of all RNA secondary structures of length $n$ with exactly $k$ base pairs can be presented as the set of noncrossing poor set partitions (a poor set partition is a set partition with each block has at most two elements). A formula for $R(n, k)$ is obtained by Schmitt and Waterman [314] in terms of the Narayana numbers.

**Definition 7.91** *Given $d \geq 1$, a set partition $B_1/B_2/\cdots/B_k \mathcal{P}_n$ is said to be d-regular, if for any two distinct elements $a, b$ in the same block, we have $|b - a| \geq d$. We denote the set of all d-regular set partitions of $[n]$ by $\mathcal{P}_n^{(d)}$ and the set of all d-regular set partitions of $[n]$ with exactly $k$ blocks by $\mathcal{P}_{n,k}^{(d)}$. Clearly, $\mathcal{P}_n = \mathcal{P}_n^{(1)}$ and $\mathcal{P}_{n,k} = \mathcal{P}_{n,k}^{(1)}$.*

A natural question that arises is: can we find an explicit formula for the number of $d$-regular set partitions of $\mathcal{P}_n$? Prodinger [278] solved this question algebraically by showing that the number of $d$-regular set partitions of $[n]$ is equal to the number of set partitions of $\mathcal{P}_{n+1-d}$, that is,

$$|\mathcal{P}_n^{(d)}| = \text{Bell}_{n+1-d}. \tag{7.29}$$

A refinement of this result is given by Yang [368, Theorem 2], who showed that

$$|P_{n,k}^{(d)}| = \text{Stir}(n+1-d, k+1-d). \tag{7.30}$$

Note that the algebraic proof of Prodinger of (7.29) can be extended to prove (7.30). Chu and Wei [77] rediscovered (7.30) by using generating function techniques, and J. Zeng found another proof by provided a recursive proof of (7.30).

Another question that can arise is: find a bijection between $\mathcal{P}_n^{(d)}$ and $\mathcal{P}_{n+1-d}$. To this question were devoted several works. Prodinger [278] gave such a bijection when $d = 2$ that he attributed to F.J. Urbanek, and he noted that Urbanek's bijection can be extended to arbitrary $d$, but he also adds that "this is more complicated to describe and therefore is omitted." Another bijection, when $d = 2$, was found by Yang [368]. The general case remained unsolved until 2004 when Chen et al [68] presented an explicit bijective explanation of (7.30) by means of a simple reduction algorithm which transforms $d$-regular set partitions of $[n]$ with exactly $k$ blocks to $(d-1)$-regular set partitions of $[n-1]$ with exactly $k-1$ blocks. In 2009, Kasraoui [156] simplified the reduction algorithm by showing the model of rook placements on the triangular board for set partitions provides an elegant and quick explanation of

(7.30). In this section we present the proof of Kasraoui [156]. At first we recall the definitions and notation that we introduced in Section 3.2.4.4. Theorem 3.56 presents a bijection $\phi$ between $\mathcal{P}_{n,k}$ and $R_{n,k}$. The bijection $\phi$ has a very nice property which maps each $d$-regular set partition in $\mathcal{P}_{n,k}$ to a rook placement in $R_{n,k}$ such that there are no rooks in the $(d-1)$ highest cells of each column. Thus, we can state the following result.

**Proposition 7.92** (see Kasraoui [156, Proposition 2.1]) *The map $\phi$ establishes a bijection between $\mathcal{P}_{n,k}^{(d)}$ and $R_{n,k}^{(d)}$. Moreover, we have $|R_{n,k}^{(d)}| = |\mathcal{P}_{n,k}^{((d))}|$.*

**Example 7.93** Let $\pi = 19/26/348/5/7$ be a set partition in $\mathcal{P}_{(4)}(9,5)$, where its corresponding rook placement (see Figure 7.2) belongs to $R^{(4)}(9,5)$.

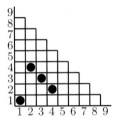

**FIGURE 7.2**: Rook placement of the set partition $19/26/348/5/7$

**Theorem 7.94** (see Chen et al. [68] and Kasraoui [156]) *Let $1 \leq d \leq k$. For all $n \geq k$, $|\mathcal{P}_{n,k}^{(d)}| = |\mathcal{P}_{n+1-d,k+1-d}|$.*

**Proof** Proposition 7.92 gives that $|R_{n,k}^{(d)}| = |\mathcal{P}_{n,k}^{((d))}|$. Thus, it suffices to prove that $|R_{n,k}^{(d)}| = |R_{n+1-d,k+1-d}|$, which is obvious. □

Actually, the above theorem shows that $|R_{n,k}^{(d)}| = |R_{n-j,k-j}^{(d-j)}|$ for all $1 \leq j \leq d-1$. Moreover, if we define

$$\psi_j : R_{n,k}^{(d)} \to R_{n-j,k-j}^{(d-j)}$$

by deleting the $j$ highest cells on each column of $R \in R_{n,k}^{(d)}$; see Figure 7.3.

It is obvious that the map $\psi_j$ establishes a bijection between the sets $R_{n,k}^{(d)}$ and $R_{n-j,k-j}^{(d-j)}$. Thus, Proposition 7.92 shows $|\mathcal{P}_{n,k}^{(d)}| = |\mathcal{P}_{n-j,k-j}^{(d-j)}|$, which implies (7.30). Note that the bijection $\psi_1$ is in fact equivalent to the reduction algorithm $\Phi$ that suggested in [68], since it can be factorized as

$$\psi_1 = \phi \circ \Phi \circ \phi^{-1}. \tag{7.31}$$

The bijection $\Phi$ can be described precisely as follows. Let $\pi$ be any set partition in $\mathcal{P}_{n,k}^{(d)}$, then construct the standard representation of $\pi' = \Phi(\pi)$ of a set partition in $\mathcal{P}_{n-1,k-1}^{(d-1)}$ from the standard representation of $\pi$ by

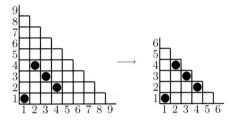

**FIGURE 7.3**: The mapping $\psi_3$

- Replacing each arc $(i,j)$ of $\pi$ by the arc $(i, j-1)$.
- Deleting the vertex $n$.

For instance, if $\pi = 19/25/36/48/7\mathcal{P}_9$, then its standard representation is given by the left side of Figure 7.4. The standard representation of $\pi' = \Phi(\pi)$ is given by the right side of Figure 7.4, which implies $\pi' = 18/247/35$.

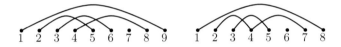

**FIGURE 7.4**: The bijection $\Phi$ maps $19/25/36/48/7$ to $18/247/35$

Now we can state the following result.

**Theorem 7.95** (see Chen et al [68]) *The map* $\Phi : \mathcal{P}_{n,k}^{(d)} \to \mathcal{P}_{n-1,k-1}^{(d-1)}$ *is a bijection.*

### 7.9.1 Noncrossing Regular Set Partitions

In order to investigate the case of noncrossing $d$-regular set partition, we present our set partitions by the standard representation. Klazar [176] investigated noncrossing $d$-regular set partitions (see [68]).

**Definition 7.96** *Denote by* $\mathrm{NC}_{n,k}^{(d)}$ *the set of noncrossing set partitions in* $\mathcal{P}_{n,k}^{(d)}$ *and the set of noncrossing set partitions in* $\mathcal{P}_{n,k}$ *by* $\mathrm{NC}_{n,k} = \mathrm{NC}_{n,k}^{(1)}$.

Let $\mathrm{NC}_{n,k}^{\leq 2,(d)}$ be the set of all poor partitions in $\mathrm{NC}_{n,k}^{(d)}$. Then Klazar proved, first with a generating function proof [176], and later bijectively [177], that

$$|\mathrm{NC}_{n,k}^{(d)}| = |\mathrm{NC}_{n-1,k-1}^{\leq 2,(d-1)}|. \tag{7.32}$$

Note that (7.32) when $d = 2$ was first obtained by Simion and Ullman [326].

**Definition 7.97** Given $\pi \mathcal{P}_n$ with its standard representation, a sequence $(i_1, j_1), (i_2, j_2), \ldots, (i_r, j_r)$ of $r$ arcs of $\pi$ is said to be an enhanced $r$-crossing if $i_1 < i_2 < \cdots < i_r \leq j_1 < j_2 < \cdots < j_r$ and said to be an $r$-crossing if $i_1 < i_2 < \cdots < i_r < j_1 < j_2 < \cdots < j_r$. Note that an $r$-crossing is just a particular enhanced $r$-crossing but the reverse is not true in general. A set partition with no enhanced $r$-crossing (respectively, $r$-crossing) is called enhanced $r$-noncrossing (respectively, $r$-noncrossing).

With this terminology, a set partition is noncrossing if and only if it is 2-noncrossing; it is poor and noncrossing if and only if it is enhanced 2-noncrossing. In particular, Klazar's [176] result can be rewritten as follows: The number of 2-noncrossing set partitions in $\mathcal{P}_{n,k}^{(d)}$ equals the number of enhanced 2-noncrossing set partitions in $\mathcal{P}_{n-1,k-1}^{(d-1)}$. Thus, it is the particular case $r = 2$ of the following result.

**Theorem 7.98** (Klazar [176]) Let $r, d \geq 2$. Then the number of $r$-noncrossing set partitions in $\mathcal{P}_{n,k}^{(d)}$ equals the number of enhanced $r$-noncrossing set partitions in $\mathcal{P}_{n-1,k-1}^{(d-1)}$.

**Proof** Let $\Phi : \mathcal{P}_{n,k}^{(d)} \to \mathcal{P}_{n-1,k-1}^{(d-1)}$ be the Chen et al. bijection as described in Theorem 7.95. In order to prove our theorem, we need to show that a set partition $\pi$ is $r$-noncrossing if and only if $\Phi(\pi)$ is enhanced $r$-noncrossing. Let $\pi \mathcal{P}_n$ and set $\pi' = \Phi(\pi)$; suppose $\pi'$ has an enhanced $r$-crossing, that is, there exists a sequence $(i_1, j1_1), \ldots, (i_r, j_r)$ of $r$ arcs of $\pi'$ such that $i_1 < i_2 < \cdots < i_r \leq j_1 < j_2 < \cdots < j_r$. By definition of $\Phi$, the pairs $(i_1, j_1 + 1), \ldots, (i_r, j_r + 1)$ are $r$ arcs of $\pi$, and they thus form an $r$-crossing of $\pi$. Therefore, we have showed that $\Phi(\pi)$ has an enhanced $r$-crossing implies that $\pi$ has an $r$-crossing, which is equivalent to $\pi$ is $r$-noncrossing implies that $\Phi(\pi)$ is enhanced $r$-noncrossing. Similarly, the converse can be verified. □

Note that the model of rook placement can be used to prove the above theorem, but the proof would be heavier and longer than the suggested proof. Also, in [68], Chen, Deng, and Du presented a reduction algorithm which transforms $d$-regular set partitions of $[n]$ to $(d-1)$-regular set partitions of $[n-1]$. They showed that the algorithm preserves the noncrossing property.

### 7.9.2 Nonnesting Regular Set Partitions

The results of the pervious subsection can be extended to the case of nonnesting $d$-regular set partitions.

**Definition 7.99** Given $\pi \mathcal{P}_n$ with its standard representation, a sequence $(i_1, j_1), (i_2, j_2), \ldots, (i_r, j_r)$ of $r$ arcs of $\pi$ is said to be an $r$-nesting if $i_1 < i_2 < \cdots < i_r \leq j_r < \cdots < j_2 < j_1$. A set partition with no $r$-nesting is called

$r$-onnesting. Denote by $\mathrm{NN}_{n,k}^{(d)}$ the set of nonnesting set partitions in $\mathcal{P}_{n,k}^{(d)}$ and set of nonnesting set partitions in $\mathcal{P}_{n,k}$ by $\mathrm{NN}_{n,k} = \mathrm{NN}_{n,k}^{(1)}$.

It is well known that

$$|\mathrm{NN}_{n,k}| = |\mathrm{NC}_{n,k}| = \mathrm{Nar}_{n,k} = \frac{1}{k}\binom{n}{k-1}\binom{n-1}{k-1},$$

where $\mathrm{Nar}_{n,k}$ is the $(n,k)$-th Narayana number.

**Theorem 7.100** (Chen, Deng, and Du [68],) *Let $r, d \geq 2$. Then, for any $1 \leq j \leq d-1$, the number of $r$-nonnesting set partitions in $\mathcal{P}_{n,k}^{(d)}$ equals the number of $r$-nonnesting set partitions in $\mathcal{P}_{n-j,k-j}^{(d-j)}$.*

**Proof** This theorem can be obtained easily by using similar arguments as in the proof of Theorem 7.98. Instead, we give another proof where we will use the model of rook placement, which leads to a quick and elegant proof. It is obvious that the correspondence $\phi : \mathcal{P}_{n,k} \to \mathrm{R}_{n,k}$ sends any $r$-nesting of a set partition $\pi$ to a *NE-chain* of length $r$ in the rook placement $\phi(\pi)$ which is a sequence of $r$ rooks in $\phi(\pi)$ such that any rook in the sequence is strictly above and to the right of the preceding rook in the sequence. It follows that the number of $r$-nonnesting set partitions in $\mathcal{P}_{n,k}^{(d)}$ equals the number of rook placements in $\mathrm{R}_{n,k}^{(d)}$ whose length of longest NE-chain is at most $r-1$. Applying the bijection $\psi_j : \mathrm{R}_{n,k}^{(d)} \to \mathrm{R}_{n-j,k-j}^{(d-j)}$, which obviously preserves the length of longest NE-chain, leads to the required result. $\square$

Consequently, the above theorem for $j = d-1$ gives that the number of $r$-nonnesting set partitions in $\mathcal{P}_{n,k}^{(d)}$ equals the number of $r$-nonnesting set partitions in $\mathcal{P}_{n-d+1,k-d+1}$. In particular, when $r = 2$, we obtain the following result.

**Corollary 7.101** (Chen, Deng, and Du [68],) *The cardinality of $\mathrm{NN}_{n,k}^{(d)}$ is given by*

$$|\mathrm{NN}_{n,k}^{(d)}| = \mathrm{Nar}_{n-d+1,k-d+1} = \frac{1}{k-d+1}\binom{n-d+1}{k-d}\binom{n-d}{k-d}.$$

*Moreover, the number of d-regular nonnesting set partitions of $[n]$ is the Catalan number $\mathrm{Cat}_{n-d+1}$.*

## 7.10  Distance Restrictions

In this section, we introduce restrictions to the distances among the elements in each block of set partition of $A$, where $A$ is given by a set of $n$ nonnegative integers. Throughout this section we fix $A = \{x_1, x_2, \ldots, x_n\}$.

**Definition 7.102** *We say that a set partition of $A$ has distance greater than $\ell$ if the distance between any two elements in the same block is at least $\ell$, that is, $|i - j| \geq \ell$ for any two different elements $x_i, x_j$ in the same block.*

*We say that a set partition of $A$ has distance that differs from $\ell$ if the distance between any two elements in the same block is not equal $\ell$, that is, $|i - j| \neq \ell$ for any two different elements $x_i, x_j$ in the same block.*

**Example 7.103** *Let $A = \{1, 3, 4, 5\}$, then we have 15 set partitions of $A$ which are*

$\{1\}, \{3\}, \{4\}, \{5\}$; $\{1\}, \{3\}, \{4, 5\}$; $\{1\}, \{3, 4\}, \{5\}$; $\{1\}, \{3, 5\}, \{4\}$; $\{1\}, \{3, 4, 5\}$;
$\{1, 3\}, \{4\}, \{5\}$; $\{1, 3\}, \{4, 5\}$; $\{1, 4\}, \{3\}, \{5\}$; $\{1, 4\}, \{3, 5\}$; $\{1, 5\}, \{3\}, \{4\}$;
$\{1, 5\}, \{3, 4\}$; $\{1, 3, 4\}, \{5\}$; $\{1, 3, 5\}, \{4\}$; $\{1, 4, 5\}, \{3\}$; $\{1, 3, 4, 5\}$.

*Among these set partitions, five have distance greater than two and there are seven set partitions have distance differing from two.*

**Theorem 7.104** *(see Chu and Wei [77, Theorem 1]) For all $n \geq \max\{1, \ell\}$, the number of set partitions of $A$ having distance greater than $\ell$ is given by*

$$S_\ell(n, k) = \sum_{j=0}^{k-\ell} \frac{(-1)^{k-\ell-j}}{(k-\ell)!} \binom{k-\ell}{j} j^{n-\ell}.$$

*Moreover, for all $n \geq k \geq \ell$, $S_\ell(n, k) = S_{n-\ell, k-\ell}$.*

**Proof** We proceed with the proof by the principle of inclusion and exclusion (see Theorem 2.72). Fix $n > \ell$. Let $M$ be the set of all distributions of $A$ into $k$ distinguishable blocks $M_1, \ldots, M_k$ with $|i - j| > \ell$ for any two elements $x_i, x_j$ in the same block. We denote the subset $M$ that contains of those distributions for which the block $M_k$ is empty by $B_k$. Let $S_\ell^*(n, k)$ be the number of the distributions in $M$ such that there no empty block. Then we have

$$S_\ell^*(n, k) = k! S_\ell(n, k) \text{ and } S_\ell^*(n, k) = \left| \cap_{j=1}^k \overline{B_j} \right|.$$

Let $\theta = \theta_1 \cdots \theta_j$ be any increasing sequence with terms in $[n]$. The cardinality $\left| \cap_{j=1}^k \overline{B_j} \right|$ can be found as follows: We distribute the first $\ell$ elements of $A$ into $\ell$ blocks from $k - j$ ones (excluding $M_i$ with $i$ as term of $\theta$), which equals $\binom{k-j}{\ell} \ell!$. On the other hand, we have $(k - \ell - j)^{n-\ell}$ ways to distribute the remaining $n - \ell$ elements $x_{\ell+1}, \ldots, x_n$ into the $k - j$ blocks such that $x_i$ with $\ell < i \leq n$ would not meet of the precedent $\ell$ elements $x_{i-1}, \ldots, x_{i-\ell}$. Thus,

$$\left| \cap_{i=1}^j B_{\theta_i} \right| = \binom{k-j}{\ell} \ell! (k - \ell - j)^{n-\ell}.$$

According to the principle of inclusion and exclusion, we obtain

$$S_\ell^*(n,k) = \sum_{j=0}^{k} \sum_{\theta} (-1)^j \left| \cap_{i=1}^{j} B_{\theta_i} \right|$$

$$= \sum_{j=0}^{k} (-1)^j \binom{k}{j} \binom{k-j}{\ell} \ell! (k-\ell-j)^{n-\ell}$$

$$= \frac{k!}{(k-\ell)!} \sum_{j=0}^{k-\ell} (-1)^j \binom{k-\ell}{j} (k-\ell-j)^{n-\ell}$$

$$= \frac{k!}{(k-\ell)!} \sum_{j=0}^{k-\ell} (-1)^{k-\ell-j} \binom{k-\ell}{j} j^{n-\ell},$$

which implies

$$S_\ell(n,k) = \sum_{j=0}^{k-\ell} \frac{(-1)^{k-\ell-j}}{(k-\ell)!} \binom{k-\ell}{j} j^{n-\ell},$$

as required. To complete the proof, we compare the explicit formula of $S_\ell(n,k)$ and Theorem 1.12 to obtain that $S_\ell(n,k) = S_{n-\ell,k-\ell}$ for all $n \geq k \geq \ell$. □

**Theorem 7.105** (Chu and Wei [77, Theorem 3]) *For all $n \geq \max\{1, \ell\}$, the number of set partitions of $A$ have distance differs from $\ell$ is given by*

$$T_\ell(n,k) = \sum_{j=1}^{k} \frac{(-1)^{k-j}}{k!} j^\ell (j-1)^{n-\ell}.$$

**Proof** Again, we proceed the proof by the principle of inclusion and exclusion (see Theorem 2.72). Fix $n > \ell > 0$ and $k > 1$. Let $M$ be the set of all distributions of $A$ into $k$ blocks $M_1, \ldots, M_k$ with $|i - j| \neq \ell$ for any two elements $x_i, x_j$ in the same block. Define $B_k$ to be the subset of $M$ containing of those distributions for which block $B_k$ is empty. Let $T_\ell^*(n,k)$ be the number of the distributions in $M$ with no empty block. Then we have

$$T_\ell^*(n,k) = k! T_\ell(n,k) \text{ and } T_\ell^*(n,k) = \left| \cap_{j=1}^{k} \overline{B_j} \right|.$$

Let $\theta = \theta_1 \cdots \theta_j$ be any increasing sequence with terms in $[n]$. The cardinality $\left| \cap_{j=1}^{k} \overline{B_j} \right|$ can be found as follows: We distribute the first $\ell$ elements of $A$ into $k - j$ blocks (excluding $M_i$ with $i$ as term of $\theta$), which equals $(k-j)^\ell!$. On the other hand, we have $(k - 1 - j)^{n-\ell}$ ways to distribute the remaining $n - \ell$ elements $x_{\ell+1}, \ldots, x_n$ into the $k - j$ blocks such that $x_i$ with $\ell < i \leq n$ would not meet of the precedent $\ell$ element $x_{i-\ell}$. Thus,

$$\left| \cap_{i=1}^{j} B_{\theta_i} \right| = (k-j)^\ell (k-1-j)^{n-\ell}.$$

According to the principle of inclusion and exclusion, we obtain

$$T_\ell^*(n,k) = \sum_{j=0}^{k} \sum_{\theta} (-1)^j \left| \cap_{i=1}^{j} B_{\theta_i} \right|$$

$$= \sum_{j=0}^{k} (-1)^j \binom{k}{j} (k-j)^\ell (k-1-j)^{n-\ell}$$

$$= \sum_{j=0}^{k} (-1)^{k-j} \binom{k}{j} j^\ell (j-1)^{n-\ell}.$$

which implies

$$T_\ell(n,k) = \sum_{j=0}^{k} \frac{(-1)^{k-j}}{k!} \binom{k}{j} j^\ell (j-1)^{n-\ell},$$

as required. □

---

## 7.11 Singletons

In this section, we focus on another statistic on set partitions, namely the largest singleton.

**Definition 7.106** *A singleton of a set partition is a block containing just one element. If $\{k\}$ is a singleton of a set partition, we denote it by $k$ for short.*

The next result states that the exponential generating function for the number of set partitions of $[n]$ without singletons is counted by $V_n$; see [327, Sequence A000296].

**Theorem 7.107** *The exponential generating function for the number of set partitions of $\mathcal{P}_n$ without singletons is given by $e^{e^x - x - 1}$.*

**Proof** Consider the number elements in the first block of a set partition, we obtain that

$$V_n = \sum_{j=1}^{n-1} \binom{n-1}{j} V_{n-1-j}$$

with the initial condition $V_0 = 1$. If we transform this recurrence in terms of exponential generating functions by Theorem 2.57, then $V'(x) = (e^x - 1)V(x)$, where $V(x) = \sum_{n\geq 0} V_n \frac{x^n}{n!}$. Therefore, $V(x) = e^{e^x - x + c}$. Using the initial condition $V(0) = 1$, we obtain the desired result. □

Note that Bernhart [32] has given a combinatorial interpretation for the relation $\text{Bell}_n = V_n + V_{n+1}$ which can also be obtain from $\text{Bell}(x) = \sum_{n\geq 0} \text{Bell}_n \frac{x^n}{n!} = V(x) + \frac{d}{dx}V(x)$. Theorem 7.107 together with Example 2.58 give $e^x V(x) = \text{Bell}(x)$ and $V(x) = e^{-x}\text{Bell}(x)$, which deduce that

$$\text{Bell}_n = \sum_{j=0}^{n} \binom{n}{j} V_j \text{ and } V_n \sum_{j=0}^{n} (-1)^{n-j} \binom{n}{j} \text{Bell}_j.$$

**Definition 7.108** *We denote the number of set partitions of $[n+1]$ with the largest singleton $k+1$ by $V_{n,k}$. Clearly, $V_{n,0} = V_n$ and $V_{n,n} = \text{Bell}_n$.*

**Lemma 7.109** (Sun and Wu [337, Lemma 2.1]) *The bivariate exponential generating function for $V_{n+k,k}$ is given by*

$$V(x,y) = \sum_{n,k\geq 0} V_{n+k,k} \frac{x^n}{n!} \frac{y^k}{k!} = e^{e^{x+y}-x-1}.$$

**Proof** By removing the largest singleton $k+1$ of a set partition of $[n+1]$ containing singletons, we obtain a set partition of $\{1,\ldots,k,k+2,\ldots,n+1\}$ whose largest singleton (if any) is at most $k$. Thus,

$$V_{n,k} = V_n + \sum_{j=0}^{k-1} V_{n-1,j},$$

which implies $V_{n,k} - V_{n,k-1} = V_{n-1,k-1}$, for all $n,k \geq 1$. Let $V_k(x) = \sum_{n\geq 0} V_{n+k,k} \frac{x^n}{n!}$. Clearly, $V_0(x) = e^{e^x-x-1}$ and $V_1(x) = e^{e^x-1}$. From the above recurrence relation, we derive

$$V_k(x) = V_{k-1}(x) + \frac{d}{dx} V_{k-1}(x) = (1+\partial_x)V_{k-1}(x) = \cdots = (1+\partial_x)^k V_0(x).$$

Hence,

$$V(x,y) = \sum_{k\geq} V_k(x) \frac{y^k}{k!} = \sum_{k\geq 0} \frac{y^k (1+\partial_x)^k}{k!} V(x) = e^{y+y\partial_x} V(x)$$
$$= e^y e^{y\partial_x} V(x) = e^y V(x+y) = e^{e^{x+y}-x-1},$$

as claimed. □

**Theorem 7.110** (Sun and Wu [337, Theorem 2.2]) *For all $n,k \geq 0$,*

$$V_{n,k} = \frac{1}{e} \sum_{m\geq 0} \frac{m^k(m-1)^n}{m!}.$$

**Proof** Lemma 7.109 gives

$$V(x,y) = e^{e^{x+y}-x-1} = e^{-x-1}\sum_{m\geq 0}\frac{e^{(x+y)m}}{m!}$$

$$= \frac{1}{e}\sum_{m\geq 0}\frac{1}{m!}\left(\sum_{n\geq 0}\frac{(m-1)^n x^n}{n!}\sum_{k\geq 0}\frac{m^k y^k}{k!}\right),$$

which, by comparing the coefficients of $\frac{x^n y^k}{n!k!}$, leads to the desired result. $\square$

**Theorem 7.111** (Sun and Wu [337, Theorem 2.4]) *For all $n,k,m \geq 0$,*

$$V_{n+m,m} = \sum_{j=0}^{n}(-1)^{n-j}\binom{n}{j}\text{Bell}_{m+j},$$

$$V_{n+m+k,m+k} = \sum_{j=0}^{m}\binom{m}{j}V_{n+k+j,k}.$$

**Proof** By Lemma 7.109 we have

$$V(x,y) = B(x+y)e^{-x} \text{ and } \frac{\partial^k}{\partial y^k}V(x,y) = V_k(x+y)e^y.$$

Thus, by comparing the coefficients of $\frac{x^n y^m}{n!m!}$, we can deduce our two identities. $\square$

Note that in [337, Theorem 2.4] provided combinatorial proofs for the identities in the statement of the above lemma.

We end this section by noting that Sun and Wu [337] found several identities between the numbers $V_n$ and Bell numbers $\text{Bell}_n$, where the proofs are based on the above results. Consequently, Sun and Wu [338] extended the above results to study the generating function for the number of set partitions of $[n]$ according to the largest singleton and sizes of the blocks; see exercises of this section.

---

## 7.12 Block-Connected

In this section we introduce two natural statistics on set partitions, namely, *connector* and *circular connector*, denoted respectively by con and ccon.

**Definition 7.112** Let $\pi = B_1/B_2/\cdots/B_k \mathcal{P}_{n,k}$ with $k > 1$.

(i) *A pair* $(a, a+1), a \in [n]$, *is called a (linear) connector if* $a \in B_i$ *and* $a + 1 \in B_{i+1}$, *for all* $i = 1, \ldots, k-1$.

(ii) *A pair* $(a, a+1), a \in [n]$, *is called a circular connector if* $a \in B_i$, *and* $a + 1 \in B_{i+1}$, *for all* $i = 1, \ldots, k-1$, *or* $a \in B_k$ *and* $a + 1 \in B_1$; *the pair* $(n, 1)$ *is a circular connector provided* $n \in B_k$.

*We define* $\mathrm{con}(\pi)$ *as the number of connectors in* $\pi$, *and* $\mathrm{ccon}(\pi)$ *as the number of circular connectors in* $\pi$, *see Tables 7.7 and 7.6.*

Note that the circular connectors are connectors when the blocks of a set partition are arranged on a circle.

**Table 7.6**: Number of set partitions of $[n]$ with exactly $r$ circular connectors

| $r \backslash n$ | 1 | 2 | 3 | 4 | 5 | 6 | 7 | 8 | 9 |
|---|---|---|---|---|---|---|---|---|---|
| 0 | 1 | 1 | 1 | 1 | 1 | 1 | 1 | 1 | 1 |
| 1 |   |   |   |   | 1 | 6 | 25 | 93 | 346 |
| 2 |   |   | 1 | 3 | 8 | 20 | 53 | 159 | 556 | 2195 |
| 3 |   |   |   | 1 | 4 | 15 | 61 | 267 | 1184 | 5366 |
| 4 |   |   |   |   | 2 | 14 | 68 | 295 | 1339 | 6620 |
| 5 |   |   |   |   |   | 1 | 11 | 97 | 694 | 4436 |

**Table 7.7**: Number of set partitions of $[n]$ with exactly $r$ connectors.

| $r \backslash n$ | 1 | 2 | 3 | 4 | 5 | 6 | 7 | 8 | 9 |
|---|---|---|---|---|---|---|---|---|---|
| 0 | 1 | 1 | 1 | 1 | 1 | 1 | 1 | 1 | 1 |
| 1 |   |   | 1 | 3 | 7 | 16 | 39 | 105 | 314 | 1035 |
| 2 |   |   |   | 1 | 6 | 24 | 86 | 307 | 1143 | 4513 |
| 3 |   |   |   |   | 1 | 10 | 61 | 313 | 1520 | 7373 |
| 4 |   |   |   |   |   | 1 | 15 | 129 | 891 | 5611 |
| 5 |   |   |   |   |   |   | 1 | 21 | 242 | 2161 |

Let $C_k(x, q)$ be the generating function for the number of set partitions of $\mathcal{P}_{n,k}$ according to the number of connectors, that is,

$$C_k(x, q) = \sum_{n \geq 0} \sum_{\pi \in \mathcal{P}_{n,k}} x^n q^{\mathrm{con}(\pi)}.$$

Our initial main goal can be formulated as follows.

**Theorem 7.113** *The generating function $C_k(x,q)$ is given by*

$$\frac{x^k}{1-x\sum_{i=1}^{k}\frac{1-x^i(q-1)^i}{1-x(q-1)}}\prod_{j=1}^{k-1}\left(q-1+\frac{1-x^{j+1}(q-1)^{j+1}}{1-x(q+j)+x\frac{1-x^{j+1}(q-1)^{j+1}}{1-x(q-1)}}\right).$$

**Proof** Since each set partition $\pi$ with exactly $k$ blocks can be decomposed as

$$\pi = 11\cdots 12\pi^{(2)}3\pi^{(3)}4\pi^{(4)}\cdots k\pi^{(k)},$$

where $\pi^{(i)}$ is a word over the alphabet $[i]$, we obtain

$$C_k(x,q) = \frac{x^k q}{1-x}W_k(x,q)\prod_{j=2}^{k-1}BW_j(x,q),$$

where $BW_k(x,q)$ is the generating function for the number of $k$-ary words $\pi$ of size $n$ with $con(k\pi(k+1)) = s$ and $W_k(x,q)$ the generating function for $k$-ary words of size $n$ according to the statistic $con$, that is,

$$BW_k(x,q) = \sum_{n\geq 0}\sum_{\pi\in[k]^n}x^n q^{con(k\pi(k+1))}$$

$$W_k(x,q) = \sum_{n\geq 0}\sum_{\pi\in[k]^n}x^n q^{con(\pi)}.$$

Exercise 7.6 gives an explicit formula for the generating function $BW_k(x,q)$. Thus,

$$C_k(x,q) = \frac{x^k q}{1-x}W_k(x,q)\prod_{j=2}^{k-1}\left(q-1+\frac{1-x^{j+1}(q-1)^{j+1}}{1-x(q-1)}W_j(x,q)\right).$$

Therefore, Exercise 7.7 completes the proof. □

**Example 7.114** *For instance, Theorem 7.113 for $k = 2, 3$, gives that the generating function $C_2(x, q)$ is given by*

$$\frac{x^2 q}{(x^2 q - x^2 + 2x - 1)(x - 1)},$$

*and the generating function $C_3(x, q)$ is given by*

$$\frac{x^3 q(xq - x - q)}{(x^3 q^2 - 2x^3 q + x^3 + 2x^2 q - 2x^2 + 3x - 1)(x^2 q - x^2 + 2x - 1)(x - 1)}.$$

Our next main theorem can be formulated as follows. Let $CC_k(x,q) = \sum_{n\geq 0}\sum_{\pi\in P_{n,k}}x^n q^{ccon(\pi)}$ be the generating function for the number of set partitions of $P_{n,k}$ according to the number $ccon$.

**Theorem 7.115** We have

$$CC_k(x,q) = \frac{x^k q}{1-x} \prod_{j=2}^{k-1} \left( q - 1 + \frac{1 - x^{j+1}(q-1)^{j+1}}{1 - x(j-1+q) + x\sum_{i=1}^{j} x^i (q-1)^i} \right).$$

$$\left[ \frac{q-1}{1-(q-1)^k x^k} + \frac{1}{(1-(k-1+q)x)(1-(q-1)x)} \right].$$

**Proof** Let $\pi$ be any set partition of $\mathcal{P}_{n,k}$; then $\pi$ can be decomposed as

$$\pi = 11 \cdots 12\pi^{(2)} 3\pi^{(3)} \cdots k\pi^{(k)},$$

where $\pi^{(j)}$ is a $j$-ary word. Thus,

$$CC_k(x,q) = \frac{x^k q}{1-x} \left( \prod_{j=2}^{k-1} BW_j(x,q) \right) BV_k(x,q|k,1),$$

where $BV_k(x,q|k,1)$ is the generating function for the number of $k$-ary words $\pi$ of length $n \geq 0$ such that $ccon(k\pi 1) = s$. With Exercise 7.8, we complete the proof. □

**Example 7.116** *For instance, Theorem 7.115 for $k = 2, 3$, yields that the generating functions $CC_2(x,q)$ and $CC_3(x,q)$ are given by*

$$\frac{q^2 x^2}{(x-1)(qx-x+1)(qx+x-1)}$$

*and*

$$\frac{qx^3(qx-x-q)^2}{(x-1)(qx+2x-1)(qx^2-x^2+2x-1)(q^2x^2-2qx^2+x^2+qx-x+1)},$$

*respectively.*

---

## 7.13 Exercises

**Exercise 7.1** *Prove the third and fifth cases of Theorem 7.82.*

**Exercise 7.2** *We have*

i. *the generating function $\sum_{n \geq k} S_\ell(n,k) x^n$ is given by*

$$\frac{x^k}{\prod_{j=1}^{k-\ell}(1-jx)}$$

**ii.** *and the numbers $S_\ell(n,k)$ are satisfy the equation*

$$x^{n-\ell} = \sum_{k=\ell+1}^{n} S_\ell(n-1, k-1)(x)_{k-\ell},$$

*where $(x)_n$ is the n-th falling factorial; see Definition 1.14.*

**Exercise 7.3** *Show that*

$$V_{n+k,k} = \sum_{i=0}^{n} \sum_{j=0}^{k} V_i \mathrm{Stir}(k,j) j^{n-i}$$

*and*

$$V_{n+k,k} = \sum_{i=0}^{n} \sum_{j=0}^{k} \mathrm{Bell}_i \mathrm{Stir}(k,j)(j-1)^{n-i}.$$

**Exercise 7.4** *A linked set partition $\pi$ of $[n]$ (see [76, 274]) is a collection of nonempty subsets $B_1, \ldots, B_k$, called blocks, of $[n]$ such that $\cup_{j=1}^{k} B_j = [n]$ and any two distinct blocks $B_i, B_j$ satisfy the following property: if $k \in B_i \cap B_j$, then either*

- $k = \min B_i$, $|B_i| > 1$ *and* $k \neq \min B_j$; *or*
- $k = \min B_j$, $|B_j| > 1$ *and* $k \neq \min B_i$.

*A linked set partition can be viewed as follows. Let $\pi$ be a linked set partition of $[n]$. First, we draw $n$ vertices on a horizontal line with points or vertices $1, 2, \ldots, n$ arranged in increasing order. For each block $B_i$ with $\min B_i = a_i$ and $|B_i| > 1$, draw an arc between the vertex $a_i$ and other vertices $b$ with $b \in B_i$ and $b > a_i$. For instance, the linear representation of the linked set partition $\{1,5,9\}, \{2,4,6\}, \{3,7\}, \{7,8\}, \{9,10\}$ is viewed in Figure 7.5.*

**FIGURE 7.5**: The linear representation of a linked set partition

1. *Find a bijection between the set of linked set partitions of $[n]$ and the set of increasing tress on $n+1$ labeled vertices (An increasing tree on $n+1$ labeled vertices is a rooted tree on vertices $0, 1, \ldots, n$ such that for any vertex $i$, $i < j$ if $j$ is a successor of $i$).*

2. *Prove that the number of linked set partitions is $n!$.*

3. A linked cycle set partition is a linked set partition where for each block the elements are arranged in a cycle. Show that the exponential generating function for the number of linked cycle set partitions of $[n]$ is given by $e^{1-\sqrt{1-2x}} - 1$ (see Sequence A001515 in [327]).

**Exercise 7.5** Following the definitions in Exercise 7.4, we are interested in the set of noncrossing linked set partitions of $\mathcal{P}_n$. The notion of noncrossing linked set partitions was introduced by Dykema [96] in the study of free probabilities. Show the following:

1. The generating function for the number of noncrossing linked set partitions of $[n]$ is given by
$$\frac{1-x-\sqrt{1-6x+x^2}}{2x}.$$

2. Find a bijection between the set of noncrossing linked set partitions of $[n]$ and the set of large Schröder paths of length $2(n-1)$. A large Schröder path of length $2n$ is a lattice path from $(0,0)$ to $(2n,0)$ consisting of up steps $(1,1)$, level steps $(2,0)$ and down steps $(1,-1)$ and never lying below the $x$-axis.

3. Find a bijection between the set of large $(3,2)$-Motzkin paths of length $n$ and the set of noncrossing linked partition of $[n+1]$. A $(\ell, \ell')$-Motzkin path of length $n$ is a lattice path from $(0,0)$ to $(n,0)$ consisting of up steps $(1,1)$, level steps $(1,0)$ and down steps $(1,-1)$ with each down step receiving one of the two colors $1, 2, \ldots, \ell'$, and each level step receiving one of the three colors $1, 2, \ldots, \ell$.

**Exercise 7.6** Show that the generating function $\sum_{n\geq 0} \sum_{\pi \in [k]^n} q^{\mathrm{con}(\pi)} x^n$ for $k$-ary words of length $n$ according to the statistic con is given by
$$W_k(x,q) = \frac{1}{1 - x\sum_{i=1}^{k} \frac{1-x^i(q-1)^i}{1-x(q-1)}}.$$

**Exercise 7.7** Show that the generating function
$$\sum_{n\geq 0} \sum_{\pi \in [k]^n} q^{\mathrm{con}(k\pi(k+1))} x^n$$

for $k$-ary words of length $n$ with $\mathrm{con}(k\pi(k+1)) = s$ is given by
$$q - 1 + \frac{1 - x^{k+1}(q-1)^{k+1}}{1 - x(k-1+q) + x\sum_{i=1}^{k} x^i(q-1)^i}.$$

**Exercise 7.8** Show that the generating function for the number of $k$-ary words $\pi$ of length $n \geq 0$ such that $\mathrm{ccon}(k\pi 1) = s$ is given by

$$\frac{(q-1)x^2}{1-(q-1)^k x^k} + \frac{x^2}{(1-(k-1+q)x)(1-(q-1)x)}.$$

**Exercise 7.9** Show that the generating function $\sum_{n\geq 0}\sum_{\pi\in\mathcal{P}_n} p^{cr(\pi)} q^{ne(\pi)} z^n$ is given by either

- $1 + \cfrac{z}{1-([1]_{p,q}+1)z - \cfrac{[1]_{p,q}z^2}{1-([2]_{p,q}+1)z - \cfrac{[2]_{p,q}z^2}{\ddots}}}$, or

- $\cfrac{1}{1-z-\cfrac{z^2}{1-([1]_{p,q}+1)z - \cfrac{[2]_{p,q}z^2}{1-([2]_{p,q}+1)z - \cfrac{[3]_{p,q}z^2}{\ddots}}}}$,

where $[r]_{p,q} = \frac{p^r - q^r}{p-q}$.

**Exercise 7.10** Show that the number of set partitions of $n$ that avoid both the subsequence patterns $\tau$ and $\tau'$ is given by $\mathrm{Fib}_{2n-2}$, where $(\tau,\tau')$ is either $(1112,1213)$, $(1122,1212)$, $(1123,1213)$, $(1123,1223)$, $(1211,1231)$, $(1212,1213)$, $(1221,1231)$, $(1222,1223)$, or $(1222,1232)$.

**Exercise 7.11** Given $\pi = \pi_1\pi_2\cdots\pi_n \in \mathcal{P}_n(111,1212)$, let $\mathrm{des}(\pi)$ denote the number of descents of $\pi$, that is, the number of indices $i$, $1 \leq i \leq n-1$, such that $\pi_i > \pi_{i+1}$, and let $\mathrm{inv}(\pi)$ denote the number of inversions of $\pi$, that is, the number of ordered pairs $(i,j)$ with $1 \leq i < j \leq n$ and $\pi_i > \pi_j$. Define the distribution polynomial $M_n(p,q)$ by $M_n(p,q) = \sum_{\pi\in\mathcal{R}_n} p^{\mathrm{des}(\pi)} q^{\mathrm{inv}(\pi)}$ with $M_0(p,q) = 1$. Show that the generating function $M(x;p,q) = \sum_{n\geq 0} M_n(p,q)x^n$ is given by

$$\cfrac{1}{1-x(1+(1-p)x) - \cfrac{px^2}{1-qx(1+(1-p)qx) - \cfrac{pq^2x^2}{1-q^2x(1+(1-p)q^2x) - \frac{pq^4x^2}{\ddots}}}}.$$

**Exercise 7.12** Prove Theorem 7.80.

**Exercise 7.13** Prove Theorem 7.81.

**Exercise 7.14** One set partition $\pi$ is said to refine another set partition $\pi'$, denoted $\pi \leq \pi'$, provided every block of $\pi$ is contained in some block of $\pi'$. Given two set partitions $\pi$ and $\pi'$, their meet, $\pi \wedge \pi'$ (respectively join, $\pi \vee \pi'$) is the largest (respectively smallest) set partition which refines (respectively, is refined by) both $\pi$ and $\pi'$. Clearly, the meet has as blocks all nonempty intersections of a block from $\pi$ with a block from $\pi'$ and the blocks of the join are the smallest subsets which are exactly a union of blocks from both $\pi$ and $\pi'$. Find the generating functions

1. for the number of pairs of set partitions of $[n]$ whose meet is the set partition $\{1\}, \{2\}, \ldots, \{n\}$?

2. for the number of pairs of set partitions of $[n]$ whose join is the set partition $\{1, 2, \ldots, n\}$?

**Exercise 7.15** Recall a crossing of a set partition is a pair of arcs $(i, j)$ and $(i', j')$ with $i < i' < j < j'$ and we can define left crossing and right crossing analogously to how it was defined for nesting arcs. Find an explicit formula for the generating function for the number of set partitions in $\mathcal{P}_n$ with exactly $k$ edges and with no right crossings.

**Exercise 7.16** The block $B$ of noncrossing set partition $\pi$ is said to be inner if there exists a block $C$ of $\pi$ such that $\min C < \min B < \max C$. Otherwise, it is said to be outer. Define the following four statistics: $isg(\pi)$ is the number of inner singletons in $\pi$, $ins(\pi)$ is the the number of inner nonsingletons in $\pi$, $osg(\pi)$ is the the number of outer singletons in $\pi$, and $ons(\pi)$ is the the number of outer nonsingletons in $\pi$. Let

$$G(a, b, c, d; t) = \sum_{n \geq 0} \sum_{\pi \in \mathcal{NCP}_n} a^{isg(\pi)} b^{ns(\pi)} c^{osg(\pi)} d^{ons(\pi)} t^n,$$

where $\mathcal{NCP}_n$ is the set of noncrossing set partitions in $\mathcal{P}_n$. Show that the generating function $G(a, b, c, d; t)$ is given by

$$\cfrac{1}{1 - ct - \cfrac{dt^2}{1 - (1+a)t - \cfrac{bt^2}{1 - (1+a)t - \cfrac{bt^2}{\ddots}}}}.$$

**Exercise 7.17** Prove Proposition 7.62.

**Exercise 7.18** Prove Proposition 7.63.

## 7.14 Research Directions and Open Problems

We now suggest several research directions which are motivated both by the results and exercises of this chapter.

**Research Direction 7.1** *In Exercise 7.4 we introduced the concepts of linked set partitions and linked cycle set partitions. In Exercise 7.5 we imposed the condition noncrossing on the set linked set partitions to obtain the set of noncrossing linked set partitions. Thus, it is naturally to ask the following question. Fix a pattern $\tau$ of length $k$ to be a linear representation of linked set partition of $[k]$. Find the number of linked set partitions of $[n]$ that avoid a pattern $\tau$ of length either three or four.*

**Research Direction 7.2** *Several research papers focused on set partitions on two or more restrictions. For instance, the study of sparse set partitions of $[n]$. A sparse set partition of $[n]$ is a set partition of $[n]$ such that for every $\in [n-1]$ the elements $i$ and $i+1$ lie in two distinct blocks. It is not hard to see that the number of sparse set partitions of $[n]$ is given by $\mathrm{Bell}_{n-1}$ the $(n-1)$-st Bell number. A set partition of $[n]$ is called abba-free if it does not happen for any four elements $i < j < k < \ell$ of $[n]$ that $i, \ell$ lie in a common block and $j, k$ in another common block. We denote the set of all sparse abba-free set partitions of $[n]$ by $\mathcal{SP}_n(abba)$. In [261], Němečeka and Klazar showed that there exists a bijection between the set of sparse abba-free set partitions of $[n]$ and the set of nonnegative words of length $n-2$ (A sequence $w = w_1w_2\cdots w_n$ is a nonnegative word if $w_i \in \{-1, 0, 1\}$ for each $i \in [n]$ and for each initial segment of a the sum of its elements is nonnegative, see [119]). In particular, they showed that the generating function for the number of sparse abba-free set partitions of $[n]$ is given by (Why?)*

$$\sum_{n\geq 0}|\mathcal{SP}_n(1221)|x^n = \frac{x}{2} + \frac{x}{2}\sqrt{\frac{1+x}{1-3x}},$$

*This result can be formulated as follows. Let $SP_n(\tau)$ be the number of standard representations of set partitions of $[n]$ that avoid the pattern*

*As consequence of this result, one can ask the following question. Find an explicit formula for the number of sparse $\tau$-free set partitions of $[n]$, where $\tau$ is any pattern from the following list:*

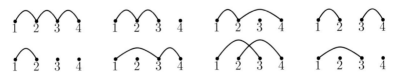

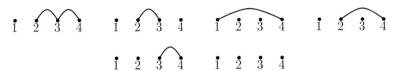

**Research Direction 7.3** *As consequence of Research Direction 7.2, we can define a* sparse set partition *as a set partition that its canonical representation form avoids the generalized pattern* 11, *there are no two consecutive equal elements. Thus, we can focus on set of sparse set partitions of* $[n]$ *that avoid a subsequence pattern* $\tau$, *that is,* $\mathcal{SP}_n(\tau)$. *Then the following question can be asked: Find an explicit formula for the number of sparse set partitions of* $[n]$ *that avoid a fixed set of patterns.*

In Tables 7.8 and 7.9 we presented the cardinality of the set $|\mathcal{SP}_n(\tau)|$ and of the set $|\mathcal{SP}_n(\tau, \tau')|$, for all subsequence set partition patterns $\tau, \tau'$ of length four.

**Table 7.8**: *Number of sparse set partitions of* $\mathcal{P}_n(\tau)$ *with* $\tau \in \mathcal{P}_4$

| $\tau, \tau' \backslash n$ | 1 | 2 | 3 | 4 | 5 | 6 | 7 | 8 | 9 | 10 |
|---|---|---|---|---|---|---|---|---|---|---|
| 1232 | 1 | 1 | 2 | 5 | 13 | 34 | 89 | 233 | 610 | 1597 |
| 1231 | 1 | 1 | 2 | 5 | 13 | 34 | 89 | 233 | 610 | 1597 |
| 1213 | 1 | 1 | 2 | 5 | 13 | 34 | 89 | 233 | 610 | 1597 |
| 1212 | 1 | 1 | 2 | 5 | 13 | 35 | 97 | 275 | 794 | 2327 |
| 1234 | 1 | 1 | 2 | 5 | 14 | 40 | 116 | 340 | 1004 | 2980 |
| 1223 | 1 | 1 | 2 | 6 | 18 | 54 | 162 | 486 | 1458 | 4374 |
| 1233 | 1 | 1 | 2 | 6 | 18 | 54 | 162 | 486 | 1458 | 4374 |
| 1211 | 1 | 1 | 2 | 5 | 14 | 42 | 135 | 459 | 1645 | 6172 |
| 1221 | 1 | 1 | 2 | 6 | 19 | 61 | 200 | 670 | 2286 | 7918 |
| 1123 | 1 | 1 | 2 | 6 | 19 | 62 | 207 | 704 | 2431 | 8502 |
| 1122 | 1 | 1 | 2 | 6 | 20 | 69 | 244 | 885 | 3295 | 12592 |
| 1222 | 1 | 1 | 2 | 6 | 19 | 63 | 221 | 817 | 3166 | 12802 |
| 1121 | 1 | 1 | 2 | 6 | 19 | 63 | 221 | 817 | 3166 | 12802 |
| 1112 | 1 | 1 | 2 | 6 | 19 | 63 | 221 | 817 | 3166 | 12802 |
| 1111 | 1 | 1 | 2 | 6 | 20 | 72 | 281 | 1188 | 5371 | 25819 |

**Table 7.9**: *Number of sparse set partitions of* $\mathcal{P}_n(\tau, \tau')$

| $\tau, \tau' \backslash n$ | 2 | 3 | 4 | 5 | 6 | 7 | 8 | 9 | 10 | 11 |
|---|---|---|---|---|---|---|---|---|---|---|
| 1111, 1234 | 1 | 2 | 5 | 13 | 30 | 59 | 90 | 90 | 0 | 0 |
| 1212, 1234 | 1 | 2 | 4 | 6 | 8 | 10 | 12 | 14 | 16 | 18 |
| 1211, 1234 | 1 | 2 | 4 | 7 | 10 | 13 | 16 | 19 | 22 | 25 |
| | | | | | | | | *Continued on next page* | | |

| | | | | | | | | | |
|---|---|---|---|---|---|---|---|---|---|
| 1211, 1232 | 1 | 2 | 4 | 7 | 11 | 16 | 22 | 29 | 37 | 46 |
| 1112, 1232 | 1 | 2 | 5 | 11 | 20 | 32 | 47 | 65 | 86 | 110 |
| 1211, 1233 | 1 | 2 | 5 | 11 | 20 | 32 | 47 | 65 | 86 | 110 |
| 1121, 1232 | 1 | 2 | 5 | 11 | 21 | 36 | 57 | 85 | 121 | 166 |
| 1213, 1221 | 1 | 2 | 5 | 11 | 21 | 36 | 57 | 85 | 121 | 166 |
| 1122, 1231 | 1 | 2 | 5 | 12 | 24 | 42 | 67 | 100 | 142 | 194 |
| 1212, 1231 | 1 | 2 | 4 | 7 | 12 | 21 | 37 | 65 | 114 | 200 |
| 1111, 1232 | 1 | 2 | 5 | 12 | 25 | 44 | 71 | 107 | 152 | 206 |
| 1221, 1234 | 1 | 2 | 5 | 12 | 24 | 43 | 71 | 110 | 162 | 229 |
| 1211, 1213 | 1 | 2 | 4 | 8 | 14 | 25 | 44 | 78 | 137 | 241 |
| 1213, 1232 | 1 | 2 | 4 | 7 | 12 | 21 | 38 | 71 | 136 | 265 |
| 1112, 1234 | 1 | 2 | 5 | 12 | 25 | 47 | 81 | 130 | 197 | 285 |
| 1121, 1234 | 1 | 2 | 5 | 12 | 25 | 47 | 81 | 130 | 197 | 285 |
| 1222, 1234 | 1 | 2 | 5 | 12 | 25 | 47 | 81 | 130 | 197 | 285 |
| 1223, 1231 | 1 | 2 | 5 | 10 | 18 | 31 | 53 | 92 | 164 | 301 |
| 1122, 1234 | 1 | 2 | 5 | 13 | 28 | 53 | 91 | 145 | 218 | 313 |
| 1213, 1222 | 1 | 2 | 5 | 11 | 21 | 38 | 68 | 121 | 214 | 377 |
| 1222, 1231 | 1 | 2 | 5 | 11 | 21 | 38 | 68 | 121 | 214 | 377 |
| 1211, 1223 | 1 | 2 | 5 | 11 | 21 | 40 | 73 | 132 | 235 | 415 |
| 1112, 1231 | 1 | 2 | 5 | 11 | 22 | 41 | 75 | 135 | 241 | 427 |
| 1211, 1231 | 1 | 2 | 4 | 8 | 16 | 32 | 64 | 128 | 256 | 512 |
| 1212, 1232 | 1 | 2 | 4 | 8 | 16 | 32 | 64 | 128 | 256 | 512 |
| 1213, 1231 | 1 | 2 | 4 | 8 | 16 | 32 | 64 | 128 | 256 | 512 |
| 1212, 1213 | 1 | 2 | 4 | 8 | 16 | 32 | 64 | 128 | 256 | 512 |
| 1231, 1232 | 1 | 2 | 4 | 8 | 16 | 32 | 64 | 128 | 256 | 512 |
| 1123, 1231 | 1 | 2 | 5 | 11 | 22 | 42 | 79 | 149 | 284 | 548 |
| 1121, 1233 | 1 | 2 | 6 | 16 | 38 | 81 | 157 | 281 | 471 | 748 |
| 1112, 1233 | 1 | 2 | 6 | 16 | 38 | 81 | 157 | 281 | 471 | 748 |
| 1213, 1233 | 1 | 2 | 5 | 11 | 23 | 47 | 95 | 191 | 383 | 767 |
| 1123, 1232 | 1 | 2 | 5 | 11 | 23 | 47 | 95 | 191 | 383 | 767 |
| 1231, 1233 | 1 | 2 | 5 | 11 | 23 | 47 | 95 | 191 | 383 | 767 |
| 1213, 1234 | 1 | 2 | 4 | 9 | 20 | 44 | 96 | 208 | 448 | 960 |
| 1231, 1234 | 1 | 2 | 4 | 9 | 20 | 44 | 96 | 208 | 448 | 960 |
| 1232, 1234 | 1 | 2 | 4 | 9 | 20 | 44 | 96 | 208 | 448 | 960 |
| 1211, 1212 | 1 | 2 | 4 | 8 | 17 | 37 | 82 | 185 | 423 | 978 |
| 1122, 1232 | 1 | 2 | 5 | 12 | 27 | 58 | 121 | 248 | 503 | 1014 |
| 1221, 1232 | 1 | 2 | 5 | 12 | 27 | 58 | 121 | 248 | 503 | 1014 |
| 1123, 1221 | 1 | 2 | 6 | 17 | 43 | 97 | 198 | 372 | 653 | 1084 |
| 1111, 1231 | 1 | 2 | 5 | 12 | 27 | 56 | 117 | 249 | 533 | 1144 |
| 1111, 1213 | 1 | 2 | 5 | 12 | 27 | 56 | 117 | 249 | 533 | 1144 |
| 1111, 1233 | 1 | 2 | 6 | 17 | 44 | 102 | 212 | 401 | 702 | 1154 |

*Continued on next page*

| | | | | | | | | | |
|---|---|---|---|---|---|---|---|---|---|
| 1121, 1231 | 1 | 2 | 5 | 12 | 27 | 59 | 128 | 278 | 605 | 1318 |
| 1223, 1232 | 1 | 2 | 5 | 12 | 28 | 64 | 144 | 320 | 704 | 1536 |
| 1223, 1234 | 1 | 2 | 5 | 12 | 28 | 64 | 144 | 320 | 704 | 1536 |
| 1212, 1223 | 1 | 2 | 5 | 12 | 28 | 64 | 144 | 320 | 704 | 1536 |
| 1212, 1233 | 1 | 2 | 5 | 12 | 28 | 64 | 144 | 320 | 704 | 1536 |
| 1213, 1223 | 1 | 2 | 5 | 12 | 28 | 64 | 144 | 320 | 704 | 1536 |
| 1232, 1233 | 1 | 2 | 5 | 12 | 28 | 64 | 144 | 320 | 704 | 1536 |
| 1123, 1212 | 1 | 2 | 5 | 12 | 28 | 64 | 144 | 320 | 704 | 1536 |
| 1233, 1234 | 1 | 2 | 5 | 12 | 28 | 64 | 144 | 320 | 704 | 1536 |
| 1121, 1213 | 1 | 2 | 5 | 12 | 28 | 64 | 145 | 328 | 743 | 1686 |
| 1123, 1233 | 1 | 2 | 6 | 16 | 39 | 89 | 194 | 410 | 849 | 1735 |
| 1221, 1231 | 1 | 2 | 5 | 12 | 28 | 65 | 151 | 351 | 816 | 1897 |
| 1112, 1213 | 1 | 2 | 5 | 12 | 28 | 65 | 151 | 351 | 816 | 1897 |
| 1122, 1213 | 1 | 2 | 5 | 12 | 28 | 65 | 151 | 351 | 816 | 1897 |
| 1212, 1221 | 1 | 2 | 5 | 12 | 28 | 65 | 151 | 351 | 816 | 1897 |
| 1222, 1232 | 1 | 2 | 5 | 12 | 28 | 65 | 151 | 351 | 816 | 1897 |
| 1123, 1211 | 1 | 2 | 5 | 12 | 28 | 67 | 159 | 382 | 917 | 2215 |
| 1222, 1233 | 1 | 2 | 6 | 16 | 40 | 96 | 224 | 512 | 1152 | 2560 |
| 1221, 1233 | 1 | 2 | 6 | 16 | 40 | 96 | 224 | 512 | 1152 | 2560 |
| 1221, 1223 | 1 | 2 | 6 | 16 | 40 | 96 | 224 | 512 | 1152 | 2560 |
| 1121, 1223 | 1 | 2 | 6 | 16 | 40 | 96 | 224 | 512 | 1152 | 2560 |
| 1223, 1233 | 1 | 2 | 6 | 16 | 40 | 96 | 224 | 512 | 1152 | 2560 |
| 1112, 1223 | 1 | 2 | 6 | 16 | 40 | 96 | 224 | 512 | 1152 | 2560 |
| 1123, 1234 | 1 | 2 | 5 | 13 | 33 | 82 | 200 | 480 | 1136 | 2656 |
| 1212, 1222 | 1 | 2 | 5 | 12 | 29 | 71 | 175 | 434 | 1082 | 2709 |
| 1112, 1212 | 1 | 2 | 5 | 12 | 29 | 71 | 175 | 434 | 1082 | 2709 |
| 1111, 1223 | 1 | 2 | 6 | 17 | 44 | 106 | 248 | 571 | 1282 | 2826 |
| 1121, 1212 | 1 | 2 | 5 | 12 | 29 | 72 | 182 | 466 | 1207 | 3158 |
| 1122, 1223 | 1 | 2 | 6 | 17 | 45 | 113 | 273 | 641 | 1473 | 3329 |
| 1111, 1212 | 1 | 2 | 5 | 12 | 29 | 73 | 189 | 497 | 1324 | 3570 |
| 1122, 1221 | 1 | 2 | 6 | 18 | 50 | 130 | 322 | 770 | 1794 | 4098 |
| 1123, 1213 | 1 | 2 | 5 | 13 | 34 | 89 | 233 | 610 | 1597 | 4181 |
| 1122, 1212 | 1 | 2 | 5 | 13 | 34 | 89 | 233 | 610 | 1597 | 4181 |
| 1123, 1222 | 1 | 2 | 6 | 17 | 45 | 115 | 287 | 706 | 1723 | 4190 |
| 1122, 1233 | 1 | 2 | 6 | 17 | 46 | 120 | 304 | 752 | 1824 | 4352 |
| 1211, 1221 | 1 | 2 | 5 | 13 | 34 | 90 | 241 | 652 | 1779 | 4889 |
| 1222, 1223 | 1 | 2 | 6 | 16 | 42 | 110 | 288 | 754 | 1974 | 5168 |
| 1112, 1221 | 1 | 2 | 6 | 17 | 45 | 116 | 298 | 770 | 2008 | 5289 |
| 1123, 1223 | 1 | 2 | 6 | 17 | 47 | 128 | 345 | 923 | 2456 | 6509 |
| 1122, 1211 | 1 | 2 | 5 | 13 | 35 | 96 | 271 | 785 | 2341 | 7174 |
| 1221, 1222 | 1 | 2 | 6 | 17 | 47 | 131 | 369 | 1047 | 2987 | 8560 |

*Continued on next page*

| | | | | | | | | | |
|---|---|---|---|---|---|---|---|---|---|
| 1121, 1123 | 1 | 2 | 6 | 17 | 48 | 136 | 388 | 1115 | 3226 | 9391 |
| 1112, 1123 | 1 | 2 | 6 | 17 | 48 | 138 | 403 | 1192 | 3564 | 10755 |
| 1121, 1221 | 1 | 2 | 6 | 18 | 52 | 149 | 431 | 1264 | 3754 | 11267 |
| 1111, 1123 | 1 | 2 | 6 | 18 | 52 | 150 | 439 | 1297 | 3852 | 11529 |
| 1122, 1123 | 1 | 2 | 6 | 18 | 53 | 156 | 462 | 1379 | 4149 | 12577 |
| 1111, 1221 | 1 | 2 | 6 | 18 | 53 | 158 | 481 | 1491 | 4688 | 14912 |
| 1112, 1122 | 1 | 2 | 6 | 18 | 54 | 163 | 498 | 1545 | 4876 | 15670 |
| 1121, 1122 | 1 | 2 | 6 | 18 | 54 | 163 | 500 | 1562 | 4977 | 16174 |
| 1211, 1222 | 1 | 2 | 5 | 13 | 37 | 112 | 363 | 1235 | 4427 | 16526 |
| 1112, 1211 | 1 | 2 | 5 | 14 | 41 | 128 | 422 | 1464 | 5316 | 20126 |
| 1121, 1211 | 1 | 2 | 5 | 14 | 41 | 129 | 426 | 1486 | 5407 | 20554 |
| 1122, 1222 | 1 | 2 | 6 | 19 | 60 | 189 | 598 | 1913 | 6216 | 20581 |
| 1112, 1222 | 1 | 2 | 6 | 17 | 49 | 148 | 474 | 1597 | 5661 | 20952 |
| 1121, 1222 | 1 | 2 | 6 | 17 | 49 | 148 | 474 | 1597 | 5661 | 20952 |
| 1111, 1211 | 1 | 2 | 5 | 14 | 41 | 130 | 434 | 1536 | 5682 | 21989 |
| 1112, 1121 | 1 | 2 | 6 | 17 | 51 | 162 | 542 | 1901 | 6961 | 26536 |
| 1111, 1122 | 1 | 2 | 6 | 19 | 60 | 194 | 645 | 2208 | 7782 | 28209 |
| 1111, 1222 | 1 | 2 | 6 | 18 | 55 | 179 | 611 | 2195 | 8207 | 31994 |
| 1111, 1121 | 1 | 2 | 6 | 18 | 57 | 193 | 691 | 2602 | 10248 | 42042 |
| 1111, 1112 | 1 | 2 | 6 | 18 | 57 | 193 | 691 | 2602 | 10248 | 42042 |

**Research Direction 7.4** We say that a set partition of $[n]$ has a type $m = (m_1, m_2, \ldots, m_p)$ if there are exactly $m_i$ blocks of size $i$, where $\sum_i i m_i = n$. Floater and Lyche [108] showed that the number of noncrossing set partitions of $[n]$ of type $m = (m_1, m_2, \ldots, m_p)$ is given by

$$\frac{p(p-1)\cdots(p-b+1)}{m_1! m_2! \cdots m_p!},$$

where $b = m_1 + m_2 \cdots + m_p$.

This result motivates the following questions: Find an explicit formula for the number of nonnesting set partitions of $[n]$ of type $m = (m_1, m_2, \ldots, m_p)$? More generally, find an explicit formula for the number of set partitions in $\mathcal{P}_n(\tau)$ of type $m = (m_1, m_2, \ldots, m_p)$, where $\tau$ is any fixed pattern.

**Research Direction 7.5** We do not know whether the bound of Corollary 7.23 is tight or whether there actually exist some more equivalences among the $(3, k)$-pairs. Note that if there more equivalences, they must involve $\tau$ of size at least 21, because for size 20 and less, we can check (with the aid of a computer) that the classes listed above are all nonequivalent. Also the additional equivalences must involve 112 (or equivalently 121) because all the pairs of the form $(122, \tau)$ are equivalent, and all the patterns of the form $(123, \tau)$ are equivalent to them as well, except for $(123, 1^k)$, which is not equivalent to any other $(3, k)$-pair. This suggests the following problem. Prove or disprove that there are equivalences among the $(3,k)$-pairs of the form $(112,\tau)$ other than those that we know about.

**Research Direction 7.6** *A set partition $\pi = B_1/B_2/\cdots/B_k \in \mathcal{P}_{n,k}$ is said to be d-smooth if $i \in B_s$ implies that*

$$i+1 \in \bigcup_{j=\max(1,s-d)}^{\min(s+d,k)} B_j.$$

*When $d = 1$, we use the term* smooth. *The study of smooth set partitions has been considered by Mansour [216]. He showed that the number of smooth partitions of $[n]$ with exactly $k$ blocks is given by*

$$\frac{4}{k+1}\sum_{i=0}^{(k-1)/2}\cos^2\tfrac{(2i+1)\pi}{2(k+1)}\left(1+2\cos\tfrac{(2i+1)\pi}{k+1}\right)^{n-1}$$
$$-\frac{4}{k}\sum_{i=0}^{k/2-1}\cos^2\tfrac{(2i+1)\pi}{2k}\left(1+2\cos\tfrac{(2i+1)\pi}{k}\right)^{n-1}.$$

*and the number of smooth set partitions of $\mathcal{P}_{n+1}$ is given by*

$$\sum_{j=0}^{n}(-1)^{n+j}\binom{n}{j}\binom{2j+1}{j+1}.$$

*Moreover, he established a bijection $\phi : SP_n \to SD_n$ between the set of smooth set partitions of $\mathcal{P}_n$ and the set of symmetric Dyck paths of semilength $2n-1$ with no peaks at even levels; see Exercise 2.19. These results motivate us to study a more general enumeration problem, namely, the set of d-smooth set partitions, which is left as a new research direction for the interested reader.*

**Research Direction 7.7** *In [219] is considered the problem of finding the number of m-gap-bounded set partitions of $\mathcal{P}_n$, that is, set partitions in which every pair of adjacent elements $a$, in a block satisfies $|a - b| \leq m$. Mansour and Munagi showed that the generating function $F_m(x)$ for the number of m-gap set partitions of $\mathcal{P}_n$ is given by $\frac{1-x}{1-2x}$, $\frac{1-2x}{1-3x+x^2}$ and $\frac{1-3x+x^2}{(1-x)(1-3x-x^2+x^3)}$, for $m = 1, 2, 3$, respectively. The question is to find an explicit formula for $F_m(x)$, for any $m \geq 1$.*

**Research Direction 7.8** *Let $\mathcal{P}_n^m$ be all set partitions of $\mathcal{P}_n$ such that the size of each block is at most $m$. Clearly, $\mathcal{P}_n^1 = \{12\cdots n\}$ and $\mathcal{P}_n^n = \mathcal{P}_n$. In this direction, we suggest to the reader to extend some of the results of this chapter in particular and the previous chapter in general by considering the same research question for $\mathcal{P}_n^m$ instead of $\mathcal{P}_n$. For instance, if $f(x)$ is the generating function for the number of set partitions of $\mathcal{P}_n^m(1212)$ (noncrossing set partitions of $\mathcal{P}_n^m$), then it is not hard to see that $f(x)$ satisfies*

$$f(x) = 1 + xf(x) + x^2f^2(x) + \cdots + x^m f^m(x).$$

Note that for $m = 2$, $f(x)$ counts the set of Motzkin paths of size $n$ (see Example 2.62), and for $m = \infty$ (that means no restriction on sizes of the blocks) the generating function $f(x)$ counts the set of Dyck paths of size $2n$ (see Example 2.61), as expected.

Another example; let $g(x)$ be the generating function for the number of set partitions of $\mathcal{P}_n^m(1112, 1212, 1213)$; compare with Theorem 7.82(1). It is not hard to verify that the generating function $g(x)$ satisfies

$$g(x) = 1 + x + \cdots + x^m + x(g(x) - 1) + 2x^2(g(x) - 1),$$

which implies that $g(x) = \frac{1 - x^2 + x^3 + \cdots + x^m}{1 - x - 2x^2}$.

**Research Direction 7.9** *Another research direction worthy of attention is to study the set partitions of $\mathcal{P}_n$ such that the size of each block equals zero modulo $m$. In this context, we denote such sets by $\mathcal{P}_n^{mod\ m}$, that is*

$$\mathcal{P}_n^{mod\ m} = \{B_1/B_2/\cdots/B_k \in \mathcal{P}_n \mid |B_i| \equiv 0 (\mod m)\ \text{for all}\ i \in [k]\}.$$

*For instance, let us consider the enumeration of the noncrossing set partitions of $\mathcal{P}_n^{mod\ m}$, that is, $\mathcal{P}_n^{mod\ m}(1212)$. Let $f(x) = \sum_{n \geq 0} |\mathcal{P}_n^{mod\ m}(1212)| x^{n/m}$ be the generating function for the number of noncrossing set partitions of $\mathcal{P}_n^{mod\ m}$. By writing each set partition $\pi$ in $\mathcal{P}_n^{mod\ m}$ as $\pi = 1\pi^{(1)} 1\pi^{(2)} \cdots 1\pi^{(s)}$ with $s \equiv 0 (\mod m)$, we obtain*

$$f(x) = 1 + xf^m(x) + x^2 f^{2m}(x) + \cdots = \frac{1}{1 - xf^m(x)},$$

*which implies*

$$f(x) = 1 + xf^{m+1}(x).$$

*Hence, the number of noncrossing set partitions of $\mathcal{P}_n^{mod\ m}$ is given by the m-ary number $\frac{1}{(m-1)n+1}\binom{mn}{n}$. It is well known that m-ary trees (a plane rooted tree such that the number of children of each node is either zero or m) are enumerated by m-ary numbers. For more information on these numbers, we refer the reader to [136]. Actually, it is not hard to define a bijection between our noncrossing set partitions of $\mathcal{P}_n^{mod\ m}$ and m-ary trees with m internal nodes (nodes with m children). This fact motivates the following suggestion: We suggest to the reader to study the set $\mathcal{P}_n^{mod\ m}$ under certain set of conditions or statistics.*

# Chapter 8

## Asymptotics and Random Set Partition

Asymptotic analysis usually involves complex analysis or probability theory or both. The idea is to study the behavior of the number of set partitions of interest, or a statistic on set partitions as $n \to \infty$. With tools from complex analysis together with our results, we obtain growth rates for the number of set partitions of $\mathcal{P}_n$ according to a statistic of interest. On the other hand, when considering set partitions as randomly selected from $\mathcal{P}_n$ (that is, all set partitions of $\mathcal{P}_n$ are equally likely to occur), we can derive results on the average or variance of a random variable which presents a statistic. In this chapter we start with presenting the basic tools from probability theory and complex analysis. However, the interested reader who wants to get deeper into asymptotics is referred to Flajolet and Sedgewick's book [107], which contains many examples and one-step references, for the technique of asymptotic analysis.

At first the asymptotics research of set partitions held interest only for the behavior of the number of set partitions. Indeed, the research for many years focused on the asymptotic behavior of the $n$-th Bell number when $n \to \infty$. More precisely, the study of the behavior of the number of set partitions of $\mathcal{P}_n$, that is, the behavior of the $n$-th Bell number has been studied in a book by Knopp [192], where he obtained only the dominant term in the asymptotic expansion of the $n$-th Bell number. Almost 30 years later, Moser and Wyman [256] used contour integration to obtain the complete asymptotic formula for the $n$-th Bell number (for a more general study, we refer the reader to [134]). Also, the study of asymptotic behavior of the number of set partitions of $\mathcal{P}_n$ was extended to include the asymptotic behavior of the number of set partitions of $\mathcal{P}_{n,k}$. Moser and Wyman [257] presented asymptotic formulas for Stirling numbers of the second kind (see also [46, 85, 86, 103, 154, 364]).

The asymptotic on set partitions does not stop with the study of the number of set partitions of $\mathcal{P}_n$ or of $\mathcal{P}_{n,k}$. Actually, the research has been extended to cover the asymptotic behavior of the average of a fixed statistic in a random set partition of $\mathcal{P}_n$, or a subset of set partitions of $[n]$ has been fixed and then studied asymptotically the number of elements of this subset when $n$ grows to infinity. For instance, Bender, Odlyzko, and Richmond [27] studied the number of irreducible set partitions of $\mathcal{P}_n$. In [191] were considered the asymptotic behavior of the number of distinct block sizes in a set partition of $\mathcal{P}_n$ (see also [356]). In [190] were considered asymptotically the mean and the variance of two parameters on set partitions of $\mathcal{P}_n$, which are related to the records of words. For other examples, see [84, 272, 308].

Other results on set partitions were obtained by means of probability theory in general and random variables in particular. For instance, Port [275] studied the connection between Bell and Stirling numbers and the moments of a Poisson random variable. A good review paper is [270], where it examines some properties of set partitions with an emphasis on a statistical viewpoint. The paper cited 58 references focusing on proofs of Dobiński's formula and some variants, equality of Bell numbers with moments of the Poisson distribution, derivation of the generating function for Stirling numbers, and how to generate a set partition at random. In [240], Mansour and Shattuck provided a formula for the probability that a geometrically distributed word of size $n$ is a set partition. Indeed, the first consideration of such a formula was derived in [266], using the Cauchy integral formula. In addition, Mansour and Shattuck extended this result by fixing the number of blocks, levels, rises, or descents. For other examples, see [253, 271, 274].

## 8.1 Tools from Probability Theory

Oftentimes we are interested in the value of a statistic for a "typical" set partition. We can compute this "typical" value by assuming that we are presented with a set partition that is randomly selected from all set partitions of $[n]$, making all set partitions of $n$ equally likely. Here, the statistic becomes a random quantity, and then we can evaluate its statistic parameters such as average and variance. If the explicit formula for the number of set partitions with a given fixed statistic is known, then we can compute the average of the statistic over all set partitions, or equivalently, the "typical" value of the statistic for a randomly selected set partition.

In order to describe this method, first, we need very basic definitions and tools from probability theory. Here, we restrict ourselves throughout to definitions for countable sample spaces, as those are the only ones that will be considered here. We refer the reader to [37, 128] for the definitions and the results for more general sample spaces.

**Definition 8.1** *The* sample space $\Omega$ *is the set of all possible* outcomes *of an experiment. A subset $E$ of $\Omega$ is called an* event. *An event occurs if any outcome $\omega \in E$ occurs. The* probability function $\mathbf{P} \colon \Omega \to [0,1]$ *satisfies*

1. $\mathbf{P}(E) \geq 0$, *for any $E \subseteq \Omega$.*
2. $\mathbf{P}(\Omega) = 1$.
3. $\mathbf{P}(E_1 \cup E_2 \cup \cdots) = \sum_i \mathbf{P}(E_i)$, *for a countable sequence $E_1, E_2, \ldots$ of pairwise disjoint sets (that is, $E_i \cap E_j = \emptyset$ for $i \neq j$).*

*Asymptotics and Random Set Partition* 381

The probability that the event $E$ occurs is defined by $\mathbf{P}(E)$.

As a consequence of the above definition, it is sufficient to specify $\mathbf{P}(\{\omega\})$ for each of the outcomes $\omega \in \Omega$ when $\Omega$ is a countable set.

**Definition 8.2** *We say that the sample space is equipped with the* uniform probability measure *if all outcomes in a finite sample space are equally likely, and we write $\mathbf{P}(\{\omega\}) = 1/|\Omega|$ for every $\omega \in \Omega$. Consequently, for any subset $E$ of $\Omega$, $\mathbf{P}(E) = \frac{|E|}{|\Omega|}$.*

We will study what happens to a statistic of a random set partition that satisfies certain set of conditions, for example, the number of blocks in the set of noncrossing set partitions of $n$ when $[n]$ is large. Thus, our sample space is the set of all set partitions with those conditions, for example, set partitions with even number of blocks, set partitions that avoid a subword pattern, and so on. Unless stated otherwise, we will assume that the sample space is equipped with the uniform probability measure.

**Definition 8.3** *A* discrete random variable *$X$ is a function from $\Omega$ to a countable subset of $\mathbb{R}$. We define its* probability mass function *by*

$$\mathbf{P}(X = m) = \mathbf{P}(\{\omega \mid X(\omega) = m\}).$$

*The set of values together with their respective probabilities characterize the* distribution *of $X$. If the set of values of $X$ is given by $\{x_1, x_2, \ldots\}$, then the probability mass function satisfies*

1. $\mathbf{P}(X = x) = 0$ *if* $x \notin \{x_1, x_2, \ldots\}$.
2. $\mathbf{P}(X = x_i) \geq 0$ *for* $i = 1, 2, \ldots$
3. $\sum_{i=1}^{\infty} \mathbf{P}(X = x_i) = 1$.

Note that it is very common to denote the random variable with a capital letter and its values with the corresponding lower-case letter.

**Example 8.4** *Let $\Omega = \mathcal{P}_4$, where we assume that all set partitions of $\mathcal{P}_4$ occur equally likely. Define $X = \text{blo}(\pi)$ for $\pi \in \mathcal{P}_4$ to be a discrete random variable that takes values $k = 1, 2, 3, 4$. The respective probabilities are computed via the set partitions of $[4]$ that have exactly $k$ blocks.*

| $k$ | 1 | 2 | 3 | 4 |
|---|---|---|---|---|
| $\omega$ | 1111 | 1112 | 1123 | 1234 |
|  |  | 1121 | 1213 |  |
|  |  | 1122 | 1223 |  |
|  |  | 1211 | 1231 |  |
|  |  | 1212 | 1232 |  |
|  |  | 1221 | 1233 |  |
|  |  | 1222 |  |  |
| $\mathbf{P}(X = k)$ | $\frac{1}{15}$ | $\frac{7}{15}$ | $\frac{6}{15}$ | $\frac{1}{15}$ |

In this particular case, we can compute the probabilities by using Theorem 1.12 (see Definition 1.11), as it presents the number of set partitions of $[n]$ with exactly $[k]$ blocks as $\text{Stir}(n,k)$.

Other examples of random variables on the sample space of set partitions $\mathcal{P}_n$ are $X$ = number of occurrences of a pattern $\tau$ or $X$ = largest singleton in a set partition $\pi$. Example 8.4 is an illustration of the following lemma.

**Lemma 8.5** *Let $\Omega$ be any finite sample space. If $\Omega$ is equipped with the uniform probability measure and $X$ is a discrete random variable on $\Omega$, then*

$$\mathbf{P}(X = t) = |\{\omega \mid X(\omega) = t\}|/|\Omega|.$$

**Definition 8.6** *Let $A$ be any countable set. For a discrete random variable $X$ which takes values $i \in A$ with probability $\mathbf{P}(X = i)$, the expected value of $f(X)$ for a (measurable) function $f$ is given by*

$$\mathbb{E}(f(X)) = \sum_{i \in A} \mathbf{P}(X = i) f(i),$$

*assuming that the sum converges. In particular, the* mean *(or average or expected value) $\mu$ and the* variance $\text{Var}(X)$ *of $X$ are given by*

$$\mu = \mathbb{E}(X) = \sum_{i \in A} i \mathbf{P}(X = i)$$

*and*

$$\text{Var}(X) = \mathbb{E}\left((X - \mu)^2\right) = \mathbb{E}(X^2) - \mu^2.$$

*The quantity $\mathbb{E}(X^k)$ is called the $k$-th* moment *of X, and $\sqrt{\text{Var}(X)}$ is called the* standard deviation *of $X$. Furthermore,*

$$\mathbb{E}(X(X-1)(X-2) \cdots (X-k+1))$$

*is called the $k$-th factorial moment of $X$.*

**Example 8.7** (Continuation of Example 8.4) Using the distribution derived in Example 8.4 for the random variable $X = \text{blo}(\sigma)$, we have that

$$\mathbb{E}(X) = \frac{1}{15} + 2 \cdot \frac{7}{15} + 3 \cdot \frac{6}{15} + 4 \cdot \frac{1}{15} = \frac{37}{15},$$

$$\mathbb{E}(X^2) = \frac{1}{15} + 4 \cdot \frac{7}{15} + 9 \cdot \frac{6}{15} + 16 \cdot \frac{1}{15} = \frac{99}{15},$$

and

$$\text{Var}(X) = \frac{91}{15} - \frac{37^2}{15^2} = \frac{116}{15^2}.$$

If all set partitions under consideration are equally likely, then a general formula for the mean and the variance of a statistic can be derived easily.

**Lemma 8.8** *Let $\Omega$ be any sample space equipped with the uniform probability measure and fix a random discrete variable $X$ on $\Omega$. Let $a_{n,t}$ be the number of set partitions $\pi \in \mathcal{P}_n$ such that $X(\pi) = t$. Then*

$$\mathbb{E}(X) = \frac{1}{|\Omega|} \sum_t t a_{n,t}. \tag{8.1}$$

**Proof** Follows immediately from Lemma 8.5 and Definition 8.6. □

Note that if we have an explicit formula for the quantity $a_{n,m}$, then (8.1) suggests a formula for the mean of a statistic $X$. To illustrate this, we will find explicit formulas for the average number of levels in all set partitions of $[n]$.

**Example 8.9** *Let $X_{\text{lev}}(n)$ be the number of levels in a random set partition of $[n]$. By Corollary 4.11, the average number of levels is given by*

$$\mathbb{E}(X_{\text{lev}}(n)) = \frac{\sum_{m=0}^{n-1} \left( m \sum_{k=1}^{n-m} \binom{n-1}{m} \text{Stir}(n-1-m, k-1) \right)}{\text{Bell}_n},$$

*which is equivalent to*

$$\mathbb{E}(X_{\text{lev}}(n)) = \frac{(n-1) \sum_{m=0}^{n-1} \binom{n-2}{m-1} \text{Bell}_{n-1-m}}{\text{Bell}_n}.$$

*On the other hand, Corollary 4.11 gives that*

$$\sum_{m=0}^{n-1} \left( \sum_{k=1}^{n-m} \binom{n-1}{m} \text{Stir}(n-1-m, k-1) \right) = \text{Bell}_n,$$

*which implies that $\sum_{m=0}^{n-1} \binom{n-1}{m} \text{Bell}_{n-1-m} = \text{Bell}_n$. Therefore,*

$$\mathbb{E}(X_{\text{lev}}(n)) = \frac{(n-1) \text{Bell}_{n-1}}{\text{Bell}_n}.$$

**Example 8.10** *Let $N_{\text{ris}}(n)$ be the number of nontrivial rises in a random set partition of $[n]$, and let $N_{\text{des}}(n)$ be the number of nontrivial descents in a random set partition of $[n]$. Then by Lemma 4.19 we have*

$$\mathbb{E}(N_{\text{ris}}(n)) = \mathbb{E}(N_{\text{des}}(n)) = \frac{\sum_{k=1}^{n} \sum_{j=2}^{k} \binom{j}{2} \sum_{i=2}^{n-k} j^{i-2} \text{Stir}(n-i, k)}{\text{Bell}_n}.$$

*Try to simplify the formula!*

**Example 8.11** Let $X_{\text{ris-des}}(n)$ be the number of "rises–descents" in a random set partition of $[n]$. By Corollary 4.15, we obtain that the average number of rises–descents is given by

$$\mathbb{E}(X_{\text{ris-des}}(n)) = \frac{\text{Bell}_{n+1} - \text{Bell}_{n-1}}{2\text{Bell}_n} - \frac{1}{2}.$$

**Example 8.12** (Dobiński's formula, see Theorem 3.2) For $a > 0$, let $X_a$ denote a random variable with Poisson distribution, $\mathbf{P}(X_a = m) = e^{-a}\frac{a^m}{m!}$ for all $m \geq 0$, so that

$$\mathbb{E}(f(X_a)) = e^{-a} \sum_{m \geq 0} f(m) \frac{a^m}{m!}.$$

By choosing $f(m) = m^n$, we obtain $e^{-a} \sum_{m \geq 0} \frac{m^n a^m}{m!} = \mathbb{E}(X_a^m)$. Therefore, Dobiński's formula is equivalent to

$$\mathbb{E}(X_a^m) = \sum_{k=1}^{n} \text{Stir}(n,k) \mathbb{E}((X_a)_{(k)}) = \sum_{k=1}^{n} \text{Stir}(n,k) a^k,$$

where we used that $E((X_a)_{(k)}) = a^k$ (prove by induction on $k$). The above formula has an interpretation in terms of Poisson process; see [271].

Now, the question is what we should do if we do not have an explicit formula for the statistic of interest, but the generating function is known. In this case, we can compute the mean and the variance of the statistic from a different but related generating function, namely, the probability generating function.

**Definition 8.13** We define the *probability generating function of a discrete random variable* $X$ by $pg_X(u) = \sum_m \mathbf{P}(X = m)u^m = \mathbb{E}(u^X)$ and the *moment generating function of a random variable* $X$ by $M_X(t) = \mathbb{E}(e^{tX}) = pg_X(e^t)$.

**Example 8.14** Let $\Omega = \mathcal{P}_n$ with $n \geq 1$ where we assume that all set partitions occur equally likely. Then $X_{\text{blo}}(n) = $ blo *is a discrete random variable (number blocks) that takes values* $m = 1, 2, \ldots, n$. The respective (nonzero) probabilities are computed via the set partitions that have $m$ blocks:

$$\mathbf{P}(X_{\text{blo}}(n) = m) = \frac{\text{Stir}(n,m)}{\text{Bell}_n}.$$

Thus, the probability generating function of $X = X_{\text{blo}}$ is given by

$$pg_X(u) = \sum_{m=1}^{n} \mathbf{P}(X(n) = m) u^m = \frac{\text{Bell}_n(u)}{\text{Bell}_n},$$

where $\text{Bell}_n(u) = \sum_{k=1}^{n} \text{Stir}(n,k) u^k$ is the $n$-th Bell polynomial[1].

---
[1] http://mathworld.wolfram.com/BellPolynomial.html

**Lemma 8.15** *The $k$-th moment and the $k$-th factorial moment of a discrete random variable $X$ from the probability generating function $pg_X(u)$ are given by*

$$\mathbb{E}(X^k) = \left.\frac{\partial^k}{\partial t^k} M_X(t)\right|_{t=0} = \left.\frac{\partial^k}{\partial t^k} pg_X\left(e^t\right)\right|_{t=0}$$

*and*

$$\mathbb{E}(X(X-1)(X-2)\cdots(X-k+1)) = \left.\frac{\partial^k}{\partial u^k} pg_X(u)\right|_{u=1},$$

*respectively.*

The next lemma describes how the generating function for the sequence $\{a_{n,m}\}_{n,m}$ is related to the probability generating function from which we can obtain the moments via differentiation.

**Lemma 8.16** *For fixed $n$, let $X$ be a statistic on the set partitions of $n$, $A(x,q) = \sum_{n,m \geq 0} a_{n,m} x^n q^m$, and $N = [x^n] A(x,1)$. Then the mean $\mu$ and the variance $\sigma^2$ of $X$ are given by*

$$\mu = \mathbb{E}(X) = \frac{1}{N}[x^n] \left.\frac{\partial}{\partial q} A(x,q)\right|_{q=1} \tag{8.2}$$

*and*

$$\sigma^2 = \mathrm{Var}(X) = \frac{1}{N}[x^n]\left.\left(\frac{\partial^2}{\partial q^2}A(x,q) + \frac{\partial}{\partial q}A(x,q) - \left(\frac{\partial}{\partial q}A(x,q)\right)^2\right)\right|_{q=1}$$

$$= \frac{1}{N}[x^n]\left.\frac{\partial^2}{\partial q^2}A(x,q)\right|_{q=1} + \mu - \mu^2. \tag{8.3}$$

**Proof** Follows immediately from Lemma 8.15 and $pg_X(q) = \frac{1}{N}[x^n]A(x,q)$. □

To illustrate the use of this lemma, we study the total statistic ris − des over all set partitions of $[n]$ with exactly $k$ blocks.

**Example 8.17** *Corollary 4.15 shows that the generating function for the number of set partitions of $[n]$ with exactly $k$ blocks according to the statistic $X_n = $ ris − des is given by*

$$F_k(x,v) = \sum_{n \geq 0} \sum_{\pi \in \mathcal{P}_{n,k}} x^n v^{X_n(\pi)}$$

$$= \frac{x^k v^{k-1}}{(1+x(1/v-1))^k \prod_{j=1}^{k} \frac{v - \frac{1}{v}\left(\frac{1+x(v-1)}{1+x(1/v-1)}\right)^j}{v - 1/v}}.$$

Hence, after simple algebraic operations (show the details!) we have that $\frac{\partial}{\partial v}F_k(x,v)|_{v=1} = \frac{(k-1)(kx-2)x^k}{2\prod_{j=1}^{k}(1-jx)}$, which, by (2.9) implies

$$\frac{\partial}{\partial v}F_k(x,v)|_{v=1} = \frac{1}{2}(k-1)(kx-2)\sum_{j\geq 1}\text{Stir}(j,k)x^j.$$

Thus, we have $\sum_{\pi \in \mathcal{P}_n} X_n(\pi) = \frac{1}{2}(\text{Bell}_{n+1} - \text{Bell}_n - \text{Bell}_{n-1})$, which, by (8.2) implies

$$\mathbb{E}(X_n) = \frac{1}{2\text{Bell}_n}(\text{Bell}_{n+1} - \text{Bell}_n - \text{Bell}_{n-1}).$$

Furthermore, if we differentiate the generating function $F_k(x,v)$ exactly twice respect to $v$, then after simple algebraic operations (show the details!) we have

$$\frac{\partial^2}{\partial v^2}F(x,v)|_{v=1}$$
$$= \frac{(k^2 - \frac{10}{3}k + 2 - \frac{1}{6}k(2k-1)(3k-7)x + \frac{1}{36}k(9k+5)(k-1)(k-2)x^2)x^k}{\prod_{j=1}^{k})1-jx)}$$
$$+ \frac{1}{3}x^k(2x-1)(x-1)\frac{\sum_{j=1}^{k}\frac{1}{1-jx}}{\prod_{j=1}^{k}(1-jx)}.$$

Try to continue!! After several not nice algebraic operations, we can obtain an explicit formula for the coefficient of $x^n$ in the above generating function. But then we need to sum over all $k = 1, 2, \ldots, n$, to derive an explicit formula for the total statistic $X_n$ on the set partitions of $[n]$. At the end, by (8.3) can be written a formula for the variance of the statistic $X_n$.

As shown in the above example, even though we can derive an explicit formula for the variance for the statistic $X_n$ on the set partitions of $[n]$, the expression is too complicated. In such cases, we will interest on asymptotic results, see below.

**Definition 8.18** *Let $X$ be a random variable. If there exists a nonnegative, real-valued function $f : \mathbb{R} \to [0, \infty)$ such that*

$$\mathbf{P}(X \in A) = \int_A f(x)dx$$

*for any measurable set $A \subset \mathbb{R}$ (a set that can be constructed from intervals by a countable number of unions and intersections), then $X$ is called* (absolutely) continuous *and $f$ is called its* density function.

**Definition 8.19** *A continuous random variable $X$ is called a* normal random variable *with mean $\mu$ and variance $\sigma^2$ if its density function has the form*

$$f(x) = \frac{1}{\sigma\sqrt{2\pi}}\exp\left[\frac{-(x-\mu)^2}{2\sigma^2}\right],$$

and we write $X \sim \mathcal{N}(\mu, \sigma^2)$. The quantities $\mu$ and $\sigma$ that appear in the density are the mean and the standard deviation of $X$. If $\mu = 0$ and $\sigma = 1$, then $X$ is called a standard normal *random variable*.

Note that if $X \sim \mathcal{N}(\mu, \sigma^2)$, then the standardized random variable $\frac{X-\mu}{\sigma}$ is a standard normal random variable, that is, $\frac{X-\mu}{\sigma} \sim \mathcal{N}(0, 1)$. Following Bender [25], we define asymptotic normality for a sequence of two indices.

**Definition 8.20** *For given a sequence $\{a_{n,k}\}_{n,k \geq 0}$ with $a_{n,k} \geq 0$ for all $k$, we associate a normalized sequence*

$$p_n(k) = \frac{a_{n,k}}{\sum_{j \geq 0} a_{n,j}}.$$

*We say that $\{a_{n,k}\}_{n,k \geq 0}$ is asymptotically normal with mean $\mu_n$ and variance $\sigma_n^2$ or satisfies a* central limit theorem *if*

$$\lim_{n \to \infty} \sup_x \left| \sum_{k \leq \sigma_n x + \mu_n} p_n(k) - \frac{1}{\sqrt{2\pi}} \int_{-\infty}^x e^{-t^2/2} dt \right| = 0,$$

*and write $a_{n,k} \stackrel{d}{\approx} \mathcal{N}(\mu_n, \sigma_n^2)$.*

## 8.2 Tools from Complex Analysis

In this section we focus on tools from complex analysis for determining asymptotic behavior of our sequences. The aim of this section that we draw on is that for a complex function $f(z)$, the location of the singularities determines the growth behavior of the coefficient of $z^n$ of the power series expansion $f(z)$ when $n \to \infty$. Thus, we will consider our generating functions as complex functions instead of formal power series, and, for example, we will use variables $z$ and $w$ instead of our variables $x$ and $q$ to indicate this viewpoint. The necessary definitions for very basic terminology and related results are provided in Appendix F. We will start by presenting the specific results (drawn primarily from Flajolet and Sedgewick [107]) that we will use later. The typical generating functions that we will consider satisfy the following result.

**Theorem 8.21** (Flajolet and Sedgewick [107, Theorems IV.5 and IV.6]) *If $f(z)$ is an analytic complex function at the origin and whose expansion at the origin has a finite radius of convergence $R$, then it has a singularity on the boundary of its disc of convergence, $|z| = R$. Furthermore, if the expansion of $f(z)$ at the origin has nonnegative coefficients, then the point $z = R$ is a singularity of $f(z)$, that is, there is at least one singularity of smallest modulus that is real-valued and positive.*

This theorem indicates how easy it is to find the dominant singularity for our generating functions – we only need to verify the analyticity of these functions along the positive real line and then find the smallest value. This singularity will determine the rate of growth, expressed using the notation $\approx$.

**Definition 8.22** *For two sequences $\{a_n\}$ and $\{b_n\}$ of real numbers, we write $a_n \approx b_n$ if $\lim_{n\to\infty} \frac{a_n}{b_n} = 1$. If $a_n \approx c^n$ for fixed $c$, then we say that $\{a_n\}$ is of exponential order $c^n$.*

**Theorem 8.23** (Exponential growth formula, Flajolet and Sedgewick [107, Theorem IV.7]) *If $f(z)$ is analytic complex function at the origin and $R$ is the modulus of a singularity nearest to the origin in the sense $R = \sup\{r \geq 0 \mid f \text{ is analytic in } |z| < r\}$, then*

$$f_n = [z^n]f(z) \approx \left(\frac{1}{R}\right)^n.$$

*Furthermore, if the expansion of $f(z)$ at the origin has nonnegative coefficients, then $R = \sup\{r \geq 0 \mid f \text{ is analytic for all } 0 \leq z < r\}$.*

If our generating function is not only analytic but also rational, then we can yield a more detailed answer for the asymptotic behavior.

**Theorem 8.24** (Rational functions, Flajolet and Sedgewick [107, Theorem IV.9]) *If $f(z)$ is an analytic rational complex function at the origin with poles at $\rho_1, \rho_2, \ldots, \rho_m$ of orders $r_1, r_2, \ldots, r_m$, respectively, then there exist polynomials $\{Q_j(z)\}_{j=1}^m$ and a positive integer $n_0$ such that for all $n > n_0$,*

$$f_n = [z^n]f(z) = \sum_{j=1}^{m} Q_j(n)\rho_j^{-n},$$

*where the degree of $Q_j$ is $r_j - 1$.*

Theorem 8.24 gives a very precise representation of the coefficients. If we interest on asymptotic behavior of the coefficients, then it is enough to consider only in the factor that dominates the growth, while we do not need to compute the polynomials $Q_j$.

**Example 8.25** Let $f(z) = \frac{1}{(1-2z)^2(1-z)^3(1-3z)}$. Then $f$ has a pole of order three at $z = 1$, pole of order two at $z = \frac{1}{2}$, and a simple pole at $z = \frac{1}{3}$. Therefore

$$f_n = Q_2(n) + Q_1(n)2^n + Q_0(n)3^n,$$

where the degree of $Q_j$ is $j$. In order to derive the asymptotic behavior of $f_n$, we look at the simple pole which it is the closest to the origin. Substituting $z = \frac{1}{3}$ everywhere except at the singularity gives that $f(z) \approx \frac{3^5}{2^3(1-3z)}$, which implies that $f_n \approx \frac{3^{n+5}}{8}$.

When we have meromorphic generating functions (see Definition F.8), then an asymptotic result that resembles the structure of the exact result for rational functions can be obtained.

**Theorem 8.26** (Meromorphic functions, Flajolet and Sedgewick [107, Theorem IV.10]) *If $f(z)$ is a meromorphic function in the closed disk $\{z \mid |z| \le R\}$ with poles at $\rho_1, \rho_2, \ldots, \rho_m$ of orders $r_1, r_2, \ldots, r_m$, respectively, and it is analytic on $|z| = R$ and at $z = 0$, then there exist $m$ polynomials $\{Q_j(z)\}_{j=1}^m$ such that*

$$f_n = [z^n] f(z) \approx \sum_{j=1}^m Q_j(n) \rho_j^{-n},$$

*where the degree of $Q_j$ is $r_j - 1$. In particular, if $f(z) = 1/h(z)$ and $\rho^*$ is the positive zero with smallest modulus and multiplicity one of $h(z)$. Then, in the neighborhood of $\rho^*$,*

$$f(z) \approx \frac{c}{1 - z/\rho^*} \quad \text{and} \quad [z^n] f(z) \approx c \cdot (\rho^*)^{-n},$$

*where $c = \frac{-1}{\rho^* \frac{d}{dz} h(z)|_{z=\rho^*}}$.*

**Example 8.27** *We illustrate the use of Theorem 8.26 by applying it to the generating function for $a_n$ the number of set partitions of $[n]$ such that each block has either one or two elements; see Example 2.13. Here we actually know an explicit formula for the coefficients, so we can see how the asymptotic result compares to the explicit one. Example 2.13 shows*

$$a_n = \text{Fib}_n = \frac{1}{\sqrt{5}} (\alpha^{n+1} - \beta^{n+1}),$$

*where $\alpha = (1 + \sqrt{5})/2$ and $\beta = (1 - \sqrt{5})/2$. Since $|\alpha| > 1$ and $|\beta| < 1$, the dominant term is the $\alpha$ term, and we obtain that*

$$a_n \approx \frac{\alpha}{\sqrt{5}} \alpha^n = 0.723607 (1.61803)^n.$$

*Now we derive the asymptotics by applying Theorem 8.26. From Example 2.13, we know that the generating function for the sequence $\{a_n\}_{n \ge 0}$ is given by*

$$A(z) = \frac{1}{1 - z - z^2} = \frac{-1}{(z - (1 - \sqrt{5})/2)(z - (1 + \sqrt{5})/2)}.$$

*Clearly, $A(z)$ is meromorphic having two poles*

$$\rho_1 = (-1 + \sqrt{5})/2 = 0.618034 = -\beta,$$
$$\rho_2 = (-1 - \sqrt{5})/2 = -1.61803 = -\alpha.$$

The pole of smaller modulus is $\rho^* = (-1+\sqrt{5})/2$. Since $h(z) = 1 - z - z^2$, we obtain that

$$\left.\frac{-1}{\rho^* \cdot h'(z)}\right|_{z=\rho^*} = \frac{1}{\rho^*(1+2\rho^*)} = 0.723607.$$

Therefore, $a_n \approx 0.723607(1.61803)^n$.

Great! Complex analysis also helps us establish asymptotic results in terms of convergence to a normal distribution (see Definition 8.20). The following proposition is a corollary to a more general result by Bender [25, Theorem 1].

**Theorem 8.28** (Bender, [25, Section 3]) *Suppose that*

$$f(z, w) = \sum_{n,k \geq 0} a_n(k) z^n w^k = \frac{g(z, w)}{P(z, w)},$$

*where*

(i) $P(z, w)$ *is a polynomial in $z$ with coefficients continuous in $w$,*

(ii) $P(z, 1)$ *has a simple root at $\rho^*$ and all other roots have larger absolute values,*

(iii) $g(z, w)$ *is analytic for $w$ near $1$ and $z < \rho^* + \epsilon$, and*

(iv) $g(\rho^*, 1) \neq 0$.

*Then*

$$\mu = -\left.\frac{r(z, w)}{\rho^*}\right|_{z=\rho^*, w=1} \quad \text{and} \quad \sigma^2 = \mu^2 - \left.\frac{s(z, w)}{\rho^*}\right|_{z=\rho^*, w=1},$$

*where*

$$r = r(z, w) = -\frac{\frac{\partial}{\partial w} P}{\frac{\partial}{\partial z} P} \tag{8.4}$$

*and*

$$s = s(z, w) = -\frac{r^2 \frac{\partial^2}{\partial z^2} P + 2r \frac{\partial}{\partial z}\frac{\partial}{\partial w} P + \frac{\partial}{\partial w} P + \frac{\partial^2}{\partial w^2} P}{\frac{\partial}{\partial z} P}. \tag{8.5}$$

*If $\sigma \neq 0$, then $a_n(k) \stackrel{d}{\approx} \mathcal{N}(n\mu, n\sigma^2)$.*

A second result is the local limit theorem.

**Theorem 8.29** (Bender [25, Theorem 3]) *Let $f(z, w)$ have power series expansion*

$$f(z, w) = \sum_{n,k \geq 0} a_n(k) z^n w^k$$

with nonnegative coefficients $a_n(k)$, and let $a < b$ be real numbers. Define

$$R(\epsilon) = \{z \mid a \leq \text{Re}(z) \leq b, |\text{Im}(z)| \leq \epsilon\}.$$

Suppose there exist $\epsilon > 0$, $\delta > 0$, a nonnegative integer $\ell$ and functions $A(s)$ and $r(s)$ such that

1. $A(s)$ is continuous and nonzero for $s \in R(\epsilon)$;

2. $r(s)$ is nonzero and has bounded third derivative for $s \in R(\epsilon)$;

3. for $s \in R(\epsilon)$ and $|z| \leq |r(s)|(1+\delta)$

$$\left(1 - \frac{z}{r(s)}\right)^\ell f(z, e^s) - \frac{A(s)}{1 - z/r(s)}$$

is analytic and bounded;

4. 
$$\left(\frac{r'(\alpha)}{r(\alpha)}\right)^2 - \frac{r''(\alpha)}{r(\alpha)} \neq 0 \quad \text{for} \quad a \leq \alpha \leq b;$$

5. $f(z, e^s)$ is analytic and bounded for

$$|z| \leq |r(\text{Re}(s))|(1+\delta) \quad \text{and} \quad \epsilon \leq |\text{Im}(s)| \leq \pi.$$

Then we have that

$$a_n(k) \approx \frac{n^\ell e^{-\alpha k} A(\alpha)}{\ell! r(\alpha)^n \sigma_\alpha \sqrt{2\pi n}} \tag{8.6}$$

as $n \to \infty$, uniformly for $a \leq \alpha \leq b$, where

$$\frac{k}{n} = -\frac{r'(\alpha)}{r(\alpha)} \quad \text{and} \quad \sigma_\alpha^2 = \left(\frac{k}{n}\right)^2 - \frac{r''(\alpha)}{r(\alpha)}.$$

To be able to specify growth rates easily, we will use *Landau's big and little O notation*.

**Definition 8.30** *We say that a function $f(x)$ is of the* order *of $g(x)$ or $O(g(x))$ as $x \to \infty$ if and only if there exists a value $x_0$ and a constant $M > 0$ such that $|f(x)| \leq M|g(x)|$ for all $x > x_0$ or, equivalently, if and only if $\limsup_{x \to \infty} |f(x)/g(x)| < \infty$. We say that $f(x)$ is $o(g(x))$ if and only if $|f(x)/g(x)| \to 0$ as $x \to \infty$, that is, $f(x)$ grows slower than $g(x)$ as $x \to \infty$.*

In general, the function $g(x)$ is a function that is "nicer" than the function $f(x)$ and consists of the terms that dominate the growth of $f(x)$ in the limit. For instance, if $f(x) = 3x^2 + x - 1$, then $f(x) = O(x^2)$ and $f(x) = o(x^n)$ for $n \geq 3$.

We end this section by introducing a wide class of functions where we can say more on the asymptotic expansion of its Laurent coefficients. More precisely, Wyman [362] introduced a class of functions for which the saddle-point method (see [86]) yields an asymptotic expansion of the Laurent coefficients when the index tends to infinity. This class is defined as follows.

**Definition 8.31** Let $f(z)$ have radius of convergence $\rho$ with $0 < \rho \leq \infty$ and always positive on some subinterval $(R, \rho)$ of $(0, \rho)$. Let

$$a(r) = \frac{r \frac{d}{dr} f(r)}{f(r)} \text{ and } b(r) = \frac{r \frac{d}{dr} f(r)}{f(r)} + \frac{r^2 \frac{d^2}{dr^2} f(r)}{f(r)} - r^2 \left( \frac{\frac{d}{dr} f(r)}{f(r)} \right)^2.$$

The function $f(z)$ is said to be H-admissible or just admissible (Hayman admissible) if it satisfies the following three conditions:

- Capture condition: $\lim_{r \to \rho} a(r) = \lim_{r \to \rho} b(r) = +\infty$.

- Locality condition: For some function $\theta_0(r)$ defined over the interval $(R, \rho)$ and satisfying $0 < \theta_0 \leq \pi$, one has

$$f(re^{i\theta}) \sim f(r) e^{i\theta a(r) - \theta^2 b(r)/2}$$

  as $r \to \rho$, uniformly in $|\theta| \leq \theta_0(r)$.

- Decay condition: Uniformly in $\theta_0(r) \leq |\theta| < \pi$

$$f(re^{i\theta}) = o\left( \frac{f(r)}{\sqrt{b(r)}} \right).$$

**Example 8.32** Let $f(z) = e^{e^z - 1}$ and $\rho = +\infty$. Clearly, the function $f(z)$ has radius of convergence $\rho$ and always positive on $(R, \rho)$ with $R > 0$. Let

$$a(r) = \frac{re^{e^r - 1} e^r}{e^{e^r - 1}} = re^r$$

and

$$b(r) = re^r + \frac{r^2 e^r e^{e^r - 1}(1 + e^r)}{e^{e^r - 1}} - r^2 e^{2r} = r(r+1)e^r.$$

Clearly, the capture condition holds. By choosing the function $\theta_0(r) = \frac{1}{re^{2r/5}}$ which defines over the interval $(R, \rho)$ and satisfying $\theta_0 \leq \pi$ (only choose $R$ too large), we have

$$f(re^{i\theta}) = f(r) - \frac{1}{2} r^2 \theta^2 \frac{d^2}{dr^2} f(r) + O(r^3 \theta^3 e^r),$$

which implies that the locality condition holds. We leave it to the reader to verify that the decay condition holds. Hence, the function $f(z)$ is admissible with $\rho = +\infty$.

Similar arguments show that the functions $e^z, e^{z+z^2/2}$ are admissible functions with $\rho = +\infty$ and the function $e^{z/(1-z)}$ is an admissible function with $\rho = 1$. But, the functions $e^{z^2}$ and $e^{z^2} + e^z$ are not admissible since they attain values that are too large when the argument of $z$ is near $\pi$.

The following theorem analyzes the asymptotic of coefficients of $H$-admissible functions.

**Theorem 8.33** (Flajolet and Sedgewick [107, Theorem VIII.4]) *Let $f(z)$ be an $H$-admissible function and $\xi = \xi(n)$ be the unique solution in the interval $(R, \rho)$ of the equation $\frac{z \frac{d}{dz} f(z)}{f(z)} = n$. The Taylor coefficients of $f(z)$ satisfy*

$$[z^n]f(z) \sim \frac{f(\xi)\xi^{-n}}{\sqrt{2\pi \left(z^2 \frac{d^2}{dz^2} \log f(z) + z \frac{d}{dz} \log f(z)\right)_{z=\xi}}},$$

*as $n \to \infty$.*

The original proof by Hayman [134] contains in addition an exact description of the coefficient $r^n[z^n]f(r)$ in Taylor expansion of $f(r)$ as $r$ gets closer to $\rho$. Using the proof of the above theorem as described in [107], we derive the following result.

**Theorem 8.34** (Flajolet and Sedgewick [107, Proposition VIII.5]) *Let $f(z)$ be an $H$-admissible function with nonnegative coefficients. Define a family of discrete random variables $X(r)$ indexed by $r \in (0, R)$ as*

$$\mathbb{P}(X(r) = n) = \frac{[z^n]f(z)r^n}{f(r)}.$$

*Then*

$$\mathbb{P}(X(r) = n) = \frac{1}{\sqrt{2\pi b(r)}} \left(e^{-\frac{(a(r)-n)^2}{2b(r)}} + \epsilon_n\right),$$

*where $a(r), b(r)$ are given in Definition 8.31 and the error term satisfies $\epsilon_n = o(1)$ as $r \to \rho$ uniformly with $\lim_{r \to \rho} \sup_n |\epsilon_n| = 0$.*

**Example 8.35** Let $f(z) = e^{z+z^2}$. It is not hard to check that the conditions of Theorem 8.34 hold. Then

$$\mathbb{P}(X(r) = n) = \frac{[z^n]f(z)r^n}{e^{r+r^2}} \sim \frac{1}{\sqrt{2\pi r(1+4r)}r^n},$$

where $r$ is a function of $n$ given implicitly by $r + 2r^2 = n$.

Hayman [134] showed general rules on the set of $H$-admissible functions, which guarantee that large classes of functions are admissible.

**Theorem 8.36** (Flajolet and Sedgewick [107, Theorem VIII.5]) *Let $f(z)$ and $g(z)$ be any two admissible functions and let $p(z)$ be any polynomial with real coefficients such that the leading coefficient is a positive number. Then the functions $f(z)g(z)$, $e^{f(z)}$, $f(z) + g(z)$, $f(z)p(z)$, and $p(f(z))$ are admissible. Moreover, if $e^{p(z)}$ has positive coefficients in its Taylor expansion, then $e^{p(z)}$ is admissible.*

Having all the necessary tools available for the asymptotic analysis, we now consider various statistics for set partitions.

## 8.3  Z-statistics

Definition 5.1 presents a statistic (see Definition 1.17) on the block representation of set partition $\pi$ as a function of the cardinalities of the blocks of $\pi$. Ones can extend this to a function of sum of interactions between any two different blocks.

**Definition 8.37** *A statistic $f$ on set partitions is said to be Z-statistic if for any set partition $\pi = B_1/B_2/\cdots/B_k$,*

$$f(\pi) = \sum_{1 \leq i \neq j \leq k} f(\mathrm{Reduce}(B_i/B_j)).$$

**Example 8.38** *In Definition 5.1 we defined a statistic $w(\pi) = \sum_{i=1}^{k}(i-1)|B_i|$ for any set partition $\pi = B_1/B_2/\cdots/B_k$. Since*

$$\sum_{1<i<j\leq k} f(\mathrm{Reduce}(B_i/B_j)) = \sum_{1\leq i<j\leq k} |B_j| = \sum_{j=1}^{k}(j-1)|B_j| = w(\pi),$$

*where $f(B/B')$ is defined to be 0 if $\min B > \min B'$ and $|B'|$ otherwise. Thus, $w$ is a Z-statistic.*

We start by showing that the computation of the mean of a Z-statistic in a random set partition is equivalent to the computation of its mean in a random set partition with exactly two blocks. In order to show that, we define the following notation.

**Definition 8.39** *Let $f$ be any Z-statistic, we define $\mathrm{total}^{f}_{\mathcal{P}_{n,2}} = \sum_{\pi \in \mathcal{P}_{n,2}} f(\pi)$ and let $\mathrm{total}^{f}_{2}(x)$ be the exponential generating function of the sequence $\{\mathrm{total}^{f}_{\mathcal{P}_{n,2}}\}_{n\geq 0}$, that is, $\mathrm{total}^{f}_{2}(x) = \sum_{n\geq 0} \mathrm{total}^{f}_{\mathcal{P}_{n,2}} \frac{x^n}{n!}$.*

**Theorem 8.40** (Kasraoui [157]) *Let $f$ be a $Z$-statistic. Then the mean of the statistic $f$ on a random set partition of $\mathcal{P}_n$ is given by*

$$\mu_n = \frac{1}{\text{Bell}_n} \left[\frac{x^n}{n!}\right] (\text{total}_2^f(x)\text{Bell}(x)), \tag{8.7}$$

*and the mean of the statistic $f$ on a random set partition of $\mathcal{P}_{n,k}$ is given by*

$$\mu_{n,k} = \frac{1}{\text{Stir}(n,k)} \left[\frac{x^n}{n!}\right] (\text{total}_2^f(x)\text{Stir}_k(x)), \tag{8.8}$$

*where* $\text{Bell}(x) = e^{e^x-1}$ *and* $\text{Stir}_k(x) = \frac{1}{k!}(e^x - 1)^k$.

**Proof** By definition of the mean, we have

$$\mu_n = \frac{\sum_{\pi \in \mathcal{P}_n} f(\pi)}{\text{Bell}_n} = \frac{\sum_{n \geq 1} \sum_{\pi \in \mathcal{P}_{n,k}} f(\pi)}{\text{Bell}_n} = \frac{1}{\text{Bell}_n} \sum_{k \geq 1} \text{Stir}(n,k)\mu_{n,k},$$

So if (8.8) holds, then

$$\mu_n = \frac{1}{\text{Bell}_n} \sum_{k \geq 1} \left[\frac{x^n}{n!}\right] (\text{total}_2^f(x)\text{Stir}_{k-2}(x))$$

$$= x \frac{1}{\text{Bell}_n} \left[\frac{x^n}{n!}\right] \left(\text{total}_2^f(x) \sum_{k \geq 1} \text{Stir}_{k-2}(x)\right)$$

$$= \frac{1}{\text{Bell}_n} \left[\frac{x^n}{n!}\right] (\text{total}_2^f(x)\text{Bell}(x)),$$

which proves (8.7). Therefore, to prove Theorem 8.40 it is sufficient to show (8.8). By definitions of mean, (8.8) can be written as

$$\sum_{\pi \in \mathcal{P}_{n,k}} f(\pi) = \sum_{j=0}^{n} \binom{n}{j} \text{Stir}(n-j, k-2)\text{total}_{j,2}^f. \tag{8.9}$$

Thus, we will show (8.9). Let $DJ$ be the set of pairs of nonempty and disjoints subsets of $\mathbb{N}$, that is, $DJ = \{(A, A') \mid A, A' \subseteq \mathbb{N}, A, A' \neq \emptyset, A \cap A' = \emptyset\}$. By definition of a $Z$-statistic, we have

$$\sum_{\pi \in \mathcal{P}_{n,k}} f(\pi) = \sum_{\pi \in \mathcal{P}_{n,k}} \sum_{A,B \in \pi} f(\text{Reduce}(A/B))$$

$$= \sum_{A,B \subseteq [n]} \sum_{\pi \in \mathcal{P}_{n,k}, A,B \in \pi} f(\text{Reduce}(A/B)) \tag{8.10}$$

$$= \sum_{A,B \subseteq [n], (A,B) \in DJ} f(\text{Reduce}(A/B)) \text{Stir}(n - |A \cup B|, k-2).$$

Now, let $C \subseteq \mathbb{N}$ such that $|C| = j$. Then

$$\sum_{(A,B)\in DJ,\, A\cup B=C} f(\text{Reduce}(A/B)) = \sum_{\pi\in\mathcal{P}_{j,2}} f(\pi) = \text{total}^{f}_{j,2}, \qquad (8.11)$$

where the first equality follows from the definition of the set $DJ$. By (8.10) and (8.11), we obtain

$$\sum_{\pi\in\mathcal{P}_{n,k}} f(\pi) = \sum_{(A,B)\in DJ,\, A,B\subseteq[n]} f(\text{Reduce}(A/B)) \text{Stir}(n-|A\cup B|, k-2)$$

$$= \sum_{C\subseteq[n]} \sum_{(A,B)\in DJ,\, A\cup B=C} f(\text{Reduce}(A/B)) \text{Stir}(n-|C|, k-2)$$

$$= \sum_{C\subseteq[n]} \text{Stir}(n-|C|, k-2) \text{total}^{f}_{|C|,2}$$

$$= \sum_{j=0}^{n} \binom{n}{j} \text{Stir}(n-j, k-2) \text{total}^{f}_{j,2},$$

which completes the proof. □

Our first example is to study the mean of the $Z$-statistic $w$ which was defined in Definition 5.1 (see Example 8.38).

**Theorem 8.41** (Kasraoui [157]) *The mean $\mu_n$ of the $Z$-statistic $w$ in a random set partition of $\mathcal{P}_n$ satisfies*

$$\mu_n = -\frac{1}{4}\frac{\text{Bell}_{n+2}}{\text{Bell}_n} + \frac{2n+1}{4}\frac{\text{Bell}_{n+1}}{\text{Bell}_n} - \frac{2n-1}{4} \qquad (8.12)$$

$$= \frac{n^2}{2\log n}\left(1 + \frac{\log\log n}{\log n}(1+o(1))\right), \qquad (8.13)$$

*as $n \to \infty$. Moreover, the mean $\mu_{n,k}$ of the $Z$-statistic $w$ in a random set partition of $\mathcal{P}_{n,k}$ satisfies*

$$\mu_{n,k} = \frac{n(k-1)}{2} - \frac{1}{2}\binom{k}{2} + \frac{n+1-k}{2}\frac{\text{Stir}(n,k-1)}{\text{Stir}(n,k)} \qquad (8.14)$$

$$= \frac{n(k-1)}{2} - \frac{1}{2}\binom{k}{2} + o(1), \qquad (8.15)$$

*as $n \to \infty$.*

**Proof** At first we compute the exponential generating function $\text{total}^{w}_2(x)$ for the sequence $\{\text{total}^{w}_{\mathcal{P}_{n,2}}\}_{n\geq 0}$, where $\text{total}^{w}_{\mathcal{P}_{n,2}} = \sum_{\pi\in\mathcal{P}_{n,2}} w(\pi)$. Since each set partition of $\mathcal{P}_{n,2}$ can be written as $11\cdots 12\sigma$ where $\sigma$ is a word over alphabet $[2]$, we obtain

$$\text{total}^{w}_{\mathcal{P}_{n,2}} = \sum_{j=1}^{n-1} j\binom{n-1}{j} = (n-1)2^{n-2}.$$

Thus,
$$\text{total}_2^W(x) = \sum_{n\geq 2} \text{total}_{P_{n,2}}^W \frac{x^n}{n!} = \frac{1}{4} + \frac{1}{4}(2x-1)e^{2x}.$$

Using the facts that $e^{2x}\text{Bell}(x) = \frac{d^2}{dx^2}\text{Bell}(x) - \frac{d}{dx}\text{Bell}(x)$ and
$$e^{2x}\text{Stir}_{k-2}(x) = \text{Stir}_{k-2}(x) + 2(k-1)\text{Stir}_{k-1}(x) + k(k-1)\text{Stir}_k(x),$$
we may write

$\text{total}_2^W(x)\text{Bell}(x)$
$$= \frac{1}{4}\text{Bell}(x) + \frac{2x-1}{4}\frac{d^2}{dx^2}\text{Bell}(x) - \frac{2x-1}{4}\frac{d}{dx}\text{Bell}(x),$$
$\text{total}_2^W(x)\text{Stir}_{k-2}(x)$
$$= \frac{x}{2}\text{Stir}_{k-2}(x) + \frac{(k-1)(2x-1)}{2}\text{Stir}_{k-1}(x) + \frac{k(k-1)(2x-1)}{4}\text{Stir}_k(x).$$

Hence, by comparing the coefficient of $\frac{x^n}{n!}$ in both sides of the above two equations, we obtain

$\left[\frac{x^n}{n!}\right](\text{total}_2^W(x)\text{Bell}(x))$
$$= -\frac{1}{4}\text{Bell}_{n+2} + \frac{2n+1}{4}\text{Bell}_{n+1} - \frac{2n-1}{4}\text{Bell}_n,$$
$\left[\frac{x^n}{n!}\right](\text{total}_2^W(x)\text{Stir}_{k-2}(x))$
$$= \frac{n}{2}\text{Stir}(n-1,k-2) + \frac{(k-1)}{2}(2n\text{Stir}(n-1,k-1) - \text{Stir}(n,k-1))$$
$$+ \frac{k(k-1)}{4}(2n\text{Stir}(n-1,k) - \text{Stir}(n,k)).$$

The last expression can be simplified by using Theorem 1.12. Division of the above expressions by $\text{Bell}_n$ gives (8.12) and (8.14). The expression (8.13) can be obtained from Corollary 8.55 and (8.12). Also, from Theorem 8.59 and (8.14), we obtain (8.15). □

Note that for given a $Z$-statistic, the proof of the above theorem describes a general process for obtaining explicit formulas for the mean $\mu_n$ and $\mu_{n,k}$ and by Corollary 8.55 and Theorem 8.59 one can obtain asymptotics expressions for the mean $\mu_n$ and $\mu_{n,k}$. For instance, we present the following result.

**Theorem 8.42** (Kasraoui [157]) *The mean $\mu_n$ of the number crossing $\text{cr}_2$ in a random set partition of $\mathcal{P}_n$ satisfies*

$$\mu_n = -\frac{5}{4}\frac{\text{Bell}_{n+2}}{\text{Bell}_n} + \frac{2n+9}{4}\frac{\text{Bell}_{n+1}}{\text{Bell}_n} + \frac{2n+1}{4} \tag{8.16}$$
$$= \frac{n^2}{2\log n}\left(1 + \frac{\log\log n}{\log n}(1+o(1))\right), \tag{8.17}$$

as $n \to \infty$. Moreover, the mean $\mu_{n,k}$ of the number crossing $cr_2$ in a random set partition of $\mathcal{P}_{n,k}$ satisfies

$$\mu_{n,k} = \frac{n(k-1)}{2} - \frac{5}{2}\binom{k}{2} + \frac{3(n+1-k)}{2}\frac{\text{Stir}(n,k-1)}{\text{Stir}(n,k)} \tag{8.18}$$

$$= \frac{n(k-1)}{2} - \frac{5}{2}\binom{k}{2} + o(1), \tag{8.19}$$

as $n \to \infty$.

**Proof** Each set partition of $\mathcal{P}_{n,2}$ can be expressed as

$$\pi^{(s)} = 1^{a_1}2^{b_1}\cdots 1^{a_s}2^{b_s}1^{a_{s+1}}$$

with $s \geq 1$, where $a_1, b_1, \ldots, a_s, b_s \geq 1$ and $a_{s+1} \geq 0$. Clearly, $cr_2(\pi^{(1)}) = 0$ and $cr_2(\pi^{(s)}) = 2k - 3 + \delta_{a_{s+1}>0}$. Thus, the generating function for the number of crossing in set partitions $\pi^{(s)}$ in $\mathcal{P}_n$ is given by

$$(2s-3)\frac{x^{2s}}{(1-x)^{2s}} + (2s-2)\frac{x^{2s+1}}{(1-x)^{2s+1}}.$$

Summing over $s \geq 2$, we obtain that (give the details)

$$F(x) = \sum_{n \geq 0} \text{total}^{cr_2}_{\mathcal{P}_{n,2}} x^n = \sum_{s \geq 2}\sum_{n \geq 4}\sum_{\pi^{(s)} \in \mathcal{P}_n} cr_2(\pi^{(s)})x^n = \frac{x^4}{(1-x)^2(1-2x)^2}.$$

This implies that the coefficient of $x^n$ in $F(x)$ is given by

$$\text{total}^{cr_2}_{\mathcal{P}_{n,2}} = (n-5)2^{n-2} + n + 1$$

for $n \geq 4$. A straightforward computation then gives

$$\text{total}^{cr_2}_2(x) = \sum_{n \geq 3} \text{total}^{cr_2}_{\mathcal{P}_{n,2}}\frac{x^n}{n!} = \frac{1}{4} + (x+1)e^x + \frac{2x-5}{4}e^{2x}.$$

Similar to the proof of the above theorem, we can write

$$\text{total}^{cr_2}_2(x)\text{Bell}(x)$$
$$= \frac{1}{4}\text{Bell}(x) + \frac{2x+9}{4}\frac{d}{dx}\text{Bell}(x) + \frac{2x-5}{4}\frac{d^2}{dx^2}\text{Bell}(x),$$
$$\text{total}^{cr_2}_2(x)\text{Stir}_{k-2}(x)$$
$$= \frac{3x}{2}\text{Stir}_{k-2}(x) + \frac{(k-1)(4x-3)}{2}\text{Stir}_{k-1}(x) + \frac{k(k-1)(2x-5)}{4}\text{Stir}_k(x).$$

Extracting the $x^n/n!$ coefficients, we obtain

$$\left[\frac{x^n}{n!}\right](\text{total}_2^{\text{cr}_2}(x)\text{Bell}(x))$$
$$= -\frac{5}{4}\text{Bell}_{n+2} + \frac{2n+9}{4}\text{Bell}_{n+1} + \frac{2n+1}{4}\text{Bell}_n,$$
$$\left[\frac{x^n}{n!}\right](\text{total}_2^{\text{cr}_2}(x)\text{Stir}_{k-2}(x))$$
$$= \frac{3n}{2}\text{Stir}(n-1,k-2) + \frac{k-1}{2}(4n\text{Stir}(n-1,k-1) - 3\text{Stir}(n,k-1))$$
$$+ \frac{k(k-1)}{4}(2n\text{Stir}(n-1,k) - 5\text{Stir}(n,k)).$$

By using Theorem 1.12 several times, we obtain

$$\left[\frac{x^n}{n!}\right](\text{total}_2^{\text{cr}_2}(x)\text{Stir}_{k-2}(x))$$
$$= \frac{(2n-5k)(k-1)}{4}\text{Stir}(n,k) + \frac{3(n+1-k)}{2}\text{Stir}(n,k-1).$$

Division of the last expression by $\text{Bell}_n$ gives (8.16) and (8.18), while their asymptotics approximations are followed easily by using Corollary 8.55 and Theorem 8.59. □

Note that the study of number crossing on set partitions has been interesting several researchers. More precisely, let $CR_{n,k}(q)$ be the generating function for the number of set partitions of $\mathcal{P}_{n,k}$ according to the number of crossings $\text{cr}_2$, that is, $CR_{n,k}(x) = \sum_{\pi \in \mathcal{P}_{n,k}} q^{\text{cr}_2(\pi)}$. Biane [35] found an explicit formula for $CR(x,q) = \sum_{n \geq k \geq 0} CR_{n,k}(q)y^k x^n$ in terms of continued fractions. Later Kasraoui, Stanton, and Zeng [158] showed that

$$CR(x,q) = \sum_{k \geq 0} \frac{(xyq)^k}{\prod_{i=1}^{k}(q^i - q^i[i]_q x + xy(1-q^i))},$$

from which they obtained

$$CR_{n,k}(q)$$
$$= \sum_{j=1}^{k}(-1)^{k-j}\left[\frac{[j]_q^n}{[j]_q!}\sum_{i=0}^{k-j}\frac{(1-q)^i}{[k-j-i]_q!}q^{\binom{k-j-i+1}{2}-kj}\left(\binom{n}{i}q^i + \binom{n}{i-1}\right)\right].$$

Another explicit formula for $CR_{n,k}(q)$ was established by Josuat-Vergés and Rubey [155]:

$$CR_{n,k}(q)$$
$$= \sum_{j=0}^{k}\sum_{i=j}^{n-k}\frac{(-1)^i q^{\binom{j+1}{2}}}{(1-q)^{n-k}}\left[\binom{n}{k+i}\binom{n}{k-j} - \binom{n}{k+i+1}\binom{n}{k-j-1}\right]\begin{bmatrix}i\\j\end{bmatrix}_q.$$

The interest in the number crossing on set partitions has been extended to study the number crossing on *circular set partitions*, that is, on the circular representation of the set partitions (see Section 3.2.2). Kasraoui [157] showed the following result (for the proof, see [157, Section 6]).

**Theorem 8.43** (Kasraoui [157]) *The mean $\mu_n$ of the number crossing $cr_2$ in a random circular set partition of $\mathcal{P}_n$ satisfies*

$$\mu_n = \frac{n}{2} \frac{\text{Bell}_{n+1}}{\text{Bell}_n} - \frac{n(4n+1)}{2} \frac{\text{Bell}_{n-1}}{\text{Bell}_n} - \binom{n}{2} \frac{\text{Bell}_{n-2}}{\text{Bell}_n}$$

$$+ \binom{n}{4} \frac{\text{Bell}_{n-4}}{\text{Bell}_n} + \frac{3n}{2}$$

$$= \frac{n^2}{2\log n} \left(1 + \frac{\log \log n}{\log n}(1 + o(1))\right),$$

*as $n \to \infty$. Moreover, the mean $\mu_{n,k}$ of the number crossing $cr_2$ in a random circular set partition of $\mathcal{P}_{n,k}$ satisfies*

$$\mu_{n,k} = \frac{n(k-1)}{2} - \frac{n(4n-5k+1)}{2} \frac{\text{Stir}(n-1, k-1)}{\text{Stir}(n,k)}$$

$$- 10\binom{n}{2} \frac{\text{Stir}(n-2, k-2)}{\text{Stir}(n,k)} + \binom{n}{4} \frac{\text{Stir}(n-4, k-2)}{\text{Stir}(n,k)}$$

$$= \frac{n(k-1)}{2} + o(1),$$

*as $n \to \infty$.*

Another interesting result with this style is counting occurrences of subsequence binary patterns.

**Theorem 8.44** (Kasraoui [157]) *Let $\tau = \tau_1 \tau_2 \cdots \tau_m$ be any binary subsequence pattern of size $m$.*

*If $\tau_1 = 1$, then the mean $\mu_n$ of the number occurrences $occ_\tau$ in a random set partition of $\mathcal{P}_n$ is given by*

$$\mu_n = -\sum_{j=0}^{m-1} p_j(n) \frac{\text{Bell}_{n+2-j}}{\text{Bell}_n} - \frac{1}{4}\left(2\binom{n}{m} - \binom{n}{m-1}\right) \frac{\text{Bell}_{n+2-m}}{\text{Bell}_n}$$

$$- \frac{1}{2}\binom{n}{m} \frac{\text{Bell}_{n+1-m}}{\text{Bell}_n} + \frac{(-1)^m}{2^{m+1}},$$

*and if $\tau_1 = 2$ then the mean $\mu_n$ of the number occurrences $occ_\tau$ in a random set partition of $\mathcal{P}_n$ is given by*

$$\mu_n = \sum_{j=0}^{m-1} p_j(n) \frac{\text{Bell}_{n+2-j}}{\text{Bell}_n} - \frac{1}{2}\binom{n}{m} \frac{\text{Bell}_{n+1-m}}{\text{Bell}_n} + \frac{(-1)^{m+1}}{2^{m+1}},$$

where
$$p_j(n) = \frac{(-1)^{m-j}}{2^{m-j+2}}\left(2\binom{n}{j} + \binom{n}{j-1}\right).$$
Moreover, we have the asymptotic approximation
$$\mu_n = \frac{n^2(\log n)^{m-2}}{2m!}\left(1 - (m-2)\frac{\log\log n}{\log n} + o\left(\frac{\log\log n}{\log n}\right)\right),$$
as $n \to \infty$.

For the proof, we refer the reader to [157]. Actually, in [157] is derived an explicit formula and asymptotic approximation for the mean $\mu_{n,k}$ of the number occurrences $occ_\tau$ in a random set partition of $\mathcal{P}_{n,k}$.

**Definition 8.45** *Two sets $B, C$ are said to* overlap *if $\min B < \min C < \max B < \max C$. Otherwise, they are said to be* nonoverlapping.

**Example 8.46** *Let $\pi = 15/26/34$ be a set partition of $\mathcal{P}_6$. then the blocks 15 and 26 are nonoverlapping, while the blocks 15 and 34 are overlapping.*

Kasraoui [157] has found the mean of the number of overlappings in set partitions, where we left the proof, which is based on the arguments of the above proofs to the reader.

**Theorem 8.47** (Kasraoui [157]) *The mean $\mu_n$ of the number overlappings in a random set partition of $\mathcal{P}_n$ satisfies*
$$\mu_n = \frac{1}{4}\frac{\text{Bell}_{n+2}}{\text{Bell}_n} + \frac{3}{4}\frac{\text{Bell}_{n+1}}{\text{Bell}_n} - frac4n + 54 - \frac{n}{2}\frac{\text{Bell}_{n-1}}{\text{Bell}_n}$$
$$= \frac{n^2}{4\log^2 n}\left(1 + \frac{2\log\log n}{\log n}(1 + o(1))\right),$$
*as $n \to \infty$. Moreover, the mean $\mu_{n,k}$ of the number of overlappings in a random set partition of $\mathcal{P}_{n,k}$ satisfies*
$$\mu_{n,k} = \frac{(k(k-1))}{4} + n(k-1)\frac{\text{Stir}(n-1,k-1)}{\text{Stir}(n,k)} - \frac{3(n+k-1)}{2}\frac{\text{Stir}(n,k-1)}{\text{Stir}(n,k)}$$
$$= \frac{k(k-1)}{4} + o(1),$$
*as $n \to \infty$.*

## 8.4 Set Partitions as Geometric Words

In this section, following [240], we provide a formula for the probability that a geometrically distributed word of size $n$ is a set partition. Indeed, the first consideration of such a formulas was derived in [266], using the Cauchy integral formula. In addition, we consider refinements of this result obtained by fixing the number of blocks, levels, rises, or descents.

**Definition 8.48** *If $0 \leq p \leq 1$, then a random variable $X$ is said to be geometric if $P(X = i) = pq^{i-1}$ for all positive integers $i \geq 1$, where $q = 1 - p$. We will say that a word $w = w_1 w_2 \cdots$ over the alphabet of positive integers is geometrically distributed if the positions of $w$ are independent and identically distributed geometric random variables.*

The study of geometrically distributed words has been a recent topic of study in enumerative combinatorics; see, for example, [52, 53, 54] and the references contained therein.

**Definition 8.49** *Let $P_n$ denote the probability that a geometrically distributed word of size $n$ is a set partition, with $P_0 = 1$.*

**Theorem 8.50** *For $n \geq 0$,*

$$P_n = \sum_{i=0}^{n} (-1)^i \binom{n}{i} (p;q)_i, \tag{8.20}$$

*where $(x;q)_i = (1-x) \cdots (1 - xq^{i-1})$ for $i \geq 1$ and $(x;q)_0 = 1$.*

**Proof** The sequence $\{P_n\}_{n \geq 0}$ satisfies the following recurrence relation

$$P_{n+1} = \sum_{j=0}^{n} p^{n+1-j} q^j \binom{n}{j} P_j \tag{8.21}$$

with $P_0 = 1$. To see this, we let the number of 1's occur exactly $n - j$ times after the first position, then each of these 1's contributes a factor of $p$, and the probability that the remaining $j$ letters form a set partition of $\{2, 3, \ldots\}$ is $q^j P_j$, by independence. Summing over all $j$ leads to the above recurrence relation.

Define the exponential generating function $G(x) = \sum_{n \geq 0} P_n \frac{x^n}{n!}$. Multiplying recurrence (8.21) by $\frac{x^n}{n!}$, and summing over $n \geq 0$, implies that $G(x)$ is a solution of the differential equation

$$\frac{d}{dx} h(x) = p e^{px} h(qx), \quad h(0) = 1. \tag{8.22}$$

We now find a second solution to (8.22) as follows. Let $F(x)$ be given as

$$F(x) = p \sum_{j \geq 0} \frac{(p;q)_j}{j!} (-x)^j.$$

It may be verified that $F(x)$ satisfies $F'(x) + F(x) = pF(qx)$, which, by multiplying by $\frac{1}{p}e^x = \frac{1}{p}e^{px}e^{qx}$, implies that

$$\frac{d}{dx}\left(\frac{1}{p}e^x F(x)\right) = pe^{px}\left(\frac{1}{p}e^{qx}F(qx)\right).$$

The final equality shows that $\frac{1}{p}e^x F(x)$ is also a solution to (8.22), and thus

$$G(x) = \frac{1}{p}e^x F(x), \tag{8.23}$$

by uniqueness of solutions. From (8.23), we obtain (8.20). □

Note that it would be interesting to have a probabilistic or a combinatorial proof of (8.20) in the sense of [29].

The above theorem can be extended to present the probability that a geometrically distributed word of size $n$ is a set partition of $\mathcal{P}_{n,k}$.

**Definition 8.51** *Fix $k \geq 1$. Define $P_{n,k}$ to be the probability that a geometrically distributed word of size $n$ is a set partition of $\mathcal{P}_{n,k}$.*

Recall the $q$-notation in Appendix C; the following theorem gives an exact formula for $P_{n,k}$.

**Theorem 8.52** *For all $n \geq k \geq 1$,*

$$P_{n,k} = \frac{p^n}{k_q!}\sum_{j=0}^{k}(-1)^j q^{\binom{j}{2}}((k-j)_q)^n \binom{k}{j}_q. \tag{8.24}$$

**Proof** First note that $P_{n,k}$ satisfies the recurrence

$$P_{n,k} = pq^{k-1}P_{n-1,k-1} + (1-q^k)P_{n-1,k}, \qquad n,k \geq 1, \tag{8.25}$$

with $P_{0,k} = \delta_{0,k}$ for $k \geq 0$. This holds since if no $k$ occurs among the first $n-1$ letters, then the last letter must be a $k$ with probability $pq^{k-1}$, and if a $k$ does occur among the first $n-1$ letters, then the last letter cannot be in $\{k+1, k+2, \ldots\}$, which has probability

$$1 - (pq^k + pq^{k+1} + \cdots) = 1 - \frac{pq^k}{1-q} = 1 - q^k.$$

Let $P_k(z) = \sum_{n \geq k} P_{n,k} z^n$. By multiplying both sides of (8.25) by $z^n$, summing over $n \geq k$, and solving for $P_k(z)$, we obtain

$$P_k(z) = \frac{pq^{k-1}z}{1-(1-q^k)z}P_{k-1}(z).$$

with $P_0(z) = 1$. By iterating, we derive

$$P_k(z) = p^k q^{\binom{k}{2}} z^k \prod_{j=1}^{k} \frac{1}{1-(1-q^j)z}. \qquad (8.26)$$

Now, we write

$$\prod_{j=1}^{k} \frac{1}{1-(1-q^j)z} = \sum_{j=1}^{k} \frac{a_{k,j}}{1-(1-q^j)z},$$

where is used partial fractions decomposition, then we obtain

$$a_{k,j} = \frac{(-1)^{k-j}(j_q)^{k-1}}{q^{\binom{j}{2}+j(k-j)}(k-1)_q!} \binom{k-1}{j-1}_q, \qquad 1 \le j \le k.$$

Then

$$[z^n](P_k(z)) = p^k q^{\binom{k}{2}} \sum_{j=1}^{k} a_{k,j}(1-q^j)^{n-k}$$

$$= \frac{p^k q^{\binom{k}{2}}}{(k-1)_q!} \sum_{j=1}^{k} \frac{(-1)^{k-j}(j_q)^{k-1}(1-q^j)^{n-k}}{q^{\binom{j}{2}+j(k-j)}} \binom{k-1}{j-1}_q$$

$$= \frac{p^k}{(k-1)_q!} \sum_{j=1}^{k} (-1)^{k-j} q^{\binom{k-j}{2}} (j_q)^{k-1}(1-q^j)^{n-k} \binom{k-1}{j-1}_q$$

$$= \frac{p^n}{(k-1)_q!} \sum_{j=1}^{k} (-1)^{k-j} q^{\binom{k-j}{2}} (j_q)^{n-1} \binom{k-1}{j-1}_q$$

$$= \frac{p^n}{k_q!} \sum_{j=0}^{k} (-1)^j q^{\binom{j}{2}} ((k-j)_q)^n \binom{k}{j}_q,$$

as required, upon changing the indices of summation and noting $\binom{k}{2} = \binom{j}{2} + \binom{k-j}{2} + j(k-j)$, $p+q=1$, and $\binom{k}{j}_q = \frac{k_q}{j_q}\binom{k-1}{j-1}_q$. □

Below, we give a table of values for $P_{n,k}$ for all $1 \le k \le n \le 3$.

**Table 8.1**: Values for $P_{n,k}$ for all $1 \le k \le n \le 3$

| $n\backslash k$ | 1 | 2 | 3 |
|---|---|---|---|
| 1 | $p$ | | |
| 2 | $p^2$ | $p^2 q$ | |
| 3 | $p^3$ | $p^3 q(q+2)$ | $p^3 q^3$ |

Theorem 8.52 was extended by Mansour and Shattuck [240] by finding the probability that a geometrically distributed word of size $n$ is a set partition of $\mathcal{P}_{n,k}$ with exactly $r$ occurrence of a fixed pattern of size two (see Research Direction 8.1).

## 8.5 Asymptotics for Set Partitions

In this section we show various asymptotics results for set partitions. In Section 8.5.1 we concentrated on the $n$-th Bell numbers and Stirling numbers $\text{Stir}(n, k)$ and estimated these numbers when $n$ is large, as well as the average of the number blocks and number of distinct block sizes in a random set partition. In Section 8.5.3 we continue our study in Section 5.3 of the statistic of the "number of records" (various types of records: strong records, weak records, and additional records) on set partitions, where we derive asymptotic results on these statistics.

### 8.5.1 Asymptotics for Bell and Stirling Numbers

As we saw earlier, the explicit formula for the $n$-th Bell number is not very useful when $n$ is large. We start by the finding an asymptotic formula for the $n$-th Bell number by stating the result of de Bruijn.

**Theorem 8.53** (de Bruijn [85, pp. 102–109]) *When $n \to \infty$,*

$$\text{Bell}_n \approx n! \frac{e^{\frac{n}{\log n} - 1 - \frac{1}{2}\log n + \frac{1}{2}\log\log n}}{(\log n - \log\log n)^{n+1/2}\sqrt{2\pi(1 + \log n - \log\log n)}}.$$

**Proof** Let $G(z) = e^{e^z}$. Clearly it is $h$-admissible function (see Example 8.32). Already, we know that the exponential generating function for the Bell numbers $\text{Bell}_n$ is given by $e^{-1}G(z)$, which implies that $\text{Bell}_n = \frac{n!}{e}[z^n]G(z)$. Theorem 8.33 shows that the saddle-point equation relative to $\frac{G(z)}{z^{n+1}}$ is $ze^z = n+1$. This famous equation admits an asymptotic solution by iteration (or "bootstrapping"): from $\log(ze^z) = \log(n+1)$ we obtain $z = \log(n+1) - \log z$, which is estimated by the recurrence relation $z = \log n - \log z$; then by iterating and estimating, we derive the solution (for full details, see de Bruijn [85, pp. 102–109])

$$z = \rho_n = \log n - \log\log n + O\left(\frac{\log\log n}{\log n}\right). \tag{8.27}$$

By Theorem 8.33, we have that $\text{Bell}_n \leq n!\frac{e^{e^{\rho_n}-1}}{\rho_n^n}$ and by estimating the solution $\rho_n$ as $\log n$ we obtain $\text{Bell}_n \leq n!\frac{e^{n-1}}{\log^n n}$, which implies that $\text{Bell}_n \leq n!$.

To complete our proof by using the saddle-point method (see Appendix F), the integration will be carried out over a circle $C$ of radius $r = \rho_n$. Then we set
$$F(z) = \log \frac{G(z)}{z^{n+1}} = e^z - (n+1)\log z,$$
and proceed to estimate the integral
$$I_n = \frac{1}{2\pi i} \int_C \frac{G(z)}{z^{n+1}} dz.$$

Now we express our quantities in terms of $r$ alone, which is possible since $n$ can be expressed as $n = re^r - 1$. We find
$$\frac{d^2}{dz^2}F(z)\big|_{z=r} = e^r + \frac{n+1}{r^2} = e^r + r^{-1}e^r = e^r(1+1/r),$$
$$\frac{d^3}{dz^3}F(z)\big|_{z=r} = e^r - \frac{2(n+1)}{r^3} = e^r - 2r^{-2}e^r = e^r(1-2/r^2).$$

The usual saddle–point heuristic suggests that the range of the saddle–point is determined by choosing a quantity $\theta_0 = \theta_0(r)$ such that $r^2 e^r \theta_0^2 \to \infty$ and $r^3 e^r \theta_0^3 \to 0$, which is possible by setting $\theta_0(r) = r^{-1}e^{-2r/5}$, for example. Now it remains to check the three conditions of the saddle-point method: tails pruning, central approximation, and tails completion, which can be completed in the usual way. Thus,
$$[z^n](G(z)) = \frac{e^{f(r)}}{\sqrt{2\pi f''(r)}}(1 + O(r^3 \theta_0^3 e^r)),$$
which implies
$$\text{Bell}_n = n! \frac{e^{e^r - 1}}{r^n \sqrt{2\pi r(r+1)e^r}}(1 + O(e^{-r/5})),$$
where $r = \rho_n$, which completes the proof. □

Several results have been derived to compare the $(n+1)$-st and $n$-th Bell numbers (see Exercises 8.1 and 8.2). For instance, using Theorem 8.53 twice, we obtain the following result.

**Corollary 8.54** *We have* $\frac{\text{Bell}_{n+1}}{\text{Bell}_n} = \frac{n}{\log n} + O\left(\frac{n \log \log n}{\log^2 n}\right)$.

More generally, we can state the following result.

**Corollary 8.55** (for example, see Kasraoui [157]) *For any integer $r$, we have, as $n \to \infty$,*
$$\frac{\text{Bell}_{n+r}}{\text{Bell}_n} = \left(\frac{n}{\log n}\right)^r \left(1 + r\frac{\log \log n}{\log n}(1 + o(1))\right).$$

**Proof** Suppose we are given an integer $d$; then by Corollary 8.54 we have

$$\frac{\text{Bell}_{n+d+1}}{\text{Bell}_{n+d}} = \left(\frac{n+d}{\log(n+d)}\right)\left(1 + \frac{\log\log(n+d)}{\log(n+d)}(1+o(1))\right).$$

Using the fact that $\log(1+x) = x(1+o(1))$ as $x \to 0$, we obtain $\log(n+d) = \log n(1 + O(\frac{1}{n\log n}))$ and $\frac{\log\log(n+d)}{\log(n+d)} = \frac{\log\log n}{\log n}(1+o(1))$, and thus

$$\frac{n+d}{\log(n+d)} = \frac{n(1+d/n)}{\log n}\left(1 + O(\frac{1}{n\log n})\right)^{-1} = \frac{n}{\log n}(1+O(1/n)).$$

Therefore,

$$\frac{\text{Bell}_{n+d+1}}{\text{Bell}_{n+d}} = \left(\frac{n}{\log n}\right)\left(1 + \frac{\log\log n}{\log n}(1+o(1))\right),$$

which implies

$$\frac{\text{Bell}_{n+r}}{\text{Bell}_n} = \prod_{d=0}^{r-1} \frac{\text{Bell}_{n+d+1}}{\text{Bell}_{n+d}}$$

$$= \left(\frac{n}{\log n}\right)^r \left(1 + \frac{\log\log n}{\log n}(1+o(1))\right)^r$$

$$= \left(\frac{n}{\log n}\right)^r \left(1 + r\frac{\log\log n}{\log n}(1+o(1))\right),$$

since $(1+x)^r = 1 + rx + o(x)$ as $x \to 0$. □

**Example 8.56** *This corollary is very useful when we would like to compute the mean of a statistic on set partitions. For instance, in Example 8.9, we showed that the average number of levels over all set partitions of $\mathcal{P}_n$ is given by $\mathbb{E}(X_{\text{lev}}(n)) = \frac{(n-1)\text{Bell}_{n-1}}{\text{Bell}_n}$. Thus, Corollary 8.54 asymptotically gives that $\mathbb{E}(X_{\text{lev}}(n)) \approx \log n$.*

*Another application for the above corollary is that Example 8.11 shows*

$$\mathbb{E}(X_{\text{ris-des}}(n)) = \frac{\text{Bell}_{n+1} - \text{Bell}_{n-1}}{2\text{Bell}_n} - \frac{1}{2},$$

*which, by Corollary 8.54, asymptotically implies $\mathbb{E}(X_{\text{ris-des}}(n)) \approx \frac{n}{2\log n}$.*

Theorem 8.53 has been extended in different directions; for instance, see [364, 347] and Exercise 8.4.

Table 8.2 shows the exact values of the number of set partitions of $[n]$ and our approximation $\text{Bell}_n^*$ as given in Theorem 8.53.

**Table 8.2**: Number of set partitions of $[n]$ and estimation

| $n$ | 50 | 100 |
|---|---|---|
| $\text{Bell}_n$ | $0.1857242688 \cdot 10^{48}$ | $0.4758539128 \cdot 10^{116}$ |
| $\text{Bell}_n^*$ | $0.7234084514 \cdot 10^{48}$ | $0.3332282627 \cdot 10^{117}$ |

Knuth [195, Page 64] compared the number $\text{Bell}_n$ and $n!$ and showed that $\frac{\text{Bell}_n}{n!} \to 0$ as $n \to \infty$. This result has been extended by Jakimczuk [144] as follows.

**Theorem 8.57** (Jakimczuk [144]) *Let $a > 0$ and $0 < h < 1$. Then*

$$\lim_{n\to\infty} \frac{\text{Bell}_n}{a^n} = \infty, \tag{8.28}$$

$$\lim_{n\to\infty} \frac{\text{Bell}_n}{n!^h} = \infty, \tag{8.29}$$

$$\lim_{n\to\infty} \frac{\text{Bell}_n}{n!} = 0. \tag{8.30}$$

**Proof** Theorem 8.53 gives

$$\log \frac{\text{Bell}_n}{a^n} = \log \text{Bell}_n - n \log a = n \log n - n \log a + o(n \log n),$$

which implies that $\lim_{n\to\infty} \log \frac{\text{Bell}_n}{a^n} = \infty$ and hence (8.28). Using the well-known Stirling formula $n! \sim \sqrt{2\pi} n^{n+1/2} e^{-n}$, we obtain

$$\log n! = n \log n - n + \frac{1}{2} \log n + \log \sqrt{2\pi} + o(1) = n \log n - n + o(n)$$

and $\log n!^h = hn \log n - hn + o(n)$. Therefore,

$$\log \frac{\text{Bell}_n}{n!^h} = (1-h) n \log n - n \log \log n - (1-h) n + o(n),$$

which implies that $\lim_{n\to\infty} \log \frac{\text{Bell}_n}{n!^h} = \infty$ and hence (8.29) for all $0 < h < 1$. If we let $h = 1$, then we have $\lim_{n\to\infty} \log \frac{\text{Bell}_n}{n!} = -\infty$ and hence (8.30). □

The number of partitions of $[n]$ into $k$ blocks is given by $\text{Stir}(n,k)$. The asymptotics of $\text{Stir}(n,k)$ were known already to Laplace (see [25, 46, 144, 249] for extensive bibliographies).

**Theorem 8.58** (Moser and Wyman [257]) *The following holds:*

$$\text{Stir}(n,k) = \frac{n!(e^r - 1)^k}{k! r^n \sqrt{2\pi k b}} \left(1 - \frac{6r^2\theta^2 + 6r\theta + 1}{12re^r}\right),$$

*where $re^r = n$, $e^r - 1 = k + \theta$ with $\theta = O(1)$, and $b = \frac{re^{2r} - r(r+1)e^r}{(e^r-1)^2}$.*

**Proof** By Theorem 3.16, we have that the exponential generating function for the number of set partitions of $\mathcal{P}_{n,k}$ is given by

$$\sum_{n \geq k} \text{Stir}(n,k) \frac{x^n}{n!} = \frac{1}{k!} (e^x - 1)^k.$$

The Cauchy integral formula gives

$$\frac{\text{Stir}(n,k)}{n!} = \frac{1}{2\pi i k!} \oint_{|z|=R} \frac{(e^z-1)^k}{z^{n+1}} dz,$$

for any $R > 0$. If we take $R = r$ and choose integer $k$ such that $e^r - 1 = k + \theta$ with $\theta = O(1)$, we derive the requested estimation. □

**Theorem 8.59** (for example, see Kasraoui [157]) *For $i \geq 0$ and $j \geq 1$, we have, as $n \to \infty$,*

$$\frac{\text{Stir}(n-i,k)}{\text{Stir}(n,k)} = \frac{1}{k^i} + O\left((1-1/k)^n\right), \quad \frac{\text{Stir}(n-i,k-j)}{\text{Stir}(n,k)} = O\left((1-j/k)^n\right).$$

**Proof** From the above theorem or Theorem 3.18 we have

$$\text{Stir}(n,k) = \frac{k^n}{k!} + O((k-1)^n) = \frac{k^n}{k!}\left(1 + O((1-1/k)^n)\right),$$

which implies

$$\frac{\text{Stir}(n-i,k-j)}{\text{Stir}(n,k)}$$
$$= \frac{(k-j)^{n-i}}{(k-j)!}\left(1 + O((1-1/(k-j))^{n-i})\right)\frac{k!}{k^n}\left(1 + O((1-1/k)^n)\right)^{-1}$$
$$= \frac{k!}{(k-j)!(k-j)^i}\left(1-j/k\right)^n\left(1 + O((1-1/k)^n)\right).$$

By considering the cases $j = 0$ and $j \geq 1$, we complete the proof. □

### 8.5.1.1 On the Number of Blocks

The number of partitions of $[n]$ into $k$ blocks is given by $\text{Stir}(n,k)$. The asymptotics of the $\text{Stir}(n,k)$ is given in Theorem 8.58. So one can ask, what is the average of the number blocks in all set partitions of $[n]$? The answer is given by our next result.

**Theorem 8.60** (Flajolet and Sedgewick [107, Section VIII.41]) *Let $X_n$ be the number blocks in a random set partition of $[n]$. Then*

$$\mathbb{E}(X_n) = \frac{n}{\log n} + \frac{n \log \log n}{\log^2 n}(1 + o(1))$$

*and*

$$\mathbb{V}(x_n) = \frac{n}{\log^2 n} + \frac{n(2\log\log n - 1 + o(1))}{\log^3 n}.$$

**Proof** The exponential generating function for the number of set partitions of $[n]$ according to the number of the blocks is given by

$$F(z,u) = \sum_{n\geq 0}\sum_{\pi \in \mathcal{P}_n} z^n u^{\text{blo}(\pi)} = e^{u(e^z-1)}.$$

Clearly, $F(z) = F(z,1) = e^{e^z-1}$ is the generating function for the number of set partitions of $[n]$. Let $F'(z) = \frac{\partial}{\partial u} F(z,u)\,|_{u=1}$, which is given by $F'(z) = e^{e^z-1}(e^z - 1)$. Thus, the mean number of blocks in a random set partition of $[n]$ can be expressed as

$$\frac{f'_n}{f_n} = \frac{[z^n]F'(z)}{[z^n]F(z)}.$$

Example 8.32 shows that the generating function $F(z)$ is an admissible function, and so is $F'(z)$ by Theorem 8.36. The saddle–point for $F(z)$ shows that $\xi$ satisfies the equation $\xi e^\xi = n$, and it is already known that $\xi = \log n - \log\log n + o(1)$. Let $a(r), b(r)$ and $a'(r), b'(r)$ be the functions as defined in Definition 8.31 for the function $F(z)$ and $F'(z)$, respectively. Then

$$\log F(z) = e^z - 1, \qquad \log F'(z) = e^z + z - 1,$$
$$a(r) = re^r, \qquad a'(r) = a(r) + r,$$
$$b(r) = r(r+1)e^r, \qquad b'(r) = b(r) + r.$$

Therefore, by estimating $f'_n$ by Theorem 8.34 with $r = \xi$, we obtain

$$f'_n = \frac{e^\xi F(\xi)}{\xi^n \sqrt{2\pi b'(\xi)}} \left( e^{-\frac{\xi^2}{2b'(\xi)}} + o(1) \right).$$

While we know the estimation of $f_n$ is given by

$$f_n = \frac{F(\xi)}{\xi^n \sqrt{2\pi b(\xi)}} (1 + o(1)).$$

Using the facts that $b'(\xi) \sim b(\xi)$ and $\xi^2$ is of smaller order than $b'(\xi)$, we derive

$$\frac{f'_n}{f_n} = e^\xi(1 + o(1)) = \frac{n}{\log n}(1 + o(1)),$$

as claimed. Similar computations give the second moment. □

The above proof can be easily simplified when we use our terminology. More precisely, the generating function $e^{q(e^x-1)}$ for the number of set partitions of $[n]$ according to the number of blocks has been computed in Theorem 3.16. Now, by Theorem 2.54 we derive

$$[x^n]\left(\frac{d}{dq}e^{q(e^x-1)}\right)\Big|_{q=1} = [x^n](e^{e^x+x-1} - e^{e^x-1})$$

$$= [x^n]\left(\frac{d}{dx}(e^{e^x-1})\right) - \text{Bell}_n$$

$$= \text{Bell}_{n+1} - \text{Bell}_n.$$

Thus, $\mathbb{E}(X_n) = \frac{\text{Bell}_{n+1}}{\text{Bell}_n} - 1$. Hence, Corollary 8.54 shows that

$$\mathbb{E}(X_n) = \frac{n}{\log n} + \frac{n \log \log n}{\log^2 n}(1 + o(1)),$$

as required.

Theorem 8.53 presents an asymptotic formula for the $n$-th Bell number in terms of the solution of the unique real solution of the equation $re^r = n$. This asymptotic formula was recovered by Moser and Wyman [256]. The function $r = r(n)$ is also known as $LambertW(n)$. The explicit form of their result is not convenient for obtaining asymptotics for the expectation and the variance, as $r$ will vary with $n$. Canfield [62], Canfield and Harper [64] made minor modifications to the proof of Moser and Wyman [256] to obtain an estimate for $\text{Bell}_{n+h}$, which holds uniformly for $h = O(\log n)$, using a single $r = r(n)$ value, as $n \to \infty$. This estimation very helpful for deriving the expectation and the variance (for example, see [84]), which provides another proof of Theorem 8.60.

### 8.5.1.2  On the Number of Distinct Block Sizes

Now, we will focus on the asymptotic behavior of the number of distinct block sizes in a set partition of $[n]$. This behavior is more complicated than other problems on set partitions. Although there is a simple generating function, the usual analytic methods for estimating coefficients fail in the direct approach, and elementary approaches combined with some analytic methods are used to obtain most of the results.

**Definition 8.61** *Let $b_n$ (respectively, $b_{n,k}$) be the number of set partitions of $\mathcal{P}_n$ (respectively, of $\mathcal{P}_{n,k}$) with distinct block sizes. Let $B(x)$ (respectively, $B(x,q)$) be the exponential generating function for the sequence $\{b_n\}_{n\geq 0}$ (respectively, $\{b_{n,k}\}_{n,k\geq 0}$), that is, $B(x) = \sum_{n\geq 0} b_n \frac{x^n}{n!}$ (respectively, $B(x,q) = \sum_{n\geq 0} \sum_{k=0}^n b_{n,k} \frac{x^n q^k}{n!}$).*

Carlitz [66] has found an explicit formula for the exponential generating function for the sequence $b_n$. Later, Wilf [356] showed that this exponential generating function is a special case of an enumerator in an exponential family for hands whose cards all have different types. In 1999, Knopfmacher et al. [191] determined the asymptotic behavior of $b_n$.

Our next result presents the explicit formula for $B(x,q)$.

**Theorem 8.62** (Carlitz [66], Wilf [356]) *The generating function $B(x,q)$ is given by*

$$B(x,q) = \prod_{m\geq 1}\left(1 + q(e^{x^m/m!} - 1)\right).$$

**Proof** Note that the number of ways of choosing $\ell_1, \ell_2, \ldots, \ell_k$ blocks of sizes $m_1, m_2, \ldots, m_k$ from the set $[n]$ when the order of the blocks is irrelevant is given by

$$b_{n,k} = \sum_{m_1 > m_2 > \cdots > m_k > 0} \frac{n!}{\prod_{i=1}^{k} \ell_i!(m_i!)^{\ell_i}},$$

where $\ell_i > 0$ and $\ell_1 m_1 + \cdots \ell_k m_k = n$. Thus

$$B(x, q) = \sum_{n \geq 0} \sum_{k=0}^{n} \sum_{m_1 > m_2 > \cdots > m_k > 0} \frac{n!}{\prod_{i=1}^{k} \ell_i!(m_i!)^{\ell_i}} \frac{x^n q^k}{n!}.$$

Exchanging the order of the sums, we obtain

$$B(x, q) = \sum_{n \geq 0} \sum_{k=0}^{n} \sum_{m_1 > m_2 > \cdots > m_k > 0} \frac{1}{\prod_{i=1}^{k} \ell_i!(m_i!)^{\ell_i}} x^n q^k$$

$$= \prod_{m \geq 1} \left(1 + q \sum_{n \geq 1} \frac{x^{mn}}{m!^n n!}\right)$$

$$= \prod_{m \geq 1} \left(1 + q(e^{x^m/m!} - 1)\right),$$

which ends the proof. $\square$

**Definition 8.63** Define $\tilde{b}_n = \sum_{k=0}^{n} k b_{n,k}$ and $\hat{b}_n = \sum_{k=0}^{n} k^2 b(n, k)$. We denote the corresponding exponential generating functions for the sequences $\{\tilde{b}_n\}_{n \geq 0}$ and $\{\hat{b}_n\}_{n \geq 0}$ by $\tilde{B}(x)$ and $\hat{B}(x)$.

Applying Lemma 8.16 on the exponential generating function $B(x, q)$, we obtain the following result.

**Corollary 8.64** We have

$$\tilde{B}(x) = B(x, 1) \sum_{m \geq 1} (1 - e^{-\frac{x^m}{m!}}),$$

$$\hat{B}(x) = B(x, 1) \left[\left(\sum_{m \geq 1}(1 - e^{-\frac{x^m}{m!}})\right)^2 + \sum_{m \geq 1}(1 - e^{-\frac{x^m}{m!}}) - \sum_{m \geq 1}(1 - e^{-\frac{x^m}{m!}})^2\right].$$

**Proof** By Theorem 8.62 and Lemma 8.16, we obtain

$$\tilde{B}(x) = \frac{d}{dq} B(x, q)|_{q=1} = B(x, q) \sum_{m \geq 1} \frac{e^{\frac{x^m}{m!}} - 1}{1 + q(e^{\frac{x^m}{m!}} - 1)}|_{q=1}$$

$$= B(x, 1) \sum_{m \geq 1} \frac{e^{\frac{x^m}{m!}} - 1}{e^{\frac{x^m}{m!}}} = B(x, 1) \sum_{m \geq 1} (1 - e^{-\frac{x^m}{m!}})$$

and similarly

$$\widehat{B}(x) = \frac{d}{dq}\left(q\frac{d}{dq}B(x,q)\right)\Big|_{q=1}$$

$$= B(x,1)\left[\left(\sum_{m\geq 1}(1-e^{-\frac{x^m}{m!}})\right)^2 + \sum_{m\geq 1}(1-e^{-\frac{x^m}{m!}}) - \sum_{m\geq 1}(1-e^{-\frac{x^m}{m!}})^2\right],$$

which ends the proof. □

Now we are ready to estimate the requested average $\tilde{b}_n/b_n$, namely, the average number of the distinct block sizes in set partitions of $[n]$.

**Theorem 8.65** *(Wilf [356]) We have $\tilde{b}_n/b_n \sim e\log n$ as $n \to \infty$.*

**Proof** At first we show that the coefficient of $z^n$ in the Taylor expansion of $f_k(z) = e^{e^z-1}(1-e^{-\frac{z^k}{k!}})$ is approximately to $\tilde{b}_n/b_n$ for $k \leq e\log n$ and is negligible for $k > e\log n$. We start by noting that the Taylor series of $e^z - 1$ and $e^z - 1 - \frac{z^k}{k!}$ are given by

$$e^z - 1 = \sum_{n\geq 1}\frac{z^n}{n!} \text{ and } e^z - 1 - \frac{z^k}{k!} = \sum_{n\geq 1, n\neq k}\frac{z^n}{n!}, \tag{8.31}$$

where both have nonnegative coefficients. For any $z \in \gamma$ (a circle around the origin),

$$\left|e^z - 1 - \frac{z^k}{k!}\right| \leq e^{|z|} - 1 - \frac{|z|^k}{k!}. \tag{8.32}$$

Define $e^{e^z-1-\frac{z^k}{k!}} = \sum_{n\geq 0}a_{n,k}z^n$. By (8.31) we have

$$0 \leq a_{n,k} \leq \frac{b_n}{n!} \text{ and } |f_k(z)| \leq f_k(|z|). \tag{8.33}$$

Now, we proceed with the critical idea of the proof. Fix $\epsilon$ any small positive real number, say $\epsilon \leq 10^{-3}$. Consider $k \leq (e-\epsilon)\log n$, where $n$ is sufficiently large (depending only on $\epsilon$). By Cauchy's theorem (see Theorem F.9), we have

$$a_{n,k} = \frac{1}{2\pi i}\int_{|z|=\rho_n}\frac{e^{e^z-1-\frac{z^k}{k!}}}{z^{n+1}}dz,$$

where $\rho_n$ is the unique positive root of $ze^z = n+1$; see (8.27). By (8.32) we obtain

$$a_{n,k} \leq \frac{1}{\rho_n^n} \max_{|z|=\rho_n} \left| e^{z-1-\frac{z^k}{k!}} \right|$$

$$= e^{-n\log \rho_n} \cdot e^{\rho_n - 1 - \frac{\rho_n^k}{k!}}$$

$$= e^{\rho_n - 1 - \frac{\rho_n^k}{k!} - n\log \rho_n}$$

$$= \frac{\sqrt{2\pi}b_n}{n!} e^{\log \rho_n + \frac{\rho_n}{2} - \frac{\rho_n^k}{k!} + o(1)}.$$

Since $\rho_n^k \geq k!\rho_n$ for all $1 \leq k \leq (e-\epsilon)\log n$ ($n$ is taken sufficiently large), we have

$$a_{n,k} \leq \frac{b_n}{n!} e^{-\frac{\rho_n}{4} + o(1)},$$

as $n \to \infty$. Therefore, the coefficient of $z^n$ in the generating function

$$\sum_{k \leq (e-\epsilon)\log n} f_k(z) = e^{e^z - 1} \sum_{k \leq (e-\epsilon)\log n} (1 - e^{-\frac{z^k}{k!}})$$

is given by $\frac{(e-\epsilon)b_n \log n}{n!}$, as $n \to \infty$. Also, by (8.33) the corresponding coefficient is at most $\frac{2\epsilon n! \log n}{n!}$ for all $k$ such that $(e-\epsilon)\log n \leq k \leq (e+\epsilon)\log n$. It remains to show the case $k \geq (e+\epsilon)\log n$. If $k \geq 100\epsilon^{-1}\log n$, then by Stirling's formula (see Theorem F.1) we have

$$\left| \frac{z^k}{k!} \right| = \frac{\rho_n^n}{k!} \leq \frac{e^k \log^k n}{k^k} \leq e^{-3k},$$

which implies

$$|f_k(z)| \leq e^{\rho_n - 1 - 2k}.$$

If $(e+\epsilon)\log n \leq n \leq 100\epsilon^{-1}\log n$, then

$$1 - e^{-\frac{z^k}{k!}} = \sum_{1 \leq m \leq 100\epsilon^{-1}k^{-1}\log n} (-1)^{m-1}\frac{z^{km}}{k!^m} + O\left(e^{-50\log n}\right).$$

Thus, the coefficient of $z^n$ in the expansion of the function $\sum_{k \geq (e+\epsilon)\log n} f_k(z)$ is given by

$$\frac{1}{2\pi i} \int_{|z|=\rho_n} \frac{\sum_{k \geq (e+\epsilon)\log n} f_k(z)}{z^{n+1}} dz$$

$$= \sum_{k \geq (e+\epsilon)\log n} \sum_{1 \leq m \leq \frac{100\log n}{\epsilon k}} \left( \frac{(-1)^{m-1}}{2\pi i k!^m} \int_{|z|=\rho_n} \frac{e^{e^z - 1}}{z^{n+1-km}} dz \right)$$

$$+ O\left(e^{e^{\rho_n} - 5\log n - n\log \rho_n}\right)$$

$$= \sum_{k \geq (e+\epsilon)\log n} \sum_{1 \leq m \leq \frac{100\log n}{\epsilon k}} \left( \frac{(-1)^{m-1}}{2\pi i k!^m} \int_{|z|=\rho_n} \frac{e^{e^z - 1}}{z^{n+1-km}} dz \right) + O\left(\frac{b_n}{n^5 n!}\right).$$

But by Cauchy's theorem, we obtain

$$\frac{(n-km)!}{2\pi i}\int_{|z|=\rho_n}\frac{e^{e^z-1}}{z^{n+1-km}}dz = b_{n-km}.$$

Now, we conclude the proof by showing that $\frac{b_{n-km}}{(n-km)!k!^m}$ is very small when compared to $\frac{b_n}{n!}$, where $(e+\epsilon)\log n \leq k \leq 100\epsilon^{-1}\log n$ and $1 \leq m \leq 100\epsilon^{-1}k^{-1}\log n$. Assume that $(e+\epsilon)\log n \leq v \leq 10^4\epsilon^{-2}\log n$, by Theorem 8.53 we have

$$\log\frac{b_{n-v}n!}{b_n(n-v)!} = -\frac{n+1}{\rho_n}+\frac{n-v+1}{\rho_{n-v}}+(n+1)\log\rho_n-(n-v+1)\log\rho_{n-v}$$
$$+\frac{\rho_n}{2}-\frac{\rho_{n-v}}{2}+o(1).$$

Thus, by (8.27) we obtain $\rho_{n-v} = \rho_n - \frac{v}{n}+O\left(\frac{v}{n\log n}\right)$, which implies

$$\log\frac{b_{n-v}n!}{b_n(n-v)!} = (n+1)\left(\frac{1}{\rho_{n-v}}-\frac{1}{\rho_n}\right)-\frac{v}{\rho_{n-v}}+(n+1)\log\frac{\rho_n}{\rho_{n-v}}$$
$$+v\log\rho_{n-v}+o(1)$$
$$= v\log\rho_n+O(1).$$

Note that $\log(k!)^m \geq mk\log k - (1+\epsilon/100)mk$ for $n$ sufficient large, thus

$$\log\frac{b_{n-km}n!}{b_n k!^m(n-km)!}$$
$$\leq km\log\rho_n-mk\log k+(1+\epsilon/100)km+O(1)$$
$$\leq km\left(\log\log n-\log(e+\epsilon)-\log\log n+1+\epsilon/100+o(1)\right)+O(1)$$
$$\leq -\epsilon km/1000 \leq -\epsilon 10^{-3}\log n.$$

Hence,

$$\sum_{k\geq(e+\epsilon)\log n}\sum_{1\leq m\leq \frac{100\log n}{\epsilon k}}\left(\frac{(-1)^{m-1}}{2\pi i k!^m}\int_{|z|=\rho_n}\frac{e^{e^z-1}}{z^{n+1-km}}dz\right) = O\left(\frac{b_n}{n!n^{\epsilon/2000}}\right).$$

Combining the results of the above three cases, namely, $1 \leq k \leq (e-\epsilon)\log n$, $(e-\epsilon)\log n \leq k \leq (e+\epsilon)\log n$, and $k \geq (e+\epsilon)\log n$, we obtain $\frac{\tilde{b}_n}{b_n} \sim e\log n$ as $n \to \infty$. □

We end this section by commenting that Wilf [356] (see preprint work of A.M. Odlyzko and L.B. Richmond: "On the Number of Distinct Block Sizes in Partitions of a Set) gave two proofs for the above theorem. The first is based on estimating the average $\frac{\tilde{b}_n}{b_n}$ by Theorem 8.53, while the second is based on the results of Hayman [134] on the coefficients of Laurent expansion of a $H$-admissible function.

## 8.5.2 On Number of Blocks in a Noncrossing Set Partition

First, we find the mean of the number of blocks of a given size $m$ in $\mathrm{NC}_{n,k}$.

**Theorem 8.66** (Arizmendi [6]) *The mean of the number of blocks of size $m$ over all set partitions of $\mathrm{NC}_{n,k}$ (see Definition 7.33) is given by*

$$\frac{n\binom{n-1-m}{k-2}}{\binom{n}{k}}.$$

**Proof** If we sum the number of blocks of size $m$ in all set partitions of $\mathrm{NC}_{n,k}$ where the $i$-th block has $r_i$ elements, then we want to calculate the following sum:

$$\sum_{\sum_{j\geq 1} r_j=k,\, \sum_{j\geq 1} jr_j=n} \frac{n!\,r_m}{(n+1-\sum_{i\geq 1} r_i)!\prod_{i\geq 1} r_i!}$$

$$= \sum_{\sum_{j\geq 1} r_j=k,\, \sum_{j\geq 1} jr_j=n} \frac{n!\,r_m}{(n+1-k)!\prod_{i\geq 1} r_i!}$$

$$= \binom{n}{k-1} \sum_{\sum_{j\geq 1} r_j=k,\, \sum_{j\geq 1} jr_j=n} \frac{(k-1)!\,r_m}{\prod_{i\geq 1} r_i!}$$

$$= \binom{n}{k-1} \sum_{\sum_{j\geq 1} r'_j=k-1,\, \sum_{j\geq 1} jr'_j=n-m} \frac{(k-1)!}{\prod_{i\geq 1} r'_i!}$$

$$= \binom{n}{k-1}\binom{n+1-k}{k-2},$$

(We leave it to the reader to explain the last equality; try to count in two ways the number of lattice paths from $(0,0)$ to $(n-1, k-1)$ with steps $(1,0)$ or $(0,1)$). Thus, by Corollary 3.53 we have

$$\frac{n\binom{n}{k-1}\binom{n-1-m}{k-2}}{\binom{n}{k}\binom{n}{k-1}},$$

which implies our result. □

As a corollary of the above theorem, we can state the mean of the number of blocks of size $m$ over all noncrossing set partitions of $\mathcal{P}_n$.

**Theorem 8.67** (Arizmendi [6]) *The mean of the number of blocks of size $m$ over all set partitions of $\mathrm{NC}_n$ (see Definition 7.33) is given by*

$$\frac{(n+1)\binom{2n-1-m}{n-1}}{\binom{2n}{n}}.$$

*In particular, the expected number of blocks of size $M$ of a noncrossing partition chosen uniformly at random in $\mathrm{NC}_n$ is asymptotically $n/2^m$, where $n \to \infty$.*

**Proof** By the proof of Theorem 8.66 and Theorem 3.52 (the number of non-crossing set partitions of $\mathcal{P}_n$ is given by $\operatorname{Cat}_n$), we can state that the mean of the number of blocks of size $m$ over all set partitions of $\operatorname{NC}_n$ is given by

$$\frac{\sum_{k=1}^{n}\binom{n}{k-1}\binom{n-1-m}{k-2}}{\frac{1}{n+1}\binom{2n}{n}} = \frac{\sum_{k=0}^{n}\binom{n}{k}\binom{n-1-m}{n-m-k}}{\frac{1}{n+1}\binom{2n}{n}}$$

$$= \frac{(n+1)\binom{2n-1-m}{n-m}}{\binom{2n}{n}},$$

as claimed. By Stirling's formula (see (F.1)) we derive that the expected number of blocks of size $M$ of a noncrossing partition chosen uniformly at random in $\operatorname{NC}_n$ is asymptotically $n/2^m$, where $n \to \infty$. $\square$

Note that the above two results have been extended by Arizmendi [6] to study the number of blocks of size $\ell m$ over all set partitions of $\mathcal{P}_{n,k}$ ($\mathcal{P}_n$).

### 8.5.3 Records

In Section 5.3 we studied the statistic of the "number of records" (various types of records: strong records, weak records and additional records)) on set partitions. Here, we continue these investigations to obtain asymptotic results on the statistic of the "number of records".

**Theorem 8.68** *The distribution of the number of additional weak records in a random set partition of $[n]$ is asymptotically Gaussian, with mean*

$$\frac{(n-1)\operatorname{Bell}_{n-1}}{\operatorname{Bell}_n} = \log n - \log\log n + O\left(\frac{\log\log n}{\log n}\right) \quad (8.34)$$

*and variance*

$$\frac{(n-1)(n-2)\operatorname{Bell}_{n-2} + (n-1)\operatorname{Bell}_{n-1}}{\operatorname{Bell}_n} - \frac{(n-1)^2\operatorname{Bell}_{n-1}^2}{\operatorname{Bell}_n^2}$$

$$= \log n - \log\log n + O\left(\frac{\log\log n}{\log n}\right).$$

**Proof** By (5.8) we obtain that the generating function for the number of set partitions of $[n]$ according to the number of additional weak records is given by $e^{e^x+qx-1}$. Thus

$$\sum_{\pi \in \mathcal{P}_n} \operatorname{addrec}(\pi) = (n-1)![x^{n-1}]\frac{d}{dq}e^{e^x+qx-1}\bigg|_{q=1}$$

$$= (n-1)![x^{n-1}]xe^{e^x+x-1} = (n-1)\operatorname{Bell}_{n-1},$$

which implies (8.34). The computation of the variance is similar (we leave it to the reader). Using Corollary 8.54, we obtain the asymptotic estimates for the moments. At the end, for the limiting distribution, consider the probability that there are exactly $\ell$ additional weak records, which equals, for $\ell = O(\log n)$,

$$\binom{n-1}{\ell}\frac{\text{Bell}_{n-1-l}}{\text{Bell}_n} = \binom{n-1}{\ell} \cdot \frac{(n-1-l)!r^{\ell+1}}{n!}\left(1+O\left(\frac{\log n}{n}\right)\right)$$

$$= \frac{r^{\ell+1}}{n\ell!}\left(1+O\left(\frac{\log n}{n}\right)\right).$$

By setting $\ell = \rho_n + t\sqrt{\rho_n}$, see (8.27), and applying Stirling's formula (see Theorem F.1) we obtain, for $t = o(\rho_n^{1/6})$,

$$\binom{n-1}{\ell}\frac{\text{Bell}_{n-1-l}}{\text{Bell}_n}$$

$$= \frac{\rho_n}{n} \cdot \exp\left(\ell + \ell\log\rho_n - \ell\log\ell - \frac{1}{2}\log(2\pi\ell)\right)\left(1+O\left(\frac{\log n}{n}\right)\right)$$

$$= \frac{\rho_n}{n}\exp\left(\ell\left(1-\log(1+t\rho_n^{-1/2})\right) - \frac{1}{2}\log(2\pi\rho_n) + O(\rho_n^{-1/2})\right)$$

$$= \frac{\rho_n}{n}\exp\left((\rho_n+t\sqrt{\rho_n})\left(1-\frac{t}{\sqrt{\rho_n}}+\frac{t^2}{2\rho_n}\right) - \frac{1}{2}\log(2\pi\rho_n) + O\left(\frac{1+t^3}{\sqrt{\rho_n}}\right)\right)$$

$$= \frac{\rho_n}{n}\exp\left(\rho_n - \frac{t^2}{2} - \frac{1}{2}\log(2\pi\rho_n) + O\left(\frac{1+t^3}{\sqrt{\rho_n}}\right)\right)$$

$$= \frac{1}{\sqrt{2\pi\rho_n}}\exp\left(-\frac{t^2}{2} + o(1)\right),$$

which completes the proof. □

Corollary 8.54 is a good tool to obtain asymptotic results for our statistics. For instance, by a similar technique as in the proof of the last theorem one can show that the mean and variance of **sumrec** asymptotically are given by $\sim \frac{n^2}{\log^2 n}$ and $\sim \frac{n^3}{2\log^3 n}$, respectively (see Theorem 5.22).

---

## 8.6 Exercises

**Exercise 8.1** *Show that*
$$\lim_{n\to\infty} \frac{\text{Bell}_{n+1}}{\text{Bell}_n} = \infty.$$

**Exercise 8.2** *Prove that* $\text{Bell}_{n+1} - \text{Bell}_n \sim \text{Bell}_{n+1}$, $\frac{\sqrt[n]{\text{Bell}_n}}{\frac{\text{Bell}_n}{\text{Bell}_{n-1}}} \to \frac{1}{e}$ *when* $n \to \infty$, *and* $\text{Bell}_{n+1} \sim e(\text{Bell}_n)^{1+1/n}$.

**Exercise 8.3** *Show that if $\omega(x)$ is the number of Bell numbers that does not exceed $x$, then $\omega(x) \sim \frac{\log x}{\log \log x}$.*

**Exercise 8.4** *As we have already showed, the exponential generating function $e^{e^x-1}$ for the number of set partitions of $[n]$ stems from the classical translation of combinatorial sets.*

1. *Show that the exponential generating function for the number of set partitions of $[n]$ with even (resp. odd) number of elements is given by $\cosh(e^z - 1)$ (resp. $\sinh(e^z - 1)$).*

2. *Define $S(z) = \cosh(e^z - 1) = \sinh(e^z - 1)$, and show that $S(z) = e^{1-e^z}$.*

3. *Prove that $S_n = a_n \cos(\alpha_n) + O(b_n)$, where $\log(b_n/a_n) = o(n/\log^k n)$ for all $k > 0$,*

$$\log(a_n/n!) = n\left(-\log\log n + \frac{\log\log n + 1}{\log n} + \frac{(\log\log n)^2 - \pi^2}{2\log^2 n}\right.$$
$$\left. + \frac{2(\log\log n)^3 - 3(\log\log n)^2 - 6\pi^2(\log\log n) + 3\pi^2}{6\log^3 n}\right.$$
$$\left. + O\left(\frac{(\log\log n)^4}{\log^4 n}\right)\right)$$

*and*

$$\alpha_n = \pi n \left(\frac{1}{\log n} + \frac{\log\log n}{\log^2 n} + \frac{(\log\log n)^2 - \log\log n - \frac{pi^3}{3}}{\log^3 n}\right.$$
$$\left. + O\left(\frac{(\log\log n)^3}{\log^3 n}\right)\right).$$

**Exercise 8.5** *A set $\mathcal{A}$ of sequences is $t$-intersection if any two elements $a, b \in \mathcal{A}$ have at least $t$ positions in common, that is, $|\{i \mid a_i = b_i\}| \geq t$. Ku and Renshaw [205] extended the study of $t$-intersections from the family of permutations to family of set partitions. In particular, they showed the following results.*

1. *Let $n \geq 2$ and let $\mathcal{A} \subset \mathcal{P}_n$ be 1-intersecting. Then $|\mathcal{A}| \leq \text{Bell}_{n-1}$ with equality if and only if $\mathcal{A}$ consists of all set partitions of $[n]$ with a fixed singleton (block of size 1).*

2. *Let $t \geq 2$ and let $\mathcal{A} \subset \mathcal{P}_n$ be $t$-intersecting. If $n \geq n_0(t)$, then $|\mathcal{A}| \leq \text{Bell}_{n-t}$ with equality if and only if $\mathcal{A}$ consists of all set partitions with $t$ fixed singletons.*

*Prove these two results.*

**Exercise 8.6** *A set partition of $[n]$ is said to be irreducible if no proper subinterval of $[n]$ is a union of blocks; see [27]. Determine the asymptotic relationship between the numbers of irreducible set partitions, set partitions without singleton blocks, and all set partitions when the block sizes must lie in some specified set.*

---

## 8.7 Research Directions and Open Problems

We now suggest several research directions which are motivated both by the results and exercises of this chapter.

**Research Direction 8.1** *Fix $\tau$ to be any subword pattern. Let $PG_\tau(n, k, r)$ denote the probability that a geometrically distributed word of size $n$ belongs to $\mathcal{P}_{n,k}$ and has exactly $r$ the subword pattern $\tau$. Define $PG_{\tau,k}(z, w) = \sum_{n,r \geq 0} P_\tau(n, k, r) z^n w^r$. Mansour and Shattuck [240] showed for all $k \geq 1$,*

$$P_{11,k}(z,w) = \prod_{j=1}^{k} \frac{\frac{pq^{j-1}z}{1-pq^{j-1}z(w-1)}}{1 - \sum_{i=1}^{j} \frac{pq^{i-1}z}{1-pq^{i-1}z(w-1)}},$$

$$P_{12,k}(z,w) = \frac{p^k q^{\binom{k}{2}} w^{k-1} z^k}{\prod_{\ell=0}^{k}(1-(1-w)pq^\ell z)^{k-\ell} \prod_{\ell=0}^{k-1}\left(1 - \sum_{j=0}^{\ell} \frac{pq^j wz}{\prod_{i=0}^{j}(1-(1-w)pq^i z)}\right)},$$

*and for $k \geq 2$,*

$$P_{21,k}(z,w) = \frac{p^k q^{\binom{k}{2}} z^k}{1-pz} \prod_{j=2}^{k} \frac{w W_j(z,w) + 1 - w}{1 + (w-1)pq^{j-1}z}$$

*with $P_{21,1}(z,w) = \frac{pz}{1-pz}$, where $W_j(z,w)$ is given by*

$$\frac{1}{W_j(z,w)} = \left[1 - \sum_{i=0}^{j-1} \frac{pq^i wz}{\prod_{\ell=0}^{i}(1-(1-w)pq^\ell z)}\right] \prod_{i=0}^{j-1}(1-(1-w)pq^i z).$$

*These results motivate the following question: Find the probability that a geometrically distributed word of size $n$ belongs to $\mathcal{P}_{n,k}$ and has exactly $r$ the subword pattern $\tau$, where $\tau$ is any fixed patten of size at least three (not necessary to be a subword pattern, it could be a subsequence pattern or a generalized pattern).*

**Research Direction 8.2** *In Section 4 we studied the generating function for the number of set partitions of $\mathcal{P}_n$ that avoid a fixed subword pattern. Also, we* obtained exact formulas for the total number of occurrences of a fixed subsword pattern $\tau$ in all set partitions of $\mathcal{P}_{n,l}$; see Corollary 4.30 when $\tau$ = peak, Corollary 4.37 when $\tau$ = valley, Corollary 4.42 when $\tau = 12$, Corollary 4.46 when $\tau = 1^\ell$, Corollary 4.23 when $\tau = 21$, Corollary 4.54 when $\tau = 12^{\ell-1}$, Corollary 4.59 when $\tau = 2^{\ell-1}$, Corollary 4.62 when $\tau = m\rho m$, Corollary 4.65 when $\tau = m\rho(m+1)$, Corollary 4.68 when $\tau = (m+1)\rho m$, Corollary 4.71 when $\tau = 121$, Corollary 4.74 when $\tau = 132$, and Corollary 4.77 when $\tau = 231$. Our research questions can be formulated as follows:

- Find asymptotically the total number $\text{total}_\tau(n)$ of occurrences of $\tau$ in all set partitions of $\mathcal{P}_n$, where $\tau$ is ones of the above subwords.

- Let $X_\tau(n)$ be the number of occurrences of $\tau$ in a random set partition of $\mathcal{P}_n$. Find asymptotically the mean $\mathbb{E}(X_\tau(n))$ and the variance $\mathbb{V}(X_\tau(n))$ of the random variable $X_\tau(n)$.

- Find asymptotically the number of set partitions of $\mathcal{P}_n(\tau)$, where $\tau$ is a subword pattern of length three.

**Research Direction 8.3** *This open problem was suggested by Adam M. Goyt on July 15, 2009. We denote the number of occurrences of the subsequence pattern $\tau \in \mathcal{P}_m$ in a set partition $\pi \in \mathcal{P}_n$ by $\nu(\tau, \pi)$. Then define*

$$\mu(\tau, k, n) = \max\{\nu(\tau, \pi) \mid \pi \in \mathcal{P}_{n,1} \cup \mathcal{P}_{n,2} \cup \cdots \cup \mathcal{P}_{n,k}\}$$

*and*

$$d(\tau, \pi) = \frac{\nu(\tau, \pi)}{\binom{n}{m}}, \ d(\tau, k, n) = \frac{\mu(\tau, k, n)}{\binom{n}{m}}.$$

*Note that Goyt showed that $d(\tau, k, n-1) \geq d(\tau, k, n)$ and $d(\tau, k, n) \geq d(\tau, k-1, n)$.*

*The packing density of the pattern $\tau$ is defined by*

$$\delta(\tau) = \lim_{n \to \infty} d(\tau, n, n) = \lim_{n \to \infty} \lim_{k \to \infty} d(\tau, k, n).$$

*For instance, Goyt proved that $\delta(121) \leq \frac{1}{2}$ and he asked, can it be shown that $12121\cdots$ is the best possible for packing $121$?*

**Research Direction 8.4** *Using Section 8.5.2, find an explicit formula for the mean of the number of blocks of size $m$ over all set partitions of $\mathcal{P}_{n,k}$ ($\mathcal{P}_n$) that avoid a pattern $\tau$, where $\tau$ any pattern of size four (five).*

**Research Direction 8.5** *Using our study of the size of the longest increasing (alternating) subsequence in set permutations, words over alphabet $[k]$ and compositions, find the mean and the variance of the size of the longest alternating subsequence (or any other patterns) in set partitions of $\mathcal{P}_n$. A sequence $\pi_1 \pi_2 \cdots \pi_n$ is said to be alternating if $\pi_1 > \pi_2 < \pi_3 > \pi_4 < \cdots$.*

**Research Direction 8.6** *Using our study of the number of distinct blocks sizes, we define the following general problem. We define the* type *of a set partition of* $\mathcal{P}_n$ *to be the integer partition of the sizes of its blocks. For example, the types of the set partitions* 111, 112, 121, 122, *and* 123 *are given by* 3, 21, 21, 21, *and* 111, *respectively. Fix a subset* $A_n$ *of integer partitions of* $n$, *and let* $\mathcal{P}(n; A_n)$ *to be the set of all set partitions of* $\pi$ *such that its type belongs to* $A_n$. *For instance, if* $A_n$ *is the set of all integer partitions with distinct parts, then* $\mathcal{P}(n; A_n)$ *is the set of all set partitions of* $\mathcal{P}_n$ *with distinct block sizes. Another example; if* $A_n$ *is the set of all integer partitions such that each part occurs an even number of times, then* $\mathcal{P}(n; A_n)$ *is the set of all set partitions of* $\mathcal{P}_n$ *such that the size of each block is even. So, our suggestion can be formulated as follows. Study the asymptotics behavior of the cardinality of the set* $\mathcal{P}(n; A_n)$ *for special types of* $A_n$? *In particular, when* $A_n$ *consists of integer partitions of* $n$ *of the form* $a_1 a_2 \cdots a_{i-1} a_i (a_i)^d a_{i+1} \cdots a_s$, *where* $a_1 > a_2 > \cdots > a_s \geq 1$ *and* $d \geq 1$.

# Chapter 9

# Gray Codes, Loopless Algorithms and Set Partitions

Algorithms and their analysis play an important role in both mathematics and computer science. In particular, many algorithms make a list of all or some of the objects of a combinatorial class (see Definition 1.31). In this chapter; we are interested in efficient object listing algorithms for such combinatorial classes. Researchers in computer science are interested in finding generating algorithms for combinatorial classes. More precisely, they seek efficient generating members in a particular combinatorial class in such a way that each member is generated exactly once. Many generating problems require sampling the members of the class. Whereas early work on combinatorics focused on counting, as we saw in the previous chapters, the computer used to list the members of the combinatorial class. However, in order to present such a listing, our generating method needs to be extremely efficient. In 1973, Ehrlich [102] suggested a new approach for generating the members of a combinatorial class in which successive elements differ in a "small change". The classic example is the binary reflected Gray code [114, 124], which gives a listing of binary words of size $n$ so that successive members differ in exactly one letter, as described in Example 9.5. Such a listing has two advantages: generation of the list of members of the combinatorial class might be faster and the consecutive members which by differ in "small changes" could differ by only small computations. Finally, Gray codes typically involve elegant recursive structures, which throw new light on the combinatorial class structure.

The concept of combinatorial Gray code was investigated in 1973 by Ehrlich [102], but formally it was presented in 1980 by Joichi and White, Dennis, and Williamson [153]. The spectrum of combinatorial classes includes *permutations* and *multiset permutations, set partitions, compositions, combinations,* and many other classes; see [312]. This area is studied and well covered by several researchers around the world. Such efficient combinatorial generation of all objects of a combinatorial class, basically involves two types of generating algorithms: *Gray code algorithms* and *loopless algorithms*. So, the branch combinatorial Gray codes in mathematics and computer science includes several interesting problems in combinatorics, graph theory, computing and group theory. On June 1988, Herbert Wilf presented an interesting talk at SIAM Conference on Discrete Mathematics in San Franciso on Generalized Gray codes, where he described some results and open problems.

Several authors have been interested in Gray codes for permutations [152, 294, 344], involutions, fixed-point free involutions [351] (in the involution $\pi_1\pi_2\cdots\pi_n$ there no $i$ such that $\pi_i = i$), derangements [18] (permutations with no fixed points), *permutations with a fixed number of cycles* [17], and set partitions [102, 165, 268, 304] (for noncrossing set partitions see [140]). These authors have been studying Gray codes for $\mathcal{P}_n$ and developed algorithms for generating $\mathcal{P}_n$ looplessly (see [102, 304] and references therein). More precisely, Kaye [165] gave a loopless implementation of a Gray code for $\mathcal{P}_n$, attributed to Knuth in [357]. This problem was posed in the book by Nijenhuis and Wilf [265] and later solved by Knuth. In the Gray code that was suggested, successive set partitions differ only in that one element moved to an adjacent block; see the second column in Table 9.1. Note that, the corresponding canonical representation forms may differ in many letters. Ruskey's [303] gave a modification of Knuth's algorithm in which one element moves to a block at most two away between successive set partitions and the corresponding canonical representation forms differ only in one letter by at most two; see the third column in Table 9.1. Later, Semba [317] constructed a set partition algorithm which is faster than the previous set partition algorithm. An efficient algorithm for generating set partitions which is even faster than Semba's algorithm has been created by ER [105]. An extension of Ehrilch's algorithm to words $a_1a_2\cdots a_n$ over nonnegative integers satisfying $a_1 \leq k$ and $a_i \leq 1+\max\{a_1,\ldots,a_{i-1},k-1\}$ has been given in [283]. Later, Mansour and Nassar [223] described algorithms, as we will discuss below, which create a Gray code with distance one; see the rightmost column of Table 9.1, and also a loopless algorithm for generating $\mathcal{P}_n$. More recently, Mansour, Nassar, and Vajnovszki [224] constructed a loopless Gray code algorithm for generating $\mathcal{P}_n$, as we will discuss below.

## 9.1 Gray Code and Loopless Algorithms

Gray codes are named after Frank Gray[1] who patented the binary reflected Gray code in 1953 for use in pulse code communications. The original idea of a Gray code was to list all the binary words of size $n$ such that successive code words differ only in one letter. This idea was generalized and given the wider definition which we shall give later as discussed above.

**Definition 9.1** *A Gray code for a combinatorial class of objects is a listing of the objects so that the transition from an object to its successor takes only a "small change" or a small number of different letters. The definition of "small change" or* small distance *depends on the particular class.*

---

[1] see http://en.wikipedia.org/wiki/Frank_Gray_(researcher)

To define the words "small change" we use the terminology of Hamming distance.

**Definition 9.2** *The Hamming distance between two finite sequences of equal size is the number of positions at which the corresponding symbols are different. A Gray code with distance $d$ is a list of sequences such that the Hamming distance between any two consecutive sequences in the list is at most $d$.*

As our first example, let us give the recursive construction of the reflected binary Gray code. Assume $L_n$ is a Gray code with distance one which lists all the binary words of size $n$, say $L_n = \pi^{(1)}, \pi^{(2)}, \ldots, \pi^{(2^n)}$ (clearly, the number of binary words of size $n$ is $2^n$). Note that $L_n$ is also a Gray code with distance one for the set of binary words of size $n$. Now we define the list $L_{n+1}$ recursively as follows:

$$L_{n+1} = \begin{cases} 0\pi^{(i)}, & i = 1, 2, \ldots, 2^n, \\ 1\pi^{(2^{n+1}+1-i)}, & i = 2^n+1, 2^n+2, \ldots, 2^{n+1}. \end{cases}$$

That is, $L_{n+1}$ is the list $0\pi^{(1)}, 0\pi^{(2)}, \ldots, 0\pi^{(2^n)}, 1\pi^{(2^n)}, \ldots, 1\pi^{(2)}, 1\pi^{(1)}$, where $m\pi$ denotes the word $\pi$ with prepending a letter $m$ before the first letter. It is obvious that $L_{n+1}$ is a Gray code with distance one for the set of binary words of size $n+1$.

**Example 9.3** *A reflected Gray code for all binary words of size $n$ with distance one, for example, is given by, 0000, 0001, 0011, 0010, 0110, 0111, 0101, 0100, 1100, 1101, 1111, 1110, 1010, 1011, 1001, and 1000.*

The essence of a combinatorial generation algorithm is to generate all the combinatorial objects in a given class, for instance, generating all binary words of size $n$. Such algorithms should be designed to use the minimum possible time and space which leads to efficient running time and space complexities, and often to generate and list the objects according to some order or property. The reflected Gray code algorithm can be implemented as described in Algorithm 1 (why?). This algorithm can be implemented in Maple as follows.

```
NEXT:=proc(n,b,j)
if j=n then
  b[j]:=1-b[j]; print(b);
else
  NEXT(n,b,j+1); b[j]:=1-b[j]; print(b); NEXT(n,b,j+1);
fi:
end;

n=4; b:=vector(n,[]);
for i from 0 to n do b[i]:=0; od:
print(b);
NEXT(n,b,1);
```

---
**Algorithm 1** Recursive Gray code for binary words of size $n$
---
1: **for** $i = 1$ to $n$ **do**
2:     set $b_i = 0$
3: **end for**
4: **print** $b_1 \cdots b_n$
5: **call** $next(1)$
6: **end**
7: **Procedure** $next(j)$
8: **if** $j = n$ **then**
9:     $b_j = 1 - b_j$
10:    **print** $b_1 \cdots b_n$
11: **else**
12:    **call** $next(j+1)$
13:    $b_j = 1 - b_j$
14:    **print** $b_1 \cdots b_n$
15:    **call** $next(j+1)$
16: **end if**
17: **end procedure**
---

Note that in the literature are given several Gray codes for generation of all the binary words of size $n$; see [312] and references therein. Also, it is natural to extend the study of Gray codes algorithms to study $m$-ary word Gray codes. It was proved in [153], using the idea of a reflected Gray code for binary words, that it is always possible to list the Cartesian product of finite sets so that the Hamming distance between any successive elements is one (see also [291, 329]). The study of Gray codes does not end with the combinatorial class words, but extends to many combinatorial classes such as permutations, subsets, combinations, compositions, integer partitions, linear extension of posets, acyclic orientations, de Burijn sequences, and set partitions. For complete details, we refer the reader to [312]. In this chapter, we are interested in the Gray codes for generation of all set partitions; see the next section.

Loopless generation has a history which dates back to 1973, as Ehrlch[2] formulated explicit criteria for the loop-free concept; this concept presents an interesting challenge in the field of combinatorial generation. The loopless algorithms must generate each combinatorial object from its predecessor in no more than a constant number of operations. Hence, each object is generated in constant time. This means that powerful programming structures such as recursion and looping cannot be used in the code for generating successive objects, although, note that we always need one loop to generate all objects of some given class. At the present time, one can find a lot of research studies which adopted the loopless idea to generate the elements of some combinato-

---
[2] see http://www.informatik.uni-trier.de/~ley/db/indices/a-tree/e/Ehrlich:Gideon.html

rial class. Consequently, the loop-free concept has been generalized and used in diverse combinatorial classes as permutations, multiset permutations, permutations with fixed number of cycles, set partitions, Dyck paths, and others.

**Definition 9.4** *An algorithm which generates the objects of a combinatorial class is said to be* loopless *or* loop-free *if it takes no more than a constant amount of time between successive objects. A loopless algorithm which generates a given combinatorial class in Gray code order is called loopless Gray code or loop-free Gray code algorithm.*

---
**Algorithm 2** Loopless Gray code for binary words of size $n$
---
1: **for** $i = 0$ to $n$ **do**
2:    set $f_i = i$
3: **end for**
4: **for** $i = 1$ to $n$ **do**
5:    set $b_i = 0$
6: **end for**
7: **print** $b_1 \cdots b_n$
8: set $i = n$
9: **while** $i > 0$ **do**
10:    set $f_n = n$, $b_i = 1 - b_i$, $f_i = f_{i-1}$, $f_{i-1} = i - 1$, $i = f_n$
11:    **print** $b_1 \cdots b_n$
12: **end while**
---

**Example 9.5** *Algorithm 2 is a loopless Gray code algorithm for generating all binary words of size $n$. It generates exactly the same output as Algorithm 1, but achieves this using only $O(1)$ operations per object. It is not a radically different algorithm from Algorithm 1; rather it is similar in that it implements Line 5 looplessly (see Lines 8–12 in Algorithm 2), which implies loopless generation for the whole algorithm. The technique it uses to achieve loopless generation is called the jumping technique, which is a technique we develop to jump immediately to the next active position in the word $b$; thus, Algorithm 2 finds the rightmost active position (rightmost letter that can be changed in the next step) in $O(1)$ instead of the linear time incurred by Algorithm 2. The structure of algorithm 2 is a typical structure of loopless algorithms in general. First, the algorithm is initialized (Lines 1–7), including the establishment of the first binary word (Line 7). Then, the "while" loop generates all binary words with $O(1)$ time between any two consecutive binary words. The loop terminates when some condition equivalent to whether the last binary word has been generated evaluates to true (Line 9). Initialization is incurred $O(n)$ time, where $n$ is the number of letters in the binary word. This is the minimum complexity possible, since each of the $n$ bits must be initialized. The generation of while loop and termination condition must both run in constant* time, which are also minimum possible time complexities. One can claim that only the loop (Line 10) is loop-free, while Ehrlich's definition of looplessness encompasses the whole algorithm. Input and output statements are typically ignored during discussion, since they are of no account and not directly referred to in the generation process. More specifically, the algorithm generates the objects regardless of whether they are printed to standard output or used in some another way.

In this chapter we present a Gary code algorithm and a loop-free algorithm for set partitions; see the next two sections.

## 9.2 Gray Codes for $\mathcal{P}_n$

We start by presenting Table 9.1, which describes the list of set partitions of $\mathcal{P}_4$ according to the lexicographic order, Knuth's Gray code, Modified Gray code of Knuth [303], Ehlich's Gray code [102], Ruskey and Savage's Gray code [283], and Mansour and Nassar's Gray code [223].

**Table 9.1**: Listing $\mathcal{P}_4$

| Lex order | Knuth's Gray code | Modified Knuth | Ehrlich's algorithm | Ruskey & Savage algorithms | | Mansour & Nassar |
|---|---|---|---|---|---|---|
| 1111 | 1111 | 1111 | 1111 | 1111 | 1111 | 1111 |
| 1112 | 1112 | 1112 | 1112 | 1112 | 1112 | 1112 |
| 1121 | 1123 | 1123 | 1122 | 1122 | 1122 | 1122 |
| 1122 | 1122 | 1122 | 1123 | 1123 | 1123 | 1121 |
| 1123 | 1121 | 1121 | 1121 | 1121 | 1121 | 1123 |
| 1211 | 1231 | 1221 | 1221 | 1221 | 1221 | 1223 |
| 1212 | 1232 | 1222 | 1223 | 1223 | 1222 | 1221 |
| 1213 | 1233 | 1223 | 1222 | 1222 | 1223 | 1222 |
| 1221 | 1234 | 1233 | 1232 | 1212 | 1233 | 1212 |
| 1222 | 1223 | 1234 | 1233 | 1211 | 1234 | 1211 |
| 1223 | 1222 | 1232 | 1234 | 1213 | 1231 | 1213 |
| 1231 | 1221 | 1231 | 1231 | 1212 | 1232 | 1233 |
| 1232 | 1211 | 1211 | 1211 | 1232 | 1212 | 1231 |
| 1233 | 1212 | 1212 | 1213 | 1231 | 1213 | 1234 |
| 1234 | 1213 | 1213 | 1212 | 1234 | 1211 | 1232 |

**Definition 9.6** *A Gray code for words is said to be* strict *if successive words differ in one letter and in that position the difference between the letters is only* ±1.

**Example 9.7** *A strict Gray code for set partitions of* [3] *is given by* $123, 122, 121, 111, 112$, *for example.*

Strict Gray codes for set partitions were investigated by Ehrilch [102], where he showed that for infinitely many values of $n$, they do not exist. But in addition he presented a Gray code algorithm in which successive set partitions differ in one letter and the letter in that position can change to 1 or the maximal element in that position change to 1. In the block representation of set partitions, this change corresponds to moving one element to an adjacent block, where the first and last blocks are considered adjacent; see the fourth column in Table 9.1.

Now, we will present our recursive construction for $\mathcal{P}_n$, which is based on ER's algorithm [105]. To do so, we define the following notation.

**Definition 9.8** *Let* $\ell_1, \ldots, \ell_n$ *be* $n$ *lists, then* $\ell_1 \circ \ell_2 \circ \cdots \circ \ell_n = \bigcirc_{i=1}^{n} \ell_i$ *denotes the concatenation of these lists. Let $m$ be an integer or integer sequence, then $\ell \cdot m$ and $m \cdot \ell$, respectively, denote the list obtained by appending $\ell$ to each sequence of $m$ on the left and on the right, respectively.*

**Example 9.9** *If $\ell$ is the list $111, 112, 121$, $\ell'$ is the list $122, 123$, and $m = 2$, then $\ell \circ \ell' = 111, 112, 121, 122, 123$ and $\ell \circ m = 1112, 1122, 1212$.*

Equation (9.1) produces $V_{n,k}$ either by appending every set partition of the list $V_{n-1,k}$ with an element $i$ to the right side of it, where $i = 1, 2, \ldots, k$, or by appending every set partition of the list $V_{n-1,k-1}$ with an element $k$ to the right side of it. Therefore, $V_{n,k} = \mathcal{P}_{n,k}$ is a list. Using the fact that $\mathcal{P}_n = \bigcup_{k=1}^{n} \mathcal{P}_{n,k}$, we obtain that $V_{n,1} \circ \cdots \circ V_{n,n} = \mathcal{P}_n$ is a list, which implies the following result.

**Lemma 9.10** *Let $n \geq k \geq 1$ and define $V_{1,1} = 1$. Then the list $V_{n,k}$ defined by*

$$V_{n,k} = \begin{cases} V_{n-1,k} \cdot 1, & k = 1 \\ (V_{n-1,k-1} \cdot k) \circ (V_{n-1,k} \cdot \bigcirc_{i=1}^{k} i), & 1 < k < n \\ V_{n-1,k-1} \cdot n, & k = n \end{cases} \quad (9.1)$$

*is the set of set partitions in $\mathcal{P}_{n,k}$. Moreover, the list $V_n = V_{n,1} \circ \cdots \circ V_{n,n}$ is the set $\mathcal{P}_n$.*

Note that the cardinality of $V_{n,k}$ is given by $\text{Stir}(n, k)$ (see Definition 1.11 and Theorem 1.12). The following example illustrates the above lemma.

**Example 9.11** *For $n = 0$ there is only the empty set partition. For $n = 1$*

we have only one set partition, namely, $V_1 = V_{1,1} = 1$. For $n = 2$, $V_{2,1} = (V_{1,1}) \cdot 1 = 11$ and $V_{2,2} = (V_{1,1}) \cdot 2 = 12$. For $n = 3$, we get

$$V_{3,1} = V_{2,1} \cdot 1 = 111,$$

$$V_{3,2} = (V_{2,1} \cdot 2) \circ (V_{2,2} \cdot \bigcirc_{i=1}^{2} i) = 112 \circ 121 \circ 122 = 112, 121, 122,$$

$$V_{3,3} = V_{2,2} \cdot 3 = 123.$$

Thus, the lists of $\mathcal{P}_n$ for $n = 2, 3$ are given by $V_2 = V_{2,1} \circ V_{2,2} = 11, 12$ and $V_3 = V_{3,1} \circ V_{3,2} \circ V_{3,3} = 111, 112, 121, 122, 123$, respectively.

Lemma 9.10 defines an iterative construction for $\mathcal{P}_n$. Its algorithmic version is Algorithm 3, and the induced order is the lexicographic order; see Exercise 9.4.

**Algorithm 3** Iterative algorithm.
1: **for** $i = 1$ to $n$ **do**
2:     **set** $\pi_i = 1$
3: **end for**
4: **set** $j = n$
5: **while** $j > 0$ **do**
6:     **print** $\pi_1 \pi_2 \cdots \pi_n$
7:     **while** $\pi_j = \max_{i \le j-1} \pi_i + 1$ **do**
8:         $j = j - 1$
9:     **end while**
10:    **increase** $\pi_j$ **by** 1
11:    **for** $i = j + 1$ to $n$ **do**
12:        **set** $\pi_i = 1$
13:    **end for**
14: **end while**

In order to explain how Algorithm 3 works, we give the following definition.

**Definition 9.12** *An index $i$ of a word $w = w_1 w_2 \cdots w_n$ on the alphabet $[n]$ is said to be* active *if $w_i < \max_{j \in [i-1]} w_j + 1$.*

First, Algorithm 3 generates the initial set partition $11 \cdots 1$, and then it generates all the other set partitions of $\mathcal{P}_n$ by repeating the following two steps:

**(1)** While $i = n$ is an active index, increase $\pi_i$ by 1.

**(2)** Otherwise, find a maximal active index $i > 1$ (if it exists), increase $\pi_i$ by 1, and then set $\pi_{i+1} = \pi_{i+2} = \cdots = \pi_n = 1$.

The C++ version of Algorithm 3 appears in Appendix H. For instance, Algorithm 3 for $n = 1, 2, 3$ gives 1; 11, 12 and 111, 112, 121, 122, 123, respectively.

Note that Algorithm 3 generates iteratively $\mathcal{P}_n$, where the generating process is uncomplicated and its implementation is simple too. This algorithm is used as the basis of our programming for finding the number of set partitions of $\mathcal{P}_n$ that satisfy certain set of conditions as described in Appendix H. On the other hand, Algorithm 3 is not efficient, since the transition from some set partition to its successors takes $O(n)$ time, and the number of letters that change from some set partition to its successors is at least one, but sometimes $n-1$ changes are required.

Algorithm 3 suggests a method for generating a random set partition of size $[n]$. To see this, we only choose any positive integer number $n$ and define our set partition $pi$ as follows: (1) set $\pi_1 = 1$ and $i = j = 1$, (2) do the following for all $i = 2, 3, \ldots, n$; in the $i$-th step we define $\pi_i$ to be any random integer number between 1 and $\min(j+1, n)$; in this case if $\pi_i = j+1$, then increase $j$ by 1. This procedure produces a random set partition of $n$ and can be coded in Maple as

```
n:=10;
pi:=vector(n,[]);
pi[1]:=1;
j:=1;
for i from 2 to n do
   dpos:=rand(1..min(j+1,n));
   pos:=dpos():
   if pos>j then j:=j+1: fi:
   pi[i]:=pos:
od:
print(pi);
```

For example, running the above algorithm (here we took $n = 10$) three times, we get three set partitions 1123224444, 1111234355 and 1234231563 of $\mathcal{P}_{10}$.

Now, our aim is to present a Gray code for generating $\mathcal{P}_n$. To do so, we need the following notation and definition.

**Definition 9.13** *Assume that $\mathcal{P}_n$ is the list $L = \pi^{(1)}, \pi^{(2)}, \ldots, \pi^{(\mathrm{Bell}_n)}$. With each set partition $\pi^{(j)} \in L$ we associate the following:*

1. $s_{i,j} = j$ for all $j \in [\max_{k \in [i-1]} \pi_k^{(j)} + 1]$, $ls_i = \max_{k \in [i-1]} \pi_k^{(j)} + 1$.

2. $fs_{i,j} = b_j$, $j \in [ls_i]$, where $b_i$ equals 1 if the $i$-th integer coordinate of the vector $s_i$ has already been used, and is 0 otherwise.

3. $ns_i$ is the number of ones in the binary array $fs_i$.

We denote the data structure $(\pi^{(j)}, s, fs)$ by $D_j$.

Now, we are interested in constructing an algorithm to generate a Gray code for $\mathcal{P}_n$ starting with the set partition $11 \cdots 1 \in \mathcal{P}_n$ and ending with the set partition $123223223\cdots \in \mathcal{P}_n$, $n \geq 3$, which differs from the known algorithms listed in Table 9.1. The algorithm works as follows: when the algorithm generates the $j$-th set partition $\pi$ in the list, namely, the data structure $D_j$ is given, we replace the letter $\pi_j$ in the maximal active position $j$ of $\pi$ with the maximal unused letter in $s[j]$. The algorithm contains one critical point, which updates the data structure $D_j$ when it generates the set partition $\pi^{(j)}$.

**Example 9.14** Let $\pi = 1111 \in \mathcal{P}_4$ with $s_1 = 1$, $s_{2,1} = s_{3,1} = s_{4,1} = 1$, $s_{2,2} = s_{3,2} = s_{4,2} = 2$, $fs_1 = 1$, $fs_{2,1} = fs_{3,1} = fs_{4,1} = 1$ and $fs_{2,2} = fs_{3,2} = fs_{4,2} = 0$. Then the maximal active position in $\pi$ is $j = 4$, and the maximal letter that does not use in position $j = 4$ is 2 (since $fs_{4,2} = 0$). Hence, the next set partition is $\pi' = 1112$, and we set $fs_{4,1} = fs_{4,2} = 1$. Again, the maximal active position in $\pi'$ is 3 and the maximal letter that does not use in this position is 2, which implies the next set partition in our list is 1122.

The implementation of our algorithm is given by Algorithm 4 as pseudo code, where the updating of the data structure $D_j$ makes the algorithm longer and complicated. For the original version of this code, we refer the reader to [223].

**Theorem 9.15** *Algorithm 4 with input $n$ generates a Gray code for $\mathcal{P}_n$ with distance 1.*

**Proof** Algorithm 4 generates at first the set partition $11\cdots 1 \in \mathcal{P}_n$ associated with the data structure $D_1$. The algorithm finds a positive active site $j$, and it replaces the letter $\pi_j$ in the maximal active position $j$ of $\pi$ with the maximal unused letter in $\{s_{j,i} \mid i\}$. Thus, the algorithm covers all the possible values of a letter in an active site starting from the rightmost active site to the leftmost active site (see Algorithm 3). Hence, by the definition of the data structure $D_j$ we find that the algorithm generates $\pi^{(j+1)}$ from $\pi^{(j)}$, where $\pi^{(j+1)}$ is a set partition of $\mathcal{P}_n$ that differs from all the previous set partitions $\pi^{(1)},\ldots,\pi^{(j)}$. Moreover, the Hamming distance between $\pi^{(j)}$ and $\pi^{(j+1)}$ is exactly one. Thus, the algorithm is a Gray code algorithm for $\mathcal{P}_n$ with distance one. □

Algorithm 4, written in C++, appears in Appendix H. For example, Algorithm 4 for $n = 1, 2, 3, 4$ gives the lists $L_n$, where

$L_1 = 1$.

$L_2 = 11, 12$.

$L_3 = 111, 112, 122, 121, 123$.

$L_4 = 1111, 1112, 1122, 1121, 1123, 1223, 1221, 1222, 1212, 1211, 1213, 1233,$ 1231, 1234, 1232.

---

**Algorithm 4** Gray code algorithm for $\mathcal{P}_n$
---
 1: **for** $i = 1$ to $n$ **do**
 2:    set $\pi_i = 1$
 3: **end for**
 4: set $s_{1,1} = 1$, $ls_1 = 1$, $fs_{1,1} = 1$, $ns_1 = 1$
 5: **for** $i = 2$ to $n$ **do**
 6:    set $s_{i,1} = 1$, $s_{i,2} = 2$, $ls_i = 2$, $fs_{i,1} = 1$, $fs_{i,2} = 0$, $ns_i = 1$
 7: **end for**
 8: set $pos = 2$
 9: **while** $pos > 1$ **do**
10:    print $\pi_1 \pi_2 \cdots \pi_n$
11:    let $pos = \max\{i \in [n] \mid ns_i \neq ls_i\}$ (if it does not exists set $pos = 0$)
12:    **if** $pos > 0$ **then**
13:       increase $ns_{pos}$ by 1
14:       let $p = \max\{\pi_i \mid i \in [pos - 1]\}$
15:       let $kp = \max\{i \mid i \in [ls_{pos}], fs_{pos,i} = 0\}$
16:       set $\pi_{pos} = s_{pos,kp}$, $fs_{pos,kp} = 1$
17:       **for** $i = pos$ to $n - 1$ **do**
18:          **if** $p < \pi_i$ **then**
19:             $p = \pi_i$
20:          **end if**
21:          set $ns_{i+1} = 1$
22:          **for** $j = 2$ to $p + 1$ **do**
23:             set $s_{i+1,j-1} = j$, $fs_{i+1,j-1} = 0$
24:          **end for**
25:          set $s_{i+1,p+1} = 1$, $ls_{i+1} = p + 1$, $fs_{i+1,p+1} = 0$
26:          **if** $\pi_{i+1} > 1$ **then**
27:             set $fs_{i+1,\pi_{i+1}-1} = 1$
28:          **else**
29:             set $fs_{i+1,p+1} = 1$
30:          **end if**
31:       **end for**
32:    **end if**
33: **end while**

$L_5 = $ 11111, 11112, 11122, 11121, 11123, 11223, 11221, 11222, 11212, 11211, 11213, 11233, 11231, 11234, 11232, 12232, 12231, 12234, 12233, 12213, 12211, 12212, 12222, 12221, 12223, 12123, 12121, 12122, 12112, 12111, 12113, 12133, 12131, 12134, 12132, 12332, 12331, 12334, 12333, 12313, 12311, 12314, 12312, 12342, 12341, 12345, 12344, 12343, 12323, 12321, 12324, 12322.

## 9.3 Loopless Algorithm for Generating $\mathcal{P}_n$

In this section, our aim is to write a loop-free Gray code for generating algorithm $\mathcal{P}_n$. Indeed, we introduce the set of **e**-set partitions or **e**-restricted growth functions (a generalization of set partitions) and give a Gray code with distance one for this set; and as a particular case, we obtain a new Gray code for set partitions. As we will see, our algorithm is a loop-free generating algorithm using classical techniques.

**Definition 9.16** *Let* $\mathbf{e} = e_1 e_2 \ldots e_n$ *be an integer sequence of size* $n$ *with* $e_1 = 0$ *and* $e_i \geq 1$ *for* $i \geq 2$. *An* **e**-set partition *of size* $n$ *is a sequence* $\pi = \pi_1 \pi_2 \ldots \pi_n$ *that satisfies* $\pi_1 = 1$ *and* $1 \leq \pi_i \leq e_i + \max\{\pi_1, \pi_2, \ldots, \pi_{i-1}\}$, *for* $2 \leq i \leq n$. *In particular, if there exists an integer* $d$ *such that* $e_2 = e_3 = \ldots = e_n = d$, *then* $\pi$ *is called the set partition of size* $n$ *and order* $d$. *Clearly, our set partitions of* $\mathcal{P}_n$ *correspond to set partitions of size* $n$ *and order* $d = 1$. *We denote the set of* **e**-*set partitions by* $\mathcal{P}_{\mathbf{e},n}$ *and the set of all set partitions of size* $n$ *and order* $d$ *by by* $\mathcal{P}_{(d),n}$ .

Our main aim is to construct a Gray code with distance 1 for $\mathcal{P}_{\mathbf{e},n}$, that is, we define a list $\mathcal{L}_{\mathbf{e},n}$ for the set $\mathcal{P}_{\mathbf{e},n}$ and we will show that $\mathcal{L}_{\mathbf{e},n}$ is a Gray code.

**Definition 9.17** *A list for a set of sequences is prefix partitioned if all sequences in the list having the same prefix are consecutive.*

To achieve the goal of this section, we construct a prefix partitioned Gray code for $\mathcal{P}_{\mathbf{e},n}$ by assigning to each position of a sequence in $\mathcal{P}_{\mathbf{e},n}$ a status: *active* or *inactive*; and initially all positions—except the leftmost one—are active. After the initialization step, the algorithm repeatedly does on the current sequence $\pi = \pi_1 \pi_2 \cdots \pi_n \in \mathcal{P}_{\mathbf{e},n}$ the following: (1) First, find the rightmost active position $i$ in $\pi$; (2) change appropriately the $\pi_i$ and output $\pi$; (3) if all prefixes of the form $\pi_1 \pi_2 \ldots \pi_{i-1} x$ have been obtained, then make position $i$ inactive and all the positions at the right of $i$ active.

Fix a prefix $\pi_1 \pi_2 \ldots \pi_{i-1}$ and $m = e_i + \max\{\pi_1, \pi_2, \ldots, \pi_{i-1}\}$. Then the algorithm sketched above will exhaust all possible values for $\pi_i \in [m]$ in an appropriate order. So, toward this end we need to define an order. Actually, we define two such orders on the set $[m]$ depending on a parameter $d \in \{1,2\}$, called *direction*. For $m \geq 2$, we define the ordering $\text{succ}_{d,m}$ on the set $[m]$ by

$$\text{succ}_{d,m}(x) = \begin{cases} m, & \text{if } x = d \text{ and } (m > 2 \text{ or } d = 1); \\ x - 1, & \text{if } x \neq d, d+1 \text{ and } x > 2; \\ 1, & \text{if } d = 2 \text{ and } (m = 2 \text{ or } x = 3). \end{cases} \quad (9.2)$$

For instance, the successive elements of the set $[m]$ are listed in $\text{succ}_{1,m}$ and $\text{succ}_{2,m}$ order as $1, m, m-1, \ldots, 2$ and $2, m, m-1, \ldots, 3, 1$, respectively.

The implementation of the above algorithm needs three auxiliary array (sequences): $d = d_1 d_2 \cdots d_n$, $m = m_1 m_2 \cdots m_n$, and $a = a_1 a_2 \cdots a_n$, where $d_i$ holds the direction of the next change of $\pi_i$, $m_i$ is given by $m_i = e_i + \max\{\pi_1, \pi_2, \ldots, \pi_{i-1}\}$, and $a_i$ is 0 or 1 according to whether $i$ is an active position or not in $\pi$, respectively. These three auxiliary array are initiated as $d_i = a_i = 1$, $m_i = e_i + 1$ for all $i$, except $a_1 = 0$.

We denote the list produced by the previous algorithm by $\mathcal{L}_{\mathbf{e},n}$. Now we give a more formal description of the above algorithm: which after the initialization stage of the auxiliary arrays as above and of $\pi$ by $11 \cdots 1$ performs

**global array**: $\pi, a, d, m, e$
**set** $\pi = 11 \cdots 1$, $d = 11 \cdots 1$, $a = 011 \cdots 1$
**set** $m_i = e_i + 1$ for all $i \in [n]$
**output** $\pi$
**while** not all $a_i$ are zeros **do**
    $i = \max_{1 \leq j \leq n}\{j \mid a_j = 1\}$
    $\pi_i = \text{succ}_{d_i, m_i + e_i}(\pi_i)$
    **if** $\pi_i = 1$ **and** $d_i = 2$ **or** $\pi_i = 2$ **and** $d_i = 1$
    **then**     $a_i = 0$
                $d_i = \pi_i$
    **endif**
    **for** $j$ from $i+1$ to $n$ **do**
        $a_j = 1$
        $m_j = \max(m_{i-1}, \pi_i)$
    **enddo**
    **output** $\pi$
**enddo**

Because of the searching of the largest index $i$ with $a_i = 1$ and of the inner loop **for**, this generating algorithm is not efficient in general. At the end of this section, we will explain how it can be implemented efficiently by a loop-free algorithm.

**Proposition 9.18** *The list $\mathcal{L}_{\mathbf{e},n}$ is a Gray code for the set $\mathcal{P}_{\mathbf{e},n}$ with distance one.*

**Proof** Fix $\pi' = \pi_1 \pi_2 \ldots \pi_j$ with $1 \leq j < n$. If there exist $\pi''$ such that $\pi'\pi'' \in \mathcal{P}_{\mathbf{e},n}$, then our algorithm produces sequences with prefix $\pi'\ell$ for all

$\ell \in \{1, 2, \ldots, e_i + \max\{\pi_1, \pi_2, \ldots, \pi_{i-1}\}\}$. Iteratively applying this fact, we have that the list $\mathcal{L}_{\mathbf{e},n}$ defined by the previous algorithm is an exhaustive list for the set $\mathcal{P}_{\mathbf{e},n}$. In addition, since a single element is changed in the current sequence in order to obtain its successor, then the list $\mathcal{L}_{\mathbf{e},n}$ is a Gray code for the set $\mathcal{P}_{\mathbf{e},n}$ with distance one, as claimed. □

Note that Walsh [352] constructs a general generating algorithm for Gray code lists $\mathcal{L}$ satisfying the following: (1) sequences with the same prefix are consecutive and (2) for each proper prefix $\pi_1 \pi_2 \cdots \pi_i$ of a sequence in $\mathcal{L}$ there are at least two values $a$ and $b$ such that $\pi_1 \pi_2 \cdots \pi_i a$ and $\pi_1 \pi_2 \cdots \pi_i b$ are both prefixes of sequences in $\mathcal{L}$. Our Gray code $\mathcal{L}_{\mathbf{e},n}$ satisfies Walsh's previous requirements and so it can be generated by a loop-free algorithm by applying his general method. Alternatively, a loop-free implementation can be obtained by using the *finished and unfinished lists* method, introduced in [211].

**Example 9.19** *By running our algorithm for* $\mathbf{e} = 0212$ *(clearly, $n = 4$), we obtain the list* $\mathcal{L}_{\mathbf{e},4}$, *namely* 1111, 1113, 1112, 1122, 1124, 1123, 1121, 1321, 1325, 1324, 1323, 1322, 1342, 1346, 1345, 1344, 1343, 1341, 1331, 1335, 1334, 1333, 1332, 1312, 1313, 1311, 1211, 1214, 1213, 1212, 1232, 1235, 1234, 1233, 1231, 1221, 1224, 1223 *and* 1222.

In conclusion, we note that [223] gave only a loop-free algorithm for generating $\mathcal{P}_n$ which is based on a special traversal of the set partition tree $T_n$, where $T_n$ is a plane tree with root vertex $11 \cdots 1 \in \mathcal{P}_n$ and if $v$ is a vertex $\pi = \pi_1 \cdots \pi_m 11 \cdots 1$ at level $m$ in $T_n$ with $\pi_m \neq 1$, then its children are the vertices $\pi_1 \pi_2 \cdots \pi_m 11 \cdots 1 k 11 \cdots 1 \in \mathcal{P}_n$, for all $k = 2, 3, \ldots, 1 + \max_{j \in [m]} \pi_j$.

---

## 9.4 Exercises

**Exercise 9.1** *Write an iterative algorithm for generating all the words of length $n$ over alphabet $\{0, 1, 2\}$. Then use your algorithm to generate all the words of length 3 over alphabet $\{0, 1, 2\}$.*

**Exercise 9.2** *Write a Gray code algorithm for generating all the words of length $n$ over alphabet $\{0, 1, 2\}$. Then use your algorithm to generate all the words of length 3 over alphabet $\{0, 1, 2\}$.*

**Exercise 9.3** *Write a Maple code for the algorithm in Exercise 9.2.*

**Exercise 9.4** *Use Algorithm 3 to list the set $\mathcal{P}_4$.*

**Exercise 9.5** *Reconstruct (as few queries as possible) an unknown set partition of $[n]$ into at most $k$ disjoint non-empty subsets (classes) by querying* subsets $Q_i \subset [n]$ such that the query returns the number of classes represented in $Q_i$.

**Exercise 9.6** *Write an algorithm for generating all the set partitions of $[n]$ with exactly $k$ blocks of sizes $\ell_1, \ldots, \ell_k$.*

## 9.5 Research Directions and Open Problems

We now suggest several research directions which are motivated by the results and exercises of this chapter.

**Research Direction 9.1** *In this chapter, we present a Gray code and loop-free algorithm for generating the all the set partitions of $\mathcal{P}_n$. On the other hand, in [140], Huemer et al. gave a Gray code for noncrossing set partitions of $\mathcal{P}_n$. In other words, this paper motivates the question of generating subsets of $\mathcal{P}_n$, where our subsets are characterized by pattern permutations. More precisely, the research question is to find a Gray code or/and loopless algorithm for generating the set partitions of $[n]$ that avoid a fixed subsequence (subword) pattern of size at least four.*

**Research Direction 9.2** *Mansour, Nassar, and Vajnovszki [224] presented a loop-free Gray code algorithm for the $\mathbf{e}$-restricted growth functions. In this context, a restricted growth function of size $n$ means a word $\pi_1 \pi_2 \cdots \pi_n$ such that $\pi_j \leq \max(\pi_1, \ldots, \pi_{j-1}) + e_j$ for all $j = 1, 2, \ldots, n$. Now, let $\tau$ be any subword pattern. For instance, $\tau = 11$. We suggest to write a loop-free Gray code algorithm for $\mathbf{e}$-restricted growth functions that avoid the pattern $\tau$.*

**Research Direction 9.3** *Let $f$ be any statistic on set partitions of $\mathcal{P}_n$. Write a generating Gary code, loop-free, or Gray code loop-free algorithm for all set partitions $\pi \in \mathcal{P}_n$ such that $f(\pi) = k$, for fixed $k$. For instance, let $f(\pi) = (-1)^{n-k}$. Write a Gray code generating algorithm for all set partitions of $\pi \in \mathcal{P}_n$ such that $f(\pi) = 1$ (that is, all even set partitions in $\mathcal{P}_n$).*

**Research Direction 9.4** *Following Exercise 3.22, find a generating Gray code (loop-free) algorithm for the set $M_{n,r}$.*

# Chapter 10

# Set Partitions and Normal Ordering

In this chapter, we are interested in one mode boson creation $a^\dagger$ and annihilation $a$ (operators) satisfying the commutation relation

$$[a, a^\dagger] = aa^\dagger - a^\dagger a = 1. \tag{10.1}$$

For the multi-mode case, we refer the reader to [225]. We are interested in the set partitions arising in the problem of normal ordering of boson expressions.

The *normal ordering* is a functional representation of boson operator functions in which all the creation operators stand to the left of the annihilation operators with the use of (10.1). This procedure leads to an operator which is equivalent to the original one but has a different functional representation; see the following sections. A standard approach to the normal ordering problem is through the *Wick's theorem*; see Section 10.3. It directly links the problem to enumerative combinatorics, that is, searching for all possible *contractions* in the boson expression and then summing up the resulting terms. This may be efficiently used for solving problems with a finite number of boson operators. But this does not help in considering problems concerning operators defined through infinite series expansions. To do so, we would have to know the recursive structure of the sequence of numbers involved, and this requires a lot of small details. Here we approach these problems using methods of advanced combinatorial analysis [82] in which combinatorial structures (such as set partitions and lattice paths) play critical roles. It has proved to be an efficient way of obtaining exact formulas for normally ordered expansion coefficients and then analyzing their properties.

Since Katriel's [161] seminal work in 1974, the combinatorial aspects of normal ordering have been received a lot of attention, considered intensively, and many different meanings have also been investigated; for example see [39, 45, 42, 43, 44, 41, 110, 162, 163, 227, 228, 248, 250, 315, 346, 361] and the references given therein. We refer the reader to Wilcox [355] for the earlier literature on normal ordering of noncommuting operators. Since the paper of Katriel and Kibler [164], the combinatorics of normal ordering of an arbitrary expression in the creation and annihilation operators $a^\dagger$ and $a$ of a single-mode $q$-boson having the commutation relations

$$[a, a^\dagger]_q = aa^\dagger - qa^\dagger a = 1, \quad [a, a] = 0, \quad [a^\dagger, a^\dagger] = 0 \tag{10.2}$$

have also been studied [40, 164, 316, 346].

This chapter is organized as follows. In the next section, we give a modern review of normal ordering and make use of a specific realization of (10.1) in terms of operators of the multiplication $X$ and derivative $D$. In Section 10.2, we will focus on normal ordering of the operator $(aa^\dagger)^n$. Here, we establish two bijections between the set of contractions of this operator and the set $\mathcal{P}_n$.

In Section 10.3, we study combinatorial aspects of the $q$-normal ordering of single-mode boson operators. It is shown how, by introducing appropriate $q$-weights for the associated "Feynman diagrams", the normally ordered form of a general expression in the creation and annihilation operators can be written as a sum over all $q$-weighted Feynman diagrams, representing Wick's theorem in the present context. In particular, we show that the conventional Stirling numbers of the second kind can also be interpreted as the number of "Feynman diagrams" of *degree* $n - k$ on the particular *operator* $(a^\dagger a)^n$.

In Section 10.4, we generalize some results of Katriel, where our generalizations are based on the number of contractions whose vertices are next to each other in the linear representation of the boson operator function. In this way, we naturally introduce a parameter that allows one to refine the set of all contractions and to specify a combinatorial statistics. In Section 10.5, we consider linear representations of noncrossing set partitions, and we define the notion of noncrossing normal ordering. Given the growing interest in noncrossing set partitions, because of their many unexpected connections (like, for example, with free probability), noncrossing normal ordering appears to be an intriguing notion. We explicitly give the noncrossing normally ordered form of the functions

$$(a^r(a^\dagger)^s)^n) \text{ and } (a^r + (a^\dagger)^s)^n,$$

plus various special cases. We are able to give bijections between noncrossing contractions of these functions, $k$-ary trees, and special sets of lattice paths.

## 10.1 Preliminaries

In this section, we present a brief introduction to the hidden physics of the normal ordering problem (see also Appendix G). Let $a^\dagger$ and $a$ be the boson creation and annihilation operators satisfying (10.1). The operators $a, a^\dagger, I$, where $I$ is the identity operator, are the generators of the *Heisenberg–Weyl algebra*. The *occupation number representation* arises from the interpretation of $a$ and $a^\dagger$ as operators annihilating and creating a particle (object) in the model. Here, the *Hilbert space* $\mathcal{H}$ of states (sometimes called *Fock space*) is generated by the number *states* $|n\rangle$, $n \geq 0$, counts the number of particles (for bosons up to infinity), and we assume the existence of a unique *vacuum state* $|0\rangle$ such that $a|0\rangle = 0$. The set of the number of states $\{|n\rangle\}_{n\geq 0}$ presents an orthogonal basis in $\mathcal{H}$, that is,

$$\langle k|n\rangle = \delta_{n,k} \text{ and } \sum_{n\geq 0}|n\rangle\langle n| = 1.$$

Thus, the operators $a$ and $a^\dagger$ can be acting on the number of states as

$$a|n\rangle = \sqrt{n}|n-1\rangle \text{ and } a^\dagger|n\rangle = \sqrt{n+1}|n+1\rangle,$$

which gives that all states can be introduced from the vacuum state through

$$|n\rangle = \frac{1}{\sqrt{n!}}(a^\dagger)^n|0\rangle$$

and the *number operator* $N$, defined by $N|n\rangle = n|n\rangle$, counting the number of particles in the model can be represented as $N = a^\dagger a$ and satisfying

$$[a, N] = a \text{ and } [a^\dagger, N] = -a^\dagger.$$

Note that the boson creation $a^\dagger$ and annihilation $a$ operators do not commute. This is the main reason for some ambiguities in the definitions of the operator functions in quantum mechanics. In order to deal with this problem, one has to define the order of the operators involved, which leads to the definition of the normally ordered form as follows.

**Definition 10.1** *A function (operator) $F(a, a^\dagger)$ can be written as a word (of possible infinite length) on the alphabet $\{a, a^\dagger\}$. We denote the normal ordering of a function $F(a, a^\dagger)$ by $\mathcal{N}[F(a, a^\dagger)]$. A contraction consists of substituting $a = \varnothing$ and $a^\dagger = \varnothing^\dagger$ in the word whenever $a$ precedes $a^\dagger$. Among all possible contractions, we also include the null contraction, that is, the contraction leaving the word as it is. We denote the set of all contractions of expression $F(a, a^\dagger)$ by $\mathcal{C}on(F(a, a^\dagger))$. The double dot operation deletes all the letters $\varnothing$ and $\varnothing^\dagger$ in the word and then arranges it such that all the letters $a^\dagger$ precede the letters $a$. Specifically,*

$$\mathcal{N}[F(a, a^\dagger)] = F(a, a^\dagger) = \sum_{\pi \in \mathcal{C}on(F(a,a^\dagger))} :\pi: . \tag{10.3}$$

**Example 10.2** *The set $Con(aa^\dagger aaa^\dagger a)$ of all contractions of the word*

$$aa^\dagger aaa^\dagger a$$

*are $aa^\dagger aaa^\dagger a$, $\varnothing\varnothing^\dagger aaa^\dagger a$, $\varnothing a^\dagger aa\varnothing^\dagger a$, $aa^\dagger \varnothing a\varnothing^\dagger a$, $aa^\dagger a\varnothing\varnothing^\dagger a$, $\varnothing\varnothing^\dagger \varnothing a\varnothing^\dagger a$, $\varnothing\varnothing^\dagger a\varnothing\varnothing^\dagger a$. The double dot operation of these contractions are $(a^\dagger)^2 a^4$, $a^\dagger a^3$, $a^\dagger a^3$, $a^\dagger a^3$, $a^\dagger a^3$, $a^2$, $a^2$, respectively. Thus, the normal ordering of the word $aa^\dagger aaa^\dagger a$ is*

$$\mathcal{N}[aa^\dagger aaa^\dagger a] = (a^\dagger)^2 a^4 + 4a^\dagger a^3 + 2a^2.$$

For more specific examples and general results on normal ordering, we refer the reader to the next sections.

Contractions can be depicted with diagrams called linear representations or Feynman diagrams.

**Definition 10.3** *Let $\pi = \pi_1 \cdots \pi_n$ be any contraction, that is, any word of length $n$ over the alphabet $\{a, a^\dagger\}$. We draw $n$ vertices, say $1, 2, \ldots, n$, on a horizontal line such that the point $i$ corresponds to the letter $\pi_i$. We represent each letter $a$ by a white vertex and each letter $a^\dagger$ by a black vertex such that a black vertex $i$ can be connected (not necessary) by an undirected edge $(i, j)$ to a white vertex $j$, where the edges are drawn in the plane above the horizontal line, if we substituted $a = \varnothing$ and $a^\dagger = \varnothing^\dagger$ in the word whenever $a$ precedes $a^\dagger$. This is the* linear representation *or* Feynman diagram *of a contraction $\pi$.*

**Example 10.4** *The linear representations of the contractions of the word*

$$aaa^\dagger a^\dagger$$

*as presented in Figure 10.1 are given by*

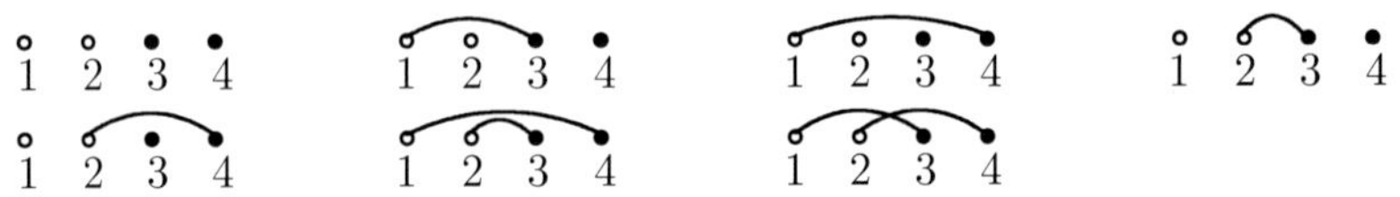

**FIGURE 10.1**: The linear representations of the contractions of the word $aaa^\dagger a^\dagger$

In this chapter, we are interested in defining some statistics on the set of contractions of a given word $F(a, a^\dagger)$ either in the deformed case (when $q \neq 1$) or in the undeformed case (when $q = 1$). To do so, we switch slightly the terminology to match that of [100] (and also [5, 35, 51]) as follows.

**Definition 10.5** *Let $S = \{\pi_1, \ldots, \pi_n\}$ be a finite linearly ordered set consisting of two types of elements, that is, there exists a "type-map" $\tau : S \to \{\mathscr{A}, \mathscr{C}\}$ which associates with each letter $\pi_i$ its type, namely $\tau(\pi_i) \in \{\mathscr{A}, \mathscr{C}\}$. We call elements $\pi_i$ with $\tau(\pi_i) = \mathscr{A}$ annihilators and elements $\pi_j$ with $\tau(\pi_j) = \mathscr{C}$ creators. We also denote by $S^+$ (respectively, $S^-$) the set of $j$ with $\tau(s_j) = \mathscr{C}$ (respectively, $i$ with $\tau(\pi_i) = \mathscr{A}$). A Feynman diagram $\gamma$ on $S$ is a partition of $S$ into one and two-element sets, where the two-element sets have the special property that the two elements are of different type (that is, contain exactly one creator and one annihilator) and where the element of type $\mathscr{C}$ is the one with larger index. We also regard $\gamma$ as a set of ordered pairs $\{(i_1, j_1), \ldots, (i_p, j_p)\}$ with $i_1 < i_2 < \cdots < i_p$, $i_k < j_k$, $i_k \neq i_l$, $j_k \neq j_l$, $\pi_{i_k} \in S^-$ and $\pi_{j_k} \in S^+$.*

Note that, in our concrete model, the set $S$ is given by the word $F(a, a^\dagger)$, the two types are given by $\mathscr{C} = a^\dagger$ and $\mathscr{A} = a$, and a Feynman diagram $\gamma$

corresponds precisely to a contraction. Indeed, the two-element sets $(i_k, j_k)$ correspond to the edges of the contraction connecting a creator $a^\dagger$ with a preceding annihilator $a$. Thus, the Feynman diagram $\gamma = \{(i_1, j_1), \ldots, (i_p, j_p)\}$ corresponds to a contraction with $p$ edges, which invites the following definition.

**Definition 10.6** *Let $\gamma$ be any Feynman diagram on $S$. We call the elements of $S$ vertices. A Feynman diagram with representation*

$$\gamma = \{(i_1, j_1), \ldots, (i_k, j_k)\}$$

*is said to have degree $k$ and the two-element sets $(i_a, j_a)$ are called edges of $\gamma$. The unpaired indices in $\gamma$ are called singletons. We denote the set of all Feynman diagrams on $S$ by $\mathscr{F}(S)$, the set of Feynman diagrams of degree $k$ by $\mathscr{F}_k(S)$, and the set of singletons of $\gamma$ by $\mathscr{S}(\gamma)$.*

**Example 10.7** *Let $F(a, a^\dagger) = \pi = aaa^\dagger aaa^\dagger a^\dagger aa^\dagger aa^\dagger a^\dagger$, and consider the Feynman diagram $\gamma = \{(1,3), (2,6), (4,9), (5,7), (8,12)\}$ of degree 5. In its linear representation in Figure 10.2, the vertices of type $\mathscr{A} = a$ are depicted by an empty circle, while the vertices of type $\mathscr{C} = a^\dagger$ are depicted by a black circle.*

**FIGURE 10.2**: The linear representation of the Feynman diagram $\gamma$.

Let $|S| = n$. Clearly, $\mathscr{F}_p(S) = 0$ for $p > \frac{n}{2}$. Given a Feynman diagram $\gamma \in \mathscr{F}_p(S)$ with $2p \leq n$, there will be $n - 2p$ singletons (in the terminology of [226, 225] these are the vertices of degree 0). Now, we introduce the double dot operation for a Feynman diagram $\gamma$ on $S$. Intuitively, it means that we omit all vertices contained in the two-element sets of $\gamma$ and order the remaining singletons in such a fashion that all creators precede the annihilators. Formally, we have the following definition.

**Definition 10.8** *Let $\gamma \in \mathscr{F}_p(S)$ and assume that $\gamma$ has $r$ singletons of type $\mathscr{C}$ and $s$ singletons of type $\mathscr{A}$, where $s = n - 2p - r$. The dot operation of $\gamma$ is given by : $\gamma := \mathscr{C}^r \mathscr{A}^s$. Using this terminology, the normal ordering of the undeformed case (10.3) can be presented as*

$$\mathcal{N}[F(a, a^\dagger)] = F(a, a^\dagger) = \sum_{\gamma \in \mathscr{F}(F(a, a^\dagger))} : \gamma :, \tag{10.4}$$

*for any word $F(a, a^\dagger)$.*

We introduce now a $q$-weight for Feynman diagrams such that we can write the normally ordered form of words $F(a, a^\dagger)$ in the $q$-boson operators in a form analogous to (10.4). For this we have to introduce some more terminology following [100].

**Definition 10.9** *Let $\gamma = \{(i_1, j_1), \ldots, (i_p, j_p)\}$ be any Feynman diagram. We say that a pair $(i_k, j_k)$ is a left crossing for $(i_m, j_m)$ if $i_k < i_m < j_k < j_m$ and we define $\mathrm{cross}_l(i, j)$ to be the number of such left crossings for $(i, j)$. We define $\mathrm{cross}(\gamma) := \sum_{(i,j) \in \gamma} \mathrm{cross}_l(i, j)$ as the crossing number of $\gamma$ (Biane [35] calls it restricted crossing number); it counts the intersections in the corresponding graph (the linear representation of $\gamma$).*

We also need to count another type of crossing.

**Definition 10.10** *Let $\gamma = \{(i_1, j_1), \ldots, (i_p, j_p)\}$ be a Feynman diagram. We say that a triple $(i_m, k, j_m)$ is a degenerate crossing if $i_m < k < j_m$, $k$ is a singleton and $(i_m, j_m) \in \gamma$. Let $\mathrm{dcross}(i, j)$ be the number of such unpaired $k$ for the edge $(i, j)$; then $\mathrm{dcross}(\gamma) := \sum_{(i,j) \in \gamma} \mathrm{dcross}(i, j)$ counts the number of such triples in $\gamma$.*

We are ready now to define the total number of crossings in a Feynman diagram as follows.

**Definition 10.11** *Let $\gamma$ be any Feynman diagram on $S$. The total number of crossings of a Feynman diagram $\gamma$ is defined by*

$$\mathrm{tcross}(\gamma) = \mathrm{cross}(\gamma) + \mathrm{dcross}(\gamma).$$

**Example 10.12** *Example 10.7 shows that the number of crossings of $\gamma$ is given by $\mathrm{cross}(\gamma) = 4$ and there are two degenerate crossings, that is, $\mathrm{dcross}(\gamma) = 2$, yielding the total crossing number $\mathrm{tcross}(\gamma) = 6$.*

Clearly, the total crossing number accounts for the "interaction" between edges (the crossings) and the "interaction" between singletons and edges (covering of singletons by edges). In addition, we need a measure which accounts for the "interaction" between singletons.

**Definition 10.13** *Let $\gamma$ be any Feynman diagram on $S = \{\pi_1, \ldots, \pi_n\}$. The length (to the right) of a singleton $\pi_k$, denoted by $l_r(\pi_k)$, of $\gamma$ is defined as follows: If the singleton $\pi_k$ is of type $\mathscr{C}$, then $l_r(\pi_k) = 0$, and if the singleton $\pi_k$ is of type $\mathscr{A}$, then $l_r(\pi_k)$ is given by the number of singletons of type $\mathscr{C}$ to the right of $\pi_k$. Then we define length of $\gamma$ to be $\mathrm{len}(\gamma) = \sum_{i=1}^{n} l_r(\pi_i)$.*

Now, after these lengthy preparations, we define the $q$-weight of a Feynman diagram.

Set Partitions and Normal Ordering    445

**Definition 10.14** We define the q-weight of a Feynman diagram $\gamma$ to be
$$\mathscr{W}_q(\gamma) = q^{\mathrm{tcross}(\gamma)+\mathrm{len}(\gamma)}. \tag{10.5}$$

**Example 10.15** Continuing Examples 10.7 and 10.12, we obtain that there is only one singleton of type $\mathscr{A} = a$ having length $l_r(\pi_{10}) = 1$, yielding the length of $\gamma$ is $\mathrm{len}(\gamma) = 1$. Therefore, the q-weight of $\gamma$ is given by $\mathscr{W}_q(\gamma) = q^{6+1} = q^7$.

We give now a general simple example on normal ordering, which is the ordering of the product $a^k(a^\dagger)^\ell$, which is in the so called *anti-normal form* (that is, all annihilation operators stand to the left of creation operators). The double dot operation readily gives $:a^k(a^\dagger)^\ell: = (a^\dagger)^\ell a^k$, while the normal ordering $\mathcal{N}$ giving the following result.

**Theorem 10.16** (See [38, Equation 2.10]) *For all $k, \ell$.*
$$a^k(a^\dagger)^\ell = \mathcal{N}[a^k(a^\dagger)^\ell] = \sum_{i=0}^{k} \binom{k}{i} (\ell)_i (a^\dagger)^{\ell-i} a^{k-i},$$
*where $(\ell)_i$ is the falling polynomial.*

**Proof** We proceed with the proof by induction on $k$. The theorem holds for $k = 0$, since it is equivalent to $(a^\dagger)^\ell = (a^\dagger)^\ell$. Assume that the theorem holds for $k$. Then
$$a^{k+1}(a^\dagger)^\ell = \mathcal{N}[a^{k+1}(a^\dagger)^\ell] = \mathcal{N}\left[a \sum_{i=0}^{k} \binom{k}{i}(\ell)_i (a^\dagger)^{\ell-i} a^{k-i}\right]$$
$$= \sum_{i=0}^{k} \binom{k}{i}(\ell)_i \mathcal{N}\left[a(a^\dagger)^{\ell-i}\right] a^{k-i}.$$

Using the fact that (See Exercise 10.3)
$$\mathcal{N}\left[a(a^\dagger)^j\right] = (a^\dagger)^j a + j(a^\dagger)^{j-1}, \tag{10.6}$$

we obtain that
$$a^{k+1}(a^\dagger)^\ell = \mathcal{N}[a^{k+1}(a^\dagger)^\ell] = \sum_{i=0}^{k} \binom{k}{i}(\ell)_i (a^\dagger)^{\ell-1-i}\left[(a^\dagger)a + (\ell-i)\right] a^{k-i}$$
$$= \sum_{i=0}^{k} \binom{k}{i}(\ell)_i (a^\dagger)^{\ell-i} a^{k+1-i} + \sum_{i=0}^{k} \binom{k}{i}(\ell)_{i+1} (a^\dagger)^{\ell-1-i} a^{k-i}$$
$$= \sum_{i=0}^{k+1} \binom{k}{i}(\ell)_i (a^\dagger)^{\ell-i} a^{k+1-i} + \sum_{i=0}^{k+1} \binom{k}{i-1}(\ell)_i (a^\dagger)^{\ell-i} a^{k+1-i}$$
$$= \sum_{i=0}^{k+1} \left(\binom{k}{i} + \binom{k}{i-1}\right)(\ell)_i (a^\dagger)^{\ell-i} a^{k+1-i}.$$

Fact 2.20(i) gives

$$a^{k+1}(a^\dagger)^\ell = \mathcal{N}[a^{k+1}(a^\dagger)^\ell] = \sum_{i=0}^{k+1} \binom{k+1}{i}(\ell)_i (a^\dagger)^{\ell-i} a^{k+1-i},$$

which completes the induction. □

The choice of the representation of the algebra can simplify the calculations. There are some common choices of the Heisenberg-Weyl algebra representation in quantum mechanics. For instance, let $X$ be the formal multiplication and $D$ be the derivative that acts in the space of (formal) polynomials, where they are defined by their acting on monomials as $Xx^n = x^{n+1}$ and $Dx^n = nx^{n-1}$, we have $[D, X] = 1$ (see (10.1)). In Example 10.18, we present an application for such representations.

## 10.2 Linear Representation and $\mathcal{N}((a^\dagger a)^n)$

This section presents a simple introduction to some methods and problems encountered later in the normal ordering form of a expression on $a$ and $a^\dagger$. We present a basic example in which all the essential techniques and methods of the proceeding sections are used, where we define and investigate Stirling and Bell numbers as solutions to the normal ordering problem [38, 161, 162, 163]. We start by considering the normal ordering form of the $n$-power (iteration) of the number operator $N = a^\dagger a$.

**Theorem 10.17** *(see [161, 162]) For all $n \geq 1$,*

$$(a^\dagger a)^n = \sum_{k=1}^{n} \text{Stir}(n,k)(a^\dagger)^k a^k.$$

**Proof** Let the normal ordering of $(a^\dagger a)^n$ be given by $\sum_{k=1}^{n} s_{n,k}(a^\dagger)^k a^k$. Then the normal ordering of $(a^\dagger a)^{n+1}$ is given by $a^\dagger a \sum_{k=1}^{n} s_{n,k}(a^\dagger)^k a^k$, which, by (10.6), implies that

$$\mathcal{N}[(a^\dagger a)^{n+1}] = \sum_{k=1}^{n+1} s_{n+1,k}(a^\dagger)^k a^k$$

$$= \sum_{k=1}^{n} s_{n,k} a^\dagger ((a^\dagger)^k a + k(a^\dagger)^{k-1}) a^k$$

$$= \sum_{k=1}^{n} s_{n,k}(a^\dagger)^{k+1} a^{k+1} + \sum_{k=1}^{n} k s_{n,k}(a^\dagger)^k a^k.$$

Therefore, $s_{n+1,k} = s_{n,k-1} + k s_{n,k}$, $s_{0,0} = 1$ and $s_{n,k} = 0$ for $k > n$ or $n > k = 0$. This and Theorem 1.12 complete the proof. □

**Example 10.18** *To present an application of the above theorem, we replace $a^\dagger$ and $a$ by $X$ and $D$ (as noted at the end of the pervious section), respectively. Then $(XD)^n = \sum_{k=1}^n \mathrm{Stir}(n,k) X^k D^k$, which implies that*

$$m^n = (XD)^n x^m = \sum_{k=1}^n \mathrm{Stir}(n,k) X^k D^k x^m = \sum_{k=1}^n \mathrm{Stir}(n,k) (m)_k,$$

*where $(m)_k = m(m-1)\cdots(m-k+1)$ is the k-falling factorial. Also, if we apply $(XD)^n$ to $e^x$, we obtain*

$$\sum_{k\geq 0} k^n \frac{x^k}{k!} = e^x \sum_{k=1}^n \mathrm{Stir}(n,k) x^k,$$

*which implies that the n-th Bell polynomial is given by*

$$\mathrm{Bell}_n(x) = \sum_{k=1}^n \mathrm{Stir}(n,k) x^k = e^{-x} \sum_{k\geq 0} k^n \frac{x^k}{k!}, \qquad (10.7)$$

*which gives another proof for Dobiński's formula, Theorem 3.2.*

We come now return to normal ordering.

**Corollary 10.19** *(see [162]) For all $n \geq 0$, $\langle z|(a^\dagger a)^n|z\rangle = \mathrm{Bell}_n(|z|^2)$. Moreover, $\langle z|e^{ta^\dagger a}|z\rangle = e^{|z|^2(e^t - 1)}$.*

**Proof** By using the properties of coherent states (see Appendix G), we conclude from Theorem 10.17 and (10.7) that diagonal coherent state matrix elements generate Bell polynomials $\langle z|(a^\dagger a)^n|z\rangle = \mathrm{Bell}_n(|z|^2)$. Then by expanding the exponential $e^{ta^\dagger a}$ and taking the diagonal coherent state matrix element, we have

$$\langle z|e^{ta^\dagger a}|z\rangle = \sum_{n\geq 0} \langle z|(a^\dagger a)^n|z\rangle \frac{t^n}{n!} = \sum_{n\geq 0} \mathrm{Bell}_n(|z|^2) \frac{t^n}{n!} = e^{|z|^2(e^t - 1)},$$

as required. □

Our aim now to give a combinatorial proof for Theorem 10.17. At first we need the following definition.

**Definition 10.20** *Let $\pi$ be a linear representation of a contraction of $(a^\dagger a)^n$ on $2n$ vertices, say $v_1, v_2, \ldots, v_{2n}$. A component of $\pi$ is a sequence of vertices $v_{i_1}, v_{i_1+1}, v_{i_2}, v_{i_2+1}, v_{i_3}, v_{i_3+1}, \ldots, v_{i_s}, v_{i_s+1}$ such that $i_1$ is minimal, $v_{i_j}$ is a black vertex, and the vertex $v_{i_j+1}$ connects the vertex $v_{i_{j+1}}$, for all $j = 1, 2, \ldots, s-1$.*

**Example 10.21** *Figure 10.3 presents the contractions of $(a^\dagger a)^3$ which ordered from left to right. The number of components of these contractions are 3, 2, 2, 1, and 2.*

**FIGURE 10.3**: The linear representations of the contractions of the word $(a^\dagger a)^3$.

**Theorem 10.22** *Let $n \geq 1$. There exists a bijection between the set of contractions of $(a^\dagger a)^n$ with $k$ components and the set $\mathcal{P}_{n,k}$.*

**Proof** Let $\pi$ be any linear representation of a contraction of $(a^\dagger a)^n$. At first, we label the vertices of $\pi$ from left to right by $v_1, v_2, \ldots, v_{2n}$ and define them as "good" vertices. Clearly, the vertex $v_{2i-1}$ ($v_{2i}$) is black (white), for all $i = 1, 2, \ldots, n$. We define a labeling function $f : \{v_1, v_2, \ldots, v_{2n}\} \to [n]$ as follows. Fix $j = 1$. Keep doing the following until all the vertices are "bad":
(1) Fix $B = v_{i_1}, v_{i_1+1}, v_{i_2}, v_{i_2+1}, \ldots, v_{i_s}, v_{i_s+1}$ as a component of $\pi$ of "good" vertices. Define $f(v) = j$ for all $v \in B$.
(2) Increase $j$ by one and define any vertex in $B$ as a "bad" vertex.
  Define $\pi' = f(v_2)f(v_4)\cdots f(v_{2n})$. For instance, let $\pi'$ be either the leftmost, one after the leftmost, ..., or rightmost contraction in Figure 10.3, then $\pi'$ is given by 123, 112, 122, 111, or 121, respectively. It is obvious that the map $\pi \mapsto \pi'$ defines a bijection between the set of contractions of $(a^\dagger a)^n$ and the set $\mathcal{P}_n$. Moreover, the maximal letter in $\pi'$ denotes the number components of a contraction $\pi$. Hence, the map $\pi \mapsto \pi'$ is a bijection between the set of contractions of $(a^\dagger a)^n$ with $k$ components and the set $\mathcal{P}_{n,k}$. □

We construct now another bijection between the set of contractions of $(a^\dagger a)^n$ and the set $\mathcal{P}_n$. In order to do that, we need the following definition.

**Definition 10.23** *Let $ConS_n$ be the set of all sequences $\pi_1\pi_2\cdots\pi_n$ of length $n$ such that $\pi_i \in \{e, i, i+1, \ldots, n\}$ and if $\pi_i, \pi_j \neq e$ then $\pi_i \neq \pi_j$. More generally, let $ConS_{n,k}$ be the set of all sequences in $ConS_n$ having exactly $k$ terms $e$.*

**Example 10.24** *The set $ConS_2$ consists of five sequences ee, e2, 1e, 2e, and 12. The set $ConS_3$ consists of 15 sequences: eee, ee3, e2e, e23, e3e, 1ee, 1e3, 12e, 123, 13e, 2ee, 2e3, 23e, 3ee, and 32e.*

Our next observation shows that the elements of $ConS_{n-1,k-1}$ are enumerated by the Stirling number $Stir(n, k)$.

**Lemma 10.25** *There exists a bijection between the set $\mathcal{C}on\mathcal{S}_{n-1,k-1}$ and the set $\mathcal{P}_{n,k}$. In particular, the cardinality of the set $\mathcal{C}on\mathcal{S}_{n-1,k-1}$ is given by* Stir(n, k).

**Proof** Let $\pi' = \pi_1\pi_2\cdots\pi_{n-1}$ be any sequence in $\mathcal{C}on\mathcal{S}_{n-1,k-1}$ and let $\pi = \pi'e$. We define the $i$th, $i = 1, 2, \ldots, k$, block of the set partition of $[n]$ by

$$B_i = \{\beta_{i1} = a_i, \beta_{i2} = 1 + \pi_{\beta_{i1}}, \ldots, \beta_{i\ell} = 1 + \pi_{\beta_{i(\ell-1)}}\},$$

where $\pi_{\beta_{i\ell}} = e$ and $a_i$ is the minimal term of the sequence $\pi$ such that $a_i \notin B_j$ for all $j = 1, 2, \ldots, i-1$ and $a_1 = 1$. For instance, if $\pi = 1e3eee$ then $B_1 = \{1, 2\}$, $B_2 = \{3, 4\}$, $B_3 = \{5\}$, and $B_4 = \{6\}$. From the above construction, we see that $\pi_1\pi_2\cdots\pi_{n-1}$ is a sequence in $\mathcal{C}on\mathcal{S}_{n-1,k-1}$ if and only if $B_1, B_2, \ldots, B_k$ is a set partition of $[n]$, as required. □

**Example 10.26** *Lemma 10.25 for $n = 2$ gives*

$ee \mapsto \{1\}, \{2\}, \{3\}; \quad e2 \mapsto \{1\}, \{23\}; \quad 1e \mapsto \{12\}, \{3\};$
$12 \mapsto \{123\}; \quad 2e \mapsto \{13\}, \{2\}.$

We present now a bijection between the set of contractions of $(a^\dagger a)^n$ and the set $\mathcal{C}on\mathcal{S}_n$. In order to do that, we recall the degree of a contraction: we say that a contraction $\pi \in \mathcal{C}on((a^\dagger a)^n)$ has degree $k$ if : $\pi := (a^\dagger)^{n-k}a^{n-k}$; see Definition 10.6.

**Theorem 10.27** *There exists a bijection between the set $\mathcal{C}on((a^\dagger a)^n)$ of contractions with degree $k$ and the set $\mathcal{C}on\mathcal{S}_{n-1,k-1}$.*

**Proof** Let $w = w_{2n}w_{2n-1}\cdots w_1$ be any contraction of $(a^\dagger a)^n$. For each $j = 1, 2, \ldots, n-1$, define $\pi_j = e$ if $w_{2j} = a^\dagger$, and $\pi_j = i$ if $w_{2j} = e^\dagger$ and $w_{2i+1} = e$, where $i$ is minimal and greater than $j$. The definition of $\pi$ implies that $w$ is a contraction of $(a^\dagger a)^n$ if and only if the sequence $\pi = \pi_1\pi_2\cdots\pi_{n-1} \in \mathcal{C}on\mathcal{S}_{n-1}$. Moreover, if $w$ is a contraction then the sequence $\pi$ has $k-1$ coordinates $e$ if and only if $k - 1 = \#\{j | w_{2j} = a^\dagger\}$, or, in other words, $\pi \in \mathcal{C}on\mathcal{S}_{n-1,k-1}$ if and only if $w$ is a contraction of degree $k$ in $\mathcal{C}on((a^\dagger a)^n)$. □

**Example 10.28** *Figure 10.3 presents the contractions of $(a^\dagger a)^3$ which are ordered from left to right. Then Theorem 10.27 maps these contractions to $ee$, $e2$, $1e$, $12$, and $2e$ in $\mathcal{C}on\mathcal{S}_2$, respectively.*

---

## 10.3 Wick's Theorem and $q$-Normal Ordering

Since Katriel and Kibler's [164] results, several authors have considered the combinatorial aspects of normal ordering arbitrary words in the creation and annihilation operators of a single-mode $q$-boson having the commutation relations

$$[a, a^\dagger]_q \equiv aa^\dagger - qa^\dagger a = 1, \quad [a, a] = 0, \quad [a^\dagger, a^\dagger] = 0; \tag{10.8}$$

see [40, 315, 316, 346].

**Definition 10.29** *We define the $q$-normal ordering of $F(a, a^\dagger)$ to be*

$$\mathcal{N}_q(F(a, a^\dagger)) = \sum_{\pi \in \mathcal{C}(F(a,a^\dagger))} q^{\deg(\pi)} : \pi :, \tag{10.9}$$

*where $\deg(\pi)$ is the degree of the Feynman diagram $\pi$.*

**Example 10.30** *The degrees of the Feynman diagrams (contractions) of the operator $aaa^\dagger a^\dagger$ are given by $0, 1, 1, 1, 1, 2, 2$; see Figure 10.1 from left to right and from top to bottom. Thus, the $q$-normal ordering of $aaa^\dagger a^\dagger$ is given by $1 + 4q + 2q^2$.*

Clearly, the operation of normal ordering $\mathcal{N}_q$ is a linear map, that is, $\mathcal{N}_q(F(a, a^\dagger) + G(a, a^\dagger)) = \mathcal{N}_q(F(a, a^\dagger)) + \mathcal{N}_q(G(a, a^\dagger))$ for any two operator functions $F(a, a^\dagger)$ and $G(a, a^\dagger)$. By (10.9), it follows that its normally ordered form $\mathcal{N}_q[F(a, a^\dagger)] = F(a, a^\dagger)$ can be written as

$$\mathcal{N}_q[F(a, a^\dagger)] = F(a, a^\dagger) = \sum_{k,\ell} C_{k,\ell}(q)(a^\dagger)^k a^\ell \tag{10.10}$$

for some coefficients $C_{k,\ell}(q)$, and the main task consists of determining the coefficients as explicitly as possible. Varvak [346] has described that the general coefficients $C_{k,\ell}(q)$ can be interpreted as $q$-rook numbers. In the next subsection, we present a different approach associated with "$q$-weighted Feynman diagrams". Note that in [5, 51, 100] very similar results have been described in slightly different situations.

Now, we can state a generalization of (10.4) to the $q$-deformed case. We refer the reader to [227] for the details of the proof.

**Theorem 10.31** *Let $F(a, a^\dagger)$ be an operator function of the annihilation and creation operators of the $q$-boson (10.8). Then the normally ordered form $\mathcal{N}_q[F(a, a^\dagger)]$ can be written with $q$-weighted Feynman diagrams and the double dot operation as*

$$\mathcal{N}_q[F(a, a^\dagger)] = F(a, a^\dagger) = \sum_{\gamma \in \mathscr{F}(F(a,a^\dagger))} \mathscr{W}_q(\gamma) : \gamma : . \tag{10.11}$$

Clearly, the case $q = 1$ reduces (10.11) to the undeformed case (10.4). Now, let us present a simple example for Theorem 10.31 before we discuss the connection between $q$-rook numbers and Stirling numbers.

**Example 10.32** Let $F(a, a^\dagger) = aa^\dagger aa^\dagger$. Since there are only two creators in the word, the Feynman diagrams can have degree at most two. The trivial Feynman diagram $\gamma$ of degree zero yields : $\gamma := (a^\dagger)^2 a^2$ and has q-weight $\mathscr{W}_q(\gamma) = q^{\text{len}(\gamma)} = q^3$. Thus, the Feynman diagram of degree zero gives the contribution $q^3 (a^\dagger)^2 a^2$. There are three Feynman diagrams of degree one, namely

$$(1,2), (1,4), (3,4)$$

and their q-weights are given by (same order) $q, q^2, q$. Thus, the Feynman diagrams of degree one give the contribution $(2q + q^2) a^\dagger a$. There is exactly one Feynman diagram of degree two, namely, $(1,2)(3,4)$ with q-weight $q^0$, which yields the contribution 1. Thus,

$$\mathcal{N}_q(aa^\dagger aa^\dagger) = q^3 (a^\dagger)^2 a^2 + (2q + q^2) a^\dagger a + 1,$$

which may also be calculated by hand from (10.8).

Now, we like to set a connection between Theorem 10.31 and Varvak's [346] results. Given a word $w = F(a, a^\dagger)$ containing $m$ creation operators $a^\dagger$ and $n$ annihilation operators $a$ (with $n \leq m$), she associates with $w$ a certain *Ferrers board* $B_w$ outlined by $w$. We denote the $k$th q-rook number of the board $B_w$ by $R_k(B_w, q)$; see Section 3.2.4.4. She shows (see [346, Theorem 6.1])

$$w = \sum_{k=0}^{n} R_k(B_w, q) (a^\dagger)^{m-k} a^{n-k}.$$

By Theorem 10.31 we first note that the set of Feynman diagrams is the disjoint union of Feynman diagrams of degree $k$, that is, $\mathscr{F}(w) = \cup_{k=0}^{n} \mathscr{F}_k(w)$, and that $\gamma \in \mathscr{F}_k(w)$ implies : $\gamma := (a^\dagger)^{m-k} a^{n-k}$, giving

$$w = \sum_{k=0}^{n} \left\{ \sum_{\gamma \in \mathscr{F}_k(w)} \mathscr{W}_q(\gamma) \right\} (a^\dagger)^{m-k} a^{n-k}.$$

Comparing the last two expressions yields the following corollary.

**Corollary 10.33** *Given a word $w = F(a, a^\dagger)$, the $k$th q-rook number of the associated Ferrers board $B_w$ equals the sum of the q-weights of all Feynman diagrams of degree $k$ on $w$, that is,*

$$R_k(B_w, q) = \sum_{\gamma \in \mathscr{F}_k(w)} \mathscr{W}_q(\gamma).$$

Let us return to the undeformed case ($q = 1$) and consider the particular word $F(a, a^\dagger) = (a^\dagger a)^n$. The same argument as above gives

$$\mathcal{N}[(a^\dagger a)^n] = \sum_{k=0}^{n} \sum_{\gamma \in \mathscr{F}_{n-k}((a^\dagger a)^n)} (a^\dagger)^k a^k.$$

Comparing this with the undeformed case of [164] shows that the conventional Stirling numbers of the second kind can also be interpreted as the number of Feynman diagrams of degree $n - k$ on the particular word $(a^\dagger a)^n$ of length $2n$, that is,
$$\text{Stir}(n, k) = |\mathscr{F}_{n-k}((a^\dagger a)^n)|.$$
Turning to the $q$-deformed situation, the same argument implies
$$\text{Stir}_q(n, k) = \sum_{\gamma \in \mathscr{F}_{n-k}((a^\dagger a)^n)} \mathscr{W}_q(\gamma) = \sum_{\gamma \in \mathscr{F}_{n-k}((a^\dagger a)^n)} q^{\text{tcross}(\gamma)+\text{len}(\gamma)}.$$

Finally, we note the following on the multi-mode boson operators. In [225] the study of combinatorial aspects of the normal ordering of multi-mode boson operators and interesting combinatorial questions were addressed. In view of $q$-Wick's theorem, it is natural to consider the analogous problem for the $q$-deformed variant of the multi-mode boson operator. However, this does not seem to be easy, as the following example shows. For concreteness we consider the deformed two-mode boson having the commutation relations
$$[a, a^\dagger]_{q_a} = 1, \quad [b, b^\dagger]_{q_b} = 1, \quad [a, b^\dagger]_{q_{ab}} = 1$$
and all other commutators vanish. Here the parameters $q_a, q_b, q_{ab}$ are arbitrary (up to this step). Let us study the simple example $F(a, a^\dagger, b, b^\dagger) = abb^\dagger$. Commuting $b$ and $b^\dagger$ using the above commutation relations yields $abb^\dagger = q_b a b^\dagger b + a$; commuting then $a$ and $b^\dagger$ implies $abb^\dagger = q_b q_{ab} b^\dagger a b + q_b b + a$. On the other hand, first commuting $a$ and $b$ and then commuting $b^\dagger$ to the left yields $abb^\dagger = q_b q_{ab} b^\dagger ba + q_{ab} a + b$. Clearly, for the two results to be the same we have to assume that $q_b = q_{ab} = 1$. A similar computation for $baa^\dagger$ shows that also $q_a = 1$. Therefore, it seems that a $q$-deformed version of the multi-mode boson considered in [225] does not exist.

## 10.4 $p$-Normal Ordering

Now let us present another generalization for the normal ordering as given in Definition 10.3.

**Definition 10.34** Let $F(a, a^\dagger)$ be any (infinite) word on the alphabet $\{a, a^\dagger\}$. We define $\mathcal{C}(F(a, a^\dagger))$ to be the multiset of all words obtained by substituting $a = e$ and $a^\dagger = e^\dagger$ whenever $a$ precedes $a^\dagger$; moreover, we replace any two adjacent letters $e$ and $e^\dagger$ with $p$. In this context, each element of $\mathcal{C}(F, a, a^\dagger)$ is called a p-contraction or just contraction. For each word $\pi$ in $\mathcal{C}(F(a, a^\dagger))$, the double dot operation of $\pi$ is defined by deleting all letters $e$ and $e^\dagger$ and arranging it such that all letters $a^\dagger$ precede the letter $a$; clearly, $: \pi := (a^\dagger)^v a^u p^w$ for some $u, v, w \geq 0$. We now define the p-normal ordering of $F(a, a^\dagger)$ to be

$$\mathcal{N}_p(F(a, a^\dagger)) = \sum_{\pi \in \mathcal{C}(F(a,a^\dagger))} : \pi : . \tag{10.12}$$

**Example 10.35** *The elements of the contractions in the set $\mathcal{C}(aaa^\dagger a^\dagger)$ are given by $aaa^\dagger a^\dagger$, $eae^\dagger a^\dagger$, $eaa^\dagger e^\dagger$, $apa^\dagger$, $aea^\dagger e^\dagger$, $epe^\dagger$, $eee^\dagger e^\dagger$, as illustrated in Figure 10.1. Thus, $\mathcal{N}_p(aaa^\dagger a^\dagger) = (a^\dagger)^2 a^2 + (p+3)a^\dagger a + p + 1$.*

Clearly, the p-normal ordering form is a generalization of the normally ordered form, namely $\mathcal{N}(F(a, a^\dagger)) = \mathcal{N}_1(F(a, a^\dagger))$. Also, the p-normal ordering form is a particular case of Definition 10.8; that is, if we define $\alpha(\pi)$ to be the number of adjacent letters $e$ and $e^\dagger$ in the Feynman diagram $\pi$, then $\mathcal{N}_p(F(a, a^\dagger)) = \sum_{\pi \in \mathcal{C}(F(a,a^\dagger))} p^{\alpha(\pi)} : \pi :$, where the double dot operation is given by Definition 10.1.

The next theorem presents a starting point of the p-normal ordering of an expression that extends Theorem 10.16.

**Theorem 10.36** *For all $\ell, k \geq 0$,*

$$\mathcal{N}_p(a^\ell (a^\dagger)^k) = \sum_{i=0}^{\ell} \left[ \binom{\ell}{i}(k-i) + (\ell - 1 + p)\binom{\ell-1}{i-1} \right] (k-1)_{i-1} (a^\dagger)^{k-i} a^{\ell-i},$$

*where $(x)_i = x(x-1) \cdots (x-i+1)$ and $(x)_{-i} = (x+1)(x+2) \cdots (x+i)$ with $(x)_0 = 1$.*

**Proof** The proof is based on the following recurrence:

$$\mathcal{N}_p(a^\ell (a^\dagger)^k) = a^\dagger \mathcal{N}_1(a^\ell (a^\dagger)^{k-1}) + (\ell - 1 + p) \cdot \mathcal{N}_1(a^{\ell-1}(a^\dagger)^{k-1}),$$

which holds immediately from the definitions. The proof is completed by Theorem 10.16. □

Before dealing with an explicit formula for $\mathcal{N}_p((a^\dagger a)^n)$, let us note the physical aspects of the p-normal ordering (for a full discussion, see [226]). In Definition 10.34, we have only mentioned the letters $a$ and $a^\dagger$ and have not interpreted these as bosonic operators. The main reason for this is very simple: the prescription (10.12) is not consistent with (10.1)! Since this is a critical point, let us discuss it explicitly. We consider the simplest nonempty case where only one contraction is involved, namely the word $aa^\dagger$. From Definition 10.34 it follows that

$$\mathcal{N}_p(aa^\dagger) =: aa^\dagger : + : pI := a^\dagger a + pI. \tag{10.13}$$

On the other hand, by (10.1) and the linearity of $\mathcal{N}_p$ we have that $aa^\dagger = a^\dagger a + I$, which implies

$$\mathcal{N}_p(aa^\dagger) = \mathcal{N}_p(a^\dagger a + I) = \mathcal{N}_p(a^\dagger a) + \mathcal{N}_p(I) = a^\dagger a + I,$$

which clearly contradicts (for the case $p \neq 1$ we are interested in) the result (10.13). This explains that the $p$-normal ordering cannot be interpreted as bosonic annihilation and creation operators! Therefore, it seems to be very difficult to extend the usual Fock representation (and the coherent states) of the bosonic operators that satisfy (10.1) and (10.3) to the letters $a$ and $a^\dagger$ that satisfy (10.12). This is the main reason why we saying *letters* $a$ and $a^\dagger$ and not operators. However, it would be interesting to see which algebraic relations are satisfied by letters $a$ and $a^\dagger$ without contradicting (10.12); further, it would be interesting to find a relation from which (10.12) could be *derived* as consequence, in analogy with the conventional case where the usual normal ordering can be derived from the canonical commutation relation.

As an example of $p$-normal ordering, we study the case $(a^\dagger a)^n$. In order to find an explicit formula for $p$-normal ordering of the word $(a^\dagger a)^n$, we need to define the following statistic on $\mathcal{P}_n$.

**Definition 10.37** *Let $Q$ be any subset of $[n]$, we say that $Q$ has a set-rise at $i$ if $i, i+1 \in Q$. The number of set-rises of $Q$ is denoted by* $\mathrm{srise}(Q)$. *Let $\pi = B_1, B_2, \ldots, B_k \in \mathcal{P}_n$, we define* $\mathrm{srise}(\pi) = \sum_{j=1}^{k} \mathrm{srise}(P_j)$.

Lemma 10.25 together with Lemma 10.27 yields the following result.

**Proposition 10.38** *There exists a bijection between the set of contractions $\mathrm{Con}((a^\dagger a)^n)$ of degree $k$ and the set of set partitions of $[n]$ with $k$ set-rises.*

We give now an explicit formula for $\mathcal{N}_p((a^\dagger a)^n)$.

**Theorem 10.39** *For all $n \geq 1$,*

$$\mathcal{N}_p((a^\dagger a)^n) = \sum_{k=0}^{n} S_p(n,k)(a^\dagger)^k a^k, \qquad (10.14)$$

*where $S_p(n,k)$ satisfies the following recurrence relation*

$$S_p(n,k) = (k-1+p)S_p(n-1,k) + S_p(n-1,k-1), \qquad (10.15)$$

*with the initial conditions $S_p(n,1) = p^{n-1}$ and $S_p(n,k) = 0$ for all $k > n$.*

**Proof** We denote the number of contractions $\pi$ of $(a^\dagger a)^n$ such that $: \pi := (a^\dagger)^k a^k p^m$ by $S(n,k;m)$. Proposition 10.38 gives that $S(n,k;m)$ equals the number of set partitions of $\mathcal{P}_n$ into $k$ blocks $B_1, \ldots, B_k$ with $m$ set-rises. In order to write a recurrence relation for the sequence $S(n,k;m)$, we consider the position of $n$ in the blocks $B_1, \ldots, B_k$. If $B_k = \{n\}$, then there are $S(n-1, k-1; m)$ such set partitions; if $n \in B_i$ and $n-1 \notin B_i$, then there are $S(n-1,k;m)$ set partitions; and if $n, n-1, \in B_i$, then the number of such set partitions equals $S(n-1,k;m-1)$. Therefore,

$$S(n,k;m) = (k-1)S(n-1,k;m) + S(n-1,k;m-1) + S(n-1,k-1;m),$$

which implies that
$$S_p(n,k) = (k-1+p)S_p(n-1,k) + S_p(n-1,k-1),$$
where $S_p(n,k)$ is defined to be $\sum_{m=0}^{n-k} S(n,k;m)p^m$. Hence,
$$\mathcal{N}_p((a^\dagger a)^n) = \sum_{k=0}^{n}\sum_{m=0}^{n-k} S(n,k;m)p^m (a^\dagger)^k a^k = \sum_{k=0}^{n} S_p(n,k)(a^\dagger)^k a^k.$$
The initial conditions can be verified directly from the definitions. $\square$

As we have seen above, the polynomials $S_p(n,k)$ satisfy (10.15). When $p=1$ this recurrence reduces to the recurrence relation of the conventional Stirling numbers which, by Theorem 1.12, gives $S_1(n,k) = \text{Stir}(n,k)$.

**Definition 10.40** Define $S_p(x;k) = \sum_{n\geq k} S_p(n,k)\frac{x^n}{n!}$.

**Theorem 10.41** For all $k \geq 1$,
$$S_p(x;k) = \int_0^x \frac{e^{pt}}{(k-1)!}(e^t-1)^{k-1}dt.$$
Moreover, the generating function $S_p(x,y) = \sum_{n\geq 1}\sum_{k=1}^{n} S_p(n,k)y^k \frac{x^n}{n!}$ is given by
$$S_p(x,y) = \int_0^x y e^{pt} e^{y(e^t-1)} dt.$$

**Proof** Multiplying (10.15) by $\frac{x^{n-1}}{(n-1)!}$ and summing over $n \geq k$, we obtain
$$\frac{d}{dx}S_p(x;k) = \frac{d}{dx}\left(\sum_{n\geq k} S_p(n,k)\frac{x^n}{n!}\right) = (k-1+p)S_p(x;k) + S_p(x;k-1),$$
with the initial condition $S_p(x;0) = 1$. Now, let us show the following claim by induction on $k$:
$$S_p(x;k) = \int_0^x \frac{e^{pt}}{(k-1)!}(e^t-1)^{k-1}dt.$$
Clearly, the claim holds for $k=1$. If we assume that the claim holds for $k-1$, then by the above recurrence we have
$$\frac{d}{dx}S_p(x;k) = (k-1+p)S_p(x;k) + \int_0^x \frac{e^{pt}}{(k-2)!}(e^t-1)^{k-2}dt,$$
which implies that
$$\frac{d^2}{dx^2}S_p(x;k) = (k-1+p)\frac{d}{dx}S_p(x;k) + \frac{e^{px}}{(k-2)!}(e^x-1)^{k-2}.$$

It is not difficult to see that the function $\frac{e^{px}}{(k-1)!}(e^x-1)^{k-1}$ is a particular solution, which shows

$$\frac{d}{dx}S_p(x;k) = \frac{e^{px}}{(k-1)!}(e^x-1)^{k-1} + \frac{d}{dx}f_k(x),$$

where $\frac{d^2}{dx^2}f_k(x) = \frac{d}{dx}f_k(x)$ with the initial conditions $f_k(0) = \frac{d}{dx}f_k(0) = 0$. Therefore $S_p(x;k) = \int_0^x \frac{e^{pt}}{(k-1)!}(e^t-1)^{k-1}$ and this completes the induction. □

Note that the above theorem for $p = 1$ gives

$$S_1(x;k) = \int_0^x \frac{e^t}{(k-1)!}(e^t-1)^{k-1}dt.$$

Also, $S_1(x,y) = e^{y(e^x-1)}$, which is the generating function for Stirling numbers of the second kind (see Example 2.58). One of the first explicit expressions for the numbers $S_p(n,k)$ was given by d'Ocagne [93].

**Theorem 10.42** For all $n \geq k \geq 1$,

$$S_p(n,k) = \frac{(-1)^{k-1}}{(k-1)!}\sum_{j=0}^{k-1}(-1)^j\binom{k-1}{j}(p+j)^{n-1}. \tag{10.16}$$

**Proof** Theorem 10.41 gives that

$$S_p(x;k) = \int_0^x \frac{e^{pt}}{(k-1)!}(e^t-1)^{k-1}dt$$

$$= \frac{(-1)^{k-1}}{(k-1)!}\sum_{j=0}^{k-1}\left((-1)^j\binom{k-1}{j}\int_0^x e^{(p+j)t}dt\right)$$

$$= \frac{(-1)^{k-1}}{(k-1)!}\sum_{j=0}^{k-1}\left((-1)^j\binom{k-1}{j}\frac{e^{(p+j)x}-1}{p+j}\right),$$

which implies that the coefficient of $x^n$ in the generating function $S_p(x;k)$ is given by

$$S_p(n,k) = \frac{(-1)^{k-1}}{(k-1)!}\sum_{j=0}^{k-1}(-1)^j\binom{k-1}{j}(p+j)^{n-1},$$

as claimed. □

By (10.16) and Theorem 10.39 we obtain a formula for $\mathcal{N}_p((a^\dagger a)^n)$.

**Theorem 10.43** For all $n \geq 1$,

$$\mathcal{N}_p((a^\dagger a)^n) = \sum_{k=0}^n \left(\frac{(-1)^{k-1}}{(k-1)!}\sum_{j=0}^{k-1}(-1)^j\binom{k-1}{j}(p+j)^{n-1}\right)(a^\dagger)^k a^k.$$

Theorem 10.43 with $p = 1$ gives Theorem 10.17. Theorem 10.39 with $p = 0$ gives the following corollary.

**Corollary 10.44** *For all $n \geq 0$,*

$$\mathcal{N}_0((a^\dagger a)^n) = \sum_{j=0}^{n} S_0(n, j)(a^\dagger)^j a^j,$$

*where $S_0(n, k)$ satisfies the following recurrence relation $S_0(n, k) = (k - 1)S_0(n - 1, k) + S_0(n - 1, k - 1)$ with initial condition $S_0(n, 1) = \delta_{n,1}$ and $S_0(n, k) = 0$, for $k > n$. Thus, $\sum_{k=1}^{n} S_0(n, k) = \text{Bell}_{n-1}$.*

## 10.5 Noncrossing Normal Ordering

The aim of this section is to introduce and study the notion of noncrossing normally ordered form. In the previous sections, we represented the contractions by a graphical representation or linear representation (see Definitions 10.3 and 10.23). In this section, we introduce a word representation called canonical sequential form (see Definition 1.3 in the case of set partitions).

**Definition 10.45** *We represent any contraction $\pi$ by a sequence $a_1 a_2 \ldots a_n$ on the set $\{1, 2, \ldots, n, 1', 2', \ldots, n'\}$. In order to define the canonical sequential form $a_1 a_2 \ldots a_n$ of $\pi$, we read the contraction $\pi$ (see Definition 10.20) from right to left*

- *If $\pi_j$ is a white (respectively, black) vertex of degree 0 (that is, incident with no edges), then replace it with $i'$ (respectively, $i$); $i$ is the smallest number not appearing in the sequence.*

- *If $\pi_j$ is a black vertex of degree 1, then replace it with $i$, where $i$ is the smallest number not appearing in the sequence.*

- *If $\pi_j$ is a white vertex of degree 1, then replace it with $i$, where $i$ is associated to the black vertex connected to $\pi_j$.*

**Example 10.46** *The canonical sequential forms of the contractions in Figure 10.1 are given by $123'4'$, $123'2$, $1213'$, $1221$, $123'1$, $1212$, and $1223'$.*

Note that we denote contractions also by enumerating the edges. For example, the contractions in Figure 10.1 are given by $\emptyset$, $(42)$, $(31)$, $(41)(32)$, $(41)$, $(42)(31)$, and $(32)$.

**Definition 10.47** Let $e = (i, j)$ and $e' = (p, q)$ be any two edges of a contraction $\pi$. We say that $e$ crosses $e'$ if they intersect with each other, or, in other words, if $i < p < j < q$ or $p < i < q < j$. In this context, we say that $e$ crosses $e'$ or $e$ and $e'$ are crossing. Otherwise, $e$ and $e'$ are said to be a noncrossing.

**Example 10.48** The rightmost contraction in the second line of Figure 10.1 presents two crossing edges, that is, $(42)(31)$ is a crossing. Any another contraction in Figure 10.1 is a noncrossing.

Using the linear representation form of contractions, the normally ordered form (see Definition 10.8) of an expression $F(a, a^\dagger)$ can be defined as follows.

**Definition 10.49** Let $\pi$ be any contraction associated with its linear representation form. Define $\mathrm{freeb}(\pi)$ (respectively, $\mathrm{freew}(\pi)$) to be the number of black (respectively, white) isolated vertices in $\pi$. The normally ordered form of $F(a, a^\dagger)$ is given by

$$\sum_{\pi \text{ is a contraction of } F(a,a^\dagger)} (a^\dagger)^{freeb(\pi)} a^{freew(\pi)}.$$

A $q$-analogue of the normally ordered form (see Definition 10.14) can be defined as follows.

**Definition 10.50** The operator $R_q$ acts on a contraction $\pi \in Con(F(a, a^\dagger))$ is defined by

$$R_q(\pi) = q^{cross(\pi)}(a^\dagger)^{freeb(\pi)} a^{freew(\pi)},$$

where $\mathrm{cross}(\pi)$ counts the number of crossing edges in $\pi$. We extend $R_q$ to a linear operator by $R_q(F(a, a^\dagger)) = \sum_{\pi \in Con(F(a,a^\dagger))} R_q(\pi)$.

Note that the operator $R_q$ is a $q$-analogue of the standard double dot operation but not a $q$-normal ordering operator as defined in Section 10.3. Namely, for a given expression $F$, $R_1(F)$ is exactly the normally ordered form of $F$.

By applying our operator $R_q$ to the expression $(a^\dagger a)^n$, we can write

$$R_q((a^\dagger a)^n) = \sum_{j=0}^{n} r_{n,j}(q)(a^\dagger)^j a^j,$$

where $r_{n,j}(q)$ is a polynomial in $q$ (clearly, $r_{n,j}(1) = \mathrm{Stir}(n, j)$; see Theorem 10.17). Here, we study the combinatorial structure of $R_0(F)$.

**Definition 10.51** A noncrossing contraction is a contraction whose edges are all noncrossing.

Figure 10.1 presents all contractions of the word $aaa^\dagger a^\dagger$, where it consists of exactly one crossing and six noncrossing contractions. With this terminology, we give the definition of noncrossing normal ordering.

**Definition 10.52** *The* noncrossing normally ordered form *of a word $w$ over alphabet $\{a, a^\dagger\}$ is defined by* $\mathcal{NC}(w) = \sum_{\pi \in \mathcal{C}on(w)} :\pi:$. *More generally, for any arbitrary expression $F(a, a^\dagger)$ (by linearity), we define*

$$\mathcal{NC}(F(a, a^\dagger)) = \sum_{\pi \in \mathcal{C}on(F(a,a^\dagger))} :\pi:. \qquad (10.17)$$

Note that $R_0(F(a, a^\dagger)) = \mathcal{NC}(F(a, a^\dagger))$. Here, we study $\mathcal{NC}((a^r(a^\dagger)^s)^n)$ (see Section 10.5.2), $\mathcal{NC}((a^r + (a^\dagger)^s)^n)$ (see Section 10.5.3) and some special cases. In Section 10.5.4, we establish bijections between sets of noncrossing contractions sets (for example, $(a + (a^\dagger)^2)^n$, $(a^r a^\dagger)^n$ and $(a(a^\dagger)^r)^n$), sets of trees and sets of lattice paths.

### 10.5.1 Some Preliminary Observations

We denote for arbitrary $q$ the corresponding normal ordering by

$$\tilde{\mathcal{N}}_q(F(a, a^\dagger)) \equiv R_q(F(a, a^\dagger))$$

such that $\tilde{\mathcal{N}}_1 \equiv \mathcal{N}$ and $\tilde{\mathcal{N}}_0 \equiv \mathcal{NC}$; see Definitions 10.50 and 10.52. Define the associated coefficients by $C_{F;k,l}(q)$, that is,

$$\tilde{\mathcal{N}}_q(F(a, a^\dagger)) = \sum_{k,l} C_{F;k,l}(q)(a^\dagger)^k a^l.$$

It is clear that $0 \leq C_{F;k,l}(0) \leq C_{F;k,l}(1)$, for any $F$ and all $k, l$.

**Example 10.53** *Let $F(a, a^\dagger) = aaa^\dagger a^\dagger$ be the example of Figure 10.1. Definition 10.3 shows that $\tilde{\mathcal{N}}_1(aaa^\dagger a^\dagger) = (a^\dagger)^2 a^2 + 4a^\dagger a + 2$. Since there is exactly one crossing contraction of degree two, namely $(42)(31)$, we have that $\tilde{\mathcal{N}}_0(aaa^\dagger a^\dagger) = (a^\dagger)^2 a^2 + 4a^\dagger a + 1$.*

Let us try to reproduce the noncrossing statistics according to Definition 10.17 by modifying the commutation relations and using the usual normal ordering process. Thus, we consider operators $b, b^\dagger$ satisfying

$$bb^\dagger = \kappa b^\dagger b + \alpha b^\dagger + \beta b + \gamma, \qquad (10.18)$$

where $\alpha$, $\beta$, and $\gamma$ are some constants (here we have assumed that the right-hand side has lower degree than the left-hand side). By this commutator, it follows

$$\tilde{\mathcal{N}}_0(bb^\dagger) \stackrel{(10.18)}{=} \tilde{\mathcal{N}}_0(\kappa b^\dagger b + \alpha b^\dagger + \beta b + \gamma) = \kappa b^\dagger b + \alpha b^\dagger + \beta b + \gamma,$$

where we have used in the second equation the fact that $\tilde{\mathcal{N}}_0$ is linear and that all summands are already normally ordered. However, if this has to be the result obtained from Definition 10.52 then $\kappa = 1 = \gamma$ and $\alpha = \beta = 0$, reproducing for $b, b^\dagger$ the conventional commutation relation. However, Example 10.53 shows that the operators $b, b^\dagger$ cannot satisfy the conventional commutation relations and Definition 10.52 simultaneously! Thus, it is not clear whether one should speak of the operators $a, a^\dagger$ as "bosonic" operators anymore.

### 10.5.2 Noncrossing Normal Ordering of $(a^r(a^\dagger)^s)^n$

Now, our goal is to study the noncrossing normal ordering of $(a^r(a^\dagger)^s)^n$.

**Definition 10.54** *We denote the set of all the linear representations of the noncrossing contractions of $(a^r(a^\dagger)^s)^n$ by $\mathcal{V}_{rs}(n)$. For each linear representation $\pi \in \mathcal{V}_{rs}(n)$, define $e(\pi)$ to be the number of edges in $\pi$. Let $A_{r,s}(x,y)$ be the generating function for the number of linear representations $\pi \in \mathcal{V}_{rs}(n)$ with exactly $m$ edges, that is, $A_{r,s}(x,y) = \sum_{n \geq 0} \sum_{\pi \in \mathcal{V}_{rs}(n)} x^n y^{e(\pi)}$.*

The noncrossing normally ordered form of $(a^r(a^\dagger)^s)^n$ is given by

$$\mathcal{NC}\left((a^r(a^\dagger)^s)^n\right) = \sum_{j \geq 0} [x^n y^j](A_{r,s}(x,y))(a^\dagger)^{sn-j} a^{rn-j}.$$

Thus, in order to find the noncrossing normally ordered form of $(a^r(a^\dagger)^s)^n$, it is sufficient to derive an explicit formula for the generating function $A_{r,s}(x,y)$. Here, we present a nonlinear system of equations whose solution gives an explicit formula for $A_{r,s}(x,y)$. Let $n \geq 1$; since any linear representation $\pi \in \mathcal{V}_{rs}(n)$ has exactly $s$ black vertices stand on the right, then we can write

$$A_{r,s}(x,y) = 1 + A_{r,s}(x,y;s), \tag{10.19}$$

where $A_{r,s}(x,y;d)$ is the generating function for all the linear representations $\pi = \pi_{n(r+s)} \ldots \pi_1 \in \mathcal{V}_{rs}(n)$ such that the canonical sequential form of $\pi$ starts with $12 \ldots d$. The following lemma presents a recurrence relation for the sequence $A_{r,s}(x,y;d)$.

**Lemma 10.55** *Let $z = x^{\frac{1}{r+s}}$. For all $t = 1, 2, \ldots, s$,*

$$A_{r,s}(x,y;t) = zA_{r,s}(x,y;t-1) + A_{r,s}(x,y) \sum_{j=1}^{r} z^{r-j} A_{r,s}(x,y;j,t)$$

*with $A_{r,s}(x,y;0) = z^r A_{r,s}(x,y)$, where $A_{r,s}(x,y;a,b)$ is the generating function for the number of linear representations $\pi \in \mathcal{V}_{rs}(n)$ such that $\pi$ starts with $b$ black vertices and ends with $a$ white vertices and there is an edge between the first black and last white vertex.*

**Proof** Clearly, we have that $A_{r,s}(x,y;0) = z^r A_{r,s}(x,y)$. Let $\pi \in \mathcal{V}_{rs}(n)$, and assume that $\pi$ starts with $d$ black vertices, that is, $\pi = \pi_{n(r+s)} \cdots \pi_d \pi_{d-1} \cdots \pi_1$

where $\pi_i$ with $1 \le i \le d$ corresponds to a black vertex. Let us write an equation for $A_{r,s}(x,y;d)$. The first black vertex corresponding to $\pi_1$ has degree either zero or one. The contribution of the first case is

$$z A_{r,s}(x,y;d-1).$$

Now, let us consider the case where the first black vertex has degree one. The contraction $\pi$ can be presented as $\pi = \pi''\beta\pi'$, where $\pi'$ starts with $d$ black vertices, ends with $j$ white vertices and there is an edge between the first black vertex of $\pi'$ and the last white vertex of $\pi'$. Moreover, $\pi'$ is followed by $r-j$ white vertices. Since we are interested in the noncrossing contractions, there are no edges from $\beta\pi'$ to $\pi''$. Therefore, the contribution of the second case gives

$$A_{r,s}(x,y) \sum_{j=1}^{r} z^{r-j} A_{r,s}(x,y;j,d).$$

By combining these cases, we derive our recurrence relation. $\square$

Lemma 10.55 shows that to derive a formula for the generating function $A_{r,s}(x,y)$, we need to find a recurrence relation for the generating functions $A_{r,s}(x,y;a,b)$.

**Lemma 10.56** *Let $z = x^{\frac{1}{r+s}}$, $a = 2, 3, \ldots, r$, and $b = 2, 3, \ldots, s$. Then*

(i) $\quad A_{r,s}(x,y;1,1) = yz^2 + xyz^2 A_{r,s}(x,y).$

(ii) $\quad A_{r,s}(x,y;a,1) = z A_{r,s}(x,y;a-1,1)$

$$+ yz^{r+2} A_{r,s}(x,y) \sum_{j=1}^{s} z^{s-j} A_{r,s}(x,y;a-1,j).$$

(iii) $\quad A_{r,s}(x,y;1,b) = z A_{r,s}(x,y;1,b-1)$

$$+ yz^{s+2} A_{r,s}(x,y) \sum_{j=1}^{r} z^{r-j} A_{r,s}(x,y;j,b-1).$$

(iv) $\quad A_{r,s}(x,y;a,b) = z A_{r,s}(x,y;a-1,b)$

$$+ yz^2 \sum_{j=1}^{a-1} z^{b-1-j} A_{r,s}(x,y;a-1,j)$$

$$+ yz^2 A_{r,s}(x,y;b-1) \sum_{j=1}^{s} z^{s-j} A_{r,s}(x,y;a-1,j).$$

**Proof** Let $\pi \in \mathcal{V}_{rs}(n)$ such that the last $a$ vertices of $\pi$ are white, the first $b$ vertices of $\pi$ are black, and there is an edge between the first black vertex and the last white vertex. The generating function for the number of such linear representations $\pi$ is given by $A_{r,s}(x,y;a,b)$. Now, let us write an equation for $A_{r,s}(x,y;a,b)$ for each of Cases (i)–(iv).

(i) If $a = b = 1$, then $\pi$ can be decomposed as either $\pi = 11$ or $\pi = 12'3'\ldots r'\pi'(d+1)(d+2)\ldots(d+s)1$. The first contribution gives $z^2 y$ and the second contribution gives $yz^{r+s+2}A_{r,s}(x,y) = yxz^2 A_{r,s}(x,y)$. Adding together these cases, we obtain (i).

(ii) If $a \geq 2$ and $b = 1$, then the degree of the vertex $v$, the one before the last white vertex (which is also a white vertex), is either zero or one. The first contribution gives $zA_{r,s}(x,y;a-1,1)$. In the second case, there exists a black vertex connected to $v$, and this contribution gives $yz^{r+2}A_{r,s}(x,y)\sum_{j=1}^{s} z^{s-j}A_{r,s}(x,y;a-1,j)$. Combining these cases, we obtain (ii).

(iii) If $b \geq 2$ and $a = 1$, then the degree of $v$, the second vertex (which is black), is either zero or one. The first contribution gives $zA_{r,s}(x,y;1,b-1)$. In the second case, there exists a white vertex connected to $v$, and this contribution gives $yz^{s+2}A_{r,s}(x,y)\sum_{j=1}^{r} z^{r-j}A_{r,s}(x,y;j,b-1)$. Adding these cases, we get (iii).

(iv) We leave it to the interested reader, which is very similar to Cases (ii)–(iii). For exact details, we refer the reader to [228]. □

As we have seen, Lemmas 10.55 and 10.56 together with (10.19) give a (nonlinear) system of equations in the variables $A_{r,s}(x,y)$, $A_{r,s}(x,y;t)$ and $A_{r,s}(x,y;a,b)$. Here, we solve this system for several interesting cases. For example, the cases of either $s=1$ or $r=1$ give the following result (prove it!).

**Theorem 10.57** *Let $r \geq 1$. Then*

$$A_{r,1}(x,y) = A_{1,r}(x,y) = (1 + xA_{1,r}(x,y))(1 + xyA_{1,r}(x,y))^r.$$

*Moreover, for all $n \geq 0$,*

$$\mathcal{NC}\left((a^r a^\dagger)^n\right) = \sum_{j=0}^{n} \frac{1}{n+1}\binom{n+1}{j+1}\binom{rn+r}{j}(a^\dagger)^{n-j}a^{rn-j},$$

$$\mathcal{NC}\left((a(a^\dagger)^r)^n\right) = \sum_{j=0}^{n} \frac{1}{n+1}\binom{n+1}{j+1}\binom{rn+r}{j}(a^\dagger)^{rn-j}a^{n-j}.$$

In Section 10.5.4, we present a combinatorial proof for this theorem. Another application of Lemmas 10.55 and 10.56 is the next result (Show that!).

**Theorem 10.58** *The generating function $A_{2,2}(x,y)$ satisfies*

$$\begin{aligned}A_{2,2}(x,y) &= 1 + x(1+y)^2 A_{2,2}(x,y) + 2xy(1 + x(1+y) + xy^2)A_{2,2}^2(x,y) \\ &\quad + x^2 y^2(x(1+y)^2 - 1)A_{2,2}^3(x,y) + x^4 y^4 A_{2,2}^4(x,y).\end{aligned}$$

### 10.5.3  Noncrossing Normal Ordering of $(a^r + (a^\dagger)^s)^n$

Now, our goal is to study the noncrossing normal ordering of $(a^r + (a^\dagger)^s)^n$.

**Definition 10.59** *We denote the set of all the linear representations of the noncrossing contractions of $(a^r + (a^\dagger)^s)^n$ by $\mathcal{W}_{rs}(n)$. For each linear representation $\pi \in \mathcal{W}_{rs}(n)$, define $w(\pi)$ and $e(\pi)$ to be the number of white vertices and edges in $\pi$, respectively. Let $B_{r,s}(x,y,z)$ be the generating function for the number of linear representations $\pi \in \mathcal{W}_{rs}(n)$ with exactly $m$ edges and $d$ white vertices, that is, $B_{r,s}(x,y,z) = \sum_{n\geq 0} \sum_{\pi \in \mathcal{W}_{rs}(n)} x^n y^{e(\pi)} z^{w(\pi)}$.*

The noncrossing normally ordered form of $(a^r + (a^\dagger)^s)^n$ is given by

$$\mathcal{NC}\left((a^r + (a^\dagger)^s)^n\right) = \sum_{i\geq 0}\sum_{j=0}^{i}[x^n y^j z^i](B_{r,s}(x,y,z))(a^\dagger)^{n-ri-j}a^{i-j}.$$

In order to find the noncrossing normally ordered form of $(a^r + (a^\dagger)^s)^n$, it is sufficient to find an explicit formula for the generating function $A_{r,s}(x,y,z)$. Again, we present a nonlinear system of equations whose solution gives an explicit formula for $B_{r,s}(x,y,z)$. At first, we write

$$B_{r,s}(x,y,z) = 1 + xz^r B_{r,s}(x,y,z) + B_{r,s}(x,y,z;s), \qquad (10.20)$$

where $B_{r,s}(x,y,z;d)$ is the generating function for all the linear representations $\pi \in \mathcal{W}_{rs}(n)$ such that the canonical sequential form of $\pi$ starts with $12\ldots d$. Applying a similar argument as in the proof of Lemma 10.55, we derive a recurrence relation for the sequence $B_{r,s}(x,y,z;d)$. Thus, we leave the proof to the reader; see [228].

**Lemma 10.60** *Let $z = x^{\frac{1}{s}}$ and $z' = x^{\frac{1}{r}}$. For all $d = 1, 2, \ldots, s$,*

$$B_{r,s}(x,y,z;d) = zB_{r,s}(x,y,z;d-1) + B_{r,s}(x,y,z)\sum_{j=1}^{r}(z'z)^{r-j}B_{r,s}(x,y,z;j,d)$$

*with the initial condition $B_{r,s}(x,y,z;0) = B_{r,s}(x,y,z)$, where $B_{r,s}(x,y;a,b)$ is the generating function for the number of linear representations $\pi \in \mathcal{W}_{rs}(n)$, such that $\pi$ starts with $b$ black vertices, ends with $a$ white vertices, and there is an edge between the first black vertex and last white vertex.*

Lemma 10.60 shows that to derive a formula for $B_{r,s}(x,y,z)$, we need to find a recurrence relation for the generating functions $B_{r,s}(x,y,z;a,b)$. We achieve this by using similar techniques as in the proof of Lemma 10.56, which implies a recurrence relation for the sequence $B_{r,s}(x,y,z;a,b)$. Thus, we leave the proof to the reader, see [228].

**Lemma 10.61** *Let $z = x^{\frac{1}{s}}$ and $z' = x^{\frac{1}{r}}$, $a = 2, 3, \ldots, r$, and $b = 2, 3, \ldots, s$.*

*Then*

(i) $B_{r,s}(x,y,z;1,b) = yzz'B_{r,s}(x,y,z;b-1)$.

(ii) $B_{r,s}(x,y,z;a,b) = yzz'B_{r,s}(x,y,z;a-1,b)$
$$+ yzz'\sum_{j=1}^{b-1} z^{b-1-j}B_{r,s}(x,y,z;a-1,j)$$
$$+ yzz'B_{r,s}(x,y,z;b-1)\sum_{j=1}^{s} z^{s-j}B_{r,s}(x,y,z;a-1,j).$$

Lemmas 10.60 and 10.61 together with (10.20) give a (nonlinear) system of equations in the variables $B_{r,s}(x,y,z)$, $B_{r,s}(x,y,z;t)$ and $B_{r,s}(x,y,z;a,b)$. For example, when $s=1$ we obtain the following result (derive that!).

**Theorem 10.62** *The generating function $B = B_{r,1}(x,y,z)$ satisfies*
$$B = 1 + x(1+z^r)B + x^2 yz^r B^r \frac{1 - x^r y^r(1+B)^r}{1 - xy(1+B)}.$$

When $r = 1$, the above theorem gives
$$B_{1,1}(x,y,z) = \frac{1 - xz - x - \sqrt{(1-x-xz)^2 - 4x^2 yz}}{2x^2 yz}.$$

Using the fact that $\frac{1-\sqrt{1-4x}}{2x} = \sum_{n\geq 0} \text{Cat}_n x^n$, we have
$$B_{1,1}(x,y,z) = \sum_{n\geq 0}\sum_{j=0}^{n}\sum_{i=j}^{n-j} \text{Cat}_j \binom{n}{i+j}\binom{i+j}{2j} x^n y^j z^i.$$

Thus, the noncrossing normally ordered form of $(a + a^\dagger)^n$ is given by
$$\mathcal{NC}\left((a+a^\dagger)^n\right) = \sum_{j=0}^{n}\sum_{i=j}^{n-j} \text{Cat}_j \binom{n}{i+j}\binom{i+j}{2j} (a^\dagger)^{n-i-j} a^{i-j}. \qquad (10.21)$$

Another application of Lemmas 10.60 and 10.61 with $r=1$ and $s \geq 1$ is that the generating function $B_{1,s}(x,y,z)$ satisfies the equation
$$B_{1,s}(x,y,z) = 1 + xzB_{1,s}(x,y,z) + xB_{1,s}(x,y,z)(1 + xyzB_{1,s}(x,y,z))^s. \qquad (10.22)$$

By Lagrange inversion formula (see Chapter 2) on the above equation, we obtain that
$$B_{1,s}(x,y,z) = \sum_{n\geq 0}\sum_{j=0}^{n}\sum_{i=j}^{n} \frac{1}{n+1}\binom{n+1}{j+1}\binom{n-j}{n-i}\binom{s(n-i)}{j} x^n y^j z^i,$$

which leads to the following result.

**Theorem 10.63** *The normally ordered form of* $(a + (a^\dagger)^s)^n$ *is given by*

$$\mathcal{NC}\left((a + (a^\dagger)^s)^n\right)$$
$$= \sum_{j=0}^{n} \sum_{i=j}^{n} \frac{1}{n+1} \binom{n+1}{j+1} \binom{n-j}{n-i} \binom{s(n-i)}{j} (a^\dagger)^{s(n-i)-j} a^{i-j}.$$

We remark that Theorem 10.63 for $s = 1$ gives

$$\mathcal{NC}\left((a + a^\dagger)^n\right)$$
$$= \sum_{j=0}^{n} \sum_{i=j}^{n-j} \frac{1}{n+1} \binom{n+1}{j+1} \binom{n-j}{n-i} \binom{n-i}{j} (a^\dagger)^{n-i-j} a^{i-j}$$
$$= \sum_{j=0}^{n} \sum_{i=j}^{n-j} \frac{n!}{j!(j+1)!(i-j)!(n-i-j)!} (a^\dagger)^{n-i-j} a^{i-j}$$
$$= \sum_{j=0}^{n} \sum_{i=j}^{n-j} \frac{(2j)!}{j!(j+1)!} \frac{n!}{(i+j)!(n-i-j)!} \frac{(i+j)!}{(2j)!(i-j)!} (a^\dagger)^{n-i-j} a^{i-j}$$
$$= \sum_{j=0}^{n} \sum_{i=j}^{n-j} \mathrm{Cat}_j \binom{n}{i+j} \binom{i+j}{2j} (a^\dagger)^{n-i-j} a^{i-j},$$

as described in (10.21).

### 10.5.4  $k$-Ary Trees and Lattice Paths

In this section, we present a bijection between our linear representations $\mathcal{V}_{rs}(n)$ and $\mathcal{W}_{rs}(n)$ with particular cases of $r$ and $s$, and different combinatorial structures such as $k$-ary trees, lattice paths, and Dyck paths (see below).

Our first aim is to show that the number of noncrossing contractions of $(a^r a^\dagger)^n$ is counted by the generalized Catalan numbers $\mathrm{Cat}_{n,k} = \frac{1}{(k-1)n+1}\binom{kn}{n}$. To do that, we consider edges whose first point is labeled by 1. We denote the edge $(j, 1)$ by $E_\pi$, and the edge with the first point $i$ by $E_i$. Clearly, $E_\pi = E_1$. Then we need to make use of the ordered set $F_\pi$ of all white vertices $j - 1, j - 2, \ldots, i + 1$, where $i$ is maximal, $j > i$ and $i$ is a black vertex.

An edge $(j, i)$ is said to *cover* the edge $(j', i')$ if and only if $j > j' > i' > i$. We have the following lemma on the structure of the noncrossing contractions of $(a^r a^\dagger)^n$, which follows immediately from the definitions.

**Lemma 10.64** *Let $\pi$ be any noncrossing contraction in the set $\mathcal{V}_{r1}(n)$. Then* the canonical subsequential form of $\pi$ is

$$\pi^{(m+1)}(a_m + s_{m+1} - 1)' \cdots (a_m + s_m)' a_1 (a_m + s_m - 1)' \cdots (a_m + s_{m-1})'$$
$$a_{m-1} \cdots (a_m + s_2 - 1)' \cdots (a_m + s_1)' a_m (a_m + s_1 - 1)' \cdots (a_m + s_0)'$$
$$\pi^{(m)} a_m \pi^{(m-1)} (a_{m-1} + r)' \cdots (a_{m-1} + 1)' a_{m-1} \cdots \pi^{(1)} (a_1 + r)'$$
$$\cdots (a_1 + 1)' a_1,$$

where $a_1 = 1$, $a_{i+1} - a_i > r$, $s_{i+1} \geq s_i \geq 0$, $\pi^{(i)} \in \mathcal{V}_{r1}(a_{i+1} - a_i - r - 1)$, $i = 1, 2, \ldots, m-1$, $\pi^{(m)}$ is either empty ($s_0 = 1$) or

$$\pi^{(m)} = (a_m + s_0 - 1)\theta(a_m + r)' \ldots (a_m + 1)'$$

with $\theta \in \mathcal{V}_{r1}(s_0 - 2 - r)$ and $\pi^{(m+1)} \in \mathcal{V}_{r1}(n - a_{m+1} - s_0 + 1)$. In other words, there exist $m$ edges, say $E_{i_1} = E_\pi, E_{i_2}, \ldots, E_{i_m}$, such that the linear representation of $\pi$ is either

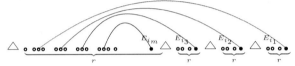

or

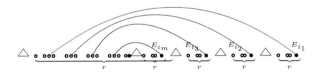

where each edge $E_{i_j}$ covers the edge $E_{i_{j+1}}$, such that the end points of the edges $E_{i_1}, \ldots, E_{i_m}$ are white vertices, $v_1, \ldots, v_m$, and there is no black vertex between $v_i$ and $v_j$, for any $i$ and $j$.

**Definition 10.65** *A $k$-ary tree is a directed tree in which each vertex has degree 0 or $k$ (for instance, see [333]). Let $\mathcal{T}_{r,n}$ be the set of $r$-ary tree with $n$ nodes. We denote by $T^1, \ldots, T^r$ the children of its root (from right to left).*

The number of $k$-ary trees with $n$ vertices is counted by the $k$-ary numbers, given by $\frac{1}{kn+1}\binom{kn+1}{n}$, for any positive integers $k$ and $n$; see [136]. We are ready now to define a bijection $\Phi$ recursively. First, the empty contraction maps to the empty $(r+1)$-ary tree, which gives the bijection $\Phi \colon \mathcal{V}_{r1}(0) \mapsto \mathcal{T}_{r+1,0}$. Define $F'_\pi = \{k_1, k_2, \ldots, k_m\}$ with $k_1 > k_2 > \ldots > k_m$ of $F_\pi$, such that the node $k_j$ is the end point of the edge $E_{i_{m+1-j}}$. Define the minimal vertex of $F_\pi$ by $k_0$. Assume we have defined the bijection $\Phi \colon \mathcal{V}_{r1}(m) \mapsto \mathcal{T}_{r+1,m}$ for all $m < n$. For $\pi \in \mathcal{V}_{r1}(n)$, according to the factorizations of the contraction $\pi$ as given by Lemma 10.64, there are two cases, for all $m \geq 1$:

If $\pi^{(m)} = \emptyset$, we define the $(k_m - k_0 + 1)$-th child of $T$ to be $T^{k_1 - k_0 + 1} = \Phi(\pi^{(m+1)})$, and the $(k_i - k_0 + 1)$-th child of $T$ to be $T^{k_i - k_0 + 1} = \Phi(\pi^{(i-1)})$, for each $i = 2, 3, \ldots, m$.

If $\pi^{(m)} \neq \emptyset$, we define the $(r+1)$-th child of $T$ to be $T^{r+1} = \Phi(\pi^{(m+1)})$, and the $(k_i - k_0 + 1)$-st child of $T$ to be $T^{k_i - k_0 + 1} = \Phi(\pi^{(i)})$, for each $i = 1, 2, 3, \ldots, m$.

When $m = 0$, we define our tree $T$ to be a $(r+1)$-ary tree with the root having only one child $T^{r+1} = \Phi(\pi^{(1)})$. Note that if there is no child $T^{r+1}$ of the root of the $(r+1)$-ary tree $T$, then this tree corresponds to a noncrossing contraction with a factorization as given by Lemma 10.64. Otherwise, the tree corresponds to a noncrossing contraction with a factorization as given by Lemma 10.64. By induction on the length of the noncrossing contractions and the unique construction of the map $\Phi$, we obtain that $\Phi$ is invertible. Therefore, we can state the following result.

**Theorem 10.66** *There is a bijection between the set of noncrossing contractions in $\mathcal{V}_{r1}(n)$ and the set $\mathcal{T}_{r+1,n}$ of $(r+1)$-ary trees.*

Let $G_r(x, y)$ be the generating function for the number of noncrossing contractions in $\mathcal{V}_{r1}(n)$ with exactly $m$ arcs, that is,

$$G_r(x, y) = \sum_{n \geq 0} \sum_{\pi \in \mathcal{V}_{r1}(n)} x^n y^{\#arcs\ in\ pi}.$$

Then the bijection $\phi$ gives that $G_r(x, y) = (1 + xG_r(x, y))(1 + xyG_r(x, y))^r$. By the Lagrange inversion formula (see Theorem 2.66), we derive

$$G_r(x, y) = \sum_{n \geq 1} \frac{x^{n-1}}{n} \sum_{i=0}^{n-1} \binom{n}{i} \binom{rn}{n-1-i} y^{n-1-i}.$$

As a consequence, we obtain the following result.

**Theorem 10.67** *The noncrossing normally ordered from of $(a^r a^\dagger)^n$ is given by*

$$\mathcal{NC}\left((a^r a^\dagger)^n\right) = \sum_{j=0}^{n} \frac{1}{n+1} \binom{n+1}{j+1} \binom{rn+r}{j} (a^\dagger)^{n-j} a^{rn-j}.$$

By using similar arguments as in the construction of the bijection $\Phi$, one obtains a bijection between the set of the noncrossing contractions of $(a(a^\dagger)^r)^n$ and the set of $(r+1)$-ary trees with $n$ nodes.

**Theorem 10.68** *The noncrossing normally ordered from of $(a(a^\dagger)^r)^n$ is given by*

$$\mathcal{NC}\left((a(a^\dagger)^r)^n\right) = \sum_{j=0}^{n} \frac{1}{n+1} \binom{n+1}{j+1} \binom{rn+r}{j} (a^\dagger)^{rn-j} a^{n-j}.$$

Now, we give a bijection $\Phi$ between the set of contractions $\mathcal{W}_{12}(n)$ of the word monomials in the expression $F(a, a^\dagger) = (a + (a^\dagger)^2)^n$ and a special set of lattice paths. These are lattice paths can be defined as follows.

**Definition 10.69** *An L-lattice path of size $n$ is a lattice path on $\mathbb{Z}^2$ from $(0,0)$ to $(3n,0)$ such that the path never goes below the $x$-axis and each of its steps is either $H = (2,1)$, $D = (1,-1)$ or $L = (1,2)$, and with no three consecutive $D$ steps (that is, there is no triple $DDD$). We denote the set of L-lattice paths of size $n$ by $\mathcal{L}_n$.*

The paths of size six are described in Figure 10.4.

**FIGURE 10.4**: The set $\mathcal{L}_2$ of L-lattices paths

Let $P \in \mathcal{L}_n$; using a first return decomposition (first return to $x$-axis), we obtain the factorization of $P$ as either

$$HDP', \ LDDP', \ LDP'HDDP'', \ HP'HDDP''D,$$
$$\text{or } LDP'LDP''HDDHDDP''',$$

where $P', P'', P'''$ are paths of smaller sizes. On the basis of this claim, the bijection $\Phi$ can be defined recursively. First, the empty contraction maps to the empty path, which gives the bijection $\Phi \colon \mathcal{W}_{12}(0) \mapsto \mathcal{L}_0$. Suppose we have defined the bijection $\Phi \colon \mathcal{W}_{12}(m) \mapsto \mathcal{L}_m$ for all $m < n$. For $\pi \in \mathcal{W}_{12}(n)$, according to the factorizations of the contraction $\pi$, there are five cases:

*(i)* The contraction $\pi$ starts with a white vertex, namely, $\pi = \pi'1' \in \mathcal{W}_{12}(n)$. We define $\Phi(\pi)$ to be the joint of the steps $HD$ and the path $P' = \Phi(\beta)$, where $\beta_i = \pi'_i - 1$ (define $i' - d = (i - d)'$), for each $i = 1, 2, \ldots, n-1$.

*(ii)* The contraction $\pi$ starts with a black vertex, namely $\pi = \pi'1 \in \mathcal{W}_{12}(n)$. We define $\Phi(\pi)$ to be the joint of the steps $LDD$ and the path $P' = \Phi(\beta)$, where $\beta_i = \pi'_i - 1$, for each $i = 1, 2, \ldots, n-1$.

*(iii)* The contraction $\pi$ starts with an arc followed by a black vertex with degree zero, that is, $\pi = \pi''1\pi'21 \in \mathcal{W}_{12}(n)$, where the letter 2 does not occur in $\pi'$. We define $\Phi(\pi)$ to be the joint of the step $LD$, the path $P' = \Phi(\beta')$, the steps $HDD$, and the path $P'' = \Phi(\beta'')$, where, for each $i$, $\beta'_i = \pi'_i - 2$ and $\beta''_i = \pi''_i - \max(2, \ell)$ such that $\ell$ is the maximal letter of $\pi'$.

*(iv)* The contraction $\pi$ starts with a black vertex of degree zero followed by an arc, that is, $\pi = \pi''2\pi'21 \in \mathcal{W}_{12}(n)$, where the letter 1 does not occur in $\pi''$. We define $\Phi(\pi)$ to be the joint of the step $H$, the path $P' = \Phi(\beta')$,

the steps $HDD$, and the path $P'' = \Phi(\beta'')$, where, for each $i$, $\beta_i' = \pi_i' - 2$ and $\beta_i'' = \pi_i'' - \max(2, \ell)$ such that $\ell$ is the maximal letter of $\pi'$.

**(v)** The contraction $\pi$ starts with two arcs, that is $\pi = \pi'''2\pi''1\pi'21 \in \mathcal{W}_{12}(n)$. We define $\Phi(\pi)$ to be the joint of the steps $LD$, the path $P' = \Phi(\beta')$, the steps $LD$, the path $P'' = \Phi(\beta'')$, the steps $HDDHDD$, and the path $P''' = \Phi(\beta''')$; for each $i$, $\beta_i' = \pi_i' - 2$, $\beta_i'' = \pi_i'' - \max(2, \ell')$ and $\beta_i''' = \pi_i''' - \max(2, \ell', \ell'')$, such that $\ell'$ and $\ell''$ are the maximal letters of $\pi'$ and $\pi''$, respectively.

Obviously, the inverse map of $\Phi$ is defined from the above cases. Thus, we obtain the following result.

**Theorem 10.70** *The map $\Phi$ is a bijection between the set of contractions in $\mathcal{W}_{12}(n)$ and the set of lattices paths in $\mathcal{L}_n$. Moreover, for any contraction $\pi \in \mathcal{W}_{12}(n)$, we have:*

**(i)** *the number of arcs in the linear representation of $\pi$ equals the number of $HDD$ in the corresponding path $\Phi(\pi)$.*

**(ii)** *the number of white vertices in the linear representation of $\pi$ equals the number of $HD$ in the corresponding path $\Phi(\pi)$.*

Let $P(x, y) = \sum_{n \geq 0} \sum_{P \in \mathcal{L}_n} x^n y^{HDD(P)}$ be the generating function for the number of paths $P$ of size $n$ in $\mathcal{L}_n$ according to the number of occurrences of the string $HDD$ in $P$. Then by the factorization of the paths in $\mathcal{P}$, we obtain that the generating function $P(x, y)$ satisfies (find the details!!)

$$P(x, y) = 1 + 2xP(x, y) + 2x^2 y P^2(x, y) + x^3 y^2 P^3(x, y),$$

which implies that $P(x, y) = B_{1,s}(x, y, 1)$. Thus, we can state the following result.

**Corollary 10.71** *The noncrossing normally ordered form $\mathcal{NC}\left((a + (a^\dagger)^2)^n\right)$ of $(a + (a^\dagger)^2)^n$ is given by*

$$\sum_{j=0}^{n} \sum_{i=j}^{n} \frac{1}{n+1} \binom{n+1}{j+1} \binom{n-j}{n-i} \binom{2n-2i}{j} (a^\dagger)^{2n-2i-j} a^{i-j}.$$

---

## 10.6 Exercises

**Exercise 10.1** *Write the contractions of the operator $aa^\dagger a^\dagger$, and draw the linear representation of each contraction.*

**Exercise 10.2** Let $\pi = aaa^\dagger a^\dagger aaa^\dagger$ be a word on boson operators associated with the Feynman diagram $\gamma = \{(1,4),(2,7)\}$. Find the degree of $\gamma$. Draw its linear representation. Find the crossing number of $\gamma$, the degenerate crossing number, and the total crossing number of $\gamma$. Find the length and the q-weight of $\gamma$.

**Exercise 10.3** Prove $\mathcal{N}\left[a(a^\dagger)^j\right] = (a^\dagger)^j a + j(a^\dagger)^{j-1}$, for all $j \geq 0$.

**Exercise 10.4** Following Example 10.32, find an explicit expression for
$$\mathcal{N}_q(aa^\dagger a^\dagger aa^\dagger a^\dagger).$$

**Exercise 10.5** Find $\mathcal{N}_p(aa^\dagger a^\dagger)$.

**Solution 10.1** By drawing the contraction of the expression $aa^\dagger a^\dagger$, we obtain
$$\mathcal{N}_p(aa^\dagger a^\dagger) = (a^\dagger)^2 a + (1+p)a^\dagger.$$

**Exercise 10.6** A 2-Motzkin path of length $n$ is a path on the plane from the origin $(0,0)$ to $(n,0)$ consisting of $U$ up steps, $D$ down steps, $L$ level steps

**FIGURE 10.5**: 2-Motzkin paths of length 2

colored black, and $L'$ level steps colored gray, such that the path does not go below the x-axis; see Figure 10.5.

Find a bijection between the set of linear representations $\mathcal{V}_{11}(n)$ and the set $\mathcal{M}_n$ of 2-Motzkin paths of length $n$.

As a consequence, show
$$\mathcal{NC}\left((aa^\dagger)^n\right) = \sum_{j=0}^n \frac{1}{n+1}\binom{n+1}{j+1}\binom{n+1}{j}(a^\dagger)^{n-j}a^{n-j}.$$

**Exercise 10.7** At first find a bijection between the set of linear representations $\mathcal{W}_{11}(n)$ and the set $\mathcal{M}_n$ of 2-Motzkin paths of length $n$. Then show
$$\mathcal{NC}\left((a+a^\dagger)^n\right) = \sum_{j=0}^n \sum_{i=j}^{n-j} c_j \binom{n}{i+j}\binom{i+j}{2j}(a^\dagger)^{n-i-j}a^{i-j},$$

as described in Theorem 10.63.

## 10.7 Research Directions and Open Problems

We now suggest two research directions, which are motivated both by the results and exercises of this chapter.

**Research Direction 10.1** *In Blasiak's thesis is described the connection between Stirling numbers of the second kind and the normal ordering form of $(aa^\dagger)^n$. Indeed, in this thesis are also described other connections between normal ordering and other types of numbers. Here we refer the reader to Section 5.4.3 in [38]. It will be interesting to extend this connection to q-normal ordering, p-normal ordering, and noncrossing normal ordering. For example, he found the normal ordering of $(a^\dagger a + a^\dagger)^n$ and $(a^\dagger e^{-a})^n$. Thus, find the q-normal ordering, p-normal ordering, and noncrossing normal ordering for the expressions $(a^\dagger a + a^\dagger)^n$ and $(a^\dagger e^{-a})^n$. Then describe the combinatorics that state on the set of the contractions of these expressions.*

**Research Direction 10.2** *The first and foremost problem consists in deriving a more physical understanding of the process of noncrossing normal ordering. As discussed in Section 10.5.1, the result of noncrossing normal ordering cannot be reproduced by the conventional normal ordering where some kind of commutation relation is assumed. Thus, the statistics resulting from the noncrossing normal ordering is a new and nontrivial phenomenon which clearly deserves closer study. In this context, it is not even clear whether one should call the operators for which the noncrossing normal ordering is applied "bosonic" anymore. Moreover, we can state two mathematical problems:*

*We say that the two edges $e = (a, b)$ and $e' = (c, d)$ are nesting if $a < c < d < b$ (that is, the edge e covers the edge $e'$) or $c < a < b < d$ (that is, the edge $e'$ covers the edge e); see [176]. Study the nonnesting normally ordered form of a given expression $F(a, a^\dagger)$.*

*Study the distribution of a given statistic on the set of normally ordered form of a given expression. For example, study the asymptotic behavior of the number of edges that cover other edges in the normally ordered form of $(a(a^\dagger)^r)^n$, when n tends to infinity.*

# Appendix A

## Solutions and Hints

## Chapter 1

**Exercise 1.1** The block representations of the set partitions of [4] are given by 1/2/3/4, 1/2/34, 1/23/4, 1/24/3, 12/3/4, 13/2/4, 14/2/3, 1/234, 12/34, 13/24, 14/23, 123/4, 124/3, 134/2, and 1234.

**Exercise 1.2** The block representations of the set partitions of [4] are given by 1111, 1112, 1121, 1122, 1123, 1211, 1212, 1213, 1221, 1222, 1223, 1231, 1232, 1233, and 1234.

**Exercise 1.3** The block representations of the set partitions of $\mathcal{P}(\{1,3,4\})$ are given by 1/3/4, 1/34, 13/4, 14/3, and 134.

**Exercise 1.4** Since 5 is an odd number, the only way a set partition of [5] can have an odd number of elements in each block is when there is an odd number of blocks, namely, one, three, or five blocks. The only set partition of [5] into one block is 12345, and the only set partition of [5] into five blocks is 1/2/3/4/5. Any set partition into three blocks must consist of two blocks with exactly one element and a third block with three elements. These are 1/2/345, 1/234/5, 1/235/4, 1/245/3, 123/4/5, 124/3/5, 125/3/4, 134/2/5, 135/2/4, and 145/2/3.

**Exercise 1.5** Applying Theorem 1.10 for $k = n_1 + n_2 + \cdots + n_m$ such that the $\ell$-th block contains $k_j$ elements, where $\ell = n_{j-1}+1, n_{j-1}+2, \ldots, n_{j-1}+n_j$, $j = 1, 2, \ldots, m$ and $n_0 = 0$, we obtain that the number of set partitions of $[n]$ with $n_j$ blocks of size $k_j$, $k_1 n_1 + \cdots + k_m n_m = n$, is given by

$$\prod_{j=1}^{m} \prod_{i=0}^{n_j-1} \binom{n - n_1 k_1 - \cdots - n_{j-1} k_{j-1} - i k_j - 1}{k_j - 1}.$$

**Exercise 1.6** By the definitions, the number of set partitions of $[n]$ such that the first block has one element, namely, the first block is $\{1\}$, is given by $\mathrm{Bell}_{n-1}$. Also, the number of set partitions of $[n]$ such that the first block has two elements, namely, the first block is $\{1, j\}$ with $2 \leq j \leq n$, is given by $(n-1)\mathrm{Bell}_{n-2}$. Thus, the required formula is given by

$$\mathrm{Bell}_{n-1} + (n-1)\mathrm{Bell}_{n-2}.$$

**Exercise 1.7**

a) He missed the pattern corresponding to 14/235, which is the top-to-bottom reflection of the second pattern.

b) The second pattern from the left in the top row is not desirable, as it decomposes.

c) We list the complete sets of partitions of the three types and underline the ones that were included in Puttenham's list. There are 10 patterns that consist of three blocks of size two:

$$\underline{13/25/46} \quad \underline{13/36/45} \quad 14/25/36 \quad 14/26/35 \quad \underline{15/23/46}$$

$$15/24/36 \quad \underline{15/26/34} \quad \underline{16/23/45} \quad 16/24/35 \quad 16/25/34$$

Thirteen pattern, consist of one block of size two and one block of size four; we just list the elements in the block of size two:

$$13 \quad 14 \quad 15 \quad 16 \quad 23 \quad 24 \quad 25 \quad \underline{26} \quad \underline{34} \quad 35 \quad 36 \quad 45 \quad 46$$

Finally, there are nine schemes with two blocks of size three; we list the first of the two blocks:

$$\underline{124} \quad 125 \quad 126 \quad 134 \quad 135 \quad 136 \quad 145 \quad 146 \quad 156$$

Overall, Puttenham only listed 9 of the 32 desirable partitions of [6].

**Exercise 1.8** The canonical form of $\pi$ is 112344223, and $\pi$ has four rises, one descent, and three levels.

**Exercise 1.9** The set of set partitions of $[n]$ with exactly two blocks contains all the set partitions of the form $11\cdots 1\pi_j\pi_{j+1}\cdots\pi_n$ such that $\pi_j = 2$, $j \geq 2$, and $\pi_m \in \{1, 2\}$ for all $m = j+1, j+2, \ldots, n$. For fixing $j \geq 2$, we have $2^{n-j}$ possibilities; thus, by adding over all values of $j$, we obtain that $\text{Stir}(n, 2) = \sum_{j=2}^n 2^{n-j} = 1 + 2 + \cdots + 2^{n-2} = 2^{n-1} - 1$, as claimed.

**Exercise 1.10** By simple induction and Theorem 1.12, we obtain

$$(\text{Stir}(n, k))^2 \geq \left(1 + \frac{3}{k}\right)\text{Stir}(n, k-1)\text{Stir}(n, k+1),$$

for all $1 \leq k \leq n$. It follows that the ration $\frac{\text{Stir}(n,k+1)}{\text{Stir}(n,k)}$ is strictly decreasing.

**Exercise 1.11** In $\mathcal{P}_5$ there are 52 set partitions. By direct checking we have 15 set partitions in $\mathcal{P}_5$ that avoid the subword pattern 11, namely, 12121, 12123, 12131, 12132, 12134, 12312, 12313, 12314, 12321, 12323, 12324, 12341, 12342, 12343, and 12345.

**Exercise 1.12** Avoiding the subsequence 11 means each block contains exactly one element; thus we have only one set partition of $[n]$ that avoids the subsequence pattern 11, namely, $11 \cdots 1 \in \mathcal{P}_n$.

**Exercise 1.13** There are several ways to write a Gray code for $\mathcal{P}_4$ with distance one. For instance, 1111, 1112, 1122, 1123, 1121, 1221, 1223, 1222, 1232, 1234, 1233, 1231, 1211, 1213, and 1212.

**Exercise 1.14** For instance, 1122, 1123, 1223, 1221, 1231, 1234, 1233, 1232, 1212, and 1213.

**Exercise 1.15** For example, we can use the following Maple code:
$with(combinat, setpartition)$;
$S := \{1, 2, 3, 4, 5\}$;
$setpartition(S, 3)$;

# Chapter 2

### Exercise 2.1

(1) The square of $n$ is just $n^2$; thus $a_n = n^2$.

(2) The sum $1 + 2 + \cdots + n = \frac{n(n+1)}{2}$, which gives that $a_n = \frac{n(n+1)}{2} = \binom{n}{2}$.

(3) To choose a subset of a set of $n$ elements, say $a_1, a_2, \ldots, a_n$, we do the following: for each element either we choose it to be one of the elements of the subset or not. That is, there are $2^n$ possibilities to choose a subset of a set of $n$ elements. Hence, $a_n = 2^n$.

(4) To choose a function $f$ from the set $\{1, 2, \ldots, n\}$ to the set $\{0, 1, 2\}$, we need to set $f(i)$ to be either 0, 1, or 2. Hence, we have $3^n$ such functions, which implies that $a_n = 3^n$.

**Exercise 2.2** (i) This property can be verified either by using the formula for $\binom{n}{k}$ or by giving a combinatorial argument. Recall that $\binom{n}{k}$ counts the number of ways to select a subset of $k$ elements from a set of $n$ elements, say, the set $[n]$. Now, any subset either does or does not contain the element $n$. If the subset contains the element $n$, then the remaining $k-1$ elements are chosen from the set $[n-1]$ in $\binom{n-1}{k-1}$ ways. If the subset does not contain the element $n$, then all $k$ elements are chosen from the set $[n-1]$, which can be done in $\binom{n-1}{k}$ ways. Since these are the only possible cases, and the two cases are disjoint, we can add the two counts to obtain the result of (i). The entries of the enclosing diagonals are of the form $\binom{n}{0}$ or $\binom{n}{n}$, and there is exactly one way to either select no or all elements from a set of $n$ objects.

(ii) The right-hand side of this formula counts the number of subsets that can be created from the set $[n]$. To obtain the left-hand side of (ii), we note that the number of $k$-element combinations (see Example 2.19) is given by $\binom{n}{k}$, which counts the number of ways to choose a set of $k$ elements from the set $[n]$. Thus, summing over all possible values of $k$, we obtain that the number of subsets that can be created from the set $[n]$ is $\sum_{k=0}^{n} \binom{n}{k}$, which completes the proof.

(iii) We proceed to the proof by induction on $n \geq k$. The claim holds for $n = k$ since $\binom{k}{k} = \binom{k+1}{k+1} = 1$. Assume that the claim holds for $n$, and let us prove it for $n + 1$: by induction hypothesis and (i) we have that

$$\sum_{i=k}^{n+1} \binom{i}{k} = \binom{n+1}{k} + \sum_{i=k}^{n} \binom{i}{k} = \binom{n+1}{k} + \binom{n+1}{k+1} = \binom{n+2}{k+1},$$

which completes the induction step.

**Exercise 2.3** Define $c_0 = 1$. Since $\xi$ is a root of $\Delta(x)$ with multiplicity $m$, then $\xi$ is also a root of $\Delta^{(s)}(x)$, and the $s$-th derivative of $\Delta(x)$ for all $s = 0, 1, 2, \ldots, m-1$. Substituting $a_n = n^i \xi^n$ into (2.2), we derive

$$\sum_{j=0}^{r} c_j (n - r + r - j)^i \xi^{n-j} = \sum_{k=0}^{i} \xi^{n-r} \binom{i}{k} (n-r)^{i-k} \sum_{j=0}^{r} (r-j)^k c_j \xi^{r-j}.$$

Let $F_k(x) = \sum_{j=0}^{r} (r-j)^k c_j x^{r-j}$. It is easy to see that $F_0(x) = \Delta(x)$ and $F_k(x) = x \cdot F'_{k-1}(x)$ for $k \geq 1$. Since $\xi$ is a root of $\Delta(x)$ with multiplicity $m$, by induction we have $F_0(\xi) = F_1(\xi) = \cdots = F_{m-1}(\xi) = 0$. Thus, $\sum_{j=0}^{r} c_j (n - r + r - j)^i \xi^{n-j} = 0$, that is, $n^i \xi^n$ is a solution of (2.2) for all $i = 0, 1, \ldots, m-1$.

**Exercise 2.4** As defined in Example 2.29, the sequence $L_n$ defined by the recurrence relation $L_n = L_{n-1} + L_{n-2}$, with initial conditions $L_0 = 2$ and $L_1 = 1$. This recurrence relation has the characteristic polynomial $\Delta(x) = x^2 - x - 1$, and hence, the same general solution is given by

$$L_n = k_1 \cdot \left(\frac{1+\sqrt{5}}{2}\right)^n + k_2 \cdot \left(\frac{1-\sqrt{5}}{2}\right)^n.$$

The initial conditions for the Fibonacci sequence give $2 = k_1 + k_2$ and $1 = k_1 \cdot \frac{1+\sqrt{5}}{2} + k_2 \cdot \frac{1-\sqrt{5}}{2}$. Using any method to solve a system of two equations, we obtain $k_1 = k_2 = 1$, and thus

$$L_n = \left(\frac{1+\sqrt{5}}{2}\right)^n + \left(\frac{1-\sqrt{5}}{2}\right)^n.$$

**Exercise 2.5** Let $a_n$ be any sequence satisfies (2.1). Assume that $p(n)$ is any solution of (2.1), that is,

$$p(n) + c_1 p(n-1) + \cdots + c_r p(n-r) = b_n \quad \text{for all } n \geq r. \tag{A.1}$$

If we replace the sequence $a_n$ by $p(n) + q(n)$, where $h(n)$ is any sequence, in (2.1), then we obtain that

$$p(n) + q(n) + c_1(p(n-1) + q(n-1)) + \cdots + c_r(p(n-r) + q(n-r)) = h_n,$$

for all $n \geq r$, which, by (A.1), implies that

$$q(n) + c_1 q(n-1) + \cdots + c_r q(n-r) = 0 \quad \text{for all } n \geq r.$$

Hence, if $q(n)$ is the general solution to the associated homogeneous recurrence relation (2.2), and $p(n)$ is any solution to (2.1), then the general solution to the nonhomogeneous linear recurrence relation as given in (2.1) has the form $p(n) + q(n)$, which completes the proof.

**Exercise 2.6** For the ordinary generating function we have

$$\sum_{n \geq 0} a_{n+k} x^n = \frac{1}{x^k} \sum_{n \geq 0} a_{n+k} x^{n+k} = \frac{1}{x^k} \sum_{n \geq k} a_n x^n$$

$$= \frac{A(x) - \sum_{j=0}^{k-1} a_j x^j}{x^k}$$

and

$$\sum_{n \geq k} a_{n-k} x^n = x^k \sum_{n \geq 0} a_n x^n = x^k A(x).$$

For the exponential generating function we have

$$\sum_{n=0}^{\infty} a_{n+k} \frac{x^n}{n!} = \sum_{n=k}^{\infty} a_n \frac{x^{n+k}}{(n+k)!}$$

$$= \underbrace{\int_0^x \int_0^{t_1} \cdots \int_0^{t_{k-1}}}_{k \text{ times}} A(t_{k-1}) dt_{k-1} \cdots dt_1 dx$$

and

$$\sum_{n=k}^{\infty} a_{n-k} \frac{x^n}{n!} = \sum_{n=0}^{\infty} a_n \frac{x^{n-k}}{(n-k)!} = \frac{d^k}{dx^k} A(x),$$

as claimed.

**Exercise 2.7** We have

$$\sum_{n \geq 0} \left( \sum_{j=0}^n a_j \right) x^n = \sum_{j \geq 0} a_j (x^j + x^{j+1} + \cdots) = \sum_{j \geq 0} a_j \frac{x^j}{1-x} = \frac{A(x)}{1-x},$$

as required.

**Exercise 2.8** By Theorem 2.55 and Example 2.34 we can state that the generating function for the sequence

$$\left\{\sum_{i=0}^{n}\binom{n}{i}\mathrm{Fib}_i\right\}_{n\geq 0}$$

is given by

$$\frac{x/(1-x)}{(1-x)(1-x/(1-x)-x^2/(1-x)^2)} = \frac{x}{(1-x)^2 - x(1-x) - x^2}$$
$$= \frac{x}{1-3x+x^2}.$$

Finding the coefficient of $x^n$ we obtain that

$$\sum_{i=0}^{n}\binom{n}{i}\mathrm{Fib}_i = \mathrm{Fib}_{2n}.$$

For a combinatorial proof see [28].

**Exercise 2.9** (i) Rewriting the recurrence relation in terms of $F(x)$ we obtain $F(x) - x = 2xF(x) + x^2F(x)$, which implies $F(x) = \frac{x}{1-2x-x^2}$.

(ii) By expanding the generating function $F(x)$ we obtain

$$F(x) = x\sum_{j\geq 0}(2x+x^2)^j = \sum_{j\geq 0}\sum_{i=0}^{j}\binom{j}{i}x^{j+i+1}2^{j-i}.$$

Comparing the coefficient of $x^n$ on both sides of the above equation yields

$$\mathrm{Pell}_n = \sum_{j\geq 0}\binom{j}{n-1-j}2^{2j-n+1} = \sum_{i\geq 0}\binom{n-1-i}{i}2^{n-1-2i}.$$

(iii) By (i) we have

$$F(x) = \frac{x}{1-2x-x^2} = \frac{\sqrt{2}}{4}\left(\frac{1}{1-(1+\sqrt{2})x} - \frac{1}{1-(1-\sqrt{2})x}\right)$$
$$= \frac{\sqrt{2}}{4}\sum_{n\geq 0}(1+\sqrt{2})^n x^n - (1-\sqrt{2})^n x^n.$$

Comparing the coefficient of $x^n$ on both sides of the above equation gives the required result.

**Exercise 2.10** We proceed to the proof by induction on $k$. Since

$$\frac{1}{1-x} = \sum_{j\geq 0}x^j,$$

then the claim holds for $k = 1$. We assume that the claim holds for $k$ and prove it for $k + 1$. The induction assumption is

$$\frac{1}{(1-x)^k} = \sum_{j \geq 0} \binom{j+k-1}{k-1} x^j,$$

which, by differentiating both sides, implies that

$$\frac{k}{(1-x)^{k+1}} = \sum_{j \geq 1} j \binom{j+k-1}{k-1} x^{j-1}.$$

Therefore,

$$\frac{1}{(1-x)^{k+1}} = \sum_{j \geq 0} \frac{j+1}{k} \binom{j+k}{k-1} x^j = \sum_{j \geq 0} \binom{j+k}{k} x^j,$$

which completes the induction.

**Exercise 2.11** Let $C$ be any drawing nonintersecting chords on a circle between $n$ points, where the points have been labeled by $1, 2, \ldots, n$. We read the points in the order $1, 2, \ldots, n$, and successively we generate the Motzkin path $M$ as follows. We read a point $j$;

if there is no point that $i$ connects to the point $j$ by a chord, we add a level step to the path $M$;

if there is a point that $i$ connects to the points $j$ by a chord such that $i > j$, then we add an up step to the path $M$;

otherwise, we add a down step to the path $M$.

For instance, the drawings of nonintersecting chords in Figure 2.1 from left to right are mapping to the following Motzkin paths: $HHHH$, $HHUD$, $HUDH$, $HUHD$, $UDHH$, $UDUD$, $UHDH$, $UHHD$, and $UUDD$. It is obvious, that we verify that our map is a bijection.

**Exercise 2.12** We denote the generating function for the number of Dyck paths in $A_n$ by $f(x)$, that is, $f(x) = \sum_{n \geq 0} |A_n| x^n$. Now let us write an equation for $f(x)$. Each nonempty Dyck path $P$ in $A_n$ can be decomposed as

$$P = UP^{(1)} UP^{(2)} \cdots UP^{(k+1)} UD^k P',$$

where $P^{(1)}, \ldots, P^{(k+1)}, P'$ are Dyck paths and $k \geq 2$. Thus, the generating function $f(x)$ satisfies

$$f(x) = 1 + x^2 f^2(x) + x^3 f^3(x) + \cdots = 1 + \frac{x^2 f^2(x)}{1 - xf(x)},$$

which implies that
$$f(x) = \frac{1+x-\sqrt{1-2x-3x^2}}{2x(1+x)}.$$

**Exercise 2.13** See [217, Theorem 2.1].

**Exercise 2.14** See [217, Equation 2.6].

**Exercise 2.15** See [217, Theorem 3.3].

**Exercise 2.16** See [217, Theorem 3.3].

**Exercise 2.17** See [88].

**Exercise 2.18** See [217] and [88].

**Exercise 2.19** Let $F(x)$ ($D(x)$) be the generating function for the number of symmetric Dyck paths (Dyck paths) of size $n$. By the definitions, any symmetric Dyck path $S$ can be written as $S = US'D$ or $S = UADS'UAD$, where $S'$ is a symmetric Dyck path and $A$ is a Dyck path. Rewriting these rules in terms of the generating function, we obtain $F(x) = 1 + xF(x) + x^2 D(x^2)F(x)$. By Example 2.61, we have $F(x) = \frac{1}{1-x-x^2C(x^2)}$, where $C(x) = \frac{1-\sqrt{1-4x}}{2x}$ is the generating function for the Catalan numbers.

**Exercise 2.20** Substitute $a_n = \text{Bell} + 2^n$ and use Theorem 1.9.

**Exercise 2.21** See [180].

**Exercise 2.22** See [209]. For more details we refer the reader to [321].

**Exercise 2.23** For example, see [209] or [321].

**Exercise 2.24** See [234].

**Exercise 2.25** See [222].

# Chapter 3
**Exercise 3.1** By induction on $n$ and (3.1), we obtain that $a_n = \text{Bell}_n$ for all $n \geq 0$.

**Exercise 3.2** By Theorem 1.16 we have $m^n = \sum_{k=1}^n \text{Stir}(n,k)(m)_k$. Multiplying by $\frac{z^m}{m!}$ and summing over all $m \geq 1$, we obtain
$$\sum_{m \geq 1} \frac{m^n z^m}{m!} = \left( \sum_{k=1}^n \text{Stir}(n,k) z^k \right) \left( \sum_{m \geq 1} \frac{z^j}{j!} \right),$$
which implies
$$\sum_{k=1}^n \text{Stir}(n,k) z^k = e^{-z} \sum_{m \geq 1} \frac{m^n z^m}{m!}.$$

Taking $z = 1$, we obtain Theorem 3.2.

**Exercise 3.3** See Anderegg [3].

**Exercise 3.4** By the fact that $\sum_{n\geq 0} \text{Bell}_n \frac{x^n}{n!} = e^{e^x - 1}$, we obtain

$$\sum_{n\geq 0} \text{Bell}_{n+1} \frac{x^n}{n!} = \frac{d}{dx}(e^{e^x - 1}) = \frac{1}{e} e^{e^x + x} = \frac{1}{e} \sum_{j\geq 0} \frac{(e^x + x)^j}{j!}$$

$$= \frac{1}{e} \sum_{j\geq 0} \sum_{i=0}^{j} \binom{j}{i} \frac{e^{ix} x^{j-i}}{j!}$$

$$= \frac{1}{e} \sum_{j\geq 0} \sum_{i=0}^{j} \sum_{k\geq 0} \binom{j}{i} \frac{i^k x^{j-i+k}}{j! k!}.$$

Comparing the coefficients of $x^n$ on both sides of the above equation, we obtain the requested identity.

**Exercise 3.5** See Constantine [83].

**Exercise 3.6** (i) Each set partition in $\text{Flatten}_{n,k}(123)$ satisfies $k = 1, 2$ and $n = 1, 2, 3$. Thus, by direct calculations we have only a set partition of $[1]$, two set partitions of $[2]$, and only a set partition of $[3]$ (namely, $13/2$), which completes this case.

(ii) A set partition $\pi$ in $\text{Flatten}_n(132)$ if and only if $\text{Flatten}(\pi) = 12\cdots n$ (Why?). If $f_n = |\text{Flatten}_n(132)|$, then by restricting the first block we obtain that the sequence $\{f_n\}_{n\geq 1}$ satisfies the following recurrence relation $f_n = f_{n-1} + f_{n-2} + \cdots + f_0$ with $f_1 = 1$, or equivalently, $f_n = 2f_{n-1}$ with $f_1 = 1$, which implies that $f_n = 2^{n-1}$ for all $n \geq 1$.

**Exercise 3.7** (i) Let $f_n(p, q)$ be the generating function for the number of set partitions of $[n]$ according to the number of singleton blocks and number of blocks, that is,

$$f_n(p, q) = \sum_{\pi \in \mathcal{P}_n} p^{\text{blo}(\pi)} q^{\text{number singleton blocks in } \pi}.$$

Using similar techniques as in the proof of Theorem 1.9 we obtain that

$$f_n(p, q) = p \sum_{j=1}^{n} \binom{n-1}{j} f_j(p, q) + pq f_{n-1}(p, q),$$

which is equivalent to

$$f_n(p, q) = p \sum_{j=0}^{n} \binom{n-1}{j} f_j(p, q) + p(q - 1) f_{n-1}(p, q).$$

Let $F(x; p, q) = \sum_{n\geq 0} f_n(p, q) x^n$. By rewriting this recurrence relation in terms of the exponential generating function by using Theorem 2.57 and Rule 2.49, we have

$$\frac{d}{dx}F(x;p,q) = pe^x F(x;p,q) + p(q-1)F(x;p,q),$$

which gives that $F(x;p,q) = e^{pe^x + p(q-1)x + pc}$, where $c$ is a constant. By the initial condition $F(0;p,q) = 1$ we obtain that $F(x;p,q) = e^{p(e^x - 1 - x + qx)}$.

(ii) Follows directly by differentiating the generating function $F(x;p,q)$ with respect to $q$ and then substituting $q = 1$.

(iii) Since the number of singleton blocks equal $\{1\}$ in set partitions of $[n]$ is equal to the number of set partitions of $[n-1]$, we see that the generating function for the number of singleton blocks in all set partitions of $[n]$, excluding any singleton blocks equal $\{1\}$, is

$$\left.\frac{d}{dq}\frac{d}{dx}F(x;p,q)\right|_{q=1} - pe^{p(e^x - 1)} = p^2 x e^x e^{p(e^x - 1)}.$$

**Exercise 3.8** Let $\pi = B_1/\cdots/B_k$ be any nonempty set partition in $\mathcal{P}_{n,k}(1/23)$. Since $\pi$ avoids $1/23$ and $1 \in B_1$, we have that $|B_i| = 1$ for all $i = 2, 3, \ldots, k$. Thus, if $k < n$, then $\pi$ has the form $12\cdots(n-k)j/n+1-k/\cdots/j-1/j+1/\cdots/n$ with $j = n+1-k, j+2-k, \ldots, n$; otherwise, clearly, we have only one set partition, namely, $1/2/\cdots/n$. Therefore, the number of set partitions in $\mathcal{P}_n(1/23)$ with exactly $k$ blocks is given by $k$. Summing over all possible values of $k$, we obtain $|\mathcal{P}(1/23)| = 1 + \sum_{k=1}^{n-1} k = \binom{n}{2} + 1$.

**Exercise 3.9** See [120, Proposition 2.13 ].

**Exercise 3.10** See [120, Propositions 4.4 and 4.6]. For instance, let us give detailed proof of the case $1/23$. By Exercise 3.8, each set partition of $[n]$ with exactly $k$ blocks that avoids $1/23$ has the form $12\cdots(n-k)j/n+1-k/\cdots/j-1/j+1/\cdots/n$ with $j = n+1-k, j+2-k, \ldots, n$. Thus, the number of even set partition of $[n]$ with exactly $k$ blocks that avoids $1/23$ is given by $k$ when $n-k$ is an even number, and $0$ otherwise. Also, the set partition $12\cdots n$ is even if and only if $n$ is an odd. Thus

$$|E\mathcal{P}_n(1/23)| = \frac{1-(-1)^n}{2} + \sum_{\substack{k=1,\, n-k \text{ even}}}^{n-1} k$$

$$= \frac{1-(-1)^n}{2} + \sum_{k=1}^{(n-1)/2}(n-2k) = \frac{1-(-1)^n}{2},$$

which, by induction, completes the proof.

**Exercise 3.11** See [335, Exercise 6.19(t)].

**Exercise 3.12** In 1970, Kreweras [203] gave this significant and surprising result. After 18 years, a simple proof is given by Liaw, Yeh, Hwang, and Chang [208].

**Exercise 3.13** See [121, Theorem 2.2].

**Exercise 3.14** Let $\mathcal{IP}_{n,j}$ be the set of all set partitions of $[n]$ that present involutions and have exactly $j$ blocks of size two. First we choose a subset of $[n]$ of size $2j$ for $j$ blocks os size two; number of choices for such subsets is $\binom{n}{2j}$. On the other hand, the number of set partitions of $[2j]$ such that each block has two elements is given by $(2j-1)!!$. Therefore, $|\mathcal{IP}_{n,j}| = \binom{n}{2j}(2j-1)!!$, which implies that $\mathcal{IP}_n = \sum_{j=0}^{n/2} \binom{n}{2j}(2j-1)!!$.

**Exercise 3.15** Let $j$ be the number of the blocks of size 2 in our involution set partitions; then $2j + k - j = n$, so $j = n - k$. Now the formula can be derived directly from Theorem 3.54.

**Exercise 3.16** See [275].

**Exercise 3.17** See [47].

**Exercise 3.18** See [57]. For extra research work, see [56, 215].

**Exercise 3.19** Immediately, followed from the definitions of $\rho(\pi)$ in Theorem 3.56.

**Exercise 3.20** Since $\text{Bell}(x) = 1 + (e^x - 1) + \sum_{r=1}^{p-1} \frac{(e^x-1)^r}{r!} + \frac{x^p}{p!} + O(x^{p+1})$, we obtain that $\text{Bell}_p \equiv 1 + 0 + 1 \pmod{p} \equiv 2 \pmod{p}$, for any prime $p$ (see [339]).

**Exercise 3.20** Since $\binom{p-1}{j} \equiv (-1)^j \pmod{p}$, we obtain

$$\sum_{j=1}^{p-1}(-1)^k \text{Bell}_j \equiv \sum_{j=1}^{p-1}\binom{p-1}{j}\text{Bell}_j \pmod{p}.$$

By Theorem 1.9 we have

$$\sum_{j=1}^{p-1}(-1)^k \text{Bell}_j \equiv \text{Bell}_p - 1 \pmod{p}.$$

Hence, by Exercise 3.20 we complete the proof (see [339]).

**Exercise 3.21** See [126].

# Chapter 4

**Exercise 4.1** An infinite number of applications of this recurrence relation derives

$$g_\ell = \alpha_\ell + \beta_\ell g_{\ell+1} = \alpha_\ell + \alpha_{\ell+1}\beta_\ell + \beta_\ell\beta_{\ell+1}g_{\ell+2}$$
$$= \alpha_\ell + \alpha_{\ell+1}\beta_\ell + \alpha_{\ell+2}\beta_\ell\beta_{\ell+1} + \beta_\ell\beta_{\ell+1}\beta_{\ell+2}g_{\ell+3}$$
$$= \cdots = \sum_{i \geq \ell} \alpha_i \prod_{j=\ell}^{i-1} \beta_j.$$

**Exercise 4.2** We proceed to the proof by induction on $n$. Clearly, the claim holds for $n = m$. Let us assume that the claim holds for all $n - 1$; then by induction hypothesis we have

$$a_n(k, m)$$
$$= a_{n-1}(k - 1, m) + (k - 1)a_{n-1}(k, m) + a_{n-1}(k, m - 1)$$
$$= \binom{n-2}{m} \text{Stir}(n - 2 - m, k - 2) + (k - 1)\binom{n-2}{m}\text{Stir}(n - 2 - m, k - 1)$$
$$+ \binom{n-2}{m-1}\text{Stir}(n - 1 - m, k - 1)$$
$$= \binom{n-2}{m}\left(\text{Stir}(n - 2 - m, k - 2) + (k - 1)\text{Stir}(n - 2 - m, k - 1)\right)$$
$$+ \binom{n-2}{m-1}\text{Stir}(n - 1 - m, k - 1).$$

By (1.12) we have

$$a_n(k, m)$$
$$= \binom{n-2}{m}\text{Stir}(n - 1 - m, k - 1) + \binom{n-2}{m-1}\text{Stir}(n - 1 - m, k - 1)$$
$$= \left(\binom{n-2}{m} + \binom{n-2}{m-1}\right)\text{Stir}(n - 1 - m, k - 1),$$

and by Fact 2.20(i) we obtain

$$a_n(k, m) = \binom{n-1}{m}\text{Stir}(n - 1 - m, k - 1),$$

which completes the induction.

**Exercise 4.3** Theorem 4.25 for $q = 0$ gives

$$W_k(x, 0) = \frac{-x\left(U_{k-1}(t) - U_{k-2}(t)\right)}{U_k(t) - U_{k-1}(t) - (1 + x)\left(U_{k-1}(t) - U_{k-2}(t)\right)},$$

which, by (D.1), completes the proof.

**Exercise 4.4** In order to give a direct proof for Corollary 4.26, it is enough to show for each $i$, $1 \leq i \leq n - 2$, that there are a total of $k^{n-3}\left(2\binom{k}{3} + \binom{k}{2}\right)$ peaks at $i$ within all the words $\pi_1 \cdots \pi_n \in [k]^n$. We consider either $\pi_i \neq \pi_{i+2}$ or $\pi_i = \pi_{i+2}$. In the former case, there are $2\binom{k}{3}k^{n-3}$ such peaks at $i$ for which $\pi_i \neq \pi_{i+2}$ within all the words of $[k]^n$, and in the later case, there are $\binom{k}{2}k^{n-3}$ peaks at $i$ for which $\pi_i = \pi_{i+2}$.

**Exercise 4.5** See Theorem 2.1 in [60].

**Exercise 4.6** Similar arguments as in the proof of Lemma 4.19 show that there are a total of $\sum_{j=2}^{k} j\binom{j}{2} f_{n,j}$ nontrivial descents at $r+1$ in which $r$ is not minimal (minimum of a block) in all the set partitions of $\mathcal{P}_{n,k}$. As in the solution of Exercise 4.4, we obtain that there are $\sum_{j=2}^{k} \binom{j+1}{3} f_{n,j}$ nontrivial descents at $r+1$, in which $r$ either goes in the same block as $r+1$ or goes in a block to the right of $r+1$, where $r$ itself is not minimal. Thus, the difference between the two sums counts all peaks at $r$ where $r+1$ is not minimal.

Now, in order to complete the proof, we will prove that the total number of peaks at $r$, where $r+1$ is minimal, is given by $\binom{k}{2}\text{Stir}(n-1,k)$. To see this, we first observe that peaks at $r$ where $r+1$ is minimal are synonymous with descents at $r+1$ where $r+1$ is minimal. To show that $\binom{k}{2}\text{Stir}(n-1,k)$ counts all descents at $i$, where $i$ is the smallest member of its block in some set partition of $\mathcal{P}_{n,k}$, first choose two numbers $a < b$ in $[k]$. Given $\pi \in \mathcal{P}_{n-1,k}$, let $j$ denote the smallest member of block that contains $b$. Increase all members of $[j+1, n-1]$ in $\pi$ by one (leaving them within their blocks) and then add $j+1$ to the block that contains $a$. This procedure produces a descent between the first element of block that contains $b$ and an element of block that contains $a$ within some set partition of $\mathcal{P}_{n,k}$.

**Exercise 4.7** See Theorem 2.2 in [59].

**Exercise 4.8** Let $C_a = C_a(x;q)$ be the generating function for the number of words $\pi$ of size $n$ over the alphabet $[a]$ according to the number of occurrences of the pattern $\tau$ in $a\pi$. Since each word $\pi$ over the alphabet $[a]$ either does not contain the letter $a$ or may be written as $\pi'a\pi''$, where $\pi'$ is a word over the alphabet $[a-1]$ and $\pi''$ is a word over the alphabet $[a]$, we have $C_a = C'_a + xC'_a C_a$, where $C'_a$ is the generating function for the number of words $\pi$ of size $n$ over the alphabet $[a-1]$ according to the number of occurrences of the pattern $\tau$ in $a\pi$. From the proof of Theorem 4.63 and the reversal operation (map each word $\pi_1 \cdots \pi_n$ to $\pi_n \cdots \pi_1$), we have $C'_a = G_a$. Hence $C_a = \frac{G_a}{1-xG_a}$. From the fact that each set partition $\pi$ of $[n]$ with exactly $k$ blocks may be expressed uniquely as $\pi = 1\pi^{(1)}2\pi^{(2)}\cdots k\pi^{(k)}$ such that each $\pi^{(i)}$ is a word over the alphabet $[i]$, we have that the generating function $P_\tau(x;q;k)$ is given by $x^k \prod_{a=1}^{k} \frac{G_a}{1-xG_a}$, which completes the proof.

**Exercise 4.9** By (4.29) and (4.30), we have from Theorem 4.66 that

$$\frac{\frac{d}{dq}P_\tau(x;q;k)|_{y=1}}{P_\tau(x,1,k)} = \sum_{a=m+1}^{k} \frac{dA_a(1)}{A_a(1)} + \frac{xdA_a(1)}{1-xA_a(1)}$$

$$= \sum_{a=m+1}^{k} x^{\ell-1}\binom{a-1}{m} + \frac{x^\ell\binom{a-1}{m+1}}{1-(a-1)x} + \frac{x^\ell\binom{a-1}{m}}{1-ax} + \frac{x^{\ell+1}\binom{a-1}{m+1}}{(1-ax)(1-(a-1)x)},$$

which implies

$$\frac{\frac{d}{dq}P_\tau(x;q;k)|_{y=1}}{P_\tau(x,1,k)} = x^{\ell-1}\binom{k}{m+1} + \sum_{a=m+1}^{k} \frac{x^\ell\binom{a}{m+1}}{1-ax},$$

which leads to the result.

We complete the proof by noting that a combinatorial proof may be given for Corollary 4.37 similar to that given above for Corollary 4.26, and we leave it to the interested reader.

**Exercise 4.10** Since the subword patterns $\tau = 2^{\ell-1}1$ and $\tau = 1^{\ell-1}2$ are equivalent on words over alphabet $[k]$ (simply replace each letter $i$ with $k+1-i$), Theorem 2.2 in [59] gives the result.

**Exercise 4.11** The problem of counting $t$-succession is introduced by Munagi [258] in 2008, where (1) and (2) have been proved in [258].

# Chapter 5

**Exercise 5.1** See [350, Equations (22) and (23)].

**Exercise 5.2** See [319, Theorem 2.2].

**Exercise 5.3** See [190].

**Exercise 5.4** See [337, Lemma 2.1 and Theorem 2.2].

**Exercise 5.5** Theorem 1.12 together with the definitions give $a(n,k;1) = \mathrm{Stir}(n,k)$. Thus, if we differentiate the recurrence relation

$$a(n,k;q) = ka(n-1,k;q) + q^n a(n-1,k-1;q)$$

and set $q = 1$, then we obtain

$$b(n,k) = kb(n-1,k) + b(n-1,k-1) + n\mathrm{Stir}(n-1,k-1)$$

with initial condition $b(n,0) = 0$ for all $n \geq 0$. Let

$$c(n,k) = k\mathrm{Stir}(n+1,k) - (n+1)\mathrm{Stir}(n,k-1)$$

and let us prove that $b(n,q) = c(n,q)$. We proceed to the proof by induction on $n, k$. Clearly, $b(n,1) = c(n,q) = 1$. Assume for $n', k'$ where $n' + k' \leq n + k$, and let us prove our claim for $n, k$. By the definitions, induction hypothesis, and Theorem 1.12 we obtain that

$$\begin{aligned}b(n,k) &= kb(n-1,k) + b(n-1,k-1) + n\mathrm{Stir}(n-1,k-1) \\ &= k(k\mathrm{Stir}(n,k) - n\mathrm{Stir}(n-1,k-1)) \\ &\quad + (k-1)\mathrm{Stir}(n,k-1) - n\mathrm{Stir}(n-1,k-2) \\ &\quad + n\mathrm{Stir}(n-1,k-1) \\ &= k\mathrm{Stir}(n+1,k) - kn\mathrm{Stir}(n-1,k-1) - \mathrm{Stir}(n,k-1) \\ &\quad - n\mathrm{Stir}(n-1,k-2) + n\mathrm{Stir}(n-1,k-1),\end{aligned}$$

which gives

$$\begin{aligned}b(n,k) &= k\text{Stir}(n+1,k) - kn\text{Stir}(n-1,k-1) - \text{Stir}(n,k-1)\\&\quad - n(\text{Stir}(n,k-1) - (k-1)\text{Stir}(n-1,k-1))\\&\quad + n\text{Stir}(n-1,k-1)\\&= k\text{Stir}(n+1,k) - (n+1)\text{Stir}(n,k-1)\\&= c(n,k),\end{aligned}$$

which completes the proof.

**Exercise 5.6** By using of the decomposition (5.7), we obtain

$$P_{\text{sumrec}}(x;q;k) = \frac{xq}{1-kx} P_{\text{sumrec}}(xq;q;k-1)$$

with the initial condition $P_{\text{sumrec}}(x;q;1) = \frac{xq}{1-x}$. This recurrence relation together with induction on $k$ give

$$P_{\text{sumrec}}(x;q;k) = \prod_{j=1}^{k} \frac{xq^{k+1-j}}{1-jq^{k-j}x}, \qquad (A.2)$$

which implies the following result:

$$P_{\text{sumrec}}(x;q;k) = \frac{x^k q^{k(k+1)/2}}{\prod_{j=1}^{k}(1-jq^{k-j}x)}. \qquad (A.3)$$

In order to determine the mean of the sumrec parameter, we differentiate the generating function $P_{\text{sumrec}}(x;q;k)$ with respect to $q$ and set $q=1$ to obtain

$$\frac{x^k \left(\frac{k(k+1)}{2} + \sum_{j=1}^{k} \frac{j(k-j)x}{1-jx}\right)}{\prod_{j=1}^{k}(1-jx)} = \frac{\frac{k(k+1)}{2} + \sum_{j=1}^{k} \frac{j(k-j)}{\frac{1}{x}-j}}{\prod_{j=1}^{k}(\frac{1}{x}-j)}. \qquad (A.4)$$

The partial fraction decomposition has the form

$$\sum_{m=1}^{k} \left(\frac{a_{k,m}}{(\frac{1}{x}-m)^2} + \frac{b_{k,m}}{\frac{1}{x}-m}\right).$$

In order to determine the coefficients $a_{k,m}$ and $b_{k,m}$, we consider the expansion of (A.4) at $\frac{1}{x} = m$ as follows:

$$\frac{\frac{k(k+1)}{2} + \frac{m(k-m)}{\frac{1}{x}-m} + \sum_{j=1,j\neq m}^{k} \frac{j(k-j)}{\frac{1}{x}-j}}{(\frac{1}{x}-m)\prod_{j=1,j\neq m}^{k}(\frac{1}{x}-m+m-j)}$$

$$= \frac{\frac{k(k+1)}{2} + \frac{m(k-m)}{\frac{1}{x}-m} + \sum_{j=1,j\neq m}^{k} \frac{j(k-j)}{\frac{1}{x}-j}}{(\frac{1}{x}-m)\prod_{j=1,j\neq m}^{k}(m-j)\left(1+\frac{\frac{1}{x}-m}{m-j}\right)},$$

which leads to

$$\frac{\frac{k(k+1)}{2} + \frac{m(k-m)}{\frac{1}{x}-m} + \sum_{j=1,j\neq m}^{k} \frac{j(k-j)}{\frac{1}{x}-j}}{(\frac{1}{x}-m)\prod_{j=1,j\neq m}^{k}(\frac{1}{x}-m+m-j)}$$

$$= \frac{(-1)^{k-m}}{(\frac{1}{x}-m)(m-1)!(k-m)!} \cdot \left(1 - \sum_{j=1,j\neq m}^{k} \frac{\frac{1}{x}-m}{m-j} + O\left((\frac{1}{x}-m)^2\right)\right)$$

$$\cdot \left(\frac{k(k+1)}{2} + \frac{m(k-m)}{\frac{1}{x}-m} + \sum_{j=1,j\neq m}^{k} \frac{j(k-j)}{m-j} + O\left(\frac{1}{x}-m\right)\right)$$

$$= \frac{(-1)^{k-m}}{(\frac{1}{x}-m)(m-1)!(k-m)!}$$

$$\cdot \left(\frac{m(k-m)}{\frac{1}{x}-m} + \frac{k(k+1)}{2} + \sum_{j=1,j\neq m}^{k} \frac{j(k-j)-m(k-m)}{m-j} + O\left(\frac{1}{x}-m\right)\right)$$

$$= \frac{(-1)^{k-m}}{(\frac{1}{x}-m)(m-1)!(k-m)!}$$

$$\cdot \left(\frac{m(k-m)}{\frac{1}{x}-m} + \frac{k(k+1)}{2} + \sum_{j=1,j\neq m}^{k} (m+j-k) + O\left(\frac{1}{x}-m\right)\right)$$

$$= \frac{(-1)^{k-m}}{(\frac{1}{x}-m)(m-1)!(k-m)!} \left(\frac{m(k-m)}{\frac{1}{x}-m} + (m+2)k - 2m + O\left(\frac{1}{x}-m\right)\right).$$

This gives

$$a_{k,m} = \frac{m(k-m)(-1)^{k-m}}{(m-1)!(k-m)!} \quad \text{and} \quad b_{k,m} = \frac{((m+2)k-2m)(-1)^{k-m}}{(m-1)!(k-m)!}.$$

Passing to the exponential generating function, we have to replace

$$\frac{1}{\frac{1}{x}-m} = \frac{x}{1-mx} = \sum_{\ell=0}^{\infty} m^\ell x^{\ell+1} \quad \text{by} \quad \sum_{\ell=0}^{\infty} \frac{m^\ell x^{\ell+1}}{(\ell+1)!} = \frac{e^{mx}-1}{m}$$

and similarly $\frac{1}{(\frac{1}{x}-m)^2}$ by $\frac{e^{mx}(mx-1)+1}{m^2}$. Furthermore, we sum over all $k$ to obtain the bivariate generating function

$$\sum_{k=1}^{\infty} p^k \sum_{m=1}^{k} \frac{m(k-m)(-1)^{k-m}}{(m-1)!(k-m)!} \cdot \frac{e^{mx}(mx-1)+1}{m^2}$$

$$+ \sum_{k=1}^{\infty} p^k \sum_{m=1}^{k} \frac{((m+2)k-2m)(-1)^{k-m}}{(m-1)!(k-m)!} \cdot \frac{e^{mx}-1}{m}$$

Interchanging the order of summation, we can simplify this as follows:

$$\sum_{m=1}^{\infty} \frac{e^{mx}(mx-1)+1}{m!} \sum_{k=m}^{\infty} \frac{(-1)^{k-m}(k-m)p^k}{(k-m)!}$$

$$+ \sum_{m=1}^{\infty} \frac{e^{mx}-1}{m!} \sum_{k=m}^{\infty} \frac{(-1)^{k-m}(km+2k-2m)p^k}{(k-m)!}$$

$$= \sum_{m=1}^{\infty} \frac{e^{mx}(mx-1)+1}{m!} \sum_{\ell=0}^{\infty} \frac{(-1)^\ell \ell p^{\ell+m}}{\ell!}$$

$$+ \sum_{m=1}^{\infty} \frac{e^{mx}-1}{m!} \sum_{\ell=0}^{\infty} \frac{(-1)^\ell (m^2+m\ell+2\ell)p^{\ell+m}}{\ell!}$$

$$= -\sum_{m=1}^{\infty} \frac{e^{mx}(mx-1)+1}{m!} \cdot p^{m+1} e^{-p}$$

$$+ \sum_{m=1}^{\infty} \frac{e^{mx}-1}{m!} \left( m^2 p^m e^{-p} - (m+2)p^{m+1} e^{-p} \right)$$

$$= pe^{p(e^x-1)} \left( pe^x (e^x - x - 1) + e^x - 1 \right).$$

as required.

**Exercise 5.7** Direct calculations lead to

$$a_{n,j} = \sum_{2 \leq i_1 < i_2 < \cdots < i_{j-1} \leq n-1} \sum_{i_j = i_{j-1}+1}^{n} x^{i_j - 1}$$

$$= \frac{x}{1-x} \sum_{2 \leq i_1 < i_2 < \cdots < i_{j-1} \leq n-1} x^{i_{j-1}-1} - \frac{x^n}{1-x} \sum_{2 \leq i_1 < i_2 < \cdots < i_{j-1} \leq n-1} 1$$

$$= \frac{x}{1-x} a_{n-1,j-1} - \frac{x^n}{1-x} \binom{n-2}{j-1}.$$

Using the initial condition $a_{n,1} = \frac{x-x^n}{1-x}$ and iterating the above recurrence, we obtain $a_{n,j} = \frac{x^j}{(1-x)^j} - \frac{x^n}{(1-x)^j} \sum_{i=0}^{j-1} \binom{n-1-j+i}{i}(1-x)^i$, as claimed.

**Exercise 5.8** Let $\pi = B_1/B_2/\cdots/B_k$ be any set partition. The proof $\widehat{\operatorname{inv}}(\pi) = \operatorname{occ}_{\widehat{1}/2}(\pi) + \operatorname{occ}_{\widehat{1}/23}(\pi)$ is similar to the proof of Proposition 5.70. The only difference here is that the minimum of a block can represent the $b$ in a dual inversion $(b, B_j)$.

**Exercise 5.9** See [349, Corollary 6.2].

**Exercise 5.10** Assume that the set partition $\pi$ in $\mathcal{P}_{n,k}$ is to contain the letter 1, exactly $i+1$, for some $i$, $0 \leq i \leq n-k$, and that exactly $j$ letters greater than 1 come after the rightmost occurrence of the letter 1 for some $j$, $0 \leq j \leq n-i-1$. If $j < n-1-i$; then we write $\pi = 1\pi'1\pi''$, where $\pi'$ is nonempty and contains $i-1$ 1s and $n-1-i-j$ letters greater than 1, and $\pi''$ is a possibly empty word of size $j$. Therefore, there are

$$\binom{n-2-j}{i-1}\text{Stir}(n-1-i, k-1)$$

such set partitions, each having a $\text{dis}_1$ value of $n-1-i-j$. By summing over all possible $i$ and $j$, we derive

$$\text{total}_{\mathcal{P}_{n,k}}(\text{dis}_1) = \sum_{i=0}^{n-k} S_{n-1-i,k-1} \sum_{j=0}^{n-2-i} \binom{n-2-j}{i-1}(n-1-i-j)$$

$$= \sum_{i=0}^{n-k} S_{n-1-i,k-1} \sum_{j=1}^{n-1-i} j\binom{i-1+j}{j}$$

$$= \sum_{i=1}^{n-k} iS_{n-1-i,k-1} \binom{n-1}{i+1},$$

as required.

**Exercise 5.11** Note that Theorem 5.54 gives $F(x,1;0,0,\ldots) = \frac{1-2x}{1-3x+x^2}$, which implies that the number of set partitions of $[n]$ with zero int value is $\text{Fib}_{2n-1}$ if $n \geq 1$.

**Exercise 5.12** Theorem 5.55 for $q=0$ gives $F(x,y;0) = 1 + \sum_{j\geq 1} \frac{y^j x^j}{(1-x)^{2j-1}}$, which implies that the number of set partitions $\pi$ in $\mathcal{P}_{n,k}$ having $\text{int}(\pi)=0$ is given by $\binom{n+k-2}{n-k}$ if $1 \leq k \leq n$. Note that

$$\sum_{k=1}^{n}\binom{n+k-2}{2k-2} = \sum_{k=0}^{n-1}\binom{2n-2-k}{k} = \text{Fib}_{2n-1},$$

as shown in Exercise 5.11.

**Exercise 5.13** By comparing Corollary 5.56 to Corollary 5.37, we complete the proof.

**Exercise 5.14** Let us write the expression in Corollary 5.56 as

$$\sum_{i=2}^{k}\left((i-1)\sum_{j=k}^{n-1} i^{n-1-j}\text{Stir}(j,k)\right) - \sum_{i=1}^{k-1}\left(\sum_{j=k}^{n-1} i^{n-1-j}\text{Stir}(j,k)\right).$$

Given $i,j$ with $2 \leq i \leq k \leq j \leq n-1$, consider those set partitions of $\mathcal{P}_{n,k}$ that can be written as in (5.18) except that now the last letter of the word $\pi'''$ must be less than $i$. Note that there are $(i-1)i^{n-1-j}$ choices for the word $\pi'''$, and $\text{Stir}(j,k)$ choices for the remaining letters $\pi'i\pi''$. Call a letter $\pi_i$ in a set partition $\pi = \pi_1\pi_2\cdots\pi_n$ *secondary* if there exists a letter to its left that is larger. The total number of such letters in all the set partition of $\mathcal{P}_{n,k}$ can be derived by finding the number of set partitions that can be expressed as in (5.18) for each $i,j$ and then summing over all possible values of $i,j$, which gives the first part of the last expression.

From the total number of secondary letters, we must subtract the total number of secondary letters that are not internal, that is, those letters less than $k$ for which there is no strictly larger letter occurring to the right. Fix $i,j$ such that $1 \leq i \leq k-1 < j \leq n-1$; consider all set partitions of $\mathcal{P}_{n,k}$ that can be written as

$$\pi = \pi' i \rho, \tag{A.5}$$

where $\pi' \in \mathcal{P}_{j,k}$ and $\rho \in [i]^{n-1-j}$. Finding the number of set partitions that can be written as in (A.5) for each $i,j$ and then summing over all possible values yields the total number of letters within set partitions of $\mathcal{P}_{n,k}$ that are secondary but not internal. This gives the second sum in the expression above, which we subtract to obtain $\text{total}_{\mathcal{P}_{n,k}}(\text{int})$.

**Exercise 5.15** See Deodhar (1998 Ph.D. thesis).

**Exercise 5.16** See [336, Theorem 4].

**Exercise 5.17** See [372, Theorem 2.1].

**Exercise 5.18** For (1) see Theorem 2.1, for (2) see Theorem 2.2 and for (3) see Corollary 3.2 in [91].

**Exercise 5.19** See [30].

**Exercise 5.20** See [322].

**Exercise 5.21** Followed immediately from Corollary 5.97 with $p = 1$ and $q = 0$.

# Chapter 6

**Exercise 6.1** See [306].

**Exercise 6.2** See [146].

**Exercise 6.3** Let $\Lambda$ be an arbitrary $\begin{pmatrix} B & 0 \\ 0 & A \end{pmatrix}$-avoiding filling of a Ferrers diagram $\Pi$. We say that a $(i,j)$ cell of $\Lambda$ is *gray* if the subfilling of $\Lambda$ induced by the intersection of rows $i+1, i+2, \ldots, row(\Lambda)$ and columns $1, 2, \ldots, j-1$ contains a copy of $B$. Note that the gray cells form a Ferrers shape $\Pi^- \subseteq \Pi$, and that the restriction of $\Lambda$ to the cells of $\Pi^-$ is a sparse $A$-avoiding filling $\Lambda'$. From Remark 6.20 we get that the filling $\Lambda'$ can be bijectively transformed into a sparse $A'$-avoiding filling $\bar{\Lambda}'$ of $\Pi^-$, which transforms $\Lambda$ into a semi-standard $\begin{pmatrix} B & 0 \\ 0 & A' \end{pmatrix}$-avoiding filling of $\Pi$.

**Exercise 6.4** We remark that the argument of the proof fails if the matrices $\begin{pmatrix} B & 0 \\ 0 & A \end{pmatrix}$ and $\begin{pmatrix} B & 0 \\ 0 & A' \end{pmatrix}$ are replaced with $\begin{pmatrix} A & 0 \\ 0 & B \end{pmatrix}$ and $\begin{pmatrix} A' & 0 \\ 0 & B \end{pmatrix}$, respectively. Also, the argument fails if Ferrers shapes are replaced with stack polyominoes. For instance, the matrix $A = \begin{pmatrix} 1 & 0 \\ 0 & 1 \end{pmatrix}$ is Ferrers-equivalent and stack-equivalent to $A' = \begin{pmatrix} 0 & 1 \\ 1 & 0 \end{pmatrix}$, but the two matrices $\begin{pmatrix} A & 0 \\ 0 & 1 \end{pmatrix}$ and $\begin{pmatrix} A' & 0 \\ 0 & 1 \end{pmatrix}$ are not Ferrers-equivalent, and the two matrices $\begin{pmatrix} 1 & 0 \\ 0 & A \end{pmatrix}$ and $\begin{pmatrix} 1 & 0 \\ 0 & A' \end{pmatrix}$ are not stack-equivalent.

**Exercise 6.5** A set partition $\pi$ avoids the subsequence pattern $1^m$ if and only if each symbol in $\pi$ occurs at most $m-1$ times if and only if each of its blocks has a size at most $m-1$. Therefore, fixing the number elements in the first block as $j$, we get that the number of set partitions of $[n]$ that avoids the subsequence pattern $1^m$ and the first block has $j$ elements is given by $\binom{n-1}{j-1} P_{n-j}(1^m)$. Thus, by summing over all $j = 1, 2, \ldots, n$, we obtain the recurrence relation

$$P_n(1^m) = \sum_{j=1}^{m-1} \binom{n-1}{j-1} P_{n-j}(1^m)$$

which is equivalent to

$$\frac{P_n(1^m)}{(n-1)!} = \sum_{j=0}^{m-2} \frac{1}{j!} \frac{P_{n-1-j}(1^m)}{(n-1-j)!}.$$

Multiplying by $x^n$ and summing over all $n \geq 1$ we obtain that the exponential generating function $P'_{1^m}(x)$ for the sequence $\{P_n(1^m)\}_{n \geq 0}$ satisfies

$$x \frac{d}{dx}(P'_{1^m}(x) - 1) = x P'_{1^m}(x) \sum_{j=0}^{m-2} \frac{x^j}{j!},$$

which is equivalent to

$$\frac{\frac{d}{dx} P'_{1^m}(x)}{P'_{1^m}(x)} = \sum_{j=0}^{m-2} \frac{x^j}{j!}.$$

Integrating both sides of this differential equation we obtain

$$\ln(P'_{1^m}(x)) = \ln C + \sum_{j=1}^{m-1} \frac{x^j}{j!},$$

which gives $P'_{1^m}(x) = C e^{\sum_{j=1}^{m-1} \frac{x^j}{j!}}$. Using the initial condition that $P_0(1^m) = 1$, we obtain $P'_{1^m}(0) = 1$, which implies that $C = 1$. Hence, the exponential generating function for the number set partitions of $[n]$ that avoid the subsequence pattern $1^m$ is given by

$$\sum_{n \geq 0} \frac{P_n(1^m)}{n!} x^n = e^{\sum_{j=1}^{m-1} \frac{x^j}{j!}},$$

as claimed.

Solutions and Hints

**Exercise 6.6** Each nonempty set partition $\pi$ that avoids the subsequence pattern $12\cdots m$ can be written as $\pi = 1\pi^{(1)}2\pi^{(2)}\cdots j\pi^{(j)}$ with $j \leq m-1$, where $\pi^{(i)}$ is a word over alphabet $[i]$. Thus, by Example 2.42 we obtain

$$P_{12\cdots m}(x) = 1 + \sum_{j=1}^{m-1} \prod_{i=1}^{j} \frac{x}{1-ix} = \sum_{j=0}^{m-1} \frac{x^j}{(1-x)(1-2x)\cdots(1-jx)},$$

where 1 counts the empty set partition.

Each set partition $\pi$ with exactly $j \leq m-1$ blocks can be constructed by choosing $i+1$ elements in its first block, and with the rest of the elements a set partition cab be constructed with exactly $j-1$ blocks. Thus, by summing over all $i = 0, 1, \ldots, n-1$,

$$P_{n,j}(12\cdots m) = \sum_{i=0}^{n-1} \binom{n-1}{i} P_{n-1-i,j-1}(12\cdots m),$$

which is equivalent to

$$\frac{P_{n,j}(12\cdots m)}{(n-1)!} = \sum_{i=0}^{n-1} \frac{1}{i} \frac{P_{n-1-i,j-1}(12\cdots m)}{(n-1-i)!}.$$

Multiplying by $x^n$ and summing over all $n \geq 1$ we get

$$\frac{d}{dx}\left(\sum_{n\geq 0} \frac{P_{n,j}(12\cdots m)}{j!} x^n\right) = e^x \sum_{n\geq 0} \frac{P_{n,j-1}(12\cdots m)}{n!} x^n,$$

for all $j = 1, 2, \ldots, m-1$. Thus, by the initial condition $P_{12\cdots m}(0;x) = 1$ and by induction on $j$ we obtain $\sum_{n\geq 0} \frac{P_{n,j}(12\cdots m)}{j!} x^n = \frac{(e^x-1)^j}{j!}$. Hence, the exponential generating function for the number of set partitions of $[n]$ that avoid the subsequence pattern $12\cdots m$ is given by

$$\sum_{j=0}^{m-1}\sum_{n\geq 0} \frac{P_{n,j}(12\cdots m)}{j!} x^n = \sum_{j=0}^{m-1} \frac{(e^x-1)^j}{j!},$$

as claimed.

**Exercise 6.7** Let $\pi$ be an arbitrary set partition, and let $\pi^-$ denote the set partition obtained from $\pi$ by erasing every occurrence of the symbol 1, and decreasing every other symbol by 1, that is, $\pi^-$ represents the set partition obtained by removing the first block from the set partition $\pi$. Clearly, a set partition $\pi$ avoids $\sigma$ if and only if $\pi^-$ avoids $\tau$. Thus, for every $\sigma$-avoiding set partition $\pi \in \mathcal{P}_n(\sigma)$, there is a unique $\tau$-avoiding set partition $\rho \in \cup_{i=0}^{n-1}\mathcal{P}_i(\tau)$ satisfying $\pi^- = \rho$. On the other hand, for a fixed $\rho \in \mathcal{P}_i(\tau)$, there are $\binom{n-1}{i}$ set partitions $\pi \in \mathcal{P}_n(\sigma)$ such that $\pi^- = \rho$. This obtains the recurrence relation

$$P_n(\sigma) = \sum_{i=0}^{n-1} \binom{n-1}{i} P_i(\tau).$$

By multiplying both sides of this recurrence by $\frac{x^n}{n!}$ and sum for all $n \geq 1$, we obtain that

$$P_\sigma(x) - 1 = \sum_{n \geq 1} \frac{x^n}{n!} \sum_{i=0}^{n-1} \binom{n-1}{i} P_i(\tau)$$

$$= \int_0^x \sum_{n \geq 1} \frac{t^{n-1}}{(n-1)!} \sum_{i=0}^{n-1} \binom{n-1}{i} P_i(\tau) dt$$

$$= \int_0^x \sum_{n \geq 0} \frac{t^n}{n!} \sum_{i=0}^{n} \binom{n}{i} P_i(\tau) dt$$

$$= \int_0^x \sum_{n \geq 0} \sum_{i=0}^{n} \frac{t^i}{i!} P_i(\tau) \frac{t^{n-i}}{(n-i)!} dt$$

$$= \int_0^x \left( \sum_{i \geq 0} \frac{t^i}{i!} P_i(\tau) \right) \left( \sum_{k \geq 0} \frac{t^k}{k!} \right) dt = \int_0^x P_\tau(t) e^t dt,$$

which is equivalent to

$$P_\sigma(x) = 1 + \int_0^x P_\tau(t) e^t dt,$$

as required.

**Exercise 6.8** In order to show the last claim, note that Theorem (6.55) can be written as

$$P_{n-1}(\tau) = \sum_{i=0}^{n-1} (-1)^i \binom{n-1}{i} P_{n-i}(\sigma).$$

The other claims follow directly from Theorem 6.55.

**Exercise 6.9** Theorem 6.71 and Theorem 6.72 give all the subsequence patterns from the set $T$ are equivalent. Thus, it remains to compute the exponential generating function for the number of set partitions of $[n]$ that avoid the subsequence pattern $\tau = 12^m$. The formula for $F(x)$ follows directly from Fact 6.31 and Theorem 6.55.

**Exercise 6.10** See [146].

**Exercise 6.11** See [146].

**Exercise 6.12** Only the third recurrence is nontrivial. We prove it by presenting a bijection between $\mathcal{D}_{n,k}$ and the disjoint union $\bigcup_{j=k-1}^{n-1} \mathcal{D}_{n-1,j}$. Assume that $k$ and $n$ are fixed, with $n \geq 2$ and $k \leq n$. Take a Dyck path $P \in \mathcal{D}_{n,k}$. By erasing the last up-step and the last down-step of $D$, we get a Dyck path $P' \in \mathcal{D}_{n-1,j}$, where $j \geq k-1$. Conversely, given a Dyck path $P' \in \mathcal{D}_{n-1,j}$ with $j \geq k-1$, we insert a down-step at the end of $P'$ and then insert an up-step into the resulting path immediately before its last $k$ down-steps. This inverts our mapping.

**Exercise 6.13** Each Dyck path $D$ that ends with an up-step followed by $k$ down-steps can be decomposed as $D = D^{(1)}uD^{(2)}u\cdots D^{(k)}ud^k$, where $u$ denotes the up-step, $d$ denotes the down-step, and $D^{(i)}$ any Dyck path. If $f(x)$ is the generating function for the number of Dyck paths of size $2n$ that end with an up-step followed by $k$ down-steps, then $f(x)$ satisfies $f(x) = x^k C^k(x)$, where $C(x) = \frac{1-\sqrt{1-4x}}{2x}$ is the generating function for the number Dyck paths of size $2n$. By the Lagrange inversion formula we obtain the $x^n$ coefficient in $f(x)$ equals $\frac{k}{n}\binom{2n-k-1}{n-1}$, which completes the proof.

**Exercise 6.14** See [366], where Yan proved that the set of set partitions of $[n+1]$ avoiding the subsequence pattern 12312 (12321) is in one-to-one correspondence with Schröder paths of size $2n$ without peaks at even level. Note that, as a consequence, Yan refined the enumeration of set partitions avoiding the subsequence pattern 12312 (12321) according to the number of blocks by reducing the problem to find the number of Schröder paths without peaks at even levels according to the number of peaks.

**Exercise 6.15** First note that the only places where a set partition of $\mathcal{P}_{n,k}(1\text{-}11)$ can have a 2-level are just after the first occurrences of letters. So, to form a set partition of $\mathcal{P}_{n,k}(1\text{-}11)$ having exactly $j$ 2-levels, first choose a set partition $\pi \in \mathcal{P}_{n-j,k}(11)$, which, as is well-known, $P_{n-j,k}(11) = \text{Stir}(n-j-1, k-1)$, and then choose a subset $S$ of $[n]$ with $j$ elements. Now insert a copy of the letter $i$ just after the first occurrence of $i$ within $\pi$ for each $i \in S$ to obtain a set partition of $\mathcal{P}_{n,k}(1\text{-}11)$ having exactly $j$ 2-levels.

**Exercise 6.16** See [231].

**Exercise 6.17** See [231].

**Exercise 6.18** See [231].

**Exercise 6.19** Note first that members of $P_{n,k}(2'\text{-}1\text{-}2'')$ must be of the form $a_1\cdots a_k b_{k-1}\cdots b_1$, where $a_i$ and $b_i$ denote, respectively, a nonempty and possibly empty string of the letter $i$. As for $\tau = 2'\text{-}2''\text{-}1$, note that members of $P_{n,k}(2'\text{-}2''\text{-}1)$ must be of the form $\pi = 1w^{(1)}2w^{(2)}\cdots kw^{(k)}$, where each $w^{(i)}$ is either a (possibly empty) string of the letter $i$ or a nonempty string of the letter $i-1$ followed by a (possibly empty) string of the letter $i$. In terms of generating functions, each word $w^{(i)}$, $2 \leq i \leq k$, then contributes

$$\frac{1}{1-x} + \left(\frac{x}{1-x}\right)\left(\frac{1}{1-x}\right) = \frac{1}{(1-x)^2},$$

which completes the second case.

**Exercise 6.20** Suppose $\pi = 1w^{(1)}2w^{(2)}\cdots kw^{(k)} \in P_{n,k}(2\text{-}1'\text{-}1'')$, where each $w^{(i)}$ is $i$-ary. Then either $w^{(k)}$ consists of a (possibly empty) string of the letter $k$, or $w^{(k)} = \alpha j \beta$, where $\alpha$ and $\beta$ are possibly empty strings of the letter $k$

and $j \in [k-1]$. In the latter case, each $w^{(i)}$, $j+1 \leq i \leq k-1$, can then only contain the letter $i$, because if not, and $w^{(i)}$ contained $t$ with $t < i$, then $i$-$t$-$j$ would be an occurrence of $\tau$. Thus, in terms of generating functions, we have

$$Q_\tau(x,y) - 1 = \frac{xy}{1-x} Q_\tau(x,y) + (Q_\tau(x,y) - 1) \sum_{r \geq 1} \left(\frac{x}{1-x}\right)^{r+1} y^r$$

$$= \frac{xy}{1-x} Q_\tau(x,y) + (Q_\tau(x,y) - 1) \left(\frac{x^2 y}{(1-x)(1-x-xy)}\right),$$

which yields the result.

**Exercise 6.21** See [229].

**Exercise 6.22** See [168].

**Exercise 6.22** See [147].

# Chapter 7

**Exercise 7.1** See [236].

**Exercise 7.2** (i) By Theorem 7.104 and (2.9) we have

$$\sum_{n \geq k} S_\ell(n,k) x^n = \sum_{n \geq k} S_{n-\ell, k-\ell} x^n = \sum_{n \geq k-\ell} S_\ell(n, k-\ell) x^{n+\ell}$$

$$= x^\ell \cdot \frac{x^{k-\ell}}{\prod_{j=1}^{k-\ell}(1-jx)} = \frac{x^k}{\prod_{j=1}^{k-\ell}(1-jx)}.$$

(ii) By Theorem 7.104 and Theorem 1.16 we have

$$x^{n-\ell} = \sum_{k=0}^{n-\ell-1} S_{n-\ell-1,k}(x)_{k+1} = \sum_{k=0}^{n-1-\ell} S_\ell(n-1, k+\ell)(x)_{k+1}$$

$$= \sum_{k=\ell+1}^{n} S_\ell(n-1, k-1)(x)_{k-\ell}.$$

**Exercise 7.3** See [337, Theorem 2.7].

**Exercise 7.4** See [76].

**Exercise 7.5** For (1) see [96], for (2) see [365], and for (3) [74].

**Exercise 7.6** See [221, Lemma 3.1 ].

**Exercise 7.7** See [221, Section 4].

**Exercise 7.8** See [221, Lemma 4.1].

**Exercise 7.9** See [277] and [160].

**Exercise 7.10** See [238].

**Exercise 7.11** See [232].

**Exercise 7.12** See [235].

**Exercise 7.13** See [239].

**Exercise 7.14** A more general problem has been solved by Pittel [272] and Canfield [63].

**Exercise 7.15** See [367].

**Exercise 7.16** See [369].

**Exercise 7.17** See [149].

**Exercise 7.18** See [149].

# Chapter 8

**Exercise 8.1** See either [183, Proposition 2.6], [195, Equation (30), Page 69] or [144, Theorem 10].

**Exercise 8.2** See [144, Corollary 11].

**Exercise 8.3** See [144, Example 14].

**Exercise 8.4** See [308, Theorem 1].

**Exercise 8.5** See [205, Theorems 1.7 and 1.8].

**Exercise 8.6** See [27].

# Chapter 9

**Exercise 9.1** The iterative algorithm can be described by the following code.

```
1:  set π₀ = π₁ = ··· = πₙ = 0, j = n
2:  while j > 0 do
3:      print π₁π₂···πₙ
4:      if πⱼ < 2 then
5:          set πⱼ = πⱼ + 1
6:          set πᵢ = 0 for all i = j + 1, j + 2, . . . , n
7:          set j = n
8:      else
9:          set j = j - 1
10:     end if
11: end while
```

The algorithm for $n = 3$ gives 000, 001, 002, 010, 011, 012, 020, 021, 022, 100, 101, 102, 110, 111, 112, 120, 121, 122, 200, 201, 202, 210, 211, 212, 220, 221, 222.

**Exercise 9.2** We modify Algorithm 1 as follows: we replace Lines 10–13 by the following lines:

```
 1: for i = 1 to n do
 2:     set b_i = 0
 3: end for
 4: print b_1 ··· b_n
 5: call next(1)
 6: end
 7: Procedure next(j)
 8: if j = n then
 9:     for i = 1 to 2 do
10:         b_j = 1 + b_j
11:         if b_j = 3 then
12:             set b_j = 0
13:         end if
14:         print b_1 ··· b_n
15:     end for
16: else
17:     call next(j + 1)
18:     for i = 1 to 2 do
19:         b_j = 1 + b_j
20:         if b_j = 3 then
21:             set b_j = 0
22:         end if
23:         print b_1 ··· b_n
24:         call next(j + 1)
25:     end for
26: end if
27: end procedure
```

The Modification of Algorithm 1 for $n = 3$, as described above, gives the following Gray code: 000, 001, 002, 012, 010, 011, 021, 022, 020, 120, 121, 122, 102, 100, 101, 111, 112, 110, 210, 211, 212, 222, 220, 221, 201, 202, 200.

**Exercise 9.3** For example, the Maple code can be written as

```
NEXT:=proc(n,b,j) local i:
if j=n then
  for i from 1 to 2 do
    b[j]:=b[j]+1:
    if b[j]=3 then b[j]:=0: fi:
    print(b);
  od:
else
  NEXT(n,b,j+1);
  for i from 1 to 2 do
    b[j]:=b[j]+1:
    if b[j]=3 then b[j]:=0: fi:
    print(b);
```

```
   NEXT(n,b,j+1);
   od:
 fi:
 end;
 n:=3:
 b:=vector(n,[]):
 for i from 1 to n do b[i]:=0: od:
 print(b); NEXT(n,b,1);
```

**Exercise 9.4** Algorithm 3 for $n = 4$ gives the following list: 1111, 1112, 1121, 1122, 1123, 1211, 1212, 1213, 1221, 1222, 1223, 1231, 1232, 1233, 1234.

**Exercise 9.5** This problem has been considered by Grebinski and Kucherov [125]. Also, they considered a variant of the problem where a representative of each class should be found without necessarily reconstructing the whole partition.

**Exercise 9.6** The problem has been solved by Hankin and West [133]. Moreover, they presented three examples of the algorithm in use: one from bioinformatics, one from multiprocessor scheduling, and one from forensic science.

# Chapter 10

**Exercise 10.1** The contractions of the operator $aa^\dagger a^\dagger$ are given by $aa^\dagger a^\dagger$, $\varnothing\varnothing^\dagger a^\dagger$, and $\varnothing a^\dagger \varnothing^\dagger$. The linear representation of these contractions are given by $\underset{1\ 2\ 3}{\circ\ \bullet\ \bullet}$, $\underset{1\ 2\ 3}{\overgroup{\circ\ \bullet}\ \bullet}$, and $\underset{1\ 2\ 3}{\overgroup{\circ\ \bullet\ \bullet}}$, respectively.

**Exercise 10.2** The degree of $\gamma$ is 2. After drawing the linear representation of $\gamma$ (we leave that to the reader) we obtain that $\text{cross}(\gamma) = 1$, $\text{dcross}(\gamma) = 4$, and $\text{tcross}(\gamma) = 5$. Also, $\text{len}(\gamma) = 0$. Hence $\mathscr{W}_q(\gamma) = q^{5+0} = q^5$.

**Exercise 10.3** We proceed to the proof by induction on $j$. The claim holds for $j = 0$ since it is equivalent to $a = a$. Assume that the theorem holds for $j$. Then

$$\mathscr{N}\left[a(a^\dagger)^{j+1}\right] = \mathscr{N}\bigg(((a^\dagger)^j a + j(a^\dagger)^{j-1})a^\dagger\bigg),$$

which is equivalent to

$$\mathscr{N}\left[a(a^\dagger)^{j+1}\right] = \mathscr{N}((a^\dagger)^j aa^\dagger + j(a^\dagger)^j).$$

Using (10.1) we obtain that

$$\mathscr{N}\left[a(a^\dagger)^{j+1}\right] = (a^\dagger)^{j+1}a + (j+1)(a^\dagger)^j,$$

which completes the induction.

**Exercise 10.4** Let $F(a, a^\dagger) = aa^\dagger a^\dagger aa^\dagger a^\dagger$. Since there are only two creators in the word, the Feynman diagrams can have degree at most two. The trivial Feynman diagram $\gamma$ of degree zero yields : $\gamma := (a^\dagger)^4 a^2$ and has $q$-weight

$\mathscr{W}_q(\gamma) = q^{\text{len}(\gamma)} = q^6$. Thus, the Feynman diagram of degree zero gives the contribution $q^6(a^\dagger)^4 a^2$. There are six Feynman diagrams of degree one, namely,

$$(1,2),(1,3),(1,5),(1,6),(4,5),(4,6).$$

and their $q$-weights are given by (same order) $q^2, q^3, q^4, q^5, q^3, q^4$. Thus, the Feynman diagrams of degree one give the contribution $(q^2 + 2q^3 + 2q^4 + q^5)(a^\dagger)^3 a$. There are exactly six Feynman diagrams of degree two, namely,

$$(1,2)(4,5),(1,2)(4,6),(1,3)(4,5),(1,3)(4,6),(1,5)(4,6),(1,6)(4,5),$$

and their $q$-weights are given by (same order) $1, q, q, q^2, q^3, q^2$. Thus, the Feynman diagrams of degree one give the contribution $(2 + 2q + 2q^2 + q^3)(a^\dagger)^2$. Hence,

$$\mathcal{N}_q(aa^\dagger a^\dagger aa^\dagger a^\dagger)$$
$$= q^6(a^\dagger)^4 a^2 + (q^2 + 2q^3 + 2q^4 + q^5)(a^\dagger)^3 a + (2 + 2q + 2q^2 + q^3)(a^\dagger)^2,$$

which may also be calculated by hand from (10.8).

**Exercise 10.5** See [228, Section 5.3].

**Exercise 10.6** See [228, Section 5.3].

# Appendix B

## Identities

Touchard's identity (see http://en.wikipedia.org/wiki/Jacques_Touchard) is given by the following theorem.

**Theorem B.1** *For all* $n \geq 1$,

$$\text{Cat}_n = \sum_{k=0}^{(n-1)/2} \binom{n-1}{2k} 2^{n-1-2k} \text{Cat}_k.$$

The Riordan number $\text{Riord}_n$ is well known to count, among other things, Dyck paths of size $2n$ with no maximal sequence of contiguous downsteps of size one (see [327, Sequence A005042]).

**Theorem B.2** *For all* $n \geq 1$,

$$\sum_{k=0}^{n} \binom{n}{k} 2^k \text{Riord}_{n-k} = \sum_{j=0}^{n} \binom{n}{j} \text{Cat}_j.$$

**Proof** It is well known that $\text{Riord}_n = \sum_{j=0}^{n} (-1)^{n-j} \binom{n}{j} \text{Cat}_j$. This leads to

$$\sum_{k=0}^{n} \binom{n}{k} 2^k \text{Riord}_{n-k} = \sum_{k=0}^{n} \sum_{j=0}^{n-k} \binom{n}{k} 2^k (-1)^{n-k-j} \binom{n-k}{j} \text{Cat}_j$$

$$= \sum_{j=0}^{n} \text{Cat}_j \left( \sum_{k=0}^{n-j} \binom{n}{k} 2^k (-1)^{n-k-j} \binom{n-k}{j} \right)$$

$$= \sum_{j=0}^{n} \binom{n}{j} \text{Cat}_j \left( \sum_{k=0}^{n-j} \binom{n-j}{k} 2^k (-1)^{n-k-j} \right)$$

$$= \sum_{j=0}^{n} \binom{n}{j} \text{Cat}_j (2-1)^{n-j}$$

$$= \sum_{j=0}^{n} \binom{n}{j} \text{Cat}_j,$$

as claimed. $\square$

# Appendix C

# Power Series and Binomial Theorem

In the following, $q$ denotes a complex number with $|q| < 1$, and some notation of Gasper and Rahman[112] will be used.

**Definition C.1** *Let $a$ be any complex number and $n \in \mathbb{N} \cup \{\infty\}$. Define $[a]_q = \frac{1-q^a}{1-q}$ and $[n]_q = 1 + q + q^2 + \cdots + q^n = \frac{1-q^n}{1-q}$. Also, $(a;q)_0 = [a]_{0;q} = 1$ with*

$$(a;q)_n = \prod_{m=0}^{n-1}(1-aq^m), \quad [a]_{n;q} = \prod_{m=0}^{n-1}[a-m]_q, \quad n \in \mathbb{N} \cup \{\infty\},$$

*and for $a_1, \ldots, a_k$ complex numbers,*

$$(a_1, \ldots, a_k; q)_n = (a_1; q)_n \cdots (a_k; q)_n, \quad n \in \mathbb{N} \cup \{\infty\}.$$

*We define the $q$-factorial as $[n]_q! = [n]_{n;q}$ with $[0]_q! = [0]_q = 1$.*

For instance, $[3]_q = 1 + q + q^2$ and $[3]_q! = (1+q)(1+q+q^2)$.

**Definition C.2** *We define the $q$-binomial coefficient as $\binom{n}{j}_q = \frac{[n]_{j;q}}{[j]_q!}$ to be the $q$-analog of the binomial coefficients, and the $q$-exponent function as $e_q(x) = \sum_{j \geq 0} \frac{x^n}{[j]_q!}$ to be the $q$-analog of $e^x$.*

For instance, $\binom{n}{0}_q = 1$, $\binom{n}{1}_q = [n]_q$, and $\binom{n}{j}_q = \binom{n}{n-j}$. By the above definition we can state the following relation.

**Theorem C.3** *For all $n \geq 2j \geq 0$,*

$$\binom{n-j}{j}_q = \binom{n-1-j}{j-1}_q + q^j \binom{n-1-j}{j}_q.$$

Theorem C.3 shows the following corollary.

**Corollary C.4** *Let $f_{n,j} = \frac{d}{dq}\binom{n}{j}_q|_{q=1}$. Then $f_{n,j} = \binom{n}{2}\binom{n-2}{j-1}$ for all $j = 0, 1, \ldots, n$.*

503

**Proof** From the fact that $\binom{n}{j}_q = q^j \binom{n-1}{j}_q + \binom{n-1}{j-1}_q$, we have $f_{n,j} = f_{n-1,j} + f_{n-1,j-1} + j\binom{n-1}{j}$. Define $F_n(t) = \sum_{j=0}^{n} f_{n,j} t^j$. Then $F_n(t) = (1+t)F_{n-1}(t) + (n-1)t(1+t)^{n-2}$ with $F_0(t) = F_1(t) = 0$. Therefore, by induction on $n$, we obtain $F_n(t) = \binom{n}{2} t(1+t)^{n-2}$, which obtains $f_{n,j} = \binom{n}{2}\binom{n-2}{j-1}$ for all $j = 0, 1, \ldots, n$. □

For any real number $\alpha$,

$$(1+t)^\alpha = \sum_{n \geq 0} \binom{\alpha}{n} t^n, \tag{C.1}$$

where $\binom{\alpha}{0} = 1$ and $\binom{\alpha}{n} = \frac{\alpha(\alpha-1)\cdots(\alpha-(n-1))}{(n-1)!}$ for all $n \geq 1$. For $q$-analog of this relation, we state the $q$-binomial theorem.

**Theorem C.5** *(Andrews [4]) For all $n \geq 0$,*

$$\frac{t^n}{(1-t)(1-tq)\cdots(1-tq^n)} = \sum_{m \geq 0} \binom{m}{n}_q t^m$$

and

$$(1+t)(1+qt)\cdots(1+q^{n-1}t) = \sum_{j=0}^{n} q^{\binom{j}{2}} \binom{n}{j}_q t^j.$$

The $q$-Stirling numbers of the second kind, described by Gould [118], arise as generating functions for the distribution of certain statistics on set partitions, much as the $q$-analogue of $n!$, $[n]_q! = [1]_q[2]_q\cdots[n]_q$, is the generating function for the distribution of certain statistics on permutations such as number inversions and major index. Milne [252], and Garsia and Remmel [111] have already given such statistics. All their statistics reflect quite naturally the recursion for $q$-Stirling numbers. There are two related $q$-Stirling numbers of the second kind $\text{Stir}_q(n,k)$ and $\widetilde{\text{Stir}}_q(n,k)$ (see [118]). We define them both by the following recurrences.

**Definition C.6** *We define $\text{Stir}_q(n,k)$ to be*

$$\begin{cases} q^{k-1}\text{Stir}_q(n-1,k-1) + [k]_q\text{Stir}_q(n-1,k), & 1 \leq k \leq n, \\ 1, & n = k = 0, \\ 0, & \text{otherwise}. \end{cases} \tag{C.2}$$

*and $\widetilde{\text{Stir}}_q(n,k)$ to be*

$$\begin{cases} \widetilde{\text{Stir}}_q(n-1,k-1) + [k]_q\widetilde{\text{Stir}}_q(n-1,k), & 1 \leq k \leq n, \\ 1, & n = k = 0, \\ 0, & \text{otherwise}. \end{cases} \tag{C.3}$$

Clearly,
$$\text{Stir}_q(n,k) = q^{\binom{k}{2}}\widetilde{\text{Stir}}_q(n,k). \tag{C.4}$$

The $p,q$-Stirling numbers of the second kind have a similar recurrence.

**Definition C.7** *We define* $\text{Stir}_{p,q}(n,k)$ *to be*
$$\begin{cases} p^{k-1}\text{Stir}_{p,q}(n-1,k-1) + [k]_{p,q}\text{Stir}_{p,q}(n-1,k), & 1 \le k \le n, \\ 1, & n = k = 0, \\ 0, & \text{otherwise.} \end{cases} \tag{C.5}$$

*where* $[k]_{p,q} = p^{k-1} + p^{k-2}q + \cdots + pq^{k-2} + q^{k-1}$.

# Appendix D

# Chebychev Polynomials of the Second Kind

**Definition D.1** *Chebyshev polynomials of the second kind are defined by*
$$U_j(\cos\theta) = \frac{\sin((j+1)\theta)}{\sin\theta}$$
*for $j \geq 0$.*

**Proposition D.2** *The $m$-th Chebyshev polynomial of the second kind, $U_m(x)$, is a polynomial of degree $m$ with integer coefficients that satisfies the recurrence relation*
$$U_{m+1}(x) = 2xU_m(x) - U_{m-1}(x) \tag{D.1}$$
*with initial conditions $U_0(x) = 1$ and $U_1(x) = 2x$.*

**Proof** By Definition D.1 we have that $U_0(\cos\theta) = 1$ and $U_1(\cos\theta) = \sin(2\theta)/\sin(\theta) = 2\cos\theta$, which implies that $U_0(x) = 1$ and $U_1(x) = 2x$. By using the trigonometric identity
$$\sin((m+2)\theta) = \sin((m+1)\theta)\cos\theta + \cos((m+1)\theta)\sin\theta$$
we obtain that
$$U_{m+1}(x) - xU_m(x) = \cos((m+1)\theta). \tag{D.2}$$
On the other hand, by the trigonometric identity
$$\cos((m+1)\theta) = \cos(m\theta)\cos\theta - \sin(m\theta)\sin\theta$$
together with (D.2), we have that
$$U_{m+1}(x) - xU_m(x) = x(U_m(x) - xU_{m-1}(x)) - U_{m-1}(x)(1 - x^2),$$
which implies that $U_m(x)$ satisfies (D.1). By induction on $m$ we can show that $U_m(x)$ is a polynomial of degree $m$ with integer coefficients, as required. □

**Example D.3** *The first terms of the sequence $\{U_m(x)\}_{m \geq 0}$ are*
$U_0(x) = 1,$ $\quad U_1(x) = 2x,$
$U_2(x) = 4x^2 - 1,$ $\quad U_3(x) = 8x^3 - 4x,$
$U_4(x) = 16x^4 - 12x^2 + 1,$ $\quad U_5(x) = 32x^5 - 32x^4 + 6x.$

*Using (D.1) one can extend the definition of the Chebyshev polynomials of the second kind to all $m \in \mathbb{Z}$ to get that $U_{-1}(x) = 0$, $U_{-2}(x) = -1$ and so on.*

Chebyshev polynomials were invented for the needs of approximation theory but are also widely used in various other branches of mathematics including algebra, combinatorics, and number theory (see Rivlin [293]).

**Fact D.4** *For all $k \geq 0$,*

$$U_k(1) = k+1 \text{ and } \frac{d}{dq}U_k(1-a+aq)\mid_{q=1} = 2a\binom{k+2}{3}.$$

**Proof** By (D.1) we have

$$U_k(1) = 2U_{k-1}(1) - U_{k-2}(1)$$

with $U_0(1) = 1$ and $U_1(1) = 2$. By induction on $k$ we derive that $U_k(1) = k+1$.
Again by (D.1) we have

$$U_k(1-a+aq) = 2(1-b+bq)U_{k-1}(1-a+aq) - U_{k-2}(1-a+aq)$$

with $U_0(1-a+aq) = 1$ and $U_1(1-a+aq) = 2-2a+2aq$. Thus, if we define $a_k = \frac{d}{dq}U_k(1-a+aq)\mid_{q=1}$, then

$$a_k = 2aU_{k-1}(1) + 2a_{k-1} - a_{k-2}$$

with $a_0 = 0$ and $a_1(1-a+aq) = 2a$. Again, by $U_k(1) = k+1$ and induction on $k$ we derive that $a_k = 2a\binom{k+2}{3}$, as required. $\square$

**Proposition D.5** *Let $a_n$ be any sequence given by $a_n = \frac{A+Ba_{n-1}}{C+Da_{n-1}}$ with $a_0 = 1$ such that $\alpha = BC - AD \neq 0$. Then for all $n \geq 0$,*

$$a_n = \frac{A\left(\frac{A+B}{\sqrt{\alpha}}U_{n-1}(t) - U_{n-2}(t)\right)}{\sqrt{\alpha}\left(\frac{A+B}{\sqrt{\alpha}}U_n(t) - U_{n-1}(t)\right) - B\left(\frac{A+B}{\sqrt{\alpha}}U_{n-1}(t) - U_{n-2}(t)\right)},$$

*where $t = \frac{B+C}{2\sqrt{\alpha}}$ and $U_m$ is the $m$-th Chebyshev polynomial of the second kind.*

**Proof** We proceed by induction on $n$. Since $U_0(t) = 1$, $U_{-1}(t) = 0$ and $U_{-2}(t) = -1$, the lemma holds for $n = 0$. Let

$$g_n = \frac{A+B}{\sqrt{\alpha}}U_{n-1}(t) - U_{n-2}(t).$$

Assume the lemma holds for $n$, and let us prove it for $n+1$:

$$a_{n+1} = \frac{A+Ba_n}{C+Da_n} = \frac{A+B\frac{Ag_n}{\sqrt{\alpha}g_{n+1}-Bg_n}}{C+D\frac{Ag_n}{\sqrt{\alpha}g_{n+1}-Bg_n}}$$

$$= \frac{A\sqrt{\alpha}g_{n+1}}{C\sqrt{\alpha}g_{n+1} - (BC-DA)g_n} = \frac{Ag_{n+1}}{Cg_{n+1} - \sqrt{\alpha}g_n}.$$

Using (D.1) we get that $g_n = \frac{B+C}{\sqrt{\alpha}} g_{n+1} - g_{n+2}$, which implies

$$a_{n+1} = \frac{A g_{n+1}}{C g_{n+1} - (B+C) g_{n+1} + \sqrt{\alpha} g_{n+2}} = \frac{A g_{n+1}}{\sqrt{\alpha} g_{n+2} - B g_{n+1}},$$

which completes the induction step. $\square$

# Appendix E

## Linear Algebra and Algebra Review

**Definition E.1** *A group $(G, +)$ is a set $G$ with a binary operation $+$ on $G$ that satisfies:*

1. *Closure: For all $a, b \in G$, $a + b \in G$.*
2. *Associativity: For all $a, b, c \in G$, $(a + b) + c = a + (b + c)$.*
3. *Identity element: There exists an (unique) element $e \in G$ such that for all $a \in G$, $e + a = a + e = a$.*
4. *Inverse element: For each $a \in G$, there exists an element $b \in G$ such that $a + b = b + a = e$, where $e$ is the identity element.*

*If the group operation $+$ is commutative so that for all $a, b \in G$, $a + b = b + a$, then $G$ is said to be an* abelian *or* commutative *group.*

**Example E.2** *The set of integers $\mathbb{Z}$ together with the usual addition is a commutative group with identity element $e = 0$, and inverse elements $-a$. $\mathbb{Z}$ together with the usual multiplication is not a group, as there is no inverse element in $\mathbb{Z}$ for every $a \in \mathbb{Z}$.*

**Example E.3** *The set of square matrices $A$ with $\det(A) \neq 0$ of size $n \times n$ together with the usual matrix multiplication is a group. However, the group is not commutative because for the matrices*

$$A = \begin{bmatrix} 1 & 2 \\ -3 & 4 \end{bmatrix} \quad \text{and} \quad B = \begin{bmatrix} 0 & 2 \\ -1 & 1 \end{bmatrix}$$

*we have that*

$$A \cdot B = \begin{bmatrix} -2 & 4 \\ 4 & -2 \end{bmatrix} \neq B \cdot A = \begin{bmatrix} -6 & 8 \\ -4 & 2 \end{bmatrix}.$$

**Definition E.4** *A* ring *is an abelian group $(R, +)$ equipped with a second binary operation $*$ such that for all $a, b, c \in R$,*

1. $a * (b * c) = (a * b) * c$.
2. $a * (b + c) = (a * b) + (a * c)$.

3. $(a+b)*c = (a*c) + (b*c)$.

If there also exists a multiplicative identity in the ring, that is, an element $e'$ such that for all $a \in R$, $a*e' = e'*a = a$, then the ring is said to be a ring with unity.

**Example E.5** *The set of rational numbers $\mathbb{Q}$ together with regular addition and multiplication is a ring, but it does not have a unity.*

# Appendix F

## Complex Analysis Review

In this section, we present the basic definitions and theorems that are used in the asymptotic analysis, which can be found in any reference text (see for example [1, 246]).

**Definition F.1** *A complex function $f(z)$, defined over a region $\Re$, is said to be* analytic *at a point $z_0 \in \Re$ if, for some open disc centered at $z_0$ and contained in $\Re$, $f(z)$ can be represented by a convergent power series expansion*

$$f(z) = \sum_{n \geq 0} c_n (z - z_0)^n.$$

*The function $f(z)$ is said to be* analytic *in a region $\Re$ if and only if it is analytic at every point in $\Re$. If $f$ is analytic at $z_0$, then there is a disc (of possibly infinite radius) such that $f(z)$ converges for $z$ inside the disc and diverges for $z$ outside the disc. This disc is called the* disc of convergence *and its radius is the* radius of convergence,*, denoted by $R(f; z_0)$.*

**Example F.2** *Let $a$ be any positive real number. Consider the complex function $f(z) = 1/(1-az)$ which is defined in $\mathbb{C}\backslash\{\frac{1}{a}\}$. It is analytic at $z = 0$ since it represents the series $\sum_{n \geq 0} a^n z^n$ which converges for $|z| < \frac{1}{a}$ and $R(f; 0) = \frac{1}{a}$.*

**Definition F.3** *A function $f(z)$ defined over a region $\Re$ is called* complex-differentiable *or* holomorphic *at $z_0$ if*

$$\lim_{\delta \to 0} \frac{f(z_0 + \delta) - f(z_0)}{\delta}$$

*exists, that is, the limit is independent of the way in which $\delta$ tends to $0$ in $\mathbb{C}$. A function is* holomorphic *in the region $\Re$ if and only if it is holomorphic for every $z_0 \in \Re$.*

**Theorem F.4** (*Basic Equivalence Theorem*) [107, Theorem IV.1] *A function is analytic in a region $\Re$ if and only if it is holomorphic on $\Re$. Furthermore, if $f(z)$ is an analytic function in $\Re$, then $f(z)$ has derivatives of all orders and the derivatives can be obtained through term-by-term differentiation of the series expansion of $f(z)$.*

514                    Combinatorics of Set Partitions

**Definition F.5** *A curve $\gamma$ in a region $\Re$ is a continuous function which maps $[0,1]$ into $\Re$. A curve is called* simple *if the mapping $\gamma$ is one-to-one, and* closed *if $\gamma(0) = \gamma(1)$. A closed curve is a* loop *if $\gamma$ can be continuously deformed to a single point within $\Re$.*

**Example F.6** *The circle $\gamma$ centered at the origin and has radius 1 is a closed simple curve since it is given by the continuous function $f(x) = e^{2\pi x i} = \cos(2\pi x) + i \sin(2\pi x)$ for all $x \in [0,1]$.*

**Theorem F.7** *(Cauchy's coefficient formula, [107, Theorem IV.4]) Let $f(z) = \sum_{n \geq 0} f_n z^n$ be an analytic complex function in a region $\Re$ containing the origin 0 and let $\gamma$ be any simple loop around 0, in $\Re$ that is positively oriented. Then*

$$f_n = [z^n]f(z) = \frac{1}{2\pi} \int_\gamma \frac{f(z)}{z^{n+1}} dz.$$

**Definition F.8** *A function $h(z)$ is said to be* meromorphic *at $z_0$ if and only if it can be represented as $f(z)/g(z)$ for $z$ in a neighborhood of $z_0$ with $z \neq z_0$, where $f(z)$ and $g(z)$ are analytic at $z_0$. If $h(z)$ is meromorphic, then it has the following representation $h(z) = \sum_{n \geq -M} h_n(z-z_0)^n$ near $z_0$. If $h_{-M} \neq 0$ and $M \geq 1$, then we say that $h(z)$ has a* pole of order $M$ *at $z_0$. A pole of order one is called a* simple pole. *The coefficient $h_{-1}$ is called the* residue *of $h(z)$ at $z = z_0$ and is denoted by $\mathrm{Res}[h(z); z = z_0]$. A function is* meromorphic *in a region $\Re$ if and only if it is meromorphic at every point of the region.*

The most important results in complex analysis is Cauchy's residue theorem, which relates global properties of a function (the integral along closed curves) to local properties (residues at poles).

**Theorem F.9** *(Cauchy's residue theorem, [107, Theorem IV.3]) Let $h(z)$ be any meromorphic function in the region $\Re$, and let $\gamma$ be any simple loop in $\Re$ along which the function $h(z)$ is analytic. Then*

$$\frac{1}{2\pi} \int_\gamma h(z) dz = \sum_r \mathrm{Res}[h(z); z = r],$$

*where the sum is over all poles $r$ of $h(z)$ enclosed by $\gamma$.*

**Example F.10** *Let $\gamma$ be the circle of radius one around the origin in the complex plane which is positively oriented. Then by Cauchy's residue theorem, we have*

$$\int_\gamma \frac{e^z}{(1-2z)(2-z)} dz = 2\pi \mathrm{Res}\left[\frac{e^z}{(1-2z)(2-z)}; z = \frac{1}{2}\right].$$

**Definition F.11** *Let $f(z)$ be an analytic complex function defined over the interior region determined by a simple closed curve $\gamma$, and let $z_0$ be a point of the bounding curve $\gamma$. If there exists an analytic function $f^*(z)$ defined over some open set $\Re^*$ containing $z_0$ such that $f^*(z) = f(z)$ in $\Re \cap \Re^*$, then $f$ is* analytically continuable *at $z_0$. If $f$ is not analytically continuable at $z_0$, then $z_0$ is called a* singularity *or* singular point*. If $f(z)$ is analytic at 0, then a singularity that lies on the boundary of the disc of convergence is called a* dominant singularity*.

**Theorem F.12** *(Rouché's theorem) Let the functions $f(z)$ and $g(z)$ be analytic in a region $\Re$ containing the interior of the simple closed curve $\gamma$. Assume that $|g(z)| < |f(z)|$ on $\gamma$. Then $f(z)$ and $f(z) + g(z)$ have the same number of zeros inside the interior domain bounded by $\gamma$. Equivalently, $f(\gamma)$ and $(f+g)(\gamma)$ have the same winding number.*

**Definition F.13** *Let $f(x)$ be a function defined over the positive real numbers. Then the* Mellin transform *of $f$ is defined as the complex function $f^*(s)$ where*

$$f^*(s) = \int_0^\infty f(x) x^{s-1} dx.$$

The Mellin transform is used in many areas. We will use it for the asymptotic analysis of sums of the form

$$G(x) = \sum_k \lambda_k g(\mu_k x).$$

These types of sums are often called *harmonic sums*. In this context, the $\lambda_k$ are the *amplitudes*, the $\mu_k$ are the *frequencies*, and $g(x)$ is the *base function*.

For full details on Saddle-point method, we refer the reader to [107, Section VIII].

**Theorem F.14** *Consider the integral $\int_a^b e^{f(z)} dz$, where the $e^{f(z)}$ is an analytic function depending on a large parameter and $a, b$ lie in opposite valleys across a saddle-point $\xi$, which is a root of the saddle point equation $f'(\xi) = 0$. Assume that the contour $C$ connecting $a$ to $b$ can be split into $C = C' \cup C''$ in such a way that the following conditions are satisfied:*
*(i) Tails are negligible in the sense $\int_{C'} e^{f(z)} dz = o\left(\int_C e^{f(z)} dz\right)$.*
*(ii) A central approximation hold, that is, along $C'$ the equation*

$$f(z) = f(\xi) + \frac{1}{2} f''(\xi)(z-\xi)^2 + O(\eta_n),$$

*is valid with $\eta_n \to 0$ as $n \to \infty$, uniformly with respect to $z \in C'$.*
*(iii) Tails can be completed back, that is,*

$$\int_{C'} e^{\frac{1}{2} f''(\xi)(z-\xi)^2} dz \sim \pm i e^{-i\phi/2} \int_{-\infty}^\infty e^{-|f''(\xi)| x^2/2} dx = \pm i e^{-i\phi/2} \sqrt{\frac{2\pi}{|f''(\xi)|}},$$

where $f''(\xi) = e^{i\phi}|f''(\xi)|$ and $i^2 = -1$.
Then one has,
$$\frac{1}{2i\pi}\int_a^b e^{f(z)}dz \sim \pm\frac{e^{f(\xi)}}{\sqrt{2\pi f''(\xi)}}.$$

The saddle-point method can used to provide complete asymptotic expansions. The idea is still to localize the main contribution in the central region but take into account corrections that come from tails, see (iii) in the above theorem. As an illustration of this general method, we refer the reader to [107, Section VIII].

A famous and absolutely asymptotic formula is Stirling's approximation or "Stirling's formula":

$$\Gamma(a+1) = a\Gamma(a) \sim a^a e^{-a}\sqrt{2\pi a}\left(1 + \frac{1}{12a} + \frac{1}{288a^2} - \frac{139}{51840a^3} + \cdots\right). \quad (\text{F.1})$$

It is valid for (large) positive real number $a$, and more generally for all $a \to \infty$ in $|\arg(a)| < \pi - \delta$ with $\delta > 0$. This formula is usually established by appealing to method of Laplace applied to integral representation for function $\Gamma$, by Euler–Maclaurin summation, or by Mellin transforms, see [107].

# Appendix G

## Coherent States

In this section, we define and review several properties of coherent states (see [175, 373]). For the proofs, see [174]. Let $a$ and $a^\dagger$ be the annihilation and creation operators of the harmonic oscillator. Set $N = aa^\dagger$ to be the number operator, then

$$[N, a^\dagger] = a^\dagger, \quad [N, a] = -a, \quad [a^\dagger, a] = -\mathbf{1},$$

where $\mathbf{1}$ is the identity operator. Let $\mathcal{H}$ be a *Fock space* generated by $a$ and $a^\dagger$, and $\{|n\rangle \mid n \in \mathbb{N} \cup \{0\}\}$ be its basis. The actions of the annihilation and creation operators on $\mathcal{H}$ are given by

$$a|n\rangle = \sqrt{n}|n-1\rangle, \quad a^\dagger|n\rangle = \sqrt{n+1}|n+1\rangle, \quad N|n\rangle = n|n\rangle,$$

where $|0\rangle$ is a normalized vacuum satisfies $a|0\rangle = 0$ and $\langle 0|0\rangle = 1$. Thus, the states $|n\rangle$ are given by $|n\rangle = \frac{(a^\dagger)^n}{\sqrt{n!}}|0\rangle$, where they satisfy the orthogonality and completeness conditions

$$\langle m|n\rangle = \delta_{m,n} \text{ and } \sum_{n\geq 0} |n\rangle\langle n| = \mathbf{1}.$$

For the normalized state $|z\rangle \in \mathcal{H}$ for $z \in \mathbb{C}$, the following conditions are equivalent: (i) $a|z\rangle = z|z\rangle$ and $\langle z|z\rangle = 1$, (ii) $|z\rangle = e^{-\frac{1}{2}|z|^2} \sum_{n\geq 0} \frac{z^n}{\sqrt{n}} |n\rangle = e^{-\frac{1}{2}|z|^2} e^{za^\dagger}|0\rangle$, and (iii) $|z\rangle = e^{za^\dagger - \bar{z}a}|0\rangle$, where $\bar{z}$ is the conjugate of the complex number $z$. In order to prove (iii) from (ii), we need the famous elementary Baker–Champbell–Hausdorff formula $e^{A+B} = e^{-\frac{1}{2}[A,B]}e^A e^B$, where $[A, [A, B]] = [B, [A, B]] = 0$, see [174].

**Definition G.1** *The state $|z\rangle$ that satisfies $a|z\rangle = z|z\rangle$ and $\langle z|z\rangle = 1$ is called* coherent state.

The important feature of coherent states is the following partition of unity:

$$\frac{1}{\pi}\int_\mathbb{C} |z\rangle\langle z| d^2z = \sum_{n\geq 0} |n\rangle\langle n| = \mathbf{1}.$$

Note that $\langle z|w\rangle = e^{-\frac{1}{2}(|z|^2+|w|^2)+\bar{z}w}$, which implies that $|\langle z|w\rangle| = e^{-\frac{1}{2}|z-w|^2}$ and $\langle w|z\rangle = \overline{\langle z|w\rangle}$, so $|\langle z|w\rangle| < 1$ if $z \neq w$.

Define $A(z) = e^{za^\dagger - \bar{z}a}$ to be operator on $\mathbb{C}$. Then $A(z)$ is a unitary operator (we call this a displacement coherent operator). This operator satisfies $A(z)A(w) = e^{z\bar{w}=\bar{z}w}A(w)A(z)$ and $A(z+w) = e^{-\frac{1}{2}(z\bar{w}-\bar{z}w)}A(z)A(w)$, for all $z, w \in \mathbb{C}$. Actually, this operator satisfies several properties. For instance, it satisfies

The matrix elements of $A(z)$ are

(i) $\quad n \leq m \langle n|A(z)|m\rangle = e^{-\frac{1}{2}|z|^2}\sqrt{\frac{n!}{m!}}(-\bar{z})^{m-n}L_n^{(m-n)}(|z|^2),$

(ii) $\quad m \leq n \langle n|A(z)|m\rangle = e^{-\frac{1}{2}|z|^2}\sqrt{\frac{m!}{n!}}z^{n-m}L_m^{(n-m)}(|z|^2),$

where $L_k^{(\alpha)}(x)$ is the $k$th associated Laguerre polynomial which is defined by $L_k^{(\alpha)}(x) = \sum_{j=0}^{k}(-1)^j\binom{k+\alpha}{k-j}\frac{x^j}{j!}$. Note that $L_k^{(0)}$ is the standard $k$th Laguerre polynomial and these are related to the main diagonal elements of $A(z)$. These polynomials satisfy

$$\frac{e^{-\frac{xt}{1-t}}}{(1-t)^{\alpha+1}} = \sum_{k \geq 0} L_k^{(\alpha)}(x)t^k$$

and

$$\int_0^\infty e^{-x}x^\alpha L_k^{(\alpha)}(x)L_m^{(\alpha)}(x)dx = \frac{\Gamma(\alpha+n+1)}{n!}\delta_{n,m},$$

for $|t| < 1$ and $\Re(\alpha) > -1$, where $\Gamma(z)$ is the $\Gamma$ function.

The projection on coherent state $|z\rangle$ is given by $|z\rangle\langle z|$ which equals $\mathcal{N}(e^{-(a-z)^\dagger(a-z)})$, where $\mathcal{N}$ means the normal ordering. Note that $|z\rangle\langle w| \neq \mathcal{N}(e^{-(a-z)^\dagger(a-w)})$ for $z, w \in \mathbb{C}$, and $z \neq w$.

Note that coherent states are widely used in quantum optics [115, 175] as well as in other areas of physics [174]. In this context we especially exploit the property of $a|z\rangle = z|z\rangle$. It is because for an operator $F(a, a^\dagger)$, which is in the normal form $F(a, a^\dagger) = \mathcal{N}(F(a, a^\dagger))$, its coherent state matrix elements may be readily presented as

$$\langle z|F(a, a^\dagger)|z'\rangle = \langle z|z'\rangle F(z', \bar{z}).$$

Also for the double dot operation, it immediately yields

$$\langle z| : G(a, a^\dagger) : |z'\rangle = \langle z|z'\rangle G(z', \bar{z}).$$

Note that for the general operator none of these formulae hold. But there is a very useful property which is true: if for an arbitrary operator $F(a, a^\dagger)$ we have

$$\langle z|F(a, a^\dagger)|z'\rangle = \langle z|z'\rangle G(z', \bar{z}),$$

then the normally ordered form of $F(a, a^\dagger)$ is given by $\mathcal{N}(F(a, \bar{a})) =: G(a, a^\dagger) :$.

# Appendix H

## C++ Programming

This chapter contains C++ programs that mentioned in the text. In particular, those programs to generate the set $\mathcal{P}_n$ and those programs to count the number of set partitions that satisfy certain set of conditions.

### Recursive Algorithm for Generating $\mathcal{P}_n$

Our first programs present a recursive algorithm for generating the set of partitions of $[n]$, which it implements in the following C++ code.

```c
#include <stdio.h>
#define MAXSIZE 13

void main()
{
    int SetPartition[MAXSIZE],n=4;
    int i,max,j=n;

    while(j>0)
    {
        for(i=1;i<=n;i++) SetPartition[i]=1;
        printf("\n");

        max=0;
        for(i=1;i<j;i++)
            if (SetPartition[i]>max) max=SetPartition[i];

        while(SetPartition[j]==max+1)
        {
            j--; max=0;
            for(i=1;i<j;i++)
                if (SetPartition[i]>max) max=SetPartition[i];
        }
        if (j>0)
        {
            SetPartition[j]++;
            for(i=j+1;i<=n;i++) SetPartition[i]=1;
            j=n;
        }
    }
}
```

Note that the above algorithm can be used to generate any subset of set partitions of $\mathcal{P}_n$. For instance, if we interest to generate all the set partitions $\pi$ of $\mathcal{P}_n$ such that lev$(\pi) = 0$ ($\pi$ does not contain levels), then we replace

519

Lines 11–12 in the above algorithm by the following lines with adding a new variable after line 7, namely int flag=1;.

```
for ( i=2;i<=n; i++)
    if (SetPartition[i]==SetPartition[i-1]) flag=0;

if(flag==1)
{
    for(i=1;i<=n;i++) SetPartition[i]=1;
    printf("\n");
}
```

In more general, if we replace the lines 5–6 in the above procedure by any other conditions, we can generate the required subset of $\mathcal{P}_n$.

## Gray Code for Generating $\mathcal{P}_n$

The next of our programs is a program that present a Gray code generating algorithm for the set partitions of $\mathcal{P}_n$, see [223].

```
#include <stdio.h>
#define MAXSIZE 13
int SetPartition[MAXSIZE],n=4;

void main()
{
    int i,j,p,pos,kp,tt;
    int s[MAXSIZE][MAXSIZE], ls[MAXSIZE],
        fs[MAXSIZE][MAXSIZE], ns[MAXSIZE];
    SetPartition[1]=1; s[1][1]=1; ls[1]=1; fs[1][1]=1; ns[1]=1;

    for (i=2;i<=n; i++)
    {
        SetPartition[i]=1; s[i][1]=1; s[i][2]=2;
        ls[i]=2; fs[1][1]=1; fs[i][2]=0; ns[i]=1;
    }
    pos=2;
    while(pos>1)
    {
        for(i=1;i<=n;i++) SetPartition[i]=1;
        printf("\n");

        pos=n;
        while(ns[pos]==ls[pos]) pos--;
        if (pos>0)
        {
            p=0;
            for (i=1;i<=pos-1;i++)
                if(SetPartition[i]>p) p=SetPartition[i];
            kp=ls[pos];
            while(fs[pos][kp]==1) kp--;
            ns[pos]++;
            SetPartition[pos]=s[pos][kp]; fs[pos][kp]=1;
            for (i=pos;i<=n-1;i++)
            {
                if(p<pi[i]) p=SetPartition[i];
                ns[i+1]=1;
```

```
38            for(j=2;j<=p+1;j++)
39            {
40               s[i+1][j-1]=j; fs[i+1][j-1]=0;
41            }
42            s[i+1][p+1]=1; ls[i+1]=p+1; fs[i+1][p+1]=0;
43            if(SetPartition[i+1]>1)
44                fs[i+1][SetPartition[i+1]-1]=1;
45            else
46                fs[i+1][p+1]=1;
47         }
48      }
49   }
50 }
```

## Flattened Set Partitions

This program computes the number of set partitions $\pi$ of $[n]$ for $n = 2, 3, \ldots, 12$ such that Flatten$(\pi)$ avoids a pattern entered by the user. The user needs to entry the size of the pattern between three and seven and the pattern. The routine vpatterns$d$(), where $d = 3, 4, 5, 6, 7$, check if the Flatten$(\pi)$ avoids the pattern $\tau$ of size $d$. The program also contains one routine that checks whether a pattern occurs: test. The routine builtsp recursively creates all the set partitions of $[n]$ and increases the count if the set partition avoids the pattern $\tau$. After the computation, the output is presented on the screen.

```
 1 #include <stdio.h>
 2 #define MAXSIZE 50
 3 int n,SetPartition[MAXSIZE],pattern[MAXSIZE],tau[MAXSIZE],lentau;
 4 long num1=0;
 5
 6 int order(int a,int b)
 7 { if (a==b) return 0;
 8   else if(a>b) return 1;
 9        else return -1;
10 }
11
12 int test(int k,int *p1,int *p2)
13 { int i;
14   for (i=0;i<k;i++)
15      if (order(p1[i],p1[k])!=order(p2[i],p2[k])) return 1;
16   return 0;
17 }
18
19 int vpatterns3()
20 { int cur[3],i1,i2,i3;
21     for(i3=0;i3<n-2;i3++){
22       cur[0]=SetPartition[i3];
23       for(i2=i3+1;i2<n-1;i2++){
24         cur[1]=SetPartition[i2];
25         if (test(1,cur,tau)) continue;
26           for(i1=i2+1;i1<n;i1++){
27             cur[2]=SetPartition[i1];
28             if (!test(2,cur,tau)) return 0;
29         }
30       }
```

```
31      }
32      return 1;
33  }

35  int vpatterns4()
36  { int cur[4], i1, i2, i3, i4;
37      for(i4=0; i4<n-3; i4++){
38          cur[0]=SetPartition[i4];
39          for(i3=i4+1; i3<n-2; i3++){
40              cur[1]=SetPartition[i3];
41              if (test(1,cur,tau)) continue;
42              for(i2=i3+1; i2<n-1; i2++){
43                  cur[2]=SetPartition[i2];
44                  if (test(2,cur,tau)) continue;
45                  for(i1=i2+1; i1<n; i1++){
46                      cur[3]=SetPartition[i1];
47                      if (!test(3,cur,tau)) return 0;
48                  }
49              }
50          }
51      }
52      return 1;
53  }

55  int vpatterns5()
56  { int cur[5], i1, i2, i3, i4, i5;
57      for(i5=0; i5<n-4; i5++){
58          cur[0]=SetPartition[i5];
59          for(i4=i5+1; i4<n-3; i4++){
60              cur[1]=SetPartition[i4];
61              if (test(1,cur,tau)) continue;
62              for(i3=i4+1; i3<n-2; i3++){
63                  cur[2]=SetPartition[i3];
64                  if (test(2,cur,tau)) continue;
65                  for(i2=i3+1; i2<n-1; i2++){
66                      cur[2]=SetPartition[i2];
67                      if (test(3,cur,tau)) continue;
68                      for(i1=i2+1; i1<n; i1++){
69                          cur[3]=SetPartition[i1];
70                          if (!test(4,cur,tau)) return 0;
71                      }
72                  }
73              }
74          }
75      }
76      return 1;
77  }

79  int vpatterns6()
80  { int cur[6], i1, i2, i3, i4, i5, i6;
81      for(i6=0; i6<n-5; i6++){
82          cur[0]=SetPartition[i6];
83          for(i5=i6+1; i5<n-4; i5++){
84              cur[1]=SetPartition[i5];
85              if (test(1,cur,tau)) continue;
86              for(i4=i5+1; i4<n-3; i4++){
87                  cur[2]=SetPartition[i4];
```

```cpp
            if (test(2,cur,tau)) continue;
            for(i3=i4+1;i3<n-2;i3++){
                cur[3]=SetPartition[i3];
                if (test(3,cur,tau)) continue;
                for(i2=i3+1;i2<n-1;i2++){
                    cur[4]=SetPartition[i2];
                    if (test(4,cur,tau)) continue;
                    for(i1=i2+1;i1<n;i1++){
                        cur[5]=SetPartition[i1];
                        if (!test(5,cur,tau)) return 0;
                    }
                }
            }
          }
        }
      }
    }
    return 1;
}

int vpatterns7()
{ int cur[7],i1,i2,i3,i4,i5,i6,i7;
    for(i7=0;i7<n-6;i7++){
        cur[0]=SetPartition[i7];
        for(i6=i7+1;i6<n-5;i6++){
            cur[1]=SetPartition[i6];
            if (test(1,cur,tau)) continue;
            for(i5=i6+1;i5<n-4;i5++){
                cur[2]=SetPartition[i5];
                if (test(2,cur,tau)) continue;
                for(i4=i5+1;i4<n-3;i4++){
                    cur[3]=SetPartition[i4];
                    if (test(3,cur,tau)) continue;
                    for(i3=i4+1;i3<n-2;i3++){
                        cur[4]=SetPartition[i3];
                        if (test(4,cur,tau)) continue;
                        for(i2=i3+1;i2<n-1;i2++){
                            cur[5]=SetPartition[i2];
                            if (test(5,cur,tau)) continue;
                            for(i1=i2+1;i1<n;i1++){
                                cur[6]=SetPartition[i1];
                                if (!test(6,cur,tau)) return 0;
                            }
                        }
                    }
                }
            }
        }
    }
    return 1;
}

void patterns()
{ if(lentau==3) num1=num1+vpatterns3();
    if(lentau==4) num1=num1+vpatterns4();
    if(lentau==5) num1=num1+vpatterns5();
    if(lentau==6) num1=num1+vpatterns6();
    if(lentau==7) num1=num1+vpatterns7();
```

```
145  }
146
147  void builtsp(int l)
148  {
149      int j,i,vv,mvv,i1,i2,i3;
150      if (l==n)
151      {
152  mvv=0;
153      for(i1=0;i1<n;i1++)
154          if(pattern[i1]>mvv) mvv=pattern[i1];
155  i3=0;
156  for(i1=1;i1<=mvv;i1++)
157  {
158      for(i2=0;i2<n;i2++)
159          if(pattern[i2]==i1) {SetPartition[i3]=i2+1; i3++;}
160  }
161      patterns();
162  }
163      else
164      {
165  vv=0;
166      for(i=0;i<l;i++)
167          if(pattern[i]>vv) vv=pattern[i];
168      for(j=1;j<=vv+1;j++)
169  {
170  pattern[l]=j; builtsp(l+1);
171  }
172      }
173  }
174
175
176  int main()
177  {
178      int i,j,s=0;
179      printf("Give me the length (3-7) of the pattern : ");
180      scanf("%d",&lentau);
181      printf("Write a pattern as sequence (example 3 1 2) : ");
182      for(i=0;i<lentau;i++) scanf("%d",&tau[i]);
183
184      printf("The pattern tau equals ");
185      for(i=0;i<lentau;i++) printf("%d",tau[i]);
186
187      for(n=2;n<13;n++)
188          {
189              num1=0; for(i=0;i<n;i++) pattern[i]=0;
190              builtsp(0);
191              printf("\n | Flatten_%d(tau)|=%d" ,n,num1);
192          }
193      printf("\n");
194      return(1);
195  }
```

## Classification Subword Patterns

This program computes the number of set partitions $\pi$ of $[n]$ for $n = 1, 2, \ldots, 12$ such that $\pi$ avoids a pattern entered by the user. The user needs

# C++ Programming 525

to entry the size of the pattern between three and seven and the pattern. The routine vpatterns$d$(), where $d = 3, 4, 5, 6, 7$, check if the $\pi$ avoids the subword pattern $\tau$ of size $d$. The program also contains one routine that checks whether a pattern occurs: test. The routine builtsp recursively creates all the set partitions of $n$ and increase the count if the set partition avoids the pattern $\tau$. After the computation, the output is presented on the screen.

```
1  #include <stdio.h>
2  #define MAXSIZE 50
3  int n,word[MAXSIZE],tau[MAXSIZE],lentau;
4  long num1=0;
5
6  int order(int a,int b)
7  { if (a==b) return 0;
8    else if(a>b) return 1;
9         else return -1;
10 }
11
12 int test(int k,int *p1,int *p2)
13 { int i;
14   for (i=0;i<k;i++)
15     if (order(p1[i],p1[k])!=order(p2[i],p2[k])) return 1;
16   return 0;
17 }
18
19 int vpatterns3()
20 { int curword[3],i1,i2,i3;
21   for(i3=0;i3<n-2;i3++){
22     curword[0]=word[i3];
23     i2=i3+1; curword[1]=word[i2];
24     if (test(1,curword,tau)) continue;
25     i1=i2+1; curword[2]=word[i1];
26     if (!test(2,curword,tau)) return 0;
27   }
28   return 1;
29 }
30
31 int vpatterns4()
32 { int curword[4],i1,i2,i3,i4;
33   for(i4=0;i4<n-3;i4++){
34     curword[0]=word[i4];
35     i3=i4+1; curword[1]=word[i3];
36     if (test(1,curword,tau)) continue;
37     i2=i3+1; curword[2]=word[i2];
38     if (test(2,curword,tau)) continue;
39     i1=i2+1; curword[3]=word[i1];
40     if (!test(3,curword,tau)) return 0;
41   }
42   return 1;
43 }
44
45 int vpatterns5()
46 { int curword[5],i1,i2,i3,i4,i5;
47   for(i5=0;i5<n-4;i5++){
48     curword[0]=word[i5];
49     i4=i5+1; curword[1]=word[i4];
```

```
50        if (test(1,curword,tau)) continue;
51        i3=i4+1; curword[2]=word[i3];
52        if (test(2,curword,tau)) continue;
53        i2=i3+1; curword[2]=word[i2];
54        if (test(3,curword,tau)) continue;
55        i1=i2+1; curword[3]=word[i1];
56        if (!test(4,curword,tau)) return 0;
57      }
58      return 1;
59  }
60
61  int vpatterns6()
62  { int curword[6],i1,i2,i3,i4,i5,i6;
63      for(i6=0;i6<n-5;i6++){
64        curword[0]=word[i6];
65        i5=i6+1; curword[1]=word[i5];
66        if (test(1,curword,tau)) continue;
67        i4=i5+1; curword[2]=word[i4];
68        if (test(2,curword,tau)) continue;
69        i3=i4+1; curword[3]=word[i3];
70        if (test(3,curword,tau)) continue;
71        i2=i3+1; curword[4]=word[i2];
72        if (test(4,curword,tau)) continue;
73        i1=i2+1; curword[5]=word[i1];
74        if (!test(5,curword,tau)) return 0;
75      }
76      return 1;
77  }
78
79  int vpatterns7()
80  { int curword[7],i1,i2,i3,i4,i5,i6,i7;
81      for(i7=0;i7<n-6;i7++){
82        curword[0]=word[i7];
83        i6=i7+1; curword[1]=word[i6];
84        if (test(1,curword,tau)) continue;
85        i5=i6+1; curword[2]=word[i5];
86        if (test(2,curword,tau)) continue;
87        i4=i5+1; curword[3]=word[i4];
88        if (test(3,curword,tau)) continue;
89        i3=i4+1; curword[4]=word[i3];
90        if (test(4,curword,tau)) continue;
91        i2=i3+1; curword[5]=word[i2];
92        if (test(5,curword,tau)) continue;
93        i1=i2+1; curword[6]=word[i1];
94        if (!test(6,curword,tau)) return 0;
95      }
96      return 1;
97  }
98
99  void patterns()
100 { if(lentau==3) num1=num1+vpatterns3();
101   if(lentau==4) num1=num1+vpatterns4();
102   if(lentau==5) num1=num1+vpatterns5();
103   if(lentau==6) num1=num1+vpatterns6();
104   if(lentau==7) num1=num1+vpatterns7();
105 }
106
```

```
107  void builsp(int l)
108  {
109    int j,i,vv;
110    if (l==n) patterns();
111    else
112    {
113      vv=0;
114      for(i=0;i<l;i++)
115         if(word[i]>vv) vv=word[i];
116      for(j=1;j<=vv+1;j++)
117    {
118      pword[l]=j; builsp(l+1);
119    }
120    }
121  }
122
123  int main()
124  {
125    int i,j,s=0;
126    printf("Give me the length (3-7) of the pattern :");
127    scanf("%d",&lentau);
128    printf("Write a pattern as sequence (example 3 1 2) :");
129    for(i=0;i<lentau;i++) scanf("%d",&tau[i]);
130
131    printf("The pattern tau equals ");
132    for(i=0;i<lentau;i++) printf("%d",tau[i]);
133
134    for(n=1;n<13;n++)
135    {
136       num1=0;  for(i=0;i<n;i++) word[i]=0;
137       builtsp(0);
138       printf("\n | P_%d(tau)|=%d",n,num1);
139    }
140    printf("\n");
141    return(1);
142  }
```

## Number of Set partition that Avoid a Fixed Subsequence Pattern

This program computes the number of set partitions $\pi$ of $[n]$ for $n = 1, 2, \ldots, 12$ such that $\pi$ avoids any subsequence pattern of a given size. The user needs to set the size of the pattern between three and seven by setting the variable lenp. The routine vpatternsd(), where $d = 3, 4, 5, 6, 7$, check if the $\pi$ avoids the subsequence pattern $\tau$ of size $d$. The program also contains one routine that checks whether a pattern occurs: test. The routine builtsp recursively creates all the set partitions of $[n]$ and increase the count if the set partition avoids the pattern $\tau$. After the computation, the output is presented on the screen.

```
1  #include <stdio.h>
2  #define MAXSIZE 50
3  int n,word[MAXSIZE],p[MAXSIZE];
4  int nlet[MAXSIZE];
5  long num1=0;
```

```
6  int listp[1000][7], nlistp=0;
7  FILE *fout;
8
9  int order(int a, int b)
10 { if (a==b) return 0;
11    else if(a>b) return 1;
12            else return -1;
13 }
14
15 int test(int k, int *p1, int *p2)
16 { int i;
17    for (i=0;i<k;i++)
18      if (order(p1[i],p1[k])!=order(p2[i],p2[k])) return 1;
19    return 0;
20 }
21
22 int vpatterns3()
23 { int curword[3];
24    int i1,i2,i3;
25    for(i3=0;i3<n-2;i3++){
26      curword[0]=word[i3];
27      for(i2=i3+1;i2<n-1;i2++){
28        curword[1]=word[i2];
29        if (test(1,curword,p)) continue;
30        for(i1=i2+1;i1<n;i1++){
31          curword[2]=word[i1];
32          if (!test(2,curword,p))
33            return 0;
34        }
35      }
36    }
37    return 1;
38 }
39
40 int vpatterns4()
41 { int curword[4];
42    int i1,i2,i3,i4;
43    for(i4=0;i4<n-3;i4++){
44      curword[0]=word[i4];
45      for(i3=i4+1;i3<n-2;i3++){
46        curword[1]=word[i3];
47        if (test(1,curword,p)) continue;
48        for(i2=i3+1;i2<n-1;i2++){
49          curword[2]=word[i2];
50          if (test(2,curword,p)) continue;
51          for(i1=i2+1;i1<n;i1++){
52            curword[3]=word[i1];
53            if (!test(3,curword,p))
54              return 0;
55          }
56        }
57      }
58    }
59    return 1;
60 }
61
62 int vpatterns5()
```

```cpp
63  {   int curword[5];
64      int i1,i2,i3,i4,i5;
65      for(i5=0;i5<n-4;i5++){
66          curword[0]=word[i5];
67          for(i4=i5+1;i4<n-3;i4++){
68              curword[1]=word[i4];
69              if (test(1,curword,p)) continue;
70              for(i3=i4+1;i3<n-2;i3++){
71                  curword[2]=word[i3];
72                  if (test(2,curword,p)) continue;
73                  for(i2=i3+1;i2<n-1;i2++){
74                      curword[3]=word[i2];
75                      if (test(3,curword,p)) continue;
76                      for(i1=i2+1;i1<n;i1++){
77                          curword[4]=word[i1];
78                          if (!test(4,curword,p))
79                              return 0;
80                      }
81                  }
82              }
83          }
84      }
85      return 1;
86  }

88  int vpatterns6()
89  {   int curword[6];
90      int i1,i2,i3,i4,i5,i6;
91      for(i6=0;i6<n-5;i6++){
92          curword[0]=word[i6];
93          for(i5=i6+1;i5<n-4;i5++){
94              curword[1]=word[i5];
95              if (test(1,curword,p)) continue;
96              for(i4=i5+1;i4<n-3;i4++){
97                  curword[2]=word[i4];
98                  if (test(2,curword,p)) continue;
99                  for(i3=i4+1;i3<n-2;i3++){
100                     curword[3]=word[i3];
101                     if (test(3,curword,p)) continue;
102                     for(i2=i3+1;i2<n-1;i2++){
103                         curword[4]=word[i2];
104                         if (test(4,curword,p)) continue;
105                         for(i1=i2+1;i1<n;i1++){
106                             curword[5]=word[i1];
107                             if (!test(5,curword,p))
108                                 return 0;
109                         }
110                     }
111                 }
112             }
113         }
114     }
115     return 1;
116 }

118 int vpatterns7()
119 {   int curword[7];
```

```
120    int i1,i2,i3,i4,i5,i6,i7;
121    for(i7=0;i7<n-6;i7++){
122        curword[0]=word[i7];
123        for(i6=i7+1;i6<n-5;i6++)
124  {        curword[1]=word[i6];
125           if (test(1,curword,p)) continue;
126           for(i5=i6+1;i5<n-4;i5++){
127               curword[2]=word[i5];
128               if (test(2,curword,p)) continue;
129               for(i4=i5+1;i4<n-3;i4++){
130                   curword[3]=word[i4];
131                   if (test(3,curword,p)) continue;
132                   for(i3=i4+1;i3<n-2;i3++){
133                       curword[4]=word[i3];
134                       if (test(4,curword,p)) continue;
135                       for(i2=i3+1;i2<n-1;i2++){
136                           curword[5]=word[i2];
137                           if (test(5,curword,p)) continue;
138                           for(i1=i2+1;i1<n;i1++){
139                               curword[6]=word[i1];
140                               if (!test(6,curword,p))
141                                   return 0;
142                           }
143                       }
144                   }
145               }
146           }
147       }
148   }
149   return 1;
150 }
151
152 void builtsp(int l,int lenp)
153 { int j,i,vv;
154     if ((l==n))
155     {
156       if(word[0]==1)
157       {
158       if(lenp==3) num1=num1+vpatterns3();
159       if(lenp==4) num1=num1+vpatterns4();
160       if(lenp==5) num1=num1+vpatterns5();
161          if(lenp==6) num1=num1+vpatterns6();
162       if(lenp==7) num1=num1+vpatterns7();
163       }
164     }
165     else
166     {
167     vv=0; for(i=0;i<l;i++) if(word[i]>vv) vv=word[i];
168         for(j=1;j<=vv+1;j++)
169     {
170       nlet[j]++; word[l]=j;
171       builtwsp(l+1,lenp);
172       nlet[j]--;
173     }
174     }
175 }
176
```

```
177  void builtpatterns(int l,int lenp)
178  { int i,j,v;
179    if (l==lenp)
180    {
181        for(i=0;i<lenp;i++) listp[nlistp][i]=p[i];
182      nlistp++;
183    }
184    else
185    {
186    v=0; for(i=0;i<l;i++) if(p[i]>v) v=p[i];
187      for(j=1;j<=(v+1);j++)
188    {
189    p[l]=j;
190    builtpatterns(l+1,lenp);
191    }
192    }
193  }
194
195  int main()
196  {
197    int   lenp=7;
198    fout=fopen("fpart7.txt","w");
199    builtpatterns(0,lenp);
200    int i,j,i1,j1,j2,ii,sw;
201    int ncc=0;
202
203  for(i1=0;i1<nlistp;i1++)
204  {
205    for(i=0;i<lenp;i++) p[i]=listp[i1][i];
206        printf("patterns "); fprintf(fout,"patterns ");
207        for(i=0;i<lenp;i++) {printf("%d",p[i]); fprintf(fout,"%d",p[i]);}
208    printf(" : "); fprintf(fout," : ");
209
210        for(n=7;n<13;n++)
211          {
212            num1=0; num2=0;
213            for(i=0;i<n;i++) {word[i]=0; nlet[i]=0;}
214            builtwords(0,lenp,lenq);
215            printf("%7d ",num1); fprintf(fout,"%7d ",num1);
216      }
217    ncc++;
218      printf("\n");
219      fprintf(fout,"\n");
220  }
221
222    fclose(fout);
223    return(1);
224  }
```

## Number of Set Partition that Contain Exactly Once a Fixed Subsequence Pattern

This program computes the number of set partitions $\pi$ of $[n]$ for $n = 1, 2, \ldots, 12$ such that $\pi$ contains exactly once a fixed subsequence pattern of a given size. The user needs to set the size of the pattern between three and seven by setting the variable lenp. The routine c1patterns$d$(), where $d = 3, 4, 5, 6, 7$, check if the $\pi$ contains exactly once the subsequence pattern $\tau$ of size $d$. The program also contains one routine that checks whether a pattern occurs: test. The routine builtsp recursively creates all the set partitions of $[n]$ and increase the count if the set partition contains exactly once the subsequence pattern $\tau$. After the computation, the output is presented on the screen and on a file.

Note that if we change Line 173 in this program from $if((word[0] == 1)/*\&\&(sparse() == 1)*/)$ to $if((word[0] == 1)\&\&(sparse() == 1))$, we obtain a new program that computes the number of set partitions $\pi \in \mathcal{P}_n$ for $n = 1, 2, \ldots, 12$ such that $\pi$ avoids the subword word pattern 11 and contains exactly once a fixed subsequence pattern of a given size.

```
1  #include <stdio.h>
2  #define MAXSIZE 50
3  int n,NK;
4  int word[MAXSIZE],p[MAXSIZE];
5  long num1=0;
6  int listp[100000][10],nlistp=0,nl=0;
7  int nlet[MAXSIZE];
8
9  FILE *fout; /*file contains the output*/
10
11 int order(int a,int b)
12 { if (a==b) return 0;
13    else if(a>b) return 1;
14         else return -1;
15 }
16
17 int test(int k,int *p1,int *p2)
18 { int i;
19    for (i=0;i<k;i++)
20       if (order(p1[i],p1[k])!=order(p2[i],p2[k])) return 1;
21    return 0;
22 }
23
24 int c1patterns3()
25 { int curword[4];
26    int i1,i2,i4,acc=0;
27    for(i4=0;i4<n-2;i4++){
28       curword[0]=word[i4];
29       for(i2=i4+1;i2<n-1;i2++){
30          curword[1]=word[i2];
31          if (test(1,curword,p)) continue;
32          for(i1=i2+1;i1<n;i1++){
33             curword[2]=word[i1];
34             if (!test(2,curword,p))
35                {acc++; if(acc>1) return 0;}
36          }
37       }
38    }
39    if (acc==1) return 1;
40    else return 0;
41 }
42
43 int c1patterns4()
```

```cpp
44  { int curword[4];
45      int i1,i2,i3,i4,acc=0;
46      for(i4=0;i4<n-3;i4++){
47          curword[0]=word[i4];
48          for(i3=i4+1;i3<n-2;i3++){
49              curword[1]=word[i3];
50              if (test(1,curword,p)) continue;
51              for(i2=i3+1;i2<n-1;i2++){
52                  curword[2]=word[i2];
53                  if (test(2,curword,p)) continue;
54                  for(i1=i2+1;i1<n;i1++){
55                      curword[3]=word[i1];
56                      if (!test(3,curword,p))
57          {acc++; if(acc>1) return 0;}
58                  }
59              }
60          }
61      }
62      if(acc==1) return 1;
63      else return 0;
64  }
65
66  int c1patterns5()
67  { int curword[5];
68      int i1,i2,i3,i4,i5,acc=0;
69      for(i5=0;i5<n-4;i5++){
70          curword[0]=word[i5];
71          for(i4=i5+1;i4<n-3;i4++){
72              curword[1]=word[i4];
73              if (test(1,curword,p)) continue;
74              for(i3=i4+1;i3<n-2;i3++){
75                  curword[2]=word[i3];
76                  if (test(2,curword,p)) continue;
77                  for(i2=i3+1;i2<n-1;i2++){
78                      curword[3]=word[i2];
79                      if (test(3,curword,p)) continue;
80                      for(i1=i2+1;i1<n;i1++){
81                          curword[4]=word[i1];
82                          if (!test(4,curword,p))
83          {acc++; if (acc>1) return 0;}
84                      }
85                  }
86              }
87          }
88      }
89      if(acc==1) return 1;
90      else return 0;
91  }
92
93  int c1patterns6()
94  { int curword[6];
95      int i1,i2,i3,i4,i5,i6,acc=0;
96      for(i6=0;i6<n-5;i6++){
97          curword[0]=word[i6];
98          for(i5=i6+1;i5<n-4;i5++){
99              curword[1]=word[i5];
100             if (test(1,curword,p)) continue;
```

```
101        for(i4=i5+1;i4<n-3;i4++){
102           curword[2]=word[i4];
103           if (test(2,curword,p)) continue;
104           for(i3=i4+1;i3<n-2;i3++){
105              curword[3]=word[i3];
106              if (test(3,curword,p)) continue;
107              for(i2=i3+1;i2<n-1;i2++){
108                 curword[4]=word[i2];
109                 if (test(4,curword,p)) continue;
110                 for(i1=i2+1;i1<n;i1++){
111                    curword[5]=word[i1];
112                    if (!test(5,curword,p))
113       {acc++; if(acc>1) return 0;}
114                    }
115                 }
116              }
117           }
118        }
119     }
120     if (acc==1) return 1;
121     else return 0;
122  }
123
124  int c1patterns7()
125  { int curword[7];
126     int i1,i2,i3,i4,i5,i6,i7,acc=0;
127     for(i7=0;i7<n-6;i7++){
128        curword[0]=word[i7];
129        for(i6=i7+1;i6<n-5;i6++){
130           curword[1]=word[i6];
131           if (test(1,curword,p)) continue;
132           for(i5=i6+1;i5<n-4;i5++){
133              curword[2]=word[i5];
134              if (test(2,curword,p)) continue;
135              for(i4=i5+1;i4<n-3;i4++){
136                 curword[3]=word[i4];
137                 if (test(3,curword,p)) continue;
138                 for(i3=i4+1;i3<n-2;i3++){
139                    curword[4]=word[i3];
140                    if (test(4,curword,p)) continue;
141                    for(i2=i3+1;i2<n-1;i2++){
142                       curword[5]=word[i2];
143                       if (test(5,curword,p)) continue;
144                       for(i1=i2+1;i1<n;i1++){
145                          curword[6]=word[i1];
146                          if (!test(6,curword,p))
147                 {acc++; if(acc>1) return 0;}
148                       }
149                    }
150                 }
151              }
152           }
153        }
154     }
155     if (acc==1) return 1;
156     else return 0;
157  }
```

```
158
159  int sparse()
160  {
161    int i;
162    for(i=0;i<n-1;i++) if (word[i]==word[i+1]) return(0);
163    for(i=0;i<n;i++) printf("%d",word[i]);
164    return(1);
165  }
166
167  void builtsp(int l,int lenp)
168  { int j,i,vv,vv1;
169    if (l==n)
170    {
171      if(word[0]==1)
172      {
173        if((lenp==3)) {num1=num1+c1patterns3();}
174        if((lenp==4)) {num1=num1+c1patterns4();}
175        if((lenp==5)) {num1=num1+c1patterns5();}
176        if((lenp==6)) {num1=num1+c1patterns6();}
177        if((lenp==7)) {num1=num1+c1patterns7();}
178      }
179    }
180    else
181    {
182    vv=0; for(i=0;i<l;i++) if(word[i]>vv) vv=word[i];
183        for(j=1;j<=vv+1;j++)
184    {
185        nlet[j]++;
186        word[l]=j;
187        builtsp(l+1,lenp);
188        nlet[j]--;
189    }
190    }
191  }
192
193  void builtpatterns(int l,int lenp)
194  { int i,j,v;
195    if (l==lenp)
196    {
197          for(i=0;i<lenp;i++) listp[nlistp][i]=p[i];
198      nlistp++;
199    }
200    else
201    {
202    v=0; for(i=0;i<l;i++) if(p[i]>v) v=p[i];
203        for(j=1;j<=(v+1);j++)
204    {
205      p[l]=j;
206      builtpatterns(l+1,lenp);
207    }
208    }
209  }
210
211  int main()
212  {
213      int lenp=4, max_n=12;
214
```

```
215    fout=fopen("sppart_one4.txt","w");
216    builtpatterns(0,lenp);
217
218    int i,j,i1,j1,j2,ii,sw;
219    int ncc=0;
220
221    for(i1=0;i1<nlistp;i1++)
222       {
223  printf("%d)",ncc+1);
224  for(i=0;i<lenp;i++) p[i]=listp[i1][i];
225      printf("patterns "); fprintf(fout,"patterns ");
226      for(i=0;i<lenp;i++) {printf("%d",p[i]); fprintf(fout,"%d",p[i
               ]);}
227  printf(" :");
228  fprintf(fout," : ");
229
230      for(n=1;n<max_n;n++)
231         {
232            num1=0;
233            for(i=0;i<n;i++) {word[i]=0; nlet[i]=0;}
234            builtwords(0,lenp);
235            printf("%5d ",num1); fprintf(fout,"%5d ",num1);
236         }
237      printf("\n"); fprintf(fout,"\n");
238  }
239   fclose(fout);
240   return(1);
241 }
```

# Appendix I

## Tables

In Chapter 6, we classify the Wilf-equivalence classes for subsequence patterns of sizes four, five, six and seven. Tables 6.3, 6.4, 6.5, and 6.6 provide the complete listing of all Wilf-equivalence classes, with/without the ones that consist of a single pattern. The program above was used to compute values of the sequence $\{P_n(\tau)\}$. Since large values of $n$ are time-consuming to compute, we display the values of $P_n(\tau)$ for small $n$ and when $\tau$ is a subsequence pattern of size six. For patterns of size seven, we refer the reader to the home page of the author.

**Table I.1**: The numbers $P_n(\tau)$ where $\tau$ is a subsequence pattern of size six and $n = 6, 7, \ldots, 11$

| $\tau$ | $P_n(\tau)$ for $n = 6, 7, \ldots, 11$ |
|---|---|
| 123415, 123425, 123435, 123445, 123451, 123452, 123453, 123454, 123455, 123456 | 202, 855, 3845, 18002, 86472, 422005 |
| 123414 | 202, 856, 3867, 18286, 89291, 445879 |
| 123413, 123424 | 202, 856, 3867, 18288, 89348, 446801 |
| 123134, 123143 | 202, 856, 3867, 18289, 89375, 447219 |
| 123241 | 202, 856, 3868, 18312, 89684, 450407 |
| 123142, 123314 | 202, 856, 3868, 18312, 89684, 450408 |
| 123124, 123145, 123214, 123234, 123243, 123245, 123324, 123341, 123342, 123345, 123412, 123421, 123423, 123431, 123432, 123434, 123441, 123442, 123443 | 202, 856, 3868, 18313, 89711, 450825 |
| 123144, 123244, 123344 | 202, 856, 3869, 18340, 90135, 455917 |
| 121342 | 202, 857, 3888, 18555, 92027, 470221 |
| 122314 | 202, 857, 3889, 18578, 92339, 473499 |
| 122341 | 202, 857, 3889, 18578, 92341, 473559 |
| 121324 | 202, 857, 3889, 18579, 92369, 474015 |
| 121334, 121343, 122334, 122343 | 202, 857, 3890, 18605, 92767, 478726 |
| 121345, 122345 | 202, 857, 3891, 18628, 93074, 481845 |
| 123141 | 202, 857, 3891, 18628, 93082, 482103 |
| 123242 | 202, 857, 3891, 18628, 93084, 482169 |
| Continued on next page | |

| $\tau$ | $P_n(\tau)$ for $n = 6, 7, \ldots, 11$ |
|---|---|
| 121344, 122344 | 202, 857, 3891, 18630, 93135, 482921 |
| 123114, 123224, 123334, 123343, 123411, 123422, 123433, 123444 | 202, 857, 3891, 18630, 93136, 482957 |
| 121234, 122134 | 202, 858, 3908, 18801, 94448, 491234 |
| 123132 | 202, 858, 3909, 18821, 94686, 493433 |
| 123213 | 202, 858, 3909, 18822, 94712, 493834 |
| 122313 | 202, 858, 3910, 18844, 95008, 497017 |
| 121332 | 202, 858, 3910, 18845, 95037, 497499 |
| 123123, 123312, 123321 | 202, 858, 3910, 18846, 95058, 497753 |
| 123231 | 202, 858, 3910, 18847, 95086, 498215 |
| 121323 | 202, 858, 3910, 18847, 95087, 498248 |
| 122331 | 202, 858, 3911, 18871, 95434, 502205 |
| 121341 | 202, 858, 3911, 18872, 95455, 502471 |
| 121314 | 202, 858, 3911, 18872, 95460, 502606 |
| 121233, 122133 | 202, 858, 3911, 18872, 95461, 502640 |
| 112342 | 202, 858, 3911, 18873, 95485, 502977 |
| 122342 | 202, 858, 3911, 18873, 95486, 503008 |
| 122324 | 202, 858, 3911, 18874, 95511, 503380 |
| 112324 | 202, 858, 3911, 18874, 95513, 503436 |
| 112334, 112343 | 202, 858, 3912, 18897, 95828, 506812 |
| 112345 | 202, 858, 3912, 18900, 95904, 507935 |
| 112344 | 202, 858, 3912, 18900, 95909, 508102 |
| 121134, 122234 | 202, 859, 3929, 19077, 97377, 518804 |
| 112234 | 202, 859, 3930, 19096, 97599, 520871 |
| 112332 | 202, 859, 3930, 19100, 97700, 522415 |
| 112323 | 202, 859, 3930, 19100, 97700, 522417 |
| 123313 | 202, 859, 3931, 19115, 97828, 523144 |
| 123131 | 202, 859, 3931, 19115, 97828, 523161 |
| 123133 | 202, 859, 3931, 19115, 97831, 523233 |
| 123113 | 202, 859, 3931, 19116, 97852, 523508 |
| 121313 | 202, 859, 3931, 19117, 97872, 523757 |
| 121132 | 202, 859, 3931, 19117, 97882, 523978 |
| 121312 | 202, 859, 3931, 19118, 97898, 524139 |
| 121223, 121232, 121322, 122123, 122132, 122213, 122231, 122312, 122321, 123112, 123122, 123212, 123221, 123223, 123233, 123323, 123331, 123332 | 202, 859, 3931, 19119, 97921, 524460 |
| 121123 | 202, 859, 3931, 19120, 97945, 524813 |
| 121213, 123121, 123232 | 202, 859, 3931, 19120, 97947, 524870 |
| 122323 | 202, 859, 3931, 19120, 97947, 524871 |
| Continued on next page ||

# Tables 539

| $\tau$ | $P_n(\tau)$ for $n = 6, 7, \ldots, 11$ |
|---|---|
| 121231 | 202, 859, 3931, 19120, 97948, 524896 |
| 121321 | 202, 859, 3931, 19121, 97972, 525251 |
| 112341 | 202, 859, 3931, 19122, 97987, 525337 |
| 112314 | 202, 859, 3931, 19122, 97992, 525474 |
| 112233 | 202, 859, 3931, 19123, 98023, 526040 |
| 122131 | 202, 859, 3932, 19139, 98173, 527077 |
| 122113 | 202, 859, 3932, 19141, 98222, 527792 |
| 121331 | 202, 859, 3932, 19142, 98242, 528050 |
| 122311, 123211, 123311, 123322 | 202, 859, 3932, 19142, 98246, 528141 |
| 122332 | 202, 859, 3932, 19144, 98296, 528903 |
| 121333, 122333 | 202, 859, 3932, 19145, 98321, 529292 |
| 121133, 122233 | 202, 859, 3932, 19146, 98345, 529646 |
| 112312 | 202, 860, 3948, 19308, 99685, 539558 |
| 112132 | 202, 860, 3948, 19310, 99730, 540193 |
| 112123 | 202, 860, 3948, 19310, 99730, 540195 |
| 112134 | 202, 860, 3949, 19327, 99908, 541664 |
| 112313 | 202, 860, 3949, 19327, 99914, 541829 |
| 112321 | 202, 860, 3949, 19330, 99990, 542997 |
| 112223, 112232, 112322 | 202, 860, 3949, 19330, 99993, 543077 |
| 112213 | 202, 860, 3949, 19331, 100010, 543248 |
| 112231 | 202, 860, 3949, 19332, 100031, 543522 |
| 112331 | 202, 860, 3950, 19350, 100240, 545542 |
| 112133 | 202, 860, 3950, 19354, 100332, 546867 |
| 112333 | 202, 860, 3950, 19354, 100338, 547019 |
| 111223, 111232 | 202, 861, 3964, 19488, 101434, 555332 |
| 111234 | 202, 861, 3966, 19516, 101662, 556533 |
| 121131 | 202, 860, 3954, 19434, 101338, 557307 |
| 121311 | 202, 860, 3954, 19434, 101342, 557389 |
| 121113, 122223, 122232, 122322, 123111, 123222, 123333 | 202, 860, 3954, 19434, 101350, 557570 |
| 111233 | 202, 861, 3966, 19523, 101837, 559225 |
| 112131 | 202, 861, 3970, 19599, 102778, 568978 |
| 112113 | 202, 861, 3970, 19599, 102786, 569157 |
| 112311 | 202, 861, 3970, 19600, 102802, 569328 |
| 111213 | 202, 862, 3984, 19731, 103869, 577539 |
| 111231 | 202, 862, 3984, 19733, 103905, 577981 |
| 111123 | 202, 863, 3996, 19837, 104726, 584231 |
| 121212, 122121 | 202, 863, 3999, 19880, 105134, 587479 |
| 121221 | 202, 863, 3999, 19880, 105135, 587501 |
| 122112 | 202, 863, 3999, 19881, 105150, 587670 |
| 112122 | 202, 863, 3999, 19882, 105176, 588067 |
| Continued on next page | |

| $\tau$ | $P_n(\tau)$ for $n = 6, 7, \ldots, 11$ |
|---|---|
| 121122 | 202, 863, 3999, 19883, 105188, 588178 |
| 112212 | 202, 863, 3999, 19883, 105192, 588263 |
| 122211 | 202, 863, 3999, 19885, 105226, 588686 |
| 112221 | 202, 863, 3999, 19885, 105233, 588828 |
| 111222 | 202, 863, 3999, 19889, 105314, 589939 |
| 121222, 122122, 122212, 122221 | 202, 863, 4001, 19917, 105594, 592404 |
| 121112 | 202, 863, 4001, 19918, 105614, 592671 |
| 112112 | 202, 863, 4001, 19918, 105614, 592676 |
| 121121 | 202, 863, 4001, 19918, 105618, 592756 |
| 111212, 112121, 121211 | 202, 863, 4001, 19919, 105636, 592976 |
| 112222 | 202, 863, 4002, 19938, 105878, 595579 |
| 122111 | 202, 863, 4002, 19939, 105886, 595601 |
| 111221 | 202, 863, 4002, 19939, 105893, 595738 |
| 112211 | 202, 863, 4002, 19939, 105895, 595781 |
| 111122 | 202, 863, 4002, 19939, 105901, 595895 |
| 111112, 111121, 111211, 112111, 121111, 122222 | 202, 864, 4020, 20150, 107964, 614574 |
| 111111 | 202, 869, 4075, 20645, 112124, 648649 |

In Chapter 7, we classify the Wilf-equivalence classes of noncrossing set partitions for subsequence patterns of sizes at most six. Tables 7.3 and 7.4 provide the complete listing of all Wilf-equivalence classes. Our program was used to compute values of the sequence $\{P_n(1212, \tau)\}$. Since large values of $n$ are time-consuming to compute, we display the values of $P_n(1212, \tau)$ for small $n$.

**Table I.2**: The numbers $P_n(1212, \tau)$ where $\tau$ is a subsequence pattern of size four and $n = 4, 5, \ldots, 12$

| $\tau$ | $P_n(1212, \tau)$ for $n = 4, 5, \ldots, 12$ |
|---|---|
| 1234 | 13, 31, 66, 127, 225, 373, 586, 881, 1277 |
| 1223, 1233, 1123 | 13, 33, 81, 193, 449, 1025, 2305, 5121, 11265 |
| 1231 | 13, 33, 82, 202, 497, 1224, 3017, 7439, 18343 |
| 1213, 1232, 1221, 1122 | 13, 34, 89, 233, 610, 1597, 4181, 10946, 28657 |
| 1222, 1112 | 13, 35, 96, 267, 750, 2123, 6046, 17303, 49721 |
| 1121, 1211 | 13, 35, 97, 275, 794, 2327, 6905, 20705, 62642 |
| 1111 | 13, 36, 104, 309, 939, 2905, 9118, 28964, 92940 |

**Table I.3**: The numbers $P_n(1212, \tau)$ where $\tau$ is a subsequence pattern of size five and $n = 6, 7, \ldots, 12$

| $\tau$ | $P_n(1212, \tau)$ for $n = 6, 7, \ldots, 12$ |
|---|---|
| 12345 | 116, 302, 715, 1549, 3106, 5831, 10352 |
| 12234, 12334, 12344, 11234 | 119, 334, 902, 2351, 5945, 14660, 35408 |
| 12341 | 119, 336, 927, 2527, 6870, 18717, 51155 |
| 12233, 11233, 11223 | 121, 354, 1021, 2901, 8130, 22513, 61713 |
| 12343, 12134, 12324 | 121, 355, 1032, 2973, 8496, 24111, 68017 |
| 12331, 12231 | 121, 355, 1033, 2986, 8594, 24674, 70757 |
| 12342, 12314 | 121, 356, 1044, 3057, 8948, 26192, 76674 |
| 12332, 12333, 12133, 11123, 12213, 12321, 12223, 11232 | 122, 365, 1094, 3281, 9842, 29525, 88574 |
| 11122, 12221, 11222 | 123, 374, 1147, 3538, 10958, 34042, 105997 |
| 11231, 12311 | 122, 367, 1117, 3441, 10720, 33727, 107012 |
| 12113, 12322, 12232, 11213 | 123, 375, 1157, 3603, 11304, 35683, 113219 |
| 12211, 11221 | 123, 375, 1158, 3615, 11393, 36209, 115940 |
| 12131 | 123, 376, 1168, 3678, 11716, 37688, 122261 |
| 12222, 11112 | 124, 384, 1210, 3865, 12482, 40677, 133572 |
| 11121, 11211, 12111 | 124, 385, 1221, 3939, 12886, 42648, 142544 |
| 11111 | 125, 393, 1265, 4147, 13798, 46476, 158170 |

**Table I.4**: The numbers $P_n(1212, \tau)$ where $\tau$ is a subsequence pattern of size six and $n = 7, 8, \ldots, 12$

| $\tau$ | $P_n(1212, \tau)$ for $n = 7, 8, \ldots, 12$ |
|---|---|
| 123456 | 407, 1205, 3313, 8398, 19691, 43022 |
| 122345, 123345, 123445, 123455, 112345 | 411, 1263, 3750, 10721, 29571, 79009 |
| 123451 | 411, 1266, 3799, 11152, 32313, 93228 |
| 123344, 112344, 112234, 122334, 122344, 112334 | 414, 1302, 4035, 12274, 36626, 107331 |
| 123245, 123454, 123435, 121345 | 414, 1304, 4063, 12497, 37960, 114013 |
| 123441, 123341, 122341 | 414, 1304, 4065, 12530, 38265, 116103 |
| 123452, 123415 | 414, 1306, 4094, 12766, 39695, 123324 |
| 123425, 123145, 123453 | 415, 1317, 4163, 13090, 40957, 127603 |
| 123444, 122234, 123334, 111234 | 415, 1318, 4173, 13150, 41242, 128801 |
| 112233 | 416, 1327, 4219, 13315, 41637, 128941 |
| 123244, 121344, 122343, 112324, 112343, 121334 | 416, 1328, 4233, 13430, 42362, 132827 |
| 122331 | 416, 1328, 4234, 13446, 42509, 133846 |
| 123443, 122134, 123324 | 416, 1329, 4247, 13544, 43071, 136568 |
| Continued on next page | |

| $\tau$ | $P_n(1212,\tau)$ for $n=7,8,\ldots,12$ |
|---|---|
| 123431, 123241 | 416, 1329, 4247, 13545, 43088, 136732 |
| 123342, 123314, 123442, 122314 | 416, 1329, 4248, 13560, 43217, 137570 |
| 123421, 112342, 123144 | 416, 1330, 4261, 13658, 43782, 140347 |
| 112341, 123411 | 415, 1321, 4218, 13533, 43704, 142170 |
| 111233, 122333, 111223, 122233, 112223, 112333 | 417, 1340, 4321, 13939, 44916, 144484 |
| 121343, 122133, 112332, 123432, 122231, 123321, 123214, 123331 | 417, 1341, 4334, 14041, 45542, 147798 |
| 122324, 123343, 112134, 123224, 121134, 123433 | 417, 1342, 4347, 14142, 46155, 151014 |
| 122311, 112231, 123311, 112331 | 417, 1342, 4349, 14173, 46434, 152931 |
| 122342, 112314, 123114, 123422 | 417, 1343, 4361, 14257, 46883, 154962 |
| 122213, 123332, 111232, 121333 | 418, 1352, 4410, 14463, 47605, 157084 |
| 123141, 121341 | 417, 1344, 4375, 14372, 47614, 158961 |
| 122321, 121133, 112133, 112322, 112232, 123221 | 418, 1353, 4422, 14553, 48145, 159924 |
| 122113, 112213, 123322, 122332 | 418, 1353, 4423, 14567, 48263, 160702 |
| 123211, 112321 | 418, 1353, 4423, 14568, 48279, 160853 |
| 123333, 111123, 122223 | 418, 1354, 4434, 14643, 48688, 162809 |
| 121314, 123242 | 418, 1354, 4435, 14655, 48776, 163324 |
| 121331, 122131 | 418, 1354, 4436, 14670, 48908, 164220 |
| 111222 | 419, 1362, 4476, 14821, 49338, 164875 |
| 112221, 122211 | 419, 1363, 4488, 14914, 49921, 168094 |
| 111122, 112222, 122221 | 419, 1364, 4498, 14980, 50280, 169836 |
| 111231, 123111, 112311 | 418, 1356, 4461, 14859, 50036, 170096 |
| 111213, 122232, 123222, 112113, 122322, 121113 | 419, 1365, 4511, 15081, 50901, 173181 |
| 122111, 112211, 111221 | 419, 1365, 4512, 15096, 51032, 174064 |
| 121311, 112131, 121131 | 419, 1366, 4524, 15183, 51532, 176587 |
| 122222, 111112 | 420, 1375, 4576, 15431, 52603, 180957 |
| 111211, 112111, 121111, 111121 | 420, 1376, 4588, 15521, 53144, 183825 |
| 111111 | 421, 1385, 4642, 15795, 54418, 189454 |

# Appendix J

# Notation

The table below lists the notation. The table contains an alphabetical listing of named quantities. Greek letters occur according to their sound. We use "gf", "sps", and "sp" as an abbreviation for generating function, set partitions, and set partition, respectively. Note that in rare instances we use the same symbol for two different objects. This is a result of following the common notation in the literature. In each case, the two different notation do not occur in the same chapter.

| Notation | Definition |
|---|---|
| $a, a^\dagger$ | Boson creation and annihilation operators |
| $a^j$ | The word $aa \cdots a$ of size $j$ |
| $\{a_n\}_{n\geq 0}$ | Sequence |
| $A^n$ | The set of all words of size $n$ over alphabet $A$ |
| $A^*$ | The set of all the words over the alphabet $A$ of all sizes |
| $B_1/B_2/\cdots/B_k$ | Block representation of a sp |
| $B_j$ | Blocks of a sp |
| $\text{Bell}_n$ | The $n$-th Bell number |
| $\text{Bell}_{n;q}$ | The $n$-th $q$-Bell number |
| $\text{blo}(\pi)$ | The number of blocks in $\pi$ |
| $\times$ | Cartesian product |
| $\overline{C}$ | Complement |
| $\text{Col}_i$ | The set of all cells in column $i$ |
| $\text{Cat}_n$ | The $n$-th Catalan number |
| $\text{Cat}(x)$ | The gf for the sequence $\{\text{Cat}_n\}_{n\geq 0}$ |
| $\mathcal{C}on(F(a, a^\dagger))$ | The set of all contraction of an operator |
| $\text{cross}_l(i, j)$ | The number of such left crossings for $(i, j)$ |
| $\text{cr}_k(\pi)$ | The number of $k$-crossings of $\pi$ |
| $D$ | The differential operator |
| $\text{dcross}(i, j)$ | The number of such unpaired $k$ for the edge $(i, j)$ |
| $\Delta$ | The difference operator |
| $\Delta_q$ | The $q$-difference operator |
| $\delta_C$ | A function equals 1 if $C$ holds and 0 otherwise |
| $\Delta(x)$ | The characteristic polynomial |
| $\text{des}(\pi)$ | Number of descents in $\pi$ |

543

| | |
|---|---|
| $\frac{\partial}{\partial x_i} A(\vec{x})$ | The derivative of $A(\vec{x})$ with respect to $x_i$ |
| $e_q(x)$ | The $q$-exponential function |
| $e = (x, y)$ | An edge connects $x$ and $y$ |
| $f : A \to B$ | A function from $A$ to $B$ |
| $F(a, a^\dagger)$ | Operator |
| $\mathscr{F}(S)$ | The set of all Feynman diagrams on $S$ |
| $\mathscr{F}_k(S)$ | The set of Feynman diagrams of degree $k$ |
| $\text{Fib}_n$ | The $n$-th Fibonacci number |
| $\text{Fib}(x)$ | The gf for the sequence $\{\text{Fib}_n\}_{n \geq 0}$ |
| $\text{Flatten}(\pi)$ | The flattened sp of $\pi$ |
| $\text{Flatten}_n(\tau)$ | The set of all sps in $\mathcal{P}_n$ that avoid the pattern $\tau$ |
| $\text{Flatten}_{n,k}(\tau)$ | The set of all sps in $\mathcal{P}_{n,k}$ that avoid the pattern $\tau$ |
| $\mathcal{I}_n$ | set of involutions of $[n]$ |
| $\int A(\vec{x}) dx_i$ | The integral of $A(\vec{x})$ with respect to $x_i$ |
| $\cap$ | Intersection |
| $\text{inv}(\pi)$ | The number of inversions in $\pi$ |
| $\widehat{\text{inv}}(\pi)$ | The number of dual inversions in $\pi$ |
| $[k]$ | Set of integers from 1 to $k = \{1, 2, \ldots, k\}$ |
| $[k]^n$ | Set of words of size $n$ over alphabet $[k]$ |
| $L(\pi)$ | The set of leftmost of letters of $\pi$ |
| $\mathbb{L}[[\vec{x}]]$ | The set of *formal power series* or *generating functions* in $\vec{x} = (x_1, \ldots, x_k)$ |
| $\text{len}(\gamma)$ | The size of $\gamma$ |
| $\text{lev}(\pi)$ | Number of levels in $\pi$ |
| $\text{Luc}_n$ | $n$th Lucas number |
| $M \stackrel{F}{\sim} M'$ | $M$ and $M'$ are Ferrers-equivalent |
| $M \stackrel{F}{\sim} M'$ | $M$ and $M'$ are stack-equivalent |
| $\text{maj}(\pi)$ | The major index of $\pi$ |
| $\widehat{\text{maj}}(\pi)$ | The dual major index of $\pi$ |
| $\text{Mot}_n$ | $n$th Motzkin number |
| $[n]_q$ | The $q$-number |
| $n!$ | $n$ factorial $= n(n-1) \cdots 1$ |
| $[n]_q!$ | The $q$-analog of the factorial |
| $n!!$ | $n$ double factorial $= n(n-2)(n-4) \cdots$ |
| $NC_n$ | The set of noncrossing sps of $[n]$ |
| $\text{ne}_k(\pi)$ | The number of $k$-nestings of $\pi$ |
| $\binom{n}{k}$ | $= \frac{n!}{k!(n-k)!}$ |
| $\begin{bmatrix} n \\ j \end{bmatrix}$ | The $q$-binomial coefficient |
| $\mathbb{N}$ | Set of natural numbers $= \{1, 2, 3, \ldots\}$ |

# Notation

| | |
|---|---|
| $\mathbb{N}_0$ | $\mathbb{N} \cup \{0\}$ |
| $\mathcal{N}[F(a, a^\dagger)]$ | Normal ordering of an operator |
| strongrec | Number of strong records in $\pi$ |
| weakrec | Number of weak records in $\pi$ |
| addrec | Number of additional records in $\pi$ |
| $\mathrm{occ}_\tau(\pi)$ | Number occurrences of the pattern $\tau$ in $\pi$ |
| $:\pi:$ | Double dot operation |
| $\mathcal{P}(S)$ | The set of all sps of $S$ |
| $\mathcal{P}_n$ | The set of all sps of $[n]$ |
| $p_n$ | The number of all sps of $[n]$ |
| $\mathcal{P}_{n,k}$ | The set of all sps of $[n]$ with exactly $k$ blocks |
| $\mathcal{P}_n(\tau)$ | The set of all $\tau$-avoiding sps of $[n]$ |
| $\mathcal{P}_n(T)$ | The set of all sps of $[n]$ that avoid each pattern in $T$ |
| $\mathcal{P}_{n,k}(\tau)$ | The set of all $\tau$-avoiding sps of $[n]$ with exactly $k$ blocks |
| $\mathcal{P}_{n,k}(T)$ | The set of all sps of $[n]$ with exactly $k$ blocks that avoid each pattern in $T$ |
| $P_\tau(x;q)$ | The gf for the number of sps of $[n]$ according to the number occurrences of pattern $\tau$ |
| $P_\tau(x,y;q)$ | The gf for the number of sps of $[n]$ with exactly $k$ blocks according to the number occurrences of pattern $\tau$ |
| $P_\tau(x,y;q\|\theta_1\cdots\theta_m)$ | The gf for the number of sps $\pi = \theta_1\cdots\theta_m\pi'$ of $[n]$ with exactly $k$ blocks according to the number occurrences of pattern $\tau$ |
| $P_\tau(x;q;k)$ | The gf for the number of sps of $[n]$ with exactly $k$ blocks according to the number occurrences of pattern $\tau$ |
| $P_\tau(x;q;k\|\theta_1\cdots\theta_m)$ | The gf for the number of sps $\pi = \theta_1\cdots\theta_m\pi'$ of $[n]$ with exactly $k$ blocks according to the number occurrences of pattern $\tau$ |
| $R(\pi)$ | The set of rightmost of letters of $\pi$ |
| $\mathbb{R}[x]$ | The set of all polynomials in single variable $x$ |
| $\mathrm{Reduce}(\pi)$ | Reduced form of $\pi$ |
| $\mathrm{ris}(\pi)$ | Number of rises in $\pi$ |
| $RL$-minima | Right-to-left minima |
| $\mathrm{R}_{n,k}$ | The set of all rook placements of $n-k$ rooks on the $n$-triangular shape |
| $Row_j$ | The set of all cells in row $j$ |
| $\mathrm{Sign}(\pi) = (-1)^{n-k}$ | The sign of a sp $\pi$ of $n$ with exactly $k$ blocks |
| $\mathcal{S}_n$ | Set of permutations of $[n]$ |

| | |
|---|---|
| $ss_\Lambda(M)$ | The number semi-standard fillings of $\lambda$ that avoid $M$ |
| $\sigma + k$ | The sequence $(\sigma_1 + k)\cdots(\sigma_n + k)$, where $\sigma = \sigma_1\cdots\sigma_n$ |
| $\mathrm{srec}(\pi)$ | The sum over the positions of all records in $\pi$ |
| $\mathrm{Stir}(n,k)$ | Stirling number of the second kind |
| $\mathscr{S}(\gamma)$ | The set of singletons of $\gamma$ |
| $\mathcal{T}(S)$ | The generating tree whose vertices at level $n$ are the set partitions of $\mathcal{P}_n(S)$ |
| $\mathrm{tcross}(\gamma)$ | Total number crossings of $\gamma$ |
| $\tau \sim \nu$ | $\tau$ and $\nu$ are called Wilf-equivalent |
| $\tau \sim_s \nu$ | $\tau$ and $\nu$ that are strong Wilf-equivalent |
| $\theta_{p,q}$ | Denotes the sequence $2^p 12^q$ |
| $\vartheta_{p,q}$ | Denotes the sequence $1^p 21^q$ |
| $U_n(t)$ | $n$th Chebyshev polynomial of the second kind |
| $\cup$ | Union |
| $\mathscr{W}_q(\gamma)$ | The $q$-weight of $\gamma$ |
| $\lfloor x \rfloor$ | The largest integer which is small or equal $x$ |
| $(z)_n$ | Falling polynomial $z(z-1)\cdots(z-n+1)$ |
| $\mathbb{Z}$ | Set of integers $= \{\ldots, -3, -2, -1, 0, 1, 2, 3, \ldots\}$ |

# Bibliography

[1] L.V. Ahlfors. *Complex Analysis*. McGraw-Hill, New York, 3rd edition, 1978.

[2] L. Alonso and R. Schott. *Random generation of trees*. Kluwer Academic Publishers, Boston, MA, 1995. Random generators in computer science.

[3] F. Anderegg. Solutions of problems: Algebra: 129. *Amer. Math. Monthly*, 9(1):11–13, 1902.

[4] G.E. Andrews. *The theory of partitions*. Cambridge Mathematical Library. Cambridge University Press, Cambridge, 1998. Reprint of the 1976 original.

[5] M. Anshelevich. Partition-dependent stochastic measures and $q$-deformed cumulants. *Doc. Math.*, 6:343–384, 2001.

[6] O. Arizmendi. Statistics of blocks in $k$-divisible non-crossing partitions. http://arxiv.org/pdf/1201.6576.pdf, 2012.

[7] D. Armstrong. Generalized noncrossing partitions and combinatorics of Coxeter groups. *Mem. Amer. Math. Soc.*, 202(949):x+159, 2009.

[8] D. Armstrong and S.-P. Eu. Nonhomogeneous parking functions and noncrossing partitions. *Electron. J. Combin.*, 15(1):Research Paper 146, 12, 2008.

[9] D. Armstrong, C. Stump, and H. Thomas. A uniform bijection between nonnesting and noncrossing partitions. arXiv:1101.1277v1, 2011.

[10] C.A. Athanasiadis. On noncrossing and nonnesting partitions for classical reflection groups. *Electron. J. Combin.*, 5:Research Paper 42, 1998.

[11] C.A. Athanasiadis, T. Brady, and C. Watt. Shellability of noncrossing partition lattices. *Proc. Amer. Math. Soc.*, 135(4):939–949, 2007.

[12] C.A. Athanasiadis and V. Reiner. Noncrossing partitions for the group $D_n$. *SIAM J. Discrete Math.*, 18(2):397–417, 2004.

[13] E. Babson and E. Steingrímsson. Generalized permutation patterns and a classification of the Mahonian statistics. *Sém. Lothar. Combin.*, 44:Article B44b, 2000.

[14] J. Backelin, J. West, and G. Xin. Wilf-equivalence for singleton classes. *Adv. Appl. Math.*, 38(2):133–148, 2007.

[15] C. Banderier, M. Bousquet-Mélou, A. Denise, P. Flajolet, D. Gardy, and D. Gouyou-Beauchamps. Generating functions for generating trees. *Discrete Math.*, 246(1–3):29–55, 2002. Formal power series and algebraic combinatorics (Barcelona, 1999).

[16] E. Barcucci, A. Bernini, L. Ferrari, and M. Poneti. A distributive lattice structure connecting Dyck paths, noncrossing partitions and 312-avoiding permutations. *Order*, 22(4):311–328 (2006), 2005.

[17] J.-L. Baril. Gray code for permutations with a fixed number of cycles. *Discrete Math.*, 307(13):1559–1571, 2007.

[18] J.-L. Baril and V. Vajnovszki. Gray code for derangements. *Discrete Appl. Math.*, 140(1–3):207–221, 2004.

[19] G. Baróti. Calcul des nombres de birecouvrements et de birevêtements d'un ensemble fini, employant la méthode fonctionnelle de Rota. In *Combinatorial theory and its applications, I (Proc. Colloq., Balatonfüred, 1969)*, pages 93–103. North-Holland, Amsterdam, 1970.

[20] C. Bebeacua, T. Mansour, A. Postnikov, and S. Severini. On the X-rays of permutations. In *Proceedings of the Workshop on Discrete Tomography and its Applications*, volume 20 of *Electron. Notes Discrete Math.*, pages 193–203. Elsevier Sci. B. V., Amsterdam, 2005.

[21] H.W. Becker. Planar thyme schemes. *Bull. Amer. Math. Soc.*, 58:39, 1952.

[22] H.W. Becker and J. Riordan. The arithemtic of bell and stirling numbers. *Amer. J. Math.*, 70:385–394, 1934.

[23] E.T. Bell. Exponential numbers. *Amer. Math. Monthly*, 41(7):411–419, 1934.

[24] E.T. Bell. Exponential polynomials. *Ann. Math. (2)*, 35(2):258–277, 1934.

[25] E.A. Bender. Central and local limit theorems applied to asymptotic enumeration. *J. Combinatorial Theory Ser. A*, 15:91–111, 1973.

[26] E.A. Bender. Partitions of multisets. *Discrete Math.*, 9:301–311, 1974.

[27] E.A. Bender, A.M. Odlyzko, and L.B. Richmond. The asymptotic number of irreducible partitions. *European J. Combin.*, 6(1):1–6, 1985.

[28] A.T. Benjamin, A.K. Eustis, and S.S. Plott. The 99th Fibonacci identity. *Electron. J. Combin.*, 15(1):Research Paper 34, 13, 2008.

[29] A.T. Benjamin and J.J. Quinn. *Proofs that really count*, volume 27 of *The Dolciani mathematical expositions*. Mathematical Association of America, Washington, DC, 2003. The art of combinatorial proof.

[30] C. Bennett, K.J. Dempsey, and B.E. Sagan. Partition lattice $q$-analogs related to $q$-Stirling numbers. *J. Algebraic Combin.*, 3(3):261–283, 1994.

[31] N. Bergeron, C. Reutenauer, M. Rosas, and M. Zabrocki. Invariants and coinvariants of the symmetric groups in noncommuting variables. *Canad. J. Math.*, 60(2):266–296, 2008.

[32] F.R. Bernhart. Catalan, Motzkin, and Riordan numbers. *Discrete Math.*, 204(1–3):73–112, 1999.

[33] D. Bessis and R. Corran. Non-crossing partitions of type $(e, e, r)$. *Adv. Math.*, 202(1):1–49, 2006.

[34] D. Bessis and V. Reiner. Cyclic sieving of noncrossing partitions for complex reflection groups. rXiv:0701792v1, 2007.

[35] P. Biane. Some properties of crossings and partitions. *Discrete Math.*, 175(1–3):41–53, 1997.

[36] P. Biane, F. Goodman, and A. Nica. Non-crossing cumulants of type B. *Trans. Amer. Math. Soc.*, 355(6):2263–2303, 2003.

[37] P. Billingsley. *Probability and Measure*. Wiley Series in Probability and Mathematical Statistics. John Wiley & Sons, New York, 3rd edition, 1995. A Wiley-Interscience Publication.

[38] P. Blasiak. Combinatorics of boson normal ordering and some applications. Ph.D. thesis, University of Paris VI and Polish Academy of Sciences, Krakow, Poland, arXiv:quant-ph/0507206, 2005.

[39] P. Blasiak, A. Horzela, K.A. Penson, G.H. Duchamp, and A.I. Solomon. Boson normal ordering via substitutions and Sheffer-type polynomials. *Phys. Lett. A*, 338(2):108–116, 2005.

[40] P. Blasiak, A. Horzela, K.A. Penson, and A.I. Solomon. Deformed bosons: Combinatorics of normal ordering. *Czechoslovak J. Phys.*, 54(11):1179–1184, 2004.

[41] P. Blasiak, A. Horzela, K.A. Penson, A.I. Solomon, and G.H. Duchamp. Combinatorics and boson normal ordering: A gentle introduction. *Amer. J. Phys.*, 75(7):639–646, 2007.

[42] P. Blasiak, K.A. Penson, and A.I. Solomon. The boson normal ordering problem and generalized Bell numbers. *Ann. Comb.*, 7(2):127–139, 2003.

[43] P. Blasiak, K.A. Penson, and A.I. Solomon. The general boson normal ordering problem. *Phys. Lett. A*, 309(3–4):198–205, 2003.

[44] P. Blasiak, K.A. Penson, and A.I. Solomon. Combinatorial coherent states via normal ordering of bosons. *Lett. Math. Phys.*, 67(1):13–23, 2004.

[45] P. Blasiak, K.A. Penson, A.I. Solomon, A. Horzela, and G.H. Duchamp. Some useful combinatorial formulas for bosonic operators. *J. Math. Phys.*, 46(5):052110, 6, 2005.

[46] W.E. Bleick and P.C.C. Wang. Asymptotics of Stirling numbers of the second kind. *Proc. Amer. Math. Soc.*, 42:575–580, 1974.

[47] M. Bóna. Partitions with $k$ crossings. *Ramanujan J.*, 3(2):215–220, 1999.

[48] M. Bóna. *Combinatorics of Permutations*. Discrete Mathematics and its Applications (Boca Raton). Chapman & Hall/CRC, Boca Raton, FL, 2004.

[49] M. Bóna and R. Simion. A self-dual poset on objects counted by the Catalan numbers and a type-B analogue. *Discrete Math.*, 220(1–3):35–49, 2000.

[50] M. Bousquet-Mélou and G. Xin. On partitions avoiding 3-crossings. *Sém. Lothar. Combin.*, 54:Article B54e, 2005/07.

[51] M. Bożejko, B. Kümmerer, and R. Speicher. $q$-Gaussian processes: noncommutative and classical aspects. *Comm. Math. Phys.*, 185(1):129–154, 1997.

[52] C. Brennan. Value and position of large weak left-to-right maxima for samples of geometrically distributed variables. *Discrete Math.*, 307(23):3016–3030, 2007.

[53] C. Brennan and A. Knopfmacher. The distribution of ascents of size $d$ or more in samples of geometric random variables. In *2005 International Conference on Analysis of Algorithms*, Discrete Math. Theor. Comput. Sci. Proc., AD, pages 343–351. Assoc. Discrete Math. Theor. Comput. Sci., Nancy, 2005.

[54] C. Brennan and A. Knopfmacher. The first and last ascents of size $d$ or more in samples of geometric random variables. *Quaest. Math.*, 28(4):487–500, 2005.

[55] A. Burstein. *Enumeration of words with forbidden patterns*. Ph.D. thesis, University of Pennsylvania, 1998.

[56] A. Burstein. Restricted Dumont permutations. *Ann. Comb.*, 9(3):269–280, 2005.

[57] A. Burstein, S. Elizalde, and T. Mansour. Restricted Dumont permutations, Dyck paths, and noncrossing partitions. *Discrete Math.*, 306(22):2851–2869, 2006.

[58] A. Burstein and T. Mansour. Words restricted by patterns with at most 2 distinct letters. *Electron. J. Combin.*, 9(2):#R3, 2002/03.

[59] A. Burstein and T. Mansour. Counting occurrences of some subword patterns. *Discrete Math. Theor. Comput. Sci.*, 6(1):1–11, 2003.

[60] A. Burstein and T. Mansour. Words restricted by 3-letter generalized multipermutation patterns. *Ann. Comb.*, 7(1):1–14, 2003.

[61] D. Callan. Pattern avoidance in "flattened" partitions. *Discrete Math.*, 309(12):4187–4191, 2009.

[62] E.R. Canfield. Engel's inequality for Bell numbers. *J. Combin. Theory Ser. A*, 72(1):184–187, 1995.

[63] E.R. Canfield. Meet and join within the lattice of set partitions. *Electron. J. Combin.*, 8(1):Research Paper 15, 2001.

[64] E.R. Canfield and L.H. Harper. A simplified guide to large antichains in the partition lattice. In *Proceedings of the Twenty-fifth Southeastern International Conference on Combinatorics, Graph Theory and Computing (Boca Raton, FL, 1994)*, volume 100, pages 81–88, 1994.

[65] L. Carlitz. $q$-Bernoulli numbers and polynomials. *Duke Math. J.*, 15:987–1000, 1948.

[66] L. Carlitz. Set partitions. *Fibonacci Quart.*, 14(4):327–342, 1976.

[67] L. Carlitz. Some numbers related to the Stirling numbers of the first and second kind. *Univ. Beograd. Publ. Elektrotehn. Fak. Ser. Mat. Fiz.*, 544–576:49–55, 1976.

[68] W.Y.C. Chen, E.Y.P. Deng, and R.R.X. Du. Reduction of $m$-regular noncrossing partitions. *European J. Combin.*, 26(2):237–243, 2005.

[69] W.Y.C. Chen, E.Y.P. Deng, R.R.X. Du, R.P. Stanley, and C.H. Yan. Crossings and nestings of matchings and partitions. *Trans. Amer. Math. Soc.*, 359(4):1555–1575, 2007.

[70] W.Y.C. Chen, N.J.Y. Fan, and A.F.Y. Zhao. Partitions and partial matchings avoiding neighbor patterns. *European J. Combin.*, 33(4):491–504, 2012.

[71] W.Y.C. Chen, I.M. Gessel, C.H. Yan, and A.L.B. Yang. A major index for matchings and set partitions. *J. Combin. Theory Ser. A*, 115(6):1069–1076, 2008.

[72] W.Y.C. Chen, T. Mansour, and S.H.F. Yan. Matchings avoiding partial patterns. *Electron. J. Combin.*, 13(1):Research Paper 112, 17 pp. (electronic), 2006.

[73] W.Y.C. Chen, J. Qin, C.M. Reidys, and D. Zeilberger. Efficient counting and asymptotics of $k$-noncrossing tangled diagrams. *Electron. J. Combin.*, 16(1):Research Paper 37, 8, 2009.

[74] W.Y.C. Chen and C.J. Wang. Noncrossing linked partitions and large $(3,2)$-motzkin paths. arXiv:1009.0176v1.

[75] W.Y.C. Chen and D.G.L. Wang. Singletons and adjacencies of set partitions of type $B$. *Discrete Math.*, 311(6):418–422, 2011.

[76] W.Y.C. Chen, S.Y.J. Wu, and C.H. Yan. Linked partitions and linked cycles. *European J. Combin.*, 29(6):1408–1426, 2008.

[77] W. Chu and C. Wei. Set partitions with restrictions. *Discrete Math.*, 308(15):3163–3168, 2008.

[78] A. Claesson. Generalized pattern avoidance. *European J. Combin.*, 22(7):961–971, 2001.

[79] M. Cohn, S. Even, Jr.K. Menger, and P.K. Hooper. Mathematical Notes: On the number of partitionings of a set of $n$ distinct objects. *Amer. Math. Monthly*, 69(8):782–785, 1962.

[80] L. Comtet. Birecouvrements et birevêtements d'un ensemble fini. *Studia Sci. Math. Hungar.*, 3:137–152, 1968.

[81] L. Comtet. Nombres de Stirling généraux et fonctions symétriques. *C. R. Acad. Sci. Paris Sér. A-B*, 275:A747–A750, 1972.

[82] L. Comtet. *Advanced Combinatorics*. Dordrecht Holland. The Netherlands, 1974.

[83] G.M. Constantine. Identities over set partitions. *Discrete Math.*, 204(1–3):155–162, 1999.

[84] E. Czabarka, P.L. Erdös, V. Johnson, A. Kupczok, and L.A. Székely. Asymptotically normal distribution of some tree families relevant for phylogenetics, and of partitions without singletons. arXiv:1108.6015v1, Preprint.

[85] N.G. de Bruijn. *Asymptotic methods in analysis*. Bibliotheca Mathematica. Vol. 4. North-Holland Publishing, Amsterdam, 1958.

[86] N.G. de Bruijn. *Asymptotic methods in analysis*. Dover Publications, New York, 3rd edition, 1981.

[87] A. de Mier. $k$-noncrossing and $k$-nonnesting graphs and fillings of Ferrers diagrams. *Combinatorica*, 27(6):699–720, 2007.

[88] E.Y.P. Deng, T. Mansour, and N. Mbarieky. Restricted set partitions. Preprint, 2011.

[89] R.S. Deodhar and M.K. Srinivasan. An inversion number statistic on set partitions. In *Electron. Notes Discr. Math.*, volume 15, pages 84–86. Elsevier, Amsterdam, 2003.

[90] N. Dershowitz and S. Zaks. Ordered trees and noncrossing partitions. *Discrete Math.*, 62(2):215–218, 1986.

[91] J.S. Devitt and D.M. Jackson. The enumeration of covers of a finite set. *J. London Math. Soc. (2)*, 25(1):1–6, 1982.

[92] G. Dobiński. Summirung der reihe $\sum \frac{n^m}{n!}$ für $m = 1, 2, 3, 4, 5, \ldots$. *Grunert's Archiv*, 61:333–336, 1877.

[93] M. d'Ocagne. Sur Une Classe de Nombres Remarquables. *Amer. J. Math.*, 9(4):353–380, 1887.

[94] D. Drake and J.S. Kim. $k$-distant crossings and nestings and matchings and partitions. In *21st International Conference on Formal Power Series and Algebraic Combinatorics (FPSAC 2009)*, Discrete Math. Theor. Comput. Sci. Proc., AK, pages 349–360. Assoc. Discrete Math. Theor. Comput. Sci., Nancy, 2009.

[95] D. Dumont. Interprétations combinatoires des nombres de Genocchi. *Duke Math. J.*, 41:305–318, 1974.

[96] K.J. Dykema. Multilinear function series and transforms in free probability theory. *Adv. Math.*, 208(1):351–407, 2007.

[97] P.H. Edelman. Chain enumeration and noncrossing partitions. *Discrete Math.*, 31(2):171–180, 1980.

[98] P.H. Edelman. Multichains, noncrossing partitions and trees. *Discrete Math.*, 40(2–3):171–179, 1982.

[99] P.H. Edelman and R. Simion. Chains in the lattice of noncrossing partitions. *Discrete Math.*, 126(1–3):107–119, 1994.

[100] E.G. Effros and M. Popa. Feynman diagrams and Wick products associated with $q$-Fock space. *Proc. Natl. Acad. Sci. USA*, 100(15):8629–8633, 2003.

[101] R. Ehrenborg and M.A. Readdy. The Möbius function of partitions with restricted block sizes. *Adv. Appl. Math.*, 39(3):283–292, 2007.

[102] G. Ehrlich. Loopless algorithms for generating permutations, combinations, and other combinatorial configurations. *J. Assoc. Comput. Mach.*, 20:500–513, 1973.

[103] C. Elbert. Strong asymptotics of the generating polynomials of the Stirling numbers of the second kind. *J. Approx. Theory*, 109(2):198–217, 2001.

[104] L.F. Epstein. A function related to the series for $e^{e^x}$. *J. Math. Phys. Mass. Inst. Tech.*, 18:153–173, 1939.

[105] M.C. Er. A fast algorithm for generating set partitions. *The Comput. J.*, 31:283–284, 1988.

[106] A. Fink and B. Iriarte Giraldo. Bijections between noncrossing and nonnesting partitions for classical reflection groups. *Port. Math.*, 67(3):369–401, 2010.

[107] P. Flajolet and R. Sedgewick. *Analytic combinatorics*. Cambridge University Press, Cambridge, 2009.

[108] M.S. Floater and T. Lyche. Divided differences of inverse functions and partitions of a convex polygon. *Math. Comp.*, 77(264):2295–2308, 2008.

[109] D. Foata. On the Netto inversion number of a sequence. *Proc. Amer. Math. Soc.*, 19:236–240, 1968.

[110] K. Fujii and T. Suzuki. A new symmetric expression of Weyl ordering. *Modern Phys. Lett. A*, 19(11):827–840, 2004.

[111] A.M. Garsia and J.B. Remmel. Q-counting rook configurations and a formula of Frobenius. *J. Combin. Theory Ser. A*, 41(2):246–275, 1986.

[112] G. Gasper and M. Rahman. *Basic hypergeometric series*, volume 35 of *Encyclopedia of mathematics and its applications*. Cambridge University Press, Cambridge, 1990. With a foreword by Richard Askey.

[113] I.M. Gessel. A $q$-analog of the exponential formula. *Discrete Math.*, 40(1):69–80, 1982.

[114] E.N. Gilbert. Gray codes and paths on the $n$-cube. *Bell System Tech. J*, 37:815–826, 1958.

[115] R.J. Glauber. Coherent and incoherent states of the radiation field. *Phys. Rev. (2)*, 131:2766–2788, 1963.

[116] N. Glick. Breaking records and breaking boards. *Amer. Math. Monthly*, 85(1):2–26, 1978.

[117] K. Goldberg, M. Newman, and E. Haynsworth. *Combinatorial analysis. Handbook of mathematical functions*, (eds. M. Abramowitz and I. A. Stegun). National Bureau of Standards/Dover Publ., New York, 1964.

[118] H.W. Gould. The $q$-Stirling numbers of first and second kinds. *Duke Math. J.*, 28:281–289, 1961.

[119] D. Gouyou-Beauchamps and G. Viennot. Equivalence of the two-dimensional directed animal problem to a one-dimensional path problem. *Adv. Appl. Math.*, 9:334–357, 2003.

[120] A.M. Goyt. Avoidance of partitions of a three-element set. *Adv. Appl. Math.*, 41(1):95–114, 2008.

[121] A.M. Goyt and L.K. Pudwell. Avoiding colored partitions of lengths two and three. arXiv:1103.0239v1, Preprint.

[122] A.M. Goyt and B.E. Sagan. Set partition statistics and $q$-Fibonacci numbers. *European J. Combin.*, 30(1):230–245, 2009.

[123] R.L. Graham, D.E. Knuth, and O. Patashnik. *Concrete mathematics*. Addison-Wesley Publishing Company Advanced Book Program, Reading, MA, 1989. A foundation for computer science.

[124] F. Gray. Pulse code communications. *U.S. Patent 2632058*, 1953.

[125] V. Grebinski and G. Kucherov. Reconstructing set partitions. Preprint, 1997.

[126] M. Griffiths. Generalized near-Bell numbers. *J. Integer Seq.*, 12(5):Article 09.5.7, 12, 2009.

[127] R.P. Grimaldi. *Discrete and combinatorial mathematics: An applied introduction*. Addison Wesley, Amsterdam, 4th edition, 2003.

[128] G.R. Grimmett and D.R. Stirzaker. *Probability and random processes*. Oxford University Press, New York, 3rd edition, 2001.

[129] H. Gupta. A new look at the permutations of the first $n$ natural numbers. *Indian J. Pure Appl. Math.*, 9(6):600–631, 1978.

[130] J. Haglund, N. Loehr, and J.B. Remmel. Statistics on wreath products, perfect matchings, and signed words. *European J. Combin.*, 26(6):835–868, 2005.

[131] T. Halverson and T. Lewandowski. RSK insertion for set partitions and diagram algebras. *Electron. J. Combin.*, 11(2):Research Paper 24, 2004/06.

[132] J.M. Hammersley. *A few seedlings of research*, volume 1. Berkeley/Los Angeles, University of California Press, 1972.

[133] R.K.S. Hankin and L.J. West. Set partitions in r. *J. Stat. Soft.*, 23:1–12, 2007.

[134] W.K. Hayman. A generalisation of Stirling's formula. *J. Reine Angew. Math.*, 196:67–95, 1956.

[135] S. Heubach, S. Kitaev, and T. Mansour. Avoidance of partially ordered patterns in compositions. *Pure Math. Appl. (PU.M.A.)*, 17(1–2):123–134, 2006.

[136] S. Heubach, N.Y. Li, and T. Mansour. Staircase tilings and $k$-Catalan structures. *Discrete Math.*, 308(24):5954–5964, 2008.

[137] S. Heubach and T. Mansour. Enumeration of 3-letter patterns in compositions. In *Combinatorial number theory*, pages 243–264. deGruyter, 2007.

[138] S. Heubach and T. Mansour. *Combinatorics of compositions and words. Discrete mathematics and its applications* (Boca Raton). CRC Press, Boca Raton, FL, 2010.

[139] Q.-H. Hou and T. Mansour. The kernel method and systems of functional equations with several conditions. *J. Comput. Appl. Math.*, 235(5):1205–1212, 2011.

[140] C. Huemer, F. Hurtado, M. Noy, and E. Omaña-Pulido. Gray codes for non-crossing partitions and dissections of a convex polygon. *Discrete Appl. Math.*, 157(7):1509–1520, 2009.

[141] C. Ingalls and H. Thomas. Noncrossing partitions and representations of quivers. *Compos. Math.*, 145(6):1533–1562, 2009.

[142] M. Ishikawa, A. Kasraoui, and J. Zeng. Euler-Mahonian statistics on ordered partitions and Steingrímsson's conjecture—a survey. In *Combinatorial representation theory and related topics*, RIMS Kôkyûroku Bessatsu, B8, pages 99–113. Res. Inst. Math. Sci. (RIMS), Kyoto, 2008.

[143] M. Ishikawa, A. Kasraoui, and J. Zeng. Euler-Mahonian statistics on ordered set partitions. *SIAM J. Discrete Math.*, 22(3):1105–1137, 2008.

[144] R. Jakimczuk. Integer sequences, functions of slow increase, and the bell number. *J. Integer Seq.*, 14(5):Article 11.5.8, 2011.

[145] V. Jelínek, N.Y. Li, T. Mansour, and S.H.F. Yan. Matchings avoiding partial patterns and lattice paths. *Electron. J. Combin.*, 13(1):Research Paper 89, 12 pp. (electronic), 2006.

[146] V. Jelínek and T. Mansour. On pattern-avoiding partitions. *Electron. J. Combin.*, 15(1):Research paper 39, 52, 2008.

[147] V. Jelínek and T. Mansour. Wilf-equivalence on $k$-ary words, compositions, and parking functions. *Electron. J. Combin.*, 16(1):Research Paper 58, 9, 2009.

[148] V. Jelínek and T. Mansour. Matchings and partial patterns. *Electron. J. Combin.*, 17(1):Research Paper 158, 30, 2010.

[149] V. Jelínek, T. Mansour, and M. Shattuck. On multiple pattern avoiding set partitions. Preprint, 2011.

[150] W.P. Johnson. A $q$-analogue of Faà di Bruno's formula. *J. Combin. Theory Ser. A*, 76(2):305–314, 1996.

[151] W.P. Johnson. Some applications of the $q$-exponential formula. In *Proceedings of the 6th Conference on Formal Power Series and Algebraic Combinatorics (New Brunswick, NJ, 1994)*, volume 157, pages 207–225, 1996.

[152] S.M. Johson. Generating of permutations by adjacent transposition. *Math. Comput.*, 17:282–285, 1963.

[153] J.T. Joichi, Dennis E. White, and S.G. Williamson. Combinatorial Gray codes. *SIAM J. Comput.*, 9(1):130–141, 1980.

[154] C. Jordan. *Calculus of finite differences*. Chelsea, New York, 2nd edition, 1950.

[155] M. Josuat-Vergès and M. Rubey. Crossings, Motzkin paths and moments. *Discrete Math.*, 311(18–19):2064–2078, 2011.

[156] A. Kasraoui. $d$-regular set parittions and rook placements. *Sémin. Loth. Combin.*, 62:Article B62a, 2009.

[157] A. Kasraoui. Average values of som z-parameters in a random set partition. *Electron. J. Combin.*, 18(1):Research Paper 228, 2011.

[158] A. Kasraoui, D. Stanton, and J. Zeng. The combinatorics of Al-Salam–Chihara $q$-Laguerre polynomials. *Adv. Appl. Math.*, 47(2):216–239, 2011.

[159] A. Kasraoui and J. Zeng. Distribution of crossings, nestings and alignments of two edges in matchings and partitions. *Electron. J. Combin.*, 13(1):Research Paper 33, 2006.

[160] A. Kasraoui and J. Zeng. Euler-Mahonian statistics on ordered set partitions. II. *J. Combin. Theory Ser. A*, 116(3):539–563, 2009.

[161] J. Katriel. Combinatorial aspects of boson algebra. *Lett. Nuovo Cimento*, 10:565–567, 1974.

[162] J. Katriel. Bell numbers and coherent states. *Phys. Lett. A*, 273(3):159–161, 2000.

[163] J. Katriel. Refined Stirling numbers: enumeration of special sets of permutations and set-partitions. *J. Combin. Theory Ser. A*, 99(1):85–94, 2002.

[164] J. Katriel and M. Kibler. Normal ordering for deformed boson operators and operator-valued deformed Stirling numbers. *J. Phys. A*, 25(9):2683–2691, 1992.

[165] R. Kaye. A gray code for set partitions. *Information Processing Lett.*, 5(6):171–173, 1976.

[166] J.S. Kim. Bijections on two variations of noncrossing partitions. *Discrete Math.*, 311(12):1057–1063, 2011.

[167] J.S. Kim. Chain enumeration of $k$-divisible noncrossing partitions of classical types. *J. Combin. Theory Ser. A*, 118(3):879–898, 2011.

[168] J.S. Kim. Front representation of set partitions. *SIAM J. Discrete Math.*, 25(1):447–461, 2011.

[169] J.S. Kim. New interpretations for noncrossing partitions of classical types. *J. Combin. Theory Ser. A*, 118(4):1168–1189, 2011.

[170] S. Kitaev. Partially ordered generalized patterns. *Discrete Math.*, 298(1–3):212–229, 2005.

[171] S. Kitaev. Introduction to partially ordered patterns. *Discrete Appl. Math.*, 155(8):929–944, 2007.

[172] S. Kitaev. *Patterns in Permutations and Words*. Springer, USA, 2011.

[173] S. Kitaev and T. Mansour. Partially ordered generalized patterns and $k$-ary words. *Ann. Comb.*, 7:191–200, 2003.

[174] J.R. Klauder and B.-S. Skagerstam. *Cherent states*. World Scientific, Singapore, 1985.

[175] J.R. Klauder and E.C.G. Sudarshan. *Fundamentals of quantum optics*. W. A. Benjamin, New York-Amsterdam, 1968.

[176] M. Klazar. On *abab*-free and *abba*-free set partitions. *European J. Combin.*, 17(1):53–68, 1996.

[177] M. Klazar. On trees and noncrossing partitions. *Discrete Appl. Math.*, 82(1–3):263–269, 1998.

[178] M. Klazar. Counting pattern-free set partitions. I. A generalization of Stirling numbers of the second kind. *European J. Combin.*, 21(3):367–378, 2000.

[179] M. Klazar. Counting pattern-free set partitions. II. Noncrossing and other hypergraphs. *Electron. J. Combin.*, 7:Research Paper 34, 2000.

[180] M. Klazar. Bell numbers, their relatives, and algebraic differential equations. *J. Combin. Theory Ser. A*, 102(1):63–87, 2003.

[181] M. Klazar. Counting even and odd partitions. *Amer. Math. Monthly*, 110(6):527–532, 2003.

[182] M. Klazar. Non-$P$-recursiveness of numbers of matchings or linear chord diagrams with many crossings. *Adv. Appl. Math.*, 30(1–2):126–136, 2003. Formal power series and algebraic combinatorics (Scottsdale, AZ, 2001).

[183] M. Klazar. Counting set systems by weight. *Electron. J. Combin.*, 12:Research Paper 11, 2005.

[184] M. Klazar. On identities concerning the numbers of crossings and nestings of two edges in matchings. *SIAM J. Discrete Math.*, 20(4):960–976, 2006.

[185] M. Klazar. On growth rates of permutations, set partitions, ordered graphs and other objects. *Electron. J. Combin.*, 15(1):Research Paper 75, 22, 2008.

[186] F. Klein-Barmen. Über ein (max, min)-theorem in der Theorie der Protoverbände und Dreigespanne. *J. Reine Angew. Math.*, 205:107–112, 1960/1961.

[187] E. Knobloch. *Die mathematischen Studien von G. W. Leibniz zur Kombinatorik*. Franz Steiner Verlag GMBH, Wiesbaden, 1973. Auf Grund fast ausschliesslich handschriftlicher Aufzeichnungen dargelegt und kommentiert, Studia Leibnitiana. Supplementa. Band XI.

[188] E. Knobloch. *Die mathematischen Studien von G. W. Leibniz zur Kombinatorik: Textband*. Franz Steiner Verlag GMBH, Wiesbaden, 1976. Im Anschluss an den gleichnamigen Abhandlungsband zum ersten Mal nach den Originalhandschriften herausgegeben, Studia Leibnitiana. Supplementa, Vol. XVI.

[189] A. Knopfmacher and T. Mansour. Record statistics in integer compositions. In *Discrete Math. and Theor. Computer Sci., Proceedings AK*, volume AK, pages 527–536, 2009.

[190] A. Knopfmacher, T. Mansour, and S. Wagner. Records in set partitions. *Electron. J. Combin.*, 17(1):Research Paper 109, 2010.

[191] A. Knopfmacher, A.M. Odlyzko, B. Pittel, L.B. Richmond, D. Stark, G. Szekeres, and N.C. Wormald. The asymptotic number of set partitions with unequal block sizes. *Electron. J. Combin.*, 6:Research Paper 2, 1999.

[192] K. Knopp. *Theory and Application of Infinite Series.* Blackie & Son Ltd., London, Glasgow, 1928.

[193] D.E. Knuth. *The art of computer programming. Volume 3.* Addison-Wesley Publishing Co., Reading, Mass.-London-Don Mills, Ont., 1973. Sorting and searching, Addison-Wesley Series in Computer Science and Information Processing.

[194] D.E. Knuth. *The art of computer programming. Volume 1.* Addison-Wesley Publishing Co., Reading, Mass.-London-Amsterdam, 2nd edition, 1975.

[195] D.E. Knuth. *The art of computer programming. Vol. 4, Fasc. 3.* Addison-Wesley, Upper Saddle River, NJ, 2005. Generating all combinations and partitions.

[196] D.E. Knuth. *The art of computer programming. Vol. 4, Fasc. 4.* Addison-Wesley, Upper Saddle River, NJ, 2006. Generating all trees—history of combinatorial generation.

[197] I. Kortchemski. Asymptotic behavior of permutation records. *J. Combin. Theory Ser. A*, 116(6):1154–1166, 2009.

[198] T. Koshy. *Fibonacci and Lucas numbers with applications.* Pure and Applied Mathematics (New York). Wiley-Interscience, New York, 2001.

[199] C. Krattenthaler. Permutations with restricted patterns and Dyck paths. *Adv. Appl. Math.*, 27(2–3):510–530, 2001. Special issue in honor of Dominique Foata's 65th birthday (Philadelphia, PA, 2000).

[200] C. Krattenthaler. The $M$-triangle of generalised non-crossing partitions for the types $E_7$ and $E_8$. *Sém. Lothar. Combin.*, 54:Article B541, 2005/07.

[201] C. Krattenthaler. Growth diagrams, and increasing and decreasing chains in fillings of Ferrers shapes. *Adv. Appl. Math.*, 37(3):404–431, 2006.

[202] C. Krattenthaler and T.W. Müller. Decomposition numbers for finite Coxeter groups and generalised non-crossing partitions. *Trans. Amer. Math. Soc.*, 362(5):2723–2787, 2010.

[203] G. Kreweras. Une famille d'identités mettant en jeu toutes les partitions d'un ensemble fini de variables en un nombre donné de classes. *C. R. Acad. Sci. Paris Sér. A-B*, 270:A1140–A1143, 1970.

[204] G. Kreweras. Sur les partitions non croisées d'un cycle. *Discrete Math.*, 1(4):333–350, 1972.

[205] C.Y. Ku and D. Renshaw. Erdős-Ko-Rado theorems for permutations and set partitions. *J. Combin. Theory Ser. A*, 115(6):1008–1020, 2008.

[206] V. Lakshmibai and B. Sandhya. Criterion for smoothness of Schubert varieties in $Sl(n)/B$. *Proc. Indian Acad. Sci. Math. Sci.*, 100(1):45–52, 1990.

[207] N.Y. Li and T. Mansour. An identity involving Narayana numbers. *European J. Combin.*, 29(3):672–675, 2008.

[208] S.C. Liaw, H.G. Yeh, F.K. Hwang, and G.J. Chang. A simple and direct derivation for the number of noncrossing partitions. *Proc. Amer. Math. Soc.*, 126(6):1579–1581, 1998.

[209] E.H. Lieb. Concavity properties and a generating function for Stirling numbers. *J. Combinatorial Theory*, 5:203–206, 1968.

[210] N. Lindquist and G. Sierksma. Extensions of set partitions. *J. Combin. Theory Ser. A*, 31(2):190–198, 1981.

[211] J.M. Lucas, D. Roelants van Baronaigien, and F. Ruskey. On rotations and the generation of binary trees. *J. Algorithms*, 15(3):343–366, 1993.

[212] P.A. MacMahon. *Combinatory analysis. Vol. I, II (bound in one volume)*. Dover Phoenix Editions. Dover Publications, Mineola, NY, 2004. Reprint of *An introduction to combinatory analysis* (1920) and *Combinatory analysis. Vol. I, II* (1915, 1916).

[213] R. Mamede. A bijection between noncrossing and nonnesting partitions of types A and B. In *21st International Conference on Formal Power Series and Algebraic Combinatorics (FPSAC 2009)*, Discrete Math. Theor. Comput. Sci. Proc., AK, pages 597–610. Assoc. Discrete Math. Theor. Comput. Sci., Nancy, 2009.

[214] T. Mansour. Restricted 1-3-2 permutations and generalized patterns. *Ann. Comb.*, 6(1):65–76, 2002.

[215] T. Mansour. Restricted 132-Dumont permutations. *Australas. J. Combin.*, 29:103–117, 2004.

[216] T. Mansour. Smooth partitions and Chebyshev polynomials. *Bull. Lond. Math. Soc.*, 41(6):961–970, 2009.

[217] T. Mansour and N. Mbarieky. Partitions of a set satisfying certain set of conditions. *Discrete Math.*, 309(13):4481–4488, 2009.

[218] T. Mansour and A.O. Munagi. Enumeration of partitions by rises, levels and descents. In *Permutation Patterns, St. Andrews 2007*, volume 376, pages 221–232. Cambridge University Press, Cambridge, 2007.

[219] T. Mansour and A.O. Munagi. Enumeration of gap-bounded set partitions. *J. Autom. Lang. Comb.*, 14(3–4):237–245 (2010), 2009.

[220] T. Mansour and A.O. Munagi. Enumeration of partitions by long rises, levels, and descents. *J. Integer Seq.*, 12(1):Article 09.1.8, 2009.

[221] T. Mansour and A.O. Munagi. Block-connected set partitions. *European J. Combin.*, 31(3):887–902, 2010.

[222] T. Mansour, A.O. Munagi, and M. Shattuck. Recurrence relations and two-dimensional set partitions. *J. Integer Seq.*, 14(4):Article 11.4.1, 2011.

[223] T. Mansour and G. Nassar. Gray codes, loopless algorithm and partitions. *J. Math. Model. Algorithms*, 7(3):291–310, 2008.

[224] T. Mansour, G. Nassar, and V. Vajnovszki. Loop-free Gray code algorithm for the e-restricted growth functions. *Inform. Process. Lett.*, 111(11):541–544, 2011.

[225] T. Mansour and M. Schork. On the normal ordering of multi-mode boson operators. *Russian J. Math. Phy.*, 15:50–61, 2008.

[226] T. Mansour, M. Schork, and S. Severini. A generalization of boson normal ordering. *Phys. Lett. A*, 364(3–4):214–220, 2007.

[227] T. Mansour, M. Schork, and S. Severini. Wick's theorem for $q$-deformed boson operators. *J. Phys. A*, 40(29):8393–8401, 2007.

[228] T. Mansour, M. Schork, and S. Severini. Noncrossing normal ordering for functions of boson operators. *Internat. J. Theoret. Phys.*, 47(3):832–849, 2008.

[229] T. Mansour and S. Severini. Enumeration of $(k, 2)$-noncrossing partitions. *Discrete Math.*, 308(20):4570–4577, 2008.

[230] T. Mansour and M. Shattuck. Counting peaks and valleys in a partition of a set. *Journal Int. Seq.*, 13:Article 10.6.8., 2010.

[231] T. Mansour and M. Shattuck. Avoiding type $(1, 2)$ or $(2, 1)$ patterns in a partition of a set. Preprint, 2011.

[232] T. Mansour and M. Shattuck. Pattern avoiding partitions and motzkin left factors. *Central Europ. J. Math.*, 9:1121–1134, 2011.

[233] T. Mansour and M. Shattuck. Pattern avoiding partitions, sequence a054391, and the kernel method. *Appl. Applied Math.: An Intern. J.*, 6:397–411, 2011.

[234] T. Mansour and M. Shattuck. A recurrence related to the bell numbers. *Integers*, 11:#A67, 2011.

[235] T. Mansour and M. Shattuck. Restricted partitions and generalized catalan numbers. *Pure Math. Appl.*, 22(2):239–251, 2011.

[236] T. Mansour and M. Shattuck. Restricted partitions and $q$-pell numbers. *Central Europ. J. Math.*, 9(2):346–355, 2011.

[237] T. Mansour and M. Shattuck. Avoiding type $(1,2)$ or $(2,1)$ patterns in a partition of a set. *Integers*, 12:A20, 2012.

[238] T. Mansour and M. Shattuck. Free rises and restricted partitions. To appear, 2012.

[239] T. Mansour and M. Shattuck. Pattern avoiding set partitions and catalan numbers. Discr. Math., To appear, 2012.

[240] T. Mansour and M. Shattuck. Set partitions as geometric words. *Australas. J. Combin.*, 53:31–39, 2012.

[241] T. Mansour, M. Shattuck, and S. Wagner. Enumerating set partitions by the number of positions between adjacent occurrences of a letter. *Appl. Ana. Discr. Math.*, 4(2):284–308, 2010.

[242] T. Mansour, M. Shattuck, and S.H.F. Yan. Counting subwords in a partition of a set. *Electron. J. Combin.*, 17(1):Research Paper 19, 21, 2010.

[243] T. Mansour and B.O. Sirhan. Counting $l$-letter subwords in compositions. *Discrete Math. Theor. Comput. Sci.*, 8(1):285–297, 2006.

[244] E. Marberg. Crossings and nestings in colored set partitions. Preprint, arXiv:1203.5738v1.

[245] E. Marberg. Actions and identities on set partitions. *Electron. J. Combin.*, 19:Research Paper 28, 2012.

[246] J.E. Marsden and M.J. Hoffman. *Basic complex analysis*. W. H. Freeman and Company, New York, 2nd edition, 1987.

[247] J. McCammond. Noncrossing partitions in surprising locations. *Amer. Math. Monthly*, 113(7):598–610, 2006.

[248] M.A. Méndez, P. Blasiak, and K.A. Penson. Combinatorial approach to generalized Bell and Stirling numbers and boson normal ordering problem. *J. Math. Phys.*, 46(8):083511, 8, 2005.

[249] V.V. Menon. On the maximum of Stirling numbers of the second kind. *J. Combinatorial Theory Ser. A*, 15:11–24, 1973.

[250] V.V. Mikhaĭlov. Ordering of some boson operator functions. *J. Phys. A*, 16(16):3817–3827, 1983.

[251] F.L. Miksa, L. Moser, and M. Wyman. Restricted partitions of finite sets. *Canad. Math. Bull.*, 1:87–96, 1958.

[252] S.C. Milne. Restricted growth functions, rank row matchings of partition lattices, and $q$-Stirling numbers. *Adv. Math.*, 43(2):173–196, 1982.

[253] J.A. Mingo and A. Nica. Crossings of set-partitions and addition of graded-independent random variables. *Internat. J. Math.*, 8(5):645–664, 1997.

[254] M. Mishna and L. Yen. Set partitions with no $k$-nesting. Preprint, arXiv:1106.5036v1, 2011.

[255] C. Montenegro. The fixed point non-crossing partition lattices. Preprint, 1993.

[256] L. Moser and M. Wyman. An asymptotic formula for the Bell numbers. *Trans. Roy. Soc. Canada. Sect. III. (3)*, 49:49–54, 1955.

[257] L. Moser and M. Wyman. Stirling numbers of the second kind. *Duke Math. J.*, 25:29–43, 1957.

[258] A.O. Munagi. Extended set partitions with successions. *European J. Combin.*, 29(5):1298–1308, 2008.

[259] A.N. Myers and H.S. Wilf. Left-to-right maxima in words and multiset permutations. *Israel J. Math.*, 166:167–183, 2008.

[260] R. Natarajan. A bijection between certain non-crossing partitions and sequences. *Discrete Math.*, 286(3):269–275, 2004.

[261] J. Němeček and M. Klazar. A bijection between nonnegative words and sparse *abba*-free partitions. *Discrete Math.*, 265(1–3):411–416, 2003.

[262] A. Nica. Crossings and embracings of set-partitions and $q$-analogues of the logarithm of the Fourier transform. In *Proceedings of the 6th Conference on Formal Power Series and Algebraic Combinatorics (New Brunswick, NJ, 1994)*, volume 157, pages 285–309, 1996.

[263] A. Nica and I. Oancea. Posets of annular non-crossing partitions of types $B$ and $D$. *Discrete Math.*, 309(6):1443–1466, 2009.

[264] A. Nica and R. Speicher. A "Fourier transform" for multiplicative functions on non-crossing partitions. *J. Algebraic Combin.*, 6(2):141–160, 1997.

[265] A. Nijenhuis and H.S. Wilf. *Combinatorial algorithms*. Academic Press [Harcourt Brace Jovanovich Publishers], New York, 1975. Computer Science and Applied Mathematics.

[266] K. Oliver and H. Prodinger. Words coding set partitions. *Appl. Anal. Discrete Math.*, 5(1):55–59, 2011.

[267] F. Oravecz. Symmetric partitions and pairings. *Colloq. Math.*, 86(1):93–101, 2000.

[268] M. Orlov. Efficient generation of set partitions. http://www.informatik.uni-ulm.de/ni/Lehre/WS04/DMM/Software/partitions.pdf, 2002.

[269] A. Panayotopoulos and A. Sapounakis. On Motzkin words and non-crossing partitions. *Ars Combin.*, 69:109–116, 2003.

[270] J. Pitman. Some probabilistic aspects of set partitions. *Amer. Math. Monthly*, 104(3):201–209, 1997.

[271] B. Pittel. Random set partitions: asymptotics of subset counts. *J. Combin. Theory Ser. A*, 79(2):326–359, 1997.

[272] B. Pittel. Where the typical set partitions meet and join. *Electron. J. Combin.*, 7:Research Paper 5, 2000.

[273] A.D. Polyanin, V.F. Zaitsev, and A. Moussiaux. *Handbook of first order partial differential equations*, volume 1 of *Differential and Integral Equations and Their Applications*. Taylor & Francis Ltd., London, 2002.

[274] M. Popa. Non-crossing linked partitions and multiplication of free random variables. In *Operator theory live*, volume 12 of *Theta Ser. Adv. Math.*, pages 135–143. Theta, Bucharest, 2010.

[275] D.N. Port. Circular numbers and $n$-set partitions. *J. Combin. Theory Ser. A*, 83(1):57–78, 1998.

[276] Y. Poupard. Étude et dénombrement parallèles des partitions non-croisées d'un cycle et des découpages d'un polygone convexe. *Discrete Math.*, 2(3):279–288, 1972.

[277] S. Poznanović and C. Yan. Crossings and nestings of two edges in set partitions. *SIAM J. Discrete Math.*, 23(2):787–804, 2009.

[278] H. Prodinger. On the number of Fibonacci partitions of a set. *Fibonacci Quart.*, 19(5):463–465, 1981.

[279] H. Prodinger. A correspondence between ordered trees and noncrossing partitions. *Discrete Math.*, 46(2):205–206, 1983.

[280] H. Prodinger. Combinatorics of geometrically distributed random variables: left-to-right maxima. In *Proceedings of the 5th Conference on Formal Power Series and Algebraic Combinatorics (Florence, 1993)*, volume 153, pages 253–270, 1996.

[281] H. Prodinger. Records in geometrically distributed words: sum of positions. *Appl. Anal. Discrete Math.*, 2(2):234–240, 2008.

[282] G. Puttenham. *The Arte of English Poesie*. The Kent State University Press, London, 1970. A facsimile reproduction of the 1906 reprint published by A. Constable and Co., Ltd.

[283] D.J. Rasmussen and C.D. Savage. Hamilton-connected derangement graphs on $S_n$. *Discrete Math.*, 133(1–3):217–223, 1994.

[284] N. Reading. Chains in the noncrossing partition lattice. *SIAM J. Discrete Math.*, 22(3):875–886, 2008.

[285] N. Reading. Noncrossing partitions, clusters and the Coxeter plane. *Sém. Lothar. Combin.*, 63:Article B63b, 2010.

[286] N. Reading. Noncrossing partitions and the shard intersection order. *J. Algebraic Combin.*, 33(4):483–530, 2011.

[287] V. Reiner. Non-crossing partitions for classical reflection groups. *Discrete Math.*, 177(1–3):195–222, 1997.

[288] A. Rényi. Théorie des éléments saillants d'une suite d'observations. *Ann. Fac. Sci. Univ. Clermont-Ferrand No.*, 8:7–13, 1962.

[289] M. Rey. Algebraic constructions on set partitions. In *Proceedings of the 19th Conference on Formal Power Series and Algebraic Combinatorics (Nankai University, Tianjin, China, 2007)*, 2007.

[290] B. Rhoades. Enumeration of connected Catalan objects by type. *European J. Combin.*, 32(2):330–338, 2011.

[291] D. Richards. Data compression and Gray-code sorting. *Inform. Process. Lett.*, 22(4):201–205, 1986.

[292] J. Riordan. *An introduction to combinatorial analysis*. Wiley Publications in Mathematical Statistics. John Wiley & Sons, New York, 1958.

[293] Th. Rivlin. *Chebyshev polynomials*. Pure and Applied Mathematics (New York). John Wiley & Sons, New York, 2nd edition, 1990.

[294] D. Roelants van Baronaigien and F. Ruskey. Generating permutations with given ups and downs. *Disc. Appl. Math.*, 36:57–65, 1992.

[295] D.G. Rogers. Ascending sequences in permutations. *Discrete Math.*, 22(1):35–40, 1978.

[296] M.H. Rosas and B.E. Sagan. Symmetric functions in noncommuting variables. *Trans. Amer. Math. Soc.*, 358(1):215–232, 2006.

[297] G.-C. Rota. The number of partitions of a set. *Amer. Math. Monthly*, 71:498–504, 1964.

[298] G.-C. Rota. On the foundations of combinatorial theory. I. Theory of Möbius functions. *Z. Wahrscheinlichkeitstheorie und Verw. Gebiete*, 2:340–368, 1964.

[299] D. Rotem. On a correspondence between binary trees and a certain type of permutation. *Information Processing Lett.*, 4(3):58–61, 1975/76.

[300] D. Rotem. Stack sortable permutations. *Discrete Math.*, 33(2):185–196, 1981.

[301] M. Rubey. Increasing and decreasing sequences in fillings of moon polyominoes. *Adv. in Appl. Math.*, 47(1):57–87, 2011.

[302] M. Rubey and C. Stump. Crossings and nestings in set partitions of classical types. *Electron. J. Combin.*, 17(1):Research Paper 120, 19, 2010.

[303] F. Ruskey. Combinatorial generation. http://www.cs.sunysb.edu/ algorith/implement/ruskey/implement.shtml.

[304] F. Ruskey and C. Savage. Gray codes for set partitions and restricted growth tails. *Australas. J. Combin.*, 10:85–96, 1994.

[305] B.E. Sagan. A maj statistic for set partitions. *European J. Combin.*, 12(1):69–79, 1991.

[306] B.E. Sagan. Pattern avoidance in set partitions. *Ars Combin.*, 94:79–96, 2010.

[307] M. de Sainte-Catherine. Couplages et pfaffiens en combinatoire, physique et informatique. Ph.D thesis, University of Bordeaux I, 1983.

[308] B. Salvy. Even-odd set partitions, saddle-point method and wyman admissibility. *Rapport De Recherche Inria*, page 14pp., 2006.

[309] J.P.O. Santos and A.V. Sills. q-Pell sequences and two identities of V. A. Lebesgue. *Discrete Math.*, 257(1):125–142, 2002.

[310] A. Sapounakis, I. Tasoulas, and P. Tsikouras. Counting strings in Dyck paths. *Discrete Math.*, 307(23):2909–2924, 2007.

[311] A. Sapounakis and P. Tsikouras. On the enumeration of noncrossing partitions with fixed points. *Ars Combin.*, 73:163–171, 2004.

[312] C. Savage. A survey of combinatorial Gray codes. *SIAM Rev.*, 39(4):605–629, 1997.

[313] D. Savitt. Polynomials, meanders, and paths in the lattice of noncrossing partitions. *Trans. Amer. Math. Soc.*, 361(6):3083–3107, 2009.

[314] W.R. Schmitt and M.S. Waterman. Linear trees and RNA secondary structure. *Discrete Appl. Math.*, 51(3):317–323, 1994.

[315] M. Schork. On the combinatorics of normal ordering bosonic operators and deformations of it. *J. Phys. A*, 36(16):4651–4665, 2003.

[316] M. Schork. Normal ordering $q$-bosons and combinatorics. *Phys. Lett. A*, 355(4–5):293–297, 2006.

[317] I. Semba. An efficient algorithm for generating all partitions of the set $\{1, 2, \cdots, n\}$. *J. Inform. Process.*, 7(1):41–42, 1984.

[318] L.W. Shapiro and C.J. Wang. A bijection between 3-Motzkin paths and Schröder paths with no peak at odd height. *J. Integer Seq.*, 12(3):Article 09.3.2, 9, 2009.

[319] M. Shattuck. Bijective proofs of parity theorems for partition statistics. *J. Integer Seq.*, 8(1):Article 05.1.5, 2005.

[320] M. Shattuck. Recounting the number of rises, levels, and descents in finite set partitions. *Integers*, 10:179–185, 2010.

[321] M. Sibuya. Log-concavity of Stirling numbers and unimodality of Stirling distributions. *Ann. Inst. Statist. Math.*, 40(4):693–714, 1988.

[322] R. Simion. Combinatorial statistics on noncrossing partitions. *J. Combin. Theory Ser. A*, 66(2):270–301, 1994.

[323] R. Simion. Combinatorial statistics on type-B analogues of noncrossing partitions and restricted permutations. *Electron. J. Combin.*, 7:Research Paper 9, 2000.

[324] R. Simion. Noncrossing partitions. *Discrete Math.*, 217(1–3):367–409, 2000. Formal power series and algebraic combinatorics (Vienna, 1997).

[325] R. Simion and F.W. Schmidt. Restricted permutations. *European J. Combin.*, 6(4):383–406, 1985.

[326] R. Simion and D. Ullman. On the structure of the lattice of noncrossing partitions. *Discrete Math.*, 98(3):193–206, 1991.

[327] N.J.A. Sloane. *The on-line encyclopedia of integer sequences.* Published electronically at http://www.research.att.com/~njas/sequences/, 2009.

[328] R. Speicher. Multiplicative functions on the lattice of noncrossing partitions and free convolution. *Math. Ann.*, 298(4):611–628, 1994.

[329] M.B. Squire. Two new gray codes for acyclic oreintations. Technical Report (North Carolina State University), 94–104, 1994.

[330] A.J. Stam. Generation of a random partition of a finite set by an urn model. *J. Combin. Theory Ser. A*, 35(2):231–240, 1983.

[331] Z. Stankova and J. West. A new class of Wilf-equivalent permutations. *J. Algebraic Combin.*, 15(3):271–290, 2002.

[332] R.P. Stanley. Catalan addendum. http://www-math.mit.edu/~rstan/ec/catadd.pdf.

[333] R.P. Stanley. *Enumerative combinatorics. Vol. 1*, volume 49 of *Cambridge Studies in Advanced Mathematics*. Cambridge University Press, Cambridge, 1997. With a foreword by Gian-Carlo Rota, Corrected reprint of the 1986 original.

[334] R.P. Stanley. Parking functions and noncrossing partitions. *Electron. J. Combin.*, 4(2):Research Paper 20, 1997. The Wilf Festschrift (Philadelphia, PA, 1996).

[335] R.P. Stanley. *Enumerative combinatorics. Vol. 2*, volume 62 of *Cambridge Studies in Advanced Mathematics*. Cambridge University Press, Cambridge, 1999. With a foreword by Gian-Carlo Rota and appendix 1 by Sergey Fomin.

[336] E. Steingrímsson. Statistics on ordered partitions of sets. http://combinatorics.cis.strath.ac.uk/einar/papers/.

[337] Y. Sun and X. Wu. The largest singletons of set partitions. *European J. Combin.*, 32(3):369–382, 2011.

[338] Y. Sun and Y. Wu. The largest singletons in weighted set partitions and its applications. *Discrete Math. Theor. Comput. Sci.*, 13(3):75–85, 2011.

[339] Z.-W. Sun and D. Zagier. On a curious property of Bell numbers. *Bull. Aust. Math. Soc.*, 84(1):153–158, 2011.

[340] G. Szekeres and F.E. Binet. On Borel fields over finite sets. *Ann. Math. Statist.*, 28:494–498, 1957.

[341] H. Thomas. Tamari lattices and noncrossing partitions in type $B$. *Discrete Math.*, 306(21):2711–2723, 2006.

[342] J. Touchard. Propriétés arithmétiques de certain nombers recurrents. *Ann. Soc. Sci. Bruxelles A*, 53:21–31, 1933.

[343] J. Touchard. Nombres exponentiels et nombres de Bernoulli. *Canad. J. Math.*, 8:305–320, 1956.

[344] H.F. Trotter. Algorithm 115, permutations. *Comm. ACM*, 5:434–435, 1962.

[345] I. Tweddle. *James Stirling's Methodus differentialis*. Sources and Studies in the History of Mathematics and Physical Sciences. Springer-Verlag London Ltd., London, 2003. An annotated translation of Stirling's text.

[346] A. Varvak. Rook numbers and the normal ordering problem. *J. Combin. Theory Ser. A*, 112(2):292–307, 2005.

[347] A.M. Vershik and Yu.V. Yakubovich. Asymptotics of the uniform measure on simplices, and random compositions and partitions. *Funktsional. Anal. i Prilozhen.*, 37(4):39–48, 95, 2003.

[348] N.Y. Vilenkin. *Combinatorics*. Academic Press, New York-London, 1971.

[349] M.L. Wachs and D. White. $p,q$-Stirling numbers and set partition statistics. *J. Combin. Theory Ser. A*, 56(1):27–46, 1991.

[350] C.G. Wagner. Partition statistics and $q$-Bell numbers ($q = -1$). *J. Integer Seq.*, 7(1):Article 04.1.1, 2004.

[351] T. Walsh. Gray codes for involutions. *J. Combin. Math. Combin. Comput.*, 36:95–118, 2001.

[352] T. Walsh. Generating Gray codes in $O(1)$ worst-case time per word. In *Discrete mathematics and theoretical computer science*, volume 2731 of *Lecture Notes in Comput. Sci.*, pages 73–88. Springer, Berlin, 2003.

[353] J. West. Permutations with forbidden subsequences and stack sortable permutations. Ph.D. thesis, M.I.T., 1990.

[354] D. White. Interpolating set partition statistics. *J. Combin. Theory Ser. A*, 68(2):262–295, 1994.

[355] R.M. Wilcox. Exponential operators and parameter differentiation in quantum physics. *J. Mathematical Phys.*, 8:962–982, 1967.

[356] H.S. Wilf. Three problems in combinatorial asymptotics. *J. Combin. Theory Ser. A*, 35(2):199–207, 1983.

[357] H.S. Wilf. *Combinatorial algorithms: An update*, volume 55 of *CBMS-NSF Regional Conference Series in Applied Mathematics*. Society for Industrial and Applied Mathematics (SIAM), Philadelphia, PA, 1989.

[358] H.S. Wilf. *Generatingfunctionology*. Academic Press, San Diego, 2nd edition, 1994.

[359] G.T. Williams. Numbers generated by the function $e^{e^x-1}$. *Amer. Math. Monthly*, 52:323–327, 1945.

[360] R.J. Wilson. The Möbius function in combinatorial mathematics. In *Combinatorial Mathematics and its Applications (Proc. Conf., Oxford, 1969)*, pages 315–333. Academic Press, London, 1971.

[361] W. Witschel. Ordered operator expansions by comparison. *J. Phys. A*, 8:143–155, 1975.

[362] M. Wyman. The asymptotic behaviour of the Laurent coefficients. *Canad. J. Math.*, 11:534–555, 1959.

[363] G. Xin and T.Y.J. Zhang. Enumeration of bilaterally symmetric 3-noncrossing partitions. *Discrete Math.*, 309(8):2497–2509, 2009.

[364] Yu.V. Yakubovich. Asymptotics of random partitions of sets. *Zap. Nauchn. Sem. S.-Peterburg. Otdel. Mat. Inst. Steklov. (POMI)*, 223(Teor. Predstav. Din. Sistemy, Kombin. i Algoritm. Metody. I):227–250, 341, 1995.

[365] S.H.F. Yan. From $(2,3)$-Motzkin paths to Schröder paths. *J. Integer Seq.*, 10(9):Article 07.9.1, 2007.

[366] S.H.F. Yan. Schröder paths and pattern avoiding partitions. *Int. J. Contemp. Math. Sci.*, 4(17-20):979–986, 2009.

[367] S.H.F. Yan and Y. Xu. On partitions avoiding right crossings. arXiv:1110.0904v1, 2011.

[368] W. Yang. Bell numbers and $k$-trees. *Discrete Math.*, 156(1–3):247–252, 1996.

[369] F. Yano and H. Yoshida. Some set partition statistics in non-crossing partitions and generating functions. *Discrete Math.*, 307(24):3147–3160, 2007.

[370] T. Yano. —. *Sugaku Seminar*, 34(11):58–61, 1995.

[371] T. Yano. —. *Sugaku Seminar*, 34(12):56–60, 1995.

[372] D.W.K. Yeung, E.L.H. Ku, and P.M. Yeung. A recursive sequence for the number of positioned partitions. *Int. J. Algebra*, 2(1–4):181–185, 2008.

[373] W.M. Zhang, D.H. Feng, and R. Gilmore. Coherent states: Theory and some applications. *Rev. Modern Phys.*, 62(4):867–927, 1990.

[374] H. Zhao and Z. Zhong. Two statistics linking Dyck paths and non-crossing partitions. *Electron. J. Combin.*, 18(1):Paper 83, 12, 2011.

[375] E. Zoque. A basis for the non-crossing partition lattice top homology. *J. Algebraic Combin.*, 23(3):231–242, 2006.

# Index

## Key words

e-set partition, 432

Algorithm, 27
   loop-free, 27
   loopless, 27
Alphabet, 3
Alternating, 419
Amplitudes, 513
Analytically continuable, 513
Arrangement, 39
   number, 40
   recurrence, 39
Asymptotic behavior, 62, 384–386, 513
Avoiding, 13
   at level, 254
   subsequence, 224
   substring, 22
   subword, 22

Bell number, 4, 5, 9, 10, 32, 34, 57, 75, 76, 78, 81, 101, 102, 114, 174, 219, 361, 444, 455, 478
   $q$-, 78, 80
   asymptotic, 390, 403, 404, 406
   explicit formula, 76, 77, 83
   extension of, 102
   generating function, 56, 73
   recurrence relation, 34, 56, 57, 77
Bell polynomial, 445
Bijection, 32, 33, 62, 72, 73, 100, 101, 113, 191, 317, 446
Binomial coefficient, 38, 71
   $q$-, 79, 501
Binomial theorem, 38

$q$-, 79
Block, 1
Boson annihilation, 437–439, 447
Boson creation, 437–439, 447
Boson normal ordering, 27

Canonical form; see set partition, 2, 455
Catalan number, 14, 59, 60, 63, 67, 92–94, 96, 100, 262, 264, 344, 415, 499
   generating function, 63
Cauchy
   coefficient formula, 512
   Residue Theorem, 512
Cell, 224
   0-, 226
   1-, 226
Central limit theorem, 385
Characteristic polynomial, 39–42
Chebyshev polynomial, 72, 119, 121, 122, 124–127, 156, 505, 506
Cluster, 255
   $k$-high, 255
   $k$-low, 255
   extra-high $k$-high, 259
   extra-low, 261
Combination, 38
   $k$-element, 38
   number, 38
Combinatorial class, 26
Commutation relation, 437
Compatible
   H-, 258
   L-, 258
Composition, 22, 307, 308
   avoiding substring, 22

avoiding subword, 22
containing substring, 22
containing subword, 22
Containing
　at level, 254
　subsequence, 224
　substring, 22
　subword, 22
Contraction, 27, 437, 439, 445
　$p$-, 450
　component, 445
　noncrossing, 438
Convergence
　disc of, 511
　radius of, 385, 511
Convolution, 51
Cramer's Rule, 243
Crossing
　degenerate, 442
　left, 442
　restricted, 442
　total, 442
Curve, 512
　closed, 512
　loop, 512
　simple, 512

Decreasing chain, 250
Density function, 384
Descent, 21, 22, 105, 106, 111
　$\ell$-, 105
　nontrivial, 115
　terminator, 91
　trivial, 115
Diagram, 224
　cell, 224
　columns, 225
　convex, 236
　filling, 225
　rows, 225
Differential operator, 54
Distributing $n$ elements, 9
Dobiński's formula, 21, 75, 78, 445
Dominate, 307
Double dot operation, 439, 441

Drop, 111
Dyck path, 15, 25, 59, 63, 72, 73, 94, 97, 262, 264, 265, 374, 477, 478
　generating function, 59, 63
　recurrence, 59
　symmetric, 73

Equivalence relation, 9
Event, 378
　occurrence, 378
Expected value
　nontrivial descents, 381
　nontrivial rises, 381
　number of levels, 381
Explicit formula, 33, 36
Exponent function
　$q$-, 501
Exponential growth, 386
Exponential number, 8

Factorial, 33, 38
　$q$-, 79, 501
　double, 40, 88
Fall, 21
Falling polynomial, 5, 77, 443
Ferrers shape, 225
Ferrers-equivalent, 228, 229, 233, 236
Ferres diagram, 225
Ferres-equivalent, 227
Feynman diagram, 438, 440
　$q$-weight, 442, 443
　degree $k$, 441
　length, 442
　singleton, 441
Fibonacci number, 25, 34, 42–47, 71, 87–89, 91, 207, 215, 276, 306, 346, 387, 476, 488
　explicit formula, 42, 47
　generating function, 45–47, 52
　identity, 71
　recurrence, 34
　shifted, 35, 36
Filling, 225
　0-1, 226

falling, 250
matrix avoiding, 227
matrix containing, 227
semi-standard, 226
sparse, 226
standard, 227
zero column, 226
zero row, 226
Fock space, 438
Formal power series, 45, 50
  addition, 51
  equality, 51
  multiplication, 51
  subtraction, 51
Frequencies, 513
Function, 32
  $h$-admissible, 391, 392
  analytic, 385–389, 511, 513
  base, 513
  holomorphic, 511
  injective, 32
  meromorphic, 387, 512
  one-to-one, 32
  onto, 32
  order of, 389
  surjective, 32
Fundamental Theorem of Algebra, 41

Gap-bounded, 374, 375
Gaussian coefficient, 79
Generalized Catalan number, 344
Generalized pattern, 201, 273
  23-1, 70
  avoidance of, 273
  size three, 274
  three letter, 201
Generating function, 39, 45, 53–57, 475
  algebraic, 52
  derivative, 52
  expansion, 72
  exponential, 45, 47, 48, 50, 58, 76, 81, 86, 475
  integral, 52
  moment, 382
  of shifted sequence, 53, 54
  ordinary, 45, 48, 50, 58, 475
  probability, 382, 383
  rational, 52
Generating tree, 67, 69, 70, 315
  $k$-word, 68
  Fibonacci, 68
  set partition, 68, 69
Genji-ko, 1, 3
Genji-mon, 1, 7
Geometric distribution, 400
Geometric series, 46
Golden ratio, 42
Gorozayemon, S., 3
Gouyou-Beauchamps, D., 246, 369
Gray code, 26, 29, 422
  distance, 26, 423
  loop-free, 425
  loopless, 425
  strict, 426
Group, 509
  abelian, 509
  identity element, 509
Guess and check, 39

Hamming distance, 423
Harmonic number, 197
Harmonic sum, 513
Heisenberg–Weyl algebra, 438, 444
Hilbert space, 438
Hybrid
  $k$-, 256

Increasing chain, 250
Index, 31
Initial condition, 34, 42
Integer partition, 309
  2-free, 309
  Sequence A027336, 309
Involution, 33, 391
Iteration, 40

Kernel method, 335

Lagrange inversion formula, 62, 63
Landscape word, 255

compatible, 255
Lattice Path
    $L$-, 466
Lattice path, 58–60, 438, 466
    down-step, 58
    horicontal-step, 58
    level-step, 58
    up-step, 58
Left-to-right maxima, 91
Level, 21, 22, 105, 106, 111
    $\ell$-, 105
Linear extension, 287
Loop-free, 425
Loopless, 425
Lucas number, 35, 42, 43, 71
    explicit formula, 42
    recursion, 35

Maple code, 29, 34, 37, 40, 42–44, 46, 48, 50, 55, 59, 61, 423, 429, 434, 473, 496
Matching, 208
Matrix, 232
    0-1, 226
    $k$-, 267
    anti-identity, 227
    Ferres-equivalent, 227
    identity, 227
    semi-standard, 267
    semicanonical, 267
    sparse, 267
Mellin transform, 513
Methodus Differentialis, 6
Moon polyomino, 236
Motzkin number, 60, 61
Motzkin path, 25, 60, 62, 72, 247, 477
    coloured, 247, 468
    generating function, 60
    recurrence, 60

Narayana number, 97, 352, 356
Nesting
    $r$-, 356
Noncrossing, 456
    $k$-, 24

$r$-, 355
    enhanced, 355
Nonintersecting chord, 477
Nonnesting
    $k$-, 24
    $r$-, 356
Nonoverlapping, 399
Normal ordering, 437, 439, 441
    $(a^\dagger a)^n$, 444, 446
    $(aa^\dagger)^n$, 438
    $(a^\dagger a)^n$, 452
    $a^k(a^\dagger)^\ell$, 443
    $p$-, 450
    $q$-, 448
    $q$-case, 438
    noncrossing, 438, 455, 457
Normal random variable, 384
    standard, 385
Number
    $q$-, 79

Occurrence
    separated, 258
One-to-one correspondence; see bijection, 32
Operator, 76, 80, 438, 439
    difference, 78
    linear, 76, 80
    number, 439
    shift, 78
Order
    exponential, 386
Overlap, 399

Pascal's triangle, 38
Pattern, 22, 23
    $(3,k)$-pair, 307
    $(m+1)\rho m$, 149
    1-2-4, 257
    1-3-4, 257
    $1'$-$1''2\cdots 2$, 291
    $1'$-$3$-$1'2''$, 295
    $1'$-$2$-$1''$, 288, 293
    $1(\tau+1)$, 241
    1/2/3, 88

# Notation

1/23, 88, 100
11, 105, 106, 111, 117, 136
111, 135, 136, 150
112, 150
1122, 265
1123, 262
$11\cdots 1$, 128, 134
11-1, 283
11-2, 283
12, 105, 106, 111, 117, 132
$12'2''$, 288
121, 70, 151–153
12112, 266, 267
$121^{m-1}$, 249
12212, 266, 267
123, 88, 100, 150
12312, 247
$12312^p 42^q$, 260
$1232^p 142^q$, 260
$1232^p 412^q$, 259
$1232^p 42^q 1$, 259
$12343^p 13^q$, 261
$123^{p+1} 13^q 4$, 261
$123^{p+1} 143^q$, 261
$123^{p+1} 413^q$, 261
$12\cdots k1$, 244
$12\cdots k12$, 246
$12\cdots k12\cdots k$, 235
$12\cdots kk\cdots 21$, 235
$12\cdots m$, 234
$12\cdots \ell$, 128, 131
$12\cdots k(\overline{\theta}_{p+q,0}+k)\theta$, 254
$12\cdots k12\cdots k(\tau+k)$, 231, 236
$12\cdots k\theta_{p+q,0}(\tau+k)$, 252
$12\cdots k\theta_{p,q}(\tau+k)$, 252
$12\cdots kk(k-1)\cdots 1(\tau+k)$, 231, 236
$12\cdots k(\overline{\theta}_{p,q}+k)\theta$, 254
$12\cdots k(\tau+k)\theta_{p,q}$, 252
$12\cdots k(\tau+k)k(k-1)\cdots 1$, 236
$12\cdots k(\tau+k)(\theta+1)$, 232
$12\cdots k(\tau+k)12\cdots k$, 236
$12\cdots k(\tau+k)\nu$, 233
$12\cdots k(\tau+k)\theta$, 228, 232, 233
$12\cdots k(\tau+k)\theta_{p+q,0}$, 252
$12\cdots k(\theta+1)(\tau+k)$, 232
$12\cdots k(k+1)12\cdots k$, 240
$12\cdots k12\cdots k(k+1)$, 240
$12\cdots k\theta(\tau+k)$, 232
$12\cdots (k+1)$, 248
$12\cdots (k-1)dk$, 248
$12\cdots (k-1)kd$, 248
12 | 3, 202
12-1, 279, 283
12-2, 278, 279, 283
12-3, 278, 283
$12^k 12^{m-k}$, 250
$12^m$, 249, 250
$12^{\ell-1}$, 138, 139
$12^{p+1} 12^q 32^{r+1}$, 257
$12^{p+1} 12^q 32^r$, 254, 256
$12^{p+1} 32^q 12^r$, 254, 256
$12^{p+2} 12^q 32^r$, 257
13/2, 88, 206
13/2, 123, 88, 100
132, 100, 153, 154
13-2, 283, 298
$1\cdots 1\text{-}1$, 278
1-11, 274, 276
1-12, 274, 277
$1\text{-}1\cdots 1$, 275, 278
1-21, 274, 277
1-22, 274, 276
1-23, 274, 277
$1\text{-}2\cdots 2$, 276
1-32, 274, 277
$1\text{-}32\cdots 2$, 277
$1^\ell$, 135
$1^j 21^{m-j}$, 249
$1^k 21^{m-k}$, 249, 250
$1^m$, 234
$1^{\ell-1} 2$, 138
$2'\text{-}2''1$, 291
$2'\text{-}2''1\cdots 1$, 290
$2'\text{-}1\cdots 12''$, 291
$2'\text{-}2''1\cdots 1$, 291
21, 105, 106, 111, 117, 137
$21'1''$, 290
211, 150
212, 150

213, 147, 150
21-1, 279, 283
21-2, 279, 283
21-3, 278, 283
$21^{\ell-1}$, 141
221, 141
22-1, 281, 283–285, 298
231, 91, 92, 155, 156
23-1, 70, 279, 283, 284, 298
2-11, 274, 275
2-12, 274, 277
2-13, 274
2-21, 274, 277
2-31, 274
$2^{\ell-1}1$, 141
312, 91, 150
31-2, 283, 298
321, 91, 94, 150
32-1, 282, 283, 285, 286, 298
3-12, 274, 275
3-21, 274, 275
$\ell$-descent, 105
$\ell$-level, 105
$\ell$-rise, 105
$\ell \cdots 21$, 128, 137
$\ell$-$\tau'$, 274
$\sigma' = 12 \cdots k(\tau + k)\theta'$, 228
$\tau(\overline{\theta}_{p+q,0} + k)$, 253
$\tau(\overline{\theta}_{p,q} + k)$, 253
$\tau$-$\ell$-$\tau$, 292
$\tau$-$\ell$-$\nu$, 294
$\tau$-$\ell$, 278
$ba_1a_2 \cdots a_{m-1}$, 289
$m\rho m$, 143
$m\rho(m+1)$, 145
avoiding, 86, 224
binary, 248
contained in, 273
containing, 86, 224
generalized, 273, 287
multi, 288
nonsubowrd, 22
nonsubword, 23
occurrence, 86
partially ordered, 287

peak, 288
POGP, 287
primary occurrence of subword, 145
rise-descent, 114
rise-level, 115
set subpartition, 85
shuffle, 288
size even, 270
size five, 266, 268
size four, 265
size seven, 270
size six, 268, 269
strong Wilf-equivalent, 223
subsequence, 23
subword, 22
subword $(m+1)\rho m$, 149
subword 111, 136
subword $21^{\ell-1}$, 141
subword $2^{\ell-1}1$, 141
subword 312, 149
subword peak, 122
subword primary occurrence, 148
ternary, 254
three letter, 201
type, 273, 274
valley, 126, 127, 288
Wilf-equivalent, 223
Pattern avoidance
in compositions, 13
in permutations, 13
in words, 13
Peak, 247
Pell number, 71, 345–349, 351, 476
explicit formula, 71
generating function, 71
Perfect matching, 40
number, 40
recurrence relation, 40
Permutation, 21, 22, 33, 100
$k$-element, 38
avoiding substring, 22
avoiding subword, 22
containing substring, 22
containing subword, 22

Dumont, 101
　generating function, 47
　number, 38
　sortable, 221
Plane tree, 67
　binary, 68
　child, 67
　internal node, 67
　labeled, 67
　node, 67
　parent, 67
　root, 67
　succession rule, 67
Pole, 386, 387
　multiplicity, 386, 387
　of order $M$, 512
　simple, 512
Principle of inclusion and exclusion, 63, 65, 66
Probability (mass) function, 378, 379
Probability measure
　uniform, 379, 381

Random variable
　$k$-th factorial moment, 380
　$k$-th moment, 380
　average, 380
　continuous, 384
　discrete, 379, 380, 382, 383
　distribution, 379
　expected value, 380, 381
　mean, 82, 380–383
　standard deviation, 380
　variance, 380–383
Recurrence relation, 33, 34, 36–39, 42, 44, 73, 88, 91, 99, 109, 110, 156, 182, 215, 506
　$P$-recursive, 39, 40
　basic solution, 41
　characteristic polynomial, 41
　constant coefficients, 39
　explicit solution, 40
　general solution, 41–44
　homogeneous, 39, 41, 43, 44
　linear, 39, 42, 48, 49

　nonhomogeneous, 39, 41, 43, 44, 48–50, 475
　order, 39
　particular solution, 43, 44
Reduce form, 21
Residue, 512
Rhyme scheme, 9
Right-to-left minima, 91
Ring, 50, 509
　with unity, 510
Riordan number, 499
Rise, 21, 22, 105, 106, 111
　$\ell$-, 105
　nontrivial, 115, 133
　trivial, 115, 133
RNA sequence, 352
Rook number
　$q$-, 448
Rook placement
　column northwest inversion vector, 168
　column southwest inversion vector, 168
　inversion, 168
　row northwest inversion vector, 168
Rouché's theorem, 513

Sample space, 378
Schröder path, 247
　$UH$-free, 247
Semi-standard filling
　Ferres-equivalent, 227
Sequence, 22, 23, 31, 36, 62
　$k$ indices, 50
　$k$-, 266
　$k$-semicanonical, 266
　$m$-term, 23, 31, 224
　avoiding substring, 22
　avoiding subword, 22
　containing subword, 22
　finite, 31
　index, 31
　left-shadow, 238
　Log-concave, 73

rank, 263
right-shadow, 239
subsequence, 31
substring containing, 22
term, 31, 71
Sequence A000045, 34
Sequence A000108, 13, 60, 221, 265
Sequence A000110, 5, 34, 76
Sequence A000129, 71, 103, 345
Sequence A000296, 359
Sequence A001006, 61
Sequence A001515, 365
Sequence A001680, 265
Sequence A005424, 103
Sequence A005425, 265
Sequence A005773, 103
Sequence A007051, 265
Sequence A008277, 5
Sequence A025242, 344
Sequence A027336, 309, 312
Sequence A054391, 305, 337
Sequence A124302, 323
Sequential form; see set partition, 2
Set
    cartesian product, 63
    complement, 63
    difference, 64
    disjoint, 64
    intersection, 63
    union, 63
Set partition, 1–3, 22, 34
    $11\cdots 1$, 135
    $12^{\ell-1}$, 140, 143
    212, 144
    $T$-avoiding, 24
    $\ell$-descents, 128
    $\ell$-levels, 128
    $\ell$-rises, 128
    $d$-Fibonacci, 352
    $d$-regular, 352
    $k$-distant crossing, 247
    $k$-noncrossing, 235
    $k$-nonnesting, 235
    $k$-shuffle, 255
    $m\rho(m+1)$, 147

$m\rho m$, 145
almost-poor, 44
avoiding, 90, 95
avoiding substring, 22
avoiding subword, 22
block, 1
block connected, 306
block representation, 75, 84, 85, 165
Canonical representation, 95
canonical representation, 2, 3, 23, 29, 75, 84, 105
Carlitz, 113
chunk representation, 248
circular connector, 362
circular representation, 84, 89
closer, 213
colored, 101
complement, 86
connector, 362
containing subword, 22
crossing, 14
distance differs from $\ell$, 357
distance greater than $\ell$, 357
even, 88
exponential generating function, 57
flattened, 84, 90
flattened representation, 75, 84
geometric distribution, 400, 401
Graphical representation, 194
graphical representation, 75, 84, 95
involution, 101
irreducible, 16
layered, 87
left shadow, 238
left-dominated, 238
left-dominates, 237
left-dominating, 237
leftmost letters, 167
line diagram, 84, 89
nesting, 14
noncrossing, 14, 62, 96, 101, 352, 354, 355

noncrossing cycle, 100
noncrossing involution, 101
nonnesting, 14, 356
number, 4, 28
number blocks, 82
number descents, 138
number levels, 112
number peaks, 121, 123
number rises, 133
number valleys, 125, 126
odd, 88
one crossing, 101
opener, 213
peak, 121
peak-avoiding, 122
poor, 44, 352
random, 26
regular, 15, 25, 354, 356
resricted, 62
restricted, 32, 35, 39, 44, 66, 72, 73, 101
right shadow, 239
right-dominated, 239
right-dominates, 239
right-dominating, 239
rightmost letters, 167
rook placement, 75, 84, 85, 98, 99
sequential representation, 2
sign, 88
singleton, 32, 100, 213, 480
smooth, 15, 374
sparse, 369
standard representation, 75, 84, 85, 96
statistic, 21, 165
strong poor, 37
subsequence avoiding, 23
subsequence containing, 23
substring containing, 22
transient, 213
valley-avoiding, 127
Shuffle
    $k$-, 255
Singleton, 359
    largest, 360

Singularity, 386, 513
Singularity; see also pole, 385, 386
Stack polyomino, 226, 237, 250
    content, 237
    Ferres-equivalent, 227
Stack-equivalent, 231, 232, 236, 237
State, 438
    coherent, 445
    number, 438
    vacuum, 438
Statistic, 21, 23, 235, 383
    1-distance, 189–191
    1-internal, 198, 199
    111, 135
    121, 153
    132, 154
    $Z$-, 392–395, 398
    al, 206
    ccon, 361
    cnw(r), 168, 170
    con, 361
    cr, 206
    $cr_2$, 212, 395, 398
    $cr_k$, 206
    csw(r), 168, 170
    $int_m$, 194
    lbig, 167, 169, 205
    lsmall, 167, 169, 205, 206
    l$big$, 170
    l$small$, 170
    r$big$, 170
    r$small$, 170
    ne, 206
    $ne_k$, 206
    $occ_\tau$, 116
    $occ_{12}$, 132
    $occ1^\ell$, 134
    rbig, 167, 205, 206
    rnw(r), 168
    rsmall, 167, 205
    $\tilde{w}$, 165
    $m$-distance, 190
    $p$-major index, 208, 209, 211, 212
    $w$, 165
    $w^*$, 165

number of positions to the right of the rightmost 1, 192
additional weak record, 173, 176, 177, 415
alignment, 205, 214
block, 382, 407
crossing, 205, 208, 214
descent, 138, 204
distance, 183
distinct block, 409, 411
dual inversion, 204, 205
dual major, 165
dual major index, 204, 205
internal, 194, 197, 216
inversion, 203, 208
left-to-right maxima; see record, 173
level, 381, 405
major index, 204
nesting, 205, 214
nonsubword, 22, 164
nontrivial descent, 115
nontrivial rise, 115
outstanding element; see record, 173
overlapping, 399
peak, 121
position of the rightmost of the letter 1, 191
record, 173
rise-descent, 381–383, 405
strong record, 173, 174
subword, 22
sum of positions of records, 178
sum of positions of weak records, 180
total of distances, 185, 188
trivial descent, 115
trivial rise, 115
valley, 125–128
weak record, 173
Stirling number, 5, 6, 8, 10, 11, 29, 38, 66, 73, 76, 81, 100, 112, 114, 115, 123, 128, 135, 136, 138, 140, 143, 145, 148, 153, 156, 427, 438, 444, 450
$p, q$-, 169, 171, 503
$q$-, 78, 80, 164, 166, 172, 203, 214, 216, 448, 450, 502
asymptotic, 406, 407
explicit formula, 82
generating function, 56
recurrence, 5
recurrence relation, 56
Strong Wilf-equivalent, 223
Strongly tight Wilf-equivalence, 158
Subsequence, 31
Substring, 22
avoiding, 22
containing, 22, 219
Subword, 22
4-letter pattern, 158, 159
5-letter pattern, 159, 160
avoidance of, 22
containing, 219
contined in, 22
nontrivial, 116
primary occurrence, 139, 140

Taylor series, 46
Tight Wilf-equivalence, 158
Touchard's identity, 93, 499
Tree, 67
$k$-ary, 464
$k$-ary, 438
ordered, 317
Triangular Board, 98
Triangular board, 168

Wick's theorem, 27, 437, 438
Wilf-equivalent, 223, 228, 229, 231–233, 236, 240, 241, 248–250, 252–254, 257, 259–261
Winding number, 513
Word, 3, 22, 33, 51
122 ⋯ 2, 156
22 ⋯ 21, 157
$k$-ary, 3, 33
avoiding substring, 22

avoiding subword, 22
binary, 3, 248
containing substring, 22
containing subword, 22
long level, 133
long rise, 129
number peaks, 118–120
number valleys, 120, 123–125
size, 3
ternary, 3, 36
Word-statistic, 105

$e$-set partition, 432

# Names of Authors

Ahlfors, L.V., 511
Al-Karaji, 38
Alonso, L., 332
Anderegg, F., 10, 479
Andrews, G.E., 309, 502
Anshelevich, M., 440, 448
Arima, Y., 5
Arizmendi, O., 17
Armstrong, D., 14, 317
Athanasiadis, C.A., 317

Bóna, M., 70, 95, 163, 164, 200, 221, 223, 481
Babson, E., 222, 273
Backelin, J., 227, 233
Banderier, C., 246
Baril, J.-L., 422
Baróti, G., 217
Becker, H.W., 8, 10, 96
Bell, E.T., 5, 8, 10
Bender, E.A., 10, 15, 16, 26, 217, 377, 385, 388, 406, 417, 495
Benjamin, A.T., 401, 476
Bennett, C., 489
Bergeron, N., 323
Bernhart, F.R., 360
Bernoulli, D., 43
Bessis, D., 317

Biane, P., 317, 397, 440, 442
Billingsley, P., 378
Binet, F.E., 10
Blasiak, P., 19, 27, 437, 443, 444, 448, 469
Bleick, W.E., 16, 377, 406
Bousquet-Mélou, M., 14, 246
Bożejko, M., 440, 448
Brady, T., 317
Brennan, C., 400
Burstein, A., 13, 14, 105, 144, 146, 221, 274, 275, 277, 278, 290, 292, 481–484

Callan, D., 15, 84, 89–92, 94
Canfield, E.R., 10, 409, 495
Carlitz, L., 10, 113, 166, 409
Catalan, E.C., 60
Chang, G.J., 15, 97
Chen, W.Y.C., 13–15, 165, 208, 209, 211–214, 265, 306, 313, 353–356, 365, 494
Chu, W., 15, 25, 66, 306, 352, 357, 358
Claesson, A., 274
Cohn, M., 9, 10, 83
Comtet, L., 10, 12, 81, 83, 217, 437
Constantine, G.M., 479
Corran, R., 317
Czabarka, É., 16, 378, 409

d'Ocagne, M., 8, 10, 454
de Bruijn, N.G., 16, 377, 390, 403
de Mier, A., 227, 233
de Moivre, A., 43
Dempsey, K.J., 489
Deng, E.Y.P., 14, 15, 214, 265, 306, 313, 353–356, 478
Denise, A., 246
Deodhar, R.S., 13
Dershowitz, N., 15, 97, 317
Devitt, J.S., 12, 217, 489
di Bruno, F., 203
Dobiński, G., 8, 10, 75, 76
Drake, D., 265
Du, R.R.X., 14, 15, 214, 265, 306, 313, 353–356

Duchamp, G.H.E., 437, 448
Dykema, K.J., 366, 494

Edelman, P.H., 317
Effros, E.G., 440, 448
Ehrenborg, R., 317
Ehrlich, G., 18, 26, 27, 421, 422, 424, 426, 427
Elbert, C., 16, 377
Elizalde, S., 14, 481
Epstein, L.F., 9
ER, M.C., 19, 422, 427
Erdös, P.L., 16, 378, 409
Euler, L., 60
Eustis, A.K., 476
Even, S., 9, 10, 83

Fan, N.J.Y., 265
Flajolet, P., 174, 246, 377, 385–387, 391, 392, 407, 511, 513, 514
Floater, M.S., 373
Foata, D., 213
Fujii, K., 27, 437, 448

Gardy, D., 246
Garsia, A.M., 12, 13, 169, 170, 502
Gasper, G., 501
Gessel, I.M., 12, 13, 165, 167, 203, 208, 209, 211–213
Gilbert, E.N., 421
Glick, N., 164, 173
Goldberg, K., 8
Gorozayemon, S., 3
Gould, H.W., 11, 12, 164, 166, 502
Gouyou-Beauchamps, D., 246, 369
Goyt, A.M., 12, 14, 15, 87–89, 164, 201–207, 305, 419, 480, 481
Graham, R.L., 5
Gray, F., 421, 422
Grebinski, V., 19, 497
Griffiths, M., 481
Grimaldi, R.P., 44
Grimmett, G.R., 378
Gupta, H., 210

Haglund, J., 265

Halverson, T., 318
Hammersley, J.M., 13, 221
Hankin, R.K.S., 497
Harper, L.H., 409
Hayman, W.K., 16, 377, 391, 413
Haynsworth, E., 8
Heubach, S., 13, 31, 70, 95, 105, 113, 118, 119, 163, 164, 200, 221–223, 274, 287
Hoffman, M.J., 511
Honda, T., 7
Hooper, P.K., 9, 10, 83
Horzela, A., 27, 437, 448
Hou, Q.-H., 343
Huemer, C., 19, 422, 435
Hurtado, F., 19, 422, 435
Hwang, F.K., 15, 97

Ishikawa, M., 13, 216

Jackson, D.M., 12, 217, 489
Jakimczuk, R., 16, 406, 495
Jelínek, V., 14, 24, 25, 222, 232, 264, 305, 307, 312, 318, 320, 326, 489, 494, 495
Johnson, V., 16, 378, 409
Johnson, W.P., 12, 167
Johson, S.M., 203, 422
Joichi, J.T., 421, 424
Jordan, C., 8, 16, 377
Josuat-Vergès, M., 15, 332, 397

Kasraoui, A., 13, 15, 16, 213, 214, 216, 247, 306, 314, 316, 317, 353, 393–395, 397–399, 494
Katriel, J., 19, 27, 437, 438, 444, 445, 447, 450
Kaye, R., 19, 422
Kibler, M., 19, 27, 437, 447, 450
Kim, J.S., 15, 242, 247, 265, 494
Kitaev, S., 13, 222, 287, 288
Klazar, M., 14, 15, 52, 84, 213, 214, 222, 265, 313, 354, 355, 369, 478, 495
Klein-Barmen, F., 10, 82
Knobloch, E., 8

Knopfmacher, A., 13, 16, 164, 173, 174, 378, 400, 409
Knopp, K., 16, 377
Knuth, D.E., 3, 5, 8, 10, 13, 19, 164, 173, 221, 406, 495
Kortchemski, I., 173
Koshy, T., 35
Krattenthaler, C., 94, 214, 227, 235, 314
Kreweras, G., 12, 17, 96, 97, 262, 266, 317
Ku, C.Y., 15, 417, 495
Ku, E.L.H., 15, 216, 489
Kucherov, G., 19, 497
Kupczok, A., 16, 378, 409
Kümmerer, B., 440, 448

Lakshmibai, V., 221
Leibniz, G., 8
Lewandowski, T., 318
Liaw, S.C., 15, 97
Lieb, E.H., 478
Lindquist, N., 15
Loehr, N., 265
Lucas, J.M., 434
Lyche, T., 373

MacMahon, P.A., 208
Mamede, R., 317
Mansour, T., 13–16, 19, 20, 22, 24, 25, 27, 31, 70, 84, 85, 95, 105, 113, 118, 119, 128, 133, 136, 137, 144, 146, 157, 163, 164, 173, 174, 180, 192, 200, 221–223, 232, 242, 245, 264, 265, 274, 275, 277, 278, 287, 290, 292, 305, 307, 312, 318, 320, 324, 326, 331, 337, 340, 342–345, 378, 399, 402, 418, 422, 426, 435, 437, 441, 450, 451, 478, 481–484, 489, 493–495, 498
Marberg, E., 16, 18
Marsden, J.E., 511
Matsunaga, Y., 4, 5, 75
Mbarieky, N., 15, 478

Menger, Jr., 9, 10, 83
Menon, V.V., 10, 406
Mikhaĭlov, V.V., 27, 437
Miksa, F.L., 14
Milne, S.C., 11, 12, 21, 23, 75, 78, 81, 163, 165, 167, 217, 502
Mingo, J.A., 17, 378
Mishna, M., 14, 237
Montenegro, C., 317
Moser, L., 14, 16, 377, 406, 409
Motzkin, T., 61
Moussiaux, A., 178
Munagi, A.O., 13, 15, 22, 85, 105, 133, 136, 137, 478, 484, 494
Murasaki, L., 1
Myers, A.N., 164, 173
Méndez, M. A., 27

Nassar, G., 19, 422, 426, 435
Natarajan, R., 15, 317
Newman, M., 8
Nica, A., 17, 317, 378
Nijenhuis, A., 422
Noy, M., 19, 422, 435
Němeček, J., 15, 369

Oancea, I., 317
Odlyzko, A.M., 16, 26, 377, 378, 409, 413, 417, 495
Oliver, K., 378, 400
Omaña-Pulido, E., 19, 422, 435
Orlov, M., 19, 422

Panayotopoulos, A., 15
Pascal, B., 38
Patashnik, O., 5
Penson, K.A., 27, 437, 448
Pitman, J., 16, 76, 378
Pittel, B., 16, 17, 378, 382, 409, 495
Plott, S.S., 476
Polyanin, A.D., 178
Popa, M., 17, 365, 378, 440, 448
Port, D.N., 16, 378, 481
Poupard, Y., 15, 96, 266
Poznanović, S., 314–316, 494

Prodinger, H., 15, 25, 96, 173, 306, 352, 353, 378, 400
Pudwell, L.K., 481
Puttenham, G., 7

Qin, J., 265
Quinn, J.J., 401

Rényi, A., 164, 173
Rahman, M., 501
Rasmussen, D.J., 422
Readdy, M.A., 317
Reidys, C.M., 265
Reiner, V., 10, 317
Remmel, J.B., 12, 13, 169, 170, 265, 502
Renshaw, D., 15, 417, 495
Reutenauer, C., 323
Rey, M., 318
Richards, D., 424
Richmond, L.B., 16, 26, 377, 378, 409, 413, 417, 495
Riordan, J., 8–10
Rivlin, T., 506
Roelants van Baronaigien, D., 422, 434
Roger, D.J., 13
Rogers, D.G., 221
Rosas, M., 323
Rota, G.-C., 9, 10, 21, 75–78, 80, 163, 317
Rotem, D., 13, 221
Rubey, M., 15, 227, 236, 237, 317, 332, 397
Ruskey, F., 19, 422, 426, 434

Sagan, B.E., 12, 14, 84, 86, 87, 206, 207, 222, 234, 323, 489
Sainte-Catherine, M. de, 209
Saka, M., 5, 6, 8, 10, 11, 75, 76
Salvy, B., 16, 378, 495
Sandhya, B., 221
Santos, J.P.O., 349
Sapounakis, A., 13, 15, 344
Savage, C., 19, 27, 421, 422, 424, 426
Schmidt, F.W., 13, 221

Schmitt, W.R., 352
Schork, M., 20, 27, 437, 441, 450, 451, 498
Schott, R., 332
Sedgewick, R., 174, 377, 385–387, 391, 392, 407, 511, 513, 514
Seki, T., 3
Semba, I., 19, 422
Severini, S., 14, 20, 27, 242, 245, 265, 437, 441, 451, 494, 498
Shapiro, L.W., 247
Shattuck, M., 12–15, 22, 25, 84, 85, 105, 128, 157, 164, 180, 192, 305, 312, 318, 320, 324, 326, 331, 337, 340, 342, 344, 345, 378, 399, 402, 418, 478, 484, 493–495
Sibuya, M., 478
Sierksma, G., 15
Sills, A.V., 349
Simion, R., 13, 15, 217, 221, 317, 355, 489
Sirhan, B., 105
Sloane, N.J.A., 5, 34, 60, 61, 71, 103, 221, 234, 309, 312, 337, 344, 345, 365
Solomon, A.I., 27, 437, 448
Speicher, R., 317, 440, 448
Squire, M.B., 424
Srinivasan, M.K., 13
Stam, A.J., 19
Stankova, Z., 227
Stanley, R.P., 14, 44, 60, 61, 84, 98, 214, 265, 313, 317, 347, 464, 480
Stanton, D., 397
Stark, D., 16, 378, 409
Steingrímsson, E., 216, 222, 273
Stirling, J., 5, 6, 10, 75
Stirzaker, D.R., 378
Stump, C., 14, 317
Sun, Y., 13, 15, 215, 360, 361, 484, 494
Sun, Z.-W., 481
Suzuki, T., 27, 437, 448
Székely, L.A., 16, 378, 409

Szekeres, G., 10, 16, 378, 409

Tasoulas, I., 344
Thomas, H., 14, 317
Touchard, J., 8, 10
Trotter, H.F., 422
Tsikouras, P., 13, 344
Tweedle, I., 6

Ullman, D., 15, 317, 355
Urbanek, F.J., 353

Vajnovszki, V., 19, 422, 435
Varvak, A., 27, 437, 448
Vershik, A.M., 405
Viennot, G., 369
Vilenkin, N.Y., 44

Wachs, M., 12, 13, 169, 170, 172
Wachs, M.L., 164, 217, 487
Wagner, C.G., 12, 23, 166, 484
Wagner, S., 13, 16, 84, 164, 174, 180, 192, 378
Wallis, J., 8
Walsh, T., 422, 433
Wang, C.J., 15, 247, 494
Wang, P.C.C., 16, 377, 406
Waterman, M.S., 352
Watt, C., 317
Wei, C., 15, 25, 66, 306, 352, 357, 358
West, J., 67, 68, 221, 227, 233
West, L.J., 497
White, D., 12, 13, 164, 169, 170, 172, 217, 421, 424, 487
Wilcox, R.M., 27, 437
Wilf, H.S., 16, 26, 27, 53–55, 76, 164, 173, 377, 409, 411, 413, 421, 422
Williams, G.T., 9, 10
Williamson, S.G., 421, 424
Wilson, R.J., 317
Witschel, W., 27, 437
Wormald, N.C., 16, 378, 409
Wu, S.Y.J., 15, 365, 494
Wu, X., 13, 15, 360, 361, 494
Wu, Y., 15, 215, 484

Wyman, M., 14, 16, 377, 390, 406, 409

Xian, J., 38
Xin, G., 14, 227, 233
Xu, Y., 15, 495

Yakubovich, Yu.V., 16, 377, 405
Yan, C., 314–316, 494
Yan, C.H., 13–15, 165, 208, 209, 211–214, 265, 313, 365, 494
Yan, S.H.F., 14, 15, 22, 105, 157, 247, 265, 493–495
Yang, A.L.B., 13, 165, 208, 209, 211–213
Yang, W., 13, 352, 353
Yano, F., 13, 495
Yano, T., 3
Yeh, H.G., 15, 97
Yen, L., 14, 237
Yeung, D.W.K., 15, 216, 489
Yeung, P.M., 15, 216, 489
Yoshida, H., 13, 495

Zabrocki, M., 323
Zagier, D., 481
Zaitsev, V.F., 178
Zaks, S., 15, 97, 317
Zeilberger, D., 265
Zeng, J., 13, 213, 214, 216, 247, 314, 316, 317, 352, 397, 494
Zhang, T.Y.J., 14
Zhao, A.F.Y., 265
Zhao, H., 15, 217
Zhong, Z., 15, 217